GAUTAM SINGH

VLSI Handbook

Other McGraw-Hill Reference Books of Interest

Handbooks

Avallone and Baumeister • MARKS' STANDARD HANDBOOK FOR MECHANICAL ENGINEERS
Benson • AUDIO ENGINEERING HANDBOOK
Benson • TELEVISION ENGINEERING HANDBOOK
Coombs • PRINTED CIRCUITS HANDBOOK
Coombs • BASIC ELECTRONIC INSTRUMENT HANDBOOK
Croft and Summers • AMERICAN ELECTRICIANS' HANDBOOK
Fink and Beaty • STANDARD HANDBOOK FOR ELECTRICAL ENGINEERS
Fink and Christiansen • ELECTRONIC ENGINEERS' HANDBOOK
Harper • HANDBOOK OF ELECTRONIC SYSTEMS DESIGN
Harper • HANDBOOK OF THICK FILM HYBRID MICROELECTRONICS
Harper • HANDBOOK OF WIRING, CABLING, AND INTERCONNECTING FOR ELECTRONICS
Hicks • STANDARD HANDBOOK OF ENGINEERING CALCULATIONS
Inglis • ELECTRONIC COMMUNICATIONS HANDBOOK
Juran and Gryna • QUALITY CONTROL HANDBOOK
Kaufman and Seidman • HANDBOOK OF ELECTRONICS CALCULATIONS
Kurtz • HANDBOOK OF ENGINEERING ECONOMICS
Stout • MICROPROCESSOR APPLICATIONS HANDBOOK
Stout and Kaufman • HANDBOOK OF MICROCIRCUIT DESIGN AND APPLICATION
Stout and Kaufman • HANDBOOK OF OPERATIONAL AMPLIFIER CIRCUIT DESIGN
Tuma • ENGINEERING MATHEMATICS HANDBOOK
Williams • DESIGNER'S HANDBOOK OF INTEGRATED CIRCUITS
Williams and Taylor • ELECTRONIC FILTER DESIGN HANDBOOK

Other

Antognetti and Massobrio • SEMICONDUCTOR DEVICE MODELING WITH SPICE
Antognetti • POWER INTEGRATED CIRCUITS
Elliott • INTEGRATED CIRCUITS FABRICATION TECHNOLOGY
Hecht • THE LASER GUIDEBOOK
Mun • GaAs INTEGRATED CIRCUITS
Siliconix • DESIGNING WITH FIELD-EFFECT TRANSISTORS
Sze • VLSI TECHNOLOGY
Tsui • LSI/VLSI TESTABILITY DESIGN

For more information about other McGraw-Hill materials, call 1-800-2-MCGRAW in the United States. In other countries, call your nearest McGraw-Hill office.

VLSI Handbook

Silicon, Gallium Arsenide, and Superconductor Circuits

Joseph Di Giacomo Editor in Chief
Department of Electrical Engineering
Villanova University

McGraw-Hill Publishing Company
New York St. Louis San Francisco Auckland
Bogotá Hamburg London Madrid Mexico
Milan Montreal New Delhi Panama
Paris São Paulo Singapore
Sydney Tokyo Toronto

Library of Congress Cataloging-in-Publication Data

VLSI handbook

1. Integrated circuits—Very large scale.
integration—Handbooks, manuals, etc. I. Di Giacomo,
Joseph J.
TK7874.V563 1989 621.381'73 88-12985
ISBN 0-07-016903-9

1234567890 DOC/DOC 89321098

ISBN 0-07-016903-9

The editors for this book were Daniel A. Gonneau and David E. Fogarty and the production supervisor was Dianne Walber. It was set in Times Roman. It was composed by the McGraw-Hill Publishing Company Professional & Reference Division composition unit.

It was printed and bound by R. R. Donnelley & Sons Company.

Contents

Part 3 VLSI Fabrication

11. VLSI Processing

T. C. Smith

12. VLSI Manufacturing Technology

T. C. Smith

13. VLSI Lithography

Yefim Bukhman

Part 4 VLSI Technology Selection

14. Basic Circuit Technologies

Bhadrik Dalal

15. Bipolar Gallium Arsenide Heterojunction Integrated Injection Logic

D. A. Roberts, N. A. Schmitz, and J. D. Watkins

16. Superconductor Digital ICs

T. Van Duzer

Part 5 VLSI Design Methodology

17. Hierarchical Design

Edward L. Hepler

18. Simulation

Edward L. Hepler

19. Layout Verification

William J. Haydamack

20. Topology Verification

William J. Haydamack

Appendix

Lawrence J. Kovacs

Contributors

Vishwani D. Agrawal AT&T Bell Laboratories, Murray Hill, N.J. (CHAP. 8)

Daniel I. Amey Du Pont Electronics, Wilmington, Del. (CHAPS. 23 AND 24)

William Blood, Jr. ASIC Applications, Motorola, Inc., Chandler, Ariz. (CHAP. 3)

Yefim Bukhman Motorola Inc., Mesa, Ariz. (CHAP. 13)

Tzoyao Chan ASIC Memory Design, LSI Logic Corporation, Milpitas, Calif. (CHAP. 7)

Vincent J. Coli Advanced Micro Devices, Sunnyvale, Calif. (CHAP. 4)

Jerry L. Da Bell International Microelectronic Products, San Jose, Calif.; formerly Gould, Inc., Semiconductor Division, Pocatello, Idaho (CHAP. 2)

Bhadrik Dalal Amdahl Corporation, Sunnyvale, Calif. (CHAPS. 5, 6, AND 14)

Z. J. Delalic Department of Electrical Engineering, Temple University, Philadelphia, Pa. (CHAP. 32)

Joseph Di Giacomo (EDITOR IN CHIEF) Department of Electrical Engineering, Villanova University, Villanova, Pa. (CHAP. 1)

Curt F. Fey Xerox Microelectronics Center, Xerox Corporation, El Segundo, Calif. (CHAPS. 25 and 26)

Mark T. Fuccio Daisy Systems Corporation, Mountain View, Calif. (CHAPS. 21 AND 22)

L. J. Gallace GE Ceramics, Inc., Chattanooga, Tenn. (CHAP. 27)

William J. Haydamack Valid Logic Systems, Inc., San Jose, Calif. (CHAPS. 19 AND 20)

Edward L. Hepler Villanova University, Villanova, Pa. (CHAPS. 17 AND 18)

Herbert L. Hinstorff Daisy Systems Corporation, Mountain View, Calif. (CHAPS. 21 AND 22)

Andrew T. Jennings Unisys Corporation, Paoli, Pa. (CHAP. 30)

R. W. Keyes IBM T. J. Watson Research Center, Yorktown Heights, N.Y. (CHAP. 31)

Lawrence J. Kovacs Unisys Corporation, Blue Bell, Pa. (APPENDIX)

Anton Mavretic Department of Electrical, Computer and Systems Engineering, Boston University, Boston, Mass. (CHAP. 29)

Emory B. Michel GE Solid State, Findlay, Ohio (CHAP. 28)

Ernest Millham IBM Corporation, Manassas, Va. (CHAP. 10)

Raymon Oberly Rhinebeck, N.Y.; retired, IBM Corporation (CHAP. 10)

Demetris E. Paraskevopoulos Microelectronics Center, Xerox Corporation, El Segundo, Calif. (CHAPS. 25 AND 26)

Sudhaker M. Reddy University of Iowa, Iowa City, Iowa (CHAP. 8)

D. A. Roberts Texas Instruments, Dallas, Tex. (CHAP. 15)

N. A. Schmitz Texas Instruments, Dallas, Tex. (CHAP. 15)

Richard M. Sedmak Self-Test Services, Ambler, Pa. (CHAP. 9)

T. C. Smith Bipolar Technology Center, Motorola Semiconductor, Mesa, Ariz. (CHAPS. 11 AND 12)

T. Van Duzer Department of Electrical Engineering and Computer Sciences and the Electronics Research Laboratory, University of California, Berkeley, Calif. (CHAP. 16)

J. D. Watkins Texas Instruments, Dallas, Tex. (CHAP. 15)

Patrick Yin Structured Product Development, LSI Logic Corporation, Milpitas, Calif. (CHAP. 7)

Preface

Traditionally, very-large-scale integration (VLSI) circuits were designed and developed by integrated circuit specialists who work for large semiconductor firms. However, with the advent of standard design approaches, standard processes, and user-friendly computer-aided design (CAD) tools, the design skills can now be acquired by more people at higher steps in the VLSI design methodology process. This Handbook is in every sense a practical manual that provides both a specific in-depth understanding of VLSI concepts and also a broad coverage of VLSI issues.

The purpose of this Handbook is to describe and explain the VLSI design methodology process from the customer's point of view. By doing this, the customer will gain a broad and general understanding of the three major tasks involved in VLSI: design, test, and fabrication. Each task is discussed in sufficient depth to allow the customer to project costs, schedules, and resources required to perform a VLSI design. It should be clear that all personnel, including design engineers, CAD operators, and management, will benefit from this Handbook. The tasks involved in VLSI design cover many disciplines and involve much detail coordination. This Handbook clearly illustrates the relationship between and among design, CAD, and factory interfaces.

I would like to express my grateful appreciation to the contributing authors of this Handbook. They have given graciously of their time and talent to provide VLSI customers with very useful and readable information.

Joseph Di Giacomo

About the Editor in Chief

Joseph Di Giacomo is a Senior Member of the IEEE. In addition, he is an Assistant Professor at Villanova University, where he also serves as chairman of the Electrical Engineering Research Committee, working in conjunction with corporations on advanced technology issues.

P • A • R • T • 1

VLSI DESIGN

CHAPTER 1
DESIGN METHODOLOGY

Joseph Di Giacomo
Department of Electrical Engineering
Villanova University
Villanova, Pennsylvania

1.1 DEFINITIONS

Initially the classification of integrated circuit (IC) complexity was by gate count only. The gate counts usually followed some multiple of 10. Thus,

- *Small-scale integration (SSI):* 1 to 30 gates
- *Medium-scale integration (MSI):* 10 to 300 gates
- *Large-scale integration (LSI):* 100 to 3000 gates
- *Very-large-scale integration (VLSI):* 1000 to 30,000 gates

However, there are other ways to classify levels of ICs, such as pin counts, feature size, chip size, and functionality. Feature size refers to the linewidths, line spacings, and transistor geometry dimensions. Another way to classify circuits is by input-output (I/O) pin counts, because there is a direct relationship between pin counts and logic complexity. The relationship varies somewhat, depending on whether the circuitry is average logic, memory, or specialized control. A third way to classify circuits is by chip area. Sometimes circuits get classified by functional complexity.

It is also important to understand how VLSI designs can be implemented. One can implement an LSI design by using full-custom, gate arrays, macrocell arrays, or standard cells. Gate arrays and standard cells are classified as semicustom approaches. A custom LSI circuit design approach involves the placement of all interconnections, circuits, and devices by a handcrafted interactive approach with a computer-aided design (CAD) system. Since the design process is interactive, the overall circuit can be optimized for performance and density. This optimization gives the highest design time but the lowest production cost. In addition, the design development cycle is relatively long.

A *semicustom* IC is a subset of a full-custom circuit that reduces design cost, minimizes development time, eliminates the need for custom-circuit design teams, and increases the probability of success. Such a circuit can be defined, designed, and implemented with a set of standard structures, such as basic gates, logic cell macro building blocks, or other predefined structures.

The *gate array* is a standard array of many gate circuits diffused into a silicon chip. The circuit designer provides the semiconductor manufacturer with an interconnection metallization pattern that converts these basic gates into functional custom circuits.

A *macrocell array* is an array of cells containing a number of unconnected transistors. Stored within a computer are the specifications for creating patterns that transform the unconnected devices within each cell into SSI-MSI logic functions, called *macros*. To create a VLSI design, the designer selects the appropriate macros and creates the necessary cell interconnecting pattern.

Automated standard cells are cells that are available from a cell library. These standard cells are basic logic components designed as a customized cell with all processing levels being unique. Most cell libraries contain cells that have at least one common denominator so that they can be placed together in rows or columns, much like variable-height or variable-width building blocks, then interconnected to create a semicustom function with no technical limit to the overall complexity.

In general, gate arrays are considered to be more cost effective at low complexities and low volume, while standard cells are better at high volumes and complexities.

1.2 DESIGN CYCLES

At least five major design cycles are involved in the design of a VLSI chip:

- System design cycle
- Internal design cycle
- CAD design cycle
- Vendor fabrication cycle
- Product evaluation cycle

There are two ways to handle these cycles. One approach is to use a single VLSI designer who is involved in the design from start to finish. This approach is usually used when there are very few designs to do, where the designs are of relatively low density, and the designer is experienced and skilled with each of the above cycles. The other approach is to create a formal VLSI design organization where the tasks are individually broken out and assigned to specialists in each category. Figure 1.1 is an example of this. This approach is used when an organization has many designs to do, when the designs are very complex, and when there is a lack of experienced people.

Table 1.1 shows the design cycle description. In the system design cycle the system designer prepares an initial VLSI specification and submits it to the de sign group. Items to be discussed would be critical performance areas, block diagram, functional description, and terminal characteristics. The components group takes the preliminary information and develops the final VLSI specification. In the internal design cycle the design group reviews the design for power conditions, size of VLSI chip, and any initial placement. A design decision must be made involving the choice of technology (bipolar, MOS) and design technique (custom, semicustom). If the design were semicustom, the system group starts the design with the library of semicustom cells selected. The test group also begins to generate a test vector list. It is the program manager's responsibility to

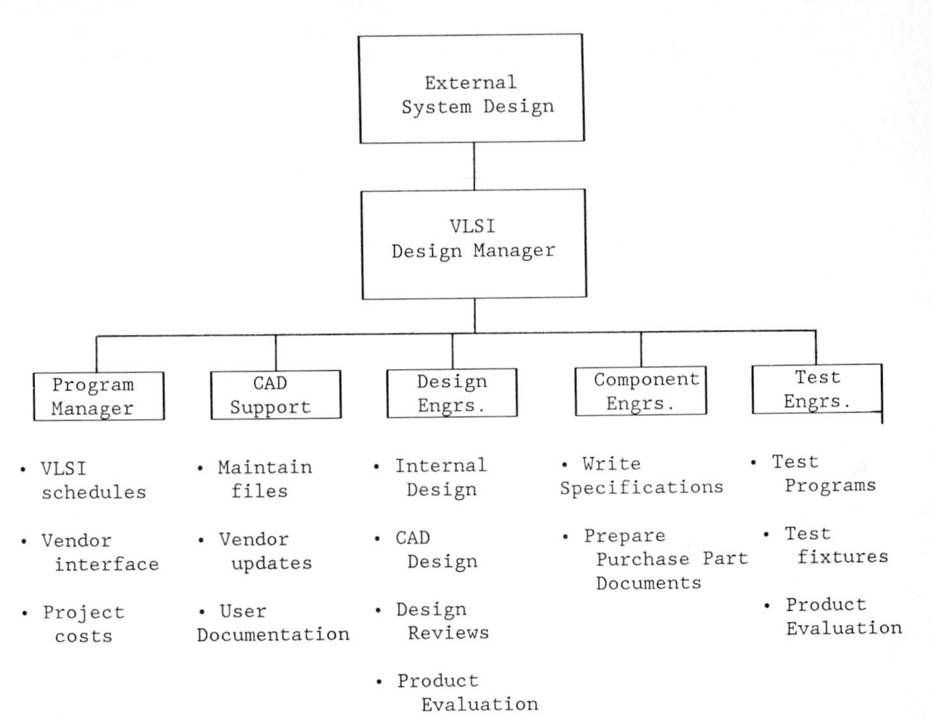

FIG. 1.1 Example of VLSI organizational chart.

plan the resources, set internal and external schedules, and maintain liaison with the semiconductor vendor.

In the CAD design cycle, the design engineer is responsible for entering the design into the CAD system. The test engineering group monitors any changes or additions to the design files, which could affect the test program. They are also responsible for validating the test tape.

TABLE 1.1 Design Cycle

Cycle	Responsible group	Affected group	Control documentation
System design cycle	Systems	Design Components	VLSI specification
Internal design cycle	System Design	Test Program manager	Design checklist VLSI schedules
CAD design cycle	Design	Test	CAD design files
Vendor FAB cycle	IC vendor	Program manager	Status reports
Product evaluation cycle	Design Test	System IC vendor	Test output data

In the vendor development cycle the semiconductor vendor has a series of tasks to accomplish to produce prototype units. The program manager is responsible for interfacing with the vendor and for handling any schedule problem.

In the product evaluation cycle the design group, using the test programs, evaluates the VLSI prototype units to verify that they meet the specifications. The design group is also responsible for signing off the design for production release.

1.3 DESIGN CYCLE METHODOLOGY

Within each of the design cycles discussed (system, internal, CAD, vendor fab, product evaluation) an explicit methodology is followed. However, the tasks to be performed are divided between the customer and the vendor. The resultant VLSI design comes about by the mutual cooperation of the two parties.

When a customer chooses a technology approach, such as gate arrays, the vendor has already developed a design methodology for that technology. In fact, a separate design methodology exists for each approach (custom, gate array, macrocells, standard cells). By *design methodology* we mean that the vendor is providing the technology, the library of elements, the customer (user) interface, the CAD software, and a definitive set of design steps to complete the VLSI function. The user interface should be easy to use, have design flexibility, and give access to all design files. The CAD software gives quick turnaround time and ensures the accuracy of the design. The technology library should give a wide choice of circuits and gate densities, and lead to minimum-cost VLSI devices. In general, as the design methodology goes from gate array to custom, the design flexibility and development costs increase, but the VLSI chip cost decreases.

In the system design cycle the first task that needs to be completed is the development of a preliminary system specification. This specification includes block diagrams, functional descriptions of subsystem elements, timing diagrams, I/O requirements, power requirements, packaging requirements, and other interface requirements. During this phase, partitions of the system are made to define separate and independent logic entities that will make the overall VLSI task more manageable.

The next task is to perform the logic design and definition of each of the subsystem elements, which includes providing circuit buffering to account for loading requirements, selecting I/O buffers, defining critical logic paths, and modifying logic to enhance the testability of the VLSI chip. In addition, formal logic design minimization techniques are used to achieve optimum gate usage. If a design is done in the customer's notation, it may have to be converted into the vendor's notation. Further, some high-level simulation can be started to verify the logic design and its timing verification, and to start test pattern development. Table 1.2 shows this methodology. These tasks are primarily customer-oriented with little interface with the vendor.

In the internal design cycle, the tasks become more circuit-oriented. Mainly, the customer determines the pinout configuration and package type, calculates die sizes or gate counts, power dissipation, and the number of voltage and ground leads required. Other parameters include fanout, race conditions, critical path, and possible noise immunity considerations. At this time some initial placement may be considered for critical path elements. A firm choice of technology (bipolar, MOS) and design technique (custom, semicustom) will be made. Table 1.3 shows this methodology.

TABLE 1.2 System Design Cycle Methodology

Design cycle	Customer tasks	Vendor tasks
System design cycle	System specification Block diagrams Functional description Timing diagrams I/O requirements Power requirements Packaging requirements Interface requirements System partitioning Logic design Buffering Critical logic paths Testability Logic conversion Simulation Functional Test patterns	Provides technology options Describes user interface Provides cell libraries Provides packaging options

In the CAD design cycle the tasks become user-interface-oriented. The VLSI design schematic must be entered into a database, the logic netlist must be generated, the logic design must be validated, and all timing simulations must be performed. In addition, a physical database must be generated to perform placement and routing of the VLSI design. It is at this point (prior to CAD input) that development agreements, as to number of designs, schedules, and costs, are signed by customer and vendor. Table 1.4 shows the CAD design cycle methodology.

A factor that must be considered is what CAD equipment and what CAD software support programs are needed. The various VLSI design approaches (custom, gate array, macrocells, standard cells) all require similar equipment and software support programs. The CAD programs required are

TABLE 1.3 Internal Design Cycle Methodology

Design cycle	Customer tasks	Vendor tasks
Internal design cycle	Pinouts Package Die sizes Gate counts Power dissipation Power pins Fan-out Race conditions Critical path Noise immunity Select technology Design technique	Provides technical and cost data Provides application support Establishes interface procedure

TABLE 1.4 CAD Design Cycle Methodology

Design cycle	Customer tasks	Vendor tasks
CAD design cycle	Enter design Build CAD files Perform simulations Generate test sequences Run fault grading Design review per design manual Runs ac simulation for placement and routing chosen Signs off VLSI design	Provides equipment Provides software Provides design manual Development costs Vendor/customer responsibilities Runs placement and routing

- Defined cell libraries or standard cells
- Logic simulator
- Fault simulator
- Circuit simulator
- Placement and routing
- Interface programs to plotters, graphics, and tape

The customer can get access to the equipment and software in several ways. Among these are

- Local access to a vendor design center
- Remote access to a vendor host computer
- Workstation(s)
- Customer computer with vendor-licensed software

The method chosen depends on the number of VLSI designs to be done, the manpower available to do the designs, and the customer's ability to fund capital expenditures.

In the vendor fabrication design cycle and product evaluation cycle, the customer supplies a pattern-generation tape as an output of the CAD system. The vendor takes the tape and makes any remaining masks that are required. From

TABLE 1.5 Vendor Fabrication Cycle

Design cycle	Customer task	Vendor tasks
Vendor fabrication cycle	Provides pattern-generation output tape	Makes masks Processes parts
Product evaluation cycle	Provides test list Evaluates parts Approves or disapproves of parts for production	Probes and tests parts Packages parts

the masks prototype parts are made and tested. The customer provides the test routine generated by the CAD system and a prototype probe card to measure the device. The customer evaluates the parts and approves or disapproves them for production release. Table 1.5 shows the vendor fabrication cycle.

1.4 ORGANIZATION OF THE BOOK

Table 1.6 shows how the 32 chapters of this book are organized. The general flow of the material covers vendor considerations, user considerations, and product considerations, which are all interrelated. Following this, some special topics and application examples are given.

Vendor considerations are covered in Parts 1, 2, and 3. In Part 1, Chaps. 1 to 7 cover design (Chap. 2), semicustom design and getting started (Chap. 3), programmable devices (Chap. 4), and circuit device fundamentals, parameters, and design (Chaps. 5, 6, and 7). In Part 2 the methods and requirements of testing are covered in Chaps. 8 and 9, followed by a thorough discussion of the automatic test equipment required to test complex devices (Chap. 10). In Part 3, Chaps. 11 to 13 cover the processing and manufacturing technology required to make VLSI devices.

User considerations are covered in Parts 4, 5, and 6. In Part 4, Chaps. 14 to 16 provide the user with a background knowledge of available technologies and the process of technology selection. In Parts 5 and 6, Chaps. 17 to 22, the complete methodology of design, simulation, layout, and topology verification is covered along with a discussion of how workstations improve the interaction between the designer and the design.

Product considerations are covered in Parts 7, 8, and 9. There are many choices available in packaging (Chaps. 23 and 24). Costs, schedules, and risk of VLSI involvement are covered in Chaps. 25 and 26. Chapters 27 and 28 cover the physics of reliability, processing failure patterns, and yield analysis.

Following this, some special topics on analog circuits (Chap. 24) and emitter-coupled logic (ECL) arrays (Chap. 30) are presented along with application and their impact on electronic equipment (Chaps. 31 and 32).

The material presented in this handbook is intended to serve as a solid foundation for users of VLSI products. The references listed at the end of each chapter can supply more information.

TABLE 1.6 Organization of Book

Vendor considerations (parts 1, 2 and 3)	User considerations (parts 4, 5, and 6)	Product considerations (parts 7, 8, 9)	Special topics applications (parts 10, 11)
Design	Selection	Packaging	Analog circuits
Test	Methods	Pricing	ECL arrays
Fabrication	Tools	Reliability	Computers
			Applications

CHAPTER 2
VLSI CUSTOM DESIGN

Jerry L. Da Bell
International Microelectronic Products
San Jose, California
formerly
Gould, Inc., Semiconductor Division,
Pocatello, Idaho

2.1 INTRODUCTION

When the discussion of which design style to use for the next integrated circuit turns to a custom solution, the client often develops an acute case of custom-chip phobia or one of the other syndromes commonly associated with developing a chip from scratch. For some, having heard of the horrors only vicariously, the disease is simply empathetic; for others, having experienced an unsuccessful development that might have seriously limited their career potential, it is real. This chapter outlines principles for developing custom chips, which, if followed, will cure custom-chip phobia and allow your system to have the technical and economic advantages of a custom circuit without the perceived risk to your career.

In this chapter we will define a custom circuit and qualitatively discuss when it is appropriate to use one. We will develop a strategy for selecting vendors, developing the specification and the project plan, and managing the project to meet the plan. The tasks that need to be done by you and the vendor will be defined. You will learn that many of the tools and procedures used for semicustom design are being applied to custom chips too. The result is that design and layout spans are much shorter than they were a few years ago, and layout errors are now uncommon. The real risk with a custom chip is getting what you want when you want it. Only the system engineer knows what the chip must do, and it is your responsibility to communicate those requirements to the vendor. Of course, you know when you want the chip, so now the task becomes one of making sure the vendor can do what you want in the time available. You must communicate exactly what you want, and then check to see that the vendor has the ability and the capacity to do it. You and the vendor must manage the development to achieve the goals. Let's start by defining a custom chip and setting some guidelines for making the decision to use one. We will also contrast custom chips to standard-cell and gate-array chips.

A custom VLSI chip has features on it that are designed especially for you. Consequently, it may cost more and may take longer to develop than a semicustom standard cell or gate array, but its recurring cost, the piece price, will generally be less than either of the others. The decision to use a custom design

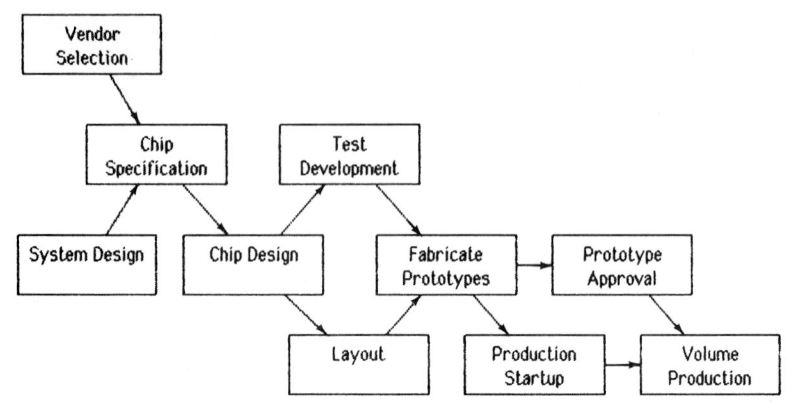

FIG. 2.1 Chip development tasks.

has been traditionally based on volume; custom chips are usually smaller than semicustom and thus cost less. If you are buying a lot of them, a small difference in piece price could make a lot of difference to your total cost. Any time you can design special circuitry that makes a significant reduction in chip area, your part will cost less. Work with your vendor to establish the regions where each design style is most effective.

Time-to-market is now such a strong factor that semicustom devices are often used when a custom chip would be a cheaper solution or give better performance. If product lifetime is long enough, a custom chip could be started after the design is frozen and phased into the early production. Several situations can only be solved with a custom chip. Analog functions generally fall into the can't-be-done-any-other-way category, as do special interface requirements and physical-space limitations.

Another reason for using custom is to protect your design from being copied by a competitor. Custom chips take longer to copy than SSI or MSI implementations, and there are techniques for making them *very* difficult to copy. Since custom chips will take you longer to develop, you will have to trade off your design time for the increased security. Once you have decided that a custom chip would meet your needs, just how do you go about getting it?

Figure 2.1 shows the top-level program evaluation and review technique (PERT) chart for the development of a typical integrated circuit. With a custom-chip development you could draw the boundary between the tasks you perform and those you will have the vendor do almost anywhere in the diagram. We will assume that you intend to specify the chip and have the vendor do the design. The critical tasks for you are vendor selection, system design, chip specification, and prototype approval.

2.2 VENDOR SELECTION

The strategy for vendor selection can be applied whether you develop just one or several chips per year. You may have already learned that maintaining good relationships with your vendors is an expensive process. By carefully selecting a small set of capable and reliable vendors, you can minimize the total cost and maximize the probability of successful projects. You should make selection of the vendor a personal experience; if you do not, the chip development will probably become very personal. The essence of the strategy is to define the capabilities the vendor needs; select two or three vendors through a series of interviews, presentations, test quotes, and factory visits; then make the selected vendors full partners in your development activities; and keep them honest by doing an annual best-value-in-the-industry survey. In discussing these issues, we will assume that you are in a high-volume situation; the principles apply to lower volume and semicustom too, but you might not want to go into as much detail. You will have to think through your situation and make a checklist of the things that are important to you. Vendor selection requires a lot of work on your part. You have to pay the price sooner or later; the best time is before you start the chip development.

What capabilities does the vendor need to have?—briefly, people, methodology, tools and equipment, capacity, and critical mass. People are the most important resource. Meet the vendor's key people in all areas of the factory. Do they have the experience, education, and skill to do your work? Have they ever done it before? How much turnover is there? How are you treated as a customer? These are the people who will design and build your chip; will they—can they—go the extra mile to make your design successful?

Investigate the vendor's methodologies, principles, rules, and procedures used in all phases of the factory. Are they memorized, documented, and incorporated in software? How are they enforced? What happens when a mistake is made? How does the vendor troubleshoot a problem? Are the procedures current? Are they sufficient? What are the statistics for turnaround time, success rate, performance to schedule? If you get answers such as, "that depends on the particular situation," it might mean that a procedure is nonexistent; dig deeper. Can the vendor train you regarding various aspects of design, particularly on the interface between the two companies? You need to have some guidelines for estimating chip size, development cost, and piece price to help you make architecture decisions. What are the quality assurance and reliability procedures? Will the vendor accommodate your special situation? Hundreds of questions need to be thought out, asked, and answered to your satisfaction. Designing, building, and delivering a chip are very complex activities, and you need to know how the potential vendor does it.

What kinds of tools and equipment does the vendor have? Those tools are intimately tied to the vendor's methodology, and the first thing to learn is how accurate they are. Just how well does simulation match the actual performance of the chip? Absolute accuracy is not practical, so to evaluate the risks, you need to understand where the compromises are. The system also needs to be tied together so that it minimizes manual checking. It should guarantee, in a liberal sense, that the chip will behave like the simulation. The usual design scenario is schematic capture, logic simulation, layout, and checking. If the logic simulation was complete and the systems engineer reviewed it thoroughly, then the logic should represent the desired behavior of the chip. If the layout is driven from the schematic, then the layout should represent the logic that represents the desired behavior. Now the layout is checked against the schematic to guarantee the correct layout. Thus it is often said that the chip is "correct by construction," but there is a loophole! The models that are used for simulation do not represent the actual configuration of transistors that were placed on the chip; simulation of a large chip would be impractical if the models were literal. There are two sets of models, simulation and layout. Without a rigorous means to ensure that the simulation, layout, and checking models are congruent, your chip just might be "incorrect by construction," one of the prime causes of custom-chip phobia.

The next thing to consider is how effective are the vendor's tools. Three things need to be optimized: design time and cost, chip area or piece price, and performance. Of course, you want to minimize the time required to develop the chip, but that usually conflicts with chip area and development cost. A sloppy layout would contribute to a fast development, but the piece price would be high. Look for hardware accelerators to reduce simulation turnaround time. How long does a simulation take? How long does it take to make a logic change and follow it through to a checked layout. Check out the placement-and-routing tools by looking at some actual chips designed with them. Is there a lot of wasted space? What is the ratio of active area (transistors and localized interconnection) versus total chip area? How long does it take to do a layout? Performance is optimized by carefully balancing the accuracy of the simulation models versus simulation time and chip area, so find out how the balance was achieved. Finally, does the design system have sufficient capacity to handle your chips along with chips from the vendor's other customers without excessive queueing? It will probably take some finessing to get the vendor to reveal any deficiencies in this area. Check all of the areas, not just design. Make sure there are enough testers and engineers available to handle the load. Many of the custom chips now have analog functions; does the vendor have analog testers?

Since you will probably depend on a single source as you start production, make sure that the vendor has the capacity to produce your chip and will be there to do it. Find out what percentage of the vendor's production capacity your circuit will require; it should be relatively small. Check to see what the capacity-limiting factors are. Will the vendor have to buy new equipment to produce your chip, or hire and train additional employees? In other words, does the vendor have *critical mass,* meaning human, physical, technical, and financial resources to build the quantity of chips you need when you need them?

This discussion should help you reduce the potential vendors to a small number that meet your requirements. Select two or three, perhaps with complementing capabilities, who will be given the opportunity to quote on all of your chips. These vendors will become your partners, in a sense. You will know enough about them to be confident that they can design and produce the chips you intend to build. When it comes to a specific chip, request budgetary quotes early in the design cycle and award the development to the vendor giving you the best value. Then work very closely with the vendor while designing the system and specifying the chip. You may want to hire the vendor as a consultant during the initial stages of development to guide your chip-related decisions. Taking advantage of your system knowledge and the vendor's chip knowledge in this manner will produce a superior system when all factors are considered.

You may be concerned about being treated fairly when you award the development so early. For a certain period of time, say a year or so, you will know enough about your vendors and the chip market to stay out of trouble, and in the short term they will keep each other honest. Over the longer term, you should conduct best-value-in-the-industry reviews. The review, which is an abbreviated version of your vendor selection procedure, will keep you abreast of new developments and provide information to help negotiate prices. As long as their performance suits you, there is no reason to change vendors, but when the need arises, information from the periodic survey will help you select a new vendor.

After selecting a few first-class vendors, you can confidently proceed with designing the chip, knowing that you will be able to make the best possible decisions regarding the chips needed for your systems, and you have a vendor that can help you.

2.3 SPECIFICATION

In the semicustom world, the specification is pretty much a formality. You have already done all of the system and logic design, everything is standard, so all you really have to do is document your choices, such as input and output levels, signal types, voltage, and operating temperature. The semicustom chip is really defined not by the specification but by the netlist and test vectors. If you think about it, those two documents are the only things the vendor can "see"; therefore, they specify the chip.

In the custom world the specification takes on much more importance and can take much longer to write. It now becomes the only means to communicate to the vendor what you want built. It is also the means for the vendor to acknowledge the capability of building what you want. During the specification phase, you should do enough work to be sure that what you want is what you specify, and the vendor should investigate the feasibility of your requests. You should work together to identify the risk areas and do enough analysis to be confident that the detail design will be successful. It is another balancing act. There is no sense in delaying a major decision until design, but it also does not make sense to design the chip during specification.

Specify only what you need. Understand what your error budgets and tolerances are, then stick to them. Overspecifying will just cost design time and chip size, so avoid it. The other important thing to realize is that the vendor needs a specification for the chip and how it will be tested, not how it will be used in the system. Do not specify things such as, "the power dissipation in an output must be less than . . ." unless you want the vendor to limit the power by design and test for it. Less design time will be wasted if you approach the specification by writing only what you expect to be tested. Some things, such as input capacitance, are not practical to test on a 100 percent basis, but if the parameter is important to control it could be tested with a special setup on a statistical basis. Another way to handle items that are not testable is to specify design goals and have the vendor demonstrate by simulation or calculation that the goals will be met.

When writing the specification, take the time to figure out what the test strategy will be. If you can figure out how to test the chip at this phase, the design will proceed smoothly; if you can't, rethink the architecture or there will be trouble later. What should be included in the specification? There are standard items such as references and absolute maximum ratings, but we will just review the items necessary for design and testing.

We are assuming that the vendor will design the logic, based on your specification. The first issue is the function you want the chip to perform. If a picture is worth a thousand words, flow diagrams,[1] state diagrams,[2] and block diagrams are worth 10,000—and at least that many dollars. Use these graphics elements liberally in the specification. They are much less ambiguous than words, and you can convey your requirements much faster. Break the design into major blocks and specify the function and interface for each block, using the appropriate diagram and a few words. Use state tables and truth tables to supplement and clarify the other diagrams.

Avoid the temptation to say, "Make this block work just like the standard product." It seems an easy way to write a specification, but you are setting yourself up for extra expense and problems. You usually do not want or need exactly what the standard product does, and you may find out that it does not do what you thought. Again, take the time to figure out what you need and specify it.

Once the functional requirements are clearly defined, it is time to specify timing relationships. Keep testing in mind here too. Timing diagrams are essential. Specify the timing relationships for each chip input and output. Outputs should be specified relative to inputs, not other outputs. It is easy enough in the lab to hang a scope probe on two outputs and observe the relative timing, but it is hard to do on a tester. The test has to be repeated many times while the strobe is adjusted to find the first output, and similarly for the second output. Then the tester has to calculate the time difference. It can be done, but it will add to the cost of the development and the testing cost of the chips. Try to use a single reference, such as the system clock.

Make the timing specifications single-ended; that is, specify a maximum or minimum delay, but not both. Double-ended specifications make design and testing expensive. If the situation is unavoidable, use sequential logic to guarantee it. The circuit elements of MOS processes have very wide variations between the fastest and slowest delays, depending on temperature, supply voltage, and normal process variations. Only count on a minimum delay if you can allow it to be zero.

You also need to define the ac load you expect the chip to see. The tester will have 50 to 100 pF inherently, and more can be added to meet your needs. If your load is less than the inherent load of the tester, and if you are pushing the performance limits of the process, you might want to specify the lighter load and derate the timing to be compatible with the tester.

Next come the static requirements of the circuit. Specify the high and low levels of the input and output voltages and the dc load the outputs will have to drive. Specifying standard levels such as TTL will help reduce the cost of the development and maybe of the chip. In general, minimize the number of different specifications. This is also the place to specify the dc or standby power consumption. The power will vary, depending on the conditions of the inputs and outputs. The minimum power will be achieved when the inputs are at the appropriate supply level, so that power is not being dissipated in a pull-up resistor or the input buffer with the dc loads disconnected. If these conditions do not meet your needs, work with the vendor; one advantage of custom is being able to get what you want.

You will probably want to specify the package type and pin sequence. The pin sequence has some effect on chip size, so ask the vendor to let you know during layout if the chip can be made smaller by changing the sequence. This is also the time to make a preliminary estimate of the number of power supply pins available. The size of the power buses inside the chip is a function of power consumption, the number of inputs and outputs, and how many of them switch simultaneously, the amount of circuitry on the chip, and the number of power pins. You can trade off the width of the buses by increasing the number of power pins. Generally speaking, it is less expensive to add power pins, if enough pins are available in the package, than to increase the bus sizes. The vendor will do the calculations, but you have control over the number of pins that are committed to power.

Taking the time to clearly specify the chip's function, timing, and interface will increase the probability of a successful development. Your confidence should also increase because you know exactly what you want; the vendor knows too, and is convinced that it can be done.

2.4 CHIP DESIGN

After the lengthy job of writing the specification, it is time to do some designing.* Actually the first steps are to review the specification, confirm the chip-size estimate, and write a project plan. If the specification was developed cooperatively with the vendor, the review will be a formality. Budgetary size estimates were made early in the project, and now a detailed estimate should be made, given the approved specification. This estimate will be used as an area budget to keep the chip on track during design and layout. The need for a project plan is obvious but often overlooked. Make it as detailed as necessary to allow the project manager to identify problems early, make sure long-lead items are started in time, identify resource requirements, and, in general, control the project. We will now design the chip.

Logic design is the first step. The actual techniques will not be covered here, but some general guidelines will be given. First, establish a systematic method for doing the design and documenting it. By doing so, you probably will not have to make this chip a career. Changes will be required at some time during the life of the chip, and it will have to be handed off to test and product engineers as the design matures. By using a systematic approach of creating and documenting the design, you can teach somebody else how to care for it. The engineer has to balance a lot of variables during design, such as area, delay, and power, and it is a good idea to build budgets for them. For example, when it is time to decide whether to build a new cell that saves a few transistors or nanoseconds, the tradeoffs between chip size and design span can be made more easily and accurately.

Remember that the chip has to be tested by a tester that can only do precisely what you tell it and that only has access to the package pins. As each piece of logic is drawn, the designer should ask, How will I test this? how much test time will it take? will it require a special setup? It will be very helpful to have a general understanding of the tester capabilities. The next question to ask is, Where will the test vectors be generated? There is really only one good place to get them, and that is from computer simulation of the logic. Although they can be captured from a breadboard or a working chip or written by hand, all of these sources will lead to excessive debug time and recurring problems during normal manufacturing. The vectors have to predict the behavior of the chip exactly, and only simulation can do that.

Many logic simulators are available that model the behavior of an MOS chip quite well. They can do that for an entire chip with thousands of equivalent gates by abstracting the gate behavior to just a few states (high, low, unknown, etc.) and several strengths for each of those states (driving, resistive, high impedance, etc.). Timing has been added to the simulators by delaying a change of state proportionally to the delay of real element. The delay is calculated by a simple RC model where R is the source impedance of the driving gate and C is the load presented by the wiring between the gates being driven and their input impedances.

*This chapter is written as a guide for the user of a VLSI chip; however, the sections on design will be addressed to the person doing the work.

The model is very simple, and since the load is considered to be purely capacitive, it cannot account for resistive wiring. If the wiring resistance is appreciable, circuit-level simulation will probably be required to analyze the timing of the path correctly. The point is, understand the simulator's limitations and use it appropriately.

Developing the models for all of the cells used in a chip is a lengthy process, so vendors often base even a custom design on existing library elements. That way most of the cells can be built and characterized without affecting your schedule; only the special ones for your chip will contribute to the development span. Cell generators are now being used to generate a new cell very quickly, once you specify various parameters, such as delay, load, cell height, and optional functions.

That explains where the simulator gets the models for the cells, but what about the netlist, or model for the chip? It is created as part of schematic capture. There are dozens of capture systems that are available commercially, based on the full range of computers from personal to mainframe, with corresponding prices. The benefit of automating this and other aspects of the design work is not so much that the original work is faster than could be done by hand, but that the machines do not make many mistakes and they minimize the time required to make a change. It is just not practical to draw the logic and build a netlist by hand.

The simulation must do two things: it must show that the chip behaves properly (it does what you want), and it must define precisely how the chip will be tested. Think through the various modes of operation and make sure the chip does what you expect. Be especially cautious at the boundaries. Check to see that switching between modes works properly. Then do some analysis to show that the chip does nothing it should not do. It is a common mistake to ignore the negative side of the question.

After you are certain that the behavior of the chip is correct under nominal conditions, it is time to develop the test patterns. The functional patterns can be used as the starting point. The goals now are to minimize the length of the patterns while testing virtually all of the circuit elements, and to ensure that the timing is correct over the expected range of operating conditions. As you look for ways to shorten the pattern length, you will probably find situations that could benefit from adding test modes. Breaking counters into short pieces, opening feedback loops in state machines, providing direct access to ROMs, PLAs, and RAMs are some of the ways to shorten the test patterns.

Set the waveforms up to look just like the tester by adding test circuitry around the chip to mimic the tester or by simply constraining the types of inputs and their timing. Figure 2.2 shows examples of waveforms allowed by Sentry testers in use today. Verify the performance of the chip over temperature, voltage, and process by simulating it at fast-speed and slow-speed conditions. Check also at the extremes of fast-rising and slow-falling parameters, and vice versa. This is a very good technique to point out any races or hazardous design configurations. Compare all of the resulting patterns at the strobe point, the point in the tester timing cycle where it makes its go-no-go decision. If all of these simulations do not match bit for bit at the strobe point, something is wrong. There are three options: change the logic to behave correctly; correct the patterns to exercise the logic properly; or ignore the differences by masking them out. Be cautious in using the latter option because the chips will behave just like the simulation. Do not forget to develop patterns to check input and output levels and power consumption. Figure 2.3 shows a scheme to verify the input levels by adding a small amount of circuitry and using a simple pattern. It might also cost a pin for the output, but it is well spent compared with the amount normally spent on

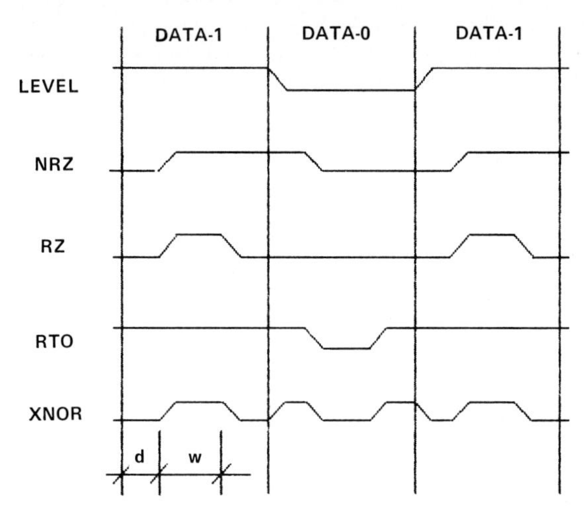

FIG. 2.2 Tester waveforms.

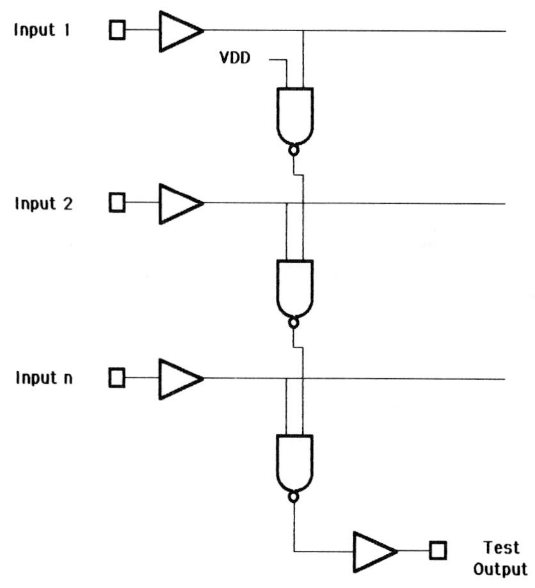

FIG. 2.3 Testing input levels.

debugging the test program. The delay used for the circuit elements must be estimated as a function of the expected wiring capacitance and of the input capacitance of the gates being driven. After layout, the actual load will be extracted and the simulation will be rerun for comparison.

The last comment about chip design is that frequent reviews of design work are extremely important. The computer tools help us design chips faster and more accurately than ever before, but there are still lots of things that depend on the person operating the tool. Reviews with peers, management, test engineering, manufacturing, and the customer all help to make sure the chip will work the first time out. These reviews will also minimize rework during layout.

2.5 LAYOUT

Layout is the process of translating schematic symbols into their physical representations. Dramatic changes have taken place in the last decade in layout technology. Where several layout designers once bent over drafting tables with pencils and erasers for months, drawing each layer of each transistor and wire, now a single person can do a chip that is larger and much more complex in a matter of days or weeks. Instead of being experts at two-dimensional drafting, layout designers now need to be expert computer operators too. For certain well-defined architectures, machines can even take a functional description and create a customized layout automatically. Instead of weeks of tedious checking by hand for design-rule errors, computers now exhaustively check very large chips in a matter of minutes or hours. What was once the most costly and error-prone task in the development of an integrated circuit has become one of the least expensive and almost error-free.

The tasks in creating the layout for a custom chip parallel those of designing the plans for a house. The first task is to create a plan for the chip by understanding the relationships between the large blocks of the architecture. Just as an architect decides the relative placement of the rooms in a house, based upon the traffic patterns of the client, the layout designer breaks the chip into blocks, looks at their relative sizes and the wiring between them, and decides where they should be placed to minimize area. That puts an additional requirement on the logic designer to draw the logic such that it can be used as a guide for the physical design of the chip. The layout designer estimates the size of the blocks by figuring the number of transistors times the area per transistor. The vendor will have rules of thumb for random logic, ordered logic like flip-flops, and ROM, RAM, etc. Successive refinements of the plan will use more accurate and detailed estimates based on actual sketches.

The design engineer will also place engineering constraints on the plan. Clock distribution is carefully specified to avoid slow transitions due to excessive load or unexpected resistance. Sensitive chip pads, like the input of a crystal oscillator, will be located to avoid noise coupling from a strong output. Input and output pads might be placed on separate power buses to prevent output transients from causing spurious transitions on inputs. Any unusual configurations that could cause the chip to be sensitive to electrostatic discharge or latch-up will be considered during planning.

To draw the plan, the layout designer needs to have an estimate of the size and shape of the cells that will be used to make up the blocks. Custom designs often use semicustom components to save time and reduce risk of error. In this case the size and shape of the appropriate standard cells would be used to plan the blocks. The planning process is interactive; that is, the layout designer and the design engineer provide information to each other that helps optimize the final chip. The planning process helps the design engineer see where the logic can be changed to save area and increase performance.

After the first pass at the plan, it becomes obvious where special cells can be built to save substantial area. Standard cells are modified to perform new functions, and special cells are drawn where the tradeoff between design time and cost, and chip size is favorable. The cells will be given thorough checking to minimize checking delays later. Specialized compilers for cells and datapaths can also be used here to create the custom cells. The design engineer can also get an estimate of the load due to wiring between blocks of circuitry, and the simulation files are updated to account for this new information.

When all of the cells are complete, the chip is assembled. Just as standard cells were used to speed cell development, *automatic placement and routing* (APAR), the same technique used for standard-cell chips, is often used for custom too. After doing all of the automatic work, the design can be transferred to a graphics editor, such as Calma,* for final adjustments to optimize area. The next two tasks of the layout phase, back annotation and checking, usually proceed in parallel.

All of the assumptions made during chip design that are affected by layout have to be checked in addition to the layout-specific issues. Constraints regarding power and clock distribution and noise coupling are analyzed by hand. Most of the other issues can be checked automatically. Dracula,† a suite of checking software by ECAD, is used by many companies for layout checking. It has several component programs for specific checks: ERC, for electrical rules check, looks for global problems like shorts between V_{ss} and V_{dd}, or transistor sources not connected to the power supply; DRC, the design rule checking component, checks all of the spacing, width, and overlap rules; continuity is verified with LVS, or layout versus schematic, where a netlist derived from the schematic is compared with one deduced from the layout. The final component of the checking software is LPE, for layout parameter extraction, which plays a key role in back annotation.

During prelayout simulation, the load on each node was assumed to be the input capacitance of each cell of the fan-out and an estimate of the wiring capacitance. These assumptions are verified by extracting the actual value of the capacitance from the layout, using LPE, and annotating the netlist (hence back annotation) with this new information. The simulations are rerun and compared with the prelayout results. Any differences have to be resolved to ensure predicted performance.

All of this checking, using computer-aided design tools, often leads to the misconception that the design will be error-free by definition. We have already discussed differences between expected and actual performance due to incorrect or incomplete simulation. Checking also has a serious weakness in most design systems: the models used for simulation and continuity checking are usually not identical. For example, a flip-flop could be simulated with a behavioral model accurately modeling the timing and function of the cell, but making no assertions about its actual structure. The flip-flop could be constructed using NAND gates, transmission gates, or NOR gates without affecting the simulation model. LVS needs a precise model of the flip-flop showing the exact topology of the cell, each transistor, and its wiring. The potential problem is now clear. If the models are not identical, how do we know that they are equivalent? We don't! Only exhaustive checking with a lot of manual analysis can ensure that they describe the same cell. This is one of the strongest reasons for basing a custom design on components from a standard library. Those components have the same problem, but effort was spent during their construction and not in the critical path of this chip

*Calma is a trademark of General Electric Company.
†Dracula is a trademark of ECAD, Inc.

to ensure accuracy of their models. New generations of design systems are striving to achieve the utopian "common database," but much more development work is needed before the IC designer's paranoia will be cured.

The last step in the layout area is to prepare the database for making reticles, or the masks used in fabricating the wafers. A *reticle* is a quartz-glass plate with a metal coating, like a mirror. Creating the pattern on the reticle plate is similar to processing the wafers themselves. Photoresist is put onto the glass plate and exposed to define the pattern by an electron beam (E-beam). The beam has a very small spot-size, only a fraction of a micron square, and its position is controlled very accurately. The layout database must be fractured, rasterized, and formatted by specialized software so that the E-beam machine can paint each layer one spot at a time.

Thus, layout, which started by converting the schematic drawing into a plan drawn hundreds of times larger than the actual chip, finally results in millions of bits of data used to make an actual reticle for each layer of the chip. Thus, one of the two physical artifacts of the design process is complete. The other artifact is the test program, and, as we learned early in the chapter, it has been considered throughout the entire design process and will be completed by the time the wafers are finished.

2.6 TEST

Most people do not realize how important testing is for developing a successful circuit. More time is spent and more frustration suffered on debugging the test program and transferring the project to manufacturing than in any other phase of development. We spent several paragraphs talking about the relationships between the specification, logic simulation, and testing, and Chaps. 8 to 10 are devoted to testing, so this section will deal with a few practical testing issues and ideas to cope with them.

Yesterday's testers are being stretched to their limits by today's chips. Chips are routinely designed to operate at 25 MHz or more; outputs drive over 100 pF and rise and fall in a few nanoseconds; inputs can respond to a glitch only a couple of nanoseconds wide. The capabilities of the tester have become a critical issue. Now a little timing skew between channels, the transition time and resolution of the timing generators, a little extra load, a little inductance in the wiring are all significant. Each of these issues could mean a few days on the tester debugging the test, and you are paying for those days in both money and schedule. The installed base of testers represents a very large capital investment that is still being depreciated. Testers from the present generation cost several times more than the installed ones but have no corresponding increase in throughput, so it is unlikely that new testers will immediately come to the rescue. With the ever increasing competition in the semiconductor business, vendors will get very creative in using their existing equipment. You need to know what you are buying and how you can save money while still getting a quality product. We will discuss the test philosophy being applied to fight noise, reflections, and tester speed.

Probably the most common source of noise is from powerful outputs switching large loads very quickly. The average current through a 100-pF capacitor, switching 5 V in 5 ns is 100 mA ($i = C\,dv/dt$). If the source impedance is just 1 Ω, the supply can bounce 100 mV, not enough to hurt, but if there are 8 or 16 outputs switching simultaneously, large noise pulses will be generated. The pulse can cause the output levels to fail if the tester samples the outputs during the noise. A simple solution to this problem is to reduce the test frequency and move the strobes out to the end of the period, allowing the signal plenty of time to reach steady state. (This solution also covers outputs that move across tester periods due to normal process variations.) Acceptable ac performance can be inferred by adding a test to measure the delay of a path through the chip to see that it is less than that predicted by simulation: the inference being that if the measured path meets the simulation, then the rest of the paths will too. With this solution, the noise still occurs, but has been avoided and debug time has been saved.

A much more serious problem is when the noise pulse is coupled into an input that recognizes it as a valid logic level and puts the chip into an unexpected state. The only way to solve this problem is to eliminate the noise or the coupling capacitance, or to relax the input levels. The noise can be reduced in several ways. A resistor can be added to the test fixture in series with the output which limits the current. Decoupling the supplies as close to the device under test (DUT) as possible will help too, but the physical configuration of autohandlers usually means that the decoupling capacitance cannot be closer than a few inches. Coupling capacitance between signal wires can be minimized by careful wiring of the test fixture. Adding capacitance to inputs can also be used to attenuate the signal near the DUT. Finally, the input voltage levels can be relaxed to compensate for the noise once the source has been identified to be the test fixturing.

Transmission line effects are also present to plague the test engineer. Just as plucking a tightly stretched string will cause a wave and its reflection to travel up and down the string, chip outputs can cause a similar effect on the tester. Now impedance matching between the chip, the test fixture, and the tester is necessary to prevent good parts from being thrown away. Diodes can be used to clamp the reflections, and sometimes ferrite beads can be used to tune the transmission line. Do not be reluctant to put a scope on the setup to observe the signals. Generally, the chip output signals should have a smooth exponential rise or fall; if not, find out why. There are several good references describing the problem and suggesting solutions.[3]

The last problem area is the speed at which testers can operate accurately. Most of the installed testers operate in the range of 10 to 20 MHz, with timing resolution of around 5 ns, timing generators that can rise or fall in about 5 ns, and round-trip delays between the DUT and the tester of a few nanoseconds. Development snags can be avoided by considering the effect of the tester specifications during specification of the chip and taking active measures to resolve them.

A generalized test strategy can be formulated from the solutions already presented:

1. Do functional testing at a low frequency, which allows the chip to reach steady state before strobing and is well within the limits of the tester. Use relaxed signal levels without loads, to compensate for and minimize noise. Take proactive measures to limit noise and reflections during the functional test.
2. Check ac performance by comparing the delay of a representative path through the chip with the delay predicted by the simulation.
3. Measure signal levels and drive capability statically.

As far as practical, the test should represent the operating conditions of the system, but it is not rational to expect the chip to compensate for noise that occurs outside its boundaries. You need to ensure that the chip behaves as it was designed to behave. Both the chip designer and the system designer need to be aware of system requirements and take active measures to satisfy them. The printed circuit board will look like a transmission line; power supplies have finite impedance; there is coupling capacitance between traces. If your system needs the chip to be impedance matched with it, then specify it, but do not match impedances by throwing away parts that behave as they were designed to behave.

The test program, the second physical artifact of the design process, is now ready. The next step is debugging. Actually there are several things being debugged simultaneously: the test program, the chip, the tester, the test hardware. All of these things will be untried at the time first silicon is ready. The concepts presented so far are intended to minimize the surprises; following them will make debugging easy.

2.7 PROTOTYPES AND PRODUCTION

There are still several steps left before you turn the chip over to the buyers for routine purchasing. The prototypes have to be evaluated in the system and characterized, and you might need to perform reliability testing. It is usually quite easy to tell whether the chip works in the system; the real question is whether the worst-case chip will work in the worst-case system under the worst-case conditions. Your first samples will be from only one or two wafers and probably from a single fabrication run; in other words, they represent only a small portion of the total population of parameters. The test program will give go–no-go information about chip performance over the extremes of temperature, voltage, loading, and timing, but, since the processing will be "nominal," both you and the vendor should take time to see how close the chip comes to the limits. If a nominal chip has sensitivities under these conditions, you should find out why. Resolve any questions now rather than when you have a lot of production committed.

It is also a good idea to start production up slowly. Everything in the design methodology is directed toward making sure every chip that passes the test will work in any system, but designing and building chips and systems is a very complicated process with lots of people, software, and machines involved. Risk will be minimized if you can control the start-up to allow you to get parts from several different manufacturing lots and install and evaluate them in systems representing the range of your production. It is often believed that reliability testing removes the risk of start-up. Actually, it only removes one of the risks—that chips which pass the production test will fail early in life under adverse environmental conditions; it says nothing about process sensitivities, races, functionality, timing. It is still prudent to get some production history by ramping up gradually.

2.8 SUMMARY

This chapter has covered several key concepts that can be used to make the development of a custom circuit go smoothly. Select a vendor who has demonstrated capability to do the kind of chip you need. Then develop a working relationship that allows your needs and the vendor's capability to be balanced. The specification is especially important to a custom chip. By working with the vendor when writing it, you will develop an early appreciation for the areas of risk in the design and you can take appropriate action to reduce the risk. Overspecification will cost time and money, so specify only what you need. Chip usage specifications should be separated from the test specification for the chip. Conduct regular reviews during the design process, and make sure you understand the simulation and the way the chip will be tested. It is often not cost-effective to test everything on every part. A streamlined production test supplemented with periodic sampling plans can get production started faster and make sure it flows smoothly. Take design measures to compensate for noise in both the test environment and the system. The test can only throw parts away; it cannot change them. Finally, get to know the part before starting volume production.

A custom chip can add value to your system that often cannot be achieved in any other way. The ideas presented in this chapter are intended to serve as background for the more detailed information in the following chapters.

REFERENCES

1. Tom DeMarco, *Structured Analysis and System Specification,* Prentice-Hall, Englewood Cliffs, N.J., 1979.
2. William I. Fletcher, *An Engineering Approach to Digital Design,* Prentice-Hall, Englewood Cliffs, N.J., 1980.
3. William R. Blood, Jr., *MECL System Design Handbook,* Motorola Inc., 1983.

CHAPTER 3
DESIGN INTERFACES FOR ASIC PRODUCTS

William Blood, Jr.
ASIC Applications
Motorola, Inc.
Chandler, Arizona

3.1 WHY ASIC?

The application specific integrated circuit (ASIC) interface starts with a need. Design engineers face a problem meeting system target goals. ASIC products can provide system solutions that would not otherwise be possible with standard off-the-shelf semiconductor circuits.

Cost is the most obvious ASIC advantage. ASIC parts can reduce semiconductor component costs. Normally the per gate price of an ASIC circuit will be less than the equivalent gate price of SSI and MSI complexity standard ICs. More important, ASIC parts reduce overall system costs. Fewer circuit boards, smaller power supplies, fewer connectors, lower assembly costs, and smaller inventories all directly reduce the expense of building a product. Lower system costs can translate to higher profit margins. Perhaps most important, lower system costs can open up new markets where it was not previously possible to address a market need in a cost-effective manner.

Smaller system size is a significant benefit for many products. Compared to using SSI or MSI complexity parts, ASIC arrays or standard cells can replace 100 or more parts with a single package. What previously required a rack of circuit boards in a cabinet may now fit in a small box on a desk. A related benefit is lower power. ASIC circuits can operate with less power dissipation than the circuits they replace. Obvious advantages are smaller, lighter, lower-cost power supplies. A less obvious advantage is reduced cooling costs. The number of fans may be reduced, or possibly the need for forced air can be eliminated.

System performance can be another big ASIC feature. ASIC products will generally outperform the products they replace. However, system speed is more than just circuit speed. Signals in high-speed systems can spend more time in circuit board runs interconnecting IC packages than they do inside the semiconductor circuits. Significant performance gains can be achieved by reducing the number of interconnects on a circuit board and doing the logic interconnect within an ASIC product.

Reliability is always important. Mechanical interconnections are major contributors to reliability problems. ASIC circuits can reduce interconnections by a ratio of 10 to 1 or more. Mechanical interconnections include wire bonds between a semiconductor chip and package, package connections to sockets or circuit boards, circuit board connections to backplanes, and cables to other system parts.

Other ASIC benefits are less apparent. Security can be a key feature. In some markets the competition will try to copy a design and introduce an equivalent product without paying development costs. It is easy to copy a system that uses all off-the-shelf components. An ASIC product with a proprietary personalization can be very difficult to copy or replace.

3.2 HOW TO GET STARTED

Getting started is a selection process; the best ASIC circuit and the ASIC supplier must be selected. Figure 3.1 shows a flow diagram covering the complete ASIC interface. Although designed around the Motorola ASIC development flow it should be sufficiently general to cover most suppliers.

The upper left corner in Fig. 3.1 shows the first step: contact a local ASIC representative. Do not restrict initial selection to only one supplier. Contact several companies and get a bigger picture of what is available. Generally it is important to meet with a technical representative from the prospective supplier. This can be done through a field applications engineer (FAE) or a factory technical specialist. The technical visit should cover, among other things, products available today, planned future products, technologies used, development interfaces, and manufacturing facilities. Sales people should also be able to provide budgetary prices for nonrecurring engineering (NRE), CAD charges, and production orders.

Technical meetings with suppliers result in a large database that must be reduced to a final decision. Figure 3.2 lists the major ASIC technology options. Each has advantages and disadvantages. Sometimes the choice is obvious; sometimes the selection is more difficult. Also, most ASIC suppliers specialize in only one or two of the choices listed in the figure. A few companies offer most or all of the technologies. If there is any doubt about the best technology for a given application, talk to a company that offers a wide range of ASIC technologies. For example, it will be difficult to get an unbiased opinion of bipolar array features from a company that provides only CMOS.

Within the choices of Fig. 3.2, CMOS offers the lowest cost, lowest power, and runs slower than BiMOS (sometimes called BiCMOS) or bipolar. Emitter-coupled logic (ECL) offers the highest performance, but does so at the expense of high power dissipation and higher prices. Generally it is desirable to go with the lowest-performance technology that meets performance requirements.

CMOS offers the biggest choice of suppliers and ASIC products. Figure 3.2 shows two basic CMOS categories: gate arrays and standard cells. Gate arrays represent a class of products that have a fixed semiconductor diffusion mask set and are customized only with metal patterns. Standard cells, which include silicon compilers, represent a class of products that are personalized with all mask layers.

Several factors influence the decision between gate arrays or standard cells. Volume is a key issue. Standard cells generally have longer development cycles and higher one-time NRE charges. Standard cells also tend to have lower per unit

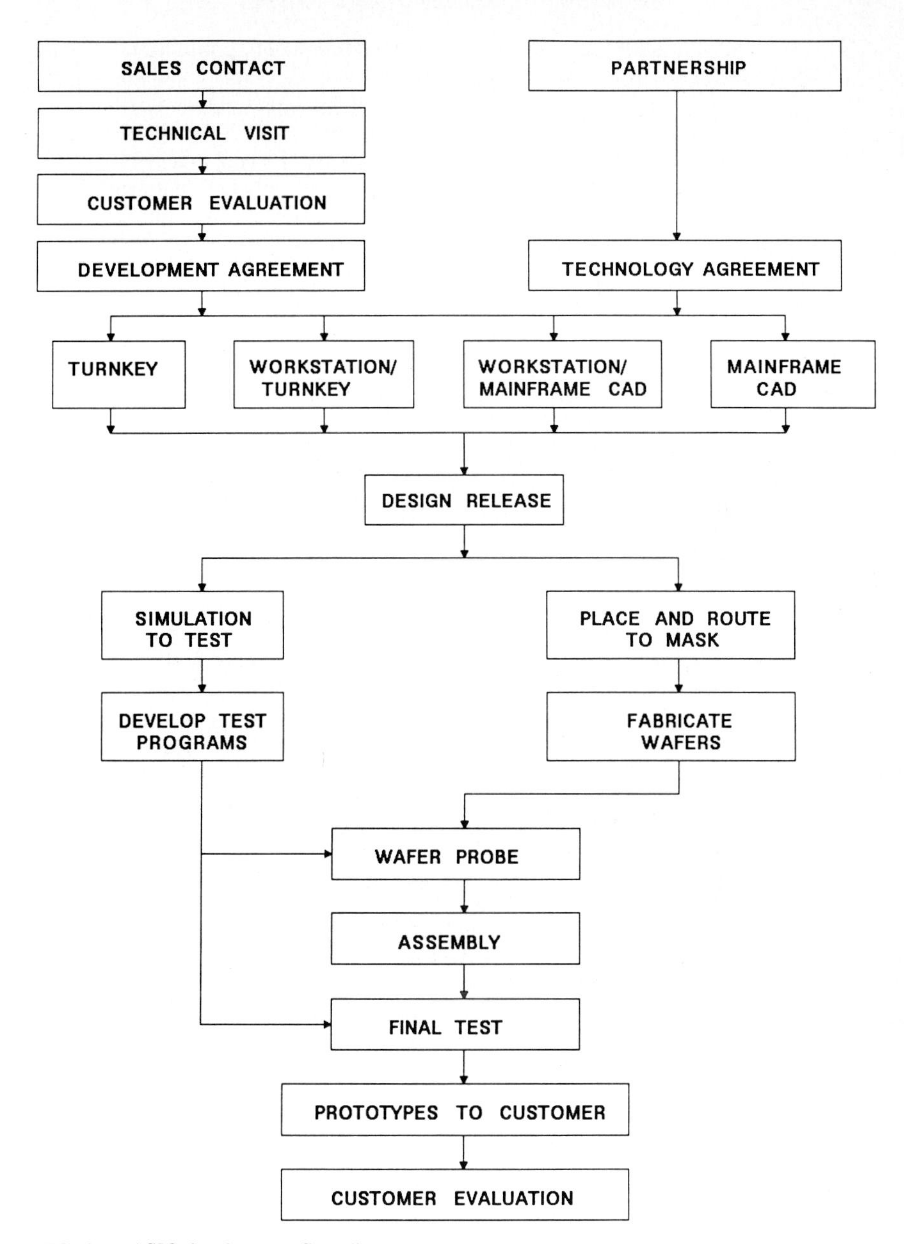

FIG. 3.1 ASIC development flow diagram.

CMOS GATE ARRAYS

CMOS STANDARD CELLS

BIMOS GATE ARRAYS

BIPOLAR TTL ARRAYS

BIPOLAR ECL ARRAYS

FIG. 3.2 Major ASIC technologies.

prices for high-volume production orders. Standard cells also offer more design flexibility through extensive cell libraries, which can include RAM, ROM, analog, and microprocessor cores. In general, the right decision is one that presents the most cost-effective answer and meets system performance goals. ASIC costs include both NRE and production. Possible rework due to design changes also has to be factored into NRE costs.

BiMOS is an emerging technology that combines many of the advantages of CMOS and bipolar. The technology seems to be moving in two directions. One is to add bipolar output drivers to an otherwise CMOS array. The other is a full BiMOS array that merges bipolar transistors into internal array cells to speed up performance. Again, it becomes a cost-performance tradeoff. Arrays with internal BiMOS offer more performance but higher cost. In addition to better drive and performance, BiMOS can add features such as combined ECL-TTL I/O interface and, eventually, high-performance analog.

ECL arrays offer the highest-performance digital logic available today. Gallium arsenide arrays offer similar performance, but at much lower densities and at higher per gate prices. However, there is a price for leadership performance, and for ECL ASIC the price is higher power dissipation and higher component costs than for CMOS or BiMOS. Process improvements continue to raise the performance levels of CMOS circuits. Similar technology improvements apply to ECL, and new products continue to be introduced that offer higher levels of density and performance.

3.3 SELECT A SUPPLIER

One of the more difficult parts of an ASIC interface is the choice of a supplier. There are many ASIC suppliers, and most of them have distinct areas of competence. There is no magic formula for choosing a supplier, but Fig. 3.3 lists several items that should be considered. Most important is the supplier's ASIC product offering. An ASIC company, no matter how good, that offers only CMOS will not be of much value for an application that needs bipolar performance. Evaluate a supplier's product offering and concentrate on those having products that meet system requirements.

Second in importance possibly is the ASIC design interface. The spectrum of available options ranges from supplying a logic schematic and receiving finished ICs to complex CAD systems that give a user complete control of an ASIC design. A common ASIC interface is through engineering workstations. Does the ASIC supplier support libraries and design interfaces for the workstation of choice? There is no right or wrong ASIC interface. Generally, for a person who occasionally does an ASIC part type, the simpler interface is better. For a project that requires several ASIC part types or pushes ASIC parts to their performance limits, the more sophisticated CAD interfaces have advantages. Specifics of ASIC interfaces are covered in more detail in Sec. 3.5. However, it is important that a potential supplier at least be able to provide a means to do simulation as a tool for logic development.

Manufacturing capability and capacity should be an important consideration.

PRODUCT OFFERING
DESIGN INTERFACE
CAD TOOLS
MANUFACTURING
QUALITY AND RELIABILITY
SERVICE
FUTURE PLANS
HISTORY
POSITION IN THE MARKET
DEVELOPMENT TURNAROUND TIME
ALTERNATIVE SOURCES

FIG. 3.3 Supplier evaluation parameters.

Does a supplier own wafer fabrication lines and packaging, or are designs farmed out to a silicon foundry? While neither the in-house nor foundry approach is a guarantee of success or failure, the company that has direct control over wafer fab and packaging can respond faster to potential yield or reliability problems.

Quality and reliability are key issues in the semiconductor industry. Replacing anything in the field is expensive and can lead to customer dissatisfaction. Get copies of any available reliability reports. Find out what life testing has been performed, and look at the test results. How does a company handle returned part failures? What programs are in place to ensure quality throughout assembly and test operations? It pays to ask questions about quality. If satisfactory answers cannot be made available, the company in question may not be giving sufficient attention to product quality and reliability.

ASIC is a service-related business. Developing ASIC array or standard-cell products is more complex than buying off-the-shelf semiconductor circuits through a sales representative. Does the ASIC supplier have local technical representation, such as FAEs? If not, how convenient is it to interface with the factory for technical assistance? Find out who are the technical contacts and how they can be reached. Documentation is another form of technical support. Can you get a full set of design manuals for the ASIC products of interest? Better documentation at the start will mean fewer questions or problems during the design.

Other decision matrix parameters are probably less critical, but can be valuable toward leading to a final selection. What are a company's future ASIC products plans? Do they line up with expected system needs? In general, there are advantages to finding an ASIC supplier, developing a business interface, and operating more as partners than in a conventional user-supplier relationship. An ASIC supplier's history may be important. How long have they been producing ASIC products, and what is the success rate? Chances for success are better with someone who has developed several thousand part types than with a company who has done only a few. A good idea is to ask for references at local companies with which it is possible to get an independent opinion.

Position in the market or market share may be a decision factor. A company that is a major factor in the market may offer more stability than a company that has recently entered the market or than one that has been around a while but has not captured a significant market share. Market share may not be an overwhelming factor. It is possible that someone newly entering the market may have a technical advantage that offsets the higher risk.

The last two items in Fig. 3.3 tend to command more attention than they deserve. Gate array turnaround time between design and finished silicon can be five weeks or less, and these times continue to be improved. In the life of a project, a week or two in getting first silicon is much less important than getting silicon that works right the first time. Standard-cell-type ASIC parts generally have longer turnaround times because the process requires a complete set of custom masks. However, with a standard-cell turnaround time of 10 weeks or less, it is still possible to get product in a reasonable amount of time.

Alternative sources tend to be more of a security issue than a practical necessity. An alternative source allows you to develop and get parts from another supplier if the first has problems. In practice, ASIC tends to be a sole-source product. Dividing ASIC business between two suppliers for a particular part can mean twice the NRE charges and higher unit purchase prices. Higher purchase prices occur because each supplier has only half the production volume. While having an alternative source can be a valuable asset, perhaps more important is the supplier's manufacturing resources. Determine the backup plan for a serious disaster in a wafer fabrication or assembly area.

3.4 DEVELOPMENT AGREEMENTS

Development agreements vary greatly between ASIC suppliers. They range from a simple purchase order to a complex formal contract. A more complex development agreement can mean increased delay starting an ASIC program while details of the agreement are discussed and understood. However, the more complex agreement up front can reduce misunderstandings later, especially if the project does not progress as expected.

Figure 3.4 lists subjects that can be covered in a development agreement. Most important is a description of the specific array and pricing. Array descrip-

ARRAY TYPE AND PRICING
PRODUCTION COMMITMENTS
PAYMENT TERMS
CAD RATES AND BILLING
STATEMENT OF WORK
DEVELOPMENT TERMS AND CONDITIONS
SCHEDULING AND PROTOTYPE DELIVERY
TERMINATION LIABILITY

FIG. 3.4 Development agreement topics.

tion covers details of the product, including array size, technology, and package. Pricing includes NRE, CAD charges, and rework. Rework is especially important to cover possible design changes at various points during the development cycle.

Some ASIC suppliers include production commitments in a development agreement. Others prefer to keep development and production as separate agreements. Combining a production order in the development agreement can mean lower NRE charges. On the other side, if there is a possibility the project could be delayed, canceled, or seriously modified, then there are advantages to avoiding any production commitments during the development phase.

Other development agreement items are designed to avoid later misunderstandings. For example, what are the terms for payment? An ASIC supplier may want 30 percent of the NRE charges when receiving a design and 70 percent when shipping prototypes. CAD charges can be a less obvious expense. Some ASIC suppliers include all CAD charges in the NRE price. Others require the user to log onto a computer for part of the development work. Associated CAD charges can be billed monthly. Will the supplier set a limit on CAD charges, or are they left open-ended? CAD charges should also cover any software the user is expected to purchase in support of an ASIC design. All charges associated with developing the ASIC part type should be covered in the development agreement.

A statement of work describes who does what. The ASIC customer will normally be expected to supply at least a schematic and a set of test vectors. In other cases the supplier will want simulation runs on either a workstation or a larger computer. Finally, the ASIC supplier may provide CAD tools and have the user do a complete design, including simulation, place and route, and design rule checks. In any case the development agreement should clearly define the tasks and responsibilities of both parties toward completing the design. Terms and conditions are related and cover items such as warranty rights, confidential information, liabilities, and ownership. Ownership is a key item since it covers such things as who owns the design, CAD database, fabrication mask sets, and wafer processing information.

The final two items of Fig. 3.4 cover possible problems. Normally a development agreement will cover delivery times for prototypes based on design release dates from the customer. It is helpful to understand any penalties or liabilities that result from the prototypes being late, due to either ASIC supplier problems or customer design problems. At times an ASIC part under development will be canceled at the customer's request. What are the liabilities when a development agreement is terminated? A more detailed agreement will cover termination charges as a function of where the circuit is in the development cycle. For example, there may be no charge prior to going to the mask shop, but there may be full NRE charges after silicon has been produced. It is also possible to have some percentage of the NRE charge after masks have been made, but prior to silicon.

The preceding ASIC interface description is the most common and covers a customer choosing an ASIC product from a wide available selection. Figure 3.1 shows an alternative path titled partnership. Larger companies may want to work with a leading ASIC supplier to codevelop a next-generation ASIC product. Both companies will work together to define the product and share resources to bring the new circuit to market.

There are advantages to both parties in such a partnership arrangement. The ASIC user drives technology in a direction that fits future system needs. The company also has an early knowledge of the product and can start design before general introduction. The ASIC supplier can get a high-technology product to market with fewer resources. The final product may be technically superior, combining the skills of a system house with those of an ASIC semiconductor company. The decision path is considerably different for a partnership agreement than for buying an existing ASIC product. Also, a partnership can require a much more comprehensive technology agreement that covers semiconductor design and/or processing rules.

3.5 ENGINEERING AN ASIC DESIGN

After selecting an ASIC product and supplier, it is time for the design development phase. Development engineering is the step that translates a design from customer specifications to a database ready for manufacturing and testing. Figure 3.1 illustrates four basic paths for ASIC circuit development. Different ASIC suppliers will offer variations of the four fundamental flows, and some may offer only one or two options. However, there are valid reasons why all four exist, and each meets a market need.

Turn-key design is the simplest interface. The ASIC user provides a product definition, commonly a logic schematic and a specification that defines package, pinouts, performance limits, and so on. The user may also supply waveforms showing proper operation. With this information the ASIC supplier engineers the circuit and develops test programs. A variation of the full turn-key is to work through an independent design center (IDC) to contract out the design. In either case the ASIC user is free from any part of the actual design conversion that is semiconductor specific.

Full turn-key has the advantage of simplicity, but it is most prone to error. Details missing or overlooked in the specification document can lead to a part that does not meet system requirements. Also, since the semiconductor company does not have the user's knowledge of what the part must do, decisions can be made by the supplier that will lead to a less than optimum design. Design reviews add some security to the turn-key approach. By having design reviews after simulation and after place and route, physical-design-related problems may be uncovered prior to building silicon.

A combination of the user doing front-end design on an engineering workstation and the supplier doing the remaining development is perhaps the most common ASIC interface. Here the user obtains software for a commonly available workstation. Software can include some or all of the following: schematic capture graphics symbols of every macro function in the ASIC product library, simulation models that accurately represent macro functionality, simulation models that accurately represent macro performance, design rule check programs, testability analysis programs, and translators to the supplier's preferred format.

Engineering workstations have several advantages with respect to an ASIC design. The stations support a schematic capture interface. Schematic capture is the process of electronically drawing and storing a schematic diagram. Circuit designs captured and electronically stored can be copied and modified far more easily than hand-drawn schematics. More important, once electronically captured, the schematic capture database can be used with simulation for verifying design accuracy. Operating off an electronically captured database minimizes a major source of errors. Another benefit of the engineering workstation schematic capture is the ability to enter information in a hierarchical manner. Hierarchical design is a method of system design in which the overall system is first planned in terms of its most general parts, each part being assigned a major function. Each major part is then considered in terms of a system of smaller components that perform the assigned function. Each smaller component can be broken down into an assortment of more specific components, and so on, down to the lowest logic primitives. Hierarchically designed circuits can be easier to simulate and debug than flat designs. Simulation can be performed on circuit sections, and problems can be uncovered by moving up and down the hierarchy rather than by examining many pages of schematics.

Figure 3.5 shows the flow pattern for design on an engineering workstation. Design starts with schematic capture using workstation software and graphics symbols provided by the ASIC supplier. Schematic capture is executed by calling up macros from the ASIC product library and logically interconnecting them with drawn lines. It is also possible to assign net names and circuit input and output names. After the schematic is entered, test vectors are entered in the workstation format, usually waveforms or a data file. Functional simulation follows using design information from the captured schematic and simulation library macro models supplied by the ASIC company. Simulation is the first major design checkpoint where logic is checked against desired functionality. Simulation tends to be an iterative process where problems are uncovered, design changes are made, and simulation is repeated. It can be viewed as the ASIC equivalent of a breadboard, where the design is tested before committing to a finished product. The advantages of doing this debug on a workstation or computer under the user's control is obvious. Changes can be made and incorporated much faster than trying to work through an ASIC supplier and having repeated design reviews.

Supplying schematic capture data and simulation results from the workstation to the ASIC supplier is one possible transfer point. The ASIC supplier takes this as the starting point and completes the rest of the design in a turn-key fashion. Remaining design functions include physical design with place-and-route programs, design rule checks, generating test programs from simulation results, and ac simulation. The interface works, but ac simulation performed by the supplier is a potential weak link. Performance goals that are not met can result in an expensive and time-consuming redesign. Design reviews or customer approval is still required for physical design parameters and ac simulation.

Continuing with the Fig. 3.5 workstation flow, one sees that it is possible to extract design information with respect to fan-out and estimated metal lengths. By using simulation models that accurately reflect macro performance limits, simulations can be run that do a reasonably accurate performance analysis. The flow diagram shows ac simulation as another decision point. If a design does not meet system performance requirements, it is possible to make design changes to the initial schematic and thus improve performance. Changes are made while in the workstation environment under the user's control. Normally this is preferable to making changes at a design review with the ASIC supplier. Chances for an ASIC part not working properly the first time decrease as more of the work is performed on a CAD system under the user's control. Another possible transfer point occurs after ac simulation, where the ASIC supplier takes the workstation database, including schematic capture, functional simulation, and ac simulation, and completes the design.

An alternative path would be to have the ASIC supplier do place and route in house, then provide the user a file of actual metal lengths or load capacitances. This information can be fed back into the workstation in a process called *back annotation*. AC simulation can be performed with actual metal delay numbers giving more accurate results. Back annotation is not available from all ASIC suppliers, but it provides several valuable benefits. AC simulation results more ac-

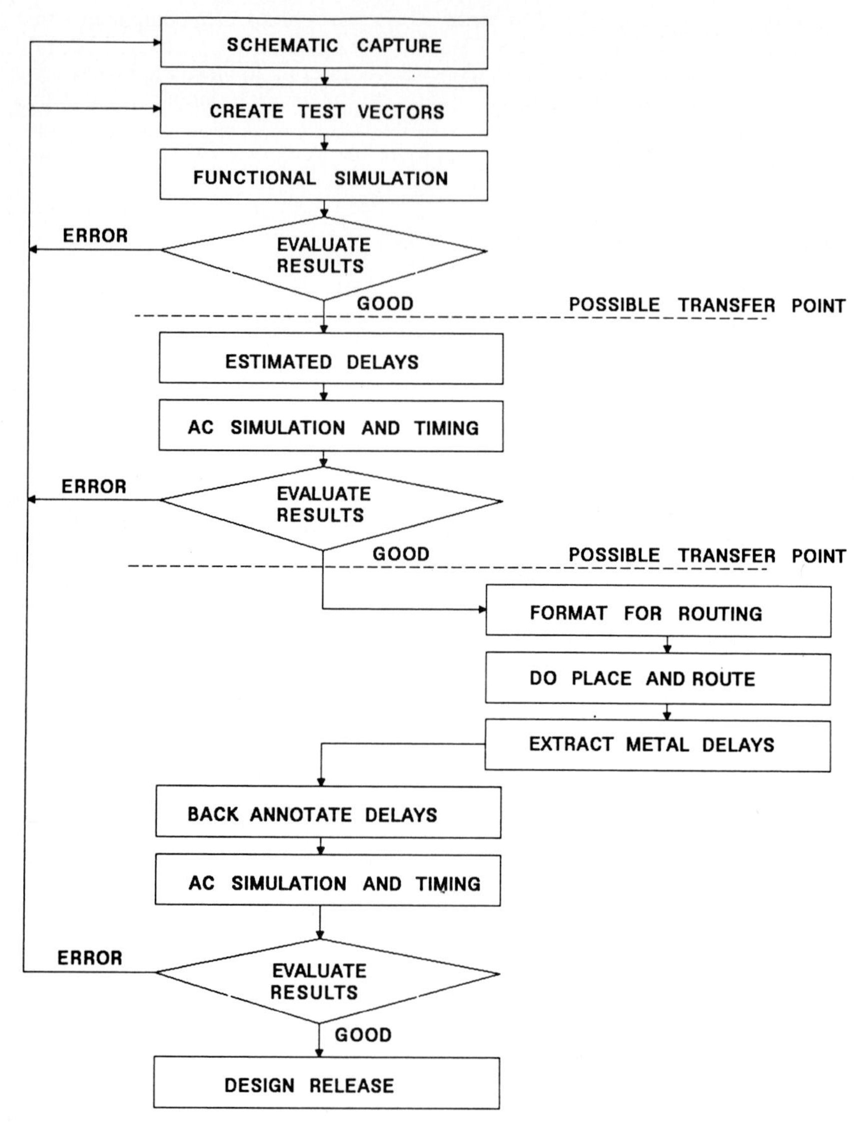

FIG. 3.5 Workstation development flow.

curately reflect actual product performance. Perhaps more important, it is possible for the customer to know product performance specifications earlier in the design cycle. System design can proceed without having to wait for the ASIC supplier to provide performance limits.

The next path in Fig. 3.1 shows ASIC circuit design being carried out on a combination of engineering workstation and mainframe computer. Here the schematic capture and simulation is performed on the workstation. Design and simulation results are translated into a format acceptable to the larger computer. The computer performs physical design with place-and-route software, does all design rule checks, performs a testability analysis, and does final ac simulation with actual metal delays. The more common interface is to use a computer owned by the ASIC supplier. Computers are available by going to an ASIC design center or over phone lines through a timeshare arrangement. Although less common, place and route can sometimes be done on an engineering workstation or on a customer-owned computer. With sophisticated CAD tools available to the ASIC user, development time is reduced and design alternatives can be evaluated earlier in the design cycle, thereby leading to a more optimized final part.

The fourth design path in Fig. 3.1 shows the entire design done on a mainframe computer. The computer can be located at an ASIC design center, accessed over phone lines from a customer's site, or be customer-owned. Figure 3.6 shows the flow diagram for doing a complete ASIC gate-array design over phone lines talking to a mainframe computer owned by the ASIC supplier. Notice that all programs are available to do the complete design, including functional and performance simulation, physical design place and route, design rule checks, and testability analysis. The user has complete control over all phases of the ASIC design and has final performance specifications prior to the circuits being built. The interface is all electronic. There is no paper sent between ASIC supplier and user. There are no design review meetings or formal design approvals. A design release on the mainframe computer signifies customer approval of the design.

The combination of engineering workstations and mainframe computers, or mainframe computers alone, can provide powerful ASIC design tools. They offer programs that thoroughly check a design and provide an excellent chance for first-pass success. Using an ASIC supplier-owned computer is a tradeoff between not buying special hardware and/or software versus paying timeshare computer charges. ASIC development costs can be minimized by doing much of the design on an engineering workstation, especially the schematic capture and functional simulation. Engineering workstations continue to get more powerful and capable, but ASIC arrays also continue to grow in size. Complex problems of physical design, testability analysis, test vector generation, and comprehensive design rule checks may still require the power of a larger computer, especially for large arrays.

The preceding information has demonstrated a wide range of ASIC interfaces. It is natural to ask which is best. There is no best interface. The choice depends on customer skills, customer-owned computing equipment, number of ASIC part types being developed, and the preferred interface for a given ASIC supplier.

3.6 ASIC FACTORY FLOW

Details of building ASIC prototypes at the factory differ between ASIC suppliers. However, the goal is to take design information from some sort of CAD system and convert it into working, tested prototypes as quickly and efficiently as possible. Figure 3.7 shows a typical flow that could be followed by an ASIC supplier.

Data preparation pulls information out of the CAD system. Place-and-route data are converted to a form required by mask-making equipment. This is not a direct translation, because additional information, such as power distribution, must be added. Simulation results are converted to wafer probe and final test programs. Again, information must be added to simulation data for testing parame-

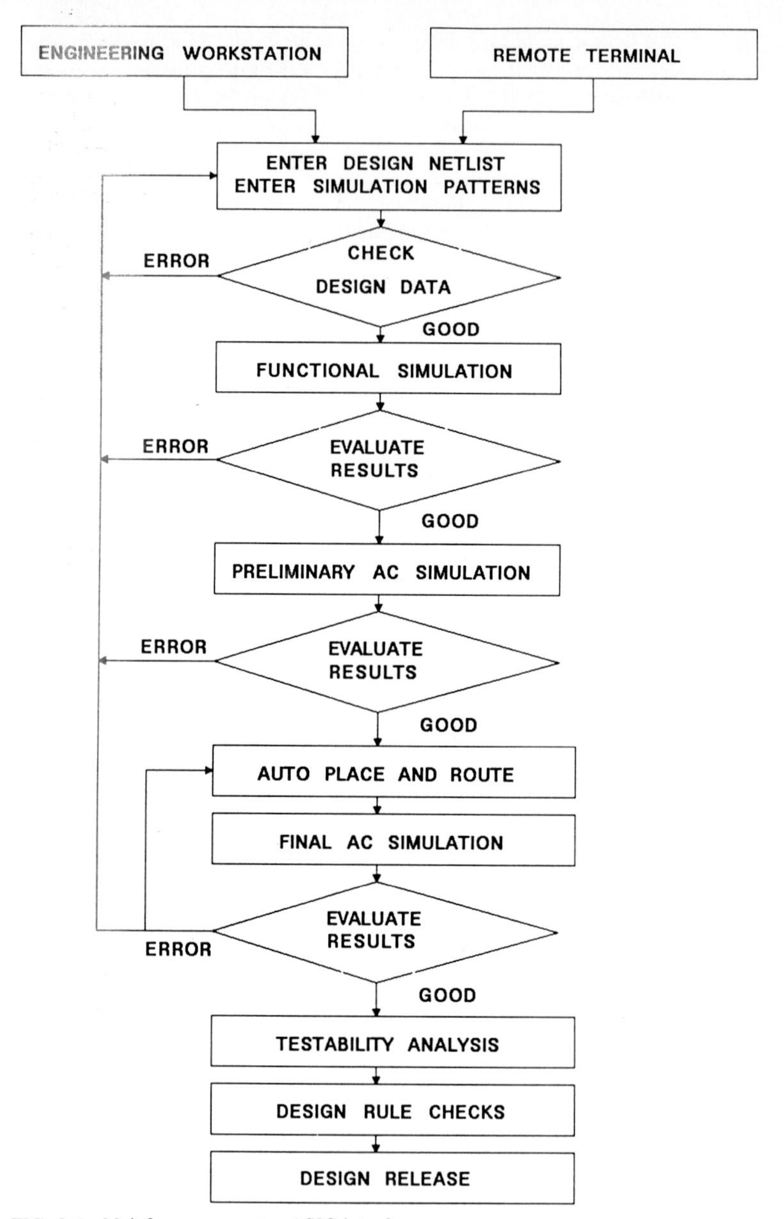

FIG. 3.6 Mainframe computer ASIC interface.

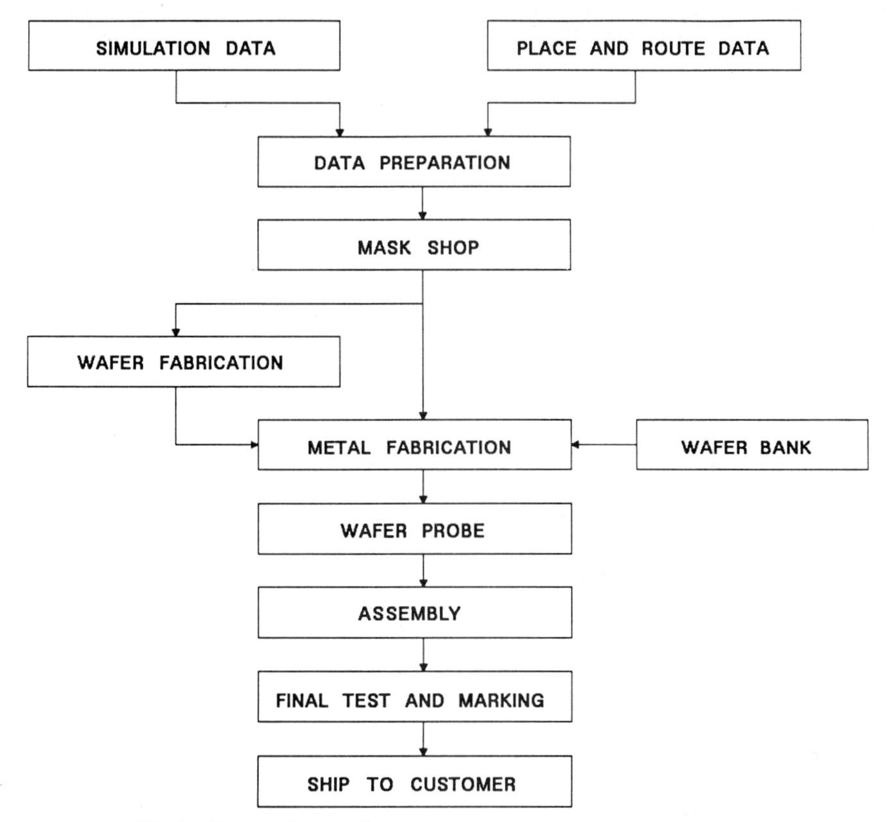

FIG. 3.7 ASIC development factory flow.

ters such as output logic levels, input thresholds, and power dissipation. Another key data preparation function is to run any final design rule checks or testability checks that may not have been included in the CAD system. Data preparation starts a complex sequence leading to prototypes being shipped. Meeting tight development schedules means careful planning and control.

The mask shop receives information based on place and route that has been modified and formatted by data preparation. Most gate-array products require three custom masks: first metal, interconnect via, and second metal. Standard cells require custom masks for all process steps, which can be 10 or more, depending on the process and number of metal layers.

Wafer fabrication builds the wafers. For gate-array products wafers can be completed without customization up to first layer metal, then banked waiting for design specific customizing masks. Standard cells require the wafer to be manufactured from the base layers through metal. Gate-array cycle times can be shorter because fewer steps are required to customize the silicon.

Wafer probe is where simulation results and silicon meet for the first time. Each die on the wafer is tested for functionality and parametric limits. Success rate at this point is heavily dependent on the design rule checks and testability analysis programs build into the CAD system. Without proper CAD tools the success rate at wafer probe and final test, even for good silicon, can be poor. Major development schedule slippages can occur when debugging test programs at wafer probe. There are various reasons why problems are experienced at wafer probe. Workstations and most other simulators have the ability to finely control input frequencies and signal timing. IC test equipment operates in a cycle mode, where inputs are applied to a circuit near the beginning of a test cycle and outputs are monitored near the end of a test cycle. While testers have some ability to stagger inputs, the resolution is not available to match the timing control of a good simulator. Software CAD tools are needed to check simulation inputs and verify that the specified patterns can be translated to meet IC test equipment requirements.

Other testability problems can exist. Simulators work in an ideal environment. If a simulator is told to change two or more inputs at the same time, the inputs change at exactly the same time. If a tester is told to change two or more package pins at the same time, there may be some variance due to tester skew. Again, it is possible to have a design that simulates perfectly, operates correctly, but fails test. Advanced ASIC CAD systems have programs that check an ASIC design and simulation inputs to verify the circuit works in the tester environment as well as with the more ideal simulation inputs.

Another important wafer probe function is diagnostics. Where all devices within a wafer lot fail, it should be possible to identify the best dies and have them packaged for additional evaluation by ASIC supplier and customer. While failure at wafer probe is possible, the goal is to have a CAD interface that catches problems during the design cycle rather than at wafer probe.

Assembly takes a good die from wafer probe and places it in IC packages. ASIC products are available in a wide range of package types including dual inline, plastic chip carriers, quad-flat packages, ceramic chip carriers, and pin-grid arrays. ASIC companies, especially those providing larger, higher-performance products, are seeing an increasing demand for unpackaged die to be put in multichip modules. In this case die from wafer probe gets placed in some form of protective carrier and sent directly to shipping, bypassing package assembly and the testing received by packaged parts.

Final test provides the most thorough device testing, including functional, parametric, and performance. In a good ASIC flow, final test should only find packaging defects. To uncover design errors with packaged parts is inefficient from both a time and cost standpoint. Marking can be tied in with final test where part number identification is put on good parts prior to shipping.

Finally, finished prototypes get shipped. Several key steps are required in the overall flow, from data input to a CAD system, to finished products shipped to customers. Each section must perform its function in a timely manner. Perhaps more important is the scheduling or tracking that keeps designs from waiting in a queue along the way. ASIC suppliers have development times down to a few weeks with a continuous push to achieve even better times.

The ASIC interface is a complicated system, but it has two things going for it. First it works. Many thousands of part types have been successfully developed and shipped to customers. Second, it is the way of the future. The advantages to a customer's end product more than compensate for the learning curve and effort required to develop ASIC circuits.

CHAPTER 4
VLSI PROGRAMMABLE DEVICES

Vincent J. Coli
Advanced Micro Devices
Sunnyvale, California

4.1 PROGRAMMABLE LOGIC DEVICES

Programmable logic devices (PLDs) are logic circuits that can be programmed by the user to perform specific logic functions. These devices offer logic designers a very attractive alternative to designing with standard 7400 series discrete logic chips. Just like any standard 7400 chip, PLDs are available off your local distributor's shelf. Many of the PLD functions have become more or less industry standard, which means they are available from many vendors and in several technologies with different speed-power-cost options. However, PLDs offer one distinct advantage that standard 7400 discrete logic can never offer: they are user-programmable. This means that the device you purchase is in a virgin state—still unprogrammed like a blank EPROM (erasable programmable read-only memory). Then in a process similar to programming an EPROM, the user programs the PLD to implement a specific logic function.

The two fundamental sections of a PLD circuit are the logic array and the macrocell. The logic array is the user-programmable section of the PLD, which consists of AND gates, OR gates, and inverters. The input signals are routed and operated upon in the logic array. The macrocell, which was added to second-generation PLDs, allows the user to customize the PLD output structure. An example of a macrocell is programmable output polarity. This macrocell is very simple; just add an extra fuse to each output outside of the logic array. The user programs this polarity fuse to configure the output polarity. As an introduction to PLDs, we discuss the logic array structures first, and then give an overview of the popular devices.

4.2 PLD ARRAY STRUCTURES

The main staple of PLDs is made with two arrays of logic gates, an AND array followed by an OR array. Specifically, the input signals to a PLD must first pass through an array of AND gates, where combinations of the input signals are formed. Each group of AND combinations is called a *minterm* in Boolean algebra or a *product line* in PLD nomenclature. Then the product lines are summed in the second array, which is an array of OR gates. The logic signal is now available to the output macrocell. Note that the polarity of the input signals is not a problem, since both the true input signal and its complement are generated by the input buffers. Also, the placement and route problem that plagues gate arrays is never a problem with PLDs, because routing is provided through the array. Therefore, 100 percent routability is always guaranteed in a PLD.

Besides being very efficient for fabrication in silicon, an AND/OR structure is quite convenient because it directly implements logic expressed in sum-of-products form. Sum-of-products form is quite natural for logic designers because, in Boolean algebra, the AND operator has hierarchy over the OR operator. Therefore the logic equations can be generated quite easily by using logic design techniques, such as Karnaugh maps and De Morgan's theorem. The concept of AND/OR logic actually has its roots in Boolean algebra, whereby any logic function can be constructed by using two levels of AND and OR gates. The catch is that the gates can potentially be very large for very dense logic. This is a problem for discrete logic but not for array-based PLDs.

There are actually three basic types of AND/OR array-based PLDs: *programmable read-only memories* (PROMs), *programmable array logic* (PAL) devices, and *programmable logic arrays* (PLAs). These structures are illustrated in Fig. 4.1. Which arrays are programmable distinguishes these three types. In a PROM, the AND array is fixed and the OR array is programmable. PAL devices are just the opposite of PROMs; a programmable AND array precedes a fixed OR array. In a PLA, the most flexible of the three types, both arrays are programmable. Unique advantages are associated with each type.

4.2.1 Programmable Read-Only Memories

PROMs and EPROMs are most often used to store fixed programs and microcode. A PROM is very easy to use in a memory application; apply an address at the inputs, and the stored data are available at the outputs after a brief memory access time delay. However, the PROM structure is also ideal for certain types of logic applications. In fact, PROMs really were used as the first PLD. In order to give PROM memories their own identity as PLDs, logic PROMs are often referred to *programmable logic elements* (PLEs).

Internally PROMs consist of a fixed AND array followed by a programmable OR array (see Fig. 4.1*a*). The fixed AND array actually consists of all combinations of minterms for each input combination. Thus PROMs can implement logic functions by programming the OR array to select the AND gate combinations (or product lines) required for a particular function. For those familiar with memory design, the fixed AND array is often called the address decoder, while the programmable OR array is used to store the memory bits. Another way of looking at this is that PROMs store the logic transfer function as a look-up table in memory.

A PROM has the unique advantage that any combination of inputs can be decoded in a PROM as long as sufficient input pins are provided, since the PROM structure provides 2^n product terms (where n is the number of inputs). The tradeoff is that PROMs are restricted in their number of input pins, because the array size must be doubled to accommodate an additional input. The arithmetic works like this: a PROM with n inputs and m outputs requires a memory size of

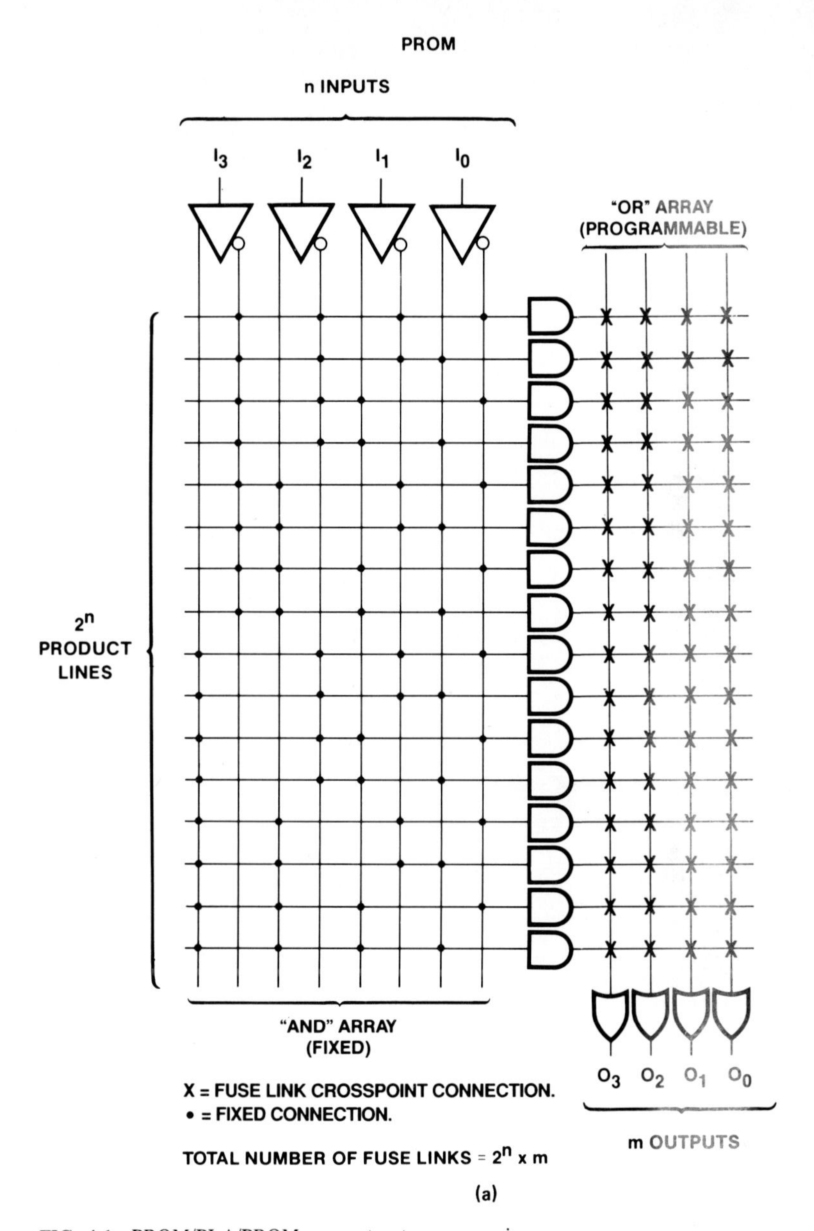

FIG. 4.1 PROM/PLA/PROM array structure comparison.

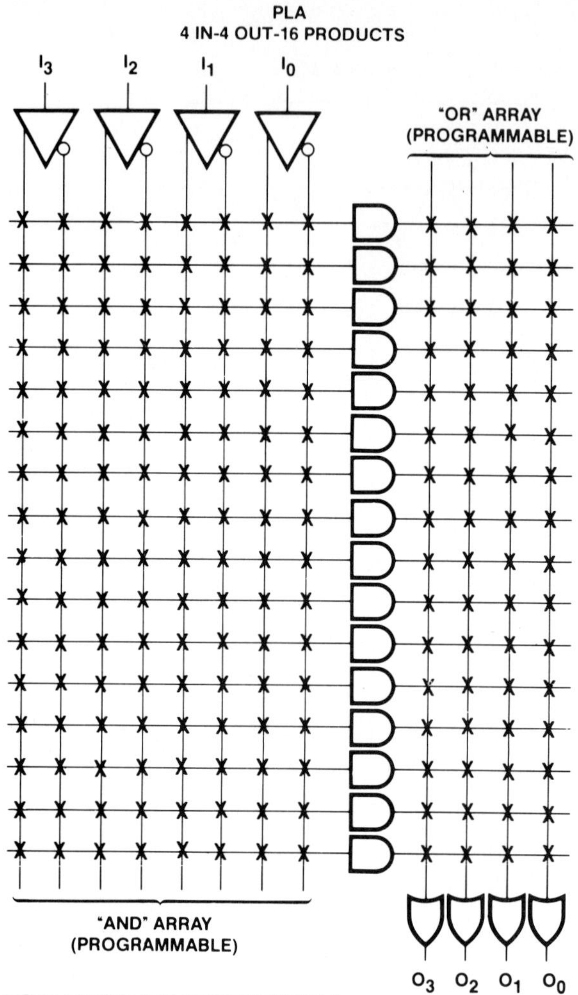

(b)

FIG. 4.1 (*Continued*)

2^n words or an OR array of 2^n lines deep by m lines wide. This translates into an OR array size of $2^n \times m$ fuse links or E-cells. For example, a PROM with 10 inputs and 8 outputs requires an OR array of $2^{10} \times 8$, or 8192 fuse locations. An eleventh input would require that the array size be doubled to 16,384. Unfortunately, the PROM structure must allocate a sufficient number of product lines for all combinations of input variables, even though a particular function may only require one product line. Cost and performance constraints limit PROMs to 13

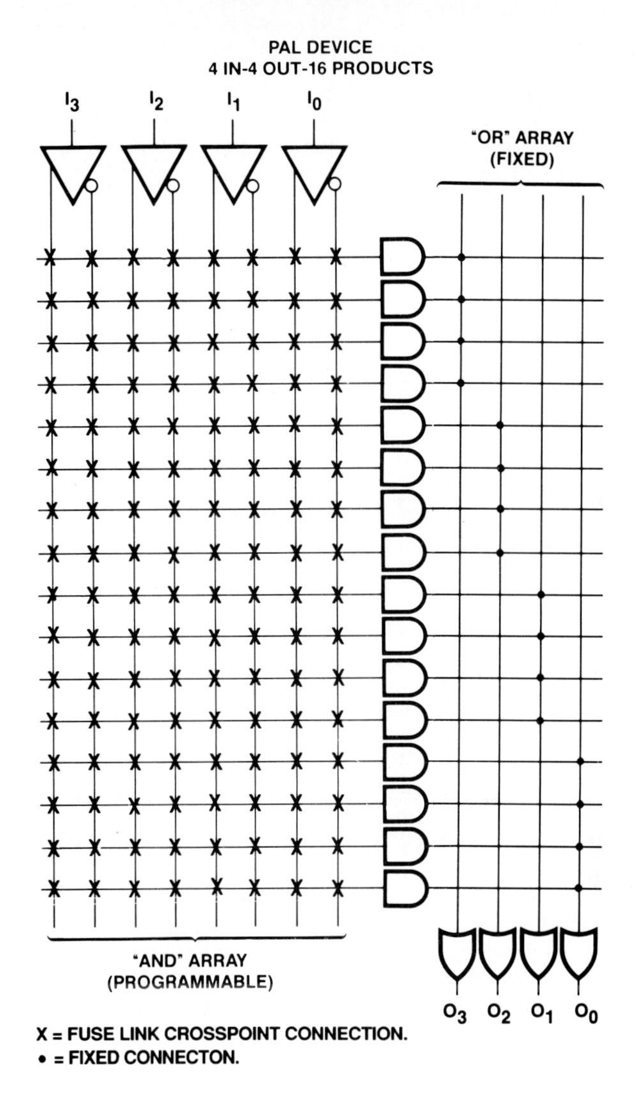

FIG. 4.1 (*Continued*)

inputs and 8 outputs. However, PROMs are very attractive for applications requiring less than 10 inputs, especially when many product lines are required.

PROMs are widely available in depths ranging from 32 locations (5 inputs) to 8192 locations (13 inputs), with either 4 or 8 outputs. A new breed of PROMs designed specifically for logic applications features either 5 or 6 inputs and 16 outputs; the large number of outputs leverages off of the PROM's abundance of product lines.

4.2.2 Programmable Logic Arrays

As stated before, the PLA structure offers the highest level of flexibility, with both arrays being programmable (see Fig. 4.1*b*). Programmability in the AND array removes the restriction found in PROMs that the AND array must be large enough to provide all possible input combinations. The reason is that different input combinations can be programmed into the product lines, and logic minimization techniques can be used to eliminate redundant combinations. Therefore almost any combination of inputs can be decoded in a PLA. Another way of looking at this is with the visual aid of a Karnaugh map. An entry of 1 in a cell is used to signify a desired input combination, 0 signifies an unwanted input combination, and "X" signifies an irrelevant or "don't care" input combination. As detailed in the Karnaugh map in Fig. 4.2, a circle drawn around the largest group of adjacent cells that contains 1s and Xs, but not 0s, defines a minimized product line.

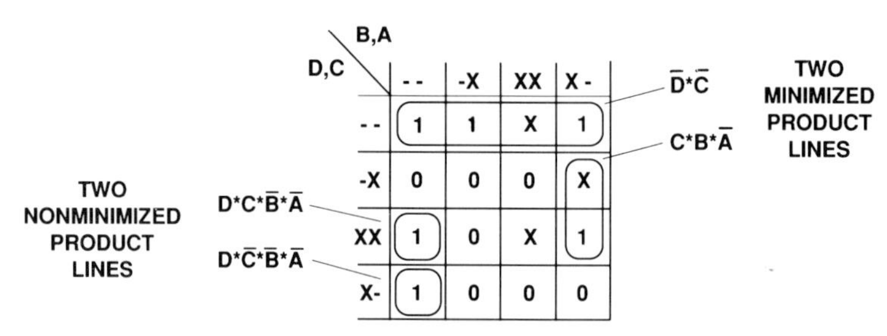

FIG. 4.2 Karnaugh map.

Programmability in the OR array offers a higher level of flexibility because product lines can be shared between outputs. For example, one product line would be saved if the same input combination (i.e., product line) were required by two outputs. Also, programmability in the OR array allows the user to assign varying numbers of product lines to each output as required.

In the world of engineering, there are always compromises. The cold facts reveal that a hefty performance and die size penalty must be realized to provide the flexibility of programming both arrays. PAL devices are generally 5 to 10 ns faster than PLAs at the same power level, and they save the silicon area required to program and verify the second array. It turns out that the flexibility of a programmable OR array is not required for most PLD applications, but it can be very useful for complex state machine and sequencer applications.

PLAs were really the first products to be offered specifically for logic applications. Due to programming limitations, early PLAs were available only in mask-programmed versions. Just like a ROM, a logic designer would indicate on the vendor's PLA AND-OR logic map where the desired connections were to be made. The vendor would then tool up a custom metal mask for the PLA to implement the customer's logic. Most of the PLAs used today are programmable by the user, just like any other PLD. However, mask-programmed PLA structures are often used in the control section of LSI-VLSI standard logic chips, such as microprocessors, and are offered in standard cell libraries.

Because of the long history of PLAs, their nomenclature can sometimes be a little confusing. Early vendors of user-programmable PLAs called their products

FPLAs to highlight their field-programmability feature and to distinguish FPLAs from factory mask-programmed PLAs. Just as ROMs and PROMs could be easily distinguished, so could PLAs and FPLAs. However, since most of the PLAs offered today are programmed by the customer, many vendors have dropped the "F" prefix and simply call them PLAs. Furthermore, PLAs designed for sequencer application are called *PLSs* (programmable logic sequencers).

4.2.3 Programmable Array Logic Devices

The structure of PAL devices is a mirror image of that of the PROM structure. Now the AND array is programmable and the OR array is fixed (see Fig. 4.1*c*). This reversal of programming arrays in PAL devices immediately relieves the input restriction associated with PROMs. Therefore, just as in PLAs, only the desired input combinations are programmed, rather than every possible combination as with PROMs.

In a PAL device, each output has an OR gate associated with it, which is the summing junction for a fixed number of product lines. Statistically there is only a limited number of product terms in any equation. Most PAL devices offer eight product lines per output.

Because the PAL structure offers a good balance of performance, flexibility, and die size, the PAL array structure has become the industry standard PLD structure. Therefore PAL and PLD are often used synonymously. Several PLD vendors add an "E" prefix to PLD, to come up with EPLD, which signifies ultraviolet-erasable PLDs. Just as there are PROMs and EPROMs, now there are PLDs and EPLDs. Also, the name *HAL* (hard array logic) refers to mask-programmed, or ROM, versions of PAL devices.

4.3 THE ORIGINAL PAL DEVICE FAMILY

The first family of PAL devices consisted of 13 members, each housed in 20-pin DIP (dual in-line) packages. Nine of these original PAL devices, often called the small PAL devices (see the first and second rows in Fig. 4.3), are strictly combinatorial arrays, while the four big brothers in the family (bottom row in Fig. 4.3) also include registered and combinatorial functions.

The small PAL devices are fairly simple; they merely consist of the programmable AND array followed by the fixed OR array, with no registers or macrocells. Each device type is distinguished by the number of inputs, number of outputs, and output polarity. The PAL10H8, for example, provides 10 inputs, 8 outputs, and active-high output polarity, while the PAL16L2 provides 16 inputs, 2 outputs, and active-low output polarity. The output drivers conform to the low-power Schottky TTL electrical characteristic for totem-pole drivers (e.g., low-level output current I_{OL} = 8 mA).

The beauty of the small PAL devices is their simplicity. It is a very straightforward process to replace a handful of SSI gates with these PAL devices. A good example of this simplicity is an address decoding application where logic generates chip selects for memory components by decoding several microprocessor addresses (see Fig. 4.4). A standard 3-to-8 decoder-demultiplexer (74S138) and additional SSI OR gates could do the job. On the other hand, a single PAL device can be programmed to perform the required decoding function with no ad-

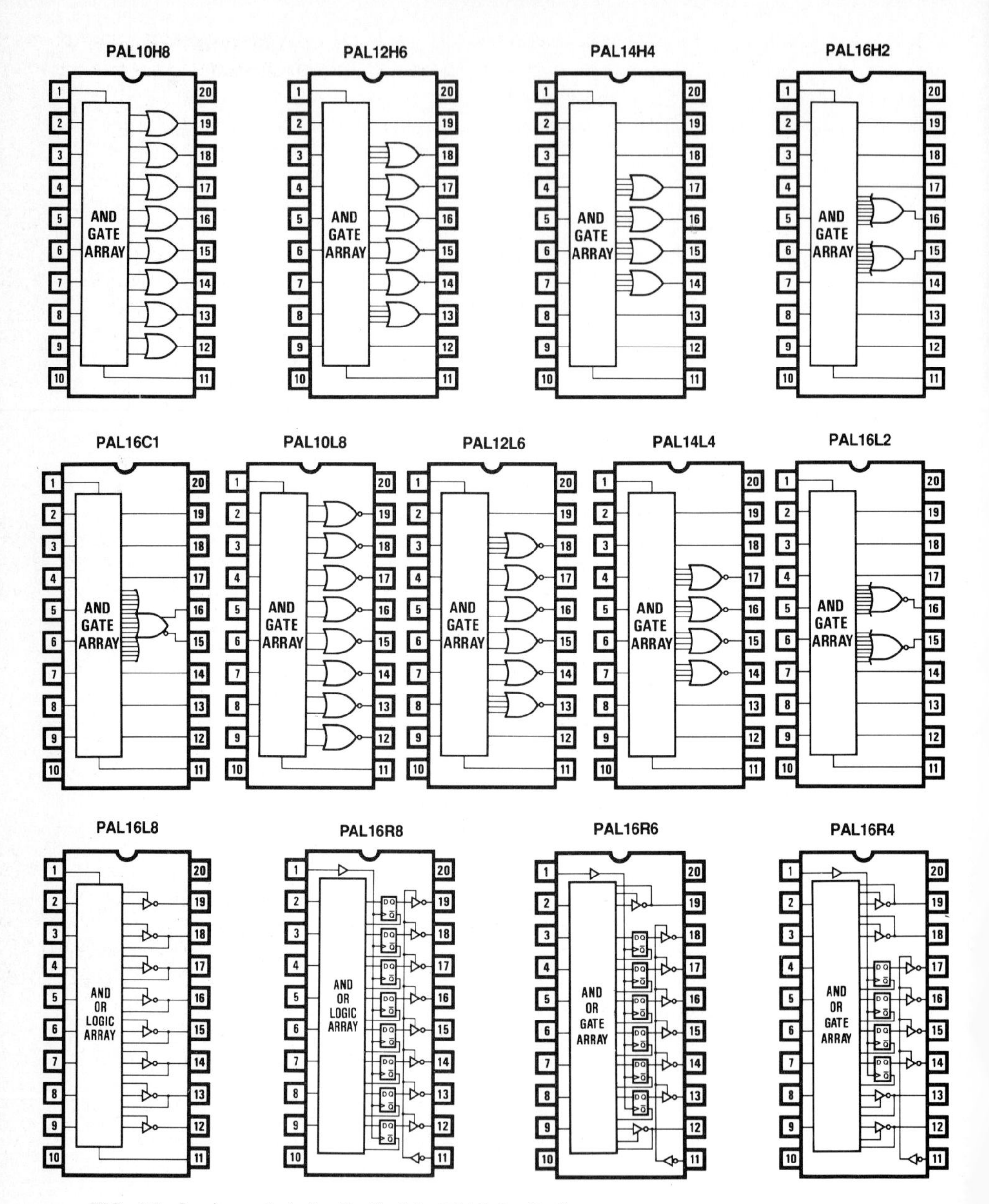

FIG. 4.3 Logic symbols for the "original PAL family."

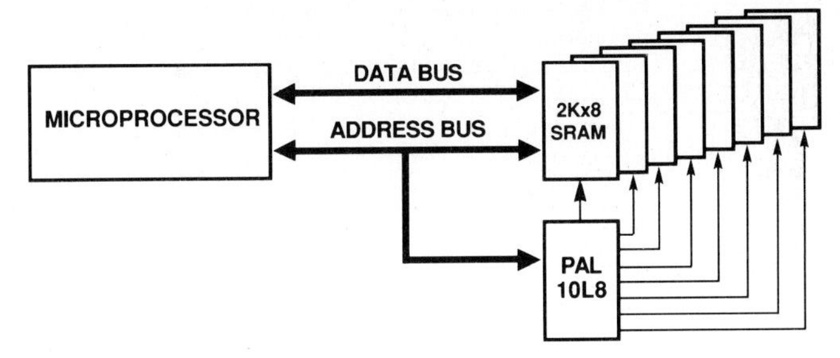

FIG. 4.4 Address decoding application for a PAL10L8.

ditional gates. All that the designer needs to do is enter the desired memory map into the PLD development software, and the software and programmer do the rest. The PAL solution is faster, uses less power, and is just one chip. Furthermore, system upgrades are easy to accommodate; just program the new memory map into another PAL device rather than changing chips and rewiring the PC board in the discrete TTL approach.

The four larger members of the original PAL device family offer a larger AND array than do previous PAL devices, and they feature a more sophisticated output structure. Each output is either combinatorial, with a programmable input-output (I/O) macrocell, or registered, with registered feedback. The PAL16L8 offers 10 inputs and 8 combinatorial outputs, while the PAL16R8 offers 8 inputs and 8 registered outputs. Actually, the remaining two members of the original family are crosses between the previous two members, differing only in the number of combinatorial versus registered outputs. The PAL16R6 offers six registered and two combinatorial outputs, while the PAL16R4 offers four of each type. Regardless of the device type, each of these PAL devices offers eight product lines per output.

4.3.1 The PAL16L8—The Most Popular PAL Device

The PAL16L8 logic diagram is shown in Fig. 4.5. Notice how the pins on the left side and bottom of the logic diagram (pins 1 to 9 and pin 11) are used for inputs, and the pins on the right (pins 12 to 19) are available as outputs. Actually six of the outputs (pins 13 to 18) are also available as inputs via the feedback line connection after the inverting output buffer. This feature, called *programmable I/O,* allows the user to program each of these six pins to be either an input or output. We will discuss how programmable I/O works in more detail later on. Now the PAL16L8 part numbering scheme should be a little more obvious; "16" signifies the maximum number of potential inputs (10 dedicated inputs and 6 programmable I/O), "8" signifies the number of outputs, and "L" signifies the output type, which is active low for this PAL part type.

In Fig. 4.5, the vertical lines running through the array, numbered 0 through 31, are the input lines. Notice that each input or I/O pin is associated with two input lines. One input line is connected to the true (or noninverted) output of the input buffer, while the other input line is connected to the complement (or inverted) output. This allows availability of both input signal polarities to the array.

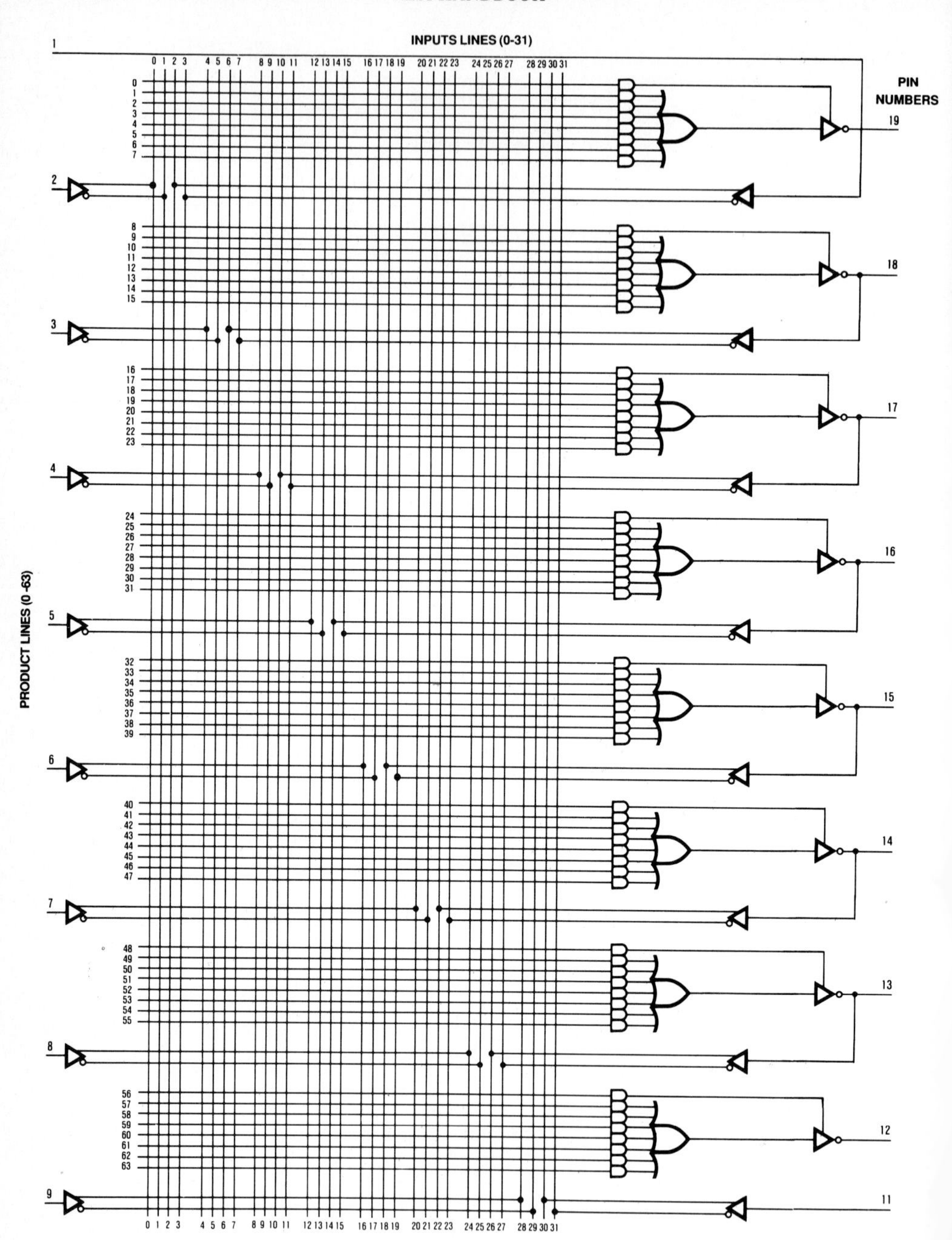

FIG. 4.5 PAL16L8 logic diagram. (Note: Fuse links not shown.)

The horizontal lines running through the array, numbered 0 through 63, are the product lines. Each of these product lines can be thought of as an AND gate with 32 inputs, which corresponds to the total number of input lines. Actually, both the true and complement of every input signal are connected via fuses to each product line before the device is programmed. This is the programmable AND array in the PAL structure. The user selects different combinations of input signals by disconnecting, via the blown fuse, the unwanted input signals in a particular product line. In total there are 2048 fuses available in this PAL device (64 product lines $\times$ 32 input lines = 2048 fuses).

Notice that each output pin has eight product lines associated with it. The lower seven product lines of each group are summed at the OR gate, while the upper product line is connected to the inverting output buffer. The lower seven product lines and the OR gate provide the sum-of-products logic power for the PAL device. The OR gate determines if any of the product lines are active, or true, and then the output buffer inverts the signal from the OR gate for the output. Note that unused product lines, or product lines with all fuses left intact, are not a problem because the logical result of each input ANDed with its complement will be false.

4.3.2 Programmable I/O

The programmable I/O upper product line associated with each output actually controls the three-state logic in the output buffer (see Fig. 4.6 for details). When this product line is active, or true, the output is enabled and the sum-of-products logic determines the output state. However, when this product line is inactive, or false, the output is disabled, with the three-state buffer in the high-impedance state. This allows the output pin to drive a three-state bus, just like a 74S240 octal buffer. Furthermore, since most PAL devices feature an output drive capability of 24 mA, they are quite handy for bus-driving applications.

The three-state product line, along with the feedback path on six of the outputs, makes the programmable I/O feature work. The pin is an input to the AND array when all the fuses in the three-state enable product line are left intact, while the pin is an output when all of the fuses are programmed. Note that a product line will always be true, regardless of input combinations, when all fuses are programmed. The programmable I/O feature lets the user allocate pins for input or output as required by the application.

An even higher level of flexibility is possible by letting the logic in the product line determine the direction of the pin. This is done by programming a condition in the product line for which the pin will be an output. This feature can be used to allocate available pins for I/O functions or to provide bidirectional transfer for operations such as shifting and rotating of data.

The flexibility of the programmable I/O feature coupled with eight product lines per output and high performance makes the PAL16L8 the most popular PAL device. The PAL16L8 is used in complex decoders, encoders, multiplexers, comparators, and replacement of SSI-MSI random logic. Another way of viewing this is that the PAL16L8 programmable AND array contains 2048 fuses. These fuses may be programmed to create almost any configuration of up to 250 AND, OR, and inverter gates, which is roughly 250 equivalent gates.

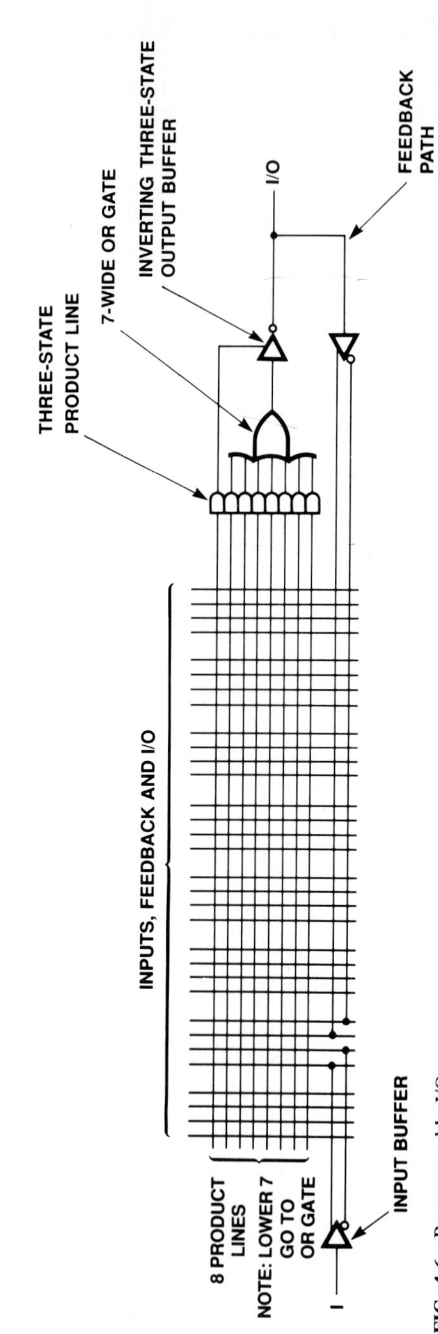

FIG. 4.6 Programmable I/O.

4.3.3 PAL Devices With Registered Outputs and Feedback

The structure of the registered PAL devices is very similar to that of the PAL16L8, except for the registered outputs and feedback. All eight of the outputs in the PAL16R8 are registered with feedback. Each of these registers is actually a *D* (data) flip-flop clocked on the rising edge (see Fig. 4.7). The clock signal, common to all eight flip-flops, comes from pin 1. Each OR gate, which sums eight product lines, is the *D* input to the flip-flop. The *Q* output from the flip-flop is available both for feedback into the PAL array and for output from the device. Either polarity of the feedback signal is available. This feedback allows the PAL device to "remember" the previous state, and it can alter its function based upon that state. This makes registered PAL devices ideal for implementing single-chip state sequencers and state machines.

The registered output structure can implement regular state sequencers such as counters and shift registers. But, more important, this structure can implement random state sequencers, or state machines, that are not available as standard functions. For example, a PAL device can be programmed as a state sequencer that executes such elementary functions as count up, count down, skip, shift, and branch. A more common application is to implement a classic state machine. In a classic state machine, the present state (or output) is a function of both the present inputs and the previous state (stored in the registers).

Registered PAL devices are ideal for implementing both Mealy and Moore types of classic state machines. The structure of the PAL16R4 closely resembles that of the Mealy type in which the four output registers store the state (Q) while the four combinatorial outputs generate the control outputs (Z), as shown in Fig. 4.8*a*. The structure of the PAL16R8 closely resembles that of the Moore type in which the output registers do double duty to store both the state and control output ($Q = Z$), as shown in Fig. 4.8*b*.

Another class of applications for registered PAL devices is that of synchronizing or pipelining combinatorial logic. Here the PAL array is used to implement combinatorial logic, while the registers ensure that the output signals change synchronously with a clock. This is required for critical timing applications where glitches or potential race conditions in the logic can create havoc for the system. Furthermore, the registers can be used to pipeline control signals or overlap delays from the PAL device to achieve higher system performance. Registered PAL devices are great for these applications because there is no penalty in chip count realized for the registers.

Each registered output also features an inverting three-state buffer controlled by an active-low enable signal on pin 11. The active-low output polarity makes these PAL devices convenient for implementing active-low control logic. Also, since each output can drive up to 24 mA of current, these PAL devices can interface directly with many control busses.

In addition to the 20-pin PAL devices, the PAL product line has been expanded to include 24-pin versions with four additional input pins. This 24-pin family consists of the combinatorial PAL20L8 and registered PAL20R8, PAL20R6, and PAL20R4. These devices offer a higher level of integration with very little additional board space, since they come in the space-saving 24-pin 0.3-in-wide DIP package.

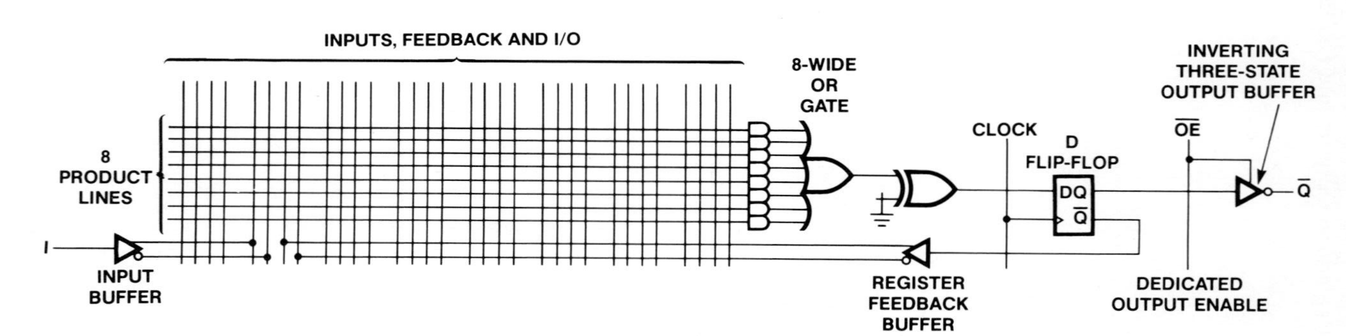

FIG. 4.7 Registered output.

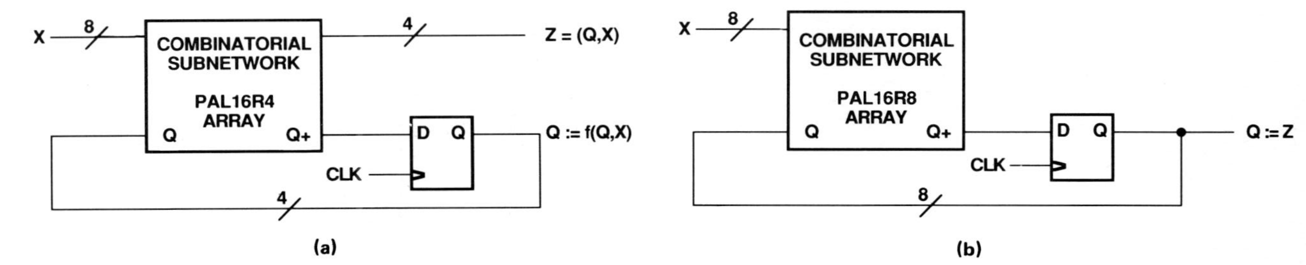

FIG. 4.8 (*a*) PAL16R4 = Mealy-type state machine; (*b*) PAL16R8 = Moore-type state machine.

4.3.4 PAL Devices With Exclusive-OR Gates

This family of four PAL devices parallels that of the PAL16L8/R8/R6/R4 family, but these devices feature an exclusive-OR function. Offered in 24-pin 0.3-in-wide DIP packages, the array size for this family is expanded to support 10 inputs and 10 outputs (either combinatorial with programmable I/O or registered with the exclusive-OR gate and feedback). All 10 of the PAL20X10 outputs are registered, while all 10 of the PAL20L10 outputs are combinatorial. The PAL20X8 with eight registered and two combinatorial outputs and the PAL20X4 with four registered and six combinatorial outputs round out the family.

The exclusive-OR gate output structure (shown in Fig. 4.9) is based upon a *D* flip-flop just as the registered PAL devices are, but now a two-input exclusive-OR connects to the *Q* input of the flip-flop. Also, each output has two OR gates and four product lines associated with it. The four product lines are segmented into two pairs. Each pair of product lines is summed at an OR gate. The two OR gates are then summed at the exclusive-OR gate.

The exclusive-OR PAL devices are handy for multifunction counter applications. For example, the PAL20X10 can implement a 10-bit up/down counter while the PAL20X8 can implement an expandable 8-bit counter (the two programmable I/Os are used for the carries). Both counters can include functions such as count enable, parallel load, preset, or clear. Or the counting sequence itself can be customized to count in BCD (binary-coded decimal) order, count by 2, count up and roll over to zero at the terminal count, or count up and then commence counting down at the terminal count.

You might ask what makes the exclusive-OR PAL devices so special for building counters, since the registered PAL devices can do counting functions also. The exclusive-OR PAL devices can build large counters quite efficiently. Ordinarily the size of a counter in a PAL device is limited by the number of product lines available, because counters, constructed using *D* flip-flops, require many product lines to implement the hold function required for counters. We briefly review how a counter works. In a binary up counter, a typical counter bit toggles (or counts up) only when all of the lesser significant bits are 1s; otherwise the counter will hold state. This process is exemplified with a 3-bit binary up counter in Fig. 4.10.

The toggle function is easy to implement in any registered PAL device; just AND the complement of the counter bit (inverted feedback from the corresponding flip-flop) with the true of all of the lesser significant counter bits. Therefore the toggle function consumes one product line regardless of the size of the counter. All that grows with counter significance is the number of inputs to the AND gate.

However, the hold function is more difficult to implement. The hold function must inhibit the counter bit from toggling when *any* of the lesser significant bits in the counter are zero. This is done by individually ANDing the true of the counter bit (noninverted feedback from the corresponding flip-flop) with each of the lesser significant counter bits. Therefore the hold function consumes one product line for each lesser significant counter bit. Thus, each additional bit in a counter requires one more product line than the previous bit does. The eighth bit, for example, requires all eight product lines available in a PAL16R8 just to implement the hold function! Fortunately, the coincidence capability of the exclusive-OR gate simplifies counter construction by enabling one product line to implement the hold function regardless of the counter stage.

Recall that an exclusive-OR gate behaves like an OR gate except when both

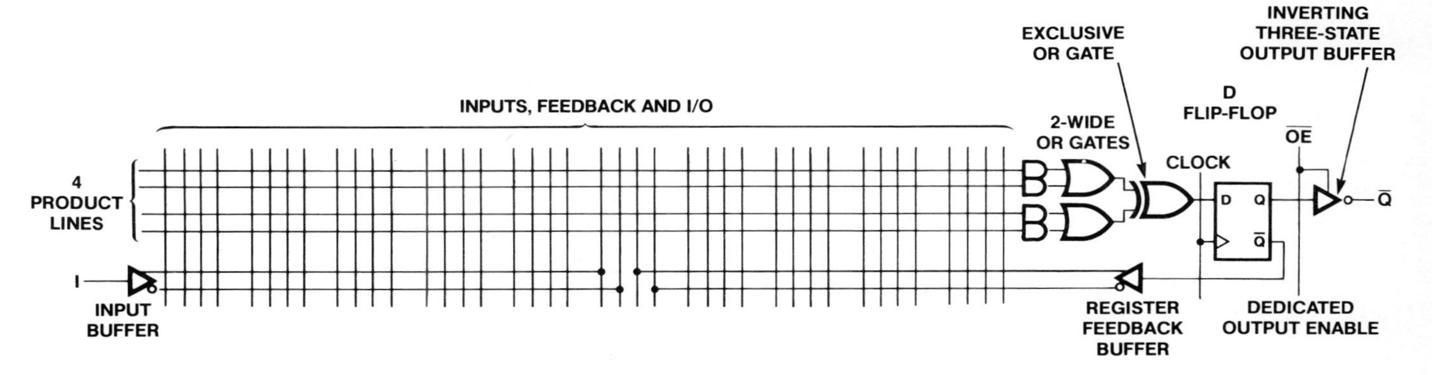

FIG. 4.9 Exclusive-OR gate output.

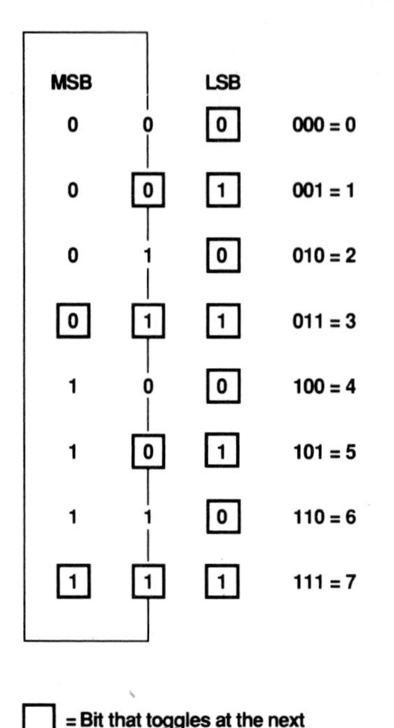

FIG. 4.10 Three-bit binary up counter example.

inputs are active simultaneously, in which case the output will be low while the OR gate would be high. (This is why the exclusive-OR gate is sometimes called the coincidence operator.) When an exclusive-OR gate is teamed up with a *D* flip-flop, something wonderful happens. The *D* flip-flop behaves like a *T* (toggle) flip-flop, whereby the state of the flip-flop will toggle on the rising edge of the clock if a set of conditions is satisfied. For a counter, this means that a particular flip-flop in the counter will toggle its state, or count, if (the condition) all of the lesser significant bits in the counter are set. If that condition is not met, then the flip-flop will not toggle and that particular bit in the counter will hold. Therefore, with this technique, only two product lines on each input to the exclusive-OR gate are required: one product line to count (or toggle), and the other product line to hold (inhibit toggle). Furthermore, only two product lines are required, regardless of the stage in the counter.

It is safe to say that *T* flip-flops are not only more efficient for counters, but they also simplify design.

Of course, the exclusive-OR PAL devices can be used in applications that require 10 outputs, but the exclusive-OR gate is not wanted. This is done by making the exclusive-OR gate behave like an OR gate using logic manipulations. Just make sure that the conditions to assert each of the four product lines are mutually exclusive or that the product line circles in a Karnaugh map do not overlap.

4.4 POPULAR PLD MACROCELL FUNCTIONS

4.4.1 What Is a Macrocell?

Until now, programmability for all PAL devices discussed was limited to the AND array. In other words, the user configures a PAL device by programming a fuse link in the AND array to perform a desired logic function. The user selected different output structures by choosing a different PAL device (e.g., PAL16R6 or PAL16R8). In this section we discuss several new PAL devices that include fuse links to control logic gates around the output structure. Now the user can select one device and program, e.g., the output to be registered or combinatorial or a *J-K* flip-flop versus a *D* flip-flop.

The four PAL devices included in this section were chosen because they are fairly representative of the new breed of PAL devices available with macrocells.

Even with this limitation, the discussion is limited to the major features of these devices. Please refer to the manufacturer's data sheets and application notes for a complete discussion of each product.

4.4.2 Programmable Output Polarity

Programmable output polarity allows the designer to select the polarity for each output. This feature can be very handy for fitting logic functions that would have otherwise exceeded available product lines, by applying De Morgan's theorem and programming the polarity fuse appropriately.

Programmable polarity is implemented by using an exclusive-OR gate situated after the OR gate and before the output buffer (for a combinatorial output) and the register (for a registered output) as shown in Fig. 4.11. The polarity fuse connects the bottom input of the exclusive-OR gate to ground when the fuse is intact. That input will be pulled high when the fuse is programmed. When the output polarity fuse is blown, the lower input to the exclusive-OR gate is high, so the output is active high. Similarly, when the output polarity fuse is left intact, the output is active low.

4.4.3 The PAL22V10—The First Advanced Macrocell

In addition to the programmable polarity feature already described, the PAL22V10 macrocell (shown in Fig. 4.12) also includes a programmable flip-flop bypass feature. Programmable flip-flop bypass enables any output to be combinatorial by bypassing the output register. This is done by setting the fuse that determines where the output multiplexer selects its data. Any combination of polarity and bypass is permissible for each output, as shown in the table in Fig. 4.12.

The PAL22V10 introduced another new feature to programmable logic, called *varied product term distribution*. This means that each pair of macrocells has a different number of product lines associated with it. There are two macros each with 8, 10, 12, 14, or 16 product terms to allow optimal use of the available product term resources.

4.4.4 PAL32VX10—High-Density 24-Pin PAL Device

The PAL32VX10 is the highest-density member of the PAL family to be housed in a 24-pin 0.3-in-wide DIP package. Often high density is linked with high-pin-count devices. However, it is possible to implement a high-density function within the I/O constraints of a space-saving 24-pin package. This is the philosophy behind the PAL32VX10 architecture. Key features like programmable flip-flops, which expand functionality, and dual independent feedback, which conserves precious I/O pins, help to achieve high density.

In addition to the familiar PAL array structure, the PAL32VX10 features 10 highly flexible I/O macrocells that are user configurable for combinatorial or registered operation. Each flip-flop is programmable to be either *J-K, S-R, T,* or *D* types for optimal design of state machines and other synchronous logic. A unique dual-feedback architecture allows I/O capability for each macrocell in both combinatorial or registered configurations, even when register feedback is

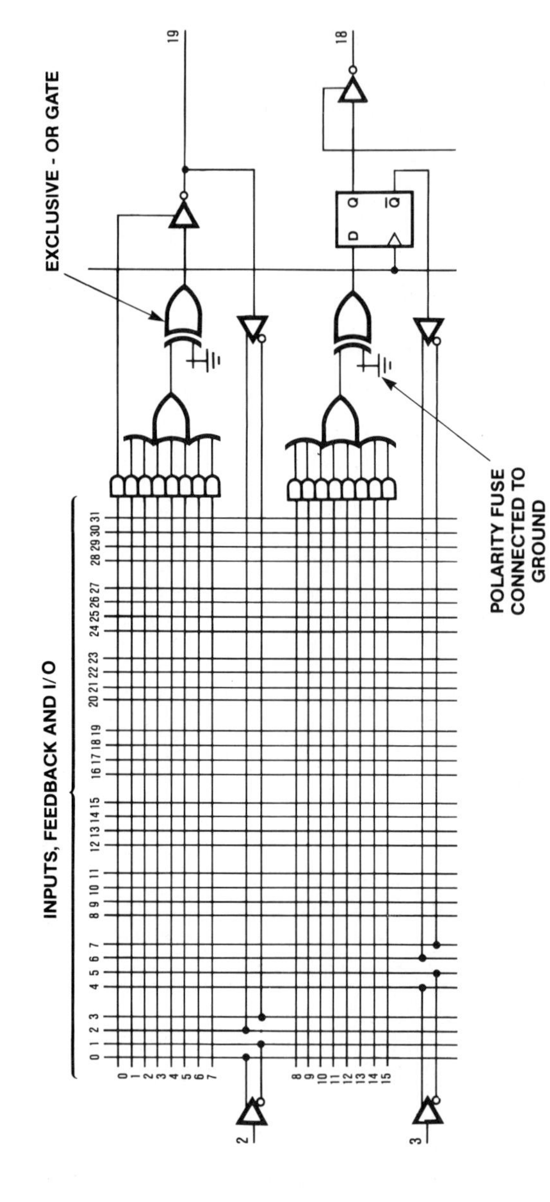

FIG. 4.11 Programmable output polarity.

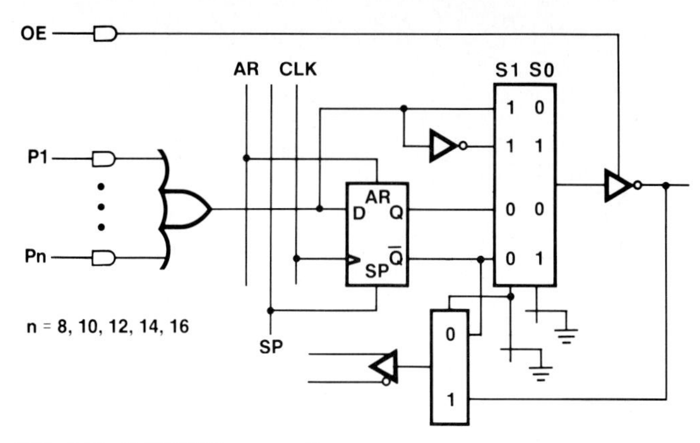

FIG. 4.12 PAL22V10 macrocell.

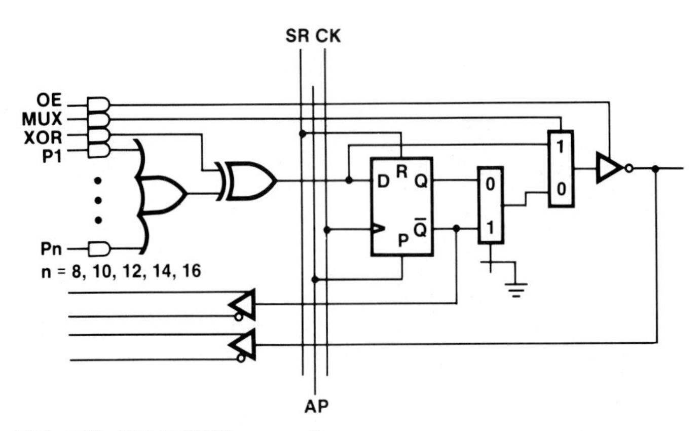

FIG. 4.13 PAL32VX10 macrocell.

present, and allows implementation of buried registers while preserving the macro input function.

The PAL32VX10 is a pin-compatible functional superset of the PAL22V10 architecture, with an improved macrocell. Details of the VX macrocell are given in Fig. 4.13. Several of the many possible VX macrocell configuration options are given in Figs. 4.14 and 4.15. Following are some significant features of the macrocell.

4.4.5 Programmable Flip-Flops (*J-K, S-R, T, D*)

Each macrocell has an exclusive-OR gate that provides the capability of emulating the operation of *J-K*, *S-R*, *T*, or *D* flip-flops. When *J-K* or *S-R* flip-flops are implemented, the available product term resources for each macrocell are shared between *J(S)* and *K(R)* functions, according to the expression

$$Q := Q \oplus (J_1 \cdots J_m) * \overline{Q} + (K_1 \cdots K_n) * Q$$

$\overline{J}$-$\overline{K}$ Flip-Flop

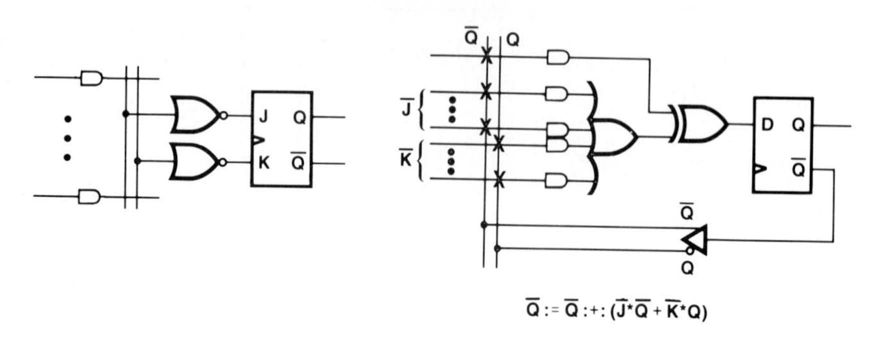

J-$\overline{K}$ Flip-Flop

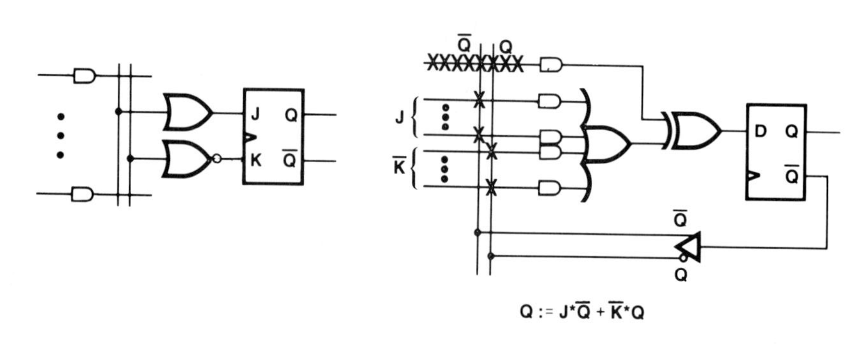

$\overline{J}$-K Flip-Flop

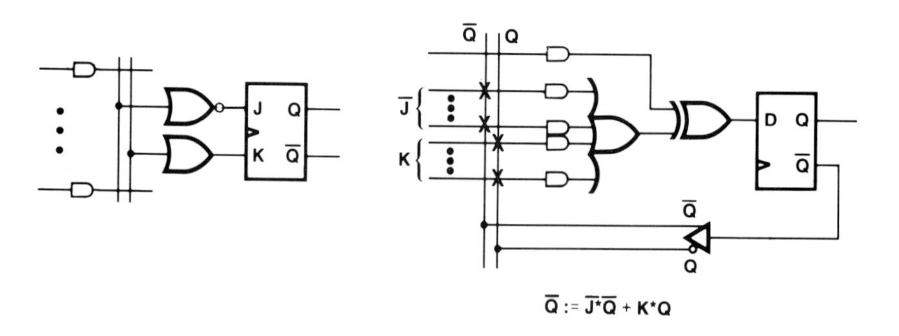

Logic Equivalent **Implementation**

FIG. 4.14 Several VX macrocell configuration options.

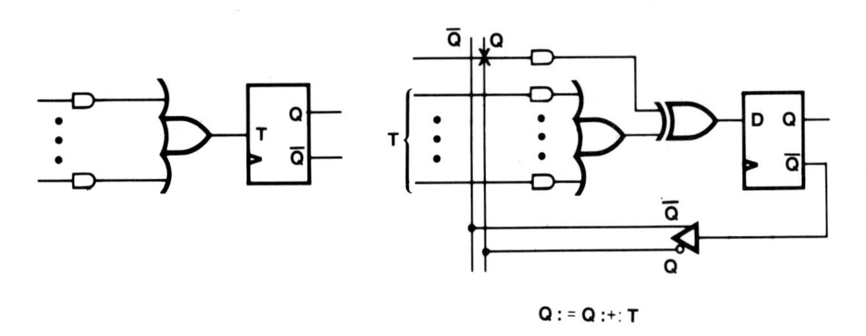

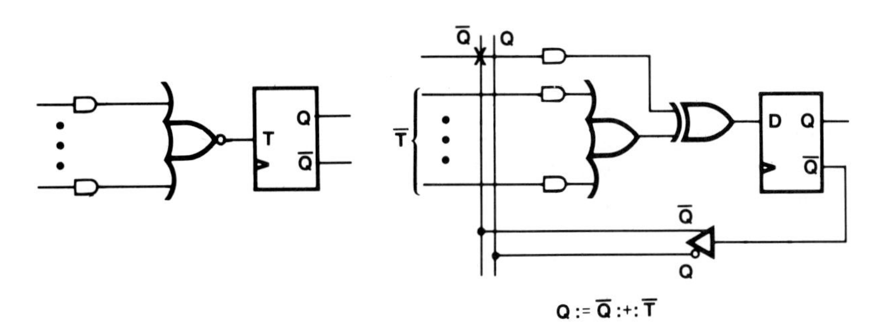

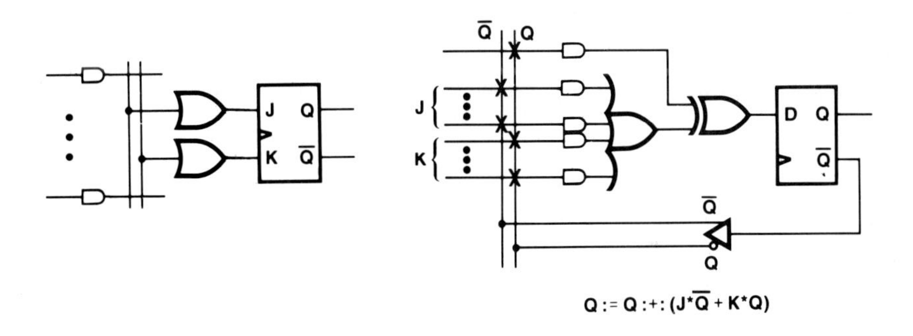

FIG. 4.14 (*Continued*)

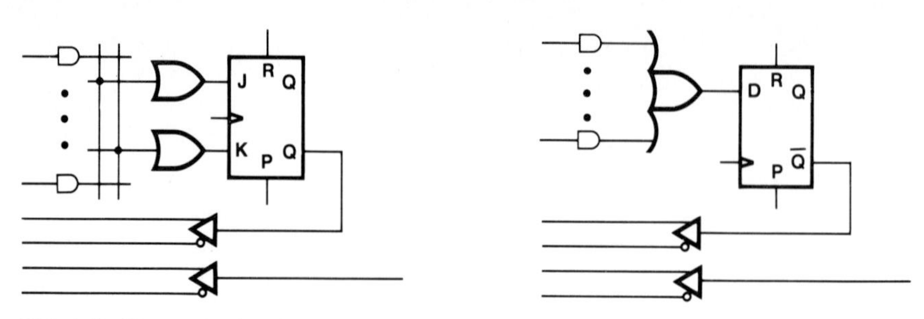

FIG. 4.15 Two examples of VX macrocells with buried registers and dedicated inputs.

where $m + n$ is the total number of product terms found in each macrocell, := denotes replaced by following the clock, and **1** denotes the exclusive-OR function operator. Several flip-flop programming options are offered in Fig. 4.14*a* and *b*.

4.4.6 Dual Independent Feedback

Each macrocell has two completely independent feedback paths. One feeds back into the array from the macrocell output pin. The second feeds back into the array from the flip-flop output. This architecture provides I/O capability for both combinatorial and registered outputs, even when register feedback is needed. In addition, it allows registers to be loaded directly from outputs and provides the capability for buried registers, while permitting use of the macrocell as an input. Two example configurations of VX macrocells with buried registers are presented in Fig. 4.15.

4.4.7 PAL20RA10—Registered-Asynchronous PAL Device

The PAL20RA10 is the industry's first registered PAL device with extensive asynchronous capabilities. Previous PAL device architectures were classified as synchronous (i.e., registered) or asynchronous (i.e., combinatorial). However, the logic content of real-world systems cannot be classified so cleanly. In reality, designers combine synchronous and asynchronous logic in the same system to obtain the most optimal solution. Therefore a programmable logic device, which also combines synchronous and asynchronous capabilities, greatly eases system design.

There are actually two members of the registered-asynchronous PAL family: PAL20RA10, which is available in a 24-pin 0.3-in-wide DIP package, and PAL16RA8, which is available in a 20-pin 0.3-in-wide DIP package. The slightly smaller PAL16RA8 has two fewer RA macrocells than does the PAL20RA10, which has 10 RA macrocells.

As with all other PAL devices, the PAL20RA10 contains a programmable AND, fixed OR array that the user can configure to implement a wide variety of logic functions expressed in sum-of-products form. However, the PAL20RA10 is unusual in that the user can configure not only the internal logic transfer function but also can implement logic functions to control the clock for each register, the output enable, and both set and reset operations for each register. In addition, the user can configure the outputs to be registered or nonregistered, active high or active low. These features are described in detail here.

4.4.8 Programmable and Hard-Wired Three-State Outputs

The top product line (see Fig. 4.16) is logically ANDed with an output enable pin ($\overline{\text{OE}}$ on pin 13) to control the three-state buffer. The output is enabled if this product line is high and the output enable pin is low; otherwise, the output is disabled. The output enable alternatives are illustrated in Fig. 4.17.

4.4.9 Programmable Clock

The second product line from the top (see Fig. 4.16) controls the clock input to the output flip-flop. The output flip-flop will receive new data from its *D* input when an input to the clock product line causes a transition from a low to a high state (i.e., a positive edge). This feature allows for asynchronous clocking for each output flip-flop, clock enables, and ripple carries.

4.4.10 Asynchronous Set-Reset

The third product line (see Fig. 4.16), when high, will cause the output flip-flop to be reset to a low state asynchronously, while the fourth product line, when high, will cause the output flip-flop to be set to a high state asynchronously. The output flip-flop will be bypassed (i.e., become combinatorial) when both of these product lines are high (i.e., reset and set are active simultaneously). The output flip-flop functions as a *D* flip-flop when both of these product lines are low.

4.4.11 Programmable Output Polarity

The output polarities are programmable to be either active high or active low. This is represented by the exclusive-OR gate before the output register. The output polarity fuse is represented as a pull down to ground. When the output polarity fuse is blown, the lower input to the exclusive-OR gate is high, so the output is active high. Similarly, when the output polarity fuse is left intact, the output is active low.

The bottom four product lines are logically summed at an OR gate. These four product lines implement the sum-of-products expression. Also, register preload is provided for logic verification via $\overline{\text{PL}}$ on pin 1. Four possible RA macrocell configurations are illustrated in Fig. 4.18. The PAL20RA10 logic diagram is given in Fig. 4.19.

4.5 NEW DEVELOPMENTS IN PLD ARCHITECTURES

Until now, all the PLDs that have been discussed employ the AND-OR array structure. Apparently this structure, coupled with the appropriate macrocells, by and large addresses PLD user needs. However, some PLD vendors are experimenting with radically new approaches in the hope of attracting new PLD users. Three new architectural developments will now be introduced.

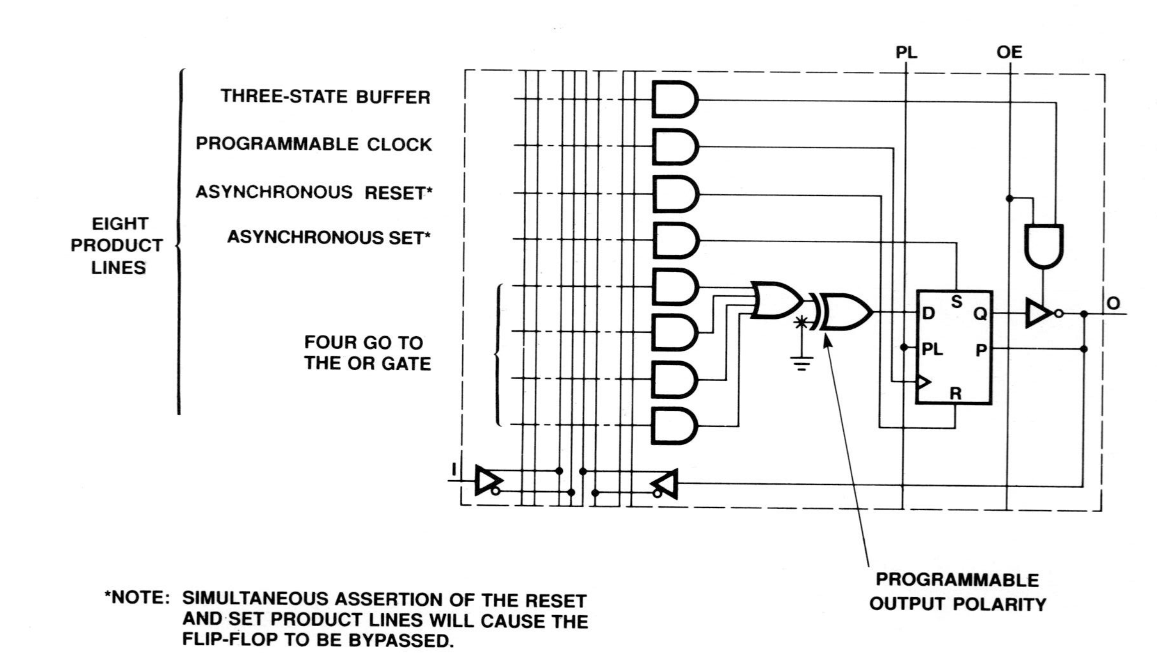
PL
OE
THREE-STATE BUFFER
PROGRAMMABLE CLOCK
ASYNCHRONOUS RESET*
ASYNCHRONOUS SET*
EIGHT PRODUCT LINES
FOUR GO TO THE OR GATE
D
S
Q
PL
P
R
O
I
PROGRAMMABLE OUTPUT POLARITY
*NOTE: SIMULTANEOUS ASSERTION OF THE RESET AND SET PRODUCT LINES WILL CAUSE THE FLIP-FLOP TO BE BYPASSED.

FIG. 4.16 RA macrocell.

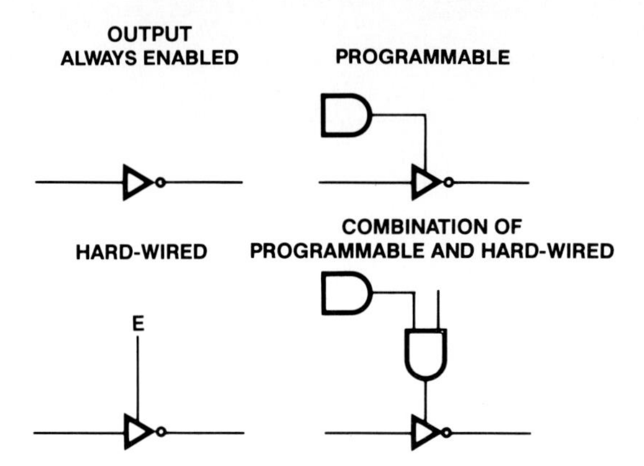

FIG. 4.17 Several output enable alternatives available in the RA macrocell.

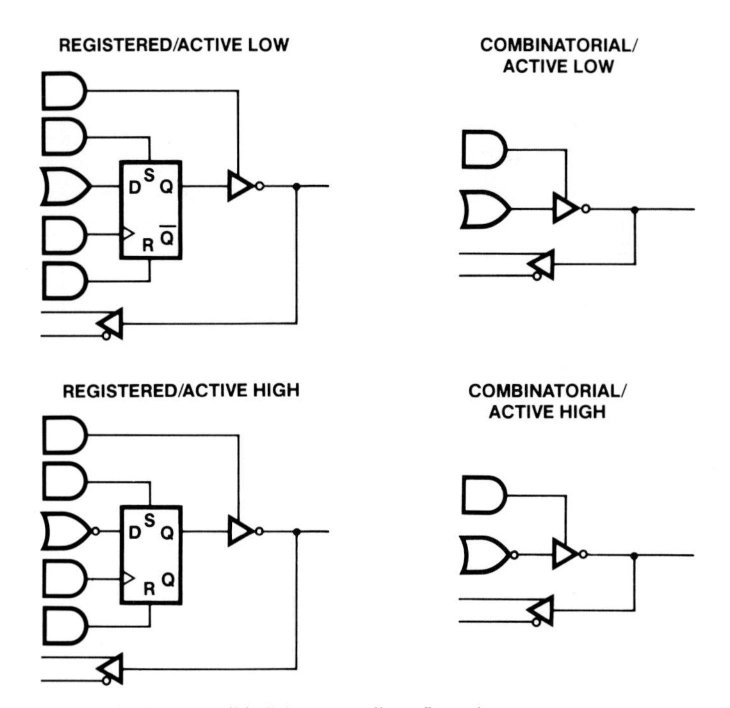

FIG. 4.18 Four possible RA macrocell configurations.

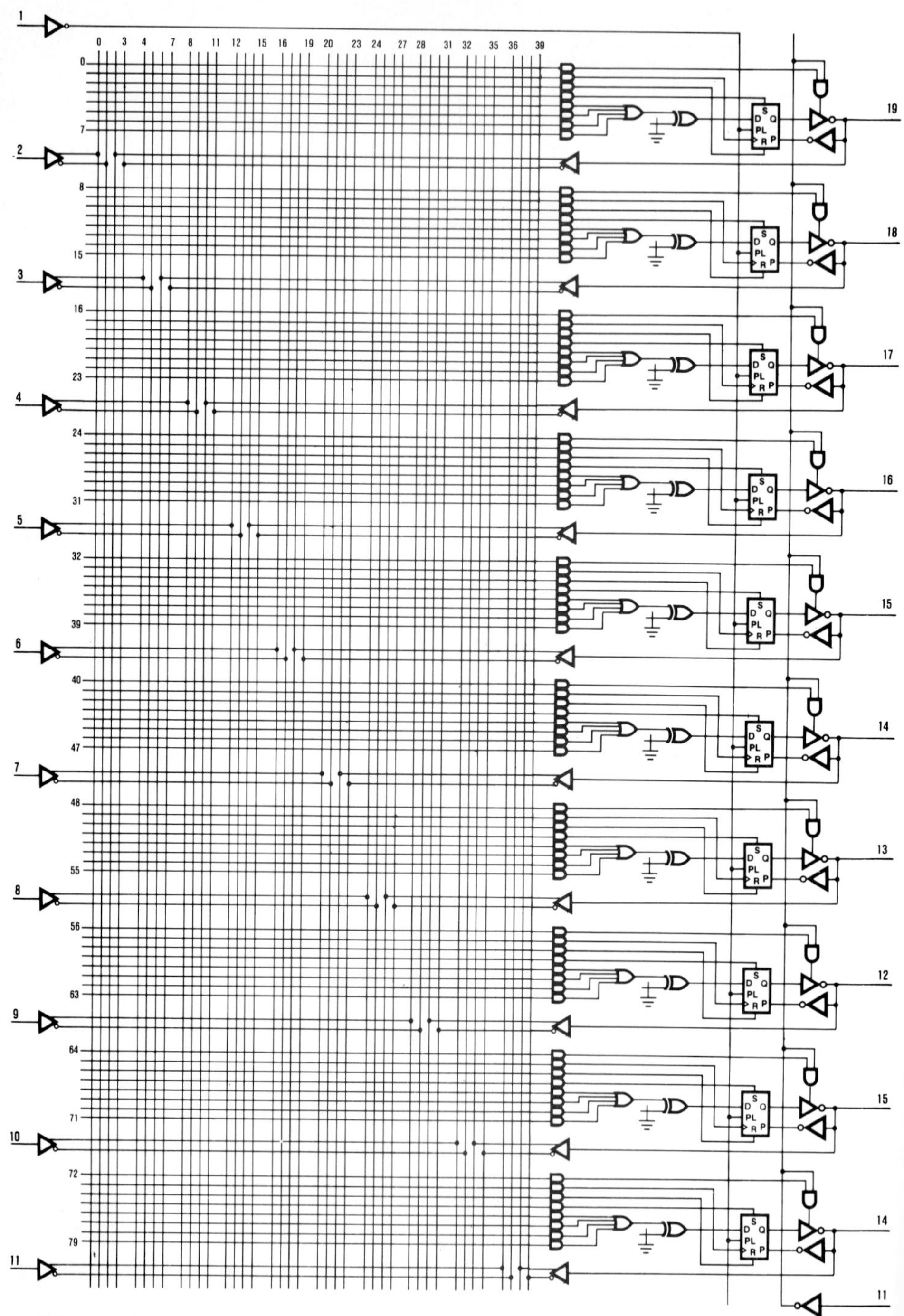

FIG. 4.19 PAL20RA10 logic diagram.

4.5.1 Folded Product Line Array Structure

The folded product line architecture is actually a simplification of the PLD AND-OR array structure. Here a single NAND gate array structure is used to implement logic functions. There is only one array, a programmable AND array, present in this structure. This array consists of a set of product lines with inverted outputs that are NAND gates. An example of such a structure is shown in Fig. 4.20. Most of these product lines feed back into the array and are hooked up to input lines, just like any other array input. This is where the term *folded product line* comes from. However, some of these product lines are not folded. Instead they hook up directly to output buffers and macrocells.

It is still possible to construct AND-OR logic structures using two levels of NAND gates (NAND-NAND logic) in the folded array structure, even though there are no OR gates present. This is done simply by feeding back one folded product line upon another. The logic equivalence of AND-OR and NAND-NAND logic can be shown with Boolean algebra. Recall that it is possible to construct sum-of-products logic, or AND-OR logic, using two levels of NAND gates. All that is required is an application of De Morgan's theorem (or bubble theory) to convert a NAND gate into an OR gate (with inverted inputs). Just push the bubble backward through the second NAND gate to create an OR gate with inverted inputs. Then the two bubbles between the first NAND gate and the OR gate (previously the second NAND gate) cancel each other. Now we have AND-OR logic!

There are basically three advantages to the folded product line array structure:

1. *Performance:* An OR gate delay will be realized in an AND-OR array-based PLD, even if only one product line is required at an output. With the folded product line approach, the equivalent OR gate delay will be realized only if two or more product lines are required for an output.
2. *Flexibility:* A user can make many of the popular macrocell functions using folded product lines. This means that the PLD vendor can focus on designing a very simple product without complicated macrocell structures.
3. *Multilevel logic:* The internal feedback of the product lines allows the user to construct as many levels of logic as are required. Note that the AND-OR restricts the user to two levels of logic.

The tradeoff for the advantages of the folded product line array structure is array size. A much larger programmable array is required to construct a given function using the folded product line approach than using the programmable AND, fixed OR array structure found in a PAL device.

4.5.2 Combined PAL and PROM Array Structure

Another new approach to PLD design is to combine two arrays, a PAL and a PROM, onto a single chip. At first glance you might think this is a PLA structure, but actually it is a little different. Take a look at the block diagram in Fig. 4.21 and recall that a PLA structure involves a programmable AND, programmable OR array. Here we actually have four arrays: the programmable AND, fixed OR PAL array structure preceding the fixed AND, programmable OR PROM array structure. Even though two separate arrays are required, the total size of the two

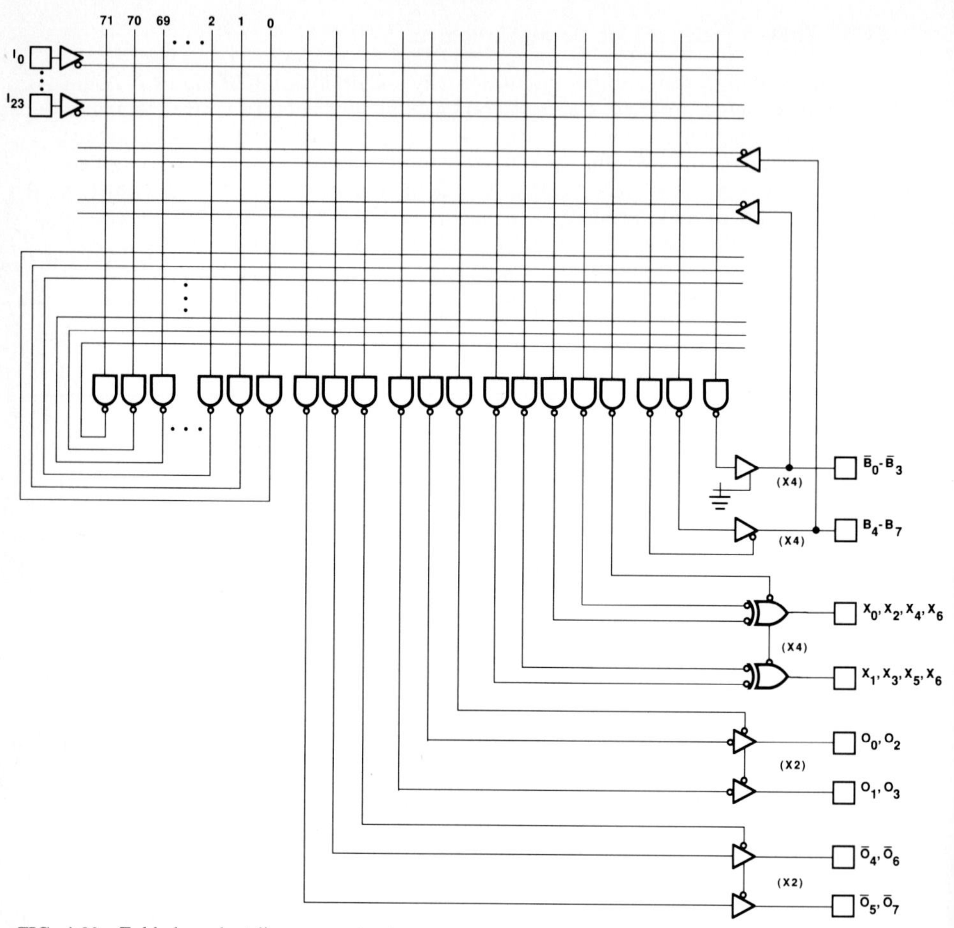

FIG. 4.20 Folded product line array structure.

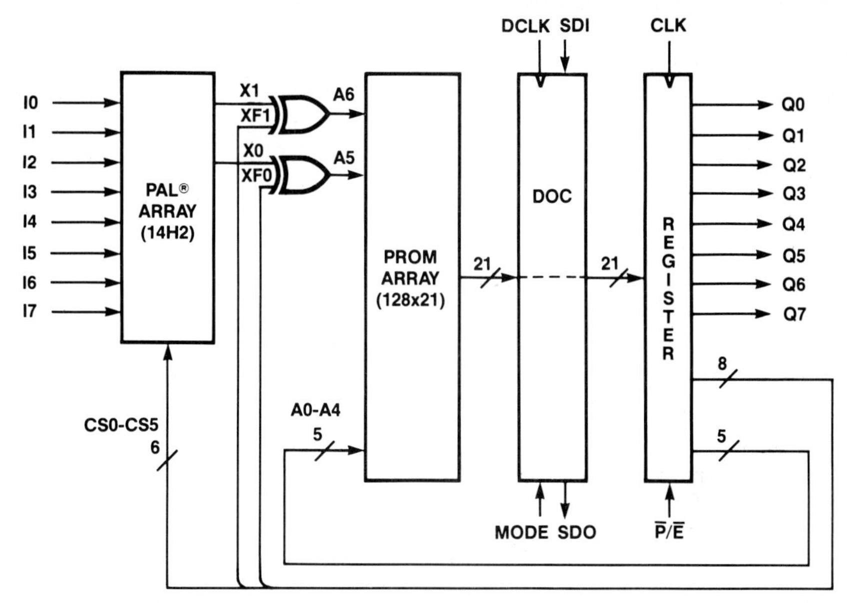

FIG. 4.21 PROSE block diagram.

arrays combined is still smaller than what would be required to achieve a similar function using a PLA structure.

Even though PLS, PAL, and registered PROM devices satisfy a wide range of state machine applications, they each are not without inherent limitations. PLS and PAL devices offer a large number of inputs and outputs, but are generally bounded by product lines in state machine applications. On the other hand, registered PROMs always have enough product lines, even for the most complex state transition sequence, but they are generally bounded by input and output limitations. The combined PAL-PROM array structure marries the best that both array structures have to offer. The PAL array, with its numerous inputs, is efficient at decoding a large number of input conditions. The PROM array, with its vast supply of product lines, is optimal for storing the state information. The combination allows a very efficient state machine with a great many inputs and state bits.

Advanced Micro Devices (AMD) has just introduced this PAL-PROM array concept in a product called PROSE (PROgrammable SEquencer). The logic diagram for this device is given in Fig. 4.22. This device boasts a state machine complexity of 128 states with 14 inputs and 8 outputs in a 24-pin package. The architecture of this product consists of a 128 × 21 (7 inputs and 21 outputs) PROM array followed by a 21-bit state register and preceded by a 14H2 (14 inputs and 2 outputs with 8 product lines each) PAL array. Eight of the 21 bits in the state register are available for output, while the remaining 13 bits are used as feedback signals (6 to the PAL array, 2 to the exclusive-OR gates, and 5 to the PROM array).

The PAL array operates on the eight conditional and six state inputs to select two control bits to the PROM. Two exclusive-OR gates between the two arrays help to minimize product terms and redundant states. Five lines feed back from

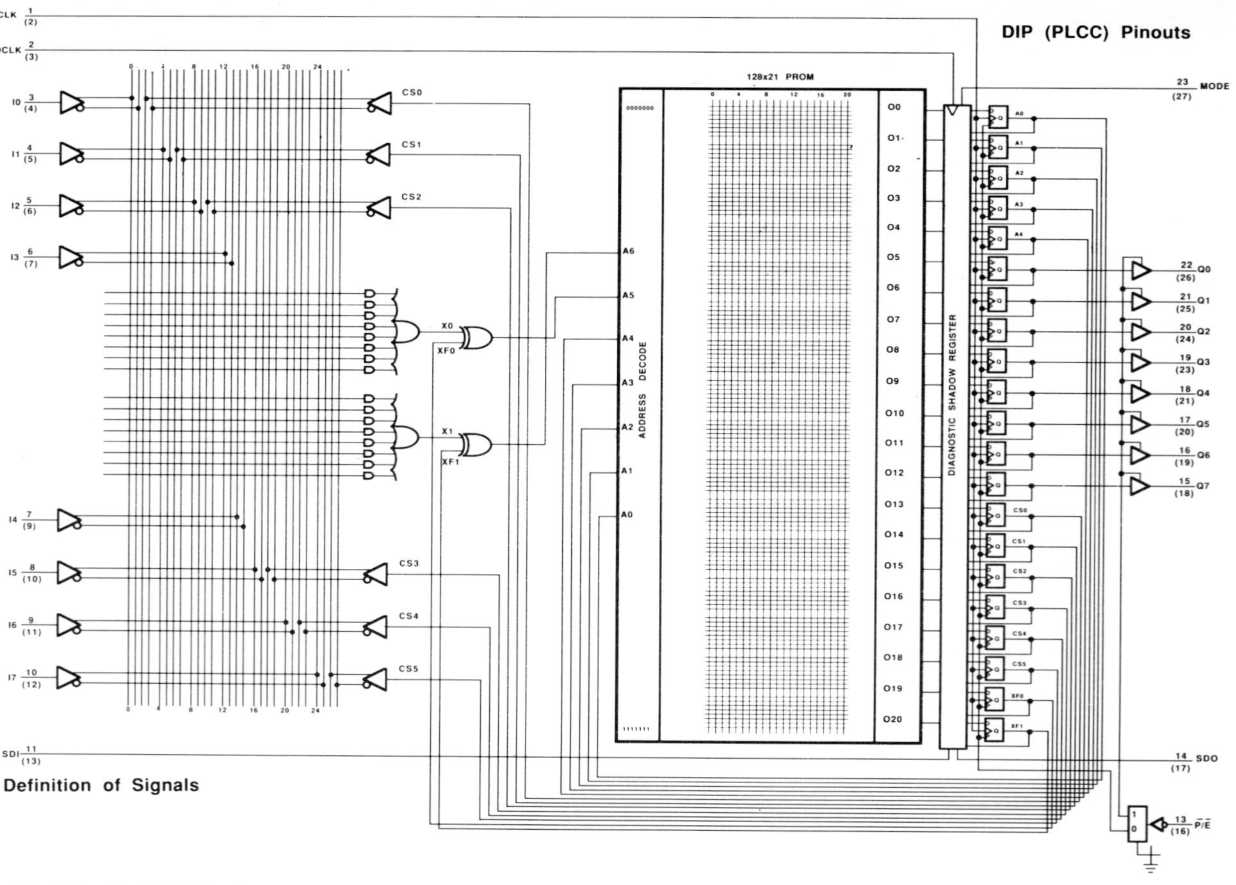

FIG. 4.22 PROSE logic diagram.

the PROM to form the primary address for the next state. The PROM stores up to 128 states of 8 outputs and 13 feedback control signals. This arrangement is optimized for four-way branching.

4.5.3 Logic Cell Array

The *logic cell array* (LCA) is a new concept in semicustom pioneered by a young company in Silicon Valley called Xilinx and alternate sourced by AMD. As you can see from the LCA structure in Fig. 4.23, this device has characteristics of gate arrays and PLDs. The LCA is like a gate array in that logic functions are synthesized by connecting together logic transistors (in groups called *configurable logic blocks,* CLBs) using a clever interconnect scheme through routing channels. Even though the LCA looks nothing like the familiar PLD AND-OR array structure, it most resembles a PLD to the user because he or she can program the chip. These characteristics make the LCA an extremely powerful logic solution realizing the density level of a small gate array (1200 to 1800 gates) and programmable by the user. Another revolutionary aspect of the LCA is its RAM-based configuration. This allows the LCA to be reconfigured by simply loading new data into the configuration RAM cells.

As we mentioned, the CLB, which is detailed in Fig. 4.24, is the functional building block of the LCA. The CLB is used to construct logic functions and as a storage element, very similar to a PLD macrocell. Each CLB is composed of five inputs (four logic inputs A, B, C, and D and a special clock input K), two outputs (X and Y), a combinatorial logic array, and a D flip-flop with asynchronous set and reset. The combinatorial logic array can generate any logic function of the four inputs, or it can generate any two independent logic functions of any three of the four inputs. The user can also determine the path through the block. For example, the flip-flop can be bypassed to construct a combinatorial logic block, or the Q output from the flip-flop can be fed back to construct a sequential logic block.

The I/O block structure is given in Fig. 4.25. Each of these I/O blocks determines the signal direction (input or output) of the pin it is associated with. If the pin is chosen to be an input, then the user has a choice of a direct, latched, or registered input buffer. If the pin is chosen to be an output, then the output buffer can be either three-state or open collector. There is a third option for the signal direction—that the I/O block configures the pin to be bidirectional, either an input or an output as determined by logic controlling the three-state buffer.

All of these CLBs and I/O blocks are hooked together using three types of interconnect resources: direct interconnect, local lines, and long lines. Three types of interconnect are available in the LCA to provide a high level of connectivity with a minimum impact on die size and yielding maximum performance. Direct interconnect, which runs in either the vertical or horizontal direction from adjacent CLBs, is the most available type of interconnect. It is ideal for connecting an output of one CLB to the input of a neighboring CLB. Also, since these lines are very short and direct, direct interconnect incurs the smallest signal delay. Local lines are the most flexible type of interconnect because they can make connections between close or distant CLBs and I/O blocks. But the price paid for this flexibility is that local lines incur slightly longer delays than direct interconnect lines, especially when making distant connections. However, long lines, which bypass the switching matrices that make the local lines so flexible, are

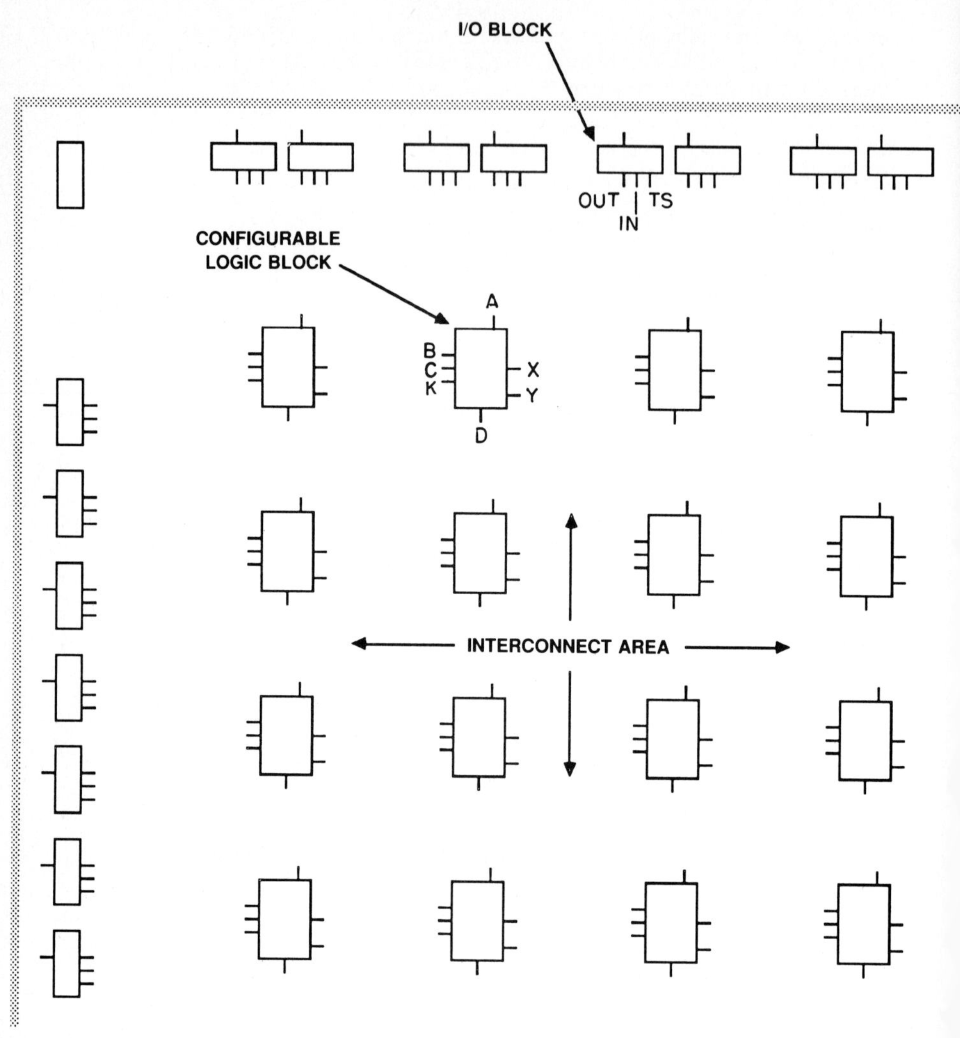

FIG. 4.23 Logic cell array structure.

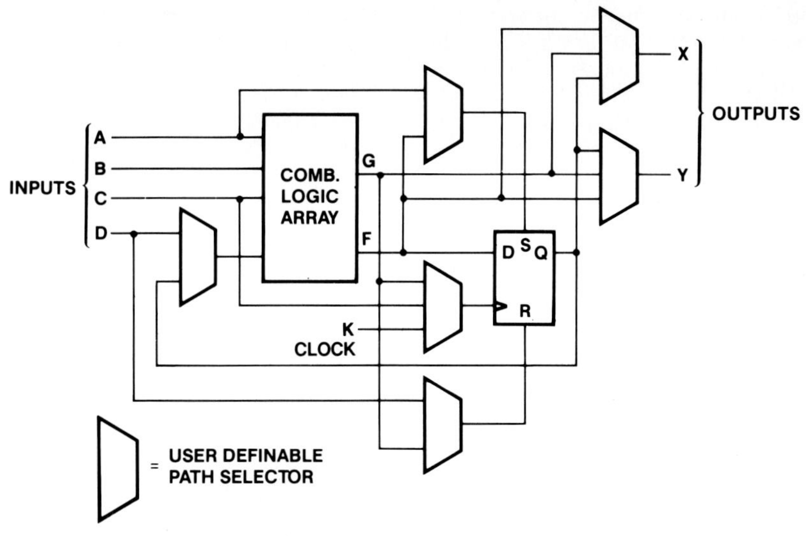

FIG. 4.24 Configurable logic block.

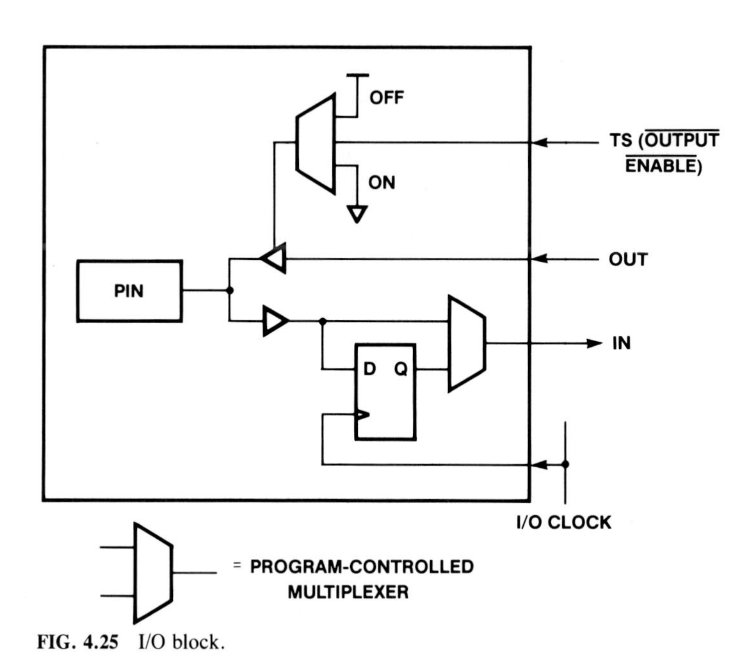

FIG. 4.25 I/O block.

ideal for making distant connections (i.e., cross chip) with a minimum delay. The tradeoff here is that long lines are the most precious interconnect resource.

There are presently two LCA products available on the market. The first is a 1200-gate device with 64 CLBs and 58 I/O blocks. The second is an 1800-gate device with 100 CLBs and 74 I/O blocks. The smaller LCA is available in either a 48-pin DIP or a 68-pin plastic chip carrier (PLCC). The larger LCA is available in an 84-pin PLCC or an 84-pin grid array (PGA).

4.6 PLD NOTATION APPENDIX

Because PLD structures are so different from ordinary TTL gates, new logic notations have been developed for PLDs to simplify their understanding. Figure 4.26 shows the logic convention adopted for a three-input AND gate. Shown on

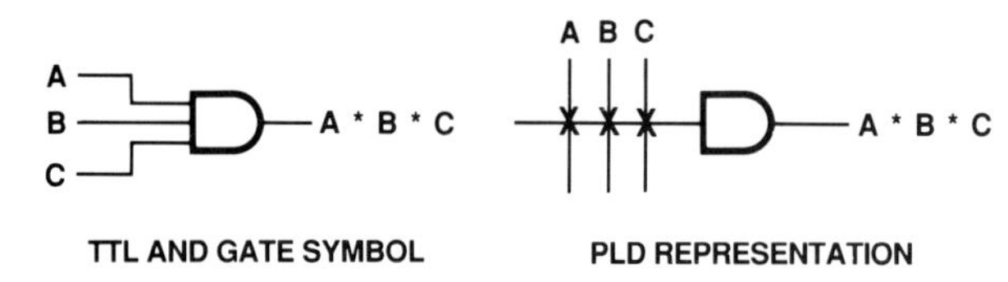

FIG. 4.26 Three-input AND gate logic representations.

the left in this figure is the TTL AND gate symbol; on the right is the equivalent PLD representation. The PLD representation for an AND gate is called a *product line*. Note that the three vertical lines are the inputs (*A*, *B*, and *C*), which are connected to the AND gate inputs through fuse links. An unprogrammed (or closed) fuse link is represented by an X at the intersection of an input line with a product line. If it was desired to disconnect one of those inputs from the AND gate, say *C*, for example, then the appropriate X would be removed from the point of intersection for the *C* input line with the product line to signify a programmed (or open) fuse link. This product line, which now implements the $A * B$ function, is shown in Fig. 4.27.

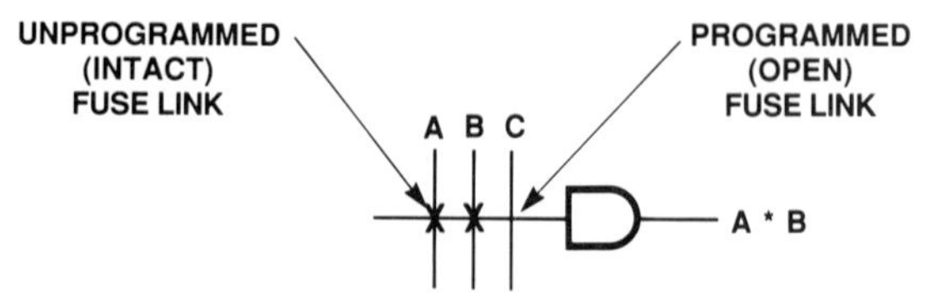

FIG. 4.27 Partially programmed product line to implement $A * B$.

Actually since every input is available to every product line in a PLD, it is convenient to show the input lines as long lines running vertically through the array. Also, there are two input lines associated with each input pin, because both input polarities are available in a PLD. Therefore the input buffer is shown

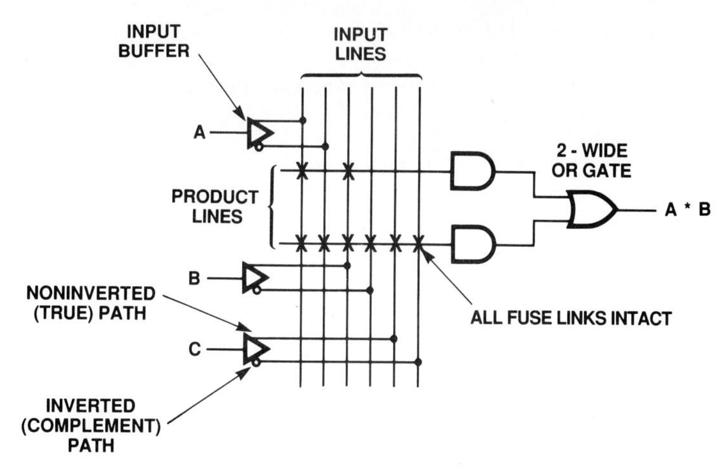

FIG. 4.28 Portion of PLD array programmed to implement $A * B$.

with both noninverted (true) and inverted (complement) output paths, which are each hard-wired connected (shown as a dot) to an input line. A portion of a PLD array illustrating the input lines and buffers is shown in Fig. 4.28. Notice that an OR gate is added to the structure. All the fuse links in the lower product line are left intact, leaving the product line in a logic low (since true inputs are ANDed with complements), while the appropriate fuse links in the upper product line are programmed to implement the $A * B$ function from Fig. 4.27.

Obviously, it is more common to implement two or more levels of logic gates, such as an AND-OR-invert circuit, in a PLD. For example, consider the following function implemented in a PAL device:

$$\overline{\text{Output}} = A * \overline{B} + \overline{A} * B$$

The standard combinatorial logic diagram for this function is shown in Fig. 4.29, with the PAL logic equivalent shown in Fig. 4.30.

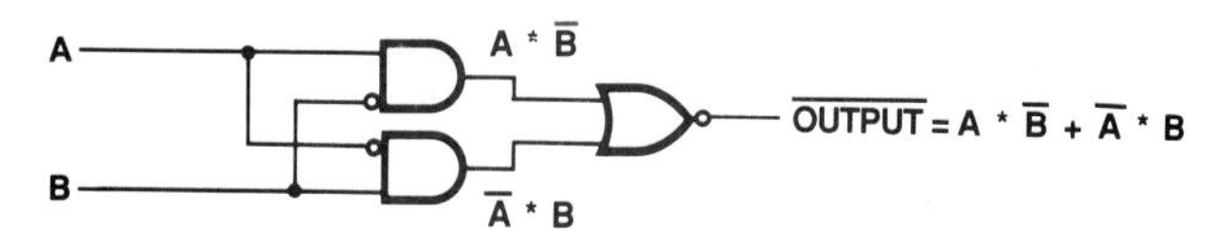

FIG. 4.29 Standard combinatorial logic diagram.

Notice the details added to Fig. 4-30, which magnify the programmed fuse link for B (since B does not appear in that product line) and the intact fuse link for $/B$ (which does appear) in the upper product line. This magnification details each fuse link and its associated diode for a bipolar PAL device. A CMOS PAL device is similar, except that an erasable cell would be substituted for the fuse link.

Even more circuit details for a bipolar PAL device implementation of the logic in Figs. 4.29 and 4.30 is given in Fig. 4.31. Notice that the logic array is a bit sideways. Now the two product lines are drawn vertically, with current sources located at the bottom, and the four input lines are drawn horizontally. Each fuse

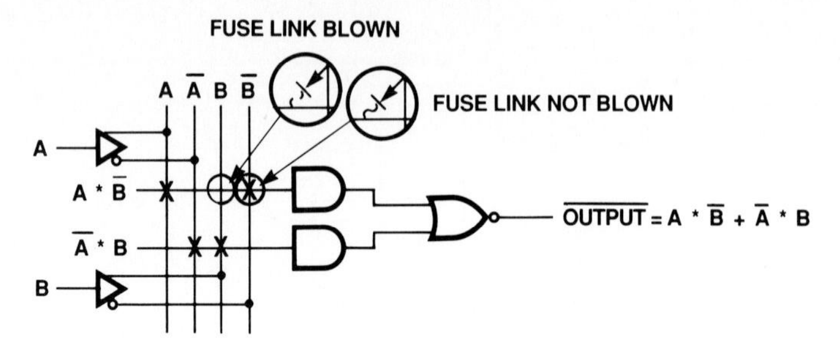

FIG. 4.30 PAL device implementation.

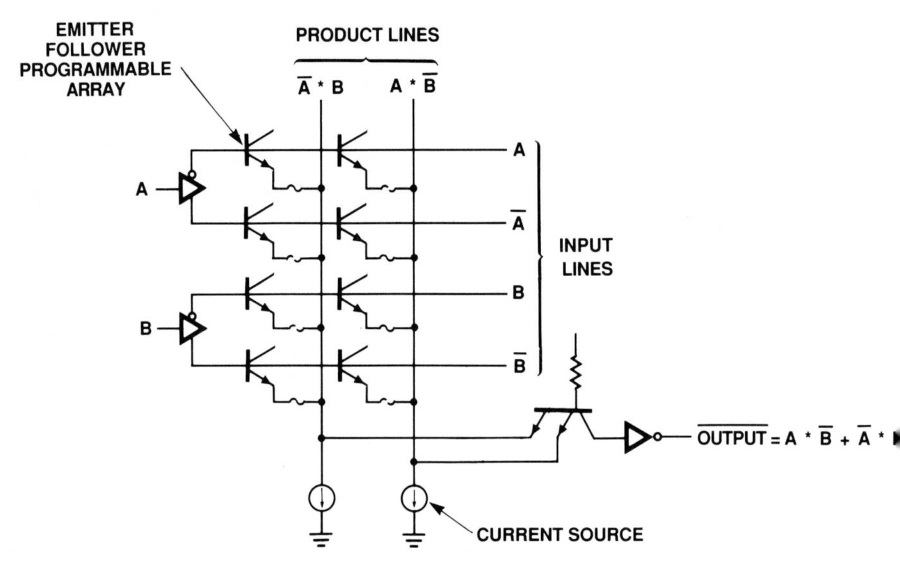

FIG. 4.31 Circuit details for PAL device implementation.

link, still located at the cross points, is associated with an *npn* transistor. A row of these *npn* transistors has a common base with an input line. The emitter of each *npn* transistor is connected to the fuse links, which are themselves connected to the product lines. This structure, a programmable array of emitter-follower transistors connected to a current source, synthesizes programmable NOR logic.

The presence of the emitter's load (i.e., intact fuse link) will make the signal level of the product line dependent upon the input signal. In turn, the absence of the emitter's load (i.e., programmed fuse link) will make the signal level of the product line independent of that input signal. The default of all fuses intact will force the product line low, while all fuses programmed will leave the product line permanently high. Obviously, a low product line summed with a high product line will force the OR gate high, but the output will be low after the inversion, as shown in Fig. 4.32.

Refer back to Fig. 4.31; the signal levels of the product lines are sensed by a

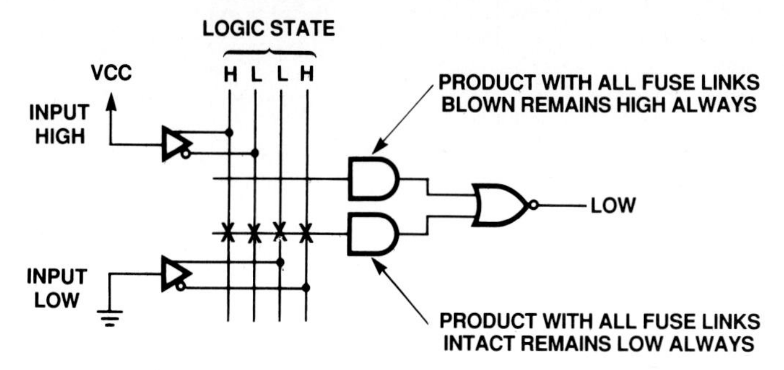

FIG. 4.32 PAL logic levels.

network of emitters in a common-base *npn* transistor associated with each output. This transistor, which implements the fixed NOR gate, is really a NAND gate.

The combination of programmable NOR gates summed by fixed AND gates might make you think the logic is reversed, but it is not. The reason is that NOR-NAND logic is equivalent to AND-NOR and AND-OR-invert logic. If you are hesitant about this logic equivalence, prove it yourself by using De Morgan's theorem or bubble theory. Or better yet, just program a PAL device!

CHAPTER 5
VLSI DEVICE FUNDAMENTALS

Bhadrik Dalal
Amdahl Corporation
Sunnyvale, California

5.1 SEMICONDUCTOR MATERIALS

Materials are electrically classified according to their ability to carry electric current as conductors, semiconductors, and insulators. This property is primarily affected by the band structure of the material.[1] Electric current conduction occurs in semiconductors due to two carriers, namely, free electrons and holes. The bandgap or energy gap between the conduction and valence bands controls the conduction properties of the material. The *bandgap* is the minimum energy required to generate free electron-hole pairs. Materials with electrical resistivity at room temperature ranging from 10 $m\Omega \cdot cm$ to 1 $G\Omega \cdot cm$ are called *semiconductors*. Good conductors have resistivities around 1 $\mu\Omega \cdot cm$. Insulators have resistivities greater than millions of gigohm-centimeter.

The most important property of a semiconductor is its ability to carry electric current. The characteristic semiconducting properties are brought about by (1) thermal excitation by phonons, (2) optical excitation by photons, (3) donor and acceptor impurities, and (4) lattice defects. Semiconductors in which electrical conduction is predominantly due to free electrons are called *n-type* semiconductors. Semiconductors in which electrical conduction is predominantly due to free holes are called *p-type* semiconductors.

5.1.1 Elemental Semiconductors

Silicon and germanium are the most important elemental semiconductors in use today. Both elements are from Group IV of the periodic table of elements. Silicon, however, is the dominant semiconductor.

Silicon. The crystal structure of silicon is the diamond structure. Each atom has 4 nearest neighbors and 12 next-nearest neighbors. The distance between two nearest neighbors is 2.35 Å. Each silicon atom has four valence electrons that provide covalent bonding with the nearest-neighbor atoms. Because of the wider energy gap of silicon, namely, 1.12 eV, it can be operated at higher temperatures than germanium can. The upper limit for operation is between 125° and 175°C. A significant processing advantage is that silicon is capable of forming a thermally grown silicon dioxide layer that can be used for surface passivation and protection of the underlying device. Silicon is an indirect-gap semiconductor. Consequently, silicon cannot be used for many electrooptical IC applications.

Germanium. Like silicon, the crystal structure of germanium is the diamond structure. The distance between two nearest neighbors is 2.45 Å. The energy gap of germanium is only 0.66 eV; consequently, it cannot be operated at higher temperatures as silicon can. The mobility of electrons and holes in germanium is higher than in silicon, resulting in higher-frequency-performance devices. However, because of the difficulty in forming a thermally grown native oxide layer for passivation, germanium has lost its appeal as a semiconducting material.

5.1.2 Compound Semiconductors

Compound semiconductors are made up of two or more elements. The elements are from Groups III and V, as in the case of gallium arsenide (GaAs) and indium phosphide (InP), or from Groups II and VI, as in cadmium telluride (CdTe). There are binary compound semiconductors, such as GaAs, and ternary compound semiconductors, such as aluminum gallium arsenide (AlGaAs). The compound semiconductors have unique properties that allow functions that cannot be performed by silicon, such as electrooptical IC applications. GaAs and related compounds are the most dominant compound semiconductors in use today.

Gallium Arsenide. Like silicon, GaAs belongs to the cubic class of crystals. The crystal structure of GaAs consists of two interpenetrating face-centered cubic sublattices, which are called the *zinc blende structure*. In GaAs, each sublattice contains atoms of only one element type. When both sublattices contain identical atoms as in the case of silicon, the lattice is called the diamond lattice. The energy gap of GaAs is 1.43 eV at room temperature.

Indium Phosphide. InP is another compound semiconductor that has been studied extensively. The energy gap of InP phosphide is 1.27 eV at room temperature.

5.1.3 Key Material Parameters

The three key material parameters for electrical conduction are the bandgap, carrier velocity, and carrier mobility.

Bandgap. As discussed earlier, the bandgap or energy gap is the most important material parameter. Materials that have a large bandgap behave like insulators. Materials with narrower bandgaps have an appreciable number of carriers in both the conduction and valence bands at room temperature, and they behave like semiconductors. In metals, the valence and conduction bands overlap, and conduction of current is almost independent of temperature.

In semiconductors, materials with larger bandgaps are capable of operation at higher temperatures. The bandgaps for germanium, silicon, GaAs, and InP are 0.66 eV, 1.12 eV, 1.43 eV, and 1.27 eV, respectively, at room temperature.[2]

Carrier Velocity. The motion of carriers in a solid is influenced by their interaction with surrounding atoms. The velocity of these carriers is proportional to the applied electric field. The mobility of carriers is the magnitude of this carrier drift velocity per unit electric field. In semiconductors, generally, the carrier velocity increases with increasing electric field. However, beyond a certain value of electric field, the velocity saturates. This is true in silicon. In GaAs, however, there are two characteristic velocities: a peak velocity and a saturated drift velocity. These velocities are separated by a region of negative differential mobility, wherein the carrier velocity actually decreases with increasing electric field.[3]

The saturated drift velocity is the primary factor in determining device switching speeds, because device operation is generally performed at high bias levels. The electron saturated drift velocity for silicon is 0.7×10^7 cm/s, whereas for GaAs it is 1.05×10^7 cm/s. The peak electron drift velocity for GaAs can be as high as 2.2×10^7 cm/s at a low electric field of 3.1 kV/cm. The electron saturated drift velocity for InP is 1.2×10^7 cm/s, whereas the peak electron drift velocity can be as high as 2.6×10^7 cm/s at an electric field of 14 kV/cm.

Carrier Mobility. As discussed earlier, the *mobility* μ of the carrier (electron or hole) is the ratio of the carrier velocity v to electric field E that is causing carrier motion. Therefore, the mobility is given by

$$\mu = \frac{v}{E}$$

The carrier mobilities are different for holes and electrons. Generally, the hole mobility is lower than the electron mobility in a given semiconductor, because the hole transport is indirect and carried out by a continuous transfer of the binding electrons. In contrast, the electron transport is direct, though it undergoes many collisions along its path.

The carrier mobilities at very low doping concentrations are essentially independent of the doping concentrations. However, at moderate doping concentrations that are typical of actual devices, the mobilities monotonically decrease with increasing doping concentrations. The room temperature mobilities for electrons and holes in silicon for low doping concentrations are 1350 $cm^2/V \cdot s$ and 480 $cm^2/V \cdot s$, respectively. The room temperature mobilities for electrons and holes in GaAs for low doping concentrations are 8000 $cm^2/V \cdot s$ and 300 $cm^2/V \cdot s$, respectively. The room temperature mobilities for electrons and holes in InP for low doping concentrations are 34,000 $cm^2/V \cdot s$ and 650 $cm^2/V \cdot s$, respectively.

5.2 DEVICE CLASSIFICATIONS

The most common semiconductor devices are two- and three-terminal devices usually referred to as *diodes* and *transistors,* respectively. The two-terminal diode is a passive device, whereas the three-terminal transistor is an active device. The key distinguishing characteristic of an active device is that of gain or amplification. The transistor is capable of achieving one or more of the following properties: current gain, voltage gain, power gain, or transconductance function.

The most common types of diodes used in ICs are the *pn*-junction diode and the Schottky diode. The Schottky diode is a metal-semiconductor rectifying contact. Among the other diodes generally used in discrete form are the tunnel diode, the backward diode, and the variable-capacitance diode. The voltage-variable-capacitance characteristic of the varactor diode can be easily realized in MOS technology and is used extensively in bootstrapping circuit applications. The two major classifications of transistors are the bipolar junction transistor (BJT) and the field-effect transistor (FET).

Electronic semiconductor devices, such as transistors, can basically be thought of as charge-control devices. These devices employ a means whereby they introduce a charge in the semiconductor and then control the movement of this charge by another means. In a FET the charge is introduced in the channel either through the mobile carriers associated with impurity atoms or through formation of an inversion layer through the application of a transverse electric field. This charge in the channel is then controlled by the application of a lateral electric field. The FET is a majority carrier device and is usually fabricated as a lateral device.

In a BJT, however, the charge is introduced or injected by means of a forward-biased *pn* junction and is collected by means of a reverse-biased junction. The movement of this charge between the two junctions is partly due to drift and partly due to diffusion. The separation of these two junctions, called the *basewidth,* is a critical dimension that has to be much less than the minority carrier diffusion length to prevent most of these carriers from recombining. The BJT is a minority carrier device and is usually fabricated as a vertical device.

5.2.1 Field-Effect Transistors

The "field-effect" is a phenomenon where the conductivity of a semiconductor is modulated by an electric field, which is applied normal to the surface of the semiconductor. The control terminal is called the *gate,* and the two terminals that connect the modulated conductor, called the *channel,* are called the *drain* and the *source.* The source and the drain are interchangeable. Field-effect devices may be either *n*- or *p*-channel, and may be operated either in enhancement or depletion modes. Because FETs are majority carrier devices, they are also referred to as *unipolar* transistors. FETs are classified as junction FETs (JFETs) or insulated-gate FETs (IGFETs), depending on the type of gate structure used. The IGFET is the most common FET structure and is commonly referred to as the *metal-oxide-semiconductor FET* (MOSFET). The JFET is a FET in which a *pn* junction is used to modulate the channel, whereas a metal-semiconductor FET (MESFET) is a FET in which a Schottky junction is used to modulate the channel.

Junction Field-Effect Transistor. The JFET was the first type of FET to be analyzed by Shockley[4] in 1952. In a JFET, the control gate electrode is a reverse-biased *pn* junction. The JFET consists of a conductive channel connected by two ohmic contacts, one acting as the source of the carriers and the other acting as the drain. The gate forms a reverse-biased rectifying junction with the channel. The width of the depletion layer associated with the *pn* junction is varied by the gate voltage. This depletion-layer-width variation results in the modulation of the channel conductance. The gate voltage at which the channel is completely pinched off by the depletion layer is called the *pinch-off voltage* V_p. The current-voltage characteristics of a depletion-mode *n*-channel JFET are described by Pierret[5] in the following drain current equations.

In the linear region (before pinch-off),

$$I_d = K\left\{V_{ds} - \frac{2}{3}(V_{bi} - V_p)\left[\left(\frac{V_{ds} + V_{bi} - V_{gs}}{V_{bi} - V_p}\right)^{3/2} - \left(\frac{V_{bi} - V_{gs}}{V_{bi} - V_p}\right)^{3/2}\right]\right\}$$

$$\text{for } 0 \le V_{ds} \le v_{ds(\text{sat})} \quad V_p \le V_{gs} \le 0$$

where K = device constant
V_{ds} = drain to source voltage
V_{gs} = gate to source voltage
V_{bi} = built-in potential of *pn* junction
V_p = pinch-off voltage

In the saturation region (post-pinch-off),

$$I_{d(\text{sat})} = K\left\{V_{gs} - V_p - \frac{2}{3}(V_{bi} - V_p)\left[1 - \left(\frac{V_{bi} - V_{gs}}{V_{bi} - V_p}\right)^{3/2}\right]\right\}$$

$$\text{for } V_{ds(\text{sat})} \ge V_{gs} - V_p$$

If the saturated drain curent at $V_{gs} = 0$ is represented by I_{dss}, the drain saturation current equation simplifies to

$$I_{d(\text{sat})} = I_{dss}\left(1 - \frac{V_{gs}}{V_p}\right)^2$$

This is the familiar *square-law* relationship for FETs.

Metal-Oxide-Semiconductor Field-Effect Transistor. The MOS structure is the basis for the MOSFET. The MOSFET is a transconductance amplifier wherein the drain current varies as a function of the gate and drain voltages. The key MOSFET device parameters are the threshold voltage V_t and the device transconductance parameter k. The device operation in the linear and saturation regions is described by Hodges and Jackson[6] in the following drain current (I_d) equations.

In the linear region,

$$I_d = \frac{k}{2}[2(V_{gs} - V_t)V_{ds} - V_{ds}^2] \qquad \text{for } V_{gs} \ge V_t \quad V_{ds} \le (V_{gs} - V_t)$$

In the saturated region,

$$I_d = \frac{k}{2}(V_{gs} - V_t)^2 \qquad \text{for } V_{ds} \ge (V_{gs} - V_t)$$

where k = device transconductance parameter
V_{gs} = the gate to source voltage
V_{ds} = drain to source voltage
V_t = threshold voltage

The process transconductance parameter k' is given by the relationship

$$k' = \frac{\mu \epsilon \epsilon_0}{t_{ox}}$$

where μ = mobility of majority carrier of device
ϵ = relative dielectric constant of gate insulator
ϵ_0 = permittivity of vacuum
t_{ox} = thickness of gate oxide

The device transconductance parameter k is given by

$$k = k' \frac{W}{L}$$

where W = width of channel and L = length of channel

Metal-Semiconductor Field-Effect Transistor. The MESFET is a FET in which a metal-semiconductor rectifying contact is used for the gate electrode. The current-voltage characteristics of a MESFET are similar to those of a JFET.

Heterostructure Field-Effect Transistor. The heterostructure FET (HFET) is similar in operation to the MESFET; however, it has superior transconductance. The HFET, usually called the high-electron-mobility transistor (HEMT),[7] goes by various names (such as MODFET, TEGFET, and SDHT). However, the principle of operation is the same. By using a superlattice heterojunction structure in which one layer is AlGaAs and the other is GaAs, and by suitably doping the two layers, high carrier mobility and velocity can be achieved. The AlGaAs layer has a higher bandgap and is doped heavily to generate electrons. The GaAs has a lower bandgap and is undoped to improve carrier mobility and velocity. The electrons from the AlGaAs layer diffuse into the lower-bandgap GaAs layer, where they are confined by the energy barrier at the heterojunction. This principle of modulating the carrier mobilities and velocities gave the transistor its name, namely, modulation-doped FET (MODFET). The current-voltage characteristics of a MODFET are similar to those of the JFET and the MESFET.

5.2.2 Bipolar Junction Transistors

There are two basic forms of BJTs: an *npn* transistor and a *pnp* transistor. These names are derived from the three semiconductor layers used to construct the transistor. The three terminals of the transistor are called the *emitter,* the *base,* and the *collector.* The base electrode is the control electrode. Unlike the FET, where the source and drain terminals are interchangeable, the emitter and collector are not generally interchangeable. There are two types of carriers responsible for current conduction in this transistor. Consequently, it is called a *bipolar* transistor. Unlike the FET, the BJT is a minority carrier device. The BJT is an active device that is capable of current gain, voltage gain, power gain, and transconductance function. The BJT can be thought of as two back-to-back *pn* junctions having a very narrow common region called the *base* region. The control electrode is the base, and the control junction is the emitter-base junction. The emitter injects majority carriers that are transported through a very narrow base region, where they become minority carriers and subsequently get collected by the collector. The BJT has four regions of operation:[8,9] the normal region, the inverse region, the cutoff region, and the saturation region. These regions are defined by the bias states of the two junctions. The equations related to the operation of the BJT are summarized below.[8]

The common-emitter current gain generally referred to as beta (β) is given by the ratio of the collector output current I_c to the base input current I_b.

$$\beta = \frac{I_c}{I_b}$$

The emitter and collector currents I_e and I_c are described by the Ebers and Moll[9] equations

$$I_e = I_{es}(e^{V_{be}/V_T} - 1) - \alpha_R I_{cs}(e^{V_{bc}/V_T} - 1)$$

$$I_c = \alpha_F I_{es}(e^{V_{be}/V_T} - 1) - I_{cs}(e^{V_{bc}/V_T} - 1)$$

where I_{es}, I_{cs} = emitter and collector junction saturation currents
V_T = thermal voltage
V_{be}, V_{bc} = bias voltages across BE and BC junctions
α_F = fraction of total current flowing across EB junction collected at CB junction
α_R = fraction of total current flowing across CB junction collected at EB junction

The voltage V_{be} in the normal region of transistor operation is approximately given by

$$V_{be} = V_T \ln\left(\frac{I_e}{I_{es}}\right) + I_b r_b + I_e r_e$$

where r_b and r_e are series base and emitter resistances. For a detailed review of the physical operation of a BJT and the equations describing its behavior, see Refs. 8, 10, and 11.

***npn* Transistor.** The *npn* transistor is the most widely used bipolar transistor today. It is a BJT with an *n*-type emitter, a *p*-type base, and an *n*-type collector. The minority carrier in the base region of an *npn* transistor is the electron, which provides higher performance because of the higher electron mobility in silicon. For this reason it is preferred in many circuit designs. The low input resistance, the high output resistance, and nearly unity common-base current gain make it possible to achieve voltage gain in this device. In the common-emitter configuration, the device provides high current gain and excellent transconductance properties.

***pnp* Transistor.** The *pnp* transistor is similar to an *npn* transistor in its principle of operation, except that the minority carrier in the base region is now the hole. Because the hole velocity and mobility are significantly lower than those of the electron, the *pnp* transistor is not used in high-frequency applications. The *pnp* transistor is usually fabricated as a lateral device, unlike the *npn* transistor, which is generally a vertical device.

Critical dimensions in a lateral device are defined by photolithographic processes, whereas those in a vertical device are determined by implantation or diffusion processes. Therefore, the critical dimensions of a vertical device are much finer in dimension and are tightly controlled. This allows the vertical device the capability of achieving very high performance over that of the lateral device. The *pnp* transistors are rarely used in digital switching circuit applications. They are extensively used in analog circuit applications.

Heterojunction Bipolar Transistor. A *pn* junction with the same semiconducting material on both sides of the junction is called a *homojunction*. However, when the two sides are made up of dissimilar semiconducting materials, the *pn* junction is called a *heterojunction*. Heterojunctions possess some unique properties, which can provide some interesting results.[12]

A heterojunction bipolar transistor (HBT) has a wider bandgap emitter. This makes it possible to achieve good emitter injection efficiency with lower emitter doping and higher base doping simultaneously. This also results in lower base-emitter parasitic capacitance and lower base resistance. Consequently, a higher performance device is realized over its conventional homojunction counterpart.

The motion of minority carriers in a BJT, as discussed earlier, results partly from diffusion and partly from drift. In an HBT, an additional effect due to the differences in the energy gap also influences carrier motion. As a result, device design and optimization for high-performance applications becomes significantly more flexible.

5.2.3 Key Device Parameters

The key device parameters for VLSI applications are the size, input capacitance, transit time, threshold voltage, and transconductance. These parameters affect the integration level, high-frequency performance, and power dissipation of the IC.

Size. One of the key features of a device is its size. The total area of a device is the sum of its intrinsic or active device area and its extrinsic or inactive device area. The active device area is determined by some critical dimension, such as the channel size in a FET, or the emitter size in a BJT. This dimension is generally determined by photolithography and is usually related to the minimum feature size. The inactive device area is determined by the necessity to electrically and physically bring the different parts of the active device to surface terminals. This enables the interconnection of devices to form circuits.

The active device area provides transistor action, and the inactive or parasitic device area provides interface to other devices to form circuits. Minimizing the parasitic device area is important to achieve high performance and high circuit integration levels.

Input Capacitance. When devices are interconnected to form circuits, their input capacitance presents itself as a load to the preceding device. This capacitance has to be charged and discharged during circuit operation.

In a FET this input capacitance is comprised of several components, such as gate to source capacitance C_{gs}, gate to drain capacitance C_{gd}, and gate to bulk capacitance C_{gb}. In a BJT the input capacitance is comprised of several components, such as depletion layer capacitances due to the emitter-base and collector-base junctions, and the diffusion capacitances associated with charge storage in the neutral base region.

As devices are scaled down to smaller and smaller feature sizes, the difference in input capacitances of the FET and BJT tends to decrease. This trend places a greater emphasis on the parasitic capacitance of the signal interconnect lines, where high-speed operation is desirable.

Transit Time. The device *transit time* is a measure of how fast the charge carriers can move from one terminal to another. In an *n*-channel FET, this transit time is a measure of how fast the majority carriers (electrons) move from the source to the drain. The movement of the carrier is primarily influenced by the electric field between the source and the drain, provided that the gate voltage is appropriately selected to form a channel of mobile charge carriers. These carriers acquire a drift velocity proportional to the electric field. However, at high electric fields the carriers experience velocity saturation. Therefore, for a FET the device transit time is directly proportional to the length of channel and inversely proportional to the carrier velocity. In an *npn* BJT this transit time is a measure of how fast the minority carriers (electrons) that are injected from the emitter traverse the neutral base region and reach the collector. The motion of these minority carriers is partly due to diffusion and partly due to drift. Because the average velocity of a diffusing carrier is proportional to the concentration gradient and the diffusion constant, the transit time across the neutral base region is proportional to the square of the length of the base region, called the basewidth. Therefore, the transit time is a nonlinear function of the basewidth. Consequently, when critical dimensions are reduced due to scaling, the BJT experiences a larger relative improvement in device transit time as opposed to the FET. However, note that device transit time is not the most significant factor in determining circuit speed in a VLSI circuit.

Threshold Voltage. The *threshold voltage* is the turn-on voltage of the transistor. A similar parameter for the normally on type of transistor is the pinch-off voltage. For a FET the threshold voltage depends on several processing parameters, such as carrier concentration in the channel and the channel thickness. The control of the manufacturing process directly affects the tolerance achievable on this threshold voltage. For a BJT the turn-on voltage depends on a fixed material parameter, namely, the energy gap of the semiconductor in the base region. The turn-on voltage also depends on the logarithm of base doping as a processing parameter.[12] Consequently, the turn-on voltage tolerance of a bipolar transistor is far superior to that of a FET. This tight tolerance permits the design of smaller logic signal voltage swings, which result in higher performance.

Transconductance. In a VLSI circuit, interconnections between devices and circuits occupy most of the chip area. These interconnects carry signals between devices and circuits. The speed at which these signals travel through the interconnects determines, to a large extent, the speed of the circuits. The interconnects mainly present parasitic capacitance loading. At finer linewidths, the interconnect can be viewed as a distributed *RC* network. The ability of the circuit to drive signals through these interconnects depends on the transconductance of the devices used to build the circuit. *Transconductance* is the ratio of the output current to the input voltage of a transistor. The larger the transconductance, the better the device is able to drive signals through the interconnect lines.

The output current of a FET varies as the square of the input voltage. However, the output current varies exponentially with input voltage for a BJT. Therefore, the BJT is widely used where high device transconductance is required, such as in high-speed digital circuit applications.

5.2.4 Interconnect Materials

The materials commonly used to provide interconnections between devices and circuits are metals, refractory metals, silicides, polycides, and polysilicon. The use of a particular material is governed by complex tradeoffs between several competing factors, such as material properties, process compatibility, device characteristics, circuit requirements, and reliability considerations.

Polysilicon. The use of polycrystalline silicon (polysilicon) films in MOS technology is widespread. This practice was a result of the shift from metal-gate *p*-channel MOS technology to self-aligned silicon-gate *n*-channel MOS technology that occurred in the mid-1970s. The application of polysilicon as gate material for MOS technology is very desirable. The use of polysilicon for interconnect tends to be limited to short local interconnects, because of the high sheet resistance of the currently available polysilicon, which ranges from 20 to 60 Ω per square. Polysilicon films can be made to be semi-insulating by controlled partial oxidation. The semi-insulating films can be used for passivation of high-voltage silicon devices.

Silicides. Silicides are used in both MOS and bipolar silicon technologies. In the bipolar technologies, the use of silicides is in the formation of Schottky barrier diodes and local interconnections. Platinum silicide is used for such applications. In the MOSFET technologies, the use of refractory metal silicides for gate material and local interconnects is becoming common as feature sizes are scaled down below 2 μm. Because of the desire to maintain a polysilicon film on top of the gate oxide,[13] the preferred solution appears to be a *polycide*: a layer of polysilicon capped with a layer of refractory metal silicide. Titanium, tantalum, molybdenum, platinum, and tungsten are among the refractory metals considered for such applications.

Refractory Metals. Tungsten and molybdenum are the two major refractory metals used for interconnections in ICs. Adhesion to silicon dioxide is generally poor,[14] though improvement can be achieved through the use of heated substrates and chemical vapor deposition techniques. The use of refractory metals is mainly confined to FET technologies.

Metals. Aluminum is predominantly used for metal interconnections in silicon ICs. Noble metals, such as platinum, gold, and silver, are occasionally used for specific applications. Gold finds extensive use in GaAs ICs. In high-performance bipolar circuits, the use of multilevel aluminum metallization is almost mandatory. Refractory metals, silicides, and polycides are too resistive for these applications. The electrical resistivity of aluminum in thin-film form is approximately 3 $\mu\Omega \cdot$ cm. Despite the extensive use of aluminum in silicon-based ICs, it suffers from electromigration and coverage problems, especially at finer linewidths. As a result, there has been considerable interest in refractory metals and silicides in recent years.

5.2.5 Key Interconnect Parameters

Interconnects must provide several important functions:[15] (1) good low-resistance contacts to the silicon and polysilicon layers of the device, (2) low-resistance interconnections between devices and circuits, (3) electromigration resistance at high current densities, (4) corrosion resistance, and (5) ease of fine-line patterning. Item (1) affects device behavior, items (3), (4), and (5) affect processing complexity, and item (2) affects circuit behavior. In VLSI system applications, the major concern appears to be with the requirement of low-resistance interconnection and fine-line patterning.

Linewidth. In the desire to provide low cost functions of ever increasing complexity, the demand to shrink device geometries remains very strong. This device miniaturization is accomplished through the scaling of minimum feature size and major process and structural enhancements. However, the demand to scale interconnects also as aggressively has not been felt in the past. This was made possible by increasing the number of levels of metal interconnect. However, with the increased use of ASICs by system designers, the demand for routing resource cannot be satisfied unless interconnect scaling is also aggressively pursued. The key feature then becomes the interconnect linewidth. Several important considerations become relevant, namely, interconnect pattern ability, electromigration resistance, and line resistance. As chip sizes grow, the long-distance interconnect length increases, and, coupled with reduced cross-sectional areas, the total line resistance begins to get excessively large.

Separation. For good interconnect density, the line separation is also generally reduced as linewidths are reduced. The sum of interconnect width and separation is called *interconnect pitch*. Scaling interconnects below about 2.5-μm pitch causes the interline capacitance to increase rapidly. This results in poor interconnect response for high-speed applications, in addition to increased crosstalk noise susceptibility. As interconnect linewidths approach submicron geometries, the line separations will have to increase somewhat to alleviate the crosstalk problem.

Distributed Resistance. As discussed earlier, the distributed resistance of the interconnect line begins to pose serious problems as linewidths approach submicron geometries. The electrical resistivity of thin films increases over that of its bulk material as linewidths approach the grain size of the interconnect material. The cross-sectional area of the line gets a nonlinear reduction with scaling because of the required accompanying thickness reduction. These two factors result in a significant increase in line resistance. As the total line resistance gets larger than the source resistance of the driver gate, which is usually very small for high performance emitter-coupled logic (ECL) bipolar circuits, the circuit performance begins to be dominated by interconnect. This phenomenon is not as noticeable in MOS technologies because of the large source resistance associated with the driver gate.

Distributed Capacitance. The distributed capacitance associated with an interconnect line reduces with scaling down to lines about 4 μm wide. As the lines become narrower, the separation also gets narrower. The interline coupling capacitance now becomes a dominant contributor to the total capacitance. At line pitch below 2.5 μm, the total capacitance actually increases per unit length. The interconnect time-constant (RC) per unit length actually gets larger, because both resistance R and capacitance C are increasing. Moreover, the total time constant of the line increases as the square of the line length, because R and C are both proportional to line length. As the device sizes are getting smaller with scaling, their current drive capability remains almost unchanged, because the increase due to smaller critical dimensions is offset by the decrease in the associated width of the device. Consequently, although the local short interconnect drive capability is somewhat improved, the global long interconnect drive capability is severely hampered. This is rapidly becoming the key factor in limiting the speed of high-performance VLSI circuits.

ACKNOWLEDGMENTS

The author would like to express his gratitude for the support and encouragement provided by Mr. Michael Clements and Mr. Bruce Beebe.

REFERENCES

1. Joseph Lindmayer and Charles Y. Wrigley, *Fundamentals of Semiconductor Devices,* Van Nostrand, Princeton, 1965, p. 199, Fig. 6-6.
2. Charles Kittel, *Introduction to Solid State Physics,* 2d ed., Wiley, New York, 1986, p. 186, Table 1.
3. Richard C. Eden et al., "The Prospects for Ultra High-Speed VLSI GaAs Digital Logic," *IEEE J. Solid-State Circuits,* **SC-14**(2):221–239 (1979).
4. W. Shockley, "A Unipolar 'Field-Effect' Transistor," *Proc. IRE,* **40:**1365–1376 (1952).
5. Robert F. Pierret, *Field Effect Devices,* Addison-Wesley, Reading, Mass., 1983, pp. 13–15.
6. David A. Hodges and Horace G. Jackson, *Analysis and Design of Digital Integrated Circuits,* McGraw-Hill, New York, 1983, pp. 50–51.
7. Hadis Morkoc and Paul M. Solomon, "The HEMT: A Superfast Transistor," *IEEE Spectrum,* **21**(2):28–35 (1984).
8. David A. Hodges and Horace G. Jackson, *Analysis and Design of Digital Integrated Circuits,* McGraw-Hill, New York, 1983, pp. 180–185.
9. J. J. Ebers and J. L. Moll, "Large Signal Behavior of Junction Transistors," *Proc. IRE,* **42:**1761–1772 (1954).
10. Raymond Warner and James N. Fordemwalt, *Integrated Circuits, Design Principles and Fabrication,* Motorola Series in Solid-State Electronics, McGraw-Hill, New York, 1965.

11. Gerold W. Neudeck, *The Bipolar Junction Transistor,* Addison-Wesley, Reading, Mass., 1983.
12. Herbert Kroemer, "Heterostructure Bipolar Transistors and Integrated Circuits," *Proc. IEEE,* **70**(1):13–25 (1982).
13. Texas Instruments, Inc., VLSI Laboratory, "Technology and Design Challenges of MOS VLSI," *IEEE J. Solid-State Circuits,* **SC-17**(3):442–448 (1982).
14. Sorab Ghandhi, *VLSI Fabrication Principles: Silicon and Gallium Arsenide,* Wiley, New York, 1983, pp. 437–439.
15. S. M. Sze, *VLSI Technology,* McGraw-Hill, New York, 1983, p. 347.

ADDITIONAL READING SUGGESTIONS

1. Robert F. Pierret, *Semiconductor Fundamentals,* Addison-Wesley, Reading, Mass., 1983.
2. Charles M. Botchek, *VLSI: Basic MOS Engineering,* Vol. I, Pacific Technical Group, Inc., Saratoga, 1983.
3. A. S. Grove, *Physics and Technology of Semiconductor Devices,* Wiley, New York, 1967.

CHAPTER 6
VLSI DEVICE PARAMETERS

Bhadrik Dalal
Amdahl Corporation
Sunnyvale, California

6.1 PHYSICAL DIMENSIONS

Devices, circuits, and interconnection networks are the basic building blocks of an IC. The physical dimensions of these building blocks influence either directly or indirectly many of the generally recognized figures of merit that are used as indicators of progress in the semiconductor industry. Device density, circuit integration level, circuit switching speed, IC power consumption, and IC reliability are some of the key figures of merit of ICs. All of these are affected by the physical dimensions of the basic devices, circuits, and interconnections.

6.1.1 Device Dimensions

Important physical dimensions of active devices are the key device dimension, the minimum feature size, and the minimum device area. The device area could be further subdivided into active device area and parasitic device area.

Key Device Dimension. The two major active device types, namely, the bipolar junction transistor (BJT) and the field-effect transistor (FET) have different carrier transport mechanisms. In the BJT, which is a minority carrier device, the basewidth is the key device dimension responsible for transistor action. The *basewidth* is a vertical dimension which is determined by successive diffusion or implantation processes. The control of basewidth to a few hundred angstroms is achievable in modern bipolar transistors having a basewidth of 1000 to 2000 Å.

In the FET, which is a majority carrier device, the channel length is the key device dimension responsible for transistor action. The channel length is a horizontal dimension determined by lithographic processes. The control of channel length to a few thousand angstroms is achievable in modern FETs having a channel length of 8000 to 20,000 Å.

Minimum Feature Size. The minimum feature size of a device or an interconnect line is determined by the line-patterning capability of lithography equipment used in the IC fabrication process. The imaging capability of an optical system depends on the wavelength of light used. The minimum feature size for production ICs is around 1 μm today. This is approaching the limit of current wafer stepper systems, which use the mercury G, H, or I lines at 436, 405, and 365 nm, respectively. To extend the maximum optical resolution beyond that provided by these current wafer steppers, researchers have used excimer laser steppers[1,2] to achieve half-micron minimum feature sizes. To achieve further reduction in minimum feature size, e-beam and x-ray imaging is necessary. The channel length of a FET and the emitter width of a BJT are around 1 μm in high-performance ICs.

Minimum Device Area. The size of a transistor is influenced by its intended application. The smallest definable transistor based solely on lithographic capabilities usually proves to be inadequate for logic switching circuits. In memory circuit applications, however, the smallest definable transistors are usually adequate. The switching transistors used in logic circuits are generally defined as separate components, whereas those used in memory circuits are often integrated with other components that make up a memory cell.

Based on a 1-μm minimum feature size, the minimum logic switching transistor areas of the FET and BJT are about 50 and 100 μm^2, respectively. This transistor area is the sum of both the active and parasitic device areas. The active area of a FET, which is the channel area, is usually 10 to 15 percent of its total device area. For a BJT, the active area is the emitter area, which represents only 4 to 8 percent of its total device area. Though the active device areas of the FET and BJT are comparable, the parasitic device overhead is significantly greater for the BJT.

6.1.2 Interconnect Dimensions

Important physical dimensions of interconnections are the minimum linewidth, minimum separation, and the thickness of the interconnect material.

Minimum Linewidth. Interconnect lines usually run over a rough surface, unlike the smooth surface over which active devices are patterned. Consequently, the minimum feature size used for interconnects is somewhat larger than that used for active devices, based on patternability considerations. In comparison with a 1-μm minimum device feature size in use today, metal interconnect lines have a minimum width of about 2 μm. Silicides, polycides, and polysilicon when used for interconnections have a width between 1 and 2 μm.

The linewidth of interconnects is also influenced by other important considerations, such as electromigration resistance and line resistance. These considerations may necessitate widening of lines for signal interconnects. For power supply and ground bus distribution, the widths are generally much wider because of noise margin considerations.

Minimum Separation. In ICs interconnect line separation is similar to the linewidth. The primary motivation is to maintain good interconnect density. In printed circuit boards, where the key considerations are to maintain a good control on the characteristic line impedance and to achieve a low crosstalk noise, the line separation is usually three to four times the linewidth. IC performance is rapidly improving to a point where these factors will become significant. Multiple metal interconnect levels will then be needed to satisfy the routing resource requirement, which is rapidly growing with increase in integration level.

Thickness. As the width of the interconnect is reduced, its thickness is also reduced to maintain good step coverage and to facilitate pattern etch. The thickness of aluminum metal lines is usually 6000 to 15,000 Å, depending on its level in a multilevel interconnect process. Two- and three-level metal interconnect processes are routinely used today in bipolar ICs, with four-level metal interconnect processes being reported[3] in the literature.

The electrical resistivity of a thin film begins to increase as the width of the line approaches the grain size of the interconnect material. The reduced cross-sectional area resulting from reduced thickness, coupled with the increase in electrical resistivity, poses a severe problem in managing the line resistance of long nets.

6.1.3 Scaling

The number of components per IC or the integration complexity of a commercially manufactured IC is perhaps the most important single measure of IC progress. Improvements in integration complexity of ICs are caused by several factors: (1) drive to reduce the cost per function, (2) drive to reduce the minimum feature size, (3) advancements in processing, (4) enhancements in device structure, and (5) innovations in circuit design techniques. The process of reducing the physical (both lateral and vertical) dimensions of both devices and interconnects is called *scaling*. The physical dimensions are scaled down by dividing them by a dimensionless constant factor S, where $S > 1$. Besides scaling physical device dimensions and doping concentrations, scaling of device threshold voltages and power supply voltages is also important, though not always possible in practice.[4]

Device and Voltage Scaling. Although device scaling approaches for FETs and BJTs are inherently different, the variables targeted for scaling are the same. These variables are the physical dimensions, supply voltages, and doping concentrations. Various MOSFET device parameters[5,6,7] are influenced through the scaling of these variables. Lateral-dimension scaling affects the device area, channel length, and channel width. Vertical-dimension scaling affects the gate oxide thickness and junction depth. Supply-voltage scaling and substrate-doping-concentration scaling are influenced by the drain to source punch-through voltage and the gate oxide breakdown. These scaling variables also influence other device parameters and effects, such as transit time, threshold voltage, transconductance, output drive current, punch-through current, hot carrier injection susceptibility, short channel effect, and narrow channel effect.

The MOSFET scaling relationships proposed by Dennard et al.[5] represent the constant-field scaling theory or the linear scaling theory. In this proposal all vertical and horizontal dimensions are reduced by the scaling factor S, the supply voltages are reduced by S, and the substrate doping concentration is increased by S. As a result, the MOSFET current is also reduced by a factor of S.

The interconnection capacitance in scaled ICs is dominated by fringing and interline coupling effects.[8] Therefore, in order to achieve increased circuit speed, the current drive capability of the MOSFET must be increased beyond that provided by linear scaling theory. To accomplish this circuit speed increase,

Chatterjee et al.[6] have proposed two approaches to non-constant-field scaling, called *constant-voltage scaling* and *quasi-constant-voltage scaling*.

In constant-voltage scaling, the supply voltage is kept constant, the dimensions and doping concentrations are scaled by S, and the gate oxide thickness is scaled by $\sqrt{S}$. Constant-voltage scaling increases the effective drive current down to a channel length of 1 μm, but falls off below a channel length of 1 μm.

In quasi-constant-voltage scaling, the supply voltage is scaled as $\sqrt{S}$, whereas all other variables, such as physical dimensions, gate oxide thickness, and doping concentrations, are scaled by S. Quasi-constant-voltage scaling increases the effective drive current down to a channel length of 0.5 μm for *n*-channel and to a channel length of 0.3 μm for *p*-channel.

Several BJT device parameters are influenced by scaling of physical dimensions, supply voltages, and doping concentrations. Lateral-dimension scaling affects the emitter size and the device area. Vertical-dimension scaling affects the basewidth, emitter depth, and epitaxy thickness. Supply-voltage scaling is influenced by punch-through voltage. Collector-doping-concentration scaling is influenced by base widening or the Kirk effect.[9] Base-doping-concentration scaling is influenced by collector to emitter punch-through.[10] These scaling variables also influence other device parameters, such as base transit time, base-emitter turn-on voltage, emitter current density, current gain, transconductance, and output drive current. BJT scaling is more complicated than MOSFET scaling because of its inherent device complexity. BJT scaling relationships have been proposed recently[11–14] that depend on whether the desired circuit delay is to be kept constant or reduced by S. In MOSFET and BJT scaling approaches, supply-voltage scaling has an impact on certain key circuit parameters, such as logic signal swing, noise margins, propagation delay, power consumption, and power delay product.

Interconnect Scaling. Two distinct types of interconnection nets are found in VLSI circuits. The first is the local interconnection net, which connects neighboring devices and circuits. This is the average type of net, which is scaled down by the same factor S by which the devices are scaled down. The other type of net is the long-distance interconnection net, which represents global interconnections between large functional blocks within the IC. This type of net gets scaled up in proportion to the increased linear dimension of the IC. Advancements in IC processing make it possible to economically fabricate ICs with increasing chip area; if the chip scaling factor is defined as S_c (where $S_c > 1$), then the long-distance net increases in proportion to S_c.

An approach to the scaling of interconnections has been proposed by Saraswat et al.[15] In this approach, the scaling of long-distance interconnections for constant local response time and constant field is carried out. This is done by scaling the long-distance interconnection directly proportional to S_c, the line resistance scaling in proportion to $S^2 S_c$, the line capacitance scaling in proportion to S_c, and the line response scaling in proportion to $(SS_c)^2$. The linewidth and separation are made equal to the minimum feature size λ, the line thickness is set at 0.25 λ, and the dielectric thickness is set at 0.35 λ.

Consequently, the average net delay reduces in inverse proportion to the device scaling factor S, whereas the long-distance net delay increases in direct proportion to the square of the product of the device scaling factor and the chip scaling factor, namely $(SS_c)^2$. This phenomenon is now being recognized as the critical limitation in the continuing improvement of future high-speed VLSI circuits.[8,16]

6.2 DEVICE MODELS AND PARAMETERS

Circuit simulation programs are generally used for nonlinear dc, nonlinear transient, and linear ac analyses. These circuit simulation programs usually have built-in device models. The most commonly used circuit simulation program for VLSI designs is the SPICE program[17] from the University of California at Berkeley. This program has undergone several updates and enhancements. The latest revision routinely used by most VLSI circuit design engineers is SPICE2G.6. However, a newer version called SPICE3[18] has been made available to selected users.

6.2.1 Device Models

The semiconductor device models for FETs and BJTs used in SPICE circuit simulation programs are listed below. The BJT model used in SPICE2 is a modified Gummel-Poon[19] model that reverts to the familiar Ebers-Moll[20] model in its simplest form. The JFET model used in SPICE2 is based on the FET model described by Schichman and Hodges.[21] Three MOSFET models are used in SPICE2. A fourth MOSFET model is available in SPICE3. The MESFET model used in SPICE3 is derived from the GaAs FET model described by Statz et al.[22]

SPICE2 BJT Model. The basic large-signal nonlinear BJT model described by Ebers and Moll[20] in 1954 is the simplest BJT model used for dc analysis. The Ebers-Moll model is described by the following model parameters: forward current gain B_F, reverse current gain B_R, saturation current I_S, and energy gap E_G. For better dc characterization, ohmic resistances for the emitter R_E, base R_B, and collector R_C terminals are added in the SPICE2 BJT model. For better transient characterization, charge storage effects need to be added. This is done by the addition of nonlinear junction-depletion capacitances for the emitter C_{JE} and the collector C_{JC} junctions, diffusion capacitances that model charge storage through forward transit time T_F and reverse transit time T_R, and a constant-substrate capacitor C_{JS}. This BJT model is based on the integral charge model described by Gummel and Poon.[19] The model reduces to the simpler Ebers-Moll model when the Gummel-Poon parameters are not specified. The Gummel-Poon model also incorporates additional features, such as basewidth modulation, variation of current gain with collector current and voltage, variation of forward transit time with collector current, and variation of device parameters with temperature. A detailed explanation of these features and the BJT model has been given by Getreu.[23]

SPICE2 JFET Model. The JFET model used in SPICE2 is derived from the FET model described by Schichman and Hodges.[21] The threshold voltage V_{TO}, device transconductance parameter β, channel length modulation parameter λ, gate junction saturation current I_S, drain ohmic resistance R_D, and source ohmic resistance R_S define the dc characteristics of the JFET. The nonlinear junction-depletion capacitances associated with the gate-source C_{GS} and the gate-drain C_{GD} *pn* junctions provide transient characterization of the JFET.

SPICE2 MOSFET Model. Three MOSFET models are provided in SPICE2. The model MOS1 is the LEVEL 1 model and is the simplest of the three models. It is based on the FET model of Schichman and Hodges.[21] The dc characteristics of the MOSFET are described by the extrapolated threshold at zero drain current V_{TO}, device transconductance parameter K_P, bulk threshold parameter γ, surface potential ϕ, and channel-length modulation parameter λ. Charge storage is modeled by several capacitors. The constant overlap capacitances associated with the gate-source, gate-drain, and gate-bulk overlaps are modeled by C_{GSO}, C_{GDO}, and C_{GBO}, respectively. The nonlinear junction-depletion layer capacitances associated with the bulk-source and bulk-drain *pn* junctions are modeled by C_{BS} and C_{BD}, respectively. The thin-oxide charge storage effects are modeled by two built-in models that use the piecewise linear voltage-dependent capacitances proposed by Meyer.[24] The LEVEL 2 and LEVEL 3 models include other second-order effects, such as subthreshold conduction, scattering limited velocity saturation, and small-size effects.[17] For a detailed understanding of these effects, Refs. 25 and 26 are recommended.

SPICE3 MESFET Model. The MESFET model used in SPICE3[18] is derived from the GaAs FET model described by Statz et al.[22] The pinch-off voltage V_{TO}, device transconductance parameter β, saturation voltage parameter α, channel length modulation parameter λ, drain ohmic resistance R_D, and source ohmic resistance R_S define the dc characteristics of the MESFET. The nonlinear junction-depletion capacitances associated with the gate-source C_{GS} and the gate-drain C_{GD} Schottky junctions provide transient characterization of the MESFET.

6.2.2 Device Parameters

BJT Parameters. The general ranges of key device parameters for a 1-μm advanced bipolar technology are as follows:[27–29]

Parameter	*Range of Values*
Emitter area	$A_e = 0.8 \times 3.3\ \mu m^2$ to $1 \times 8\ \mu m^2$
Emit.-base capacitance	$C_{JE} = 9.6$ to 18 fF
Coll.-base capacitance	$C_{JC} = 5$ to 14 fF
Coll.-sub. capacitance	$C_{JS} = 15$ to 30 fF
Gain-bandwidth product	$F_I = 12$ to 14 GHz
Base resistance	$R_B = 90$ to 500 Ω
Current gain	$B_F = 50$ to 70
Transconductance	$g_m = 5$ to 20 mS

The total device area for an advanced BJT ranges from 85 to 128 μm^2. The current density at which these transistors can be operated ranges between 200 to 670 $\mu A/\mu m^2$.

MOSFET Parameters. The typical device parameter values for a 1-μm advanced *n*-channel MOSFET technology are as follows:[30]

Parameter	*Value*
Channel length	$L = 1\ \mu m$
Channel width	$W = 7.5\ \mu m$
Oxide thickness	$T_{OX} = 250$ Å

Threshold voltage	$V_{TO} = 0.5$ V
Transconductance	$g_m = 0.57$ mS
Gate-source capacitance	$C_{GSO} = 1.7$ fF
Gate-drain capacitance	$C_{GDO} = 1.7$ fF
Gate-bulk capacitance	$C_{GBO} = 2.8$ fF
Bulk-source capacitance	$C_{BS} = 3.6$ fF
Bulk-drain capacitance	$C_{BD} = 3.6$ fF

The total device area for an advanced n-channel MOSFET ranges from 40 to 60 μm^2. The typical saturation current at which these transistors are operated ranges from 300 to 500 μA.

6.2.3 Interconnect Parameters

Two- and three-layer metal interconnect systems are generally required in high-performance bipolar VLSI circuits, whereas two-layer metal systems are generally adequate for high-performance MOS VLSI circuits. The first-layer metal is usually aluminum between 5000 and 8000 Å thick having a linewidth ranging from 1.5 to 2.0 μm with similar line separation. The line resistance ranges from 20 to 40 Ω/mm, and the line capacitance ranges from 200 to 400 fF/mm, resulting in an 8-ps time constant for 1 mm of line length. Because the time constant is proportional to the square of the line length, for nets larger than a few millimeters long, the time constant becomes the limiting factor for high-performance bipolar VLSI circuits. The second-layer metal characteristics are similar to the first metal if the process has a reasonably planar device topography; otherwise the second metal rules tend to be somewhat larger in all three dimensions.

6.3 CHIP PARAMETERS

6.3.1 Product Types

The major application of digital ICs are in computer products, telecommunication products, and consumer products. These system products utilize ICs that fall into two major IC product classifications, namely, logic IC products and memory IC products. The logic IC products are comprised of application specific integrated circuits (ASICs), standard logic circuits, custom circuits, and microprocessor circuits. The memory IC products are subdivided into major subclassifications, namely, dynamic random-access memories (DRAMs), static random-access memories (SRAMs), and programmable read-only memories (PROMs), which can be programmed through a number of ways, such as IC mask, ultraviolet light, or electrical input. Among these product types, ASICs, microprocessors, and DRAMs represent the driving force for IC technology development.

6.3.2 Integration Level

Integrated circuits have traditionally been classified as either MOS-based circuits or bipolar-based circuits. In the past few years, circuits that have both MOS and bipolar devices on the same chip, called BiMOS ICs, have emerged as a major new development. These BiMOS ICs aim to exploit the advantages of MOSFETs and BJTs without incurring the disadvantages of either.

The most dominant IC products are based on CMOS circuits. The highest-performance IC products are based on bipolar emitter-coupled-logic (ECL) circuits. Integration level is traditionally measured in the number of components per commercially manufactured IC. Other measures of integration level are the number of storage bits per IC for memory products and the number of logic gates per IC for logic products. CMOS products exhibit an integration level that is 2 to 10 times greater than that found in ECL products, whereas ECL products outperform CMOS products by a factor of 2 to 5 in high-speed applications.

The present status of integration level in various product types is as follows: DRAM products have the highest level of integration, with 1-Mbit commercial products already in production. SRAM and PROM commercial products in the 256-kbit range are becoming available. Microprocessor products with 32-bit architecture are also commercially available. ASIC products with gate complexity of about 20,000 gates are generally available commercially.

6.3.3 Process Complexity

Although process complexity is usually measured in number of process steps and mask levels, other factors that also contribute to the overall process complexity are minimum feature size, device structure, device isolation approach, and number of interconnect levels. An advanced bipolar process today uses 1-μm feature size, trench isolated extended electrode devices using double polysilicon layers having two to four layers of metal interconnect and using 14 to 18 mask levels. The process uses an epitaxial layer and an ion-implanted structure.

An advanced CMOS process today uses a 1-μm feature size, trench transistors and capacitors, devices using polysilicon gates having two layers of metal interconnect and using 12 to 14 mask levels. The process also uses epitaxial layers and ion-implanted structures. It is becoming apparent that advanced CMOS process is getting as complex as the bipolar process.

6.3.4 Area and Yield

The product type and integration level of an IC determine the chip area. Chip sizes of 1 cm^2 for microprocessor and ASIC products are common. This is true for both CMOS and bipolar products. The memory products, which are usually considerably smaller than the logic products, range from 0.3 to 0.6 cm^2 in chip area.

The yield of an IC depends on its area and defect density. Other important factors that influence yield are device type, process complexity, and parameter control.

6.3.5 Power Dissipation

The power dissipation of an IC is generally governed by its end usage. In applications where cost is the primary objective, the power dissipation tends to be usually under 1 W per IC so that cooling expense can be minimized. In forced-air-cooled applications, power dissipation of 2 to 6 W per IC is quite common. In liquid-cooled system applications, power dissipation of 8 to 12 W per IC is generally acceptable. In future, some applications could accept a power dissipation of 20 to 30 W per IC. For ICs that have a large power dissipation, considerable expense is incurred by the customer in providing adequate cooling approaches.

ACKNOWLEDGMENTS

The author would like to express his gratitude for the support and encouragement provided by Mr. Michael Clements and Mr. Bruce Beebe. The author would also like to thank Mrs. K. Bellew for her assistance in the preparation of the manuscript.

REFERENCES

1. James H. Bennewitz et al., "Excimer Laser-Based Lithography for 0.5 μm Device Technology," *IEEE IEDM Technical Digest,* December 7–10, 1986, pp. 312–315.
2. M. Sasago et al., "Half-Micron Photolithography Using a KrF Excimer Laser Stepper," *IEEE IEDM Technical Digest,* December 7–10, 1986, pp. 316–319.
3. Stewart Brenner et al., "A 10,000 Gate Bipolar VLSI Masterslice Utilizing Four Levels of Metal," *IEEE ISSCC Digest of Technical Papers,* February 23–25, 1983, pp. 152–153, 298.
4. David A. Hodges and Horace G. Jackson, *Analysis and Design of Digital integrated circuits,* McGraw-Hill, New York, 1983, pp. 112–115.
5. Robert H. Dennard et al., "Design of Ion-Implanted MOSFETs with Very Small Physical Dimensions," *IEEE J. Solid-State Circuits,* **SC-9** (5): 256–268 (1974).
6. Pallab K. Chatterjee et al., "The Impact of Scaling Laws on the Choice of n-Channel or p-Channel for MOS VLSI," *IEEE Electron Device Letters,* **EDL-1** (10): 220–223 (1980).
7. B. Hoeneisen and C. A. Mead, "Fundamental Limitations in Microelectronics. I. MOS Technology," *Solid State Electronics,* **15:** 819–829 (1972).
8. Paul M. Solomon, "A Comparison of Semiconductor Devices for High-Speed Logic," *Proc. IEEE,* **70** (5): 489–509 (1982).
9. C. T. Kirk, Jr., "A Theory of Transistor Cut-off Frequency Fall-off at High Current Densities," *IRE Transactions on Electron Devices,* **ED-9:** 164–174 (1962).
10. B. Hoeneisen and C. A. Mead, "Limitations in Microelectronics. II. Bipolar Technology," *Solid State Electronics,* **15:** 891–897 (1972).
11. Paul M. Solomon and Denny D. Tang, "Bipolar Circuit Scaling," *IEEE ISSCC Digest of Technical Papers,* February 14–16, 1979, pp. 86–87.
12. T. H. Ning et al., "Scaling Properties of Bipolar Devices," *IEEE IEDM Technical Digest,* December 8–10, 1980, pp. 61–64.
13. J. S. T. Huang, "MOS/Bipolar Technology Trade-Offs for VLSI," In *VLSI Electronics: Microstructure Science,* Vol. 9, N. Eisnpurch (ed.), Academic Press, Orlando, Fla., 1985, Chap. 1.
14. J. S. T. Huang, "Bipolar Device Scaling and Limitations," *IEEE Bipolar Circuits and Technology Meeting Proceedings,* September 11–12, 1986, pp. 67–70.
15. Krishna C. Saraswat and Farrokh Mohammadi, "Effect of Scaling of Interconnections on the Time Delay of VLSI Circuits," *IEEE Solid-State Circuits,* **SC-17** (2): 275–280 (1982).

16. James D. Meindl, "Chips for Advanced Computing," *Scientific American,* **257** (4): 78–88 (1987).
17. A. Vladimirescu et al., *SPICE Version 2G User's Guide,* Department of Electrical Engineering and Computer Sciences, University of California, Berkeley, Cal., 1981.
18. T. Quarles et al., *SPICE 3A7 User's Guide,* Department of Electrical Engineering and Computer Sciences, University of California, Berkeley, Cal., 1986.
19. H. K. Gummel and H. C. Poon, "An Integral Charge Control Model of Bipolar Transistors," *Bell Systems Technical J.,* **49:** 827–852 (1970).
20. J. J. Ebers and J. R. Moll, "Large Signal Behavior of Junction Transistors," *Proc. IRE,* **42:** 1761–1772 (1954).
21. Harold Schichman and David A. Hodges, "Modeling and Simulation of Insulated-Gate Field-Effect Transistor Switching Circuits," *IEEE J. Solid-State Circuits,* **SC-3** (3): 285–289 (1968).
22. H. Statz et al., "GaAs FET Device and Circuit Simulation in SPICE," Internal Memorandum, Raytheon Research Division, Lexington, Mass., 1985.
23. Ian Getreu, *Modeling the Bipolar Transistor,* Tektronix, Beaverton, 1976.
24. John E. Meyer, "MOS Models and Circuit Simulation," *RCA Review,* **32:** 42–63 (1971).
25. Robert F. Pierret, *Field Effect Devices,* Addison-Wesley, Reading, Mass., 1983.
26. Dieter K. Shroder, *Advanced MOS Devices,* Addison-Wesley, Reading, Mass., 1987.
27. H. Takemura et al., "Submicron Epitaxial Layer and RTA Technology for Extremely High Speed Bipolar Transistors," *IEEE IEDM Technical Digest,* December 7–10, 1986, pp. 424–427.
28. Tohru Nakamura et al., "63 ps ECL Circuits Using Advanced SICOS Technology," *IEEE IEDM Technical Digest,* December 7–10, 1986, pp. 472–475.
29. Tohru Nakamura et al., "Integrated 84 ps ECL with I^2L," *IEEE ISSCC Digest of Technical Papers,* February 22–24, 1984, pp. 152–153, 330.
30. Donald L. Fraser, "High Speed MOSFET IC Design," IEEE IEDM Short Course Lecture Notes, December 7, 1986, pp. 3.43–3.44.

CHAPTER 7
VLSI CIRCUIT DESIGN

Patrick Yin
Structured Product Development
LSI Logic Corporation
Milpitas, California

Tzoyao Chan
ASIC Memory Design
LSI Logic Corporation
Milpitas, California

7.1 INTRODUCTION

In this chapter, the fundamental characteristics of MOS transistors are reviewed. Static and dynamic MOS digital circuit operations are discussed. VLSI circuit elements such as capacitor, pass transistor, input and output buffers, memory cells, and flip-flops are then introduced.

7.2 TRANSISTOR FUNDAMENTALS

7.2.1 Basic Structures

The basic MOSFET (metal-oxide semiconductor field-effect transistor) consists of a gate electrode, a drain, and a source region. Figure 7.1 shows a diagrammatic cross section of an *n*-channel MOS transistor. The gate electrode, composed of polysilicon, is insulated from the silicon substrate by a very thin layer of silicon dioxide. The source and the drain are produced by ion implanting the region with *n*-type material for *n*-channel transistors and *p*-type material for *p*-channel transistors. The substrate, or the bulk, of the *n*-channel and the *p*-channel are lightly doped *p*-type and *n*-type materials, respectively.

When the gate electrode is open-circuited, there is no conducting path between the source and the drain. The substrate material between the drain and the source regions has high resistivity and forms two back-to-back diodes with the two regions.

When a positive voltage is applied to the gate of an *n*-channel transistor, an

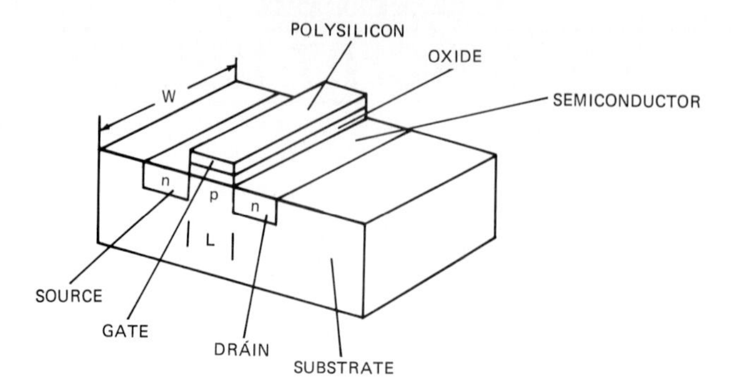

FIG. 7.1 *N*-Channel MOS structure.

electric field is established across the gate oxide. The induced negative mobile charges form the conducting channel. Electric current flows from the drain to the source, or vice versa, depending on the electric potential between the two regions. The conductance of the channel is a function of the gate-source (or the gate-drain) voltage. Similarly, a negative voltage applied to the gate of a *p*-channel MOSFET creates a conducting channel between its source and drain.

The aforementioned transistors are *enhancement* type. The gate-source voltage $|V_{gs}|$ must be equal to or greater than the threshold voltage $|V_t|$ in order to create the conducting channel.

In the *depletion* type, the channel is achieved by ion implanting the silicon surface during wafer fabrication. The channel conductance is still a function of the gate-source voltage, but a negative V_{gs} is required to deplete the channel, hence reducing the conductance to zero.

7.2.2 Device Equations

The Linear Region. The *I-V* characteristics of Fig. 7.2 show a region of almost linear dependence of the drain-source current I_{ds} on the drain-source voltage V_{ds}. In this region, where $V_{ds} < V_{gs} - V_t$,

$$I_{ds} = B\left[(V_{gs} - V_t) - \frac{V_{ds}}{2}\right] V_{ds} \tag{7.1}$$

Under this condition there is a strong inversion layer formed along the entire channel.

The Saturation Region. As V_{ds} increases, the electric field in the oxide layer at the drain end diminishes, thus reducing the inversion layers to zero at the drain end of the channel. The "pinched-off" channel produces an infinitely high resistance region. Then I_{ds} becomes essentially constant when

$$I_{ds} = \frac{B(V_{gs} - V_t)^2}{2} \qquad V_{ds} > V_{gs} - V_t \tag{7.2}$$

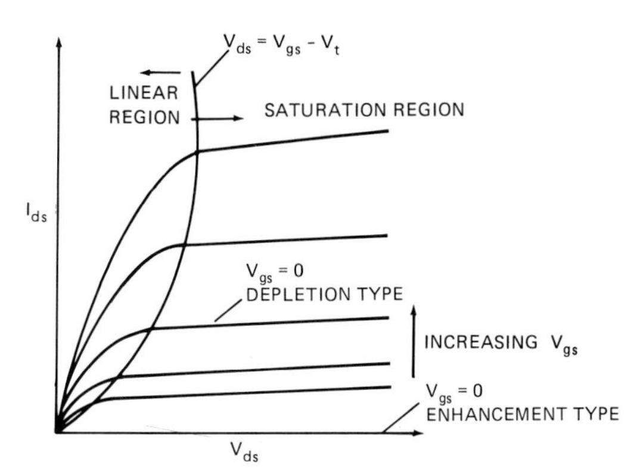

FIG. 7.2 Transfer characteristics.

But as V_{ds} increases further, the electric field at the drain end reverses direction, causing the depletion layer to grow at the expense of the conduction channel. This reduction of channel length, known as channel length modulation, causes a slight increase of I_{ds}. Therefore in the saturation region I_{ds} increases with V_{ds}, giving a finite output impedance characteristic as shown in Fig. 7.2

The Transconductance g_m

$$g_m = \frac{\Delta I_{ds}}{\Delta V_{gs}} \quad \text{for a given } V_{ds} \tag{7.3}$$

Applying this expression to the equation for the linear region, we obtain

$$g_m = BV_{ds} \tag{7.4}$$

Similarly in the saturation case the transconductance can be expressed as

$$g_m = B(V_{gs} - V_t) \tag{7.5}$$

if the effect of channel length modulation is ignored. Therefore g_m expresses the dependence of I_{ds} on V_{gs}; on the transfer characteristics (Fig. 7.2), g_m determines the spacing between the curves.

The Factor B ($= K'\ (W/L)$). The factor B has two basic components:
The Conduction Factor K'

$$K' = \frac{e_{ox}\mu}{t_{ox}} \tag{7.6}$$

where e_{ox} = permittivity of gate oxide
μ = effective surface mobility of carriers
t_{ox} = thickness of gate oxide

K' is technology dependent, that is, it depends on the process variables, such as mobility and crystal orientation. For instance, the (110) orientation has a higher value of hole mobility than the (100) orientation does. Electrons have approxi-

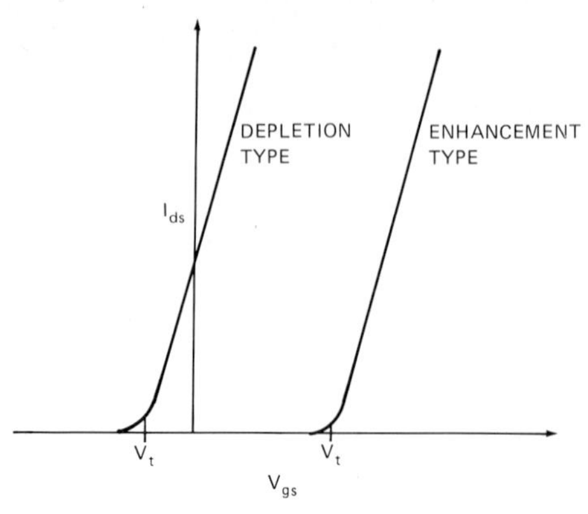

FIG. 7.3 I_{ds} versus V_{gs} characteristics.

mately three times the surface mobility values of the holes; thus n-channel transistors under the same conditions are faster than p-channel transistors by the same ratio. Furthermore, μ varies inversely with temperature. Thus MOS devices have greater propagation delays at elevated temperatures due to reduced I_{ds}. The gate oxide t_{ox} varies greatly with the fabrication process. In a 2-μm process t_{ox} is typically 45 nm.

The Geometrical Factor W/L. W and L are the "effective" channel width and length, respectively, as shown in Fig. 7.1. The term *effective* refers to the resultant dimensions after the tolerance of photomask, the effect of etching, lateral diffusion of the source and drain regions, and oxide encroachment have been taken into account. An increase in W/L increases the current drive capability of the transistor, causing a corresponding improvement in response time in a digital circuit. However, increasing W increases the physical areas of the gate, the source, the drain, and, hence, the associated capacitances. These increases in parasitic capacitances could have an adverse effect on circuit performance in higher levels of circuit integration.

7.2.3 The Threshold Voltage V_t

For an n-channel MOS transistor,

$$V_t = \phi_{gs} + \frac{Q_{ss}}{C_{ox}} + 2\phi_f + \frac{\sqrt{2e_{ox}qN_B(2\phi_f + V_{bg})}}{C_{ox}} \tag{7.7}$$

where ϕ_{gs} = polysilicon gate to silicon work function
Q_{ss} = positive fixed charges at silicon and silicon dioxide interface; value of Q_{ss} varies with crystalline orientation used
C_{ox} = gate capacitance per unit area (= e_{ox}/t_{ox})
ϕ_f = substrate Fermi potential at surface
N_B = average substrate concentration at silicon surface
V_{bg} = source-substrate voltage

From Eq. (7.7) we can see that the major portion of V_t is used to overcome the

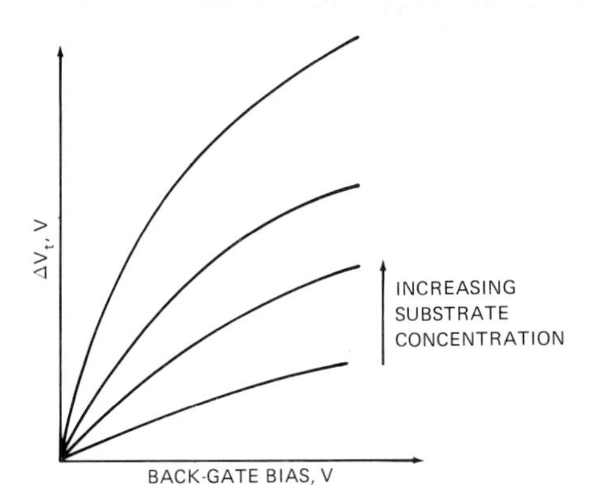

FIG. 7.4 Back bias effect.

effect of the charge created by exposed dopants at the surface, the fixed charges Q_{ss} in the gate oxide and silicon interface, and the gate to silicon work function difference ϕ_{gs}. An additional gate voltage, equal to $2\phi_f$, is responsible for the creation of the inversion layer—the conducting channel. The value of V_t can be modified by surface implant, which changes the value of N_B at the silicon surface.

The significance of V_{bg} can be seen in Fig. 7.4. Basically, the substrate can be considered as another gate, at the back of the normal gate, operating in a reverse-bias sense. For *n*-channel MOS transistors V_{bg} is typically at zero or some negative voltage. The more negative V_{bg} is, the higher V_t has to be in order to maintain channel conductance. This phenomenon of V_t dependence on V_{bg} is known as *back-gate bias* or *substrate bias* effect.

7.2.4 Static Analysis

A simple inverter is used for this discussion, which can be extended to other logic gates (i.e., NAND, NOR, and other complex gates). An inverter comprises a DRIVER, enhancement-type NMOS transistors, and a LOAD. The LOAD device could be any of the following types (see Fig. 7.5*a* to *e*):

1. A resistor
2. A saturated enhancement NMOS transistor
3. A nonsaturated enhancement NMOS transistor
4. A depletion NMOS transistor
5. A nonsaturated enhancement PMOS transistor

Figure 7.6 shows the transfer characteristics of inverters using different loads.

Resistive Load. In general, the output voltage level of the inverter is given by

$$V_{\text{out}} = V_{dd} - (I_{ds} + I_{\text{load}})R \tag{7.8}$$

where I_{ds} = driver and drain-source current and I_{load} = external dc current load.

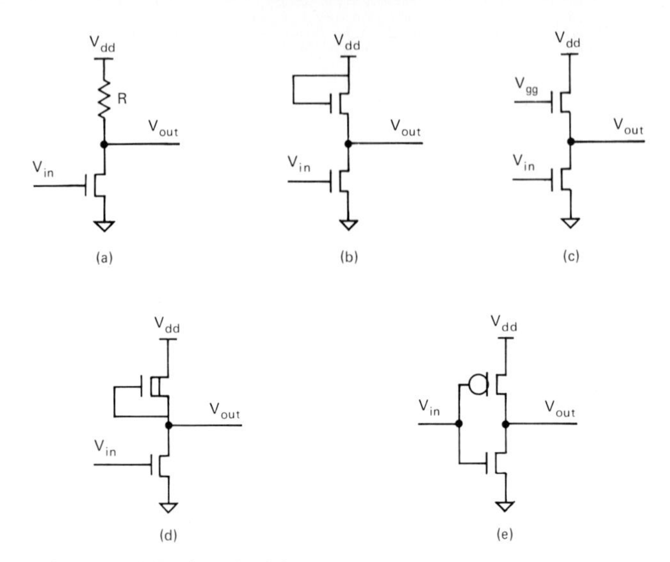

FIG. 7.5 (*a*) Resistor load inverter; (*b*) enhancement-type saturated load inverter; (*c*) enhancement-type nonsaturated load inverter; (*d*) depletion load inverter; (*e*) complementary load inverter.

When V_{in} is low (i.e., $V_{\text{in}} < V_{td}$, where V_{td} is the threshold voltage of the driver transistor), the driver transistor is off and $I_{ds} = 0$. The expression for the output voltage then becomes

$$V_{\text{out}}(H) = V_{dd} - V_l \tag{7.9}$$

where $V_l = I_{\text{load}}R$ is the voltage drop across the resistor due to external dc current loading. V_l can be adjusted to a desired value by choosing different values of R.

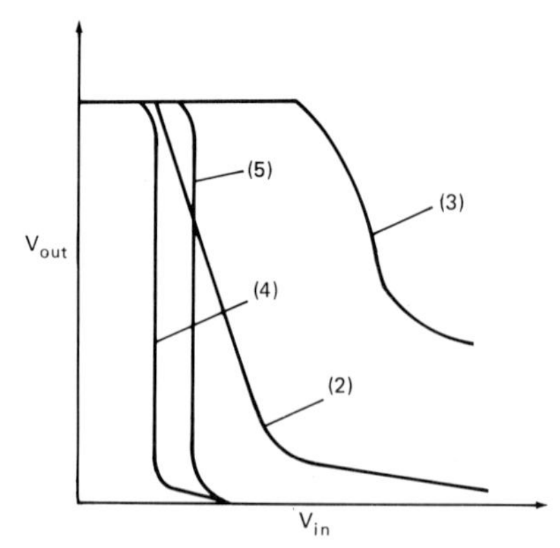

FIG. 7.6 Inverter transfer curves for devices listed on page 7.5.

When V_{in} is high (i.e., $V_{in} > V_{td}$), the driver is on, and

$$V_{out}(L) = V_{dd} - I_{ds}R \tag{7.10}$$

If the output of the inverter only drives MOS transistors, $V_l = 0$. The dc power dissipation is

$$P = I_{ds}V_{dd} \tag{7.11}$$

Saturated Enhancement NMOS Load. The gate of the load transistor is connected to the drain, making $V_{ds} = V_{gs}$. Hence the load operates in the saturation mode. When V_{in} is low, the driver is off. Then

$$V_{out}(H) = V_{dd} - V_{tl} \tag{7.12}$$

where V_{tl} = threshold voltage of the load, typically greater than V_{td} due to back-gate bias.

When V_{in} is high, the driver is on and operates in the linear region; both load and driver are conducting, and I_{ds}(load) = I_{ds}(driver). The low-level output voltage is

$$V_{out}(L) = \frac{B_l}{2B_d}\left(\frac{V_{in}^2}{V_{in} - V_{td}}\right) \tag{7.13}$$

where B_l and B_d are the β factors of the load and the driver, respectively.

The dc power dissipation is simply $I_{ds}(l)V_{dd}$, where $I_{ds}(l)$ is the drain-source current of the load device when V_{out} is low. From the equations it is evident that the output low level depends heavily on the β ratio—the ratio of the transconductance of the driver to that of the load. Figure 7.7 shows the transfer characteristics of an inverter for different values of the β ratio. The geometric factor *W/L* of the load and the driver is often used to achieve the desired β ratio.

The Nonsaturated Enhancement NMOS Load. By connecting the gate of the load transistor to V_{gg}, a voltage greater than $V_{dd} + V_{tl}$, we make the load operate in

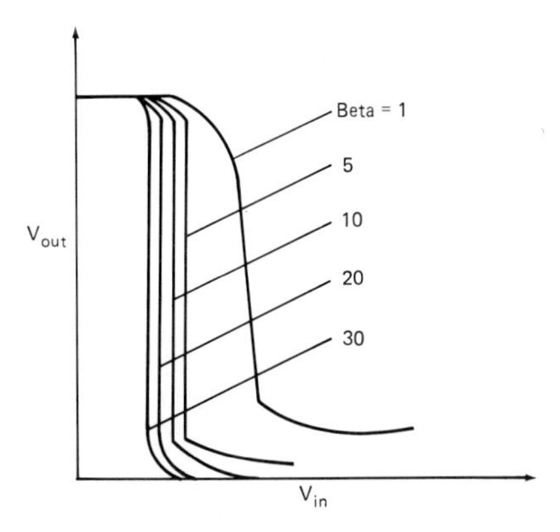

FIG. 7.7 Inverter transfer curve with different β ratio.

the linear region. A major reason for using this type of load is the higher voltage level obtainable at the output. When V_{in} is low and external dc load current is negligible,

$$V_{out}(H) = V_{dd} \tag{7.14}$$

When V_{in} is high, the output goes low and its voltage can be approximated by

$$V_{out}(L) = \frac{B_l}{B_d}\left(\frac{V_{gg} - V_{tl}}{V_{in} - V_{td}}\right)V_{dd} \tag{7.15}$$

The transfer characteristic of this type of inverter is similar to the saturated enhancement NMOS load. While V_{gg} can be used to improve the output (high) level and the transient response of the inverter, it also raises $V_{out}(L)$ and the dc power dissipation.

The Depletion NMOS Load. The depletion NMOS load is the most popular in NMOS ICs. The gate is connected to the output of the inverter as shown in Fig. 7.5*d*, making V_{gs} = O V. Since depletion transistors have a negative V_t, this type of load gives constant-current characteristics, allowing

$$V_{out}(H) = V_{dd} \tag{7.16}$$

When V_{in} is high, the driver transistor is on and

$$V_{out}(L) = \frac{B_l}{B_d}\left(\frac{V_{tl}^2}{V_{in} - V_{td}}\right) \tag{7.17}$$

For the four types of load mentioned thus far, the dc power dissipation of the inverter depends upon the transconductance of the load devices; the output level $V_{out}(L)$ is a function of the β ratio. These are all ratio-type inverters.

The Nonsaturated Enhancement PMOS Load. When a PMOS enhancement load is used as shown in Fig. 7.5*e*, we have a CMOS inverter. (C stands for the complementary devices used in the circuit.) There are many advantages in CMOS circuits, all based on the fact that the *p*-channel and *n*-channel devices are never on at the same time except during the very short transient period. The output levels do not depend on the β ratio. Thus, a full (ground to power supply) logic swing can be obtained. Consequently, excellent noise margin can be expected in most CMOS circuits. The quiescent power is practically zero. Therefore, the transient performance of the circuit is no longer constrained.

However, there are limiting factors in the transient response. The fabrication process dictates the feature sizes and layout design rules allowed in the circuit. The K' factor and the parasitic resistances and capacitances ultimately limit circuit performance. Quite often the consideration of noise generated from charging and discharging of load capacitances overrides the ac performance requirement. The noise level increases with increasing transconductances of the *p*-channel and *n*-channel devices. The negative sides of this technology are the more complicated fabrication process and the lower silicon real estate efficiency—more transistors are required to implement a logic gate, such as NAND and NOR gates, in

CMOS due to the complementary nature of the technology. The possibility of "latch-up" between parasitic *npn* and *pnp* transistors also requires a bigger spacing between *p*-channel and *n*-channel devices in the layout.

7.2.5 Noise Margin

In a system, for that matter within a highly complex IC, many sources of noise could render the system nonfunctional, giving erroneous outputs. In general, noise could be classified into ac noise and dc noise. Dc noise ranges from dc-level offsets (e.g., voltage drop on power supplies or signal lines due to dc current loading) to noise spikes whose pulse widths are longer than the propagation delay of one logic gate. On the other hand, ac noise spikes have pulse widths shorter than one logic gate delay. Noise spikes could be generated by capacitive or inductive coupling between signal lines or power supply bumps, caused by transistor charging and discharging of large load capacitances. In MOS ICs the output buffers are a great source of this kind of noise. The ac requirement of driving sizable capacitive loads (50 to 100 pF) and the dc requirement of sinking or sourcing dc current, up to 8 mA, often dictate the transconductances of the transistors used in the output buffers. Low transconductances result in high instantaneous I_{ds}, which together with the parasitic impedances of the circuit generates noise spikes. Various measures are being used by circuit and system designers to cope with this problem, such as shielding, ground planes on printed circuit boards, separate power supplies for output drivers, and slew-rate-controlled output drivers.

ICs used in a digital system typically have certain voltage levels set to represent the logic 1 and logic 0 for the input and output. Ideally we would like these to be the power supply voltages. But due to circuit and system design considerations, the voltage specifications for the logic 1 and logic 0 in practice have certain tolerances or limits within which logic states can be recognized in the system. As shown in Fig. 7.8, the output of a logic block has a minimum high voltage $V_{oh(\min)}$ and a maximum low voltage $V_{ol(\max)}$ to represent the logic 1 and logic 0, respectively. Similarly, the logic block being driven has $V_{il(\max)}$ and $V_{ih(\min)}$ as the input limits for logic 0 and logic 1, respectively. The noise margin in this case can be given as follows:

The noise margin for logic 1:

$$V_{nm_1} = V_{oh(\min)} - V_{ih(\min)} \tag{7.18}$$

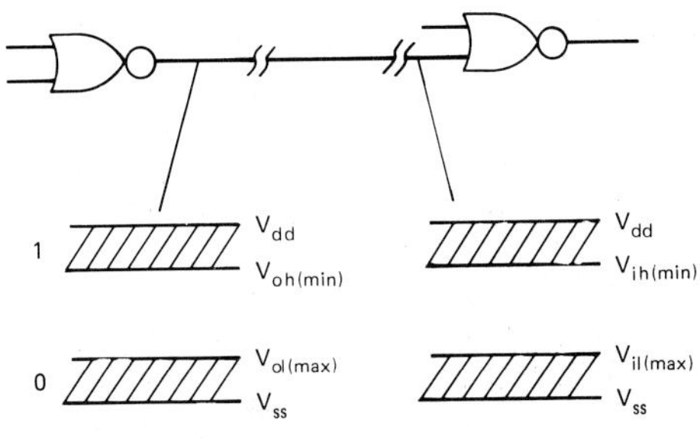

FIG. 7.8 Input and output logic levels.

The noise margin for logic 0:

$$V_{nm0} = V_{il(\max)} - V_{ol(\max)} \tag{7.19}$$

In ICs these specifications can be determined from the transferred characteristics of the logic blocks. In Fig. 7.9 points *b* and *c*, the unity gain points, are equivalent

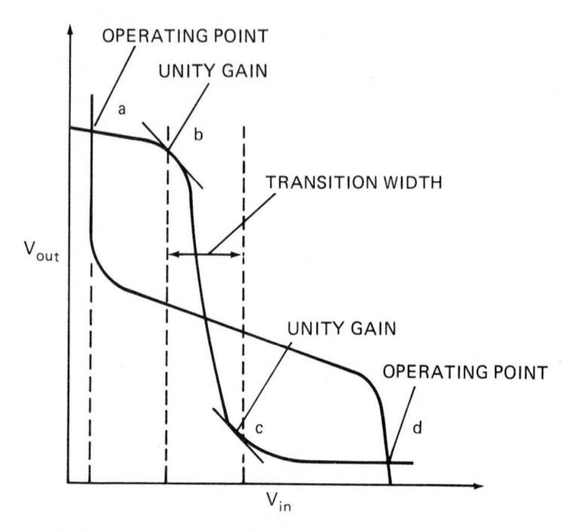

FIG. 7.9 Noise margin of inverter.

to $V_{il(\max)}$ and $V_{ih(\min)}$, respectively; between these two points is the transition region where the inverter has the highest gain. If the transfer curve is transposed and replotted on the same graph, points *a* and *d* are obtained (dashed lines in Fig. 7.9). These two points correspond to the static operating points of a logical 1 and a logical 0. In other words, when a voltage equivalent to the level at point *d* is applied to the input of the next logic block, the output of that block would be at a voltage level equal to point *a*. The noise margins of the inverter are defined:

$$V_{nm_0} = V_b - V_a \qquad V_{nm_1} = V_d - V_c \tag{7.20}$$

In practice, V_{nm_0} and V_{nm1} assume a range of values in an IC due to differences in transfer characteristic of logic blocks, process tolerances, temperature variations, loading conditions, and power supply voltage drops.

7.2.6 Transient Analysis

The transient performance of a MOS logic gate can be evaluated by calculating its output transition times T_{tlh} and T_{thl}—the low to high and high to low transition times, respectively. T_{tlh} is actually the time it takes to charge the output load capacitance from a low voltage level V_0 to a high voltage V_1; T_{thl} is the discharging time.

The major factors that govern the transient performance are the transconductances of the transistors, the power supply, and the loading condition. The process technology to a large extent is responsible for the continuing increase in IC ac performance and complexity. More advanced processes allow

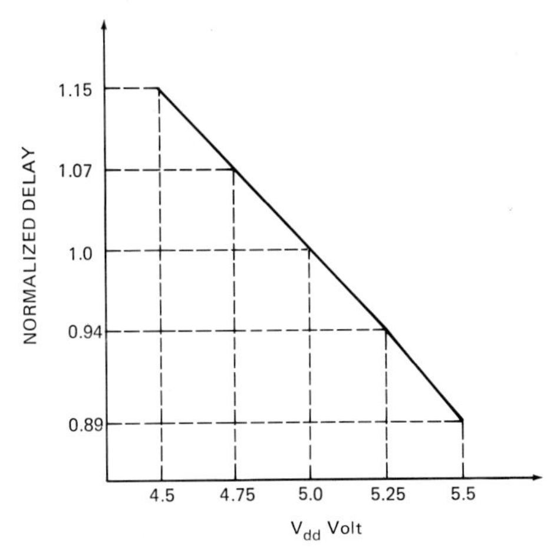

FIG. 7.10 Propagation delay versus supply voltage.

smaller feature sizes to be used in designing ICs. The reduction in feature sizes increases device transconductance, because of the smaller effective channel length, and improves loading conditions due to lower parasitic capacitances.

However, for a given IC one can expect ± 40% changes in transient performance due to process variations alone. These variations are principally due to fluctuations in V_t; the tolerance in effective channel length L_{eff} has a secondary impact, especially with today's better-controlled process technology. The ac performance of a MOS circuit is also a strong function of the power supply voltage V_{dd}. The charging and discharging current are basically the drain-source current I_{ds} of the MOS devices. I_{ds} increases with V_{dd} due to the fact that both V_{gs} and V_{ds} are a function of V_{dd}. Figure 7.10 shows a graph of the normalized delay of a CMOS gate versus V_{dd}.

Another important data point is that a CMOS IC operating at $V_{dd} = 3$ V is roughly 60% slower than at $V_{dd} = 5$ V. High operating temperature also adversely affects ac performance of MOS circuits, as shown in Fig. 7.11. I_{ds} is reduced at elevated temperatures due to reduction in the surface mobility of the carriers, notwithstanding the reduction in V_t. There is a second-order effect due to the positive temperature coefficient of the source-drain resistance and the metal interconnection resistance, which is becoming more important in VLSI, where signal lines could be very long and very narrow—a result of advanced processes. In calculating T_{tlh} and T_{thl}, we shall use the simple expression

$$t = C\frac{dV}{I}$$

where dV = voltage change
I = charging or discharging current
C = loading capacitance

The input voltage to the MOS inverter is assumed to be an ideal step function.

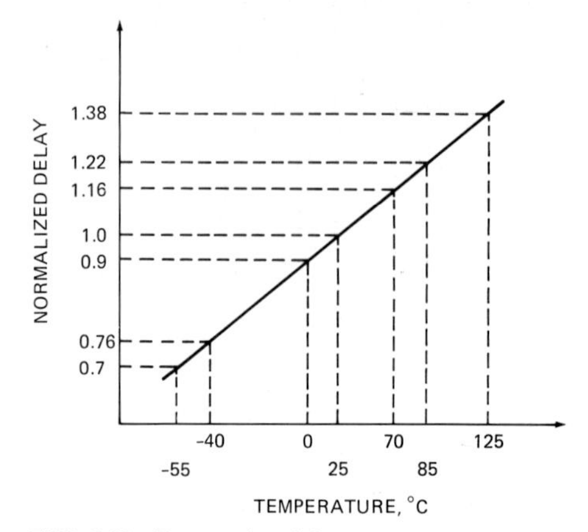

FIG. 7.11 Propagation delay versus temperature.

The Discharging Time T_{thl}. To simplify the analysis, we ignore the current coming from the load device. Then the I_{ds} of the driver transistor equals the current required to discharge the load capacitance C by a voltage swing dV.

At $t = 0$ the input and output voltages are both at V_1, and the driver transistor is on and operates in the saturation region. Therefore the saturation current equation is used for the discharging current I.

At $t = t_1$ the output is at $V' = V_1 - V_t$—the output voltage level at which the driver device crosses over to the linear operating region. Thus $dV = V_1 - V'$ is the voltage excursion after $t = t_1$, which can be written as

$$t_1 = 2\frac{CV_t}{B(V_1 - V_t)^2} \tag{7.21}$$

From t_1 onward the driver transistor operates in the linear region; hence the linear region equation should be used to calculate the second half of the discharging time t_2. By substituting the current and the final output voltage V_0 into the "time" expression, we obtain

$$t_2 = T\left\{\frac{2V_t}{V_1 - V_t} + \ln\left[\frac{2(V_1 - V_t)}{V_0} - 1\right]\right\} \tag{7.22}$$

where

$$T = \frac{C}{B(V_1 - V_t)} \tag{7.23}$$

The total discharging time T_{thl} is the sum of t_1 and t_2.

The Charging Time T_{tlh}. When the driver is off, the load capacitance C is charged from V_0 to V_1 through the load device.

Resistive Load. Use the same "time" expression $t = C\,dV/I$, where

$$I = \frac{V_{dd} - V_{\text{out}}}{R_l} \tag{7.24}$$

By substitution and solving the differential equation, we obtain

$$T_{tlh} = R_l C \ln\left(\frac{V_{dd} - V_0}{V_{dd} - V_1}\right) \tag{7.25}$$

Saturated Enhancement NMOS Load. In this case I is simply the current of the load in the saturation region. Solving the equation, we obtain

$$T_{tlh} = \frac{C(V_1 - V_0)}{B_l(V_{dd} - V_t)V_0} \tag{7.26}$$

The Nonsaturated Enhancement NMOS Load. Similarly, solving the time expression with the I_{ds} equation of the load and assuming $V_0 < 1$ V, we can approximate the charging time by

$$T_{tlh} = \frac{2Cx}{B_l V_{dd}(1 - x)} \ln\left(\frac{V_{dd} - xV_1}{V_{dd} - V_1}\right) \tag{7.27}$$

where

$$x = \frac{V_{dd}}{2(V_{gg} - V_t) - V_{dd}} \tag{7.28}$$

The Depletion NMOS Load. Using the I_{ds} equation for the saturation region with $V_{gs} = 0$, we obtain

$$T_{tlh} = \frac{2C(V_1 - V_0)}{B_l V_t^2} \tag{7.29}$$

for the charging time for the depletion load.

The CMOS Inverter. Charging the load capacitance with a PMOS transistor is similar to the discharging case discussed previously; i.e. the p-channel operates in the saturation region from $t = 0$ to t' when the output voltage level V' is such that $V_{dd} - V' = V_{\text{in}} - V_t$. From $V_{\text{out}} = V'$ to $V_{\text{out}} = V_{dd}$ the p-channel device operates in the linear region. Therefore the analysis of the charging time for CMOS inverted follows the same path as that of the discharging time.

7.3 MOS CAPACITOR

7.3.1 MOS Capacitor Characteristics

The capacitance characteristics of a MOS transistor depend on the state of the semiconductor surface. In the accumulation region, the gate forms one plate of the capacitor, and the high concentration of carriers (electrons or holes) in the substrate forms the second plate of a capacitor. The gate capacitance can be approximated by

$$C_{gb} = C_0 = \frac{e_{ox}}{T_{ox}} A \qquad \text{(in accumulation region)} \tag{7.30}$$

In the depletion region, a depletion layer is formed under the gate. This depletion capacitance is

$$C_{gb} = C_{dep} = \frac{e_{ox}}{d} A \tag{7.31}$$

where d is the depletion layer depth. The total capacitance is the capacitance of two capacitors connected in series.

$$C_{gb} = \frac{C_0 C_{dep}}{C_0 + C_{dep}} \qquad \text{(in depletion region)} \tag{7.32}$$

In the inversion region, a maximum depletion depth d_{max} exists (before surface inversion occurs). The dynamic capacitance remains the same after surface inversion.

$$C_{gb} = \frac{C_0 C_{dep(min)}}{C_0 + C_{dep(min)}} \qquad \text{(after inversion)} \tag{7.33}$$

where $C_{dep(min)}$ is the depletion capacitance with depletion layer depth of d_{max}. In ultralow-frequency operation (less than 200 Hz), the inverted layer can supply carriers fast enough to essentially shield the depletion region underneath it, making $C_{gb} = C_0$ again. However, most MOS devices are operated at much higher frequency ranges.

As shown in Fig. 7.12 in the one-transistor cell of dynamic memory (DRAM), a transistor biased in the deep accumulation region is used as the storage element. The capacitor consists of a first-level poly gate, a thin oxide layer (110 A,

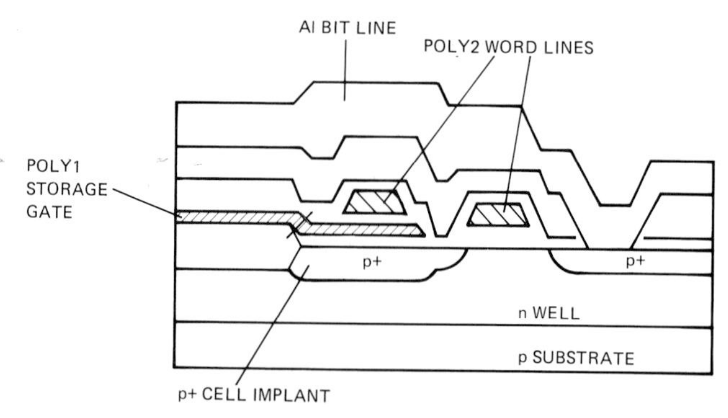

FIG. 7.12 A 1*T* dynamic memory cell.

typical), and the $p+$ diffusion region. An extra $p+$ cell implant is used here to enhance the storage capacitance by increasing the junction capacitance attached to the storage node.

7.3.2 Delay Element Capacitor

In MOS circuit design, the characteristics of the capacitors can be used to control internal timing or to enhance the circuit performance. One example is to use them as *RC* delay elements, such as the one-shot circuit in Fig. 7.13. The output is normally high because there is an odd number (three in this case) of inverters between the two inputs of the NAND gate (one input is high, the other is low). When V_{in} changes from low to high, the output will go low and remain low until V_{in} propagates through the inverter chain. The MOS capacitors control the charge and discharge rates of the internal nodes and therefore the pulse width of the clock signal output (V_{out}). This circuit should be used with extreme caution, since the generated pulse width is a function of inverter delays, which are subjected to changes across different temperatures, supply voltages, and process variations.

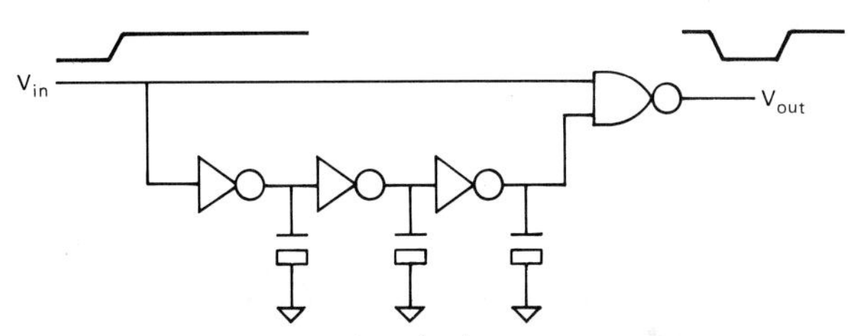

FIG. 7.13 A CMOS one-shot circuit.

7.3.3 Bootstrapping Capacitor

An NMOS delay clock circuit that uses capacitors to boost internal nodes is shown in Fig. 7.14. During the precharge cycle, clock C_p is high. Node N_A is preset to $V_{cc} - V_t$ while nodes N_B and N_C are preset to ground. In an active cycle, when clock C_1 fires, node N_C is charged to V_{cc} level (MC_1 is a bootstrapping capacitor). After the clock delay through devices M_1 and M_2, node N_A is discharged to ground. Node N_B goes toward V_{cc}, and node N_C is boosted by the depletion capacitor MC_2 to a level close to $2V_{cc}$. This allows the output device to have much higher conductance due to the boosted gate voltage on node N_C and therefore faster response on clock output C_2.

7.4 PASS TRANSISTOR

7.4.1 NMOS/CMOS Pass Transistor

An NMOS pass transistor is shown in Fig. 7.15 with its transfer characteristics. When the gate voltage is V_{cc}, the output V_2 follows the input V_1 linearly until V_2 reaches $V_{cc} - V_t$. The transistor is turned off at that time. The voltage drop at

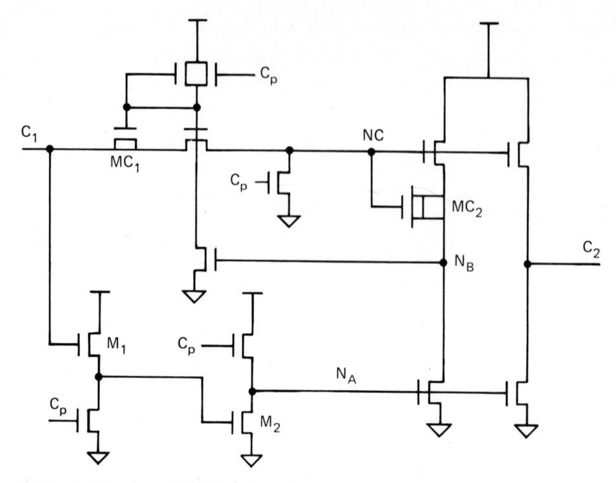

FIG. 7.14 An NMOS delay clock.

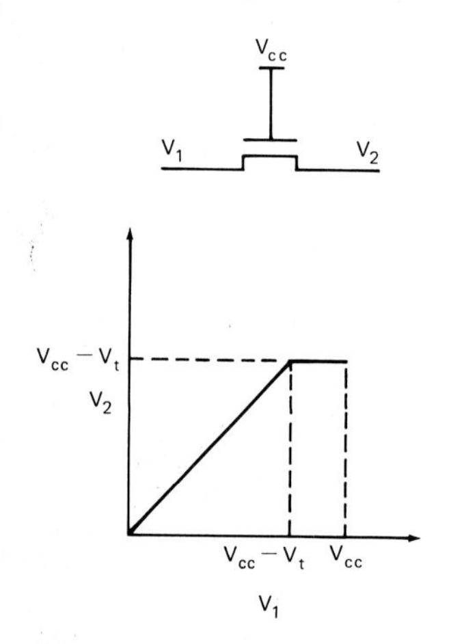

FIG. 7.15 An NMOS pass gate.

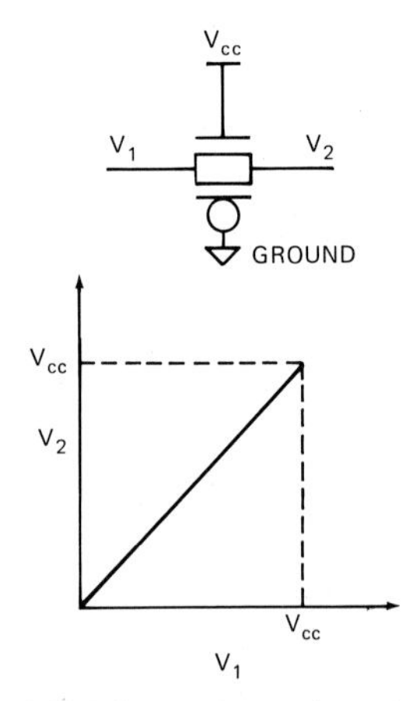

FIG. 7.16 A CMOS pass gate.

the output node may cause performance degradation in some circuits. A CMOS pass gate as shown in Fig. 7.16 should be used if full rail swing on the output is required.

7.4.2 Applications

An example of an NMOS pass gate is the transfer device between bit lines and data lines in a memory circuit, as shown in Fig. 7.17. Since both bit lines and data lines are clamped to the $V_{cc} - V_t$ level before signal separation, there is no penalty of voltage drop across the pass transistor. A CMOS shift register using CMOS pass gates as building blocks is shown in Fig. 7.18. C_1 and C_2 are com-

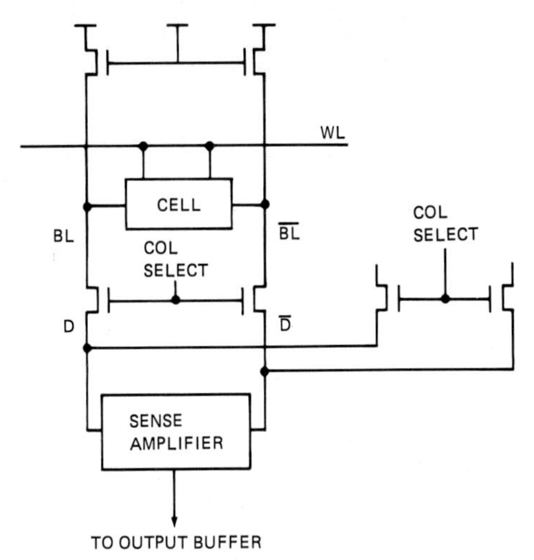

FIG. 7.17 SRAM column circuit.

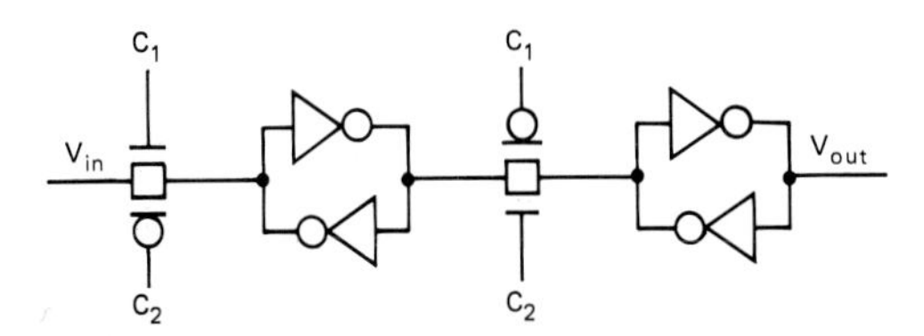

FIG. 7.18 CMOS shift register.

plementary clocks. When C_1 is high, input V_{in} is transferred to the master stage latch. When C_1 is low, V_{in} is isolated from the register circuit, and the stored signal is then transferred to the slave latch. This type of shift register is extensively used in CMOS digital circuits, such as flip-flops, counters, pointers, shift registers, and SCAN test cells.

7.5 INPUT AND OUTPUT BUFFERS

7.5.1 TriState Output Buffer

The main purpose of output buffers is to provide necessary driving currents for the printed circuit board. Standard products often have tristate buffers as outputs. A high Z state (also known as high-impedance state) is needed to avoid bus contention if several outputs are wired to the same bus, as in many modern system board designs. A very common tristate output buffer is shown in Fig. 7.19. EN is the enable control signal. A high on EN will enable the NAND and NOR gates that are connected to the p and n devices, respectively. The output state follows signal input A. In the tristate mode, EN is low and both p and n devices are turned off, leaving the output floated (high Z state). This mode allows other

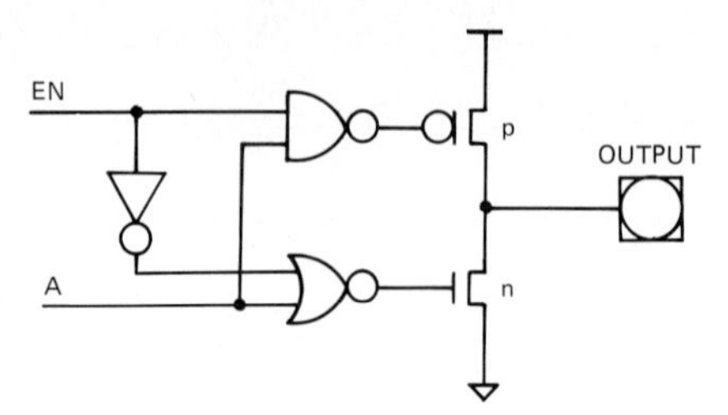

FIG. 7.19 Tristate output buffer.

devices connected to the same bus to take over the control. For example, in memory board design, the deselected memory chips will have the chip-select bar ($\overline{\text{CS}}$) high, which forces the memory outputs to be tristated. Only the selected chip (with $\overline{\text{CS}}$ low) has control of memory data output bus.

7.5.2 I/O Buffer

Common I/O (bidirectional I/O) can be implemented with a tristate output buffer. As shown in Fig. 7.20, during data input time (for example, write cycle in mem-

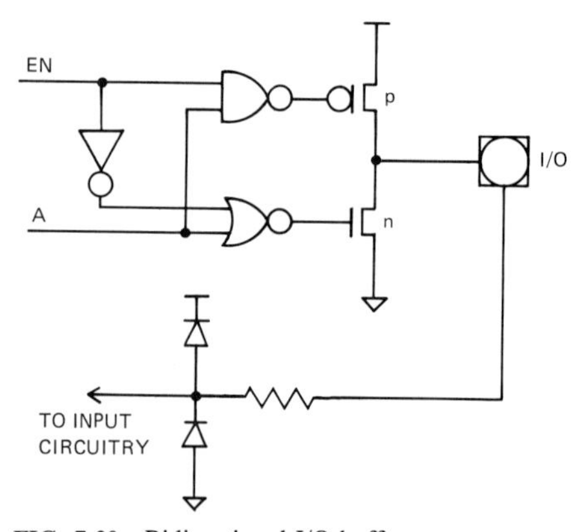

FIG. 7.20 Bidirectional I/O buffer.

ory access), EN is low, and outputs are at high-impedance states. This allows data to be passed to input circuitry. Without the tristate control, the input source will be loaded by the *p* or *n* transistors and its value may be changed, depending on the driving capability of the input clock.

7.5.3 Level Translation Buffer

The input stage usually consists of input protection devices and level translation buffers. Input protection devices are used to protect the gate oxide of the input device from rupturing due to electrostatic discharge (ESD). Level translation buffers provide the necessary interface between different logic levels (e.g., TTL, ECL).

TTL to CMOS Input Stage. Figure 7.21 shows an example of a CMOS circuit that uses a TTL level input signal. By controlling the W/L ratio of the p and n transistors, one can adjust the trip point (when output changes logic state) of the input stage to be at the center of TTL logic swing. Some input circuits have coupling capacitors between inputs and V_{dd}/V_{ss}, making the trip point independent of the internal power (V_{cc}/V_{ss}) noise on the chip.

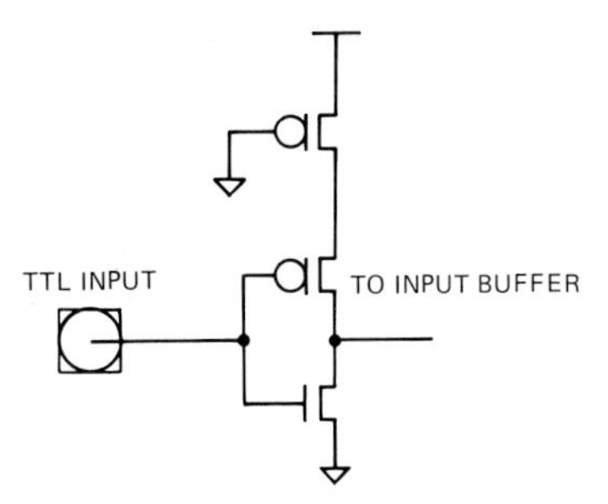

FIG. 7.21 A TTL to CMOS input stage.

ECL to CMOS Input Stage. An ECL to CMOS level input stage is shown in Fig. 7.22. It consists of three stages. The first is the differential input stage, which compares the ECL level input (centered at −1.3 V) to an on-chip reference signal of −1.3 V. A level shifter is inserted as a second stage to shift down the output voltage level of the differential amplifier to drive the third stage current mirror amplifier. A full CMOS level signal is achieved at the output of the current mirror.

CMOS to ECL Output Stage. Figure 7.23 shows a CMOS output stage with ECL level output loads. A resistor R_1 is connected between the output and a voltage supply V_{ee} of −2 V. A large p-channel transistor drives the base of the output bipolar transistor to provide the correct output high voltage and current. An output low level is supplied by resistor R_1 tied to −2 V.

7.6 MEMORY CELLS

RAM has an access time independent of the physical locations of the stored data. RAM chips are used extensively in system boards. Typical applications include address and data storage, writable control store, register files, scratch pad memory, cache memory building block, high end graphics, and a digital signal processing unit.

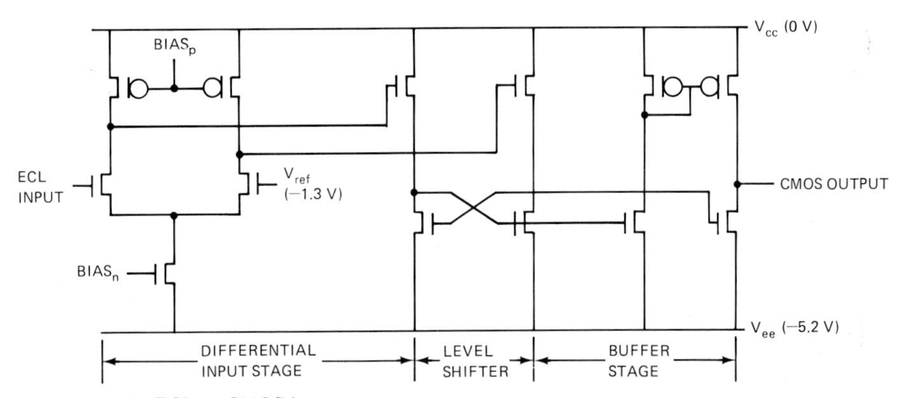

FIG. 7.22 An ECL to CMOS input stage.

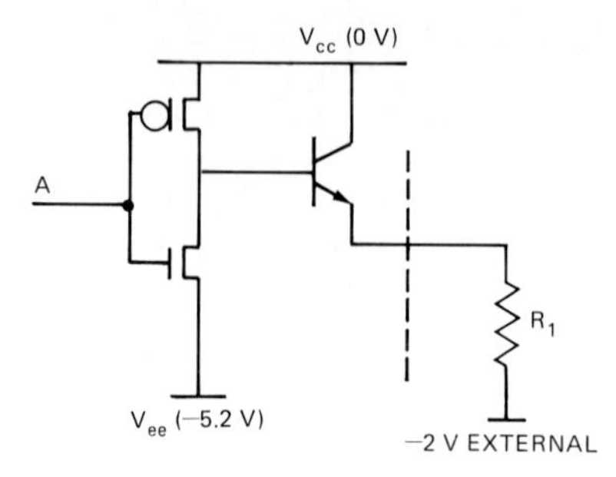

FIG. 7.23 A CMOS to ECL output stage.

7.6.1 6T SRAM Cell

Figure 7.24 shows a six-transistor static RAM (SRAM) cell. It consists of cross-coupled inverters connected to the bit lines by transfer gate transistors. In data storage mode, one of the nodes inside the transfer gate (N_1, for example) is

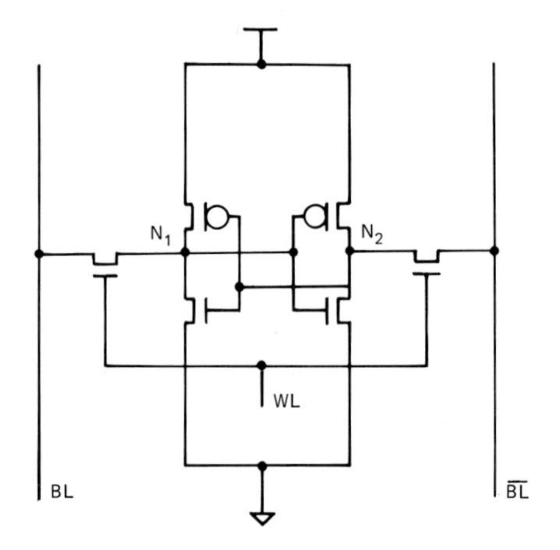

FIG. 7.24 A 6*T* SRAM cell.

latched low by the cross-coupled inverter, while the other node (N_2) is held high by the *p*-channel load device. The load transistor supplies enough currents to compensate for the leakage currents of the pull-down and transfer devices. The *W*/*L* ratio of the transfer gate and pull-down device is important for stable cell operation. In the presence of the bit-line pull-up devices (i.e., diode clamp), the pull-down device should be strong enough to hold the stored "0" to well below the threshold voltage of the cross-coupled inverters.

7.6.2 4T SRAM Cell

Figure 7.25 shows a four-transistor SRAM cell. This type of cell structure is used in most of the commercial SRAM chips because of the area advantage. It consists

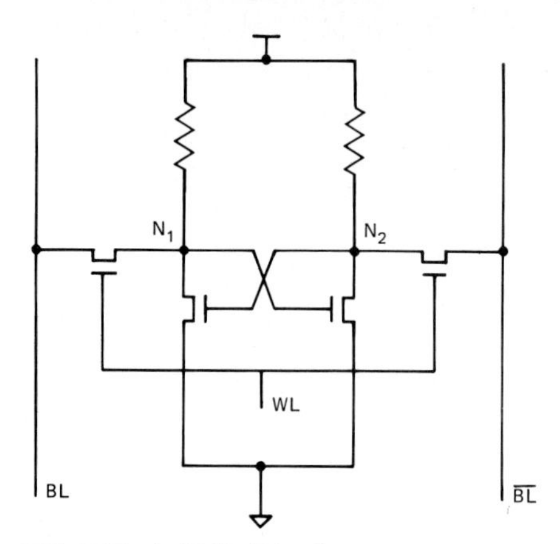

FIG. 7.25 A 4*T* SRAM cell.

of resistor-loaded cross-coupled inverters. A separate undoped polysilicon layer is used for the resistor. Typical value of the resistor runs between 100 MΩ and several gigaohms. Resistors are used here to maintain the stored "1" data. Very high resistance is required to keep the standby current at low level. In addition to the difficult control of the resistance of the undoped polysilicon, this cell is more susceptible to radiation damage because of the high resistor value.

7.6.3 Multiport Memory Cell

A two-port memory cell is shown in Fig. 7.26. Separate read and write ports are used to access the same address location. This structure is very useful in multiprocessor designs where a common memory source can be shared by two processors. An extended version is the five-port (two write, three read) RAM shown in Fig. 7.27. An inverter between the cross-coupled latch and column transfer device serves as the sense amplifier for all read bit lines. Writing of the data is

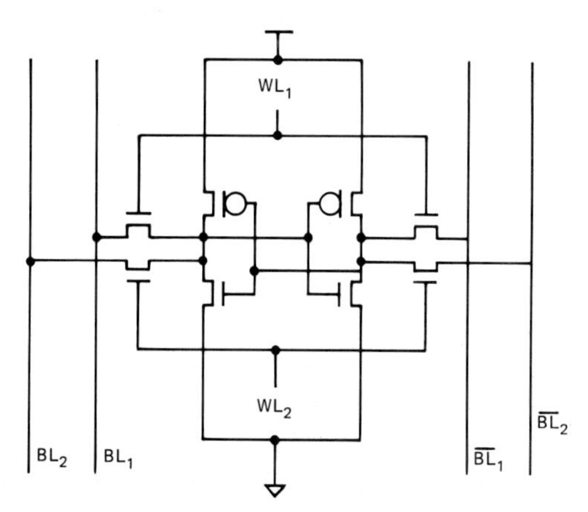

FIG. 7.26 A dual-port RAM cell.

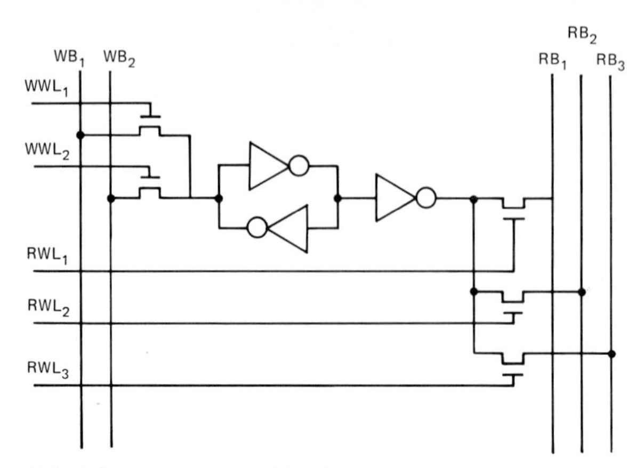

FIG. 7.27 A five-port RAM cell.

achieved by toggling a single bit line per port. It is more difficult to write a "1" into the cell because of the *n*-channel transfer gate device in the write ports. To overcome this problem, one should make the *n* feedback transistor of the memory cell latch much weaker than the *p*-channel pull-up transistor of the write driver.

7.6.4 Content Addressable Memory (CAM) Cell

A content addressable memory (CAM) cell is shown in Fig. 7.28. In addition to normal read-write operation, one can compare the stored data with the incoming bit-line data. If the incoming data matches that in the cell, the two *n*-channel transistors connected in series will pull the precharged MATCH line low. If no match occurs, the precharged line remains high. A very common application of the CAM structure is the cache memory. In a cache memory system, all data are

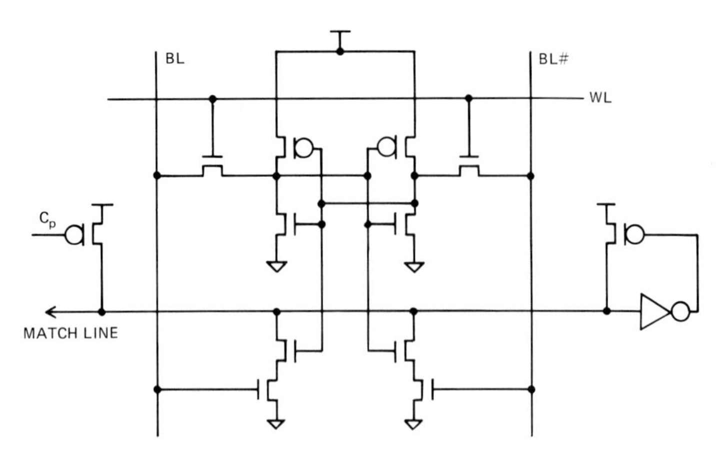

FIG. 7.28 A content addressable memory (CAM) cell.

stored in main memory, and some data are duplicated in the cache. When the processor accesses memory, it checks the cache first. If the desired data are in the cache, the processor can access the data quickly and a HIT signal will be generated. If the data are not in the cache, the cache fetches data from the main memory, a MISS signal will be generated, and the incoming data will be written back into the cache memory for subsequent access.

7.6.5 Dynamic Memory (DRAM) Cell

Dynamic memories (DRAMs) are used where massive storage of data is required. A DRAM cell is the smallest in area compared with other memory cells. A drawback of DRAMs is that they have to be constantly refreshed to maintain the stored data. It also has higher soft error rate compared to SRAM. A three-transistor dynamic memory cell is shown in Fig. 7.29. The cell stores data on the gate of storage transistors. To ensure adequate storage time (the time period when data are retained without being refreshed), capacitor C_g must be big enough to offset the leakage currents at the nodes.

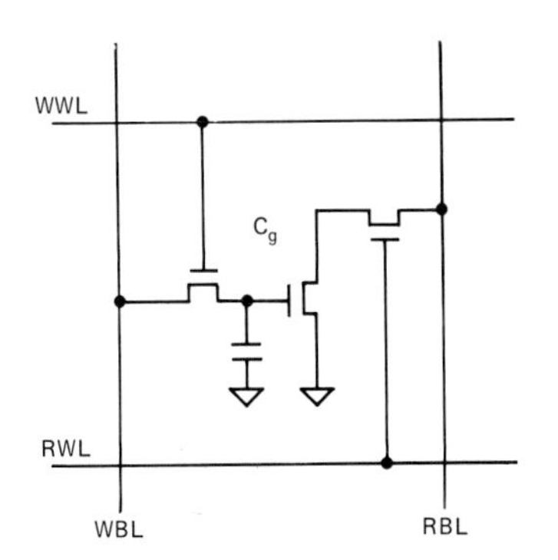

FIG. 7.29 A 3*T* dynamic memory (DRAM) cell.

One transistor memory cell is used in most of the DRAM chips. As shown in Fig. 7.30, the data are stored on a capacitor plate. Upon turning on a word line, memory cells will share charge with bit lines. A small differential signal will be generated between the paired bit lines (typically 100 mV). This signal will be amplified when sensing strobe clock SAS fires and latches the cross-coupled transistors. When DRAMs are used in a system, a separate controller is required to handle refresh timing and address multiplexing (address multiplexing is used in all commercial DRAMs to save pin counts).

Another interesting architecture is the quasi-static memory shown in Fig. 7.31. Each memory location is accessible to both normal and refresh word lines. While normal access is done by turning on and off normal word lines, background refresh is achieved at the same time. To avoid accessing the same locations by both word lines, we need some logic to let the normal word line prevail over the refresh word line when contention occurs.

7.6.6 Read Only Memory (ROM) Cell

A ROM cell is implemented with one transistor for each stored bit. A very common approach using a NOR array is shown in Fig. 7.32. During a read cycle, an "intake" transistor will pull the precharged bit line low. This will represent a 1 or 0, depending on the majority coding in actual design. Mask programming can be done by contact, diffusion (with or without the transistor diffusion regions), or ion implantation (to raise the threshold voltage of the OFF transistor). Once the programming is done, the stored data will remain permanently.

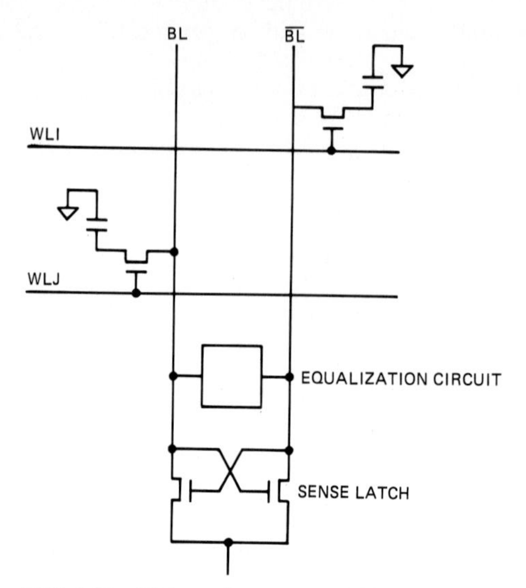

FIG. 7.30 1*T* dynamic memory (DRAM) cell.

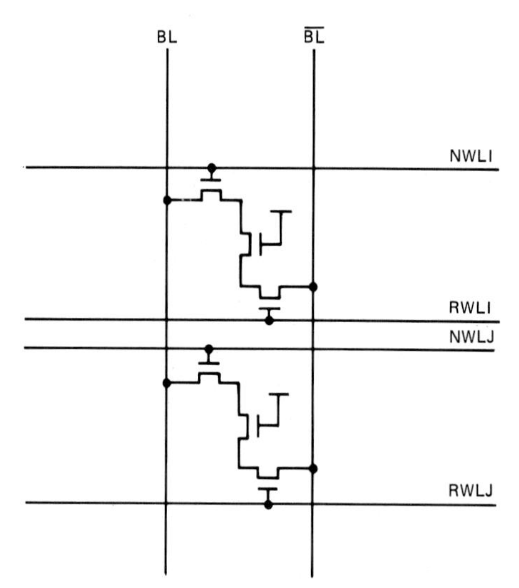

FIG. 7.31 A quasi-static memory cell.

7.7 FLIP-FLOP

7.7.1 *RS* flip-flop

The flip-flop is one of the most used elements in digital designs. It can be used for internal logic and timing control. Figure 7.33 shows an asynchronous *RS* flip-

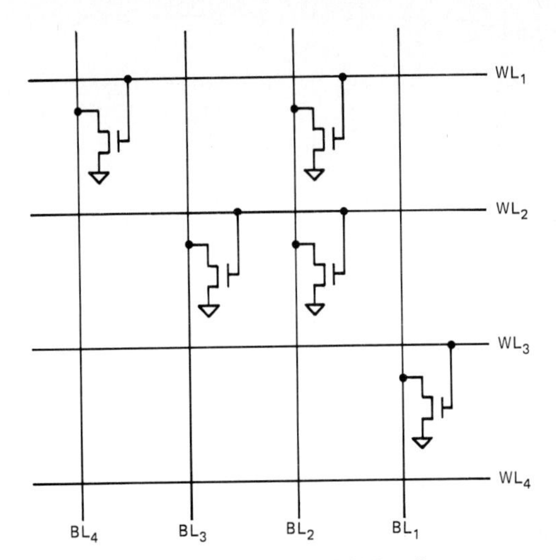

FIG. 7.32 Read-only memory (ROM) cell.

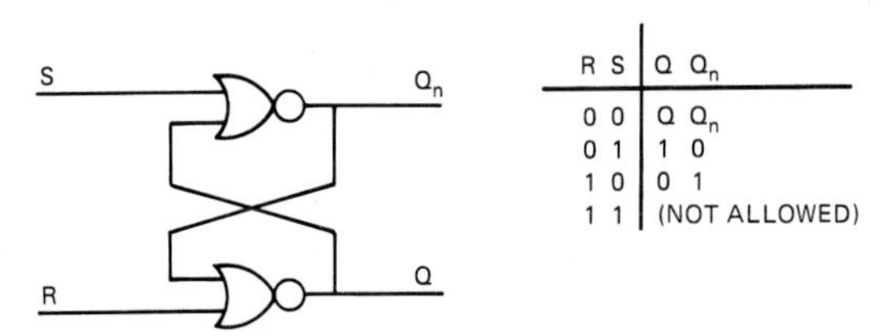

S
Qn
R
Q
R S | Q Qn
0 0 | Q Qn
0 1 | 1 0
1 0 | 0 1
1 1 | (NOT ALLOWED)

FIG. 7.33 Asynchronous flip-flop.

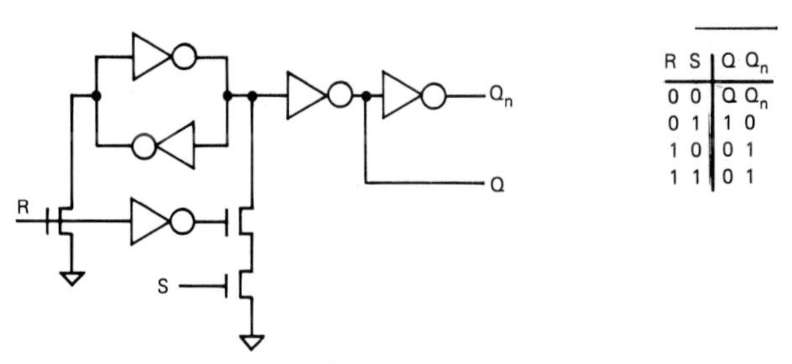

Qn
Q
R
S
R S | Q Qn
0 0 | Q Qn
0 1 | 1 0
1 0 | 0 1
1 1 | 0 1

FIG. 7.34 Latch-type flip-flop.

flop. State variable Q is set high when $R = 0$ and $S = 1$, and reset low when $R = 1$ and $S = 0$. Q is unchanged when $R = S = 0$. Note that $R = S = 1$ is an invalid input, because both output Q and Q_n will be low. A modified version is shown in Fig. 7.34, an extra inverter is used to disable the pull-down path of the two n-channel transistors in series when $R = 1$. Output Q remains low when both R and S are high ($R = S = 1$).

7.7.2 *D/JK*-Type Flip-Flop

In a clocked system, synchronous flip-flops are preferred. Figure 7.35 shows a *D*-type flip-flop together with its excitation table. A *JK* flip-flop and its excitation table are shown in Fig. 7.36. Unlike the *RS* flip-flop, the *JK* flip-flop changes state when $J = K = 1$. Another example of a *JK* flip-flop with asynchronous SET-CLEAR capabilities is shown in Fig. 7.37. Active low signal CD (clear direct) and SD (set direct) will set the *Q* state to 0 and 1, respectively.

7.7.3 Data Setup Time

A flip-flop's setup time is the minimum time the data must be stable before the active edge of the clock occurs. The hold time is the minimum time the data must

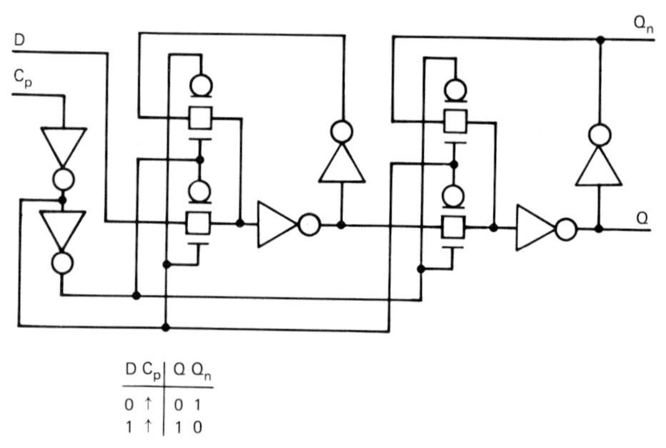

D	C_p	Q	Q_n
0	↑	0	1
1	↑	1	0

FIG. 7.35 *D*-type flip-flop.

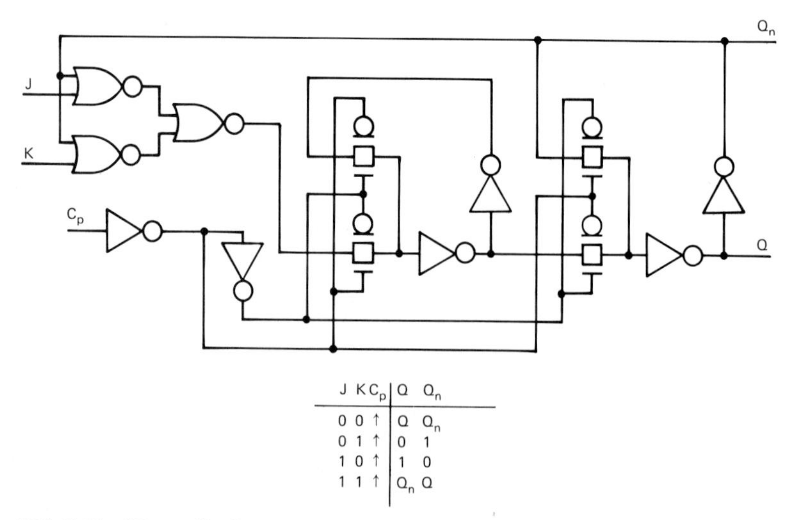

J	K	C_p	Q	Q_n
0	0	↑	Q	Q_n
0	1	↑	0	1
1	0	↑	1	0
1	1	↑	Q_n	Q

FIG. 7.36 *JK*-type flip-flop.

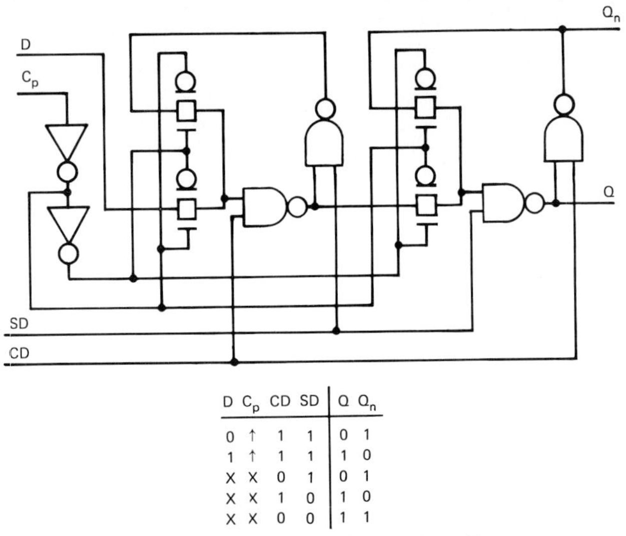

D	C_p	CD	SD	Q	Q_n
0	↑	1	1	0	1
1	↑	1	1	1	0
X	X	0	1	0	1
X	X	1	0	1	0
X	X	0	0	1	1

FIG. 7.37 *D*-type flip-flop with clear direct and set direct.

be stable after the active edge of the clock. The timing is shown in Fig. 7.38. Both setup and hold times are functions of the internal delay of the master portion of the flip-flop only. They are independent of the loading on the Q and Q_n outputs. These timings can be calculated based on the various propagation delays of clock (C_p) and data (D, for example) signals.

In Fig. 7.39 when the clock input C_p is low, transmission gate G_4 is enabled and G_3 is disabled. Any signal changes occurring on the data pin will affect the outputs of G_5 and G_6. The result is a change in signal state that is set up on input of G_3. When the clock pin becomes high, G_3 is enabled and G_4 is disabled. As a result, the signal set up on G_3's input is latched in the master and transmitted to the slave latch. Since G_4 is disabled, any data changes occurring on the data pin are blocked out.

To meet the setup time requirement, the signal at G_3's input must be stable prior to the time that G_3 is enabled. Since any changes occurring on the data pin must go through G_4, G_5, and G_6 before reaching G_3, the setup time is the sum of the propagation delays through these latter three gates. However, the clock signal must go through G_1 and G_2 before enabling G_3. The setup time for the D flip-flop therefore is equal to the sum of the G_4, G_5, and G_6 propagation delays minus the sum of the G_1 and G_2 propagation delays:

$$T_{\text{setup}} = (T_{pdG_4} + T_{pdG_5} + T_{pdG6}) - (T_{pdG_1} + T_{pdG_2}) \tag{7.34}$$

7.7.4 Data Hold Time

For the hold time requirement to be met, the data pin must not change state before G_4 is disabled (see the timing diagrams of Fig. 7.38 and Fig. 7.40). Since the clock signal must go through G_1 and G_2 before disabling G_4, the hold time is equal to the sum of the G_1 and G_2 propagation delays:

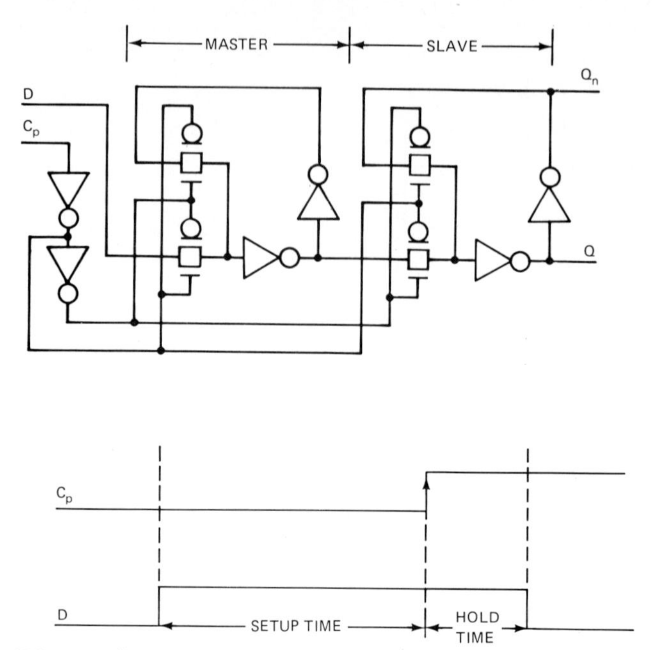

FIG. 7.38 Setup and hold time of *D* flip-flop.

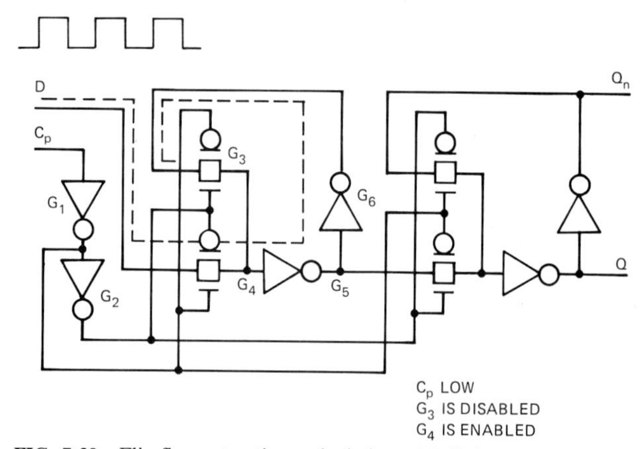

FIG. 7.39 Flip-flop setup time calculation with C_p low.

$$T_{\text{hold}} = T_{pdG_1} + T_{pdG_2} \tag{7.35}$$

7.7.5 Minimum Pulse Width

Another design consideration is the clock pulse width. As shown in Fig. 7.41, while the clock pin is low the signal from the data pin propagates through G_4, G_5,

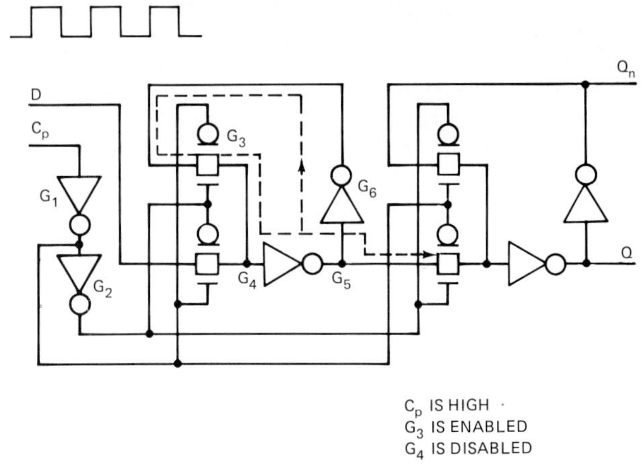

FIG. 7.40 Flip-flop hold time calculation with C_p high.

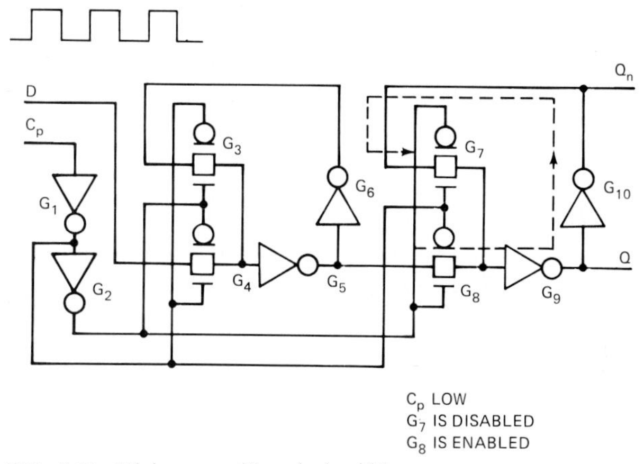

FIG. 7.41 Minimum positive clock width.

and G_6, and the signal is set up on the input of G_3. The clock signal must remain low during this period so that the master can latch in the correct data. The minimum negative pulse width is

$$T_{-pw} = T_{pdG_4} + T_{pdG_5} + T_{pdG_6} \tag{7.36}$$

As the clock is high, data are transferred from master to slave. The data signal propagates through G_8, G_9, and G_{10}, and the signal is set up on the input of G_7. The minimum positive pulse width is

$$T_{+pw} = T_{pdG_8} + T_{pdG_9} + T_{pdG_{10}} \tag{7.37}$$

The clock signal must remain high during this period, allowing the slave to latch in the correct data.

Since the master drives no external loads, the minimum negative pulse width T_{-pw} is independent of the loading on the Q and Q_n outputs. However, the slave drives external loads. If Q or Q_n drives high fan-out lines, then the propagation delay of G_9 and G_{10} increases the minimum clock pulse width for the flip-flop. This problem can be eliminated by adding inverters at Q and Q_n. The penalty is additional gate delay from the clock to Q and Q_n.

7.7.6 Other Considerations

The *RS* flip-flop in Fig. 7.33 is a level-sensitive element. It has the transparent property that the flip-flop output will follow the input changes. For edge-sensitive types of flip-flops (i.e., master-slave type), outputs change only when the clock input makes a specific transition. Subsequent changes of the data inputs have no effect until the next active transition occurs. These types of flip-flops are more often used in digital design.

From the above timing calculations, it is very obvious that clock skew is a serious concern in flip-flop design (clock skew concerns relative timing of the clock arriving at different flip-flops). A skewed clock signal will change all the timing requirements. To reduce the clock skew, keep clock line routing at a minimum, minimize the number of gates driven by the clock, and use a large buffer to drive all clock inputs. To reduce the effect of output loading on the minimum positive pulse width, buffer both Q and Q_n before using them to drive high fan-out lines.

BIBLIOGRAPHY

1. P. E. Gray and C. L. Searle, *Electronics Principles: Physics, Models, and Circuits,* Wiley, New York, 1969, pp. 313–341.
2. *COS/MOS Integrated Circuit Manual,* RCA Technical Series, CMS-271, 1972, pp. 3–33.
3. *Integrated Circuit Design Principles and Fabrication,* Motorola Series in Solid State Electronics, McGraw-Hill, New York, 1965, pp. 208–231.
4. W. N. Carr and J. P. Mize, *MOS/LSI Design and Application,* McGraw-Hill, New York, 1972, pp. 26–58.
5. A. S. Grove, *Physics and Technology of Semiconductor Devices,* Wiley, New York, 1967, pp. 243–288.
6. M. I. Elmasry, *Digital MOS Integrated Circuits: A Tutorial,* IEEE Press, New York, 1981.
7. T. Masuhara, M. Nagata, and N. Hashimoto, "A High Performance N-Channel MOS LSI Using Depletion Type Elements," *IEEE J. Solid State Circuit,* June 1972.
8. L. A. Glasser and D. W. Dubberpuhl, *The Design and Analysis of VLSI Circuits,* Anderson & Wiley, New York, 1983.
9. N. Weste and K. Eshraghian, *Principles of CMOS VLSI Design: A System Perspective,* Anderson & Wiley, New York, 1985.
10. *CMOS Macrocell Manual,* LSI Logic Corporation, 1985, Chap. 11.

P • A • R • T • 2

VLSI TEST

CHAPTER 8
FAULT MODELING AND TEST GENERATION

Vishwani D. Agrawal
AT&T Bell Laboratories
Murray Hill, New Jersey

Sudhakar M. Reddy
University of Iowa
Iowa City, Iowa

8.1 AN OVERVIEW OF VLSI TESTING

The role of testing in the VLSI device realization process is illustrated in Fig. 8.1. Test planning begins with specifications. Device specifications include functional (input-output behavior, frequency, timing, etc.), environmental (power, temperature, humidity, noise, etc.), and reliability (incoming quality, failure rate, etc.) specifications. Types of tests and test equipment are chosen to match device specifications.

Test activities are interwoven with design. Architectural design consists of partitioning of the VLSI chip into realizable functional blocks. The next step, logic design, includes several test activities. Either the logic should be synthesized in a testable form, or the synthesized logic should be analyzed (and improved) for testability. This is discussed in Chap. 9.

After logic synthesis, test vectors are generated and evaluated for their effectiveness. These topics will be discussed in this chapter. The next test activity takes place after physical design (layout, timing verification, mask generation) and fabrication (wafer processing). Using test specifications and test vectors, we develop a test program for the test equipment to be employed. Test programming and test equipment will be discussed in Chap. 10.

8.1.1 Types of Testing

Testing enters the life cycle of a VLSI device in several places. First, in the factory, device wafers and packages are subjected to *manufacturing test*. Second, users of VLSI devices conduct *acceptance test*. Third, the devices undergo test-

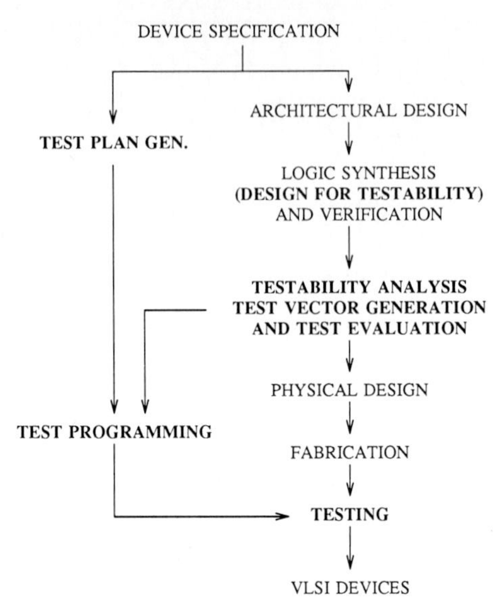

FIG. 8.1 Test functions (shown in **bold**) in VLSI device realization.

ing during the *systems test* performed on the systems (printed circuit boards, etc.) where the VLSI chips are used. Finally, the devices must be tested in the field, and that is known as the *maintenance test*.

8.1.2 Method of Testing

Testing consists of mounting chips (either wafers or packaged chips) on the automatic test equipment (ATE), applying stimuli to the input pins, and comparing the response at the output pins with the expected responses. Manufacturing tests are the most comprehensive. The same (or a subset of the same) tests are used for many other types of testing.

8.1.3 Requirements of Testing

Manufacturing tests are supposed to determine that each component (transistor, etc.) and each interconnection are fabricated correctly on the chip. These tests should thoroughly check every node in the circuit. The effect of every fault should be propagated to the circuit output. A normal requirement for these tests is a very high (95 to 100 percent) coverage of all *modeled faults* in the circuit.

The objective of acceptance testing is to ascertain the quality level of chips that have undergone the manufacturing test. Test and the coverage requirements are similar to those of manufacturing tests.

High-fault-coverage tests are also used for *in-circuit testing* of chips. This form of testing is conducted at the printed circuit board level during the manufacture or repair of boards.

8.2 FAULT MODELING

Failures occur in VLSI chips throughout their life cycle. Failures are caused by design errors, material defects, process defects, extremes in operational environment, deterioration due to length of operation or age, and so on. Phenomena causing failures can be physical or chemical in nature. However, to analyze the faulty behavior and develop techniques to detect and locate failures, we use abstract fault models. Fault models allow cost-effective development of test stimuli that will identify failed chips and, if necessary, diagnose the failure. Fault models also limit the number of tests as opposed to applying all possible inputs.

Practical fault models depend upon the chip model, the technology, and, in some cases, the particular phase in the life cycle of the chip where analysis is conducted. By chip model we mean how the chip is described. Typically, one describes a VLSI chip at the following levels: specification, behavioral, functional, logic, circuit, and layout. The translation between levels may be manual or automatic, but the complexity of the description grows in detail as we move from specification toward layout. Some of the causes of faults, which can occur at any level, are errors in specification, errors in the translation from one level to another, errors in the manufacturing process, or material failures. Fault models must mimic the effect of these errors, yet they should be easily analyzable. Compromises are often necessary to balance the complexity of a fault model necessary for accuracy against the tractability of analysis. The guiding principle in arriving at a good compromise is to model *most probable* failures. The percentage of chips found faulty in the field is used as a measure to *certify* the adequacy of the fault model used in testing. In the sequel we will describe several fault models currently in use.

8.2.1 Fault Models

Failures can be broadly classified into two classes: *permanent* and *intermittent*. Detection of intermittent failures requires repeated application of tests[1] or the use of on-line monitoring.[2] In either case, the tests applied are the ones derived for permanent failures based on the chosen model of a permanent fault.

Stuck Faults. The most commonly used fault model to represent failures in logic circuits is the *line-stuck fault*.[3] A line in a logic circuit is an input or output of a logic gate, and by fault "line *l* stuck-at-0" we mean that line *l* in the faulty circuit remains in the logic 0 state independent of the input to the circuit. Similarly, one can define the "stuck-at-1" fault for line *l*. Another class of stuck faults is the *transistor-stuck faults*.[4] A "stuck-on" fault in a transistor represents a failure that causes the transistor to be permanently turned *on*. A "stuck-open" fault represents a failure causing the transistor to remain permanently in the *open* or off state.

Since the only thing the line-stuck fault needs is a logic model, and most digital circuits, irrespective of the specific technology, can be modeled at this level, these faults are often called *logical faults* or *technology-independent faults*. In contrast, the transistor-stuck faults are specific to the MOS technology and are often referred to as *nonclassical faults*; stuck faults are regarded as the classical model.

Shorts and Opens. The diminishing feature size allows increased circuit density. Certain failures in high-density VLSI chips require fault models that are different from the stuck model. A *short or bridging fault* is defined as an electrical short circuit between two nodes that are supposed to be electrically isolated.[5] A *short* fault between nodes *A* and *B* causes both nodes to have the same voltage at all times. An *open* fault represents a failure that causes a line or *wire* in the circuit to be *broken*. An *open* fault between nodes *A* and *B*, which are normally connected, causes them to become electrically isolated.

Cross-Point Faults. Programmable logic arrays (PLAs) are often used to realize logic functions in a cost-effective and practical way. An example is shown in Fig. 8.2. A *cross point* in the PLA is the intersection of a product line with an input or

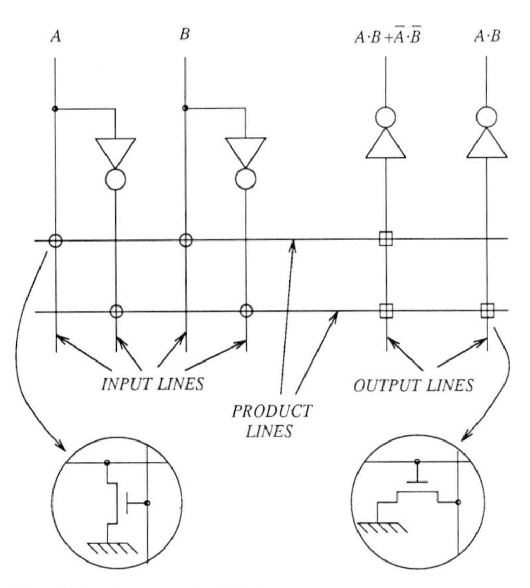

FIG. 8.2 An example PLA.

output line. At a cross point a transistor is either present or absent. A *cross-point fault* means that either a desired transistor is missing or an unintended transistor is present.[6] Correspondingly, these faults are called the *missing cross points* or the *extra cross points*. Apart from the cross-point faults, line-stuck faults on the inputs and outputs and bridging faults on adjacent lines are also modeled in PLAs.

Memory Faults. Semiconductor memories, both ROMs and RAMs, require fault models that are different from those described above. A memory is made of two parts: the address decoder and the storage cell array. To model failures in memories, one uses three classes of fault models: (a) parametric faults, (b) functional faults, and (c) pattern-sensitive faults. *Parametric* faults refer to failures that cause one or more dc or ac parameters of the memory to fall outside the specified range. Typical dc faults are unacceptable output levels, high power consumption, fan-out capability, and noise margins. Examples of ac faults are too high access time, setup time, and hold time. Functional faults model failures that influence proper storage and recall of data. These include cell-stuck (at 0 or 1) faults and stuck faults in the decoder logic.[7,8] Pattern-sensitive faults address failures that cause erroneous storage or recall of data only for certain states of the memory. For example, a faulty cell may store data correctly but may fail to hold a 0 whenever all four of its neighbor cells have 1s stored in them. In general, the pattern-sensitive fault model represents all faulty conditions under which the state of a cell is altered as a result of a certain pattern of 0s, 1s, 0-to-1 transitions, or 1-to-0 transitions in the other cells. Among the various types, the *neighborhood pattern-sensitive faults* have received the most attention.[9,10] It is generally believed that most of the probable functional faults in memories can be covered by considering *cell-stuck faults, cell-transition faults, cell-coupling faults, and multiple-access faults*.[11,12]

Functional Faults. Functional fault modeling is common for memories, but it has not been very popular for logic chips. However, it often happens that a user of a complex VLSI chip does not have access to a detailed circuit or logic level description of the chip. In such situations, *functional faults* that model failures at the register level or processor instruction level can be used.[13] Such models, developed for microprocessors, have proved effective in deriving tests that cover a high percentage of gate-level-stuck faults.[2] Functional faults have also been used for cellular logic arrays[14] and sequential machines.[15] For example, an m-input cell in a logic array may be assumed to realize any other m-input function, or the state table of a sequential machine is assumed to have erroneous entries.

Delay or Timing Faults. The fault models described essentially address faults that affect the function of the chip. To ascertain correct operation, however, one must also verify that the function is performed within the specified time limit. Failures and process variations that cause chips to produce results outside of the expected time window are modeled by *delay faults*.[16,17] *Gate-delay faults* represent failures that grossly change the signal propagation delay of every signal passing through the faulty gate. *Path-delay faults* model changes in the cumulative delay of gates on a signal path.

8.2.2 Single versus Multiple Faults

Given a fault model, one often needs to make assumptions about the number of faults occurring simultaneously in a chip. Analysis shows that tests generated to cover all single faults can have a significantly lower coverage of multiple faults.[18] However, the complexity of analyzing multiple faults and generating tests for them forces one to consider only the single faults. Perhaps the strongest point in favor of single faults is experience. Test methodology, developed on the basis of the single-fault model, has produced VLSI chips with an acceptably small fraction of field rejects.

8.2.3 Reduction of Fault Set

It is important to find a proper set of modeled faults for which tests will be generated. Considering the complexity of test generation methods, this set should be as small as possible. Toward this end, the concepts of *fault equivalence, fault dominance,* and *checkpoint faults* are useful.

Equivalence of two faults means that any test detecting one fault will also detect the other. Consider a simple example. For an AND gate, any input line stuck-at-0 is equivalent to the output line stuck-at-0. Thus testing for a line stuck-at-0 fault at the output implies that all inputs have also been checked for line stuck-at-0 faults. For the purpose of test generation, therefore, we only consider $n + 2$ faults in an n-input AND gate, such as stuck-at-1 on each input line and stuck-at-1 and stuck-at-0 on the output line. Similarly, in an n-input OR gate, we test for stuck-at-0 on each input and stuck-at-1 and stuck-at-0 on the output. Testing considerations for NAND and NOR gates are similar.

Since the faults that are equivalent are detectable by exactly the same tests, they are completely indistinguishable. In practical fault analyses (test generation and fault simulation), therefore, only one fault from a set of equivalent faults is considered. The process of eliminating all but one equivalent fault is called *equivalence fault collapsing*. This often results in considerable economy.

Fault collapsing in a logic network requires determination of equivalence between faults whose sites may be separated by several gates. In equivalence fault collapsing we only collapse the faults that are indistinguishable. If we are prepared to give up on diagnostic resolution (ability to distinguish between faults), more collapsing is possible. This is accomplished by using the concept of *fault dominance,* as explained below. In VLSI circuits, since coverage of faults rather than their exact location is the overriding consideration, *dominance fault collapsing* may be desirable.

Consider two faults f_1 and f_2. Suppose all tests for f_1 also detect f_2, but only some of the tests for f_2 detect f_1. Then f_2 is said to *dominate* f_1. This definition of dominance was given by Poage.[3] If we had to pick one, we are obviously safer to take f_1. Even though at times it may be a little harder to find a test for f_1, this test is guaranteed to cover f_2.

In an AND gate, the output stuck-at-1 dominates any input stuck-at-1. Thus if we desire dominance fault collapsing, then for an n-input AND gate we need to consider only $n + 1$ faults. One normally takes each input stuck-at-1 and any input stuck-at-0. The input stuck-at-0 fault is equivalent to the output stuck-at-0. However, the input fault is preferred because in a multilevel circuit the collapsing algorithms often select faults that are closer to the circuit inputs.

For irredundant combinational circuits, it has been found that we only need to consider stuck faults on the *checkpoints*.[19] *Checkpoints* are defined as the primary inputs and only those gate inputs that are fed by fan-out lines. If any checkpoint fault is redundant (i.e., no test exists for it), then additional faults must be considered.[20] In sequential circuits fault collapsing is often accomplished by multiple passes. More details may be found in the literature.[21–23]

8.3 TEST GENERATION

In order to determine whether a given VLSI chip performs according to specifications, one normally applies a sequence or set of test input stimuli and samples chip responses at the output. If the sampled response differs from the one expected from a good chip, then the chip is declared faulty. Obviously, the test stimuli should be so chosen that for every modeled fault at least one input will produce a response that differs from the corresponding good response.

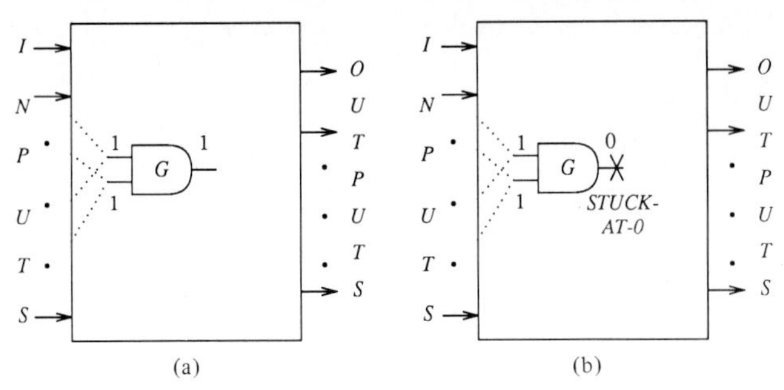

FIG. 8.3 Fault activation. (*a*) Fault-free circuit; (*b*) faulty circuit.

The test inputs (or stimuli) are derived in essentially two different ways. One way is to pick an input vector and determine the faults it detects by a fault simulator. The second approach is to determine a test vector to detect a particular fault. This is done by a test generation algorithm. In either case vectors can be accumulated to cover an adequate number of modeled faults.

The basic idea behind test generation algorithms is illustrated in Figs. 8.3 and 8.4. Let *G* be an AND gate embedded in the circuit under consideration and as-

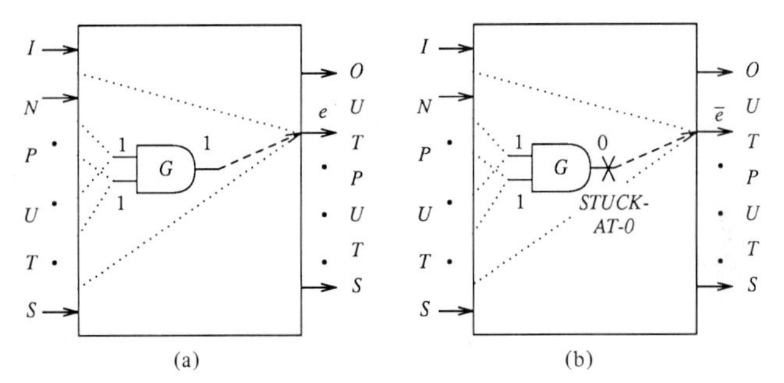

FIG. 8.4 Fault propagation. (*a*) Fault-free circuit; (*b*) faulty circuit.

sume that a test for the fault "output of *G* stuck-at-0" is desired. The test must satisfy *two* objectives. The first objective is to *activate* the fault, and the second objective is to *propagate* the effect of the fault to an observable output. Activation is achieved by ensuring that the logic state at the fault site in the fault-free circuit is different from its faulty state. For example, for the AND gate *G* in Fig. 8.3, whose output is stuck-at-0, the test must drive the output line to 1. This is done by setting both inputs of the AND gate to 1 through manipulation of states of logic within the two triangular regions bounded by dashed lines. The test inputs must also propagate the effect of this fault by ensuring that an observable output is at different logic states in the faulty and the fault-free circuits. This is achieved by sensitizing at least one path from the faulty line to an observable output, as indicated in Fig. 8.4. This time we must manipulate the circuit states within the large dashed-line triangle.

8.3.1 Test Generation for Combinational Circuits

Two basic approaches are taken to generate tests to detect faults in combinational circuits. One is *random test generation* and the other is *deterministic test generation*.

In random test generation, tests are randomly generated and the faults detected by them are determined by fault simulation. Random patterns are obtained by either software-implemented random number generators[24] or through the use of linear feedback shift registers (LFSRs). Circuit inputs can be made to assume 0 and 1 states with equal probabilities or unequal probabilities, which may also be changed as the test generation proceeds.[25,26]

Deterministic methods, commonly implemented for single line stuck faults, can be divided into the following categories: (a) tabular, (b) algebraic, (c) functional, and (d) algorithmic. Except for the functional approach, all other procedures derive tests for specific faults. The functional approach attempts to derive tests that are *implementation-independent*.

Tabular Method: In this method a table with one row for each possible input combination and columns corresponding to the outputs for the fault-free and each faulty circuit are used to derive a set of sufficient or minimal tests.

Algebraic Methods: A popular example of an algebraic method is the Boolean difference method.[27] In this method a Boolean expression representing the XOR function of the fault-free function and the faulty function of the circuit is derived. The faulty function corresponds to the fault for which a test is required. This expression, known as the *Boolean difference,* contains the primary inputs as the independent variables. All input vectors for which the Boolean difference is *true* give tests for the fault.

Functional Methods: In deriving tests from the functional description (e.g., truth table), one is interested in tests that will detect all faults of interest in any implementation. Such tests, known as *complete* or *universal,* are possible for line-stuck faults in a certain class of realizations.[28]

Gate-Level Algorithms: These methods construct tests by a gate-level analysis of the circuit. Since the gate-level description is frequently used for logic synthesis and verification, such methods are quite popular. Another reason for their widespread use is that they can be easily programmed.

Primarily, all such methods attempt to construct a vector that will *activate* the fault and *propagate* its effect to a primary output, as we mentioned earlier. Here we illustrate the principle by two examples. Consider the combinational circuit shown in Fig. 8.5. Test generation for the fault "line c stuck-at-0" is carried out in the following steps:

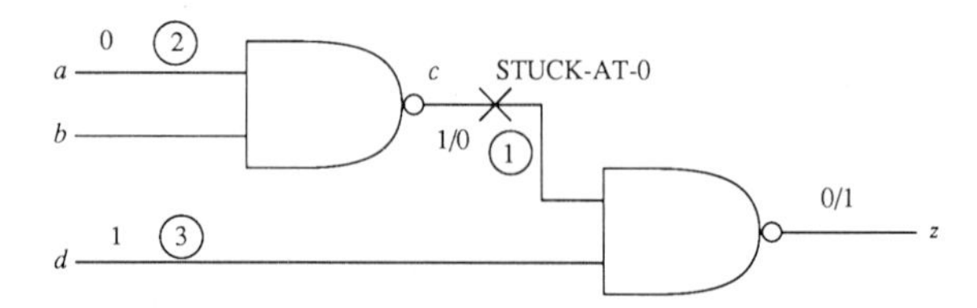

FIG. 8.5 A simple test generation example. (Circled numbers show the sequence of test generation steps.)

1. Set line c to 1. Its value, shown as 1/0, means 1 in a good circuit and 0 in a faulty circuit.
2. Justify step 1 by setting line a to 0.
3. Propagate the value of line c to output z by setting line d to 1.

The test 0X 1 is found in three steps. The value X for input b implies that the value of b could be 0 or 1 to obtain a test for the fault considered. This is the simplest case, which consists of a circuit without any reconvergent fan-out.

As a second example, consider the circuit of Fig. 8.6. Even though this circuit

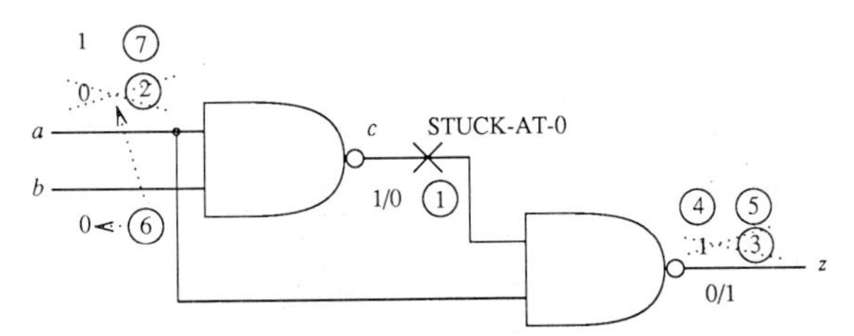

FIG. 8.6 Test generation with reconvergent fan-out. (Circled numbers show the sequence of test generation steps.)

may seem functionally trivial, derivation of a test to detect the fault "line c stuck-at-0" illustrates the main cause for the high complexity of test pattern generation.

Test generation steps for this example are as follows:

1. Set line c to 1. Its state is denoted by 1/0 as before.
2. Justify $c=1$ by setting line a to 0.
3. Carry out forward implication of $a=0$; i.e., set z to 1.
4. Propagate the state of c forward. It is impossible. Backtrack until a new choice is available.
5. Backtrack to step 3 and undo it.
6. Backtrack to step 2, use alternative choice: set $b=0$ and leave a unspecified.
7. Propagate state of c forward by setting $a = 1$.

Test "10" is found in seven steps. Although this circuit has about the same size as that in the last example, the test generation took more than twice as many steps. The added complexity is due to the reconvergent fan-out of a, which made backtracking necessary. In general, large logic circuits contain numerous reconvergent fan-outs, and the test generation complexity increases rapidly with the size of circuit.

The first comprehensive algorithm for combinational circuit test generation was the D *algorithm.*[29] This method uses five logic values, $\{0, 1, x, D, \overline{D}\}$ and generates tests by propagating a D or $\overline{D}$ from the faulty line to an observable output. The values D and $\overline{D}$ are used to represent signals that differ in the faulty and the fault-free circuits. A fault-free/faulty state 1/0 is represented by D, and the state 0/1 is represented by $\overline{D}$. The D-algorithm guarantees a test for a nonredundant fault (a fault is redundant if it does not alter the circuit function and is, therefore, undetectable by I/O testing).

A similar guarantee is offered by a more recent, and perhaps the most popular, algorithm, known as *path-oriented decision making (PODEM)*.[30] The PODEM algorithm incorporates heuristics to implicitly enumerate the circuit inputs, which, in general, result in better performance. Other extensions of these two basic algorithms are the *subscripted D-algorithm*[31] and FAN.[32]

8.3.2 Test Generation for Sequential Circuits

Sequential circuits contain memory states that are absent in combinational circuits. Functionally, the output of a sequential circuit depends on its inputs and its internal memory states. Structurally, the memory states are implemented either through *feedbacks* in logic elements or by *charge-storage elements*. A common practice in test generation is to extend the methods of combinational test generation to sequential circuits. We will illustrate this by an example.

Consider the latch in Fig. 8.7. Assume the gates have zero delay. The result of

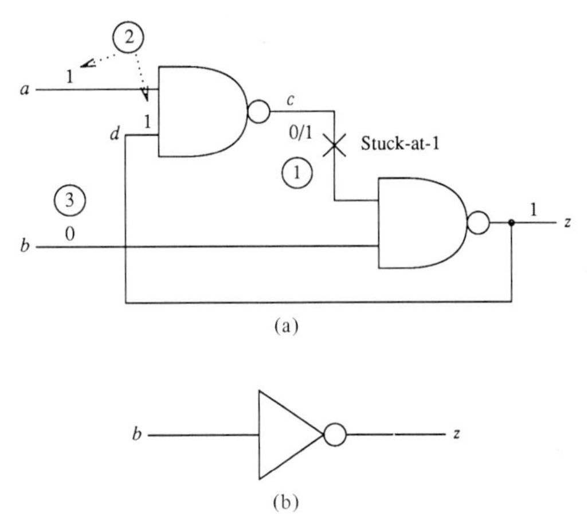

FIG. 8.7 Test generation for NAND latch. (*a*) Circled numbers show the sequence of test generation steps; (*b*) faulty function.

applying the combinational test generation approach is shown in the figure. The process stops with inputs at 10 and output at 1. Since the output is the same in both good and faulty circuits, test is not found.

Figure 8.7 also shows that the faulty function, when "line c is stuck-at-1," is $z = \bar{b}$. The fault destroys the feedback; the circuit no longer has the storing capability. The fault can be easily detected by first applying a 10 input and then following it up by 11. The first pattern will produce a 1 output, irrespective of the fault. The second pattern simply stores the state of latch. In the good circuit, the output will remain 1, but it will change to 0 in the faulty circuit.

The combinational test generation method was not able to solve this problem due to the zero-delay assumption. If we consider finite delays of gates, it is possible to apply the value of $z = 1$ to d and then change b to 1 to sensitize the path for the fault. A common practice is to cut the feedback path. This is shown in Fig.

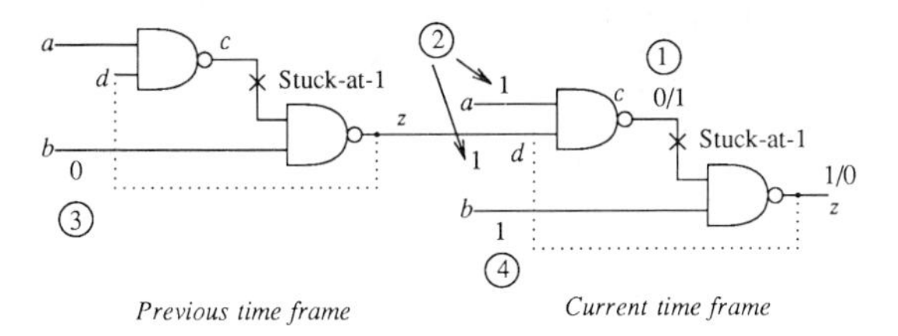

FIG. 8.8 Time frame extension of NAND latch. (Circled numbers show the sequence of test generation steps.)

8.8. A copy of the circuit is attached to generate the feedback signal *d*. The test generation, as shown in the figure, begins from the copy shown as *current time frame*. In general, any number of time frames (previous or future) can be added on either side of the current time frame. However, circuit complexity increases. Another problem occurs due to the signals left unspecified by the test generator. For example, the test in Fig. 8.8 is an 11 pattern preceded by $X0$. If we set X to 1, we get the desired test. But if X was set to 0, the test will cause a *race* in the fault-free circuit. Thus sequential circuit tests generated by such procedures require special processing to avoid timing problems.

Detailed algorithms on the procedure illustrated in the above example have been reported in the literature.[33–35] Typical program implementations have also been reported.[36–39] One common problem with automatically generated tests is the timing problem of races. This occurs because the timing is neglected by the test generators. Some of the tests, when verified by a fault simulator, are found to be ineffective. As a result, several other approaches have been considered. These include higher-level modeling approaches[40,41] simulation-based approaches[42,43] and the knowledge-based expert systems methods.[44]

In spite of all the work done so far, test generation for general sequential circuits continues to be a bottleneck in the design methodology. Complexity of gate-level test generation prevents the use of automatic methods for circuits with over 1000 or 2000 gates. The higher-level modeling approach, on the other hand, does not guarantee high fault coverage for gate-level-stuck faults. At this time the only successful automatic method is the scan design used in conjunction with an automatic test generation program.[45,46]

8.4 TEST EVALUATION

How good are the tests? Can they reject every faulty chip? If not, how many bad chips are going to pass the tests? These questions must be answered before a design can be successfully manufactured. Common methodology is to evaluate the tests for their coverage of single stuck faults. Experimental studies have shown strong correlation between fault coverage and the quality of the tested product.[47,48]

8.4.1 Evaluation Criteria

Since the coverage is assessed for the modeled faults (single stuck) and real faults may be quite different (transistor open, short, bridging, excess delay, multiple faults, etc.), even the high measured coverage cannot guarantee perfect quality. Quality is measured as the fraction of faulty chips among the tested-good chips. A reject ratio of 0.01 percent (100 defective parts per million) may be commercially acceptable, though better quality levels are required for military applications.

In practice, it is possible to obtain a 100 percent fault coverage in the combinational part of a scan-type circuit. For general sequential circuits, coverage in excess of 95 percent may be adequate. These coverages are over all stuck faults at the gate level. When higher-level modeling is used, stuck faults at the inputs and outputs of higher-level modules are analyzed. Measured coverage in such cases is higher than the gate-level coverage for the same tests.

8.4.2 Fault Simulation for General Circuits

A fault simulator is a standard CAD tool in today's VLSI design environment. Inputs to the fault simulation program are the circuit description (a gate-level or transistor-level netlist) and the input vectors. The fault simulator generates a list of collapsed faults and computes the coverage as a function of the number of input vectors. Fault simulation programs are implemented in many ways. Some of these are listed below.

Serial Fault Simulator. A serial fault simulator processes one fault at a time. It primarily requires a good circuit simulator that is run repeatedly. First, the fault-free circuit is simulated through all test vectors, and the output response is stored. A single fault is then injected by modifying the circuit description to represent the faulty behavior. Simulation is repeated, and as the output response is being computed it is compared with the stored fault-free response. If only detection is desired, the simulation can be stopped as soon as a discrepancy occurs in the response. The process is repeated over all faults. In the worst case, for n faults, the serial fault simulation requires $n + 1$ times the effort of fault-free simulation. In practice, however, if the faulty circuit simulation is stopped as soon as the fault is detected, the effort will be reduced, typically to a half or a third of the worst case. Implementation of a serial fault simulator is simple, it requires only a moderate amount of memory, and provides flexibility in circuit modeling and fault modeling. Yet it is rarely used because more efficient algorithms are available.

Parallel Fault Simulator. A natural improvement over the serial fault simulation would be to simulate several faults at the same time. The simplest such technique is *parallel fault simulation,*[49] in which, during a pass, each bit of the computer word simulates a different fault. This is possible because of the parallelism available at the level of logical instructions in all computers. Before the beginning of a pass, a set of W-1 as-yet-unsimulated faults are chosen, where W is the word size of the computer. The remaining bit is used to simulate the fault-free circuit. Gates whose inputs-outputs are directly affected by a selected fault are flagged so that when it is time to simulate such a gate, the effect of the fault can be *injected* into appropriate bits of the words representing its inputs and outputs. Thus a total of F faults can be simulated in $F/(W\text{-}1)$ passes. Parallel fault simulators give their best performance when the circuit is modeled entirely at the logic gate level.

Deductive Fault Simulator. The ultimate, in terms of reducing the number of passes, is the *deductive fault simulator.*[50] It needs just one pass for simulation independent of the number of faults simulated. The basic idea here is to associate with each line a list of just those faults that are sensitized to that line (that is, signals on the line for the normal and faulty circuits are different) by the simulated input pattern. Simulation of a gate amounts to deducing the fault list at the gate output from the input fault list. This will be illustrated for a two-input AND gate assumed to be embedded in a circuit (Fig. 8.9). The normal signals on its

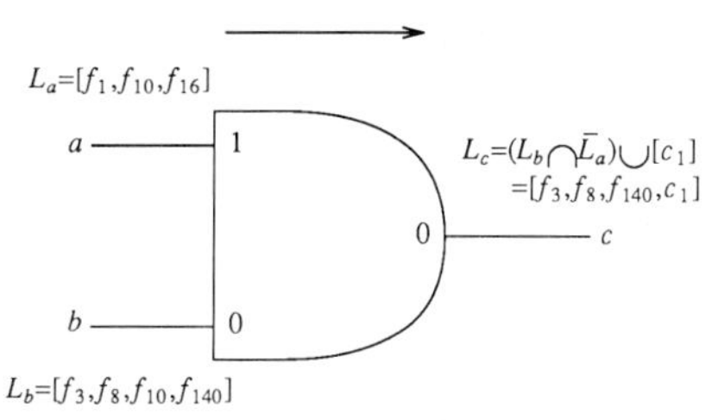

FIG. 8.9 Fault list computation in deductive fault simulation.

inputs a and b are assumed to be 1 and 0, respectively, for the pattern being simulated. We also assume that the fault list L_a associated with line a contains three single stuck-type faults denoted by f_1, f_{10}, and f_{16}. Similarly the fault list L_b contains the four faults shown. In the presence of the fault f_1 the signal value on input a will change to 0, but that on input b will stay unchanged (since f_1 is not in L_b). The output will stay at 0; hence f_1 cannot be in the fault list L_c. Indeed, no fault in L_a can occur in L_c for the same reason. Next, consider the fault f_8, which is in L_b but not in L_a. Such a fault will change the values on both lines b and c and hence must be included in L_c. Additionally, the fault "c stuck at 1" (or c_1) is in the output fault list since it complements the normal value. Thus we obtain the expression for the output fault list shown in the figure.

Compared with the parallel fault simulation, the penalties paid for a single pass in deductive simulation are (a) dynamically varying storage for fault lists associated with each line and (b) more complex processing of a gate requiring set operations on the line fault lists. In deductive simulation the computation of the fault list at a gate output must be dynamic because of its dependence on *both* the gate type and gate inputs. Less obviously, the output fault list may change even when the gate inputs do not change. In the above example, when the normal value on line a changes to 0, the value on line c is not changed, but L_c may change nevertheless. This "fault-list event" must be propagated through all the gates to which line c is an input. This observation has motivated a refinement of the deductive simulation method, which avoids unnecessary simulation of lines whose values remain unchanged under a fault.

Concurrent Fault Simulator. The basic idea behind concurrent fault simulation is quite simple. Consider how a stuck-type fault might affect the normal signal values in a circuit. In the simplest case, the faulty value may coincide with the normal value and not affect the circuit at all. Even otherwise, a typical fault affects only a very small part of a large circuit. Thus, most, if not all, of the information for simulating a fault is contained in the "good-circuit" simulation. In concurrent simulation, the good-circuit simulation is carried out in its entirety, but a faulty circuit is simulated only at gates with at least an input or the output value under the fault differing from the good-circuit values.

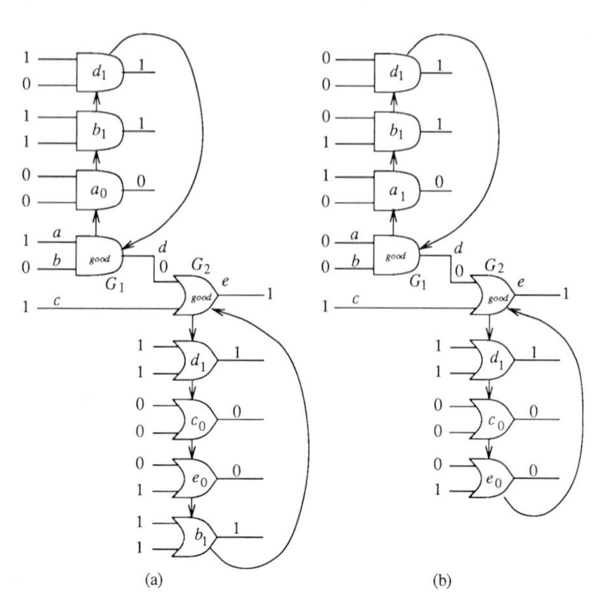

FIG. 8.10 Concurrent fault simulation example.

As an example, consider the two-gate circuit in Fig. 8.10*a* with the signals 1, 0, and 1 applied, respectively, to the inputs *a, b,* and *c*. The figure shows the true value simulation for this pattern. In addition, it shows, attached to each gate, a list of "faulty-circuit" gates that differ from the good-circuit gate in an input or the output. Often these would be attached in the form of a linked list in an implementation. The associated fault is marked inside the gate symbol. There are six single stuck-type faults associated with each of the two gates (two of which overlap since line *d* is common to both the gates). Of the six faults in G_1, only two are attached to gate G_2; the other four do not cause a change at the inputs of gate G_2. Next, suppose the input *a* was changed from 1 to 0 (Fig. 8.10*b*). Since an input of G_1 has changed, it must be resimulated with the new values. The resulting true values and the faulty gates are as shown in the figure. Notice that concurrent simulation stops at this point, since the inputs to gate G_2 have remained unchanged. In contrast, a deductive fault simulator would have processed gate G_2 also because of a fault-list event at the output of G_1. Thus, even for a simple circuit concurrent simulator will involve less processing.

The term *concurrent* is derived from the fact that each faulty gate carries enough information for independent simulation of the associated fault. This is considerably more information than just the fault index used in deductive simulation. Thus the speed is gained in concurrent simulation at the expense of additional dynamic storage per node in the circuit.

Even though we discussed concurrent simulation in terms of gate networks, the basic concept transcends the level of simulation. Some successful implementations include FMOSSIM[51] at the switch level, MOTIS[52] at a mixed level, and CHIEFS[53] with hierarchical fault simulation.

Parallel Value List Simulator. A recent fault simulation algorithm, known as the *parallel value list method,* combines the features of the parallel and the concurrent methods.[54]

Statistical Fault Analysis. Fault simulation of large circuits is a resource-intensive task both in central processor unit (CPU) time and computer memory. Fault-free simulation, on the other hand, has much lower complexity. Statistical fault analysis, known as STAFAN[55], completely avoids fault simulation. Using nodal activity generated during good-circuit simulation, STAFAN determines controllability (the degree to which the test vectors exercise circuit nodes) and observability (the likelihood of faults propagating to the output). Taking both factors into consideration, one can determine detection probabilities of individual faults and the fault coverage. The main overhead added by the analysis to good-circuit simulation is the collection of nodal activity statistics. As a result, the computation time grows only linearly with the number of circuit nodes.

8.4.3 Fault Simulation for Scan Circuits

In a scan circuit, all outputs of the combinational logic are observable. It is therefore unnecessary for a fault simulator to propagate fault effects through sequential elements. Thus, fault simulation of combinational logic would suffice. Very efficient simulators have been devised for this special situation.

Parallel Pattern Single-Fault Propagation (PPSFP). In the conventional form of a parallel fault simulator, the signal on a line is represented by a computer word. The bits in this word correspond to the states of the line for various faults. Some of these faults may be detected earlier than others. The simulation, however, must continue until all the faults in one pass are detected or until the vectors are exhausted. Thus, the faults detected earlier have to be simulated unnecessarily. To improve the efficiency, workers at IBM devised a parallel pattern simulator.[56] The simulator runs on special hardware with a 256-bit word. The combinational part of the circuit being simulated is modeled with logic gates, and the simulation is performed with just two logic states, 0 and 1. Fault simulation is similar to the serial method, but 256 vectors are processed in the time that a conventional simulator would take for one vector. This provides an improvement of two orders in performance over parallel, deductive, and concurrent simulators. Further extensions of this method allow simulation of delay faults for scan designs at a nominal 10 percent increase in computation time.[57]

Critical Path Tracing. Another efficient method of fault simulation in combinational circuits, known as *critical path tracing,*[58] completely avoids fault propagation during simulation. The simulation is performed in the true-value mode. After a vector is simulated, an analysis of criticality is performed. A line is defined as *critical* if there exists a sensitized path to a primary output. If a line is critical and its value is v, then the fault "stuck-at-$\overline{v}$" on this line is detected. The criticality analysis proceeds inward from primary outputs, which are always critical. If the output of a gate is critical, then its inputs having sensitized paths through the gate are also critical. A single pass from circuit outputs to inputs marks all critical lines in a circuit that is free from reconvergent fan-outs. This analysis, performed for each vector, will provide fault-detection data. For circuits with reconvergent fan-outs, the criticality analysis for *stem* lines (fan-out lines that produce reconverging signals) becomes more involved, and one must resort to approximations.

8.4.4 Fault Sampling

Only a fraction of faults are simulated in a sampling technique in order to estimate the fault coverage. In a method that has a direct analogy with the opinion polls, a randomly chosen sample of some fixed number of faults is simulated, and the fraction detected by the test set is used as an estimate for the fault coverage. The particular appeal of sampling techniques lies in the fact that the confidence range of the estimate depends only on the sample size and not on the population size. Many people find this result counterintuitive. However, it is the basis for the success of Monte Carlo methods[59] in solving problems involving many dimensions (e.g., n-dimensional quadrature). Also, the estimate range narrows closer to 100 percent fault coverage. For example, an estimate of 95 percent fault coverage can be accurate to within 2 percent for a sample size of 1000, while a 50 percent estimate has a 5 percent tolerance.[60]

Accuracy in a sampling experiment demands large sample size. A sample size of 1000 to 2000 faults is generally considered necessary. Simple analysis gives the following estimate (with 99 percent confidence) of fault coverage:

$$C = c \pm \frac{4.51 + Nc(1 - c)}{N}$$

where c is the fraction of covered faults in the sample, and N is the number of sampled faults (N 1000). As an example consider a sample of 2000 faults randomly picked from the fault list. If, after fault simulation with the given vectors, 1900 sample faults are found detectable, then the preceding formula gives the range of coverage as $C = 0.95 \pm 0.022$. Thus the fault coverage of the simulated vectors over the fault list is estimated to lie between 92.8 and 97.2 percent.

Considering the fact that a typical VLSI circuit can have 10,000, or even 100,000 faults, simulation of just 2000 faults will result in significant saving of computing resources. One disadvantage of the sampling approach is that it does not provide a complete list of undetected faults. Designers still consider the method useful. For instance, in the foregoing example, the sample fault simulation will give 100 undetected faults that the designer could examine to find the portions of the circuit with low detectability. On the other hand, simulation of all faults in a 100,000-fault circuit with a similar coverage would have produced a list of undetected faults with nearly 5000 entries, a number too large to examine conveniently!

8.5 TESTABILITY ANALYSIS

Fault simulation, as described in the previous section, is considered by some as a form of testability analysis. Portions of the circuit where most undetected faults are located could be considered to have poor testability. However, fault simulation requires a prior generation of test vectors. If tests were already generated, why should we be concerned about testability? Another problem with fault simulation is that it not only evaluates circuit testability, but it also evaluates the vectors. For example, fault simulation of a testable circuit with improperly generated vectors might show poor testability.

Since test generation is done after the design is completed and verified, design changes to improve testability are not desirable at that stage. The primary purpose of testability analysis is to identify problem areas before test generation, preferably in the early stages of the design. The analysis should, therefore, only be based on the circuit structure. To be useful, the effort of testability analysis should also be considerably less than the actual test generation. Test generation effort is known to increase exponentially with circuit size. Complexity of testability analysis should be close to linear.

8.5.1 SCOAP

The most common approach to testability analysis is to define and compute *controllability* and *observability* for every node (or signal-carrying line) in the circuit. The most popular testability algorithm at the present time is SCOAP (Sandia controllability and observability analysis program). SCOAP[61,62] computes six quantities for every node in the circuit. These quantities relate to the effort of controlling and observing the node. For node x, the six quantities are defined as follows:

1. Combinational zero controllability $CC^0(x)$
2. Combinational one controllability $CC^1(x)$
3. Combinational observability $CO(x)$
4. Sequential zero controllability $SC^0(x)$
5. Sequential one controllability $SC^1(x)$
6. Sequential observability $SO(x)$

The number of other nodes that must be set to either control or observe a particular node determines the combinational measures for that node. Sequential measures, on the other hand, depend on the number of time frames (e.g., clock cycles) required to control or observe a node. For all primary inputs of the circuit, both combinational controllabilities are defined as unity, and both sequential controllabilities as zero. Also, both observabilities for all primary outputs are set to zero. To understand how these measures are computed for other nodes of the circuit, consider the AND gate in Fig. 8.11. Since the gate is combinational, only combinational measures are relevant.

All node controllabilities are calculated first and used in the calculation of node observabilities. The calculation of controllabilities proceeds from input to output. In Fig. 8.11*a* the combinational one controllability of the output line c is computed as the sum of the one controllabilities of the input lines. Intuitively, this computation is obvious if controllability of a node is equated with the effort to set that node to the specified value. Thus, in order to set line c to 1, both lines a and b must be set to 1. The quantity d is added as the depth factor, again based on the intuition that the farther removed a node is from the primary input the harder it is to control. The exact value chosen for d is somewhat arbitrary; Goldstein[61] suggests $d = 1$. In Fig. 8.11*b* line c can be set to zero by setting *either a or b* to zero, whichever is easier. Thus, the minimum of the zero controllabilities for lines a and b is added to d to derive zero controllability of c.

The observability calculation starts with an *a priori* definition of observabilities for primary outputs. The observability of a gate input is defined in terms of the observability of the gate output and the controllabilities of the other inputs of the gate. Recursive application of these definitions allows the compu-

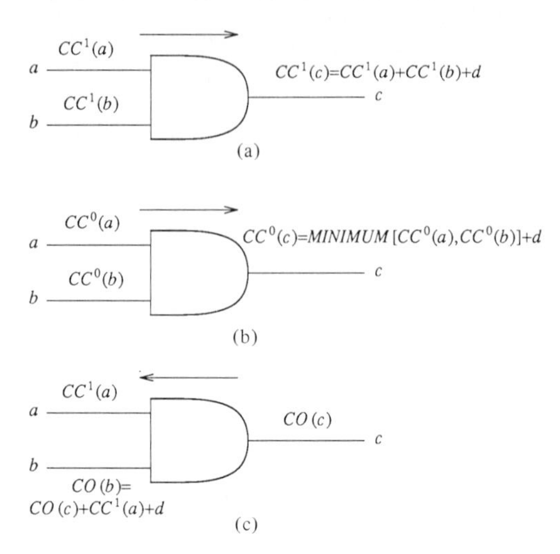

FIG. 8.11 Combinational controllabilities and observability in SCOAP.

tation of observabilities for all the nodes in the circuit. Figure 8.11*c* illustrates the calculation of observability of an input to an AND gate: for observing line *b*, one must set line *a* to 1 and observe *c*. A node in the circuit will become harder to control or observe as its corresponding measure increases.

The sequential measures are defined and computed in a similar way, but, as already mentioned, they represent the length of input sequences required to control or observe a node. Thus combinational elements add nothing to sequential measures. For a flip-flop the output sequential controllabilities will be determined by the sequential controllabilities of its data and clock inputs and a sequential depth factor.

The overall complexity of computation in SCOAP is almost linear in the number of gates, which makes it an attractive testability audit tool. Actual experience has shown the sequential measures to be less valuable than the combinational ones. Analytical approaches to relate the testability measures to data on fault coverage or fault detection have not met with much success. Studies[63] of correlation between fault detection data and testability measures show testability measures to be quite poor at predicting which individual faults will remain undetected and which will be detected. However, testability measures *can* be effective in predicting relative probabilities of detection of specific fault sets. The general information provided at the entire circuit level may be even better. Other testability methods, in the same class as SCOAP, are TMEAS,[64] TESTSCREEN,[65] CAMELOT,[66] and VICTOR.[67] These also suffer from similar inaccuracies as SCOAP.

8.5.2 Applications of SCOAP

In practice, designers use testability measures to find portions of poor controllability or observability in their circuits, as identified by large values. Proper modifications in the circuit lead to reduction of these values. The testability of the stuck-type faults on a node *x* is defined as

$$T(x \text{ stuck-at-0}) = \mathrm{CC}^1(x) + \mathrm{CO}(x)$$
$$T(x \text{ stuck-at-1}) = \mathrm{CC}^0(x) + \mathrm{CO}(x)$$

These equations reflect the requirement that to detect a stuck-type fault on a line it must be set at the opposite value *and* observed.

The solid line in Fig. 8.12*a* shows node testabilities computed by SCOAP in the form of a histogram for a complex VLSI circuit. Notice that testability represents the effort of testing; therefore, nodes with high testability values are the most difficult to test. An examination of the nodes with large testability values (above 1500) showed that they were clustered in a portion of the circuit that had poor observability. A simple modification, as shown in Fig. 8.12*b*, allows the signals to be observed at a primary output through a test multiplexer. The testability histogram of the modified circuit, shown with dotted line in Fig. 8.12*a*, clearly indicates the testability improvement.

Empirical study has shown the use of testability analysis in predicting the test length. In this work the overall circuit *testability index* is defined as[68]

$$\text{Testability index} = \log \sum_{\text{all } f_i} T(f_i)$$

The actual number of test vectors (for 90 percent fault coverage) for several circuits are shown in Fig. 8.13 as a function of their testability index.

We have given two applications of SCOAP. Testability analysis has also been used to improve the search heuristics of test generator programs.[69–71] Attempts at testability-based static analysis of fault coverage have also been quite successful for combinational circuits.[72,73] In 1984, at the International Test Conference, the panel discussing the topic *Will Testability Analysis Replace Fault Simulation?*

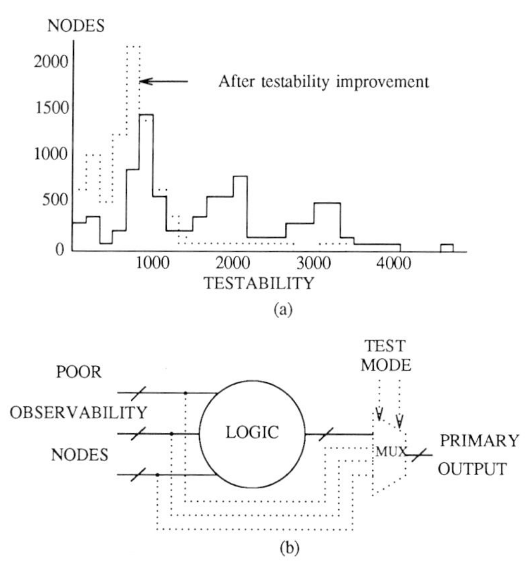

FIG. 8.12 SCOAP-guided testability improvement.

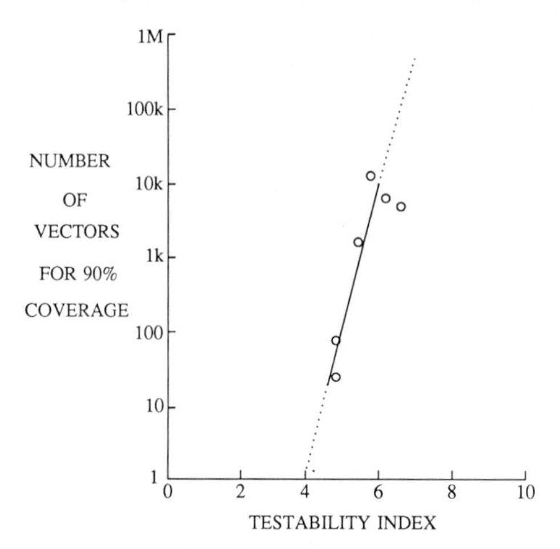

FIG. 8.13 Number of vectors versus testability index for typical VLSI circuits.

concluded with a clear identification of two problems: (1) fault simulation will be too expensive for the million-device chips of the future; (2) improvements are needed in testability analysis techniques. Use of testability analysis in guiding design for testability is described in Ref. 74.

8.5.3 Probabilistic Testability Measures

There are two problems with the SCOAP-like testability measures. First, the definition of measures (e.g., effort of controlling or observing) makes their calibration difficult. The actual test vector generation depends, in addition to the circuit complexity, on the engineer's experience, the tools, and so on. Testability measures are only effective while comparing two circuits. Second, the simplifying assumptions neglect signal correlations arising due to reconvergent fan-outs in the circuit. Since the reconvergent fan-outs are the very reasons for the difficulty of test generation, testability measures often fail to identify untestable circuitry.

Recent efforts have addressed developing testability measures based on a probabilistic definition. An example of such a measure is COP,[75] which defines controllabilities and observabilities as probabilities of controlling and observing signals in a random input environment. Exact methods for computing these probabilities, although known for some time,[76] have not become popular due to high complexity. Practical approaches include circuit partitioning to localize the reconvergent fan-out regions for exact analysis[77] or computing lower and upper bounds on the required probabilities.[78]

REFERENCES

1. S. Kamal and C. V. Page, "Intermittent Faults: A Model for a Detection Procedure," *IEEE Trans. on Computers,* **C-23:**713–719 (1974).

2. S. M. Thatte and J. A. Abraham, "Test Generation for Microprocessors," *IEEE Trans. on Computers,* **C-29:**429–441 (1980).
3. J. F. Poage, "Derivation of Optimum Tests to Detect Faults in Combinational Circuits," *Proc. Symp. on Mathematical Theory of Automata (April 1962),* Polytechnic Press, New York, 1963, pp. 483–528.
4. R. L. Wadsack, "Fault Modeling and Logic Simulation of CMOS and MOS Integrated Circuits," *Bell Syst. Tech. J.,* **57:**1449–1474 (1978).
5. K. C. Y. Mei, "Bridging and Stuck-at-Faults," *IEEE Trans. on Computers,* **C-23:**720–727 (1974).
6. D. L. Ostapko and S. J. Hong, "Fault analysis and Test Generation for Programmable Logic Arrays," *IEEE Trans. on Computers,* **C-28:**617–626 (1979).
7. J. Knaizuk, Jr. and C. R. P. Hartmann, "An Optimal Algorithm for Testing Stuck-at Faults in Random Access Memories," *IEEE Trans. on Computers,* **C-26:**1141–1144 (1977).
8. R. Nair, "Comments on an Optimal Algorithm for Testing Stuck-at Faults in Random Access Memories," *IEEE Trans. on Computers,* **C-28:**258–261 (1979).
9. J. P. Hayes, "Detection of Pattern-Sensitive Faults in Random Access Memories," *IEEE Trans. on Computers,* **C-24:**150–157 (1975).
10. D. S. Suk and S. M. Reddy, "Test Procedures for a Class of Pattern-Sensitive Faults in Semiconductor Random Access Memories," *IEEE Trans. on Computers,* **C-29:**419–429 (1980).
11. R. Nair, S. M. Thatte and J. A. Abraham, "Efficient Algorithms for Testing Semiconductor Random-Access Memories," *IEEE Trans. on Computers,* **C-27:**572–576 (1978).
12. D. S. Suk and S. M. Reddy, "A March Test for Functional Faults in Semiconductor Random Access Memories," *IEEE Trans. on Computers,* **C-30:**982–985 (1981).
13. D. S. Brahme and J. A. Abraham, "Functional Testing of Microprocessors," *IEEE Trans. on Computers,* **C-33:**475–485 (1984).
14. W. H. Kautz, "Testing for Faults in Cellular Logic Arrays," *Proc. Symp. on Switching and Automata Theory,* 1967, pp. 161–174.
15. F. C. Hennie, *Finite-State Models for Logical Machines,* Wiley, New York, 1968.
16. G. L. Smith, "Model for Delay Faults Based on Paths," *Proc. Int. Test Conf.,* Philadelphia, Pa. 1985, pp. 342–349.
17. C. J. Lin and S. M. Reddy, "On Delay Fault Testing in Logic Circuits," *Proc. Int. Conf. Comp. Aided Des. (ICCAD-86),* Santa Clara, Cal., 1986, pp. 148–151.
18. V. K. Agrawal and A. S. F. Fung, "Multiple Fault Testing of Large Circuits by Single Fault Test Sets," *IEEE Trans. on Computers,* **C-30:**855–865 (1981).
19. D. C. Bossen and S. J. Hong, "Cause-Effect Analysis for Multiple Fault Detection in Combinational Networks," *IEEE Trans. on Computers,* **C-20:**1252–1257 (1971).
20. M. Abramovici, P. R. Menon, and D. T. Miller, "Checkpoint Faults Are Not Sufficient Target Faults for Test Generation," *IEEE Trans. on Computers,* **C-35:**769–771 (1986).
21. E. J. McCluskey and F. W. Clegg, "Fault Equivalence in Combinational Logic Networks," *IEEE Trans. on Computers,* **C-20:**1286–1293 (1971).
22. D. R. Schertz and G. Metze, "A New Representation for Faults in Combinational Digital Circuits," *IEEE Trans. on Computers,* **C-21:**858–866 (1972).
23. C. W. Cha, "Multiple Fault Diagnosis in Combinational Networks," *Proc. 16th Des. Auto. Conf.,* 1979, pp. 149–155.
24. V. D. Agrawal and P. Agrawal, "An Automatic Test Generation System for Illiac IV Logic Boards," *IEEE Trans. on Computers,* **C-21:**1015–1017 (1972).
25. K. P. Parker, "Adaptive Random Test Generation," *J. Des. Automation & Fault-Tol. Comput.,* **1:**62–83 (1976).

26. H. D. Schnurmann, E. Lindbloom, and R. G. Carpenter, "The Weighted Random Test-Pattern Generator," *IEEE Trans. on Computers,* **C-24:**695–700 (1975).
27. E. F. Sellers, M. Y. Hsiao, and L. W. Bearnson, "Analyzing Errors with the Boolean Difference," *IEEE Trans. on Computers,* **C-17:**676–683 (1968).
28. S. M. Reddy, "Complete Test Sets for Logic Functions," *IEEE Trans. on Computers,* **C-22:**1016–1020 (1973).
29. J. P. Roth, W. G. Bouricius, and P. Schneider, "Programmed Algorithms to Compute Tests to Detect and Distinguish Between Failures in Logic Circuits," *IEEE Trans. on Electronic Computers,* **EC-16:**567–580 (1967).
30. P. Goel, "An Implicit Enumeration Algorithm to Generate Tests for Combinational Logic Circuits," *IEEE Trans. on Computers,* **C-30:**215–222 (1981).
31. C. Benmehrez and J. F. McDonald, "Measured Performance of a Programmed Implementation of the Subscripted D-Algorithm," *Proc. 20th Des. Auto. Conf.,* Miami Beach, Fla., 1983, pp. 308–316.
32. H. Fujiwara and T. Shimono, "On the Acceleration of Test Generation Algorithms," *IEEE Trans. on Computers,* **C-32:**1137–1144 (1983).
33. H. Kubo, "A procedure for Generating Test Sequences to Detect Sequential Circuit Failures," *NEC Res. & Dev.,* **12**:69–78 (1968).
34. R. Putzolu and J. P. Roth, "A Heuristic Algorithm for the Testing of Asynchronous Circuits," *IEEE Trans. on Computers,* **C-20:**639–647 (1971).
35. P. Muth, "A Nine-Valued Circuit Model for Test Generation," *IEEE Trans. on Computers,* **C-25:**630–636 (1976).
36. J. J. Thomas, "Automated Diagnostic Test Programs for Digital Networks," *Computer Design,* August 1971, pp. 63–67.
37. S. G. Chappell, "Automatic Test Generation for Asynchronous Digital Circuits," *Bell Syst. Tech. J.,* **53:**1477–1503 (1974).
38. R. A. Marlett, "EBT: A Comprehensive Test Generation Technique for Highly Sequential Circuits," *Proc. 15th Des. Auto. Conf.,* Las Vegas, Nev., 1978, pp. 335–339.
39. S. Mallela and S. Wu, "A Sequential Circuit Test Generation System," *Proc. Int. Test Conf.,* Philadelphia, Pa., 1985, pp. 57–61.
40. F. J. Hill and B. Huey, "A Design Language Based Approach to Test Sequence Generation," *Computer,* **10:**28–33 (1977).
41. T. Lin and S. Y. H. Su, "The S-Algorithm: A Promising Solution for Systematic Functional Test Generation," *IEEE Trans. on Computer-Aided Design,* **CAD-4:**250–263 (1985).
42. T. J. Snethen, "Simulator-Oriented Fault Test Generator," *Proc. 14th Des. Auto. Conf.* 1977, pp. 88–93.
43. V. D. Agrawal and K. T. Cheng, "Threshold-Value Simulation and Test Generation," *Testing and Diagnosis of VLSI and ULSI,* M. Sami and F. Lombardi (eds.), Martinus Nijhoff, Dordrecht, The Netherlands, 1988.
44. M. J. Bending, "Hitest: A Knowledge-Based Test Generation System," *IEEE Design & Test of Computers,* **1:**83–92 (1984).
45. V. D. Agrawal, S. K. Jain, and D. M. Singer, "Automation in Design For Testability," *Proc. Custom Integrated Circuits Conf. (CICC),* Rochester, N.Y., 1984, pp. 159–163.
46. D. Leet, P. Shearson, and R. France, "A CMOS LSSD Test Generation System," *IBM J. Res. Dev.,* **28:**625–635 (1984).
47. R. A. Harrison, R. W. Holzwarth, P. R. Motz, R. G. Daniels, J. S. Thomas, and W. H. Wiemann, "Logic Fault Verification of LSI: How it Benefits the User," *Proc. WESCON,* September 1984.
48. V. D. Agrawal, S. C. Seth, and P. Agrawal, "Fault Coverage Requirement in Production Testing of LSI Circuits," *IEEE J. Solid-State Circuits,* **SC-27:**57–61 (1982).

49. E. W. Thompson and S. A. Szygenda, "Digital Logic Simulation in a Time-Based, Table-Driven Environment, Part 2, Parallel Fault Simulation," *Computer,* **8**:38–44 (1975).

50. D. B. Armstrong, "A Deductive Method for Simulating Faults in Logic Circuits," *IEEE Trans. on Computers,* **C-21**:464–471 (1972).

51. M. D. Schuster and R. E. Bryant, "Concurrent Fault Simulation of MOS Digital Circuits," *Proc. Conf. Adv. Res. in VLSI,* Cambridge, Mass., 1984, pp. 129–138.

52. A. K. Bose, P. Kozak, C. Y. Lo, H. N. Nham, E. Pacas-Skewes, and K. Wu, "A Fault Simulator for MOS LSI Circuits," *Proc. 19th Des. Auto. Conf.,* Las Vegas, Nev., 1982, pp. 400–409.

53. W. A. Rogers and J. A. Abraham, "CHIEFS: A Concurrent, Hierarchical and Extensible Fault Simulator," *Proc. Int. Test Conf.,* Philadelphia, Pa., 1985, pp. 710–716.

54. K. Son, "Fault Simulation with the Parallel Value List Algorithm," *VLSI Systems Design,* vol. VI, December 1985, pp. 36–43.

55. S. K. Jain and V. D. Agrawal, "Statistical Fault Analysis," *IEEE Design & Test of Computers,* **2**:38–44 (1985).

56. J. A. Waicukauski, E. B. Eichelberger, D. O. Forlenza, E. Lindbloom, and T. McCarthy, "A Statistical Calculation of Fault Detection Probabilities by Fast Fault Simulation," *Proc. Int. Test Conf.,* Philadelphia, Pa., 1985, pp. 779–784.

57. J. A. Waicukauski, E. Lindbloom, B. K. Rosen, and V. S. Iyengar, "Transition Fault Simulation," *IEEE Design & Test of Computers,* **4**:32–38 (1987).

58. M. Abramovici, P. R. Menon, and D. T. Miller, "Critical Path Tracing: An Alternative to Fault Simulation," *IEEE Design & Test of Computers,* **1**:83–93, (1984).

59. Y. A. Shreider, *The Monte Carlo Method,* Pergamon Press, New York, 1966.

60. V. D. Agrawal, "Sampling Techniques for Determining Fault Coverage in LSI Circuits," *J. Digital Systems,* **5**:189–202 (1981).

61. L. H. Goldstein, "Controllability/Observability Analysis of Digital Circuits," *IEEE Trans. on Circuits and Systems,* **CAS-26**:685–693 (1979).

62. L. H. Goldstein and E. L. Thigpen, "SCOAP: Sandia Controllability/Observability Analysis Program," *Proc. 17th Des. Auto. Conf.,* Minneapolis, Minn., 1980, pp. 190–196.

63. V. D. Agrawal and M. R. Mercer, "Testability Measures—What Do They Tell Us?", *Proc. Int. Test Conf.,* Philadelphia, Pa., 1982, pp. 391–396.

64. J. Grason, "TMEAS—A Testability Measurement Program," *Proc. 16th Des. Auto. Conf.,* San Diego, Cal., 1979, pp. 156–161.

65. P. G. Kovijanic, "Single Testability Figure of Merit," *Int. Test Conf. Digest of Papers,* Philadelphia, Pa., 1981, pp. 521–529.

66. R. G. Bennetts, C. M. Maunder, and G. D. Robinson, "CAMELOT: A Computer-Aided Measure for Logic Testability," *Proc. Int. Conf. Comp. Des. (ICCC-80),* Port Chester, N.Y., 1980, pp. 1162–1165; also *IEE Proc.,* **128**(Pt. E):177–189 (1981).

67. I. M. Ratiu, "VICTOR: A Fast VLSI Testability Analysis Program," *Proc. Int. Test Conf.,* Philadelphia, Pa., 1982, pp. 397–401.

68. D. M. Singer, "Testability Analysis of MOS VLSI Circuits," *Proc. Int. Test Conf.,* Philadelphia, Pa., 1984, pp. 690–696.

69. V. D. Agrawal, S. C. Seth, and C. C. Chuang, "Probabilistically Guided Test Generation," *Proc. Int. Symp. Circ. Syst. (ISCAS),* Kyoto, Japan 1985, pp. 687–689.

70. F. Brglez, P. Pownall, and R. Hum, "Applications of Testability Analysis: from ATPG to Critical Path Tracing," Philadelphia, Pa., 1984, pp. 705–712.

71. S. Patel and J. Patel, "Effectiveness of Heuristics Measures for Automatic Test Pattern Generation," *Proc. Des. Auto. Conf.,* Las Vegas, Nev., 1986, pp. 547–552.

72. V. D. Agrawal and S. C. Seth, "Probabilistic Testability," *Proc. Int. Conf. Comp. Des. (ICCD-85),* Port Chester, N.Y., 1985, pp. 562–565.

73. F. Brglez, "A Fast Fault Grader: Analysis and Applications," *Proc. Int. Test Conf.*, Philadelphia, Pa., 1985, pp. 785–794.
74. E. Trischler, "ATWIG, An Automatic Test Pattern Generator with Inherent Guidance," *Proc. Int. Test Conf.*, Philadelphia, Pa., 1984, pp. 80–87.
75. F. Brglez, "On Testability Analysis of Combinational Networks," *Proc. Int. Symp. on Circ. Syst. (ISCAS)*, Montreal, Canada, 1984, pp. 221–225.
76. K. P. Parker and E. J. McCluskey, "Probabilistic Treatment of General Combinational Circuits," *IEEE Trans. on Computers*, **C-24:**668–670 (1975).
77. S. C. Seth, L. Pan, and V. D. Agrawal, "PREDICT—Probabilistic Estimation of Digital Circuit Testability," *Fault-Tolerant Computing Symposium (FTCS-15) Digest of Papers*, Ann Arbor, Michigan, 1985, pp. 220–225.
78. J. Savir, G. S. Ditlow, and P. H. Bardell, "Random Pattern Testability," *IEEE Trans. on Computers* **C-33:**79–90 (1984).

CHAPTER 9
BUILT-IN SELF-TEST

Richard M. Sedmak
Self-Test Services
Ambler, Pennsylvania

9.1 INTRODUCTION

There is little question that VLSI is offering a host of opportunities for significant advancements in electronic systems. However, along with the benefits comes a major bottleneck to the manufacturing of VLSI chips and the products that use them: the time and cost required for testing. Such a dilemma exists in part because leaps in integrated circuit technology are occurring at a faster rate than advancements in test technology.

Many of the popular designs for testability techniques, including the ad hoc approaches and the structured methods, such as level-sensitive scan design (LSSD), Scan/Set, and other scan path techniques, have helped to ease the difficulty of test generation and application. However, the core problem still remains: the increasing complexity and density of VLSI circuits lead to ever-worsening accessibility of internal logic nodes and to greater complexity and time associated with the generation and evaluation of test patterns. (See Chap. 8.) The problem is further complicated by the existence of some troublesome fault types, such as the stuck-open fault in CMOS,[1] which casts doubts on the very assumptions upon which many current test development tools are based. One very promising, emerging discipline that attacks not only the limited accessibility problem, but the associated test generation and application problems as well, is built-in self-test (BIST).

9.2 BASIC BIST CONCEPTS

9.2.1 Definition of BIST

BIST may be defined very simply as the capability for a unit to test itself, with little or no need for external test equipment or manual test procedures. A unit may be a chip, board, or system; and the definition implies that the testing processes of input stimulation and output response evaluation are integral to the unit being tested.[2] While BIST to some individuals connotes a chip-level capability only, the widely used definition certainly does not preclude board- or system-level implementation.

The term *built-in self-test* was coined in 1983 in conjunction with the establishment of an IEEE workshop that addresses the subject. The new term was created in order to merge the term *built-in test* (BIT), used primarily in the defense electronics industry, with the term *self-test,* embraced primarily by the commercial electronics industry. The new term *BIST* seems to be gaining support; it is beginning to appear frequently in both commercial and defense literature related to testing.

9.2.2 The Role of BIST

As depicted in Fig. 9.1, three primary subprocesses constitute a general test process: input stimulus generation, output response evaluation, and test control. The input stimulus generation subprocess involves the application of signals to the inputs of the unit under test (UUT) in such a way as to expose the symptoms of failures in the unit to a point where they can be observed, such as output lines or test points. The output response evaluation subprocess is associated with the comparison of the unit's actual test responses with the predicted, fault-free test responses of the UUT. The test control subprocess is related to the control of the input and output subprocesses, as well as the collection and processing of test results.

In conventional "external testing," shown in Fig. 9.1*a,* the three subprocesses are carried out in a piece of equipment external to the UUT, such as automatic test equipment (ATE) or a maintenance processor. Paradoxically, some members of the testing community consider the incorporation of resident ATE or mainte-

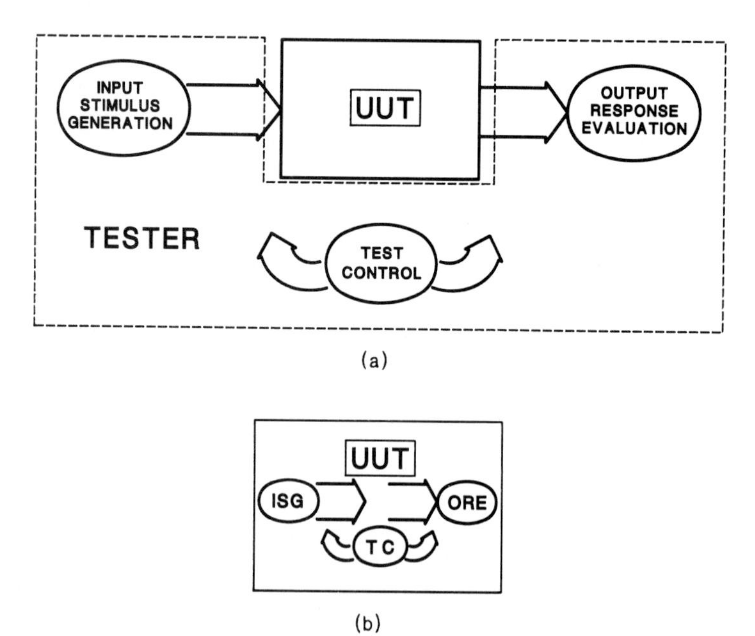

FIG. 9.1 Conventional external testing (*a*) versus BIST (*b*).

nance processors in a system as system-level BIST. That broader concept of BIST is not embraced by this author.

In a BIST environment, shown in Fig. 9.1*b*, the three subprocesses are integral to the UUT, rather than being external to it. It is from this key characteristic that the many benefits of BIST are derived, as discussed in Sec. 9.3.1. Although a total integration of the three subprocesses into the UUT is required to constitute a pure BIST capability, a hybrid approach is occasionally adopted, in which one or two of the subprocesses are integrated into the UUT, while the remaining one or two are relegated to the external tester. Such an approach is considered a partial BIST implementation by this author.

9.3 BENEFITS AND PENALTIES OF BIST

9.3.1 Benefits of BIST

The primary benefits of BIST can be summarized as follows:

1. Substantially lower test generation costs
2. Reduced storage and maintenance of test patterns
3. Simpler and less costly ATE and maintenance processors
4. Higher fault coverage, given time constraints for external test generation and application
5. Reduced test and diagnostic time and costs
6. Ability to test, cost effectively, many units in parallel
7. Commonality of test approach in all test phases
8. For some approaches, applicability to both off-line and on-line testing
9. Ease of accommodating engineering changes
10. For some approaches, independence from fault models, circuit technology, and packaging approach

Most of these benefits are derived from two causal factors:

Unlimited access to internal nodes

Ability to run at circuit speeds

Unlimited access to internal nodes is provided by virtue of the test process being moved into the UUT in a BIST environment. For example, if a BIST capability is provided within a chip, the on-chip pattern generators and response evaluators have unlimited access to any point within the chip, including such sources of testability headaches as redundant nodes, random-pattern resistant nodes, and global feedback paths.

With the test process being carried out within the UUT, the input-output pin bottleneck, associated with external testing, is eliminated. From the removal of such a critical testing barrier, substantial advantages are gained in test time and costs. Fortunately, in practice there is no need to take full advantage of the notion of unlimited access, because adequate fault coverage and test time can be achieved without resorting to the area-consuming routing of BIST elements to every node in the UUT.

The ability to run at circuit speeds is also derived from the incorporation of the test process within the UUT in a BIST implementation. Once the pattern generators and response evaluators are moved into the UUT, they are implemented using the same "fabric" from which the functional circuits are made; hence, those BIST elements can run at the same speed as the functional circuits being tested. Benefits realized from this factor include, for example, a savings in test time and an increased ability to detect timing-related faults.

The greatest payback from BIST benefits can be achieved by planning for BIST at the outset of the product development cycle and then designing and implementing it *in conjunction with the functional design,* rather than as an afterthought.

9.3.2 Penalties of BIST

The primary penalties of BIST can be summarized as follows:

1. Increased design effort and time
2. Circuit overhead
3. Pin overhead
4. Possible performance overhead
5. Yield impact
6. Reliability impact

All of the penalties associated with BIST arise from the additional hardware, firmware, or software required for a BIST implementation. The impact of those penalties can be greatly reduced by employing the following important BIST design guidelines:

1. Plan for BIST starting at the earliest stage (e.g., proposal stage) of the program.
2. Design BIST in conjunction with the functional design, not as an afterthought.
3. Employ for BIST design the same high degree of engineering cleverness and rigor that is used for the functional design.
4. Take advantage of CAD tools for the BIST design process whenever possible.
5. Incorporate the subject of BIST into peer, design, and program reviews.

9.3.3 BIST Impact on System Operational Readiness, Reliability, Maintainability, and Availability

BIST has a direct impact on a number of system attributes, including, for example, operational readiness, reliability, maintainability, and availability. A definition and discussion of each of these attributes follows.

Operational readiness: The percentage of time that a system will perform its mission correctly when activated from a powered-down or idle state

Reliability: The probability that the system will perform its mission for a specified period of time after it has entered operational mode

Maintainability: The probability that a system will be retained in, or restored to, a specified operational condition within a given period of time, when the maintenance is performed in accordance with prescribed procedures and resources[3]

Availability: The percentage of time that a system is in its correct operational state and capable of performing its function or mission

Operational Readiness. BIST aids in assessing operational readiness by providing a confidence test of the operability of the system at power-up, during idle mode, or at any other time the BIST capability is invoked on demand.

Reliability. Reliability can be improved by a BIST capability by virtue of providing a better quality test during the manufacturing process, thus weeding out marginally or completely defective components or assemblies (particularly by using BIST in conjunction with the burn-in process). BIST capabilities that have been extended to provide fault tolerance also enhance reliability by increasing the effective mean time between failures of the system. Reliability may also be negatively impacted by BIST as a result of an increase in failure rate due to the additional circuits required for its implementation.

Maintainability. Maintainability can be improved through a BIST capability by virtue of a substantial reduction in the mean time to repair, since a BIST capability may automatically detect, and in some cases, isolate the fault. Maintainability can also be somewhat negatively impacted by virtue of the need to maintain the additional circuits required for BIST. However, this effect is minimal, compared to the positive effect discussed above.

Availability. While reliability is both positively and negatively impacted by BIST, the overall impact on maintainability is positive; and the net effect on system availability is almost always positive. It can be shown algebraically that if the percentage decrease in system mean time to repair is higher than the percentage decrease in system mean time between failures, the overall impact on system availability will always be positive.

9.4 BIST ARCHITECTURAL CONCEPTS

Several notions related to the high-level design of BIST capabilities are important for understanding the development of an architectural approach to BIST. Those concepts are discussed next. (For a detailed discussion of specific examples of BIST architecture, see Ref. 46.)

9.4.1 System Aspects of BIST

As illustrated in Fig. 9.2, BIST design requires both a top-down and a bottom-up process to be performed. The top-down process is associated with the translation of the system test requirements to test requirements at lower levels of assembly. The bottom-up process relates to the normal way of designing a system. After the high-level system architecture is specified, the system is broken down into parts, and the system is actually designed by designing chips and integrating them into a system.

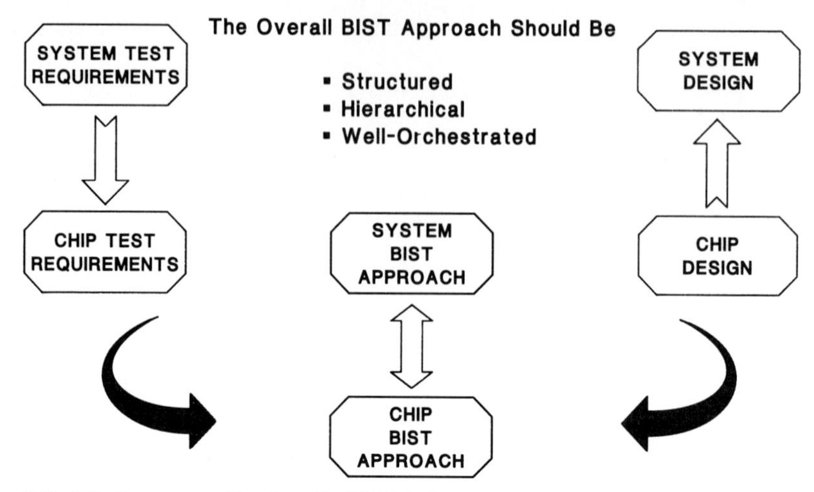

FIG. 9.2 System considerations for BIST design.

As Fig. 9.2 implies, the challenge is to design the BIST capability so that it is structured, hierarchical, and well-orchestrated, rather than ad hoc, monolithic, and loosely coupled. Instead of letting the allocation of BIST to portions of the system be left to chance or the whim of individuals in the design organization, the allocation should be performed early in the design cycle and be based on important factors such as

- Difficulty of testing via external means
- Failure rate
- Usage rate or duty cycle
- Criticality relative to the system operation

9.4.2 Building Block Philosophy

The building block philosophy is an architectural approach to BIST in which a minimum amount of BIST hardware or software, known as the *kernel,* is used to test other hardware in the system, which, if fault-free, is then "pulled in" to augment the kernel hardware, thus providing a more powerful BIST capability. The augmented BIST capability is then used to test yet additional hardware in the system, which if fault-free, is also pulled in to supplement the existing BIST hardware. This building block process progresses, with the BIST capability continuously growing in power, until all of the system required to be tested by BIST, has been tested.

9.4.3 Hierarchical BIST

If BIST is provided at the chip level, an interface, often referred to as the testability or maintenance interface, is provided to allow the results of the BIST operation at the chip level to be communicated to the board level. At the board level, interfaces between chips are tested by the BIST capability, as well as the ability of the chips to operate correctly together. The results of this BIST operation are passed on to the system level through another testability or maintenance interface at the board level. Continuing with the previous philosophy, the BIST capability is used to ensure that the interfaces between boards are operational and that the boards operate correctly together. Finally, results of this BIST process may be passed on through a system-level testability or maintenance interface to an external device such as a tester or system maintenance processor, which processes the information and determines the necessary corrective actions to be taken.

An additional benefit of the hierarchical BIST approach is that improved diagnostic resolution can often be achieved, since at a higher assembly level, an analysis can be performed of which lower-level BIST capabilities did or did not detect the same fault condition (the fault diagnostic tree concept).

9.4.4 Hardcore

The term *hardcore* relates to those portions of the system that are critical to the correct operation of the BIST capability. A failure in a hardcore portion of the system may prevent the BIST capability from starting or performing successfully. Examples of hardcore include the following:

1. BIST circuits (pattern generators, response evaluators, test controllers)
2. BIST interfacing circuits (testability or maintenance interface, visual BIST status indicators, etc.)
3. System critical functions (power, clocks, etc.)

The problem with hardcore is that while it is critical to the BIST operation, it is often very difficult to test. The frequently chosen solution of simply "trusting the hardcore" and assuming that it will not fail is not a practical solution. Additional measures must be taken to ensure the correct operation of the hardcore, such as the use of self-testing techniques within the hardcore areas or the use of an external tester to test the hardcore.

9.4.5 Centralized versus Distributed BIST

One of the degrees of freedom in implementing a BIST architecture is to choose the amount of parallelism desired in the operation of the BIST capability. The tradeoff is between test time and area overhead, and one can select a centralized architecture, distributed architecture, or one that falls between the two extremes.

As illustrated in Fig. 9.3*a*, a centralized architecture is characterized by the fact that there are only one or a few input stimulus generators (ISGs) and output response evaluators (OREs), which are multiplexed over a larger number of circuits under test (CUTs). This architecture features the lowest overhead, but suffers from being very slow because of the sequential servicing of the CUTs by the shared BIST elements.

Figure 9.3*b* shows the distributed type of architecture, which employs a multiplicity of BIST elements in order to achieve a greater degree of parallelism and, hence, higher test speeds. The penalty for the additional parallelism is a larger real estate overhead.

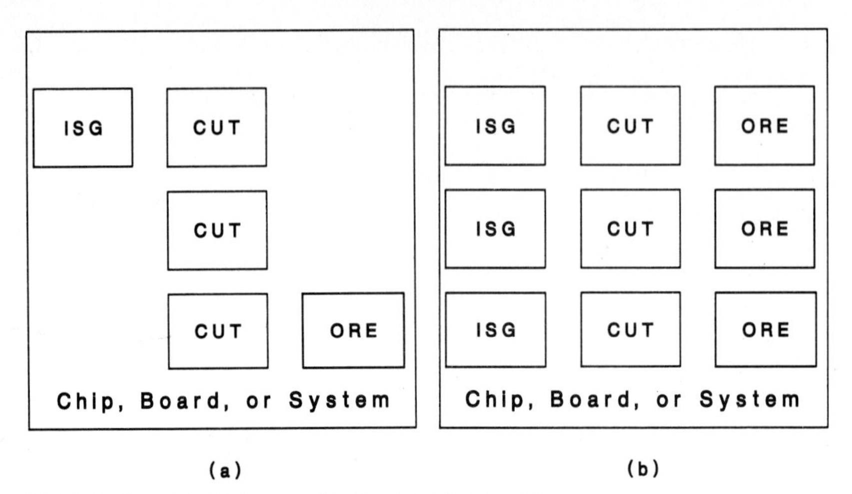

FIG. 9.3 Centralized (*a*) versus distributed (*b*) BIST architecture.

9.4.6 Separate versus Embedded BIST

The other major degree of freedom in selecting a BIST architecture is the option of choosing a separate approach rather than an embedded approach. The tradeoff in this selection process is primarily area overhead versus complexity of implementation. In the separate architecture, the BIST elements, ISGs and OREs, are separate entities from the functional circuits. As shown in Fig. 9.4*a*, the BIST elements are separate from the functional CUTs and serve no purpose other than performing the BIST function.

In Fig. 9.4*b*, the other option of an embedded approach is depicted. In this approach, the BIST elements are reconfigured from existing functional circuits, such as flip-flops and registers. During functional mode, the shared circuits

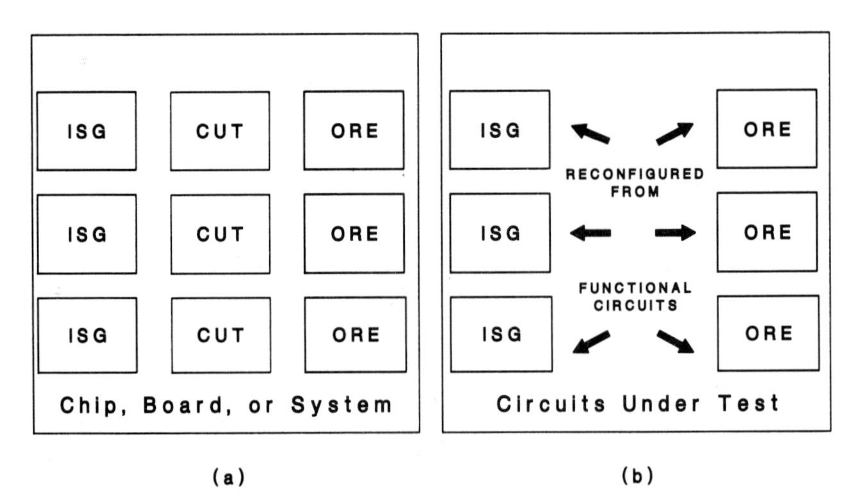

FIG. 9.4 Separate (*a*) versus embedded (*b*) BIST architecture.

would operate normally. During test mode, the functional circuits would be reconfigured to serve as ISGs or OREs. Choosing an embedded approach can save on real estate, but can result in a penalty of additional complexity in controlling the mode or role of each of the multipurpose circuits. In some cases, the added hardware required to control the multiple roles of the shared functional-BIST circuits can outweigh the real estate advantages gained by adopting the embedded architecture in the first place.

9.4.7 Scan Path Techniques for Improving Testability

One of the major breakthroughs in the area of improving testability for external testing methods is the scan path concept, pioneered in the 1970s. While the various approaches to scan path have been given different names, such as level-sensitive scan design,[4] Scan/Set,[5] and the Stanford scan path,[6] the principles and philosophies of the approaches are basically very similar. The philosophy of the scan path is that some or all of the storage elements (individual flip-flops, counters, registers, etc.) in the design are designed in such a way that they can be configured as part of a serial shift register, the serial input and output of which interfaces with the tester. By requiring such a storage element design, the problem of testing a sequential circuit is reduced to simply testing the storage elements as a serial shift register and then testing the remaining combinational logic that interfaces to the storage elements as a stand-alone combinational network.

Scan path can best be understood by relating its implementation to the Huffman model[7] for digital circuits. In Fig. 9.5, the Williams and Angell scan path approach is presented in the Huffman model form. As shown in Fig. 9.5*a*, flip-flops are designed with dual data ports, one port for normal system data and the other port for the serial scan test data. A mode control is provided for each flip-flop to select between the two ports, and the clock input is common to both

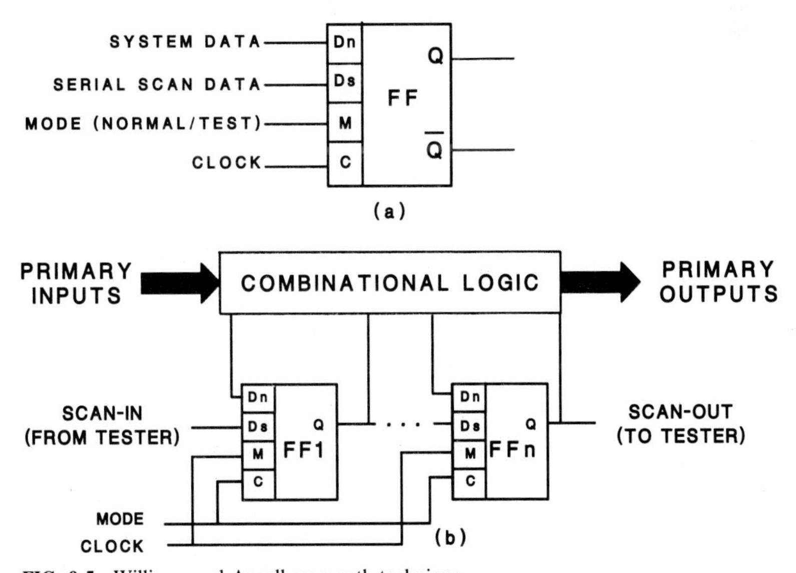

FIG. 9.5 Williams and Angell scan path technique.

modes. Figure 9.5*b* shows how the dual-port flip-flops can be configured into a serial shift register. This serial shift capability is essentially overlaid on top of the normal flip-flop connections to the combinational logic section, thereby making the structure functionally transparent to normal operation. In a practical design, however, the overhead of having all flip-flops designed to be scannable may not be tolerable, so the decision may be made to make scannable only those flip-flops that reside in those sections of the design that are most difficult to test.

The testing of any digital circuit implemented using this scan path is accomplished as follows:

1. The flip-flops are tested initially by subjecting the serial shift register to a "flush test," which is usually a simple alternating 1s and 0s pattern such as 010101...or 00110011....To apply the test, the mode control is set to the serial scan port and the pattern is supplied by the tester to the first flip-flop in the register through an input pin. Once the entire shift register has been filled with the pattern, the pattern is then serially shifted out to the tester through an output pin. The tester compares the received pattern with the input pattern to determine if there are any faults present in the storage elements.

2. The combinational logic is then tested using the previously tested shift register, which provides an enormous amount of additional controllability and observability for the tester. Before the combinational logic is tested, patterns are developed for that portion of the circuit using either manual or automatic test pattern generation techniques. However, the test pattern generation process is facilitated significantly because the points of control and observation in the circuit under test are not only the primary inputs and outputs going to pins, but also all of the combinational logic outputs (known as *pseudo-outputs*) going to the data inputs of flip-flops as well as all of the inputs to the combinational logic (known as *pseudo-inputs*) coming from the outputs of the flip-flops.

The patterns developed above are applied to the combinational logic one at a time through the primary inputs and through the serial shift register. A given pattern is allowed to propagate through the combinational logic being tested, and then the response of the combinational logic is observed by examining the primary outputs and by capturing the pseudo-outputs in the shift register, clocking it into the system data ports of the flip-flops. The contents of the shift register are then serially shifted out to the tester for evaluation, while the next input test pattern is being shifted in through the serial input pin.

In addition to its importance in improving testability in an external testing environment, scan path can be used as a foundation for implementing BIST, where the BIST pattern generators and response evaluators can be used to replace the tester mentioned in the discussion.

9.5 BIST DESIGN TECHNIQUES

The coverage of BIST design techniques that follows is divided into a discussion of functional BIST approaches and a discussion of structural approaches.

9.5.1 Functional BIST

The characteristics of functional BIST approaches are as follows:

1. The BIST design and analysis procedures are based on the functions of the unit under test, although both the function and the general structure of the UUT typically must be known.
2. The test philosophy is oriented toward exercising functions in such a way as to expose symptoms of faults in the form of erroneous UUT behavior.
3. A functional fault model (better named "error model") is assumed.
4. Test patterns applied during BIST processes are meaningful in terms of the UUT functions.

The advantages of functional BIST approaches are as follows. The BIST approach

1. Does not require gate-level knowledge of the UUT implementation.
2. May cover circuit areas not reached by structural approaches.
3. May cover some non-stuck-at-fault types.
4. Usually runs at circuit speeds.
5. Will not, in general, produce troublesome or destructive patterns.
6. Design can often take advantage of existing resources in the UUT.
7. Nucleus of tests may emerge naturally from functional testing during the design verification process.

Since the primary advantages of the functional BIST approach are that it does not require gate-level knowledge and that it can often be implemented using existing resources, the functional approach finds application most often in designs that are very sensitive to real estate overhead, designs that employ primarily off-the-shelf parts for which schematics are unavailable, or designs in which the function is fairly well known and the resulting functional test is more efficient than a structural one.

The disadvantages of functional BIST approaches include the following:

1. It may produce tests for impossible faults.
2. The degree of rigor of design and evaluation processes is questionable, particularly if the UUT is a "black box" for which little or no gate-level details are available.
3. The nonalgorithmic nature of the design process makes automated tool development difficult.

The primary disadvantage of the approach is that there is a lack of rigor associated with its application to designs for which gate-level implementations are unavailable (i.e., the testing of a "black box"). The reason that rigor is lacking is that it is only when the detailed implementation is known that the BIST designer can be assured of a thorough test and can accurately assess the fault coverage of the BIST capability.

Functional BIST Architecture and Implementation. A substantial amount of research has been performed in the area of functional BIST, primarily aimed at improving the degree of rigor associated with the design and analysis processes.[8–17] The basic functional BIST architecture is illustrated in Fig. 9.6. A processing engine driven by firmware or software is stored in a ROM or downloaded into RAM. The processing engine, which could be a microprocessor structure within

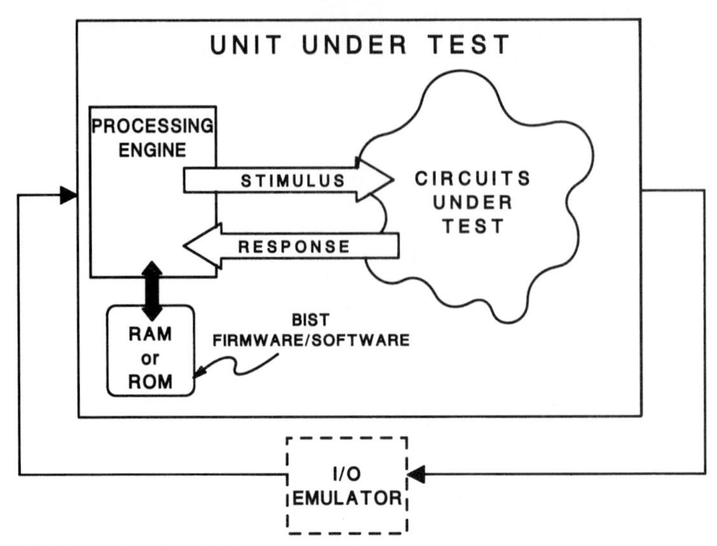

FIG. 9.6 Functional BIST architecture.

a chip or a microprocessor on a board, or even a processing unit in a system, provides the functional stimuli to exercise the circuits under test and also evaluates the responses of those circuits to the stimuli.

To test the I/O capability of the UUT, an I/O emulator is often used to emulate the I/O interface and associated handshaking activities. The emulator, which is usually implemented in the form of a specially designed piece of hardware, or which could even be the external tester itself, allows the processing engine to send stimuli out over the interface and evaluate the responses that come back into the input side.

If a bus structure is present in the UUT, as is common in many microprocessor-based designs, an architectural BIST feature known as the *wrap test* or *wraparound capability* is possible. According to Fig. 9.7, the wraparound test begins with the processing engine testing bus element 1 by functional stimuli and response evaluation over the bus. The engine then uses the bus and element 1 as a conduit through which to pass stimuli to bus element 2, the responses of which would be placed back onto the bus for subsequent evaluation by the processing engine. Such a wraparound test capability works in harmony with the building block philosophy discussed in Sec. 9.4.2. Another interesting technique shown in Fig. 9.7 is the use of the bus structure and the I/O emulator to provide the I/O test discussed previously.

A typical test flow for functional BIST is depicted in Fig. 9.8. It is based on the building block philosophy discussed previously and starts with a test of the hardcore circuits (including power, reset, the control ROM, and the processing engine itself) and follows with tests of the memory, I/O, and buses, and then tests of the remainder of the system.

9.5.2 Structural BIST

The characteristics of structural BIST approaches are as follows:

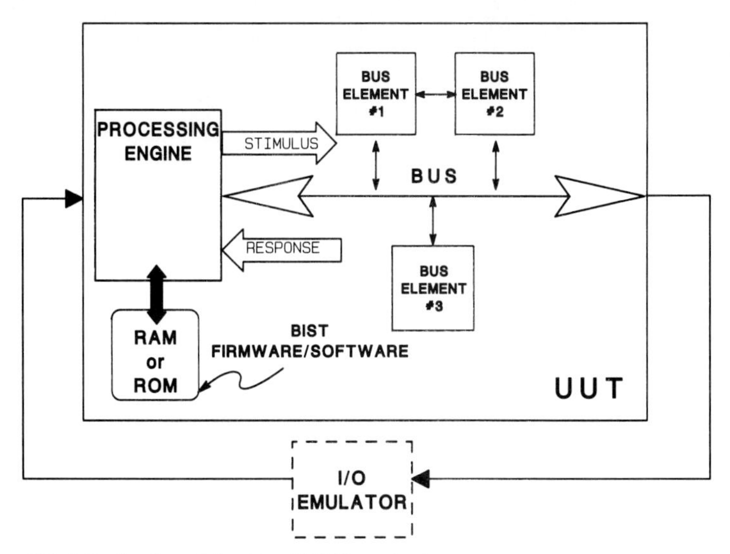

FIG. 9.7 Bus-based functional BIST architecture.

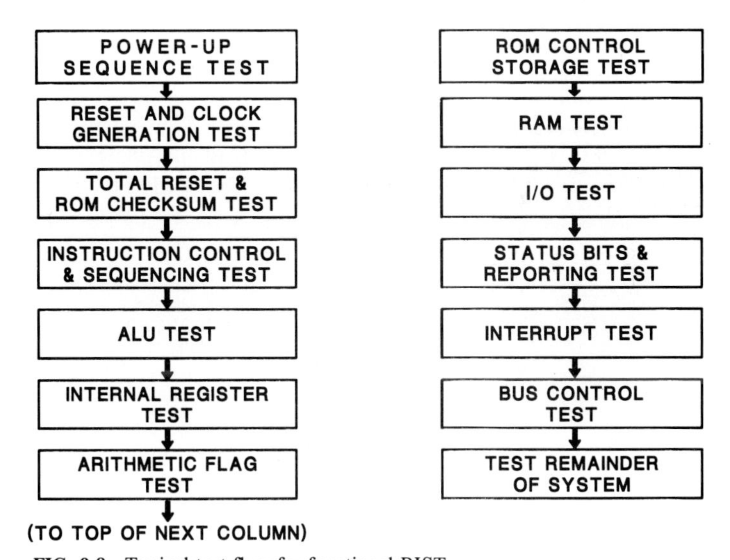

FIG. 9.8 Typical test flow for functional BIST.

1. The BIST design and analysis are based on the topology or structure of the UUT.
2. The test philosophy is oriented toward detecting and isolating faults at circuit nodes. The object is to stimulate a circuit node and then observe its state at the circuit's outputs.
3. A structural fault model (such as single stuck-at 0s and 1s) is assumed.

4. Test patterns are not generally meaningful in terms of the function of the UUT.
5. A knowledge of the UUT structure (most often gate level) is required.

The advantages of the structural BIST approach are as follows:

1. Tests for impossible faults are unlikely to be produced.
2. The degree of rigor of the design and analysis processes is very high.
3. Many automated tools are already available, and many more are emerging.

The major advantage of the structural BIST approach relates to the abstraction level to which it is applied—the gate level. At this level the degree of rigor associated with the BIST design and analysis is extremely high, since the fault sites can be determined accurately, the BIST capability can be targeted for the detection of specific faults at specific sites, and a rigorous analysis (through fault simulation) can be performed to determine the exact fault coverage of the BIST test.

The primary disadvantages of the structural BIST approaches are as follows:

1. The approach requires detailed knowledge of circuits under test.
2. The approach does not test functional behavior of the UUT.
3. Some approaches do not run at circuit speeds.
4. Troublesome or destructive patterns may be produced because of disregard for the functions taking place during the test.
5. The area penalty may be higher than for the functional approaches.

Structural BIST Architecture. The basic architecture of structural BIST approaches is shown in Fig. 9.9. The ISGs, which are usually implemented using counters or feedback shift registers, apply stimuli or patterns to the CUTs, which

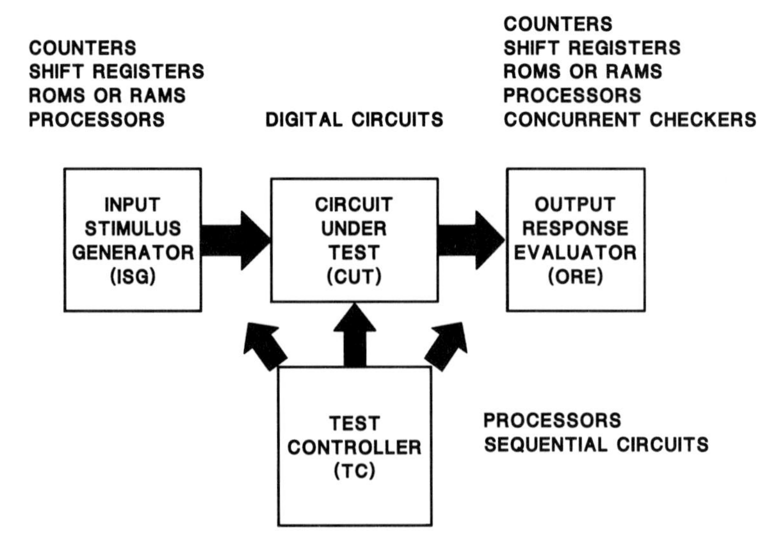

FIG. 9.9 Structural BIST architecture.

are digital circuits that can be composed of combinational logic and, in one ISG approach, sequential logic as well. The OREs evaluate the responses of the CUT and are also usually implemented in the form of counters or feedback shift registers. The test controller is typically implemented by using a processing engine, or it could be as simple as a sequential circuit that controls the starting and stopping of the BIST process as well as orchestrates the phases of a multiphase BIST test process.

Structural BIST Implementation Techniques. The methods of implementing structural BIST will be discussed by first addressing the design approaches for ISGs and then the design techniques for OREs. The design of the test controller is not treated in the discussions that follow.

Implementation Approaches for Input Stimulus Generators. The objective for input stimulus generation is to produce a sufficient set of input stimuli to force the output response of the circuit under test to be different when it is faulty than when it is fault-free. The four major approaches to input stimulus generation are as follows:

1. Stored deterministic pattern
2. Exhaustive pattern
3. Pseudoexhaustive pattern
4. Pseudorandom pattern

Each approach is illustrated in Fig. 9.10 and discussed in the sections that follow.

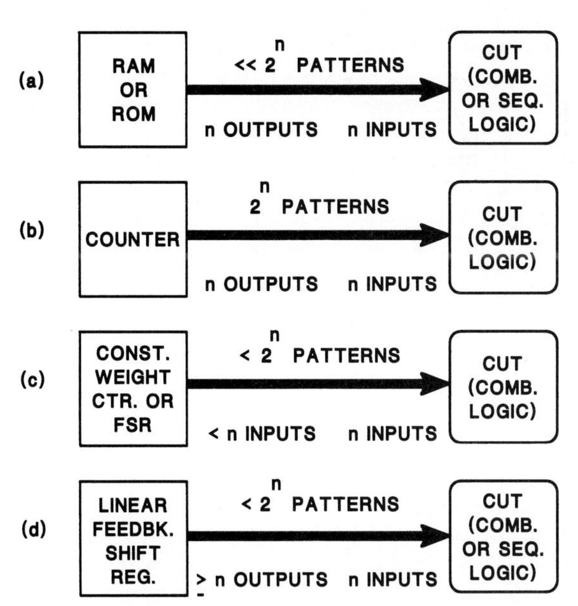

FIG. 9.10 Input stimulus generation techniques. (*a*) Stored deterministic pattern approach; (*b*) exhaustive pattern approach; (*c*) pseudoexhaustive pattern approach; (*d*) pseudorandom pattern approach.

Stored Deterministic Pattern Approach. In the stored deterministic pattern approach, as illustrated in Fig. 9.10, patterns required to test the UUT are generated deterministically off-line either manually or by an automatic test pattern generator. Those patterns are then stored in a ROM or downloaded into a RAM in the UUT. During the BIST process, the patterns are read from the memory one at a time and applied to the circuits being tested, under control of the test controller. Unfortunately this approach is not practical as the *primary* ISG method, for two reasons:

1. The approach does not provide one of the major objectives of BIST—the reduction or elimination of the need for external test pattern generation.
2. Even if the patterns could be generated cost-effectively off-line, a large memory in the UUT would still be required to store all of them.

As a result of these drawbacks, the stored deterministic pattern approach is rarely used as the primary ISG method. However, it may be used to supplement some of the other approaches to boost the fault coverage to an acceptable level, since the number of deterministic patterns required in that case would be more manageable.

Exhaustive Pattern Approach. The exhaustive pattern approach to input stimulus generation, shown in Fig. 9.10, is a brute-force method, whereby all 2^n possible bit combinations are applied to an n-input combinational circuit. This approach is extremely attractive for several reasons:

1. The design and analysis of the ISG requires no knowledge of the CUT structure or function other than the number of inputs it has and the fact that it is combinational.
2. The ISG can be implemented simply as a binary counter or a modified linear feedback shift register, a structure that will be described later in this section.
3. Fault coverage for the approach is the highest possible for any static test, being 100 percent of all (statically) detectable combinational faults. (A combinational fault is one that preserves the combinational nature of the circuit in the presence of that fault. An example of a noncombinational or sequential fault is a CMOS stuck-open fault.)[18]
4. Fault coverage is easily assessed and requires no fault simulation, as long as it is understood that any statically undetectable faults will not be detected by the approach.

The major drawbacks of this approach are that it is not applicable to sequential circuits and not practical for combinational circuits with a very large number of inputs. A realistic limit to the number of inputs that can be serviced depends on the test pattern application rate and the maximum acceptable test time. For example, a 20-input combinational circuit would require approximately 1 s to test exhaustively at a 1-MHz test application rate.

Pseudoexhaustive Pattern Approach. One approach to input stimulus generation that has almost the same benefits as the exhaustive pattern approach, but not the drawback of requiring all 2^n bit combinations, is the pseudoexhaustive pattern method, illustrated in Fig. 9.10.[19–21] The characteristics of the approach are as follows:

1. The approach is only applicable to combinational logic networks, in which no output depends on *all* inputs. (However, even for circuits that do not meet this

criterion, the approach can still be used if the circuit is first segmented or partitioned.)

2. All circuit nodes must be able to be controlled and observed at circuit inputs and outputs, respectively. Faults at nodes not meeting this requirement (e.g., redundant nodes) will not be detected by this approach.
3. The circuit must remain combinational in the presence of faults. Sequential faults, which alter the combinational behavior of the circuit (e.g., CMOS stuck-open faults), will not necessarily be detected by this approach.
4. The basis of the approach is the application of *locally* exhaustive patterns to *segments* of the circuit, not to the entire circuit at once.
5. The ISG can be implemented as a constant-weight counter or as a feedback shift register. The number of stages or output bits required in the ISG is usually less than that required for the exhaustive approach, although many BIST architectures cannot take advantage of such a reduction.
6. The fault coverage is 100 percent of all single stuck-at faults, as well as many multiple stuck-at and bridging faults (nonfeedback type).
7. As long as the initial condition is met that no output depends on all inputs, the number of patterns required will be *at most* 50 percent of the number required for a completely exhaustive test.

As indicated previously, with this approach, segments of the circuit receive an exhaustive set of patterns. In the fully exhaustive approach, sufficient patterns are applied so that the total *input* set of the CUT "sees" an exhaustive test. In contrast, the pseudoexhaustive approach uses the reverse perspective, ensuring that each *output* "sees" an exhaustive test being applied to its own input set, which will always be a subset of the total input set for the circuit (due to the assumption stated in characteristic 1 above).

A simple circuit for which pseudoexhaustive patterns have been developed is shown in Fig. 9.11. In the example, each output, *d* and *e,* "sees" an exhaustive test applied to it, yet the overall number of patterns, 4, is only half of the number that would have been required for an exhaustive test of the circuit as a whole.

While it is relatively common practice to apply locally exhaustive test patterns whenever possible to a circuit using external test or BIST, the breakthrough in the pseudoexhaustive approach is the establishment of *algorithmic* methods for identifying the subcircuits and generating the pseudoexhaustive patterns for them. One such algorithmic method is as follows:[21]

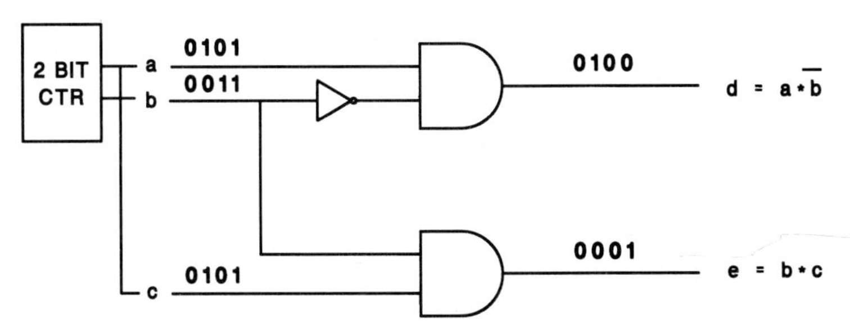

FIG. 9.11 Combinational circuit being tested with pseudoexhaustive patterns.

1. Partition the combinational network into disjoint sections, i.e., sections that do not share gates or inputs.
2. For each of the disjoint sections, perform the following steps:
   a. Generate a dependence matrix that totally characterizes the circuit in terms of the dependencies of outputs on inputs. The matrix, shown in Fig. 9.12, is an m by n matrix, with m rows representing the m outputs of the combinational network, and n columns, one for each input of the circuit. Within the matrix are 1 and 0 entries only. A 1 entry indicates that the output in that column depends on the input in that row. Dependency is determined by tracing each output back to all inputs that affect it, a procedure known as *cone tracing*. One reasonableness check of the dependence matrix is to check that it has no all-zeros columns or rows and no all-ones rows. The cone-tracing procedure and resulting dependence matrix for the sample circuit are shown in Fig. 9.13.
   b. Partition the matrix into clusters of inputs that do not affect the same output. This step is accomplished by examining the dependence matrix and initially taking pairwise combinations of columns such that *within a pair* there are zero or one 1s in each row. After one or more pairs of columns are found that meet this criterion, attempts should be made to merge additional columns (or pairs of columns) with the existing pairs such that the criterion will still be met. The final goal should be to ensure that the number of partitions is minimal. (Unfortunately, this final step is very computationally complex, and shortcuts for performing it are being sought.) One aid in seeking partitions is to realize that any all-ones column automatically becomes a single column group and cannot be merged with any other valid column. A partitioned dependence matrix for the sample circuit is shown in Fig. 9.14.
   c. Collapse each input group in the matrix to form a single equivalent input that retains the same input-output dependencies as the individual inputs in the group. Algorithmically, this step is accomplished by simply taking the inclusive-OR function of all of the columns *within* each cluster or partition

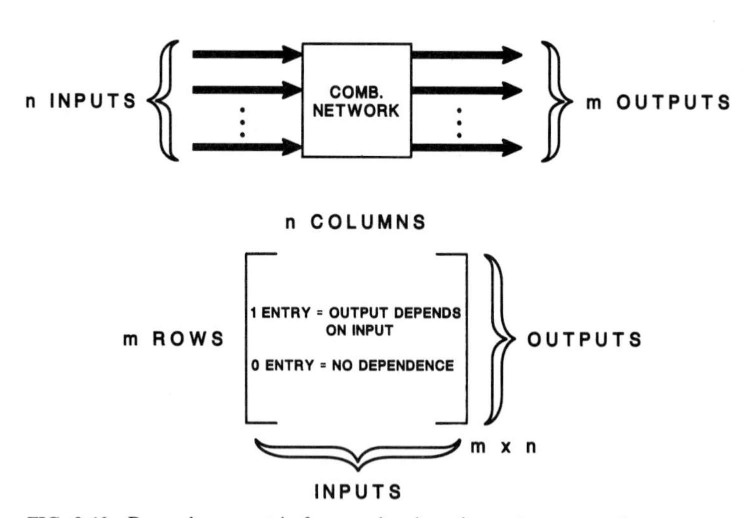

FIG. 9.12 Dependence matrix for pseudoexhaustive pattern generation.

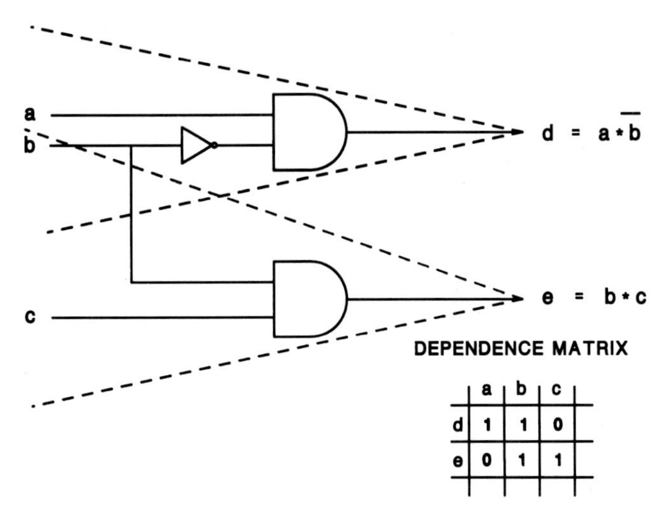

FIG. 9.13 Dependence matrix generation for the sample circuit.

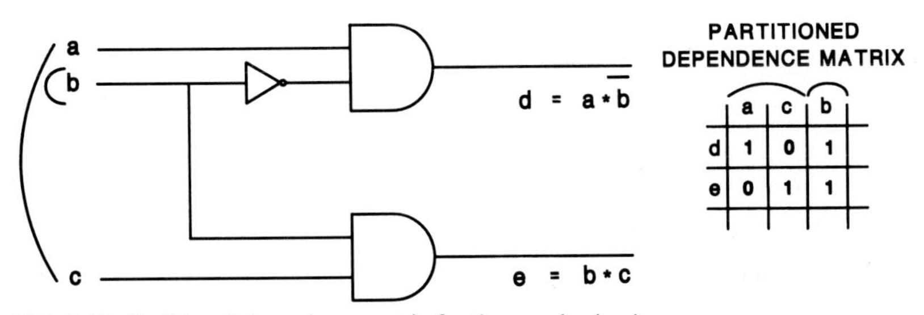

FIG. 9.14 Partitioned dependence matrix for the sample circuit.

generated in step *b*. The collapsed dependence matrix for the sample circuit is shown in Fig. 9.15.

d. Characterize the collapsed matrix in terms of two parameters: width (P), which is the number of input partitions or clusters remaining after the collapsing step above, and weight (W), which is the maximum number of input clusters upon which any output depends. Algorithmically, these two

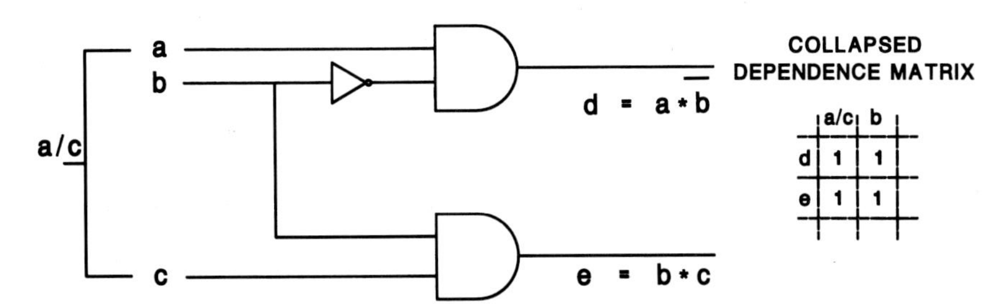

FIG. 9.15 Collapsed dependence matrix for the sample circuit.

parameters are easily determined from the collapsed dependence matrix as follows: P = the number of columns and W = the maximum number of 1s in any row. From Fig. 9.15, it can be seen that for the sample circuit, $P = W = 2$.

e. The final step is to identify the pseudoexhaustive pattern set for the circuit, based on the relative values of P and W. Three cases must be examined. However, in all three cases, the width or number of bits in each pattern is P, and the number of patterns depends on P. The three cases are as follows:

- *Case I. P = W:* Each pattern is P bits wide, and the number of patterns is 2^P. The property of the pattern set is that it consists of simply all possible combinations of P bits. (Note, however, that since P is always at least $n - 1$, then the number of pseudoexhaustive patterns is at least half of the number of fully exhaustive patterns.)
- *Case II. P = W + 1:* Each pattern is P bits wide, and the number of patterns is $2_p - 1$. The property of the patten set is that all patterns have odd parity, or all patterns have even parity. (The choice is arbitrary.)
- *Case III. P > W + 1:* Each pattern is P bits wide, and the number of patterns depends on the values of P and W. The property of the pattern set is that it consists of two or more subsets of patterns, each subset of which has patterns with a constant weight (number of 1s). In the look-up table presented in Ref. 21, for each combination of P and W, there are either one or two constant-weight multiples indicated. A constant-weight multiple indicates the constant weight for each constant-weight pattern subset described above. If two constant-weight multiples are listed, the choice between the two is arbitrary. (One constant-weight multiple is the complement of the other. Whenever W is odd, only one constant-weight multiple is shown, and it is a self-complementing one.)

A portion of a look-up table with values of P and W filled in is shown in Fig. 9.16, as is an example of a (0,3) constant-weight pattern set. The sample circuit is a Case I circuit since $P = W = 2$. Therefore, the number of patterns in the pseudoexhaustive set is 2^2 or 4, and the patterns are simply all possible combinations of 2 bits, as illustrated in Fig. 9.17.

The input stimulus generator for the pseudoexhaustive approach can be a simple binary counter or a linear feedback shift register (modified to go through the all-zeros state) for Case I circuits or a constant-weight counter or a feedback shift

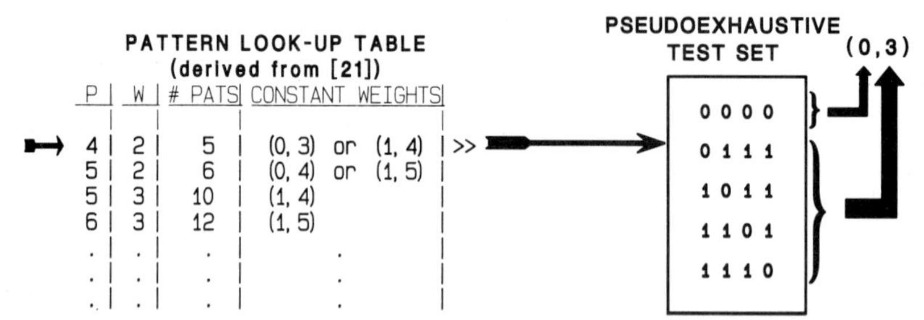

FIG. 9.16 Pseudoexhaustive test set for the case III circuits. [*Note:* (*a*, *b*) represents a constant-weight multiple, where *a* is the constant weight of one subset of patterns and *b* is the weight of the other subset.]

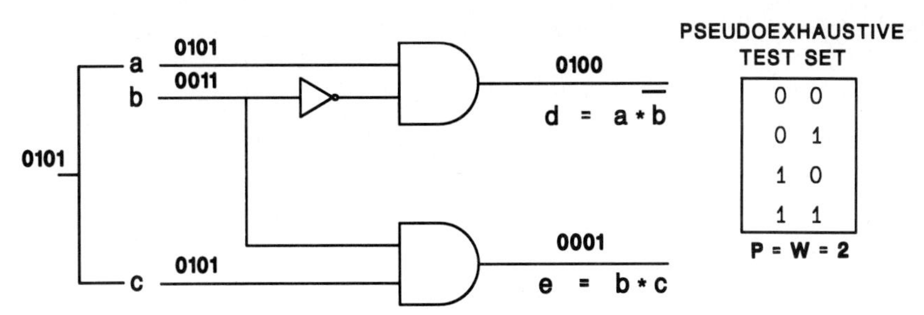

FIG. 9.17 Final pseudoexhaustive test set for the sample circuit (case I).

register for Case II and III circuits. In the situation where a circuit does not meet the requirement that no outputs depend on all inputs, the circuit must be segmented or partitioned prior to the application of the algorithm. Methods for segmentation are discussed in Ref. 19.

Pseudorandom Pattern Approach. An input stimulus generation method that has general applicability to both combinational and sequential logic is the pseudorandom pattern approach, illustrated in Fig. 9.10. The technique emerged from some studies performed on the Illiac boards,[22] which indicated that combinational logic boards could be tested effectively simply by applying pseudorandomly generated patterns, in which each bit in the pattern has approximately a 0.5 probability of being a 1 or a 0. The study also pointed out that should the fault coverage achieved by the pseudorandom method fall short of the desired level, the pseudorandom patterns could be supplemented with deterministically generated patterns (e.g., via a D-algorithm based generator). Subsequent studies[23,24] showed that the fault coverage for a circuit rises almost exponentially as the number of pseudorandom patterns applied to the circuit increases.

The characteristics of the pseudorandom pattern approach are as follows:

1. For testing a combinational circuit, the approach usually requires more patterns than the deterministic methods but fewer than the exhaustive ones.
2. The fault model assumption is usually the single stuck-at fault model.
3. The pattern generators can be implemented by using software or a hardware-implemented linear feedback shift register (LFSR).
4. Individually, the patterns produced are completely deterministic (and hence repeatable), in the sense that given the same seed and number of clock cycles for a given LFSR, the LFSR will always produce the same patterns in the same sequence. However, when the entire set of patterns is analyzed statistically, it is found to possess the following randomness properties:[25]
   - Approximately an equal number of 1s and 0s (actually, since the all-zeros pattern is not produced by the LFSR, the distribution slightly favors 1s)
   - An equal number of runs of 1s and 0s
   - A two-valued autocorrelation function

An LFSR is basically a finite-state machine consisting of flip-flops and a feedback structure between the flip-flops. The LFSR has a period, the length of which depends on the number of stages of the register and the nature of the feedback function, which in the case of an LFSR is always linear (i.e., exclusive-OR gates only). Designing an LFSR requires the selection of the proper polynomial to implement. To obtain the longest pseudorandom test sequence possible from an LFSR, we select a primitive polynomial. An LFSR implementing a primitive polynomial is called a *maximal length LFSR*. Such a pattern generator will produce a sequence of length $2^m - 1$ and will require m flip-flops, where m is the degree of the primitive polynomial. The only pattern not produced during the sequence is the all-zeros pattern. Tables listing primitive polynomials can be found in Refs. 25 and 26.

The two primary types of LFSRs are shown in Fig. 9.18. In the type 1 LFSR, shown in Fig. 9.18*a*, the feedback is derived from the outputs of various flip-flops in the LFSR and feeds back through a parity-tree-type structure to the input of the first flip-flop. The circled c_n notation indicates that there may ($c_n = 1$) or may not ($c_n = 0$) be a connection at that point, depending on which polynomial is implemented. The type 2 LFSR, depicted in Fig. 9.18*b*, uses feedback tapped off the output of only the last flip-flop and applies that feedback to the input of the first flip-flop as well as to various other intermediate flip-flops through exclusive-OR gates. Again, the c_n notation indicates that there may or may not be a connection present. If there is no connection, no exclusive-OR gate is required at that stage.

While the type 2 LFSR is always faster than the type 1 (whenever the implemented polynomial has more than three terms), there are some similarities in operation. Either LFSR can implement any primitive polynomial; and for a given polynomial, both types will produce the same set of patterns, but in a different sequence. In addition, both types of LFSRs require $T - 2$ exclusive-OR gates, where T is the number of terms in the implemented polynomial.

As indicated previously, it has been shown that the shape of the curve of fault coverage versus number of pseudorandom patterns is almost exponential, particularly for combinational logic.[22–24] In addition, the curve becomes asymptotic to some value, which depends on the pattern generator and the circuit under test. It

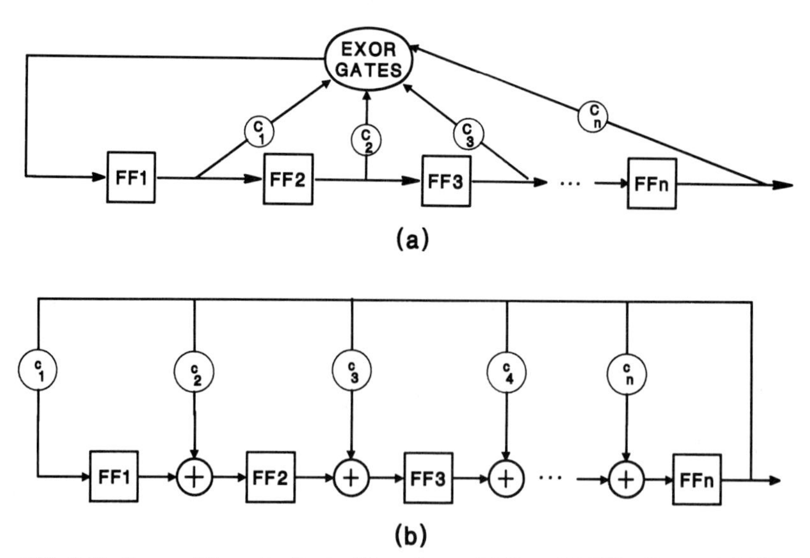

FIG. 9.18 Types of linear feedback shift registers. (*a*) Type 1 LFSR; (*b*) type 2 LFSR. (c_n = variable indicating presence or absence of connection.)

is the randomness properties, particularly the near equal distribution of 1s and 0s and the resulting high toggling rate of each bit position of a pseudorandom pattern that influences the shape of the curve and makes the LFSR attractive for stimulus generation. For a combinational circuit, the curve becomes asymptotic below 100 percent coverage if the circuit possesses any statically undetectable faults or if an m-input circuit that contains a fault requiring the all-zeros pattern for detection is being tested with an m-stage LFSR. Otherwise, the coverage will eventually reach 100 percent.

Pseudorandom patterns are not as effective for testing sequential circuits as for combinational circuits. While the coverage curve may reach 100 percent for some sequential circuits, for others it may become asymptotic below 100 percent, because it is possible that the specific *sequences* of patterns required to test the circuit fully may not even be producible by the pseudorandom pattern generator being used.

There are a number of problems that may be encountered in the use of pseudorandom patterns. While not a concern for long test sequences, it is possible to lose some of the rich randomness properties of the sequence if only a small subset of the maximal length sequence is used. This problem can be avoided by lengthening the sequence or, in some cases, by simply choosing another seed or polynomial.

A more common problem is related to a characteristic of some circuits known as *random-pattern resistance*.[27,28] Circuits that are random-pattern resistant are not tested effectively by random patterns. Examples of such circuits are wide-input gating that requires a unique, low-probability pattern to set up a required state at the output of the gating, or a cascade of gates that require a unique, low-probability pattern to sensitize an observability path to a primary output or pseudooutput. A more obvious example of random-pattern-resistant circuits is the general class of sequential circuits, which have classical controllability and observability problems.

As implied in the discussion of fault coverage, random-pattern resistance is a relative term (relative to the length of the sequence). Many combinational circuits that appear to be random-pattern resistant in a shortened sequence of an LFSR could become fully tested simply by using a longer sequence. In other cases of combinational circuits and in many cases of sequential circuits, however, the length of the sequence required may be unacceptable. In such a case, either the circuit has to be modified[27] or the outputs of the LFSR weighted to produce a bias of 1s or 0s for certain inputs to the circuit under test.[24,29,30]

A third problem relates to the fact that since pseudorandom patterns are not functional patterns (in the sense defined earlier), it is possible to force a sequential circuit into invalid, indeterminate, or destructive states, such as a state where multiple tristate drivers are active simultaneously. To eliminate the risk for all cases, precautions must be taken in the BIST design, for example, by eliminating the possibility of indeterminate or illegal states (for the first condition) and by proper termination of drivers or prevention of multiple active drivers (for the second condition).

Implementation Techniques for Output Response Evaluators. As discussed in previous sections, the objective in the input stimulus generation process is to provide a set of patterns sufficient to force detection of a high percentage of possible faults in the circuit being tested. The objective on the output side then is to evaluate the responses of the circuit to determine if one or more of those responses are incorrect. Since there will typically be hundreds of thousands or even millions of responses, it is not generally feasible for the BIST circuitry to compare each output response at each clock cycle to the expected fault-free response. It may be feasible, however, if the output responses incorporate some form of redundant data, such as parity or residue codes, and the output evaluation technique employs the associated concurrent error-checking circuits, for example, parity or residue code checkers.[31,32]

The current approach used in most BIST applications is to compress the output response stream into a word of much more manageable size, called the *signature,* which can then be compared with the predicted, fault-free, signature obtained in advance through good machine simulation or through application of the input stimuli to a known good unit. However, the challenge remains of achieving that compression without losing the hard-earned fault coverage attained during the input stimulation process.

While the performance of the compression circuit is usually measured in terms of number of erroneous bits *detected* in the output bit stream, from an overall BIST process standpoint, the real factor being measured is the ability of the compressor to *preserve* the fault detection and fault isolation coverage achieved on the input side. The preservation of detection fault coverage is jeopardized by a phenomenon known as *aliasing,* or *error masking,* which occurs in all compression processes. Aliasing is a condition whereby a faulty circuit produces the same final signature as the fault-free version of the circuit, thus giving a false indication that the circuit is good. The probability of aliasing can be reduced, but it cannot be eliminated when just one compression circuit and one input sequence are used.

There are a number of means for performing compression of output responses, including counting techniques[33–35] and parity techniques,[36–38] but the method that has proven to be most effective in terms of a low aliasing rate is signature analysis, a compression approach perhaps first discussed in Ref. 36 and used by Hewlett-Packard as a compression technique for some of their field test equipment.[39,40] The signature analysis means of compression uses the same LFSR used for pseudorandom pattern generation, but, rather than being used as an autonomous circuit, the LFSR has one or more external inputs originating from the outputs of the circuit under test.

There are several primary implementations of signature analyzers.[36] One type of implementation is the serial approach, shown in Fig. 9.19*a,* which is based on the type 2 LFSR and which can compress one circuit output at a time. To handle more than one circuit output, the multiple outputs would have to be multiplexed into the serial compressor one at a time. The other signature analyzer implementation, shown in Fig. 9.19*b* and known as a multiple-input signature register (MISR) or parallel signature analyzer (PSA), compresses multiple outputs of the circuit under test in parallel. While the MISR shown in the figure is based on the type 2 LFSR structure, both the serial and parallel analyzers can also be implemented with the type 1 LFSR structure. There are some differences between the type 1 and type 2 implementations. For example, while both signature analyzers carry out a form of polynomial division, only in the case of the type 2 LFSR is the final signature equal to the remainder after division (mod 2) of the circuit's output polynomial by the characteristic polynomial of the signature analyzer.

Aliasing in signature analyzers occurs whenever the *error polynomial* (the bit-by-bit mod-2 sum of the faulty output sequence and the fault-free output sequence) is a nonzero multiple of the divisor polynomial. A thorough analysis of the aliasing probability would require examination of the entire BIST process, since the probability depends on the input patterns applied, the nature of the circuit under test (e.g., the manner in which faults manifest themselves as errors at

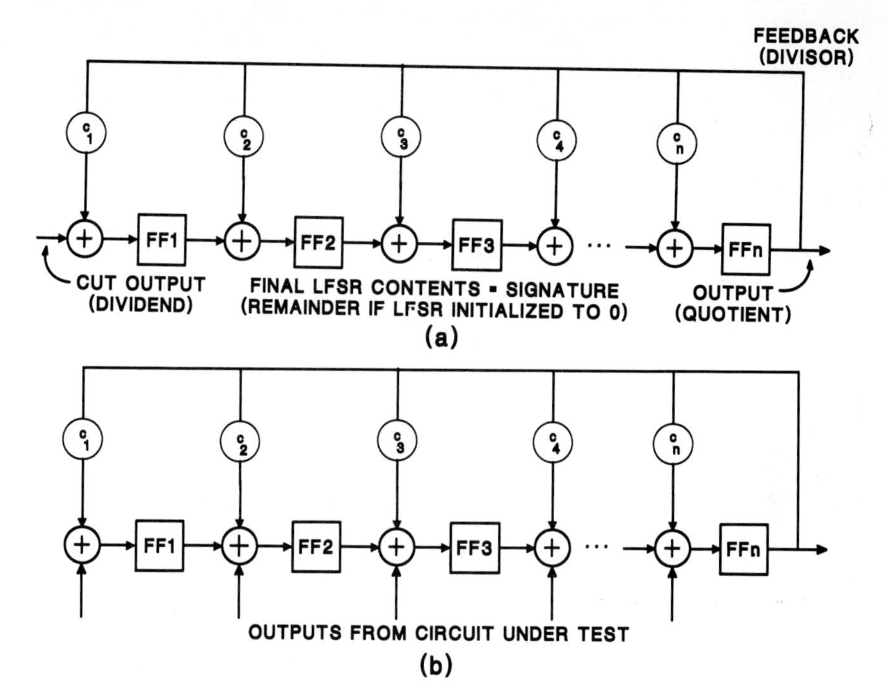

FIG. 9.19 Types of signature analyzers. (*a*) Serial signature analyzer based on type 2 LFSR; (*b*) parallel signature analyzer based on type 2 LFSR.

outputs), and the design of the signature analyzer. Such an analysis, however, implies a substantial simulation effort, which to date has not been affordable. To make the analysis manageable, many researchers have simply ignored the detailed contribution of the input patterns and the circuit under test. By making some statistical assumptions about the error polynomials, we can *estimate* the aliasing probability by examining only the contribution of the signature analyzer itself. Such an approach, used in Refs. 40 and 41, results in the following estimates of aliasing probability for serial and parallel signature analyzers:

Probability of aliasing in serial signature analyzers

$$P(A)_{\text{sss}} = \frac{2^{L-m} - 1}{2^L - 1}$$

Probability of aliasing in parallel signature analyzers

$$P(A)_{\text{psa}} = \frac{2^{L-1} - 1}{2^{L+m-1} - 1}$$

where L is the output sequence length and m is the length of the signature analyzer.

When the output sequence length is large, both equations reduce to simply $1/2^m$. Therefore, the aliasing probability can be cut in half simply by adding an-

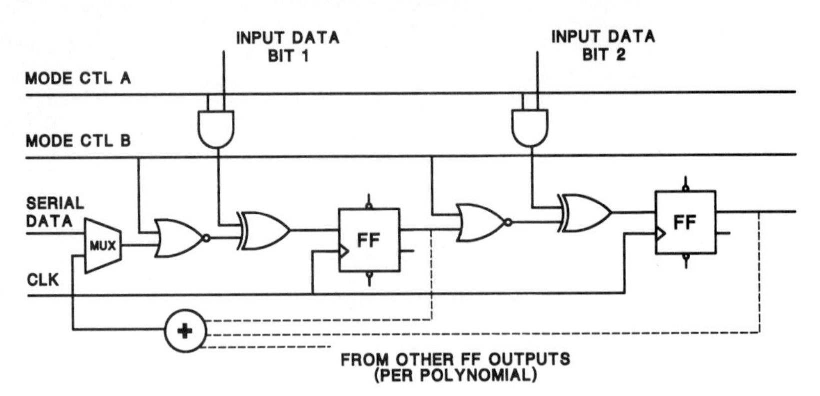

FIG. 9.20 Two cells of a built-in logic block observer.

other flip-flop stage to the signature analyzer. Additional guidelines that have been proposed and used for reducing aliasing include implementing a primitive polynomial in the signature analyzer[42,43] and using a polynomial with many terms (providing a rich feedback function). Other ideas, such as observing the final flip-flop output, modifying the circuit's outputs before compression,[44,45] and running the input patterns a second time in reverse order, have potential for reducing aliasing but have not been implemented widely because of the more significant overhead.

Combined Stimulus Generator-Response Evaluator Building Blocks. One BIST building block that is based on the recognition that the LFSR is the basis of both stimulus generator and response evaluator is the built-in logic block observer (BILBO).[47] The structure, 2-bit cells of which are illustrated in Fig. 9.20, has four modes of operation: reset, parallel load register, serial shift register, and pseudorandom-pattern generator or parallel signature analyzer. Distributed BIST architectures that employ an extensive number of BILBOs will typically provide a very high speed, high-coverage test, although the overhead may be high, particularly for the logic needed to control the modes of the BILBOs and the phases of the test process.

9.6 FAULT COVERAGE EVALUATION FOR BIST

Methods for testability and fault coverage evaluation range from estimation approaches, such as testability analysis techniques[48–51] and sampled fault simulation,[52] to the many full-fault simulation methods. A detailed discussion of these evaluation methods is presented in Chap. 8.

9.7 CONCLUSIONS

Much progress has been made in BIST technology, but several important challenges remain. Additional effort is required in several key areas, such as

1. Fault isolation methods for BIST, particularly for pseudorandom pattern–signature analysis approaches
2. An efficient, low-cost means for assessing the effectiveness of a BIST capability while the design unfolds, as well as after a detailed simulation model is available
3. BIST approaches applicable to analog designs
4. CAE tools to support BIST design and analysis
5. Well-orchestrated board- and system-level BIST approaches

As levels of integration for VLSI continue to increase and as further packaging breakthroughs occur that impede progress in conventional external testing, those approaches rooted in built-in self-test should continue to provide a cost-effective solution.

REFERENCES

1. R. L. Wadsack, "Fault Modeling and Logic Simulation of CMOS and MOS Integrated Circuits," *Bell System Tech. J.*, May–June 1978, pp. 1449–1474.
2. R. M. Sedmak, "Built-In Self-Test: Pass or Fail," *IEEE Design and Test Mag.*, April, 1985, pp. 17–19.
3. MIL-STD-1309C—"Military Standard for Definition of Terms for Test, Measurement, and Diagnostic Equipment," November 18, 1983.
4. E. B. Eichelberger and T. W. Williams, "A Logic Design Structure for LSI Testability," *Proceedings of the 14th Design Automation Conference*, 1977, pp. 462–468.
5. J. H. Stewart, "Application of Scan/Set for Error Detection and Diagnostics," *Digest of the 1977 Semiconductor Test Conference*, 1978, pp. 152–158.
6. M. J. Y. Williams and J. B. Angell, "Enhancing Testability of Large-Scale Integrated Circuits via Test Points and Additional Logic," *IEEE Trans. Computers*, January 1973, pp. 46–60.
7. D. A. Huffman, "The Synthesis of Sequential Switching Circuits," *J. Franklin Inst.*, March–April 1954, pp. 161–190, 275–303.
8. J. A. Abraham and K. P. Parker, "Practical Microprocessor Testing: Open and Closed Loop Approaches," *Proc. IEEE Computer Society 22nd Int. Conf. (COMPCON)*, February 1981, pp. 308–311.
9. J. A. Abraham and S. M. Thatte, "Fault Coverage of Test Programs for a Microprocessor," *Proc. 1979 IEEE Int. Test Conf.*, October 1979, pp. 18–22.
10. N. Annaratone and M. Sami, "An Approach to Functional Testing of Microprocessors," *Proc. 12th Symp. on Fault-Tolerant Computing*, June 1982, pp. 158–164.
11. C. Bellon, A. Liothin, S. Sadier, G. Saucier, R. Velazco, F. Grillot, and M. Issenman, "Automatic Generation of Microprocessor Test Programs," *Proc. 19th Design Automation Conf.*, 1982, pp. 566–573.
12. D. Brahme and J. A. Abraham, "Functional Testing of Microprocessors," *IEEE Trans. Computers*, **C-33** (6):475–485 (1984).
13. C. Robach and G. Saucier, "Microprocessor Functional Testing," *Proc. 1980 IEEE Int. Test Conf.*, 1980, pp. 433–443.
14. L. Shen and S. Su, "A Functional Testing Method for Microprocessors," *Proc. 14th Int. Conf. on Fault-Tolerant Computing*, 1984, 212–218.

15. T. Sridhar and J. P. Hayes, "A Functional Approach to Testing Bit-Sliced Microprocessors," *IEEE Trans. Computers,* August 1981, pp. 563–571.
16. T. Sridhar and J. P. Hayes, "Testing Bit-Sliced Microprocessors," *Digest 9th Annual Int. Symp. on Fault-Tolerant Computing,* June 1979, pp. 211–218.
17. S. M. Thatte and J. A. Abraham, "Test Generation for Microprocessors," *IEEE Trans. Computers,* **C-29**:429–441 (1980).
18. E. J. McCluskey, "Built-In Self-Test Techniques," *IEEE Design and Test Mag.,* April 1985, pp. 21–28.
19. E. J. McCluskey and S. Borzorgui-Nesbat, "Design for Autonomous Test," *Proc. 1980 Int. Test Conf.* 1980, pp. 15–21.
20. E. J. McCluskey, "Built-In Verification Test," *Proc. 1982 Int. Test Conf.,* 1982, pp. 183–190.
21. E. J. McCluskey, "Verification Testing—A Pseudoexhaustive Test Technique," *IEEE Trans. Computers,* June 1984, pp. 541–546.
22. V. D. Agrawal and P. Agrawal, "An Automatic Test Generation System for Illiac IV Logic Boards," *IEEE Trans. Computers,* **C-21**(9):1015–1017 (1972).
23. T. W. Williams and E. B. Eichelberger, "Random Patterns within a Structured Sequential Logic Design," *Proc. 1977 Int. Test Conf.,* 1977, pp. 19–27.
24. C. C. Timoc, F. Scott, K. Wickman, and L. Hess, "Adaptive Self-Test for a Microprocessor," *Proc. 1983 Int. Test Conf.,* 1983, pp. 701–703.
25. S. W. Golomb, *Shift Register Sequences,* Aegean Park Press, 1982.
26. W. W. Peterson and E. J. Weldon, *Error-Correcting Codes,* 2d ed., The Colonial Press, 1972.
27. E. B. Eichelberger and E. Lindbloom, "Random-Pattern Coverage Enhancement and Diagnosis for LSSD Logic Self-Test," *IBM J. Res. Develop.* **27**(3):265–272 (1983).
28. J. Savir, G. S. Ditlow, and P. H. Bardell, "Random Pattern Testability," *IEEE Trans. Computers,* **C-33**(1):79–80 (1984).
29. H. D. Schnurmann, E. Lindbloom, and R. G. Carpenter, "The Weighted Random Test-Pattern Generator," *IEEE Trans. Computers,* **C-24:**695–700 (1975).
30. C. Chin and E. J. McCluskey, "Weighted Pattern Generation for Built-In Self Test," Tech. Report No. 84-7, Center for Reliable Computing, Computer Systems Laboratory, Stanford University, Stanford, Ca., August 1984.
31. R. M. Sedmak, "Design for Self-Verification: An Approach for Dealing with Testability Problems in VLSI-based Designs," *Proc. 1979 IEEE Test Conf.,* October 1979, pp. 112–124.
32. R. M. Sedmak, "Implementation Techniques for Self-Verification," *Proc. 1980 IEEE Test Conf.,* November 1980, pp. 267–278.
33. K. P. Parker, "Compact Testing: Testing with Compressed Data," *Proc. 6th Annual Fault-Tolerant Computing Symp.,* June 1976, pp. 93–98.
34. J. Savir, "Syndrome-Testable Design of Combinational Circuits," *IEEE Trans. Computers,* **C-29**(6):442–451. (1980). "Corrections," *ibid.,* **C-29**(11):1012–1013 (1980).
35. J. Savir, "Syndrome-Testing of 'Syndrome-Untestable' Combinational Circuits by Maximizing the Probability of Fault Detection," *IEEE Trans. Computers,* August 1981, pp. 606–608.
36. N. Benowitz, D. F. Calhoun, G. E. Alderson, J. E. Bauer, and C. T. Joeckel, "An Advanced Fault Isolation System for Digital Logic," *IEEE Trans. Computers,* **C-24**(5):489–497 (1975).
37. W. C. Carter, "Signature Testing with Guaranteed Bounds for Fault Coverage," *Proc. IEEE Int. Test Conf.,* 1982, pp. 75–82.

38. W. C. Carter, "The Ubiquitous Parity Bit," *Digest of Papers 12th Annual Int. Symp. on Fault-Tolerant Computing,* June 1982, pp. 289–296.
39. A. Y. Chan, "Easy-to-Use Signature Analyzer Accurately Troubleshoots Complex Logic Circuits," *Hewlett-Packard J.,* May (1977), pp. 9–14.
40. R. A. Frohwerk, "Signature Analysis: A New Digital Field Service Method," *Hewlett-Packard J.,* **28**(5):2–8 (1977).
41. J. E. Smith, "Measures of the Effectiveness of Fault Signature Analysis," *IEEE Trans. Computers,* **C-29**(6):510–514 (1980).
42. T. W. Williams, W. Daehn, M. Gruetzner, and C. W. Starke, "Comparison of Aliasing Errors for Primitive and Non-Primitive Polynomials," *Proc. 1986 IEEE Int. Test Conf.,* pp. 282–288.
43. T. W. Williams, W. Daehn, M. Gruetzner, and C. W. Starke, "Aliasing Errors with Primitive and Non-Primitive Polynomials," *Proc. 1987 IEEE Int. Test Conf.,* pp. 637–644.
44. V. K. Agarwal, "Increasing Effectiveness of Built-In Testing by Output Data Modification," *Proc. 13th Annual Fault-Tolerant Computing Symp.,* 1983, pp. 227–233.
45. Y. Zorian, V. K. Agarwal, "Higher Certainty of Error Coverage by Output Data Modification," *Proc. 1984 IEEE Int. Test Conf.,* pp. 140–147.
46. E. J. McCluskey, "Built-In Self-Test Structures," *IEEE Design and Test Mag.,* April 1985, pp. 29–36.
47. B. Konemann, J. Mucha, and G. Zwiehoff, "Built-In Logic Block Observation Technique," *Digest 1979 IEEE Test Conf.,* 1979, pp. 37–41.
48. J. Savir, G. S. Ditlow, and P. H. Bardell, "Random Pattern Testability," *IEEE Trans. Computers,* **C-33:**79–90 (1984).
49. F. Brglez, "On Testability Analysis of Combinational Networks," *Proc. Int. Symp. on Circuits and Systems,* May 1984, 221–225.
50. S. K. Jain, V. D. Agrawal, "Statistical Fault Analysis," *IEEE Design and Test Mag.,* **2:**38–44 (1985).
51. S. C. Seth. L. Pan, and V. D. Agrawal, "PREDICT—Probabilistic Estimation of Digital Circuit Testability," *Proc. 1985 IEEE Fault-Tolerant Computing Symposium,* June 1985, pp. 220–225.
52. V. D. Agrawal, "Sampling Techniques for Determining Fault Coverage in LSI Circuits," *J. Digital Systems,* **5:**189–202 (1981).

CHAPTER 10
AUTOMATIC VLSI TEST EQUIPMENT

Ernest Millham
IBM Corporation
Manassas, Virginia

Raymon Oberly
Rhinebeck, New York

Many of the same disciplines are required by both the designer and the user of automatic VLSI test systems. Both must thoroughly understand the product to be tested, the testing methodologies, the cost of ownership, and the test system architecture. Both must also be aware of the many tradeoffs available—the designer-manufacturer to avoid providing features that are not needed, and the user to avoid the purchase of features which will never be used.

To understand the architecture and design of automatic VLSI test equipment, we must first review the testing requirements of the device under test (DUT) as they relate to the test system. Architecture and design are interrelated, but each has a different dependency. The test system architecture is primarily driven by the types of testing to be performed, while the actual design is mostly dependent upon cost and the technology available for implementation.

It is always necessary to insert hardware between the tester pin electronics and the DUT to adapt to the physical configuration of the DUT. This space transformer must be capable of transmitting the required electrical signals without degradation or crosstalk and typically is different for each type of DUT.

General-purpose systems, having features to allow testing of the most DUT types are usually most desirable; however, at times a special-purpose test system is more cost-effective. When there are enough parts of the same type to test, then fixturing, interfaces, and system features can be optimized, resulting in a lower purchase price and higher throughput.

10.1 VLSI TEST EQUIPMENT REQUIREMENTS

10.1.1 Interface with the Product

An automatic VLSI test system must properly communicate with the DUT, which may be chips, components, boards, subsystems, or systems. The mechanical interfaces range from single-point to multipoint probes (typically 0.010-in centers) through card and/or cable connectors. These mechanical interfaces are required to transmit signals, to and from the DUT, with minimal electrical signal degradation and interference. Electrically the interface must be capable of handling dc power to very high frequency ac stimulus and response.

Often the DUT must be tested using particular environmental conditions. Ideally the application of these conditions should not interfere with the electrical characteristics of the interface. Environmental conditions can be temperature, humidity, liquid baths, etc. When temperature is specified, the point at which the temperature is to be monitored must be clearly understood (e.g., junction, case, pin, etc.), and the test system must be capable of monitoring and maintaining the environment during testing.

For logic testing, the test system electrical interface typically consists of a driver, tristate with −4- to +8-V levels and up to 50 mA of drive capability. Also there are one or two receivers with controlled levels from −4 to +8 V and which present a total current load of less than 1 μA to the DUT. A programmable load is also often provided. Table 10.1 illustrates the typical interface specifications for automatic VLSI test equipment. The drivers are used to properly format (generate the required electrical conditions) for applying the test vectors to the DUT, while the receivers are used to convert DUT responses into digital signals for comparison to a specified format.

The tester drivers and receivers (bidirectional tester pins) must be able to be "formatted." That is, the pin function must change from one test cycle to the next with specific time relationships. A minimum of eight different formats and six unique timing edges are required.

In addition to logic signals, the interface must be capable of handling dc signals from −12 to +12 V and with 50 mA capability. Special ac signals may need to be routed to and from conventional rack-mounted instruments. Power supply pins must be capable of handling higher currents and must be correctly decoupled to keep the DUT power free of noise.

In general, the load seen by the DUT signal pins should exhibit low capacitance so that outputs from the DUT are minimally influenced by the test system. The bandpass of the entire interface must be at least 1 GHz.

Ideally the interface impedance should be a fixed value which matches the DUT ac characteristics. This matching is possible in 50-Ω systems and is generally compromised in other cases when 70-, 90-, or 100-Ω series termination of the

TABLE 10.1 Interface Requirements

	Driver	Receiver	General interface
Type	Tristate	Dual threshold	
Impedance	50-, 93-, or 100-Ω selectable (± 1%)		
Voltage	−4 to +8 V (±10 mV)	− 4 to +8V (± 10 mV)	
Current	±30 mA	< 1 μA	
Transition	1 ns/V (10–90%)	1 ns/V (10–90%)	
Perturbations	Monotonic transition (<25 mV overshoot)		
Capacitance			<30 pF
Bandwidth			>1 GHz

driver is used. The ac environment of the interface is many times compromised to best satisfy the largest range of DUTs with minimum impact on signal integrity.

10.1.2 Functional Logic

The DUTs to be tested often exceed 100,000 circuits and require over 250,000 test cycles (test vectors) for a complete test. This large number of test vectors is required since most logic design is random and highly sequential.

Functional logic testing may require that test vectors be synchronized with an on-board clock signal. This will require all tester timing to be synchronized with the DUT oscillator through a DUT terminal.

There will be cases during the functional testing of logic when the test patterns must change in response to one or more interrupts from the DUT. An example of this is the case where an unknown number of DUT cycles must be executed to establish known states for all latches prior to the execution of a predetermined test pattern set. This may be accomplished by repeatedly applying a predetermined set of stimuli while monitoring responses for a specific set of logic conditions.

Similarly, there may be redundant logic structures that are intended to be used as replacements for identical structures which are found defective during normal test. For this situation, the defective structure will be replaced (by fuse blow, voting logic or other techniques), and a modified set of test patterns will be applied for the new logic structure.

Test patterns are provided by "stuck at" test generators, where each pattern is intended to detect one or more potential circuit faults, usually with few timing requirements. Manually generated functional patterns are also used, which are directed at finding potential functional faults within each functional structure of the DUT, often with moderate timing requirements. Test patterns may also be derived from system reliability and serviceability (RAS) tests, which require stringent timing of test vectors. Total test pattern data content is typically greater than 10 M bytes for each DUT part number. See Fig. 10.1 for the relationship between circuits tested and test data requirements.

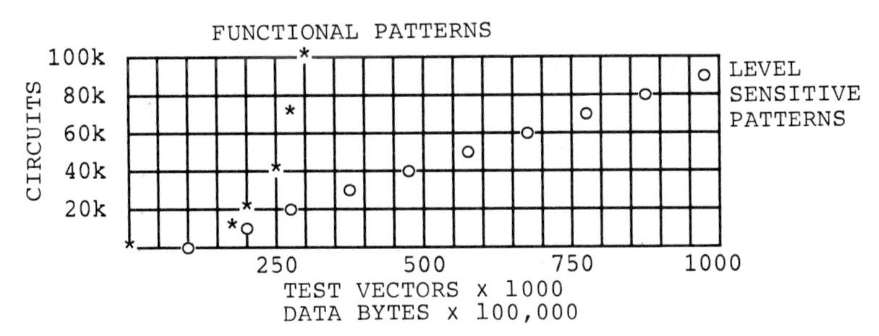

FIG. 10.1 Circuits versus test vectors.

10.1.3 Serial Scan Logic

Serial scan logic (typified by LSSD, BILBO, etc.) is a product design methodology where all latches are interconnected as one or more shift registers, while the

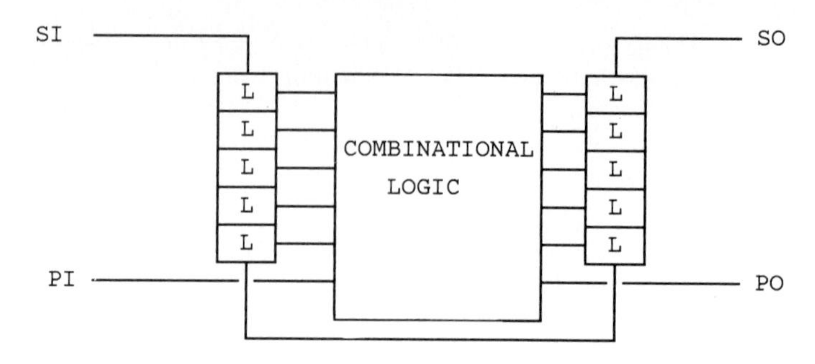

FIG. 10.2 Serial scan logic.

combinational logic remains unchanged, connecting the latches according to the functional design (e.g., see Fig. 10.2). This methodology allows all latches to be set to known states by shifting data into the shift registers. Latches driving combinational logic can then be used as pseudo inputs, while latches which are driven by combinational logic can be used as pseudo outputs. Once the latches are all set to a known state, signals are allowed to propagate through the combinational logic, forcing the latches being driven to a known state. The shift register is then shifted, and the output is examined for correct responses.

Logic designs using serial scan methodology are a boon to automatic test generation since the test generation problem is reduced to generating stuck-fault test data for combinational logic only. Test coverage for stuck faults can approach 100 percent versus the 50 to 70 percent obtainable for random sequential structures.

As shown in Fig. 10.1, the test data content can exceed 100 M bytes per part number. The number of tester cycles can easily exceed 1 million.

10.1.4 Memories

VLSI logic products will often contain on-board ROM and RAM. The testing requirements for these memories are similar to those for conventional stand-alone memory products. In general, testing will be exhaustive with stringent timing requirements and at functional cycle rates. Memory size can be up to 32 bits by 64,000 words and possibly even larger.

Dynamic on-board RAMs require periodic refresh to retain the stored data. Refresh may be indicated by an interrupt from the DUT at specific time intervals or by test data control.

Most VLSI logic products with on-board memories will be designed in such a fashion that the memory address, data, clock, and control signals are available at the DUT terminals. This control information and refresh data will be accomplished through multiplexing DUT terminals and will require logic conditioning prior to or in conjunction with testing of the array. This design methodology allows the testing of memories at functional specifications and enables the use of algorithmically generated patterns.

Some on-board memories will be electrically reconfigurable for yield enhancement. That is, a defective portion of the memory may be replaced by redundant circuitry. The testing sequence will have the ability to recognize such occurrences, to cause the defective structure to be replaced (by fuse blow, voting logic, or other methods), and to change to a sequence correct for the new structure.

10.1.5 Measurements

Logic testing alone, particularly of chips, is not generally sufficient to guarantee performance of a DUT. Other measurements are required, such as both dc and ac parametric stimulus and response, and ac analog.

Dc parametric consists of forcing voltage or current on specific terminals while measuring the results on the same or other terminals. It may often be necessary to establish the logical conditions of every DUT terminal prior to performing these tests. For instance, input leakage is measured by forcing a dc voltage and measuring the resultant current into (or out of) a DUT terminal.

Ac parametric is the measurement of time between logic signals at DUT terminals, generally with accuracies beyond the capabilities of the logic response circuits. Logical conditioning is always required in conjunction with these tests. One typical test (single-shot time measurement) is the measurement of time between two output responses, which must be accomplished for a single occurrence.

Ac analog testing requires stimulus and response characteristics outside the bounds of normal logic signals. Logic conditioning is required on other DUT terminals during analog testing. Stimulus and response characteristics cover the range of specifications normally associated with rack-mounted, digitally controlled signal generators and waveform analyzers. Some examples are the testing of coder-decoders (CODECs) analog-to-digital converters, and digital-to-analog converters.

10.1.6 Calibration

All stimulus and measurement subsystems require calibration traceable to the National Bureau of Standards to ensure tester accuracy and correlation. In those cases where recognized standards do not exist, they must be provided with the test system. Calibration should not significantly affect the overall throughput (DUTs tested per unit time) of the system.

Critical components of the system, primarily those where slight errors may cause defective DUTs to be tested as good, should be automatically self-tested prior to the testing of each DUT. Self-test should be performed as nearly as possible with the identical ac and dc conditions to be presented to the DUT during normal test.

Self-test should not impact the time available for DUT testing. Any failure of self-test must result in immediate action to correct the problem.

10.1.7 Fault Isolation

Provisions must be made for isolating faults detected on failing DUTs. On-line, test-system-directed fault isolation should have the capability for both automatic and interactive operation. For off-line fault isolation, the test system is required to collect sufficient data to allow fault isolation to the smallest possible partition and should be interfaced to popular CAD/CAM, test generation, and simulation systems. In either case, the amount of data to be handled will exceed 10 M bytes.

On-line fault isolation typically consists of a table-driven or dictionary capability to coordinate the activities of a physical probe or a noncontacting probe such as an electron beam. The probe is moved to a specific node based on an analysis of the fail data, and either the original set or a modified set of stimuli is applied while monitoring responses at the probe. This process is repeated until the fault is isolated. Once isolated, the fault can be repaired, where possible, or data can be retrieved for off-line compilation when on-line repair is not possible.

In simplest form, off-line fault isolation is a simulation of the DUT using the same stimuli as originally applied to the DUT together with the responses, detected by the test system. In more complicated form, the test system will send response data on a link to a simulator. The simulator will calculate a new set of stimuli for application to the DUT in a closed loop, so the DUT being tested effectively becomes part of the simulator.

10.2 BLOCK DIAGRAMS

10.2.1 Generic Test System

Figure 10.3 is a block diagram of the generic components common to all VLSI automatic test equipment. Physical implementation, software, and performance features are different between manufacturers of VLSI automatic test equipment, but the architecture and data flow are very similar.

The Host System. A host computer system is not required for operation of the test system. The small user with limited part numbers and manufacturing requirements may not have sufficient economic justification for the purchase of a host system. The larger user with multiple test systems and high production volumes will find that a host system is indispensable and easily cost-justified.

The host computer is generally a large data-processing system capable of manipulating large amounts of data for transfer to and from the test system and is capable of communicating with several test systems. Host systems are specific to the target test systems. Manufacturers of test systems do not normally provide the physical components of the host system, but usually do provide the software required to support their test system.

Test data are stored at the host for all DUT part numbers to be tested. Response data from the test system can be stored and manipulated by the host to provide yield analysis, off-line DUT diagnostics, quality tracking, and other functions. The host system can direct floor control operations such as product flow, scheduling, and report generation. Often the host system is used for test generation and compiling test data into the correct format for the target test system. Since there are no standards for test data formats, unique software (and often hardware) is required to communicate between the test generator/simulator and target test system.

The Test System Controller. The test system controller is a commercially available miniprocessor of sufficient power to perform many of the host functions while directing the testing operation. When used as stand-alone (no host), the re-

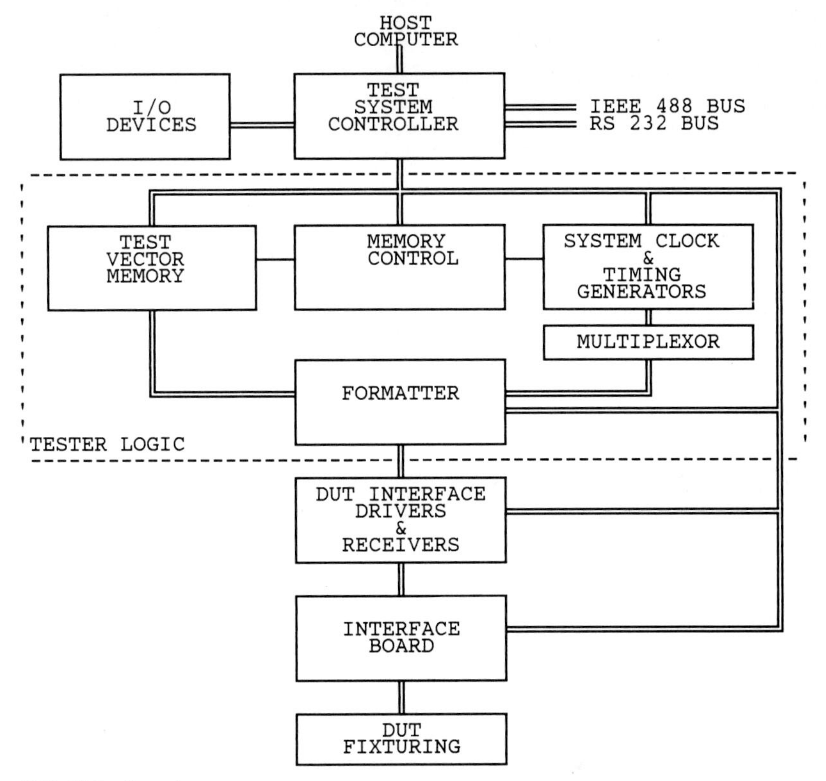

FIG. 10.3 Generic test system.

duction in product throughput can be significant due to the demands on the processor.

Usually, during initial program load (IPL), the controller performs self-test of itself followed by an exhaustive self-test of the remainder of the test system. This can take from several seconds to several minutes, depending on the completeness of the test. Results of the self-test are stored to provide maintenance records. Any failure of self-test will result in more exhaustive self-testing together with a maintenance call.

The controller is used to monitor airflow sensors, temperature gauges, power supply voltages, and other test system functions. Interrupts from any of these functions are considered catastrophic and result in system power off.

I/O Devices. Commercial I/O devices, such as terminals, printers, and plotters, are used to communicate with the test system. Sufficient magnetic storage media direct-access storage device (DASD) is provided to minimize communication with the host and/or contain enough data, when no host is used, to minimize manual intervention.

The controller communicates with the remainder of the test system via a 16- or 32-bit parallel bus. Steppers, handlers, instrumentation, and so on are connected to the RS-232 or IEEE-488 bus.

The Tester Logic. The tester logic is unique to each test system, while the architecture is generally the same for all test systems. The tester logic of Fig. 10.3 (the area inside the dashed lines) consists of the system clock, memory control, and vector memory. This is the high-speed portion of the test system, typically operating at clock rates of up to 100 MHz. Setup and test data are loaded by the controller into registers and memories. Once setup and other peripheral operations are complete and logic testing has been started, the tester logic takes control of the logic testing function, freeing the controller for other tasks. Logic testing is terminated by a pass or fail interrupt to the controller.

The System Clock and Timing Generators. System timing of the tester logic during logic testing is defined by the test system clock. The system clock determines the period of the test cycle, generally one test vector, and is used to synchronize the timing generators and memory control. In some test systems, there are up to 32 test cycle (clock) definitions, which may be switched on the fly between test cycles. The DUT test data can be set up by using different cycle rates at various points in the test. Figure 10.4 illustrates a typical method of generating tester cycles.

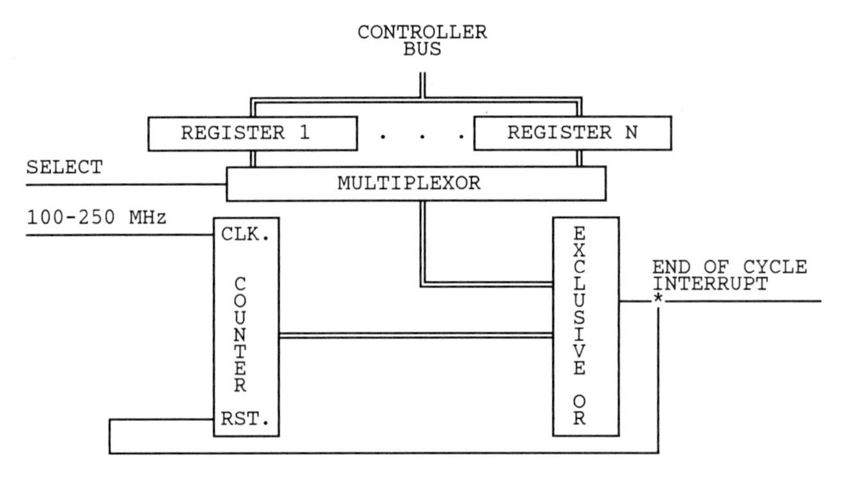

FIG. 10.4 Primitive clock and timing generator.

In primitive form, cycle and timing generators are similar in design. Data are first loaded into registers by the tester controller corresponding to the number of oscillator pulses desired in the cycle. The oscillator is enabled, allowing the counter to run until the count is equal to the value in the selected register, when an end-of-cycle pulse is generated. This pulse indicates to the memory control that the cycle is complete and resets the counter, causing the start of another count sequence. The register used for any specific cycle is selected by the memory control. Similar circuits are used to provide timing markers for the formatter to define driver and receiver timings.

The Memory Control. The memory control (Fig. 10.5) is a high-speed application-specific microprocessor with associated memory capable of operation at the fastest tester rate. The memory contains instructions and data. The memory width (word) is of a size sufficient for the specific requirement and of a depth

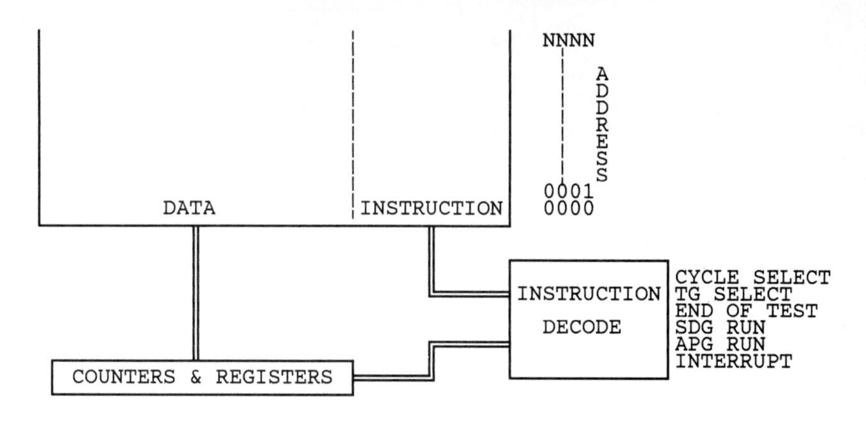

TYPICAL INSTRUCTION FORMAT

INSTRUCTION			DATA	
0000 0123	000 456	000 789	11111111112 01234567890	22222222233 12345678901
INST	CYC	TG	COUNT OR ADDRESS	COUNT OR ADDRESS

FIG. 10.5 Memory control.

(number of words) usually the same as the test vector memory. Each address of the instruction memory accesses the same address in the test vector memory. Table 10.2 lists available instructions and describes their functions.

Once started by the controller, the memory control has control of the tester until an interrupt is generated, signaling the controller for action. Interrupts are last test vector (end of test), error response, and memory control interrupts, such as selected counter equals 0. The controller, based on specific part number data, then decides what the appropriate action should be (end of test good, error, or perform other functions such as reload memories).

The Test Vector Memory. The test vector memory (Fig. 10.6) is a RAM with an effective read access time equal to the fastest cycle rate of the tester. Addressing is from the memory control. The memory is structured as 8-bit bytes for loading from the controller and in 3- or 4-bit words for use by the formatter. Each test vector is represented by one word for each tester pin (DUT interface). The number of test vectors contained varies widely by manufacturer from a low of 1000 to a high of 250,000, although the upper limit is continually growing.

Generally, the memory should be as large as possible, within cost and speed constraints, to avoid having to reload during testing. For example, if the memory supports a 256-pin interface of 3 bits per pin with 64,000 test vectors, then application time during test for all vectors applied at 1 MHz would be 6.4 ms. Reloading all 64,000 vectors at 1 MHz from an 8-bit bus would require 6.1 s (even ignoring overhead such as disk access), which is significantly longer than the time required for actual testing. There are various schemes to reduce the reload time, such as segmenting the memory and loading only where new data are required, but all methods involve compromises, and tester throughput is always sacrificed.

TABLE 10.2 Sample Memory Control Instructions

This table contains simple examples of instructions processed by the memory controller. These examples assume a 32-bit instruction and data memory. Each tester manufacturer uses a different format and number of bits. Within each instruction are data for selecting the cycle rate and timing generator set to be used during execution of that instruction.

Instruction	Data	
NOP		Data are applied from the test vector memory to the formatter and the memory increments to the next sequential address.
LOOP	COUNT	Data are applied from the test vector memory to the formatter and the memory increments to the next sequential address. The count is loaded into a counter. The address of the loop instruction is loaded into a loop register.
ENDLOOP	ADDRESS	Data are applied from the test vector memory to the formatter. The loop counter is decremented and tested for 0. If the loop counter is not 0, then the next memory address is the loop register. If the counter is 0, then the next memory address is taken from the address field.
REPEAT	COUNT	Data are applied from the test vector memory to the formatter. The count is loaded into a counter. The counter is tested during each tester cycle and if not equal to 0 this instruction and vector memory data are repeated. If equal to 0, then the next sequential memory address is used.
APG	COUNT	Data for all pins not connected to the APG are applied from the test vector memory. Data for all pins connected to the APG are applied to the formatter from the APG. The count is loaded into a counter. The counter is decremented during each tester cycle and tested for 0. When the counter equals 0, then the next sequential memory address is used.
SDG	COUNT	Data for all pins not connected to the SDG are applied from the test vector memory. Data for all pins connected to the SDG are applied to the formatter from the SDG. The count is loaded into a counter. The counter is decremented during each tester cycle and tested for 0. When the counter equals 0, then the next sequential memory address is used.
END		Data are applied from the test vector memory to the formatter. An interrupt is generated for the test system controller, indicating end of test data.

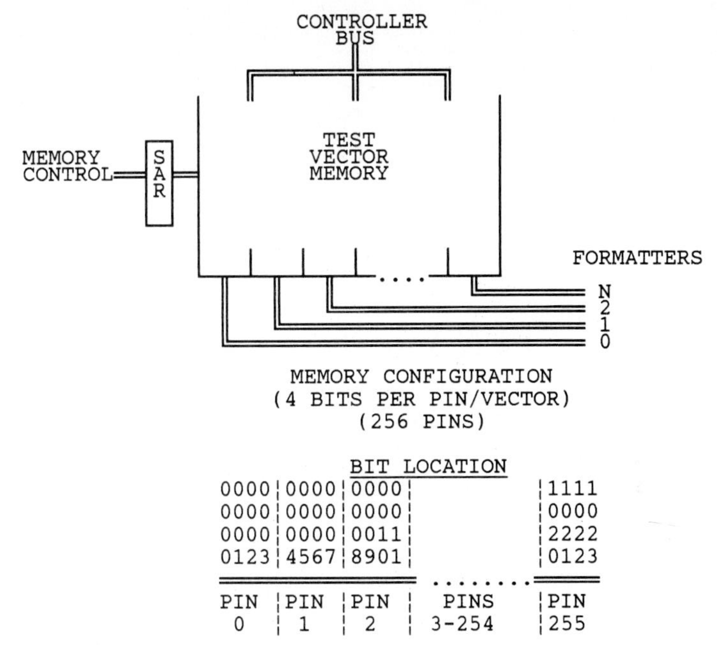

FIG. 10.6 Test vector memory.

The Formatter. In Fig. 10.7 the formatter, one for each pin in the test system, decodes the bits from the test vector memory and combines the resultant instruction with the timing generators. The output of the formatter is typically one signal to determine the logical state of the DUT interface driver and one signal to turn the driver on and off. The DUT interface receivers are compared to a formatted signal to provide an error indication to the controller when the DUT output is not as expected. Detailed in Fig. 10.7 are the formatters for driver RZ (return to zero) and receiver E0 (expect zero). The receiver format shown depicts a window compare, where an error will be indicated if the receiver output is different from the selected logic state at any time between the two timing markers. An alternative method is to allow detection of an error only during the duration of each timing marker (typically 100 ps). All other formats are similarly provided as shown in Fig. 10.8.

The timing generator inputs are timing markers (i.e., narrow pulses) of a width less than the narrowest pulse to be provided to the DUT terminals. Up to 7 of the 32 available generators may be selected for use by each formatter. Each selected timing generator is delayed by a programmable fine-trim circuit that typically has 100-ps increments. The fine trim is used to compensate for differences in line length and circuit delays.

The formatter design and function is where the largest differences exist between manufacturers. There may be as few as 2 to as many as 16 available formats, from 8 to 32 timing generators, and from 0 to 7 fine-trim circuits. Generally the more function available the easier the system is to use, and the test system has application over a broader product line.

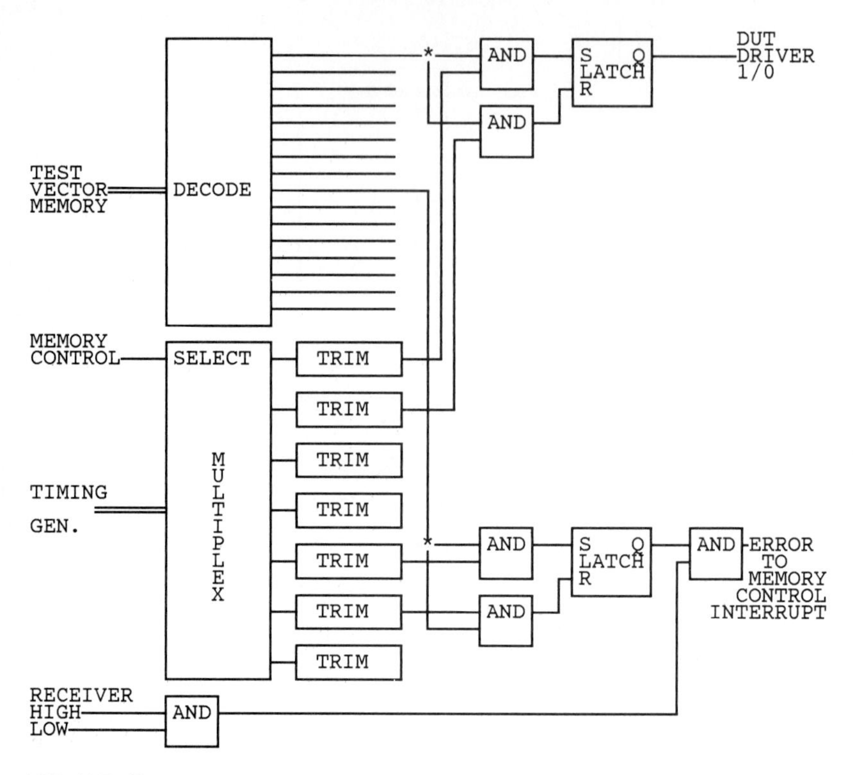

FIG. 10.7 Formatter.

The DUT Interface Circuits. The DUT interface drivers and receivers are the electrical interface between the tester and the DUT. Figure 10.9 represents a typical interface.

There are two popular methods for obtaining the dc voltage levels required to establish the logical levels out of the DUT interface drivers and the reference levels for the receivers. In the minimum case, several (up to 32) individual programmable power supplies are provided and are multiplexed to all drivers and receivers, allowing up to 32 possible drive and detect levels. In the other case (as in Fig. 10.9), individual digital-to-analog converters (DACs) are provided for each required level for each pin interface. DACs are loaded from the controller bus during setup.

The driver is comprised of three high-speed switches: one to connect the first DAC (driver high), the second to connect the second DAC (driver low), and the third to connect or disconnect the driver from the DUT. The receiver is composed of two high-speed comparators, the digital outputs of which are used as inputs to the error compare circuits in the formatter. Each comparator has a unique threshold level.

The Interface Board (Fixture). The primary function of the interface board is to provide the physical space transformation from the drivers and receivers to the DUT contactor and to allow unique DUT power supply assignments. Also pro-

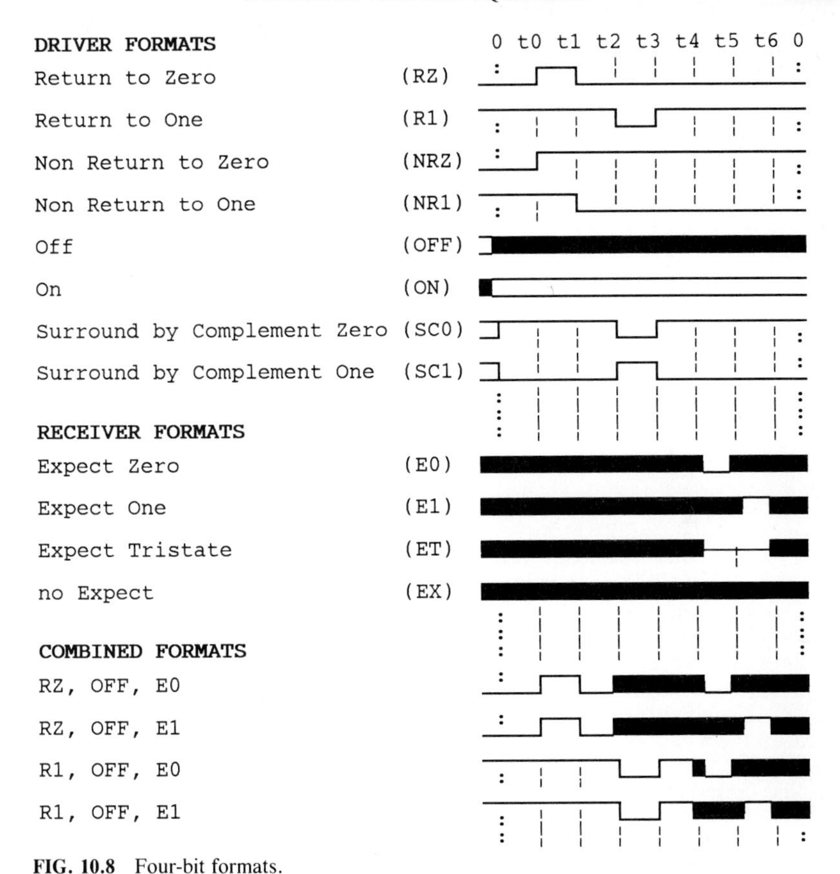

FIG. 10.8 Four-bit formats.

vided on the interface board is space for mounting DUT loads, special conversion circuits, relays, and so on. These features are controlled by a bus from the controller and are interconnected by printed circuit and/or discrete wiring. Typically, an interface board is necessary for each DUT part number with unique power or loading requirements.

In some systems the driver and receiver are not hard-wired but may be connected on the interface board. This allows unique assignment of pins as inputs or outputs, thus improving the ac characteristics as seen by the DUT.

10.2.2 Variations for Chip-Wafer

Chip-wafer testing is usually accomplished by using a wafer stepper. Packaging of the interface drivers and receivers is most critical for optimum ac performance. The package is a carrousel (cylindrical) arrangement with the drivers and receivers mounted on cards radiating from the center like spokes in a wheel. With this arrangement the pin-to-pin electrical characteristics are nearly identical, and physical line lengths to the interface board and wafer are minimal. The carrousel

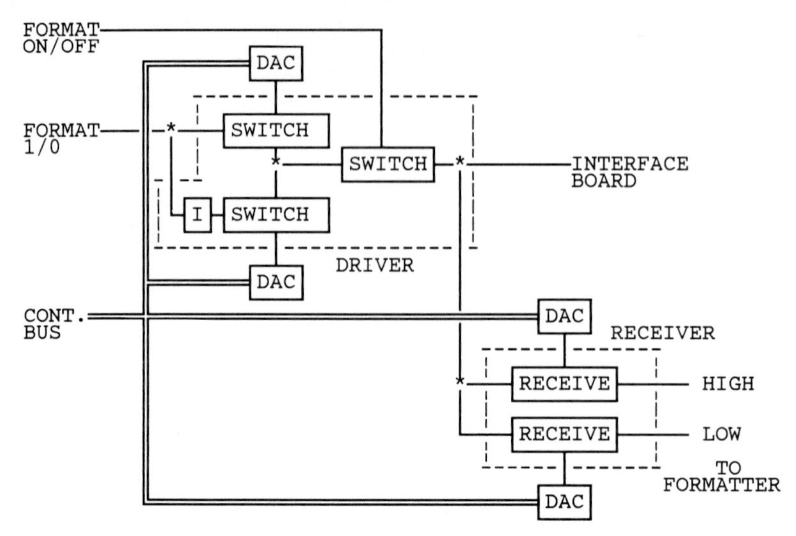

FIG. 10.9 DUT interface drivers and receivers.

is mounted on a pedestal arrangement with x, y, z, and θ adjustments for docking to the wafer stepper.

Chip-wafer testing requires faster test cycle rates, better accuracies, and higher throughputs than do other levels of the product packaging. Also, special measurements are required, such as dc parametrics (Fig. 10.10) and the ability to measure a recirculating loop frequency. Data manipulation by the controller for on-line information, such as wafer maps for chip picking and measurement data

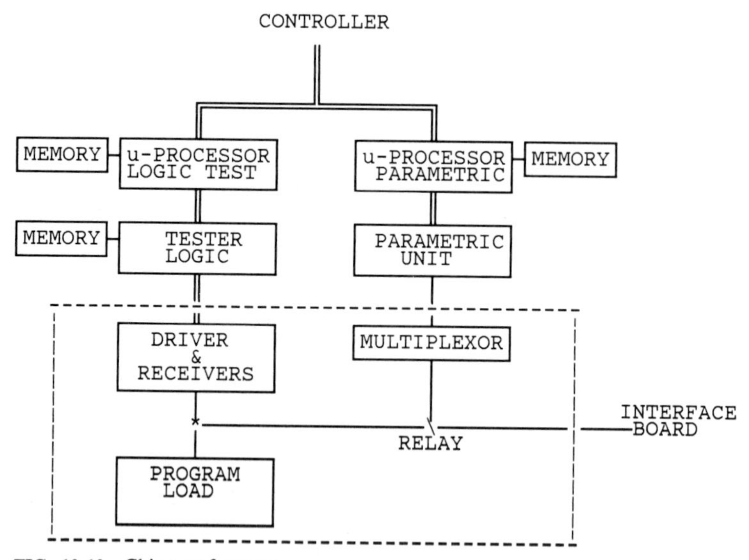

FIG. 10.10 Chip test features.

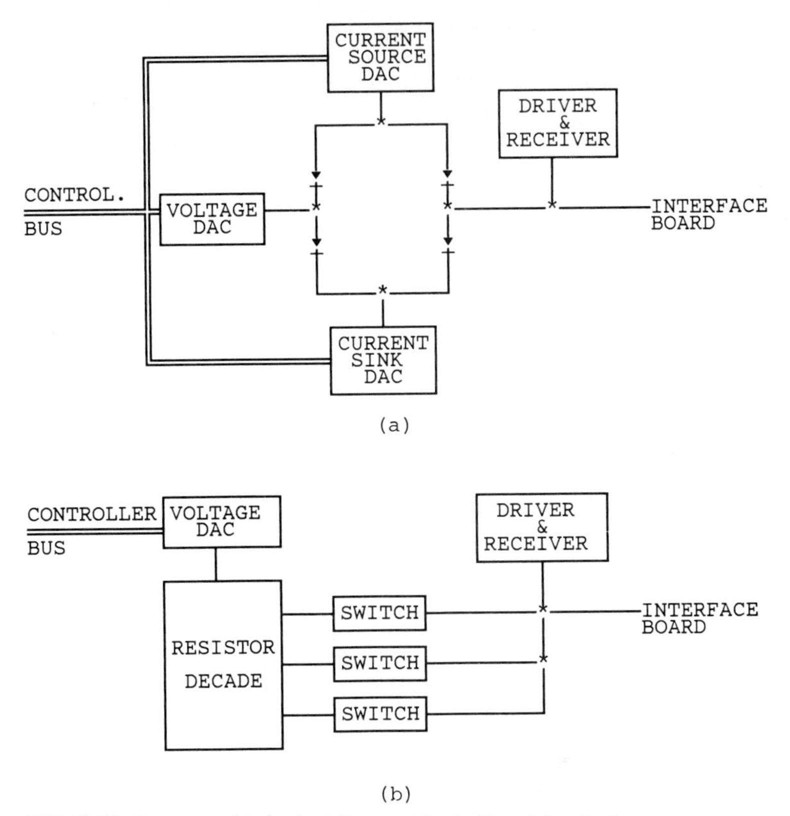

FIG. 10.11 Programmable loads: (*a*) current load; (*b*) resistive load.

for process control, is more efficiently handled by including distributed (parallel) processing, larger memories, and multiple measuring units.

A programmable load is often included, which reduces the number of interface boards required for a specific family of products. Most popular today is the quad bridge (Fig. 10.11*a*), which presents a current load to the DUT. Alternatively, a resistive load may be provided (Fig. 10.11*b*).

10.2.3 Variations for Card-Board

The most popular type of fixturing for card-board testing is the bed-of-nails approach. With this type of interface, pogo (spring loaded) contacts are arranged in a grid array to physically make contact with selected nodes internal to the DUT. The normal DUT pins are connected to the tester interface circuits through pogo pins or standard card connectors. Contact may be made on both sides of the card, but more typically only the side hidden from the operator is contacted, leaving the top side exposed for manual or automatic probing to assist in defect analysis. With the increasing use of surface-mounted devices (SMDs) on printed circuit cards, internal test points must be specially designed on the card surface to allow probe contact.

Interface driver and receiver circuits are more specialized than those required for chip testing. Accuracies, range of operation, and frequency requirements typically are less stringent. Generally, voltage rails (shared voltage sources) are used for level-setting drivers and receivers rather than DACs per pin. Some test systems use the in-circuit testing methodology where logic levels are imposed and measured on nodes internal to the DUT. In this case, the driver circuits must be capable of supplying higher currents than normal to enable backdriving of the DUT circuits. Each type of DUT circuit must be analyzed to ensure that backdriving does not damage or overstress the DUT during testing.

Test time is not particularly a critical factor, since most DUTs contain one or more defects and must be analyzed for repair. Analysis is performed by using an automatic or computer-aided manual probe, typically single point, and by dictionary search through nodes contacted by the bed of nails fixture. Large amounts of storage and computer time are required to support these analysis techniques.

10.2.4 Serial Data Generator (SDG)

DUTs designed and tested using serial scan logic require special techniques to manage the large volumes of test data required. In a serial design, a DUT has two or more scan-in and scan-out (SI and SO) pins dedicated to serial scan data while the remainder are conventional primary inputs and outputs (PIs and POs). Usually all PIs and POs change state only once for each N changes of state on the

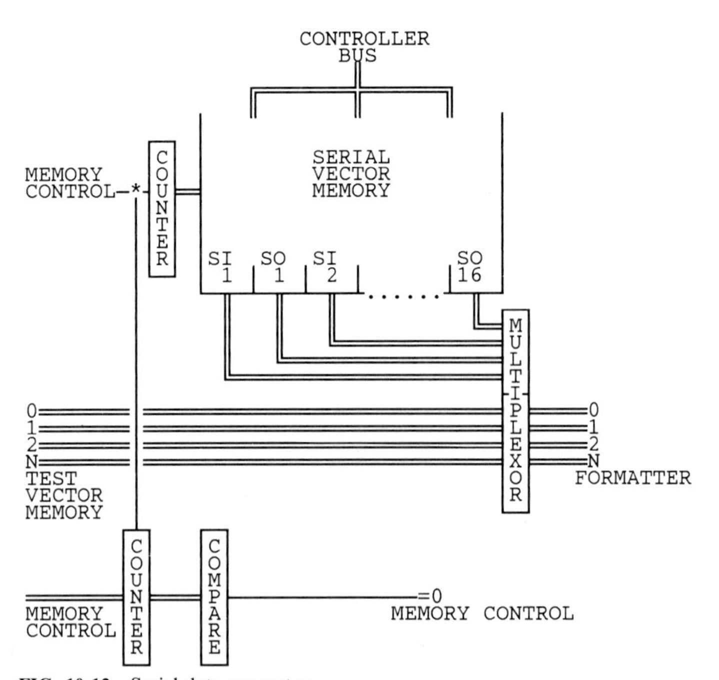

FIG. 10.12 Serial data generator.

serial scan pins. During loading and unloading of the latches (scanning of SIs and SOs) the PIs and POs do not change state. In the generic tester of Fig. 10.3, assuming a DUT with 100 pins, 2 of which are SI-SO, and 100 latches in the shift register string that require shifting 10 times, 1010 tester cycles are required to complete the test. Each tester cycle is comprised of 4 bits per pin times 100 pins for a total of 1010 by 4 by 100 bits (404,000 bits or 50,500 bytes) of data. In the SDG enhancement of Fig. 10.12, the data required for the 10 PI-PO cycles are the same (4 by 100 by 10), while the data required for the shift cycles are 4 by 2 by 100 by 10, for a total of 12,000 bits (1500 bytes). An additional 40 bytes are required for the memory control instructions. Thus, it can be seen that, for the testing of serial scan designs, about 30 times as many test vectors can be contained in a vector memory of a given size where the SDG is used.

In operation the SDG (Fig. 10.12) is first connected through the multiplexer to the SI and SO pins of the DUT; all other pins are connected as normal. Until an SDG instruction is encountered in the memory control, format data is supplied to all pin formatters from the vector memory. When an SDG instruction is encountered, the memory control stops at that address. Format data for PIs and POs are supplied from the test vector memory (the same format every cycle), while format data for the SIs and SOs are supplied from the SDG. The SDG address counter is incremented, and the SDG counter is decremented once for each tester cycle. Once the SDG counter reaches zero, the memory control increments to the next sequential address, and all pin format data are supplied by the vector memory. This process is repeated until testing is complete.

10.2.5 Algorithmic Pattern Generator (APG)

Memories typically require exhaustive testing, resulting in multiple addresses of each memory cell in the array. These memories require special techniques to manage the large volumes of test data required. Full-function APGs are complex, high-speed microprocessors of a design unique to each test equipment manufacturer.

Figure 10.13 is a simple form of a pattern generator that allows incrementing of the addresses of a memory to be tested. In operation, the outputs of the APG are first connected through the multiplexer to the DUT pins representing the address lines of the memory; all other pins are connected as normal. When an APG instruction is encountered in the memory control, pins that are connected to the APG receive formats from the APG, while all other pins (data, clock, etc.) receive formats from the test vector memory. This data flow exists until the APG counter reaches zero, at which time the memory control will advance to the next address.

Following is an example of the memory control instructions and test vector data required to write a 1 into all addresses of a 64,000-word memory of 8 bits per word, and then read all cells. The target memory for the example has 16 address pins, 8 common data-in–data-out pins, 1 read-write control pin, and 1 clock pin. With the generic tester of Fig. 10.3, the test vector memory would need to supply 26 formats (16 + 8 + 1 + 1) per test vector for 128,000 test vectors. The total test data requirement would then be 128,000 vectors by 26 pins by 4 bits per pin for a total of 13,312,000 bits (1,664,000 bytes). By using the APG feature the data requirements would be 2 vectors by 26 pins by 4 bits per pin for a total of 208 bits (26 bytes). An additional 8 bytes are required for the memory control instructions. APGs can be quite expensive and are of use only when the memory termi-

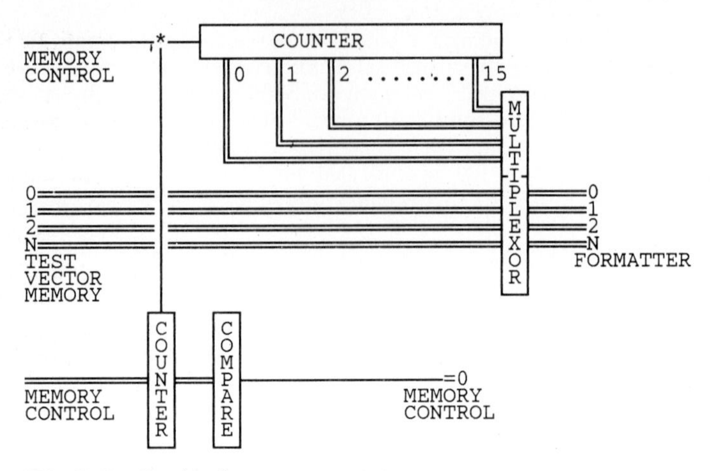

FIG. 10.13 Algorithmic pattern generator.

nals can be mapped one for one to the DUT terminals. A cost analysis is required, and careful attention must be paid to DUT design.

10.2.6 Measurement Systems

Measurement systems, required when analog types of measurements are to be made, consist of dc (see Sec. 10.2.2) and ac instrumentation. Ac instruments may be anything commercially available and are typically connected to the DUT through a 50-Ω multiplexer (Fig. 10.14). These instruments are usually rack-

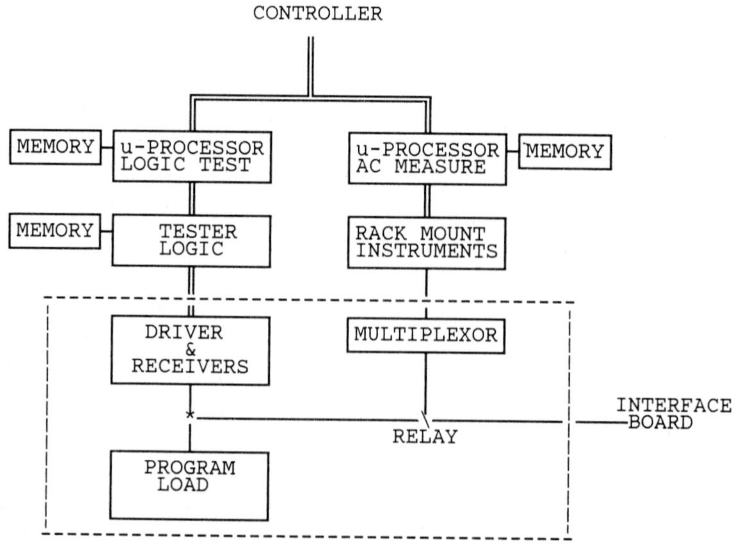

FIG. 10.14 Instrumentation.

mounted separately from the tester mainframe and are controlled by the IEEE-488 or RS-232 bus. In some cases a separate microprocessor is included to speed up operation.

10.3 TRENDS

10.3.1 Dedicated (per Pin) Resources

The per pin testers presently available are typically per pin only in the context of the interface and timing circuits (drivers, receivers, DACs, and fine timing trim). With the use of VLSI in tester design and implementation, it will be possible to provide more of the tester functions uniquely for each pin (vector memory, memory control, APG, etc.). The result of providing more per pin functions will be to further reduce test data requirements, possibly by as much as 90 percent. With reduced test data and distributed processing (possibly a microprocessor per pin), throughputs can be improved by three to five times for typical wafer test applications.

Significantly, with more per pin features, the test systems will be easier to use. Test programs, conversions between test systems of different manufacturers, and interfaces to simulators will all be simplified.

10.3.2 Pseudo-Random and Signature Analysis

Pseudo-random pattern generation for stimuli and signature analysis for response verification are becoming increasingly popular. Both will soon be included as options in most VLSI automatic test systems.

Pseudo-random patterns are provided from hardware within the test system or, sometimes, on the DUT. In either case, weighting registers and data to support them should be included in the test system to enhance testing of pseudo-random resistant logical structures. The value of pseudo-random pattern generation is to significantly reduce test pattern generation and simulation requirements. Tester memory is reduced significantly, as much as 80 percent in some cases. Implementation will consist of 16- or 32-bit linear feedback shift registers either in a parallel configuration (shared between 16 to 32 tester pins) or in a serial configuration per pin.

The use of signature analyzers in the test system can reduce the tester memory requirements by as much as 50 percent. Signature analyzers are linear feedback shift registers configured for parallel operation (shared between pins) or serial operation per pin. In either case, the shift registers will be 16 or 32 bits long.

Pseudo-random patterns and signature analysis are of particular advantage in the testing of serial scan logic. When applied to random sequential logic or tristate logic, the advantages are not nearly as significant.

10.3.3 Speed and Accuracy

The use of VLSI components in general-purpose VLSI test systems will allow cycle rates of up to 250 MHz and accuracy at the DUT terminals (wafer test) of 100 ps due to the denser packaging. The use of VLSI components allows critical circuits to be packaged closer to the DUT and enables more per pin functions. For even better speed and accuracy, there will have to be innovations in device fixturing and the tester interfaces. Accuracy improvements will be further realized by performing the ac calibration at the DUT contacts under the same voltage and timing conditions as required for testing.

10.3.4 Costs

Due to the use of VLSI components, the purchase price of VLSI automatic test systems will be about two-thirds of 1987 prices for fully configured systems. There will be more dedicated systems available at even lower cost for specific applications. Some commonality in data formats will start to emerge. This, coupled with products designed to make use of test system features, will reduce the cost of test system ownership by approximately 50 percent.

The decision to purchase a test system is a complicated procedure, particularly for a first-time user. While the development requirements are usually technology driven with little regard for manufacturing, the manufacturing requirements are more oriented to cost of ownership and may well sacrifice capabilities for throughput and reliability. It is difficult to find a common meeting ground, so development and manufacturing often follow separate paths to the detriment of the corporate balance sheet.

Given the proper motivation, development and manufacturing can usually come to terms. Doing so involves considering all the factors that contribute to cost of ownership and should be driven by those familiar with business and accounting procedures with input from the engineering and manufacturing staffs. Factors to consider beyond the initial purchase price (which may well not be the major cost contributor) are maintenance, ease of use, development and manufacturing compatibility, floor space, power, longevity, fixturing, and a host of others. Care must be taken to avoid being blinded by technology and sales hype.

10.3.5 Common Usage

Test systems architecture and design will be such that all package types can be efficiently tested with a single test system by simply changing the DUT interface. Due to the use of VLSI, it will be more cost-effective to provide all features rather than configure systems specifically for each product package type.

With the introduction of more per pin features, it will be economically and technically feasible to extend the number of pins supported by the test system far beyond present limitations. The limiting factor will be packaging of the interface electronics rather than the system architecture, although this too will improve. It will be possible, for instance, to provide systems with upward of 4000-pin capability. One tester could then support multiple interfaces, each configured for testing a different product type.

10.3.6 Calibration

Improved accuracy will require that calibration be performed under conditions nearly identical to those applied to the DUT during testing, probably with an actual (or possibly simulated) DUT. This implies that a subset of the total system calibration will be performed each time an electrical parameter is changed or at timed intervals.

Calibration standards, such as pulse integrity and timing verification at varying cycle rates, will be provided with the test systems by the manufacturer. These standards and measuring techniques will be traceable to the National Bureau of Standards. The automatic test equipment (ATE) industry should provide many of these standards and techniques.

10.3.7 Summary

Over the past two decades, roughly since the first ATE system was introduced, there has been little innovation in test system architecture. Advances have been evolutionary. The coming decade will be more revolutionary. System architecture will become more per pin oriented with significant advances in distributed processing. The purchase price for a large (250 + pins) fully featured system will decrease from $10,000 to $13,000 to about $7000 per pin as the newer system architectures are introduced.

Test systems will become easier to use, allowing product designers direct access to the tester and removing the test engineer from the loop. As a result, product design cycles will decrease, with significantly easier transition to manufacturing. Reliability of the test system will be improved, as will the time to repair when the system does fail.

Tester productivity will improve by at least a factor of 2 to 3 times in both development and manufacturing and the cost of ownership will be about half that of today.

P • A • R • T • 3

VLSI FABRICATION

CHAPTER 11
VLSI PROCESSING

T.C. Smith
Bipolar Technology Center
Motorola Semiconductor
Mesa, Arizona

VLSI circuits are fabricated by means of sequential application of various manufacturing steps that combine to form the device structures on semiconductor substrates processed in wafer form. This chapter will discuss only *front-end* processing, which consists of those steps starting with bare wafers and ending with completed circuits ready for testing in wafer form. *Back-end* processing consists of the assembly steps from backlap and die separation through packaging and final testing. Since most of the volume production of VLSI circuits is executed in silicon, the processes reviewed will be the basic building blocks used in *silicon* wafer fabrication. Processes for GaAs substrates are similar in some ways but very different in others. The commonly used processes will be discussed in terms of the capabilities that exist in a typical volume production wafer fabrication area, not in terms of ''leading-edge'' submicrometer techniques that have been used in various research laboratories around the world. The basic principles of each process are reviewed with emphasis on the parameters of interest in different product technologies. The coverage of lithography will be very general, since this is reviewed separately in Chap. 13.

This chapter briefly treats the generic silicon processes separately; Chap. 12 shows examples of how they are incorporated in process flows of both bipolar and MOS technologies. The performance of the individual devices, and therefore the circuit and product performance, are intimately connected to the design-rule-related processing capabilities of the front-end wafer fabrication facility building the ICs. Process flows, the exact details of which are usually regarded as proprietary, are made up of different combinations of sequentially applied processing steps. These steps are arranged to build the desired high-density structures in the vertical and horizontal planes with good manufacturability. A great deal of synergism exists between the design and the fabrication functions. Process characterization and control are vital to realizing the targeted design goals. Feedback of electrical results on test structures to the process engineering group helps in pinpointing processing problems and leads to solutions in complex fabrication schemes.

The technology of wafer processing involves the application of physics, chemistry, and metallurgy as well as sophisticated analytical techniques in an interdisciplinary manner. Control of contamination and particulates throughout the wafer fabrication area is crucial in order to achieve reasonable yields in the production environment. Modeling of the various processes will be discussed along with trends in processing and manufacturing. The references given will lead the reader to more in-depth treatments of specialized topics.

11.1 STARTING MATERIAL AND EPITAXY

The silicon wafers or substrates in which circuits are fabricated are themselves the product of sophisticated processing. The raw material for the preparation of single-crystal silicon is electronic-grade silicon, a polycrystalline material of high purity. The predominant method of crystal growth is Czochralski (CZ) growth, which consists of crystalline solidification of atoms from the liquid phase. Single-crystal ingots are pulled from molten silicon contained in a fused silica crucible. Many properties of the wafer, such as oxygen concentration and dopant uniformity, depend on the precise parameters and conditions of the crystal-pulling operation.

The ingots are ground to a cylindrical shape of precisely controlled diameter, and one or more flats are ground along its length. The primary flat is positioned with relation to a specific crystal direction. The smaller secondary flat(s) are used to identify the orientation and conductivity type of the wafer. The silicon slices are sawed from the ingot with an exact crystallographic orientation. For example, wafers of ⟨100⟩ and ⟨111⟩ orientation are cut "on orientation" within 0.5°, while for epitaxial use, ⟨111⟩ material is cut "off orientation" by about 3° to minimize pattern shift effects. After sawing, the wafers are lapped within a specified thickness and flatness. Following an edge-rounding treatment, the wafers are etched and polished to their final dimensions and surface finish, then cleaned, visually inspected, sorted by flatness and resistivity, and packaged.

Most MOS devices are fabricated on ⟨100⟩ material because this orientation has the lowest density of surface states. Starting material for bipolar ICs may have either ⟨111⟩ or ⟨100⟩ orientations. Common wafer sizes used in production are typically 100, 125, or 150 mm in diameter. Wafer manufacturers are providing starting wafers with lightly doped epitaxial layers on heavily doped substrates for prevention of latchup in CMOS technologies as well as wafers with backside polysilicon or backside damage for extrinsic gettering. Table 11.1 shows a comparison of the material properties and requirements for VLSI wafers.[1] Standardized tests for the various properties have been established by the American Society for Testing and Materials (ASTM) and the Semiconductor Equipment and Materials Institute (SEMI).[2,3]

Crystalline defects that can adversely affect device performance are either originally present in the substrate or are introduced by subsequent process steps. A complete description of defects and the mechanisms for their formation and movement in the substrate with a discussion of related electrical effects can be found in Chap. 2 of Ref. 1. Device quality material can be preserved by careful management of thermal processes and the use of gettering techniques, which remove harmful impurities and defects from the active device regions. For example, in fabricating MOS dynamic RAMs (DRAMs), one can make use of an intrinsic (internal to the substrate) method of forming a denuded zone on the surface of the wafer by appropriate thermal cycling of material with controlled oxygen concentrations in the bulk of the wafer.[4]

The process of epitaxial growth is one in which a thin single-crystal layer is

TABLE 11.1 Comparison of Material Properties and Requirements for VLSI

Property	Czochralski	Float zone	Requirements for VLSI
Resistivity (phosphorus) n-type, $\Omega \cdot$ cm	1–50	1–300 and up	5–50 and up
Resistivity (antimony) n-type, $\Omega \cdot$ cm	0.005–10	—	0.001–0.02
Resistivity (boron) p-type, $\Omega \cdot$ cm	0.005–50	1–300	5–50 and up
Resistivity gradient (four-point probe), %	5–10	20	< 1
Minority carrier lifetime, μs	30–300	50–500	300–1000
Oxygen, ppma	5–25	Not detected	Uniform and controlled
Carbon, ppma	1–5	0.1–1	< 0.1
Dislocation (before processing), per cm^2	$\leqslant 500$	$\leqslant 500$	$\leqslant 1$
Diameter, mm	Up to 200	Up to 100	Up to 150
Slice bow, μm	$\leqslant 25$	$\leqslant 25$	< 5
Slice taper, μm	$\leqslant 15$	$\leqslant 15$	< 5
Surface flatness, μm	$\leqslant 5$	$\leqslant 5$	< 1
Heavy-metal impurities, ppba	$\leqslant 1$	$\leqslant 0.01$	< 0.001

Source: After Wolf and Tauber.[1]

deposited on the surface of a single-crystal substrate. Silicon epitaxy is a chemical vapor deposition (CVD) process in which a batch of wafers is placed on a heated susceptor in a reactor. Deposition takes place when process gases react at the wafer surface at high temperatures (900 to 1250°C). A typical film growth rate is about 1 μm/min. The process selection and equipment capabilities are quite varied, with four compounds available for silicon sources.[5] Hydrides of the common dopants (PH_3, B_2H_6, and AsH_3) are added to control the dopant level in the growing film. The usual range of doping concentration in the epi film is 10^{14} to 10^{17} atoms/cm^3. The thickness and doping concentration of the epitaxial layer can be accurately controlled and, unlike the underlying substrate, the layer can be made oxygen- and carbon-free. Another advantage of the epitaxy technique is that the doping profile in the device structure can be controlled in a manner that is not available by diffusion or ion implantation processes. For example, retrograde doping profiles, with higher doping at the bottom of the epi layer, can be achieved. A limitation of epitaxy is that the degree of crystal perfection of the deposited layer cannot be any better than that of the substrate. Other process-related defects such as slip or impurity precipitates from contamination must be minimized.

In bipolar device technology, an epitaxial layer is commonly used to provide a high-resistivity region above a low-resistivity buried layer, which is formed by a previous diffusion or implant and drive-in process. The heavily doped buried layer serves as a low-resistance collector contact, but an additional complication arises when epitaxial layers are grown over patterned buried layer regions. A step in the wafer surface may be produced in the preepitaxial processing in order to mark its location so that subsequent layers can be aligned in relation to the pattern of the buried layer. Unfortunately, the epi deposition produces a pattern shift and distortion for which the lithographic masks must compensate. These effects are a complex function of substrate orientation, temperature, pressure, film thickness, growth rate, and silicon source.[6]

11.2 THERMAL OXIDATION

Insulating oxide films grown by thermal oxidation from the substrate silicon have characteristics that are generally superior to deposited films. High-quality oxides formed by well-controlled processes are the cornerstone of modern IC technology. Thermally grown SiO_2 is used in thicknesses ranging from less than 10 up to 1000 nm for a variety of applications: gate and capacitor dielectrics or tunneling oxides in MOS devices, oxides for isolation of individual devices, masks against diffusion or implantation, pad oxides under nitride films, and oxides for surface passivation.

In the thermal oxidation process, the surface of the wafer is exposed to an oxidizing ambient of O_2 or H_2O at elevated temperatures, usually at an ambient pressure of one atmosphere. For both dry and wet processes, the mechanism and growth kinetics of oxides whose thickness is greater than 30 nm were well described by Deal and Grove.[7] They proposed that the oxidant (O_2 or H_2O) diffuses through the existing oxide film to the SiO_2-Si interface, where the oxidation reaction occurs. As the oxide layer grows in thickness, Si is consumed and the interface moves into the Si substrate. Due to the difference in the density of SiO_2 and Si, the thickness of Si consumed is equal to 44 percent of the final oxide thickness.

In the model of Deal and Grove, a balance of the flux of oxidizing species arriving at the wafer surface with the flux diffusing through the oxide and the flux consumed at the SiO_2-Si interface leads to the equation

$$X_0^2 + AX_0 = B(t + t_0) \tag{11.1}$$

where X_0 = oxide thickness
A, B = constants
t = oxidation time
t_0 = time displacement required to account for an initial oxide thickness, X_i

Two limiting cases are of interest: one is for long oxidation times or high temperatures; the other is for very short times or low temperatures. In the former case, Eq. (11.1) reduces to

$$X_0^2 = Bt \tag{11.2}$$

where B is called the parabolic rate constant. Since B is proportional to the diffusion constant D of the oxidizing species, the parabolic regime is diffusion controlled. In the latter case, Eq. (11.1) reduces to

$$X_0 = \left[\frac{B}{A}\right](t + t_0) \tag{11.3}$$

This is the linear growth law, with the reaction at the interface controlling the rate, and $[B/A]$ is called the linear rate constant. The rate constants have been determined for common substrate orientations as a function of temperature. Figure 11.1 shows oxide thickness versus time for oxidation of Si in wet oxygen (partial pressure of H_2O of 640 torr).[8]

In addition to temperature, oxidizing ambient, pressure, and crystallographic orientation of the Si surface, other factors affect the oxidation rate, including doping levels in the Si and crystallographic damage from implantation. Chlorine impurities, which are often intentionally added to the level of a few percent in the gas phase to improve oxide properties, also affect the growth rate.[9] HCl dry oxidation cycles are typically used for growing gate oxides with improved MOS device characteristics. Wet oxidations are conveniently carried out by the pyrogenic technique in which high-purity H_2 and O_2 react inside the high-temperature furnace to form water vapor. High-pressure oxidation, at pressures up to 25 atm, provides substantially higher oxide growth rates at low temperatures. This can be used to advantage because dopant profiles from previous steps will not be broadened by diffusion effects.

Following oxidation cycles, the films are evaluated by nondestructive measurements such as ellipsometry to determine the oxide thickness and index of refraction. In addition, electrical measurements of high-frequency *C-V* characteristics, using MOS capacitor structures, are routinely made to ensure proper process control. Measurements are also made under conditions of applied bias and thermal stress in order to monitor the oxidation process to ensure that it is free of detectable contamination by mobile ionic charge, mainly due to the presence of Na^+ or K^+. Other characteristics of the oxide, such as breakdown voltage, strongly depend on the processing parameters. A full discussion of the characterization of oxides is given in Ref. 10. The standardized terminology for oxide charges associated with thermally oxidized silicon has been reported by Deal.[11] The VLSI trend toward gate oxides thinner than 20 nm presents special problems

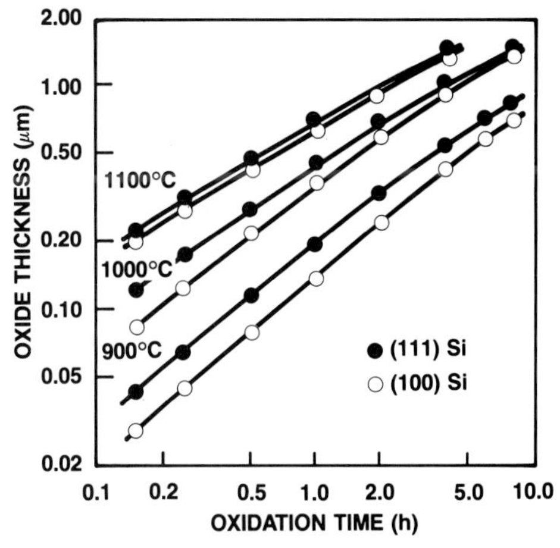

FIG. 11.1 Oxide thickness versus oxidation time for silicon in H_2O at 640 torr. (*After B. E. Deal.*[8] *Reprinted by permission of the publisher, The Electrochemical Society, Inc.*)

regarding compatibility with other processes. For example, wafer-charging phenomena in high-dose implantation or plasma etching and reactive ion etching (RIE) may lead to dielectric breakdown in MOS devices.[12]

11.3 LITHOGRAPHY

The patterns delineated on the wafers to fabricate the circuit elements and provide for interconnecting these components are created by the lithographic process. This topic is covered in greater detail in Chap. 13, but the basic steps involved will briefly be outlined here. The process consists of spin-coating photoresist on the wafer, prebaking, aligning the image on the mask or reticle to the pattern on the wafer from previous masking levels, then exposing, developing, and postbaking the pattern of the layer being printed. The subsequent step is usually an etching process or a selective doping step using ion implantation. The photoresist is then stripped prior to thermal treatments, depositions, and repatterning. This sequence is repeated for each of the layers involved, such as active area, gate definition, source and drain implants, contact, or interconnects.

Advances in circuit density have been paced by improvements in the capabilities of this patterning process, particularly with regard to achievable resolution and registration. Resolution could be defined as the minimum feature size that can be consistently exposed and developed within the required manufacturing tolerances of the product technology. Registration is a measure of the ability to align the masking pattern of the layer being printed to a previously printed pattern on the wafer. These parameters are affected by the characteristics of the photoresist system as well as by the performance of the optical system used for aligning and printing the image in the layer of photoresist.

For the majority of VLSI production applications, the lithographic process involves exposure by optical radiation (e.g., ultraviolet, UV) of positive photoresist, which is preferred because of its higher resolution capabilities. The basic requirement of photoresist is to reproduce precise patterns from the mask, which are used to selectively protect the substrate during etching or ion implantation. Optical photoresists are formulated of three basic components to optimize the tradeoffs of numerous properties that are necessary for any given layer. These consist of the matrix or resin, which serves as a binder, the sensitizer or photoactive compound, and the solvent, which keeps the resist liquid. In the case of positive resist, photoscission of sensitizer molecular chains makes the exposed regions soluble in the developer so that the areas of remaining resist correspond to the opaque features on the mask.

The materials properties of the resist include (1) mechanical and chemical properties, such as flow characteristics, thickness, adhesion and thermal stability, (2) the optical characteristics, principally the photosensitivity, contrast, and resolution, and (3) processing properties, such as process latitude, metals content, and safety considerations. Different applications require more emphasis on some properties than on others, as when etch resistance is critical but resolution is not. In some instances, reduced sensitivity to reflection from an underlying metal film over surface topography may be the most desirable property.

The pattern transfer from the mask to the wafer, which results from exposure of the photoresist, is strongly affected by the performance of the optical system of the equipment used to project the image, which is called the aligner or printer. The resolution and registration capabilities of the alignment tool are crucial to maintaining the required critical dimensions on the features being printed. The "process bias," or total offset from the "as-drawn" mask dimension to the final etched linewidth of a feature on the wafer must be characterized along with the registration actually achieved. For any photoresist-aligner system, this information is vital in determining the design rules for the overall process. Overlay accuracy, repeatability, and machine-to-machine matching are often the limiting factors in production. This is increasingly important with the trend toward increasing masking steps.

Projection printing is the principal method used for VLSI, with refracting lenses or reflecting mirrors focusing the image from the mask to the wafer. In these systems, the mask and wafer are separated by large distances. High-power mercury-vapor lamps are used as the light source. With high-quality optics, the imaging characteristics are limited not by lens aberrations but by diffraction effects. The three types of projection systems used are (1) 1:1 scanning aligners, (2) 10× or 5× reduction step-and-repeat, and (3) 1× step-and-repeat. These can be employed in a mix-and-match approach, using higher-throughput equipment for less critical alignments. Automatic alignment systems, which pick up alignment targets, are replacing human operators. One advantage of steppers is that only a portion of the wafer is exposed, with the image from a reticle (as a cluster of die) printed in each field, so wafer size and distortion are not limiting factors. However, wafer throughput generally decreases, and the sensitivity to reticle defectivity increases.

Many of the same considerations for producing high-resolution images pertain to the fabrication process for the masks and reticles used in wafer processing. The predominant method of generating these directly from computer data is by the use of electron beam writing systems. Photomasks and reticles are fabricated of glass or quartz with low thermal expansion coefficients, and the patterns are etched in a thin (100-nm) coating of chrome.

Trends in photolithography include the increased use of pellicles (transparent membranes mounted on a frame to protect the mask or reticle surface from contamination by airborne particulates), multilayer resist schemes, and the use of shorter-wavelength (deep UV) light for exposing the photoresist. The use of particulate control methods for reduction of printed defects is absolutely necessary in this processing area.

11.4 ETCHING

Following the lithography step, the pattern in the photoresist can be permanently transferred to the wafer by an etching process that selectively removes various layers. With the stringent VLSI requirements for dimensional accuracy, dry etching techniques have become the method of choice because of their superior ability to control critical dimensions reproducibly. In addition, for a well-characterized process, the bias or offset from mask dimension to the etched feature on the wafer can be predicted. Wet etching techniques are generally not suitable, since the etching is isotropic (i.e., the etching proceeds at the same rate in all directions), and the pattern in the photoresist is undercut. Dry etchers, so called because they make use of low-pressure gas discharges or plasmas, provide highly directional etching. The ratio of the etch rates of different materials is known as the selectivity of an etch process. Wet etches can be highly selective, and are used for removing thin films or blanket layers of various materials. In dry etching systems, anisotropic etching is possible, since vertical etch rates can be made high relative to etch rates in the horizontal direction.

Two of the most important characteristics of the etch process are the selectivity of the layer being etched relative to the mask and its selectivity with respect to the underlying layer on which the etching must stop. For example, when contacts are etched in SiO_2, high selectivity with respect to the underlying doped silicon substrate is required in order to avoid contact resistance problems. Because of nonuniformities in etch rate and layer thickness over all parts of all wafers in the etcher, there is a practical need for overetching (i.e., etching beyond the theoretical "endpoint" when the surface is cleared). Anisotropic etching of conformal layers over stepped topography on the surface also makes overetching necessary to remove residues along edges. The required selectivity must exceed the minimum value determined from the processing parameters. For various amounts of overetching, Fig. 11.2 shows the needed selectivity with respect to the substrate as a function of the ratio of the film thickness to the allowable substrate thickness loss.[13] In this figure, Φ_f is the fractional variation in etch rate to be expected for the process, δ is the fractional variation in film thickness, and Δ is the

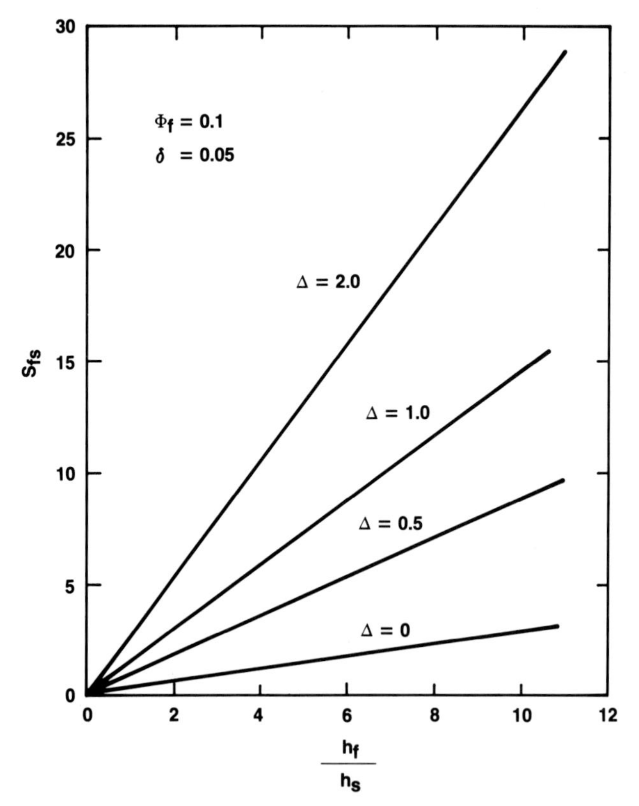

FIG. 11.2 Selectivity required as function of ratio of film thickness to substrate removed for various amounts of overetching. (*After C. J. Mogab.*[13] *Copyright 1983. Reprinted by permission of S. M. Sze.*)

fractional overetch time anticipated (e.g., $\Delta = 1.0$ corresponds to 100 percent overetch).

The low-pressure gas discharge or plasma is the source of the active species that produce etching or serve as catalysts to the etching process. Ions, atoms, and radicals are produced in the plasma at a rate determined by the discharge parameters, such as gas flow rate, pressure, excitation frequency, and power density. The goal of manipulating the complex physical and chemical mechanisms operative in the etching system is to produce proper edge profile control and reproducible critical dimensions with adequate selectivity. The feed gases themselves, such as CF_4, are virtually unreactive, but atomic fluorine and other radicals and ions generated in a discharge react spontaneously with Si-containing films to form volatile SiF_4. Anisotropy results when etching is caused by an ion-assisted reaction, from physical bombardment of the horizontal wafer surfaces by energetic ions impinging at normal incidence. Details of the various mechanisms can be found in Ref. 13.

The various films that must be patterned and etched in typical process flows each require a specific chemistry or choice of reactive gases. Several alternatives are often suitable for a given etching system. The effect of additions of O_2 to a CF_4 plasma upon the etch rates of Si and SiO_2 is shown in Fig. 11.3. Sometimes the process is tailored to produce a tapered edge profile in order to lessen step coverage problems for the next layer to be deposited. Table 11.2 summarizes typical etch gases used for various IC materials.[14]

Parallel plate plasma etching systems operate with symmetric electrodes at relatively high pressures (10^{-1} to 10^{+1} torr), with wafers placed on the grounded electrode. However, high plasma potentials can be developed, so that nonchemical mechanisms can come into play. With RIE the wafers are placed on

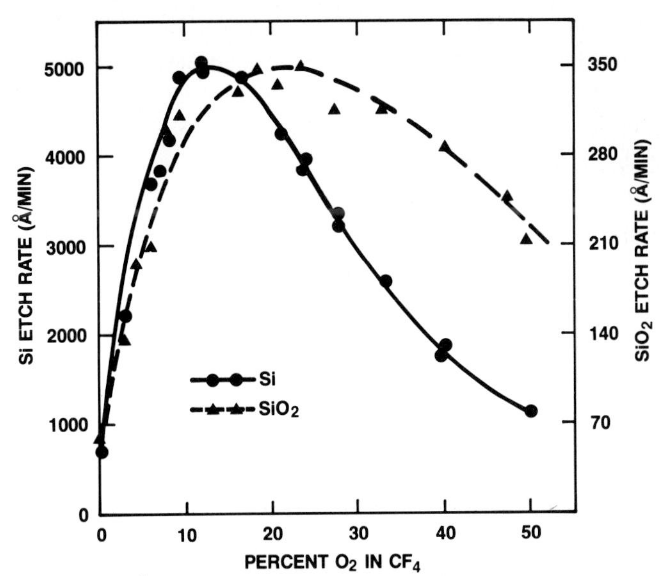

FIG. 11.3 Effect of the addition of O_2 to a CF_4 plasma upon etch rates of Si and SiO_2. (*After C. J. Mogab.*[13] *Copyright 1983. Reprinted by permission of S. M. Sze.*)

TABLE 11.2 Etch Gases Used for Various Integrated Circuit Materials

Material	Gases
Silicon (including polysilicon)	CF_4, CF_4-O_2, CF_3Cl, SF_6-Cl, $Cl_2 + H_2$, C_2ClF_5-O_2, SF_6-O_2, SiF_4-O_2, NF_3, ClF_3, CCl_3F_5, C_2ClF_5-SF_6
SiO_2	CF_4-H_2, C_2F_6, C_3F_8, CHF_3
Si_3N_4	CF_4-O_2, CF_4-H_2, C_2F_6, C_3F_8
Organic solids	O_2, $O_2 + CF_4$, $O_2 + SF_6$
Aluminum	BCl_3, CCl_4, $SiCl_4$, BCl_3-Cl_2, CCl_4-Cl_2, $SiCl_4$-Cl_2
W, WSi_2, Mo	CF_4, CF_4-O_2, C_2F_6, SF_6
$TaSi_2$	SF_6-Cl_2, CF_4-Cl_2
Au	$C_2Cl_2F_4$, Cl_2

Source: After Wolf and Tauber.[14]

the powered cathode, whose area is much less than the area of the grounded anode. This results in a high potential difference between the plasma and the wafer surface, leading to more efficient sputter etching. In addition, RIE systems operate with more complex vacuum pumps at lower pressures, ranging from about 10^{-3} to 10^{-1} torr. Various endpoint detection schemes are used wherever possible for improved run-to-run reproducibility.

The dominant systems used for plasma-assisted etching or RIE are constructed in either of two configurations: parallel electrode (planar) reactors or cylindrical batch (hexode) reactors. Components common to both of these include electrodes arranged inside a chamber evacuated to low pressures, pumping systems for maintaining proper vacuum levels, power supplies to maintain a plasma, gas inlet systems for reactant gases, and systems for controlling and monitoring process parameters, such as operating pressure and gas flow. Strong emphasis is placed on safety considerations related to the handling and pumping of hazardous or corrosive gases, some of which are by-products of the etching process.

11.5 DOPING AND DIFFUSION

Doping of silicon to alter the conductivity type in different regions of device structures has traditionally been accomplished by diffusion at high temperatures (900 to 1200°C) from chemical vapors or from a doped oxide layer on the surface of the wafers. Diffusion is the process by which impurity atoms move in the crystal lattice in the presence of a chemical gradient. With the requirement for shallow junction depths, ion implantation is the preferred doping method used in VLSI device fabrication. It also offers superior doping precision, repeatability, and uniformity. The implantation technique is discussed further in the next section. However, high-temperature treatments are necessary for annealing and activating implanted atoms. In many instances, significant diffusion or redistribution of implanted dopants may occur. Diffusion furnaces are also configured for oxidation and low-pressure chemical vapor deposition (LPCVD) processes, which are also discussed in this chapter.

The simplest description of the diffusion of dopants in the lattice, which is valid at low concentrations, is given by continuum theory with Fick's one-dimensional diffusion equation:

$$\frac{\partial C(x, t)}{\partial t} = D \frac{\partial^2 C(x, t)}{\partial x^2} \tag{11.4}$$

where C = concentration of dopant, which is a function of x and t only, atoms/cm^3
x = coordinate in direction of dopant flow
t = diffusion time
D = diffusion coefficient or diffusivity, cm^2/s or μm^2/h.

Here the diffusivity, which is a strong function of temperature, is assumed to be independent of concentration.

Two cases in thermal diffusion are commonly of interest: one in which the dopant flux at the surface is maintained constant during the entire diffusion time, and another in which the total quantity of dopant is held constant. Two-step diffusion processes make use of an initial predeposition diffusion with constant source for a short time at low temperatures followed by a drive-in phase with the dopant supply shut off.

The solution of Fick's diffusion equation with appropriate boundary conditions for the first case above is

$$C(x, t) = C_s \operatorname{erfc}\left(\frac{x}{2\sqrt{Dt}}\right) \tag{11.5}$$

where C_s is the surface concentration, erfc is the complementary error function, and $\sqrt{Dt}$ is the diffusion length. As diffusion time increases, the dopant penetrates deeper into the silicon, but the surface concentration remains constant. In the second case, where a thin layer of dopant is present at the surface either from a chemical predeposition or ion implantation, the quantity of dopant is fixed at Q_0. Here, the solution to the diffusion equation is

$$C(x, t) = C_s \exp\left(\frac{-x^2}{4Dt}\right) \tag{11.6}$$

This has the form of a Gaussian distribution where the surface concentration, $C_s = C(0, t)$ is given by

$$C_s = \frac{Q_0}{\sqrt{\pi Dt}} \tag{11.7}$$

In this drive-in diffusion, the dopant penetrates deeper into the substrate as time increases and the surface concentration decreases. Figure 11.4 is a graph of normalized concentration (C/C_s) versus normalized distance ($x/2\sqrt{Dt}$) for the complementary error function (erfc) and the Gaussian distributions.[15]

The junction depth X_j defined to be the distance from the surface at which the concentration of the dopant is equal to that of the substrate, can be calculated from the analytical expressions for the profiles in both cases. If C_B is the background or substrate concentration, the junction depth for the case of constant C_s is

$$X_j = \frac{2\sqrt{Dt}}{\operatorname{erfc}(C_B/C_s)} \tag{11.8}$$

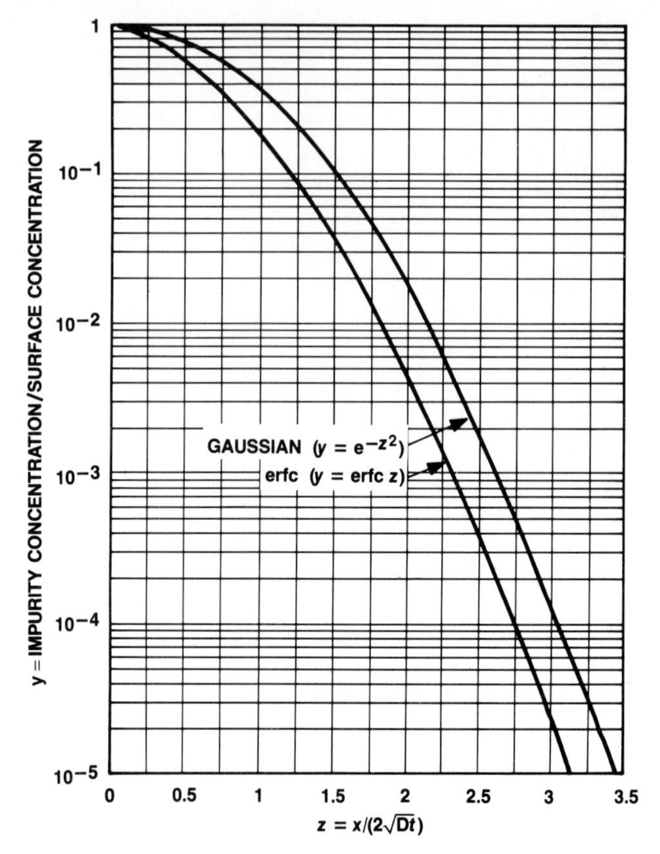

FIG. 11.4 Normalized $C(x)/C_s$ versus $x/2\sqrt{Dt}$ for gaussian and erfc profiles. (*After S. K. Ghandhi.*[15] *Copyright 1983. Reprinted by permission of John Wiley & Sons, Inc.*)

For constant Q_0 the junction is located at

$$X_j = 2\sqrt{Dt}\ \sqrt{\ln\left(\frac{C_s}{C_B}\right)} \tag{11.9}$$

This simple theory gives reasonably good estimates of the dopant profile when the density of dislocations is low and the doping concentration is less than the intrinsic carrier concentration at the diffusion temperature. At higher concentrations, D is a function of dopant concentration. In addition to temperature and concentration gradients, point defects play an important part in diffusion, and atomistic models have been developed to explain the wide variety of phenomena observed under different conditions such as oxidizing ambients. The mechanisms involve substitutional or interstitial diffusion via interactions with point defects and vacancies of different charge states. Reference 16 and its numerous references give more details of studies of these mechanisms. The damage due to ion implantation introduces vacancies and interstitials that can modify subsequent diffusion.

The temperature dependence of the diffusivity can usually be written in the form

$$D(T) = D_0 \exp\left(\frac{-E_a}{kT}\right) \tag{11.10}$$

where D_0 = frequency factor, cm^2/s
E_a = activation energy, eV
T = temperature, K
k = Boltzmann constant, 8.62×10^{-5} eV/K

Figure 11.5 graphically shows the temperature dependence of the intrinsic diffusivities of the commonly used dopants for silicon: boron, phosphorus, arsenic, and antimony.[16] Arsenic is the preferred dopant for n^+ source-drain regions of *n*-channel MOSFETS because its lower diffusivity produces shallow junctions, and it is used for bipolar emitters because it allows better control of narrow basewidths.

In device fabrication, the use of SiO_2 layers to mask diffusion is a cornerstone of silicon processing technology. Diffusivities of the common dopants are several orders of magnitude lower in SiO_2. Openings etched in the surface permit the dif-

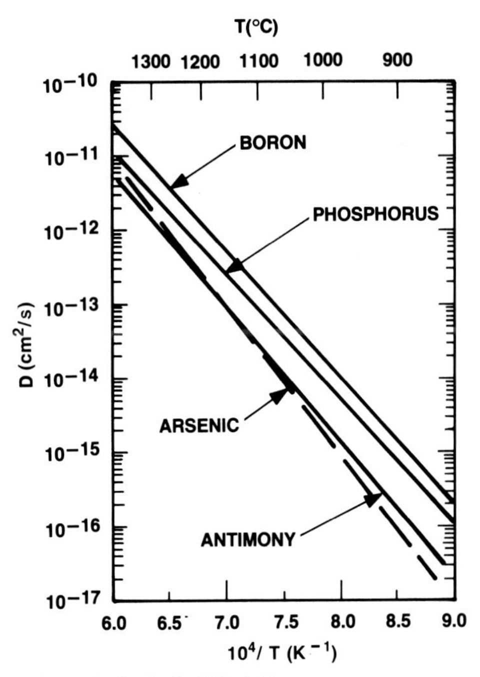

FIG. 11.5 Intrinsic diffusivities versus temperature for B, P, As, and Sb. (*After R. B. Fair.*[16] *Reprinted by permission of North-Holland Physics Publishing and the author.*)

fusion or implant to be selective. For VLSI applications, where diffusion times are short and temperatures are usually kept below 1000°C, oxide thicknesses of about 0.4 μm are sufficient to mask against boron or phosphorus diffusions. Diffusion in polycrystalline is about 10 to 100 times faster than in single-crystal silicon because grain boundary diffusion is the predominant mechanism.

Contamination control is especially important in diffusion and oxidation furnaces as well as the cleaning steps that precede the furnace runs. Constant surveillance of all processes and routine testing for mobile ionic contamination, as discussed under thermal oxidation, is imperative for maintaining furnace qualification. In-line evaluation techniques for diffused and implanted layers, which can be made on test wafers immediately following a diffusion process, consist of sheet resistance measurement by means of the four-point probe technique and junction depth measurement using the bevel-and-stain method. Analytical methods for determining dopant profiles off-line are further discussed in the section on process characterization.

11.6 ION IMPLANTATION

In VLSI processing, ion implantation is the preferred method of selectively doping the substrate, since it is a very versatile technique and provides superior control of the dopant profile. It is a low-temperature process and is compatible with other processes, such as photoresist masking. In ion implantation, the dopant atoms are introduced into an "ion source," in the vapor phase or as a gaseous compound, where electron bombardment forms a plasma consisting of electrons, positive ions, and neutrals. The positive ions are "extracted" from the source and accelerated to high energies, generally in the range of 10 to 200 keV. The output of the ion source is formed into a beam that is acted upon by electric and magnetic fields in a high-vacuum environment. The energetic ions undergo mass analysis so that only the pure dopant, such as $^{11}B^+$ (mass 11 boron, singly charged), is directed at the wafer. The beam of ions is either scanned rapidly over the surface of the wafer in an *xy* raster pattern by the action of time-varying electric fields, or the wafer is mechanically scanned in front of the stationary beam. Hybrid scanning schemes are also employed.

The net charge delivered to the wafer, determined by integrating the beam current, is a precise measure of the dose (the number of dopant atoms per unit area) incident upon the surface of the substrate. The range of doses used in VLSI processes extends from $10^{11}/cm^2$ to $10^{16}/cm^2$. In addition to this wide dynamic range, implantation provides considerable flexibility for tailoring the dopant profile by varying the ion energy. The resultant superior control of shallow doping, the high uniformity and repeatability, along with the ability to use photoresist as a masking material are the basic reasons why implantation has been widely accepted.

The profile of dopant atoms in the substrate is determined by the net effect of the nuclear and electronic energy loss processes involved in the ion-substrate interactions that take place as the ion comes to rest in the silicon. In the simplest approximation, the concentration of the dopant is given by the gaussian distribution

$$C(x) = \frac{\phi}{\sqrt{2\pi}\,\Delta R_p} \exp\left[\frac{-(x - R_p)^2}{2(\Delta R_p)^2}\right] \qquad (11.11)$$

where ϕ = dose
x = distance from the surface
R_p = mean projected range
ΔR_p = straggle or standard deviation of the distribution

In actual fact, the distributions are considerably more complex, and range statistics with higher-order moments (skewness and kurtosis), such as the Pearson-IV distribution have been calculated.[17] Figure 11.6 shows values of R_p for boron, phosphorus, and arsenic as a function of energy for amorphous silicon and thermal silicon dioxide targets.[18] For energies below 400 keV, the peak concentration of the implanted dopant is less than 1 μm deep. The depth of penetration in certain crystallographic axial or planar directions can be greater than the value of R_p given in the table due to channeling phenomena, but these undesirable effects can be counteracted by certain precautions, such as the use of screening oxide layers and control of wafer orientation with respect to the incident ion beam.[19] Figure 11.7 shows vertical straggle and transverse straggle (i.e., the straggle in a direction perpendicular to the direction of the incident ion), for B, P, and As ions in Si. In VLSI processing, the effects of lateral ion straggle and channeling are limiting factors in electrical channel length and vertical junction depth.

After the implantation process itself, the near-surface silicon substrate is in a highly disordered state as a result of nuclear interactions that displace the host atoms of the crystal. Light ions such as B produce a profile of displaced atoms with a buried peak concentration, but heavier ions such as P or As create more damage close to the surface. In some cases, an amorphous layer is formed when the concentration of displaced atoms is equal to the number of Si atoms per unit volume. In any case, a thermal "annealing" cycle is required to restore crystallinity and move the implanted dopant atoms into electrically active sites. The ef-

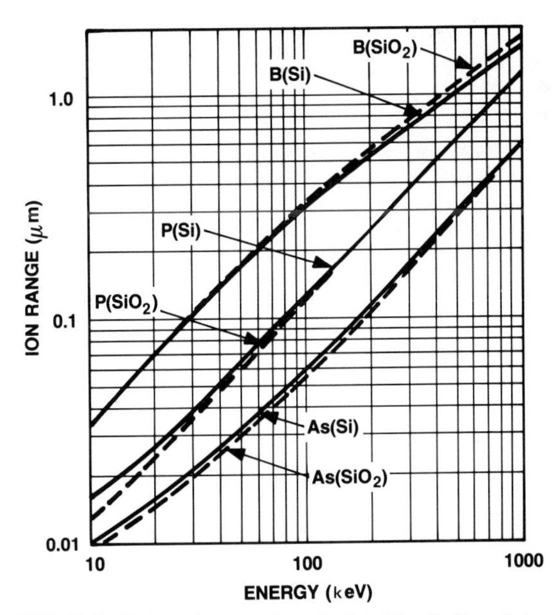

FIG. 11.6 Projected range, R_p, calculated for B, P, and As versus ion energy. (*After T. E. Seidel.*[18] *Copyright 1983. Reprinted by permission of S. M. Sze.*)

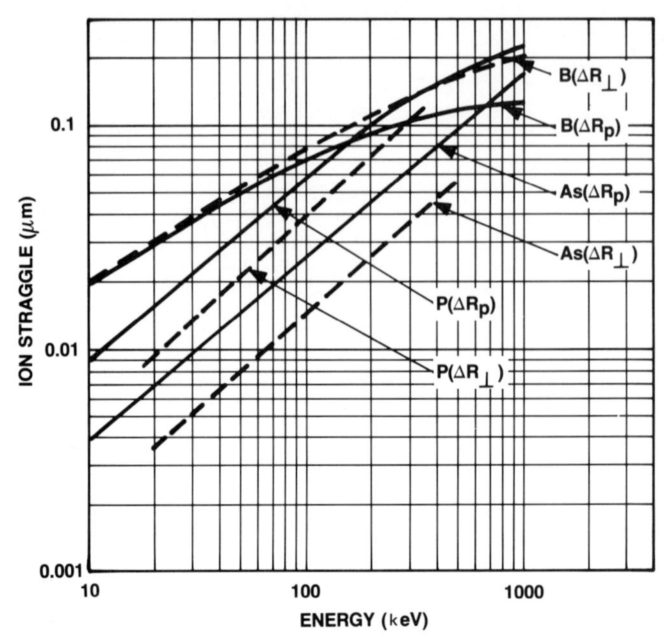

FIG. 11.7 Calculated vertical straggle ΔR_p and lateral straggle ΔR_l for the B, P, and As ions in Si. (*After T. E. Seidel.*[18] *Copyright 1983. Reprinted by permission of S. M. Sze.*)

fects of damage production are very dependent upon ion species, energy, dose, and wafer temperature during the implant, and subsequent annealing phenomena are influenced by these factors as well. Some implant profiles are significantly modified by diffusion, while others remain very close to their "as implanted" state. In any process flow, the various high-temperature processes are tailored to ensure proper activation and redistribution of dopants as required by the device considerations. Reference 20 gives an overview of the ion implantation technique, related effects of damage and annealing, and details of the equipment used to carry out this process.

11.7 DEPOSITION OF DIELECTRIC AND POLYSILICON THIN FILMS

Deposited thin films are used in fabricating ICs for providing interconnections, insulation between conducting layers, diffusion or ion implantation masks, and for passivation coating over finished devices. In addition to oxide films discussed above, which are thermally grown at high temperatures from the silicon substrate, other dielectric or polysilicon thin films are deposited in various steps at low to moderate processing temperatures. Spin-on glass, a liquid solution applied to wafers in a manner similar to photoresist, is baked at low temperatures and then densified or flowed at high temperatures, and is used as a dopant source or planarization film. Polyimide compounds are applied by spin-on techniques and baked to produce a planarized insulator between metal levels. Polyimide is also used as a final passivation layer and as an alpha particle protection layer for DRAM devices.

The most widely used deposited thin films are silicon dioxide, doped oxides, silicon nitride, and polycrystalline silicon. These films are usually applied by the CVD process, in which chemicals in the vapor phase react to form a solid film on the surface of the wafer. The energy for the reaction is supplied thermally, photochemically, or from a plasma discharge. With high wafer throughput, CVD techniques can provide a variety of high-purity films having well-controlled uniformity, composition, and structure. Good step coverage and excellent adhesion can be obtained along with good electrical properties.

CVD reactors vary in their configuration, but usually contain a reaction chamber, wafer carriers, gas inlet systems with mass flow controllers, a method of heating the wafers on which the film is being deposited, and temperature controllers. The reactor design may include a vacuum pumping system, but in any case incorporates the proper means of handling and disposing of corrosive and hazardous gases that may be reactants introduced or possibly by-products exhausted from the chamber. Safety practices associated with maintaining these processes and equipment have been published.[21] The types of reactors include atmospheric pressure (APCVD) systems, primarily used for low-temperature oxide deposition with continuous processing, LPCVD systems in a horizontal tube, hot wall configuration for depositing polysilicon, silicon nitride, and doped and undoped silicon dioxide, and plasma-enhanced (PECVD) systems. These PECVD systems use a radio-frequency (rf)-induced glow discharge and offer the advantage of lower deposition temperatures than the other CVD methods, but one potential disadvantage is damage due to ion bombardment.[22] Table 11.3 gives a comparison of the different CVD methods.[23]

To deposit a thin film for a particular application, one needs in addition to a clean wafer surface, to control process variables such as temperature, temperature gradients, total pressure, partial pressures of reactants, pumping speed, and wafer spacing. In addition to these, in plasma systems the reactor geometry, electrode materials and spacing, rf power, and frequency also affect the characteristics of the resulting films. Some of these process variables interact to produce various results in the output characteristics. A more complete discussion of process variables and film properties is given in Refs. 23 and 24. The ideal process should reproducibly yield films with well-controlled thickness and uniformity, composition, density, film stress, step coverage, electrical properties, etch rate, and other specified properties dictated by the requirements of the device structure. Measurements of the refractive index and the etch rate in buffered HF solution are often used as process control parameters to verify the quality of deposited films.

Silicon dioxide can be deposited by several different methods. The specific use or occurrence in the process flow dictates which method is used. For reflow glass, the SiO_2 is formed by reacting silane (SiH_4), oxygen, and phosphine (PH_3) during a low-temperature deposition to form phosphosilicate glass (PSG), a binary glass consisting of SiO_2 and P_2O_5. At high temperatures of 1000 to 1100°C, with 6 to 8 wt % phosphorus, the glass becomes viscous and surface tension forces produce flow over steps. By the addition of a boron dopant (B_2H_6) to form a ternary component oxide system, borophosphosilicate glass (BPSG), the reflow temperature may be reduced to as low as 700°C in a nitrogen ambient. SiO_2 can be deposited in an LPCVD reactor at about 700°C by decomposing tetraethoxysilane (TEOS), vaporized from a liquid source. This method can be

TABLE 11.3 Comparison of Different Deposition Methods

Deposition properties	Methods: Atmospheric-pressure CVD	Low-temperature LPCVD	Medium-temperature LPCVD	Plasma-assisted CVD
Temperature (°C)	300–500	300–500	500–900	100–350
Materials	SiO_2, P-glass	SiO_2, P-glass	Poly-Si, SiO_2, P-glass, Si_3N_4	SiN, SiO_2
Uses	Passivation, insulation	Passivation, insulation	Gate metal, insulation, passivation	Passivation, insulation
Throughput	High	High	High	Low
Step coverage	Poor	Poor	Conformal	Poor
Particles	Many	Few	Few	Many
Film properties	Good	Good	Excellent	Poor
Low temperature	Yes	Yes	No	Yes

Source: After Adams.[23]

used to deposit oxides over polysilicon, but not over aluminum because of the high temperature required. For oxide deposition over aluminum, where the process temperature must stay below 450°C, the APCVD reaction is generally used for its high deposition rate.

Silicon nitride (Si_3N_4) thin films are used as a mask for the selective oxidation of silicon, to produce planar device structures, and as a passivating layer, because it is an almost impervious barrier to diffusion of sodium and moisture. If the process flow allows temperatures of 700 to 800°C, an LPCVD process of reacting dichlorosilane ($SiCl_2H_2$) and ammonia (NH_3) gives very good uniformity at high wafer throughputs. For nitride passivation over metals that require low process temperatures, a PECVD technique that reacts silane (SiH_4) and ammonia (NH_3) at 200 to 400°C would be employed. In this latter case, the film tends to be nonstoichiometric and contains 10 to 30 at. % H, and the stress in the film must be carefully controlled.

Polysilicon (poly-Si or poly) is an important deposited thin film, especially in MOS processing, where it is used as a gate electrode having excellent interface properties with gate oxides and compatibility with subsequent high-temperature processing and allows for self-aligned gate processes. It is also used as a diffusion source for shallow junction devices, as high-value resistors, and as a means of filling trenches for dielectric isolation. Poly is generally deposited by thermal decomposition of silane (SiH_4) at a temperature of about 600°C. The LPCVD technique is used, mainly because of the ability of the process to economically run batches of 100 to 200 wafers with good uniformity and repeatability.

The structure and properties of LPCVD poly films are different from those of bulk single-crystal silicon.[25] As deposited, these poly films may be amorphous or polycrystalline (i.e., made up of small single crystals, about 100 nm on a side and separated by grain boundaries). After subsequent treatments at elevated temperatures, the films show a polycrystalline structure, usually exhibiting an increase in grain size. Polysilicon can be doped in situ by adding dopant gases during deposition or by diffusion or ion implantation after deposition. The rearrangement of grain boundaries and the different diffusion coefficients that affect the concentration of dopant atoms in the crystallites and on the grain boundaries have a strong effect upon the electrical properties of the poly, particularly its resistivity. The work function of the poly, which is important in determining the value of the threshold voltage of the MOS devices, depends upon the doping level in the film. The etch rate of poly in a plasma and the thermal oxidation rate also depend on the dopant concentration. Reference 23 gives more detail on these related factors.

In addition to poly and dielectric deposited films, CVD techniques are being applied to the task of forming suitable low-resistivity conductors or via filling plugs using refractory materials such as tungsten, in both selective and nonselective modes, and tungsten silicide. For VLSI applications, the emphasis on reducing the process temperatures and the need to control the generation of particles by the CVD process itself has led to the development of cold wall reactors designed for specific films.[26]

11.8 INTERCONNECTS

The applications for interconnects include ohmic contacts of metal to various regions of the silicon substrate, metal conductors connecting circuit elements, Schottky barrier diodes used as clamps in bipolar circuits, and gate interconnects in MOS circuits. The trends to achieve higher packing density and smaller chip size with increased circuit complexity will require the use of multiple levels of metallization in all product areas. Many MOS circuit designs already employ multiple layers of interconnects using polysilicon, which offers the advantage of tolerating high-temperature processing. The need to reduce *RC* time constants for high-performance MOS circuits has led to the use of low-resistivity refractory metals and their silicides in place of conventional doped polysilicon.

The range of materials used for interconnecting functional devices has expanded in the VLSI era, and the number of techniques for depositing them has increased. Aluminum alloys are the primary metallization used in both MOS and bipolar processes, and doped polysilicon, which has been commonplace in MOS devices, is now being utilized in fabricating bipolar devices. Physical vapor deposition (PVD) systems provide considerable flexibility in varying the film properties, but CVD techniques, which tend to produce conformal coatings, are used for some silicides. Planarization and via filling techniques are especially needed for multilevel metal interconnect schemes with two to four levels of metal and one to three dielectric interlayers. Barrier films are often interposed between two materials to inhibit the mixing of those materials by diffusion in the processes that follow. For instance, films of refractories such as Ti-W (with about 10 percent Ti by weight) may be used as a diffusion barrier between $PtSi_2$ and Al alloy films.

The properties of the thin films that must be controlled include mechanical and electrical characteristics and parameters related to reliability issues. Among the various alloys and related interlayer dielectric structures in use, there are tradeoffs involved with the deposition techniques and associated processes. For

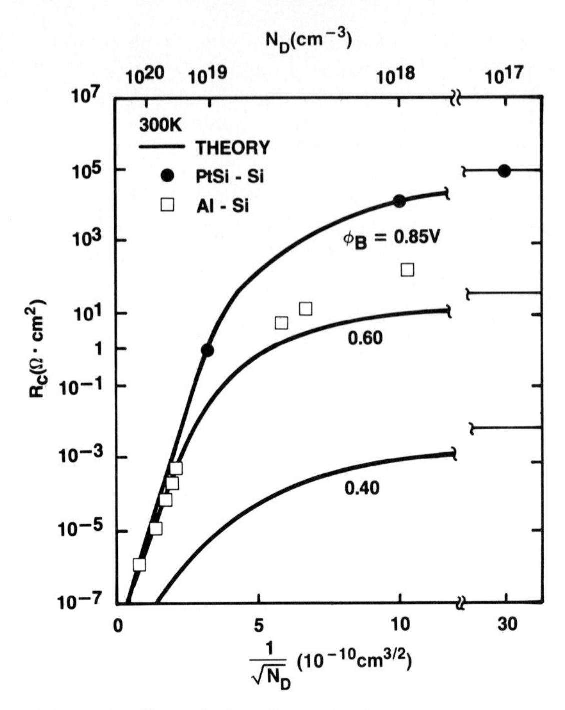

FIG. 11.8 Theoretical and measured values of specific contact resistance as a function of donor concentration and barrier height. (*After D. B. Fraser.*[27] *Copyright 1983. Reprinted by permission of S. M. Sze.*)

example, the addition of a barrier film preserves junction integrity but increases the effective resistance of the metal contact. The contact resistance is strongly affected by the surface concentration of the dopant in the substrate, as seen in Fig. 11.8, where N_D is the donor concentration and ϕ_B is the Schottky barrier height.[27] The contact resistance is also sensitive to the effectiveness of the surface cleaning treatment prior to deposition and to any impurities incorporated in the film.

Film characteristics to be monitored and controlled include thickness, resistivity, reflectivity, adhesion, internal stress, hardness, contact resistance, and step coverage. Uniformity of all of these parameters over large wafer diameters is crucial. Some of the metal characteristics, such as hardness and corrosion resistance, are related to the requirements and constraints of the assembly and packaging operations. Film thicknesses for different layers and applications vary from tens of nanometers for Schottky contacts to several micrometers in multilevel metal structures. Reflectivity and film composition must be controlled to achieve reproducible results in the patterning and etching of interconnect layers. The formation of hillocks or protrusions when subjected to subsequent thermal cycling must be suppressed. Scaling down the width and thickness of interconnecting metal reduces the cross-sectional area, thereby increasing the current per unit area. Electromigration, the mass transport of metal under the influence of dc electric fields, is a failure mechanism of concern at high current densities (up to 10^6 A/cm^2). For reliability reasons, the process must produce films having adequate resistance to corrosion and electromigration failure under normal operating conditions.

With the trends toward smaller geometries, the issues of step coverage and contact resistance are of increasing concern. The ratio of sidewall film thickness to flat surface thickness is a measure of step coverage. In contact windows or vias, this parameter depends strongly upon the aspect ratio (depth-width) of the opening in the dielectric layer. In sputtering systems, a bias may be applied to the substrate to promote redeposition on the sidewalls, or additional substrate heating may be used to improve the surface mobility of the deposited material. The pre-metal-cleaning operation, which usually involves the use of buffered HF solutions, must be carefully controlled to remove thin residual oxides from single-crystal Si and polycrystalline contacts.

The techniques used to deposit metals and silicides include PVD, where material is transferred from a source or target to the substrate in a vacuum system, and CVD, where a film is formed on the substrate by the reaction of chemicals in the vapor phase. The methods of depositing aluminum and its alloys have progressed from evaporation, in which the evaporated atoms travel in straight lines from the source to the substrate, to magnetron sputtering systems, where the directionality of the sputtered metal vapor is more random. The process of magnetron sputtering consists of bombarding a large target with ions from a magnetically confined plasma in argon gas. At a pressure of about 1 Pa (7.5 mT), collisions between the sputtered metal atoms and the argon gas molecules in the plasma result in more randomness of the direction of the arriving metal atoms in the growing film. With close-coupled targets and high deposition rates (10 to 20 nm/s), significant substrate heating can be achieved to further improve step coverage. Sputtering permits flexibility in producing alloy films with silicon added to avoid junction spiking or copper added to ensure improved resistance to electromigration failure. In concentrations of a few weight percent, these additives will increase the resistivity of the alloy film from about 2.9 $\mu\Omega \cdot$ cm (characteristic of pure Al), to about 3.4 $\mu\Omega \cdot$ cm, depending upon the exact composition and the method of deposition.

Refractory metal silicides have been successfully used in MOS device fabrication as low-resistance gate interconnects. Since the minimum sheet resistance of doped polysilicon is about 15 Ω per square, materials with lower resistivities are required to reduce *RC* time constants for more aggressive design rules. Aluminum is not a suitable replacement because, with its low melting point, all subsequent processes would have to be held to less than 500°C. Refractory metals such as W, Ta, or Mo have high melting temperatures, but their oxides are typically of poor quality. The predominant structure for replacing polysilicon as the gate electrode has been "polycide," a silicide layer on top of polysilicon, which retains the known work function of doped polysilicon and the reliable poly-Si-SiO_2 interface yet has a resistivity which is a factor of 10 lower. Methods of forming the silicide include direct metallurgical reaction with the pure metal, coevaporation or cosputtering from independent Si and metal sources, sputtering from a composite target, and CVD techniques. Considerations of the desired properties of silicides for use in IC fabrication, particularly with regard to process compatibility and device requirements, narrow down the list of candidates considerably. Murarka has summarized much of the published literature on this topic.[28] The characteristics of four silicides that have been extensively evaluated are compared in Table 11.4.

TABLE 11.4 Comparison of the Properties of the Silicides of Ti, Ta, Mo, and W

Property	$TiSi_2$	$TaSi_2$	$MoSi_2$	WSi_2
Resistivity, μΩ·cm	~25	~50–55	~90–100	~70
Formed by codeposition* of metal and Si followed by sintering	Yes	Yes	Yes	Yes
Etch rate in (10:1) BHF, Å/min	2000	100–300	Small	Small
Dry etching in				
(*a*) Plasma	Yes	Yes	Yes	Yes
(*b*) RSE	Yes	Yes	Yes	Yes
Oxidation in				
(*a*) Dry O_2	Yes	No (yes)†	Yes	Yes
(*b*) Wet O_2	Yes	Yes	Yes	Yes
Stability of the metal oxide	Good	Good	Poor, vaporizes	Poor, vaporizes
Oxidation stability of silicide on				
(*a*) Silicon	Good	Good	Good	Good
(*b*) Oxide	Good	Fair	Poor	Poor
Al-silicide interaction temperature, °C	>500	>500	>500	>500
Recommended for use on oxide	Yes	Yes‡	No	No
on silicon	Yes	Yes	Yes	Yes

*Cosputtering, sputtering from sintered target or coevaporation.
†No for Ta sintered on polysilicon to form silicide and yes for cosputtered silicide on polysilicon.
‡Not to be exposed to oxidizing ambients at high temperatures.
Source: After Murarka.[28]

11.9 SUMMARY

From the description of the processes in this chapter, it is obvious that the fabrication of VLSI circuits involves the repeated application of thin films of different types. Some of the films are conductors, but many are dielectric layers. Some of the layers are in place only temporarily, such as photoresist for etching or masking of implants. Nitride layers that are used for selective oxidation are subsequently removed. Doping of the substrate is sometimes accompanied by doping of polysilicon regions that are exposed at the same time. Oxide layers are grown from the silicon substrate as well as from poly layers. Typical process flows which integrate these individual steps are discussed in Chap. 12 for bipolar and MOS technologies.

Various process approaches to fulfilling the requirements for any particular layer are usually available, both in terms of suitable equipment and the specific details of its operation. Many different schemes may produce a similar result in some regards, but might have other effects upon the structure that existed before, or will exist after, that operation. The optimization of the individual process parameters and the resolution of the tradeoffs involved in combining processes into a manufacturable process flow is a formidable task. Planned experimentation using factorial analysis and response surface methodology are useful techniques in process integration. Subtle changes made in the process flow may have an unexpected impact on a device or circuit parameter that was previously regarded as well-characterized. Interactions between certain processing parameters as they affect device performance must be understood, and tolerances must be realistic in order to put together a robust process flow.

ACKNOWLEDGMENTS

The author is indebted to Judy Cannon for help in preparing this chapter, and to John Martin for his encouragement. Thanks also to Jack Saltich and W. J. Kitchen and to my colleagues in the Bipolar Technology Center for their support.

REFERENCES

1. S. Wolf and R. N. Tauber, *Silicon Processing for the VLSI Era,* Lattice Press, Sunset Beach, Cal. 1986, p. 27.
2. The American Society for Testing and Materials, standardized tests for VLSI-required properties, Committee F-1 on Electronics, Philadelphia, Pa.
3. Semiconductor Equipment and Materials Institute, standardized tests for VLSI-required properties, Mountain View, Cal.
4. D. Huber and J. Reffle, "Precipitation Process Design for Denuded Zone Formation in CZ-Silicon Wafers," *Solid State Technol.,* August 1983, p. 137.
5. C. W. Pearce, "Epitaxy," in *VLSI Technology,* S. M. Sze (ed.), McGraw-Hill, New York, 1983, Chap. 2, p. 66.
6. H. M. Liaw and J. W. Rose, "Silicon Vapor-Phase Epitaxy," in *Epitaxial Silicon Technology,* B. Jayant Baliga (ed.), Academic Press, Orlando, Fla, 1986, p. 46.
7. B. E. Deal and A. S. Grove, "General Relationship for the Thermal Oxidation of Silicon," *J. Appl. Phys.,* **36**:3770 (1965).
8. B. E. Deal, "Thermal Oxidation Kinetics of Silicon in Pyrogenic H_2O and 5% HCI/H_2O Mixtures," *J. Electrochem. Soc.,* **125**:576 (1978).
9. D. W. Hess and B. E. Deal, "Kinetics of Thermal Oxidation of Silicon in O_2/HCI Mixtures," *J. Electrochem. Soc.,* **124**:735 (1977).
10. F. J. Feigl, "Characterization of Dielectric Films," in *VLSI Electronics: Microstructure Science,* **6.** N. G. Einspruch and G. B. Larrabee (ed.), Academic Press, New York, 1983, chap. 3.
11. B. E. Deal, "Standardized Terminology for Oxide Charges Associated with Thermally Oxidized Silicon," *IEEE Trans. Electron Devices,* **ED-27**:606 (1980).
12. T. Watanabe and Y. Yoshida, "Dielectric Breakdown of Gate Insulator Due to RIE," *Solid State Technol.,* April 1984, p. 213.
13. C. J. Mogab, "Dry Etching," in *VLSI Technology,* S. M. Sze (ed.), McGraw-Hill, New York, 1983, chap. 8, p. 311.
14. S. Wolf and R. N. Tauber, *Silicon Processing for the VLSI Era,* Lattice Press, Sunset Beach, Cal., 1986, p. 581.
15. S. K. Ghandhi, *VLSI Fabrication Principles,* Wiley, New York, 1983, p. 143.
16. R. B. Fair, "Concentration Profiles of Diffused Dopants in Silicon," in *Impurity Doping Processes in Silicon,* F. F. Y. Wang (ed.), North-Holland, New York, chap. 7, 1981.

17. J. F. Gibbons, W. S. Johnson, and S. W. Mylroie, *Projected Range Statistics-Semiconductors and Related Materials,* 2d ed., Dowden, Hutchinson, and Ross Inc., Stroudsburg, Pa., 1975.
18. T. E. Seidel, "Ion Implantation," in *VLSI Technology,* S. M. Sze (ed.), McGraw-Hill, New York, 1983, chap. 6, p. 232.
19. N. L. Turner, M. Current, T. C. Smith, and D. Crane, "Effects of Planar Channeling Using Modern Ion Implantation Equipment," *Solid State Technol.,* February 1985, p. 163.
20. J. F. Ziegler (ed.), *Ion Implantation: Science and Technology,* Academic Press, Orlando, Fla., 1984.
21. L. C. Beavis, V. J. Harwood, and M. T. Thomas, Vacuum Hazards Manual, American Vacuum Society, New York, 1979.
22. S. V. Nguyen, "Plasma Assisted Chemical Vapor Deposited Thin Films for Microelectronics Applications," *J. Vac. Sci. Technol.,* **B(5)**:1159 (1986).
23. A. C. Adams, "Dielectric and Polysilicon Film Deposition," in *VLSI Technology,* S. M. Sze (ed.), McGraw-Hill, New York, 1983, chap. 3, p. 125.
24. W. Kern and V. S. Ban, "Chemical Vapor Deposition of Inorganic Thin Films," in Thin Film Processes, J. L. Vossen and W. Kern (eds.), Academic Press, New York, 1978, Part III, chap. 2.
25. T. I. Kamins, "Structure and Properties of LPCVD Silicon Films," *J. Electrochem. Soc.,* **127**:686 (1980).
26. C. Murray, "Process Specific Chemical Vapor Deposition," *Semiconductor International,* March 1987, p. 50.
27. D. B. Fraser, "Metallization," in *VLSI Technology,* S. M. Sze (ed.), McGraw-Hill, New York, 1983, chap. 8, p. 249.
28. S. P. Murarka, Silicides for *VLSI Applications,* Academic Press, Orlando, Fl., 1983, p. 151.

CHAPTER 12
VLSI MANUFACTURING TECHNOLOGY

T. C. Smith
Bipolar Technology Center
Motorola Semiconductor
Mesa, Arizona

The manufacture of ICs in wafer form may be regarded as an art that applies the generic process steps in a complete process flow, beginning with blank starting material and ending with finished circuits ready for testing. This chapter reviews additional topics related to wafer fabrication and gives examples of the manufacturing flows in both bipolar and MOS technologies. In both of these, the process integration is the result of considerable interaction between the product, design, device, processing, manufacturing, and reliability engineering functions. Feedback from one function to the others results in improved performance of the product and the processes. Once a basic process flow is established, whole families of products can be run utilizing that flow, with different mask sets being used for various parts. The capabilities of the wafer fab area are intimately related to the technology being run, especially regarding lithographic capability and the level of cleanliness in the manufacturing facility.

There are more similarities than differences in the manufacturing of bipolar products compared to MOS products. The required equipment and process capabilities are generally very similar. The major application of high-performance digital bipolar ICs is for high-speed memory and logic circuits in mainframe computers. Linear bipolar ICs are used for analog applications such as amplifiers and regulators. The various product requirements have led to some distinctive features of the respective process flows. For instance, the use of epitaxial layers and multilayer metallization is fairly commonplace in bipolar processing, but they have not been widely implemented in MOS device manufacture in the past. In addition, bipolar circuits often incorporate Schottky barrier diodes using metal silicides and barrier metallization schemes that are not encountered in MOS circuits. Certain trends, such as the development of BiMOS products that combine bipolar and CMOS technologies in the same circuit, are leading to more commonality of processing.[1]

Whether running in high-volume production of commodity parts or modest volumes of custom ICs, the modern facility must incorporate state-of-the-art clean-room technology and a disciplined work force indoctrinated in clean-room culture.

The factors in the environment of the manufacturing area that are closely controlled include high-purity chemicals and process gases, ultrapure deionized water, airflow, temperature, humidity, vibration, and contamination. In a typical operation, conditioned air is delivered from high-efficiency particulate air (HEPA) filters with vertical laminar flow from the ceiling to the floor. Federal standard 209B, which defines particle measurement specifications at various size ranges and sets down procedures for verification of cleanliness, is being revised to define higher levels of cleanliness. Tentative federal standard 209C has been submitted for consideration.* All personnel in the fab area wear clean-room garments that completely cover their street clothes, preserving the integrity of the clean room.

Wafers are processed in batches from one process step to another, usually in lots consisting of 20 to 50 wafers requiring identical masking at the various layers. Within a process step, the wafers may be treated as a batch, as in diffusion processes, or processed sequentially, as in aligning and exposing patterns in photoresist. Computer tracking systems are typically used for inventory control and to ensure that the overall flow of material is kept running smoothly. Complete process specifications and training of qualified operators are crucial to successful operation with high process yields. Statistical quality control methods should be used in monitoring the performance of all of those characteristics of the individual processes that can be measured on a sampling basis. In addition, proper statistical methods must be employed when analyzing problems and when making decisions about the effects of modifications to existing processes and flows.

12.1 WAFER CLEANING AND CONTAMINATION CONTROL

Wafer cleaning is an important part of many of the processes used in device fabrication, along with techniques used to monitor and control sources of contamination. Wafers may be contaminated by various films or residues, which can lead to processing problems, as well as by particulates randomly distributed on the surface, which can result in nonfunctionality of circuits and therefore produce yield loss. Chemical cleaning techniques are used to remove the films that are chemically bonded to the wafer surface. Insoluble particulate contamination is generally removed by either ultrasonic or mechanical scrubbing. In situ cleaning cycles are sometimes incorporated into reactive ion etching (RIE) processes as a "descum" step, in epitaxial reactors as an HCl preclean, or in sputtering systems as a sputter etch step for removal of native oxide films.

Chemical cleaning processes must be compatible with the materials exposed on the wafer surface, so many cleaning solutions are regarded as proprietary. For wafers with bare Si or thermal oxides on the surface being cleaned prior to high-temperature processes, a cleaning sequence similar to one formulated at RCA is used.[2] After a preliminary photoresist strip in an oxygen plasma or an oxidizing wet stripper, and a rinse in deionized water, the batch of wafers has the residual organic contaminants removed in an H_2O_2-NH_4OH-H_2O solution. Next, a dip in dilute HF removes the oxide that was chemically formed on bare Si. Then the remaining ionic contaminants are desorbed in an H_2O_2-HCl-H_2O cleaning solution, and the wafers are rinsed and dried.

*For additional information, contact The Institute of Environmental Sciences, 940 E. Northwest Hwy., Mt. Prospect, Ill. 60056.

For contamination and particulate control, it is preferable to eliminate the sources of contamination, but the processing equipment or chemicals themselves, such as photoresist residues, may be the cause of the problem. The utilization of ultrapure deionized water and ultrapure chemicals together with point-of-use filtration of all chemicals and gases in the processing area should be standard practice. During processing, wafer surfaces are constantly inspected under bright light and microscopes are used with various modes of illumination. Instrumentation is available that scans a sharply focused light beam over the surface and detects particles by collecting scattered light.[3] These systems generate wafer maps of particle locations and give a statistical breakdown by size categories for continuous monitoring of processes and equipment.

The impact of contamination and particulates upon final test yield of ICs is evident in terms of both *gross defects,* failures to achieve certain device parameters, and *random defects,* failure of the circuits in different regions of some wafers to function and meet certain test criteria. Random defects may be induced in any of the "hot process" steps [diffusion, implantation, oxidation, epitaxial film growth, or chemical vapor deposition (CVD) depositions], metallization and interlayer dielectric depositions, or in photoresist and etching operations. Failures may be related to open or shorted conductors, pinholes, contacts, emitter piping, or junction leakage. IC yield statistics have been described by different models, but any model that is convenient and fits the data for devices and test structures used as monitors can lead to successful yield management.[4]

For a given process flow, yields can be related to DA, where D is the average defect density and A is the critical area, which is the sensitive area susceptible to killing defects. The critical area may be taken to be the area of the chip or die. Figure 12.1 shows the dramatic effect upon yield of the average defect density for a large VLSI die size (0.5 cm^2 or 77,500 mil^2), according to each of several commonly used yield models. Chip areas have tended to decrease due to device scaling, which should reduce the probability of getting a defective die, but the trend toward increasing circuit complexity is making chip areas larger. In either case, the decrease in feature sizes within the die means that the size of particulates that can produce killing defects is also decreasing. For process flows with a large number of masking steps, the reduction of particles added by these and associated process steps is critical for achieving high circuit yields.

12.2 PROCESS CHARACTERIZATION AND CONTROL

In the course of developing the various process steps that are put into sequence as a process flow, process and device engineers carry out a considerable number of tests to characterize the individual steps as they interact and relate to the total series of operations. The effect of expected variations in the process parameters upon device electrical parameters and overall circuit performance is determined. Previous experience and the results of process and device modeling calculations guide the way. More work is required if the steps, or the sequence in which they are performed, are new or different from well-understood ones. Once the process is debugged and stabilized, the results obtained from parametric tests on completed wafers are correlated with suitable in-process test wafers that are run as a monitor on the individual processes. Continuous checking of these independent in-process monitors and timely corrective action ensure that problems due to

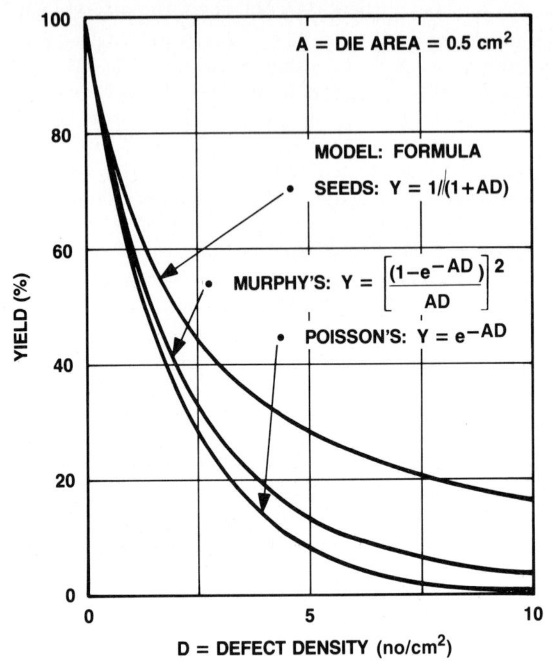

FIG. 12.1 Yield versus defect density for various models.

equipment malfunction or operator error do not develop. Otherwise, process-induced problems might not be detected until final testing at the end of the total process flow, so the manufacturing line would be full of useless product.

In the early stages of any process development effort, and later on in the course of diagnosing manufacturing problems, a wide variety of measuring equipment and analytical techniques are employed for quantitative and qualitative testing. Both extraordinary means of analysis and measuring instruments that are routinely used in evaluating process monitors are very useful in these phases of engineering work. Evaluations are performed on both patterned product wafers and blank test wafers with various films. In many VLSI wafer fab areas, the scanning electron microscope (SEM) is routinely used in photoresist and etching work to investigate the edge profiles of images developed in photoresist and details of features etched in the underlying layers. The SEM has resolution and depth-of-field capabilities that are far superior to optical microscopes. The results of physical or chemical analysis of materials are combined with pertinent electrical measurements whenever possible to confirm the modeling or determine the mechanisms operating in the process flow.

In the developmental phases, sophisticated analytical instruments such as secondary ion mass spectrometry (SIMS) may be used to determine dopant profiles as a function of depth in the surface region. Although SIMS detects the chemical presence of elements of interest down below the part-per-million levels of concentration, the electrically active dopant concentration is measured with the spreading resistance probe (SRP). In many cases, interpretation of the data by an experienced analyst is required. An overview of materials characterization techniques and suggestions for how to accomplish an effective analysis is given in Ref. 5. The principles of operation of the other commonly used instruments and their sensitivities are discussed there also.

For real-time evaluation in the fab area of the results of various processes, instrumentation is used by operators or technicians to measure properties such as thickness or sheet resistance of thin films, or linewidths of developed photoresist and etched features. Deposition and etching rates are inferred from these measurements. Optical instruments, such as ellipsometers or reflectometers, are typically used to measure the thickness and index of refraction of dielectric thin films. Four-point probes are employed to measure the sheet resistance of implanted or diffused layers in silicon substrates and deposited conducting layers. Visual inspection according to a plan for each product type at critical steps is also an important part of the monitoring of the production line. These inspections help to identify yield-limiting factors such as residues and particulates as well as a wide variety of defects in the photoresist and etching steps. As mentioned under the topic of oxidation, monitoring of capacitor test structures by *C-V* measurements with applied bias and temperature stress should be done routinely to detect possible mobile ion contamination. The reason is that sources of such contamination include not only the oxidation and diffusion furnaces but also processing chemicals, thin-film deposition systems and wafer handling. Problems with threshold shifts and surface inversion can result from uncontrolled mobile ion contamination.

Information on measurements from in-line process monitors and other measurements of equipment operating parameters are usually plotted on process control charts. The purpose is to present the information graphically to verify that the process is performing within specified limits, to observe trends, and to ensure that corrective actions are taken to keep the process in control. An example of a trend chart for a typical monitor of an ion implantation process is shown in Fig. 12.2. Methods of statistical process control are applied to these measurements to characterize the absolute value and uniformity within wafers as well as variations in results from wafer to wafer within a lot and from lot to lot. The in-line process data is correlated to electrical measurements made on automatic testing systems using test patterns that incorporate specially designed structures that give information on both device parameters and linewidth and alignment accuracy.[6] These latter results can be displayed as wafer maps as an aid to detecting and fixing systematic errors related to the processing equipment.

12.3 PROCESS MODELING AND CIRCUIT SIMULATION

To varying degrees, all VLSI circuit manufacturers use process modeling tools to characterize processes or to evaluate device structures, and they employ additional tools to simulate circuit performance. The complexity of advanced ICs and their sensitivity to process variations make this approach more cost-effective than using experimental techniques alone. When correlated to actual measurements, this modeling allows a better understanding of the interactions between process parameters and device behavior. Numerous software programs have been developed and continuously upgraded by both universities and commercial suppliers to model various aspects of the sequence from starting silicon to completed circuits. Virtually all of the process steps discussed in Chap. 11 can be

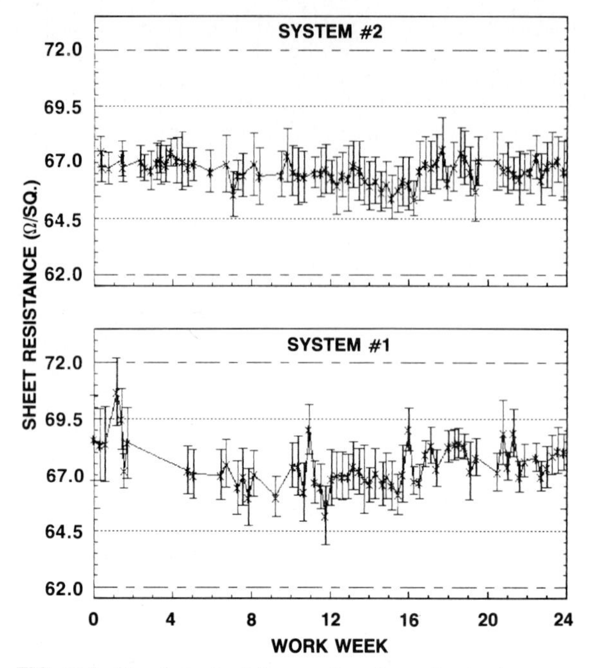

FIG. 12.2 Trend charts of line monitors for an ion implant process for two different systems.

modeled in one- or two-dimensional structures, using analytical and numerical solution techniques. The outputs from process modeling become the inputs for device simulations. Their outputs in turn provide the electrical characteristics and parameters for circuit simulations in both MOS and bipolar technologies.

Modeling or simulation programs are typically named with an acronym that describes its origin or purpose. Some of the better-known modeling programs are the processing model called SUPREM,[7] from Stanford University, and the circuit simulation program called SPICE,[8] from the University of California at Berkeley. An analysis program called PREDICT,[9] which rigorously solves coupled equations describing the processes, has been developed by the Microelectronics Center of North Carolina. These types of programs are often used in conjunction with in-house-generated computer codes for parameter extraction and analysis techniques that aid in design centering for improved performance and yield optimization. Reference 10 describes a modeling approach for CMOS VLSI circuit design that is highly automated and provides statistically relevant results.

12.4 TRENDS AND EMERGING PROCESSES

The continuing drive toward smaller dimensions in pattern definition put pressure on other processes as it relates to dimensional control. Developing trends in lithography are discussed separately in Chap. 13. Etching processes, in particular, are strongly affected by the drive toward smaller geometries, and must deal with an increasing variety of materials to be etched. All processes are affected by the attendant requirement to reduce particle contamination levels, which become more stringent as the sizes of "killer defects" also decrease. Techniques for reducing the number of particles added in handling wafers, by the chemicals and gases used or by the processing equipment itself, are being vigorously pursued. Robotic handling equipment, adapted to the specific processing tool and compatible with each process step, is incorporated into the various systems used in wafer fabrication areas.

At the same time, device scaling considerations require that vertical dimensions be reduced. This in turn impacts those processes that influence vertical dopant profiles in silicon, such as ion implantation and diffusion. Thus, the trend is to implant dopants at low energies and to limit the total time at high temperatures in order to minimize dopant redistribution and thus produce shallow junction depths. The rapid thermal processing (RTP) techniques discussed in Chap. 11 are compatible with these requirements. RTP systems, which cycle single wafers to high temperatures for about 1 min, are being incorporated into process flows whenever they can be used for special advantages.[11] Trends in diffusion technology include the use of robotic wafer cleaning and furnace loading systems for reduced particulate generation and increased use of computer controls with process monitoring capabilities.

In ion implantation technology, renewed interest in higher energies has been generated by distinctive advantages in device fabrication.[12] Other trends in ion implantation include the continuing evolution of equipment that produces higher beam currents, since higher doses correspond to greater collected charge (beam current integrated over the implanting time). Thus, for high wafer throughput, typically 100 to 200 wafers per hour for implanters, beam currents in excess of 10 mA are needed. These developments have required a better understanding of problems associated with wafer charging phenomena as they relate to the integrity of gate oxides and the operation of implanters. More emphasis has been placed on precise wafer orientation to minimize channeling artifacts and system specifications that guarantee low added particulates are now commonplace.

In all phases in the manufacture of VLSI products, from design and layout, through wafer fabrication, to assembly and testing, computer-aided manufacturing (CAM) techniques are used to optimize the designs and provide feedback on a variety of information. CAM systems are used for managing both the business and engineering aspects of the manufacturing areas. These systems provide the means of controlling inventories or work in process through lot tracking and prioritization of the flow of material in the production line. They are used for maintaining documentation such as systems of specifications and also for recording parametric and process control data. The common needs of automation and the requirement that all equipment in the processing and assembly areas be configured with a port to communicate with a host computer has led to the formulation of semiconductor industry standards for computer communication.[13] The growing complexity of the equipment and the need to ensure maximum availability of the processes will bring about the use of expert systems for maintenance and repair.

12.5 BIPOLAR IC TECHNOLOGY

The application of the various generic process steps to the fabrication of bipolar ICs is straightforward when the process building blocks are properly integrated.

With few exceptions, bipolar process flows are distinguished from those used in MOS processes only by the specifics of the device structures as they affect the device and circuit performance goals.

The cross-sectional drawings of a vertical *npn* transistor shown in Fig. 12.3*a–e* illustrate the device structure at various stages in a typical bipolar process flow. The drawings are schematic; the vertical and horizontal dimensions are not meant to be true to scale. A sequence of steps representative of those presently used in a VLSI production facility will be described, and then some of the key points and tradeoffs will be discussed. Many variations of the process sequence are possible, and additional masking layers or related optional steps

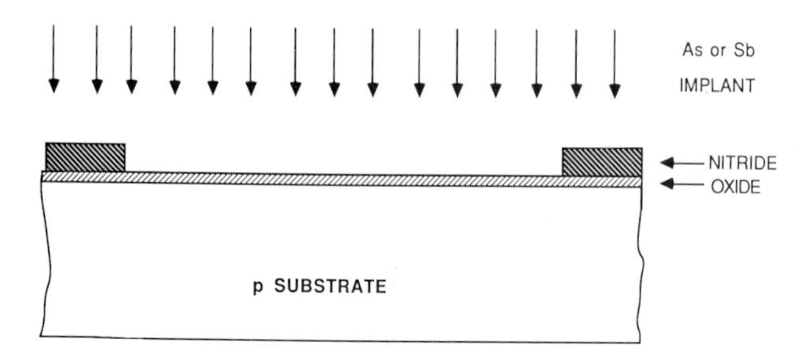

FIG. 12.3*a* Buried-layer implant.

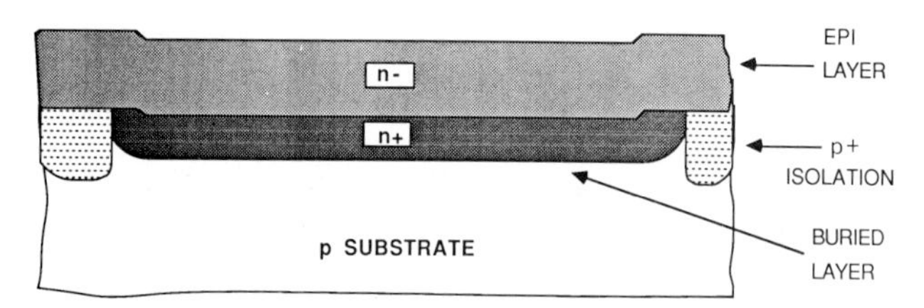

FIG. 12.3*b* Epilayer and buried layers.

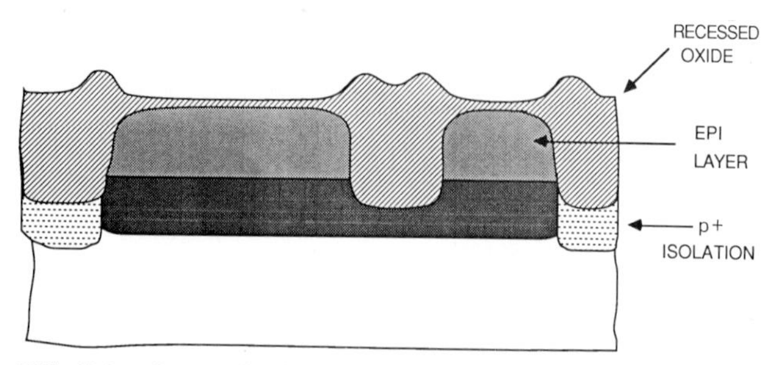

FIG. 12.3*c* Recessed oxide isolation.

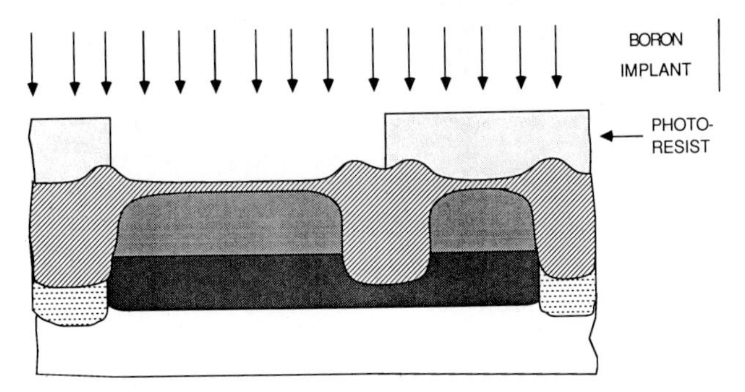

FIG. 12.3*d* Base implant.

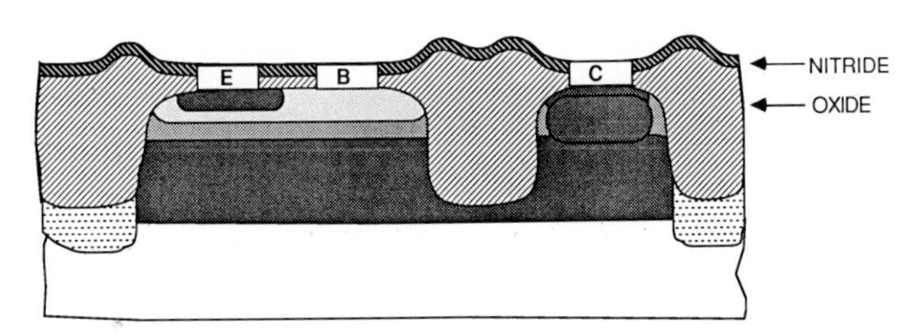

FIG. 12.3*e* Contacts.

might be added to incorporate specific product features, such as load resistors or Schottky barrier diodes. The starting wafer is processed through initial oxidation, and nitride is deposited and etched with the buried-layer pattern. The heavy $n+$ subcollector doping is usually done with an arsenic or antimony implant, as shown in Fig. 12.3*a*. After a reoxidation step, which accompanies the buried-layer drive-in, the resulting slight step in the surface provides a pattern for aligning the subsequent layer. A boron implant is performed with photoresist masking and driven in to provide $p+$ isolation between devices. This is followed by an oxide strip and deposition of an $n-$ epitaxial layer. After these steps, the region where the completed transistor will be located appears as in Fig. 12.3*b*, with the step in the surface, which was replicated through the epilayer, somewhat exaggerated.

The next sequence of processes involves pad oxide growth and nitride deposition; then photoresist is applied and patterned with the "moat" mask. Next, the nitride and oxide layers are etched together with some of the underlying epilayer. After selective oxidation, the recessed oxide structure, with its characteristic "bird's beak" and "bird's head," looks like Fig. 12.3*c*. Following nitride strip, photoresist is used to pattern the active base region, and the base implant is performed as in Fig. 12.3*d*. Nitride is deposited, and contacts for the emitter, base, and collector are etched simultaneously. The electrode structures are now self-aligned, with the geometry of the layout limited only by the minimum spacing between metal contacts. Less critical photoresist masking is now sequentially applied for implants into the various regions. For example, a phosphorus implant can be performed in the collector contact only as a "reach-through" dopant contacting the $n+$ buried layer. The emitter implant goes into both the emitter and collector contacts. The implanted dopants are annealed and redistributed as required by device performance goals. The completed transistor structure in the silicon is shown in Fig. 12.3*e*, with the emitter, base, and collector contacts indicated. Processing continues through first-metal deposition and etching, interlayer dielectric (ILD) deposition, and via etch. These deposition and etching steps are then repeated. Metallization and ILD may be required one or two more times prior to final passivation, depending upon whether the product uses two or three layers of metallization. Further details of this bipolar device fabrication sequence, with some consideration of the tradeoffs involved, are discussed below.

The starting substrate is usually lightly doped *p*-type of $\langle 100 \rangle$ or $\langle 111 \rangle$ orientation with a background doping level of about 10^{15} atoms/cm^3. The buried layer is formed under the active *npn* devices to provide a low-resistance conducting path to the collector contact. The "predep" implant is driven in to reduce the doping level from its initial high value to a concentration that is below the solid solubility limit. The desired junction depth and a final sheet resistance of about 30 Ω per square are produced. Either As or Sb is used for the buried-layer dopant, due to their low diffusivities, resulting in less up-diffusion and less autodoping in the process of epitaxial film growth. If the implant dose were increased to reduce the sheet resistance, increased up-diffusion would occur, reducing the thickness of the lighter-doped epitaxial layer. The isolation or channel-stop implant enhances the *p*-type doping level under the isolation oxide, preventing surface inversion of the substrate and increasing the breakdown voltage between epi islands. The epitaxial layer forms the collector immediately under the base region with a doping level of approximately 10^{16} atoms/cm^3. The dopant concentration affects the collector-base capacitance as well as the high-level injection performance, and proper control avoids base pushout effects.[14]

Following epitaxial film growth, a "pad oxide" of about 50 nm is grown, and a nitride layer of about 100 nm is deposited. The nitride, which is patterned by RIE or plasma etching, will act as a mask against oxidation of the epitaxial layer. The pad oxide acts as a means of stress relief to avoid defects in the crystal that could be induced in subsequent high-temperature oxidation cycles. In the selective oxidation step, the nitride-covered areas are protected, and the open areas are converted to oxide throughout the thickness of the epilayer. Lateral oxidation under the edge of the nitride produces the bird's beak, and oxidation at the corners forms the bump referred to as the bird's head.[15] If the high-pressure oxidation technique is used here, further up-diffusion from the buried layer is somewhat reduced because the oxidation process takes place at reduced temperatures.

In the following stages of the process, the total time at high temperatures is held to a minimum in order to keep device junctions shallow. After the masking nitride is stripped, a photoresist mask is applied and the base is implanted through the bottom oxide. This oxide helps to avoid ion channeling in the base profile, an effect that would be nonuniform in different parts of the wafer.[16] Since the current gain is determined by the Gummel number (i.e., the total integrated charge in the active base), the implant parameters can be tailored to give the desired gain and high-frequency cutoff characteristics.[17]

The surface is then reoxidized, and a final nitride layer, which will remain to provide protection from mobile ion contamination, is deposited. The contacts for the emitter, base, and collector are simultaneously exposed and patterned with a single mask and etched, making the next several alignments less critical. Optional masking for implanting phosphorus into an enhanced collector contact and/or a boron implant, for an inactive base with lower base sheet resistance may be employed. Arsenic is used for the emitter implant because, with its low depth of penetration and its low diffusivity, the junction depth can be made very shallow (less than 0.2 μm). A last high-temperature anneal determines the final doping profiles of the active devices. High-performance bipolar circuits may use a thin polysilicon layer in the emitter contact prior to an arsenic predep implant. This provides a higher current gain and an abrupt profile in the single-crystal emitter.[18]

The completed structure in the silicon, depicted in Fig. 12.3*e*, shows the emitter, base, and collector contacts prior to the deposition of first metal. A metal silicide may be used to incorporate a Schottky barrier diode clamp to the collector region.[19] This silicide is typically formed by sintering a sputter-deposited thin film of platinum or palladium. A thin film of titanium-tungsten is deposited as a barrier between the silicide and the aluminum alloy that follows. The last sequence of processes through multiple layers of metallization has become increasingly challenging with the trends toward decreasing linewidths and spaces. Proper cleaning of all contacts in the process flow is crucial. The interlayer dielectrics used include CVD and plasma-enhanced (PE) CVD oxides as well as polyimides. Careful tapered etching of the vias is important for achieving adequate metal step coverage. The demands upon the lithography and etching processes are great because of the severe topography of the wafer surface. Similar process steps are employed in MOS flows. Only the device structures and the circuit operating conditions differ; the basic process building blocks are the same.

12.6 MOS IC TECHNOLOGY

In the fabrication of MOS devices and circuits, the sequence of process steps is different from bipolar flows because the structures and circuit performance criteria are different. In general, active-area doping levels are lower, the use of polysilicon is commonplace, and since the devices are more sensitive to surface effects, there is more emphasis on the quality of the grown and deposited oxides. The tradeoffs between various process parameters and their effect upon device performance are different from bipolar processing. Consequently, in processes for BiMOS technologies, some compromises must be made. Figure 12.4*a–f*

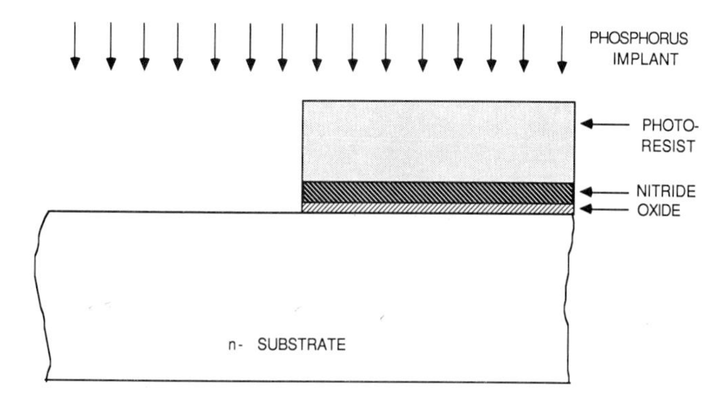

FIG. 12.4*a* *n*-Well implant.

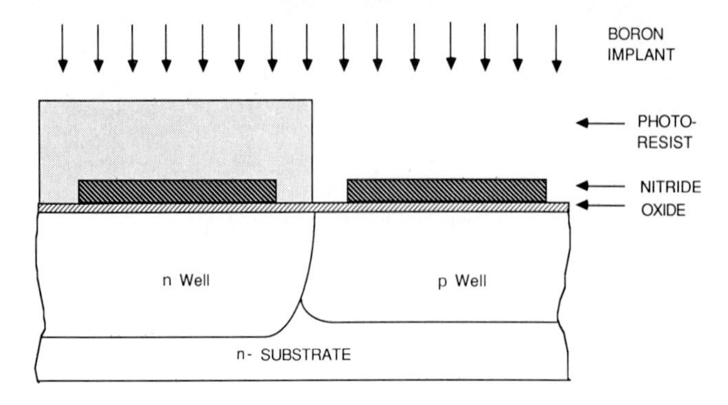

FIG. 12.4*b* Channel-stop implant.

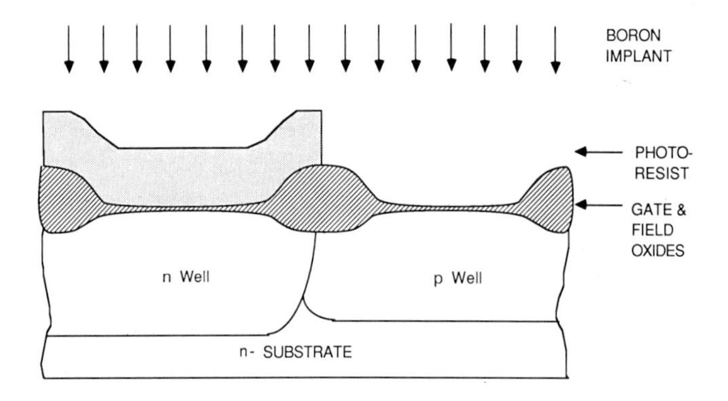

FIG. 12.4*c* NMOS V_t adjust implant.

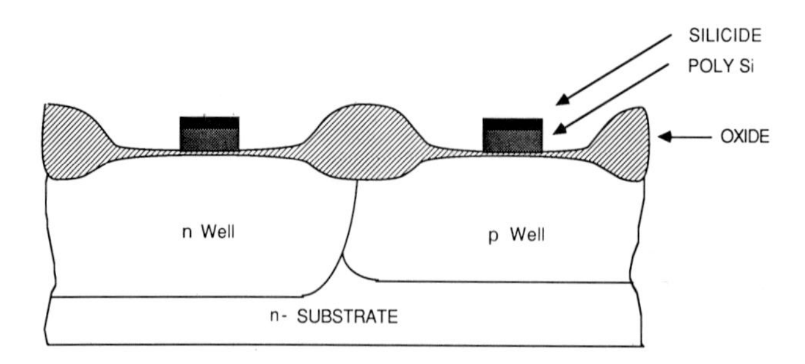

FIG. 12.4*d* Polycide structure.

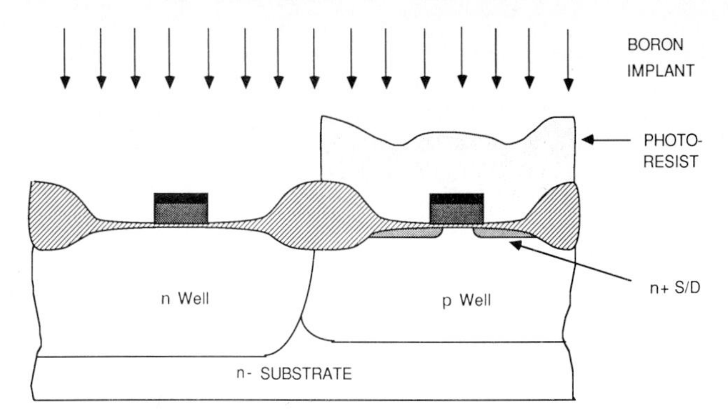

FIG. 12.4*e* p + S/D implant.

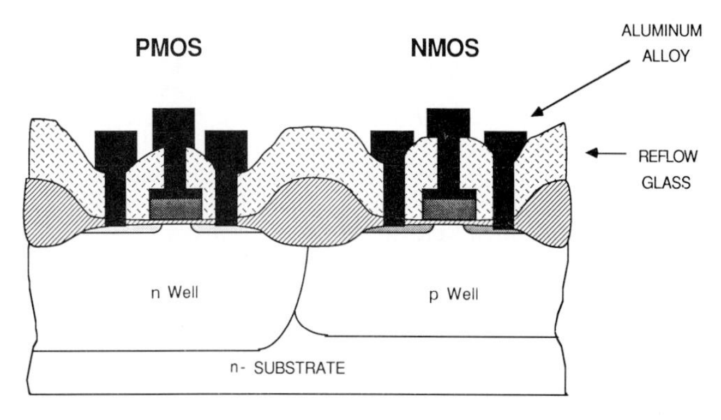

FIG. 12.4*f* Metallization.

shows cross-sectional views of portions of a CMOS circuit as it appears at various stages throughout the process flow. The process in this example implements twin wells and uses silicided gates as a low-resistivity interconnect. Again, the drawings are schematic; the vertical and horizontal dimensions are not true to scale.

The starting material could be lightly doped n or p type, possibly as an epitaxial layer deposited on a more heavily doped substrate to minimize the circuit's susceptibility to latch-up.[20] The starting substrate is oxidized to form a pad oxide, which is covered with a deposited film of silicon nitride. The pattern of the n-well region is defined in photoresist, the nitride-oxide layer is etched, and phosphorus is implanted to dope the n-well. The structure looks like Fig. 12.4*a* at that point. The photoresist is stripped, and when the surface is oxidized a thicker oxide grows on the bare silicon above the n-well regions. Next, the nitride is stripped, and boron is implanted into the p-well regions, using the thick oxide over the n-well as a self-aligned mask. The diffusion of dopants for both well regions takes place simultaneously, and the oxides are stripped. The slight step in the silicon surface that results from unequal oxide growth is used to align the next masking pattern.

A second pad oxide is grown, nitride is again deposited, and the surface is patterned with the active area photomask and etched to bare silicon in the field region. A channel-stop implant is performed, with the *n*-well active areas masked by photoresist and the *p*-well active areas masked by nitride, as shown in Fig. 12.4*b*. This produces heavier doping under the field oxide prior to the field oxidation cycle. By the LOCOS (local oxidation of silicon) process, thick oxide is grown in the field regions while nitride inhibits oxide growth in the active areas.[21] The nitride and thin oxide are stripped, gate oxide is grown, and photoresist masking is used for the low-dose implants for adjusting threshold voltages to their desired levels. Figure 12.4*c* shows boron being implanted into the *n*-channel devices, to be built in the *p* well, while the active areas of the *n* well are protected. At this point, an optional mask may be used, and the gate oxide may be etched in some regions to provide a buried contact to the substrate for the polysilicon layer, which is deposited next.

The polysilicon is doped in situ or by diffusion or implant, and a thin film of silicide is applied on top of the poly to greatly reduce the resistivity of the gate interconnect. This "polycide" gate structure is etched and, after photoresist strip, appears as in Fig. 12.4*d*. Following gate definition, the polycide is annealed, and the $n+$ source-drain (S/D) regions are implanted with arsenic while the PMOS regions are masked with photoresist. The photoresist is stripped and the $n+$ S/D implant is annealed. A lightly doped drain (LDD) structure can be achieved by implanting a lighter dose of phosphorus at this same step. When the $n+$ source-drains are annealed, the faster-diffusing phosphorus will extend under the gate electrode, reducing hot electron injection effects.[20]

The $p+$ S/D regions are implanted while the complementary $n+$ regions, which were previously implanted and driven in, are masked with photoresist, as shown in Fig. 12.4*e*. This implant may be done without masking, if the $n+$ regions can tolerate counterdoping and if the $p+$ dose is much less than the $n+$ dose. After a reflow oxide is deposited on the surface, the $p+$ S/D implant is annealed and contacts are etched prior to aluminum alloy deposition. The metal is then patterned, and the completed transistor structures prior to passivation are as shown in Fig. 12.4*f*. There are many variations of this kind of process flow used in CMOS wafer fabrication, with the choice of the specifics determined by the processing capabilities of the wafer fab area and the product performance requirements.

Trends in advanced MOS processing have been directed toward higher-density circuits, especially in memory products, with increasing use of multilayer metallization schemes as in bipolar processing. With decreasing geometries, numerous special precautions in processing must be taken to counteract short channel effects.[22] The use of trench structures for isolation and for capacitors has allowed higher density with increased processing complexity. Three-dimensional structures may become available in production as various schemes for this technology develop. An example is a Static RAM (SRAM) device that has been fabricated using laser recrystallized polysilicon for the second active layer.[23]

ACKNOWLEDGMENTS

The author is greatly indebted to Judy Cannon for her help in preparing this manuscript, and to John Martin for his continued encouragement. I would also like to thank Jack Saltich and W. J. Kitchen and my colleagues in the Bipolar Technology Center for their support.

REFERENCES

1. A. R. Alvarez et al. "Technology Considerations for High Performance BIMOS," *Int. Conf. on Computer Design: VLSI in Computers,* Rye, N.Y., 1985.
2. W. Kern, "Purifying Si and SiO_2 Surfaces with Hydrogen Peroxide," *Semiconductor International,* April 1984, p. 94.
3. C. Logan, "Analyzing Semiconductor Wafer Contamination," *Microelectronic Manufacturing and Testing,* March, 1985, p. 1.
4. C. H. Stapper, F. M. Armstrong, and K. Saji, "Integrated Circuit Yield Statistics," *Proc. IEEE,* **71**(4):453 (1983).
5. S. Wolf and R. N. Tauber, in *Silicon Processing for the VLSI Era,* Lattice Press, Sunset Beach, Cal., 1986, Chap. 17.
6. G. P. Carver, W. L. Lindholm, and T. J. Russel, "Use of Microelectronic Test Structures to Characterize IC Materials, Processes, and Process Equipment," *Solid State Technol.,* May 1985, p. 133.
7. C. P. Ho, S. E. Hansen, and P. M. Fahey, "SUPREM-III—A Program for Integrated Circuit Process Modeling and Simulation," Technical Report No. SEL84-001, Stanford Electronics Laboratories, Stanford University, Stanford, Cal., July, 1984. [SUPREM and other simulation programs are available from Technology Modeling Associates, 300 Hamilton Ave, Palo Alto, Cal. 94301.]
8. A. Vladimirescu, K. Zhang, A. R. Newton, D. O. Pederson, and A. Sangiovanni-Vincentelli, *SPICE Version 2G User's Guide,* University of California, Berkeley, August 10, 1981.
9. R. B. Fair and R. Subramanyan, "PREDICT—A New Design Tool for Shallow Junction Processes," *Proc. SPIE,* **530:**88 (1985).
10. N. Herr and J. J. Barnes, "Statistical Circuit Simulation Modeling of CMOS VLSI," *IEEE Trans. on Computer-Aided Design,* **CAD-5**(1):15 (1986).
11. S. R. Wilson, R. B. Gregory, and W. M. Paulson, "An Overview and Comparison of Rapid Thermal Processing Equipment: A User's Viewpoint," *Mat. Res. Soc. Proc.,* **52:**181 (1986).
12. T. C. Smith, "Status of IC Applications of Ion Implantation," *Nucl. Instr. & Methods in Phys. Res.,* **B21:**90 (1987).
13. J. Secrest, "Status of SECS—SEMI's Equipment Communication Standard," *Semiconductor International,* June 1987, p. 40.
14. T. H. Ning, D. D. Tang, and P. M. Solomon, "Scaling Properties of Bipolar Devices," *IEEE Int. Electron Device Meet.,* Washington, D.C., 1980, p. 61.
15. K. Y. Chiu, J. L. Moll, and J. Manoliu, "A Bird's Beak Free Local Oxidation Technology Feasible for VLSI Circuits Fabrication," *IEEE Trans. Electron Devices,* **ED-29:**536 (1982).
16. P. J. Ward, "A VLSI Compatible Bipolar Transistor Process," *Electrochemical Society VLSI Symposium,* Detroit, 1982, p. 290.
17. R. S. Payne, R. J. Scavuzzo, K. H. Olsen, J. M. Nacci, and R. A. Moline, "Fully Ion-Implanted Bipolar Transistors," *IEEE Trans. Electron Devices,* **ED-21**(4):273 (1974).
18. T. H. Ning and R. D. Isaac, "Effect of Emitter Contact on Current Gain of Silicon Bipolar Transistors," *IEEE Int. Electron Device Meet.,* Washington, D.C., 1979, p. 473.
19. L. C. Parillo, "VLSI Process Integration," in *VLSI Technology,* S. M. Sze (ed.), McGraw-Hill, New York, 1983, Chap. 11.
20. J. Y. Chen, "CMOS- The Emerging VLSI Technology," *IEEE Circuits and Devices Magazine,* March 1986, p. 16.

21. J. A. Appels et al., "Local Oxidation of Silicon and Its Applications in Semiconductor Technology," *Phillips Res. Rep.,* **25:**118 (1970).
22. L. C. Parillo et al., "A Versatile, High-Performance, Double-Level-Poly, Double-Level-Metal, 1.2 Micron CMOS Technology," *IEEE Proc. IEDM,* **86:**244 (1986).
23. Y. Inoue et al., "A Three-Dimensional Static RAM," *IEEE Electron Device Letters,* **EDL-7**(5):327 (1986).

CHAPTER 13
VLSI LITHOGRAPHY

Yefim Bukhman
Motorola, Inc.
Mesa, Arizona

VLSI fabrication requires the introduction of exactly controlled amounts of impurities into precisely defined regions of the circuit structures on the substrate material. All elements of this three-dimensional array are built by modification, addition, and removal of various insulating, conducting, or semiconducting materials at exact geometrically controlled locations. The lithographic process ultimately defines all the microcircuit elements and has the greatest technological and economical impact on VLSI performance and fabrication cost. At any given moment of IC manufacturing history, the ability to reproduce minimum feature size (*resolution*) at the exactly defined location (*registration*) determines the current state of this technology.

13.1 LITHOGRAPHIC PROCESS

The VLSI design process culminates in the generation of a CAD circuit layout, which contains information of the exact shape and position of every circuit element. This information is transferred to the starting substrate to form the desired circuit elements by what is known as the lithographic process.

The lithographic process for most VLSI circuits consists of the circuit design layout transfer into a resist material, usually an organic polymer coated on the silicon wafer. The resist film forms a conformal mask through which the underlying materials can be removed (etch processes) or impurities can be introduced (ion implantation or diffusion processes). Therefore, the lithographic process should take into account not only resolution and registration parameters associated with the resist image definition, but also the issues related to the resist material compatibility with the subsequent processing steps, such as etching and ion implantation.

13.1.1 Lithographic Strategies

While today advanced VLSI circuits are being manufactured with the minimum feature size of 1.25 to 2.0 μm, the 0.5-μm design rules will be implemented in the

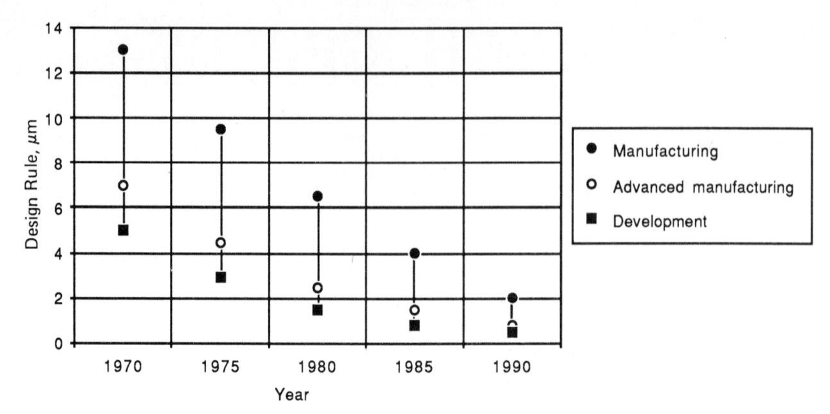

FIG. 13.1 VLSI design rules in high-volume manufacturing, advanced manufacturing, and development laboratories.

next 5 to 10 years.[1] The gap between high-volume manufacturing and advanced product development requirements, Fig. 13.1, has been reduced from 5 to 10 years to 2 to 5 years. Though several imaging techniques can be employed to achieve this lithographic performance, only optical and electron beam exposure tools are presently used in high-volume VLSI manufacturing. Both methods have considerable fundamental difficulties in achieving 0.5-μm resolution limits, and x-ray lithography may provide high-volume submicrometer manufacturing opportunities in the future.

The manufacturing sequence for a typical lithographic process begins with the generation of CAD data of the circuit design and ends with the image transfer into the substrate surface. The imaging sequence may be as short as one step for direct-write electron beam lithography, or it may include intermediate image transfer steps in optical lithography, Fig. 13.2. As device complexity increases, the resolution and registration accuracy requirements become more demanding. That is why most masks for VLSI devices are written directly by electron beam exposure tools to avoid pattern placement and dimensional errors associated with the optical replication methods. The mask for optical lithography (*photomask*) consists of a glass substrate, usually borosilicate or quartz plate, coated with about 1000 Å of chromium.[2] The pattern replication process for photomasks is very similar to that used on semiconductor substrates, and includes resist exposure, development, and etching of the chromium layer. Two types of photomasks are used in the VLSI manufacturing: *reticles* and *masks*. A reticle is a photomask printed on a surface of the

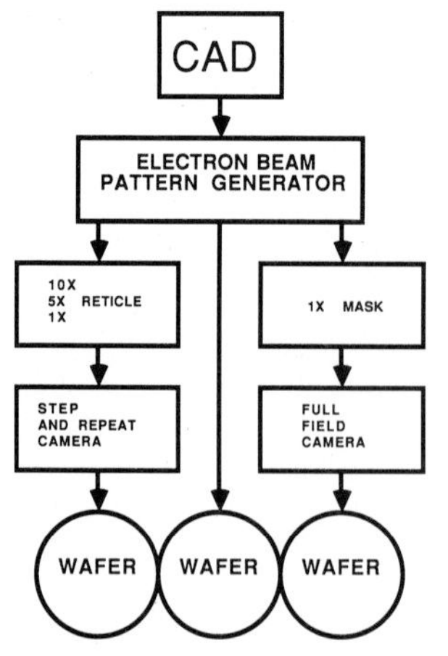

FIG. 13.2 Image transfer sequence in VLSI manufacturing.

substrate in a *step-and-repeat* fashion to expose an entire substrate area. Step-and-repeat cameras, or *steppers*, have either reduction or 1:1 projection optics. A mask is a photomask printed by a full-field optical camera over the entire substrate area without reduction.

As the design rules approach 1.0 μm minimum feature size, mask and reticle defects become very critical issues in the overall lithographic process. The photomask defects are feature placement and sizing errors, feature replication errors (i.e., added or missing information), and particle contamination. Reduction optical steppers (10× or 5×) provide considerable relief for all categories of photomask-related errors due to the simple 10× or 5× difference of the feature size at the reticle and wafer planes. For example, a 1.0-μm reticle defect will be reduced to a 0.1-μm defect at the wafer plane in a 10× stepper (0.2 μm in a 5× camera). Full-field projection cameras and 1× steppers are more sensitive to all mask-related defects. The choice of the optical cameras for VLSI lithography is usually a careful compromise between technical requirements (resolution and registration) and manufacturing considerations (productivity). Currently, steppers are offering superior resolution and registration performance but with lower productivity compared with the full-field optical cameras.

13.1.2 Resist Process Technology

The VLSI pattern formation in resist material consists of several consecutive steps and largely depends upon the particular lithographic strategy employed in manufacturing. In general, Fig. 13.3, a thin (0.5- to 4.0-μm) layer of radiation-sensitive resist is spin-coated on top of the substrate and then exposed in selected

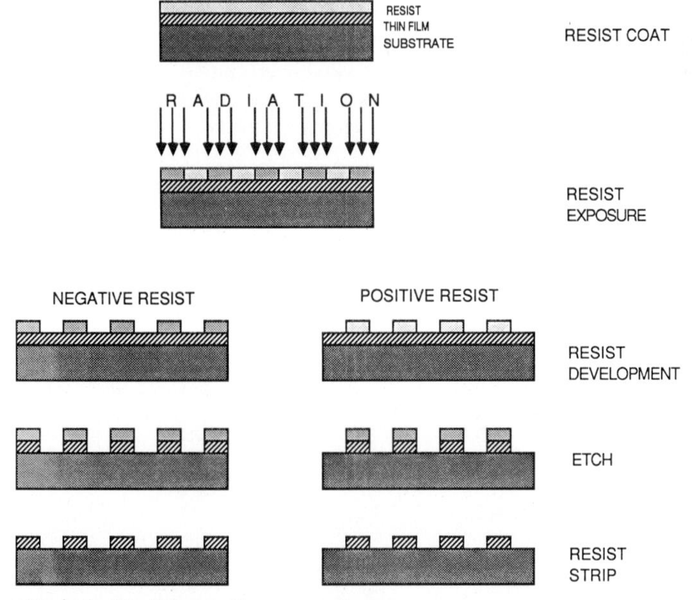

FIG. 13.3 The lithographic process diagram.

areas by that radiation. After exposure the resist material changes and, depending upon its chemical nature, becomes less soluble (*negative resist*) or more soluble (*positive resist*) in some developing solvent. The result of the resist development process is a three-dimensional relief image, which replicates the circuit layout pattern. This pattern must "resist" a subsequent etch or ion implantation process step. The remaining resist material is removed (*stripped*) from the substrate, leaving behind an etched three-dimensional structure or distribution of implanted ions. In that way, the total lithographic process can be described as a resist coat, exposure, development, and strip sequence, which must be repeated several times during VLSI fabrication process.

Two distinctive steps are responsible for the spatial definition of a pattern in a resist material: formation of a latent image by exposure to radiation, and development of that image to replicate circuit layout on the substrate surface. The choice of resist material largely depends upon the nature of exposure radiation. The lithographic method (optical, electron beam, or x-ray) also imposes physical limitations on the formation of the latent image in a resist material. Resist materials and resist processing steps are responsible for latent image development into the final three-dimensional relief image.

Sensitivity, contrast, and *resolution* are the major lithographic factors considered in resist material performance.[3] Those parameters affect throughput of a particular exposure tool (sensitivity) and determine the dimensions of the pattern replicated in resist (contrast and resolution). Resist sensitivity is commonly defined as the exposure energy per unit area (*exposure dose*) required to achieve the desired chemical changes in a resist material after development (i.e., complete removal of the positive resist, or cross-linking of the negative resist material to the required thickness in the irradiated areas). Resist sensitivity and contrast can be experimentally measured from the response curve, where normalized resist thickness (ratio between resist thickness in the exposed and unexposed areas after development) is plotted as a function of the logarithm of the exposure dose, Fig. 13.4.

For a negative resist, response curve *A*, at the exposures below threshhold dose D_n material remains completely soluble in the developer. At exposure levels several times greater than the threshhold limit, the negative resist images become almost completely insoluble in the developer solution. The sensitivity of a negative resist is measured as the exposure dose, D_{nr}, at which developed area maintains a required thickness *r*.

For a positive resist, response curve *B*, the film solubility in the developer gradually increases with the exposure dose. The sensitivity of a positive resist is defined as the exposure dose D_p required to completely dissolve resist material in the exposed area during the development process. Positive resist (γ_p) and negative resist (γ_n) contrast can be calculated from Fig. 13.4 by extrapolating linear portions of the resist response curves:

$$\gamma_p = \frac{1}{\log D_p - \log D_{p0}} = \left(\log \frac{D_p}{D_{p0}}\right)^{-1} \tag{13.1}$$

$$\gamma_n = \frac{1}{\log D_{n0} - \log D_n} = \left(\log \frac{D_{n0}}{D_n}\right)^{-1} \tag{13.2}$$

where D_{p0} is the exposure dose at which the positive resist begins to clear, and D_{n0} is the exposure dose at which the thickness of the negative resist remains

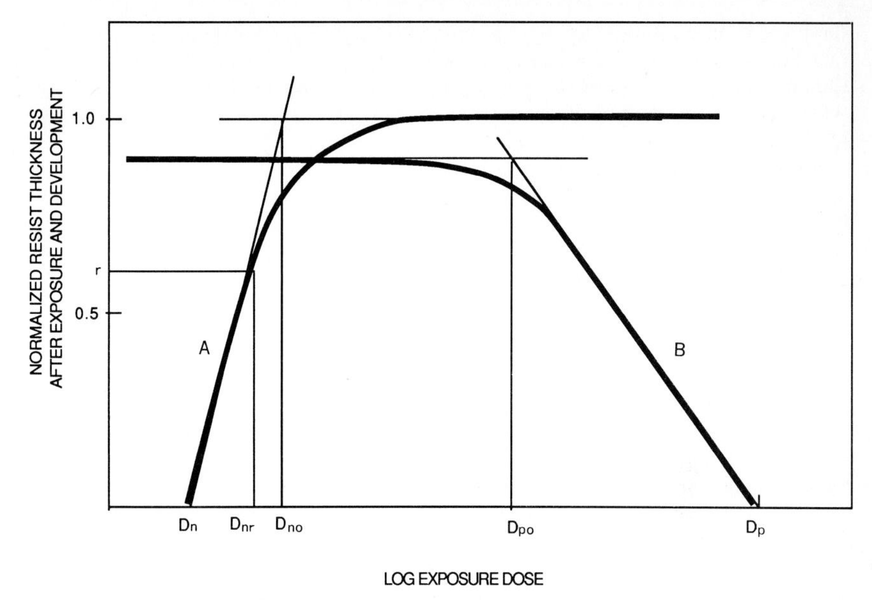

FIG. 13.4 Example of the resist exposure characteristic. (Curve *A*) Positive resist; (curve *B*) negative resist.

intact after development. The contrast has direct impact on the resolution properties (the ability to resolve minimum size images) and wall profile of a resist material after the development process.[4] Ideally, resist material should exhibit infinitely large contrast to obtain a vertical wall profile. In practice, positive organic resist materials demonstrate higher contrast ($\gamma \approx 2.0$), compared with the contrast of negative resists ($\gamma \approx 1.5$). For most negative resist materials, resolution is limited by irreversible resist deformation occurring during the development process. This is known as resist *swelling*. Positive resist materials do not show postdevelopment deformations and generally exhibit superior resolution qualities compared with negative resists. The high contrast and resolution of positive resists has made them almost the exclusive material of choice for optical VLSI lithography.[5] The resolution of a lithographic process depends ultimately upon the integrated environment in which exposure method, substrate material, resist properties, and processing parameters interrelate. These interactions determine pattern fidelity. *Optical lithography* or *photolithography,* with the radiation wavelengths around 200 to 450 nm is a predominant method of resist exposure for VLSI manufacturing. *Electron beam lithography* is used primarily in photomask fabrication processes, and *x-ray lithography* is slowly gaining acceptance as a potential method for submicrometer VLSI manufacturing.

13.2 OPTICAL LITHOGRAPHY

There are several methods of optical pattern transfer, or *printing,* of photomask images into the resist coated wafer: *contact, proximity,* and *projection* photolithography. In contact printing, the mask and wafer are brought into direct contact during the exposure process. Though intimate mask-to-wafer contact yields the best possible theoretical resolution, the defects introduced on both mask and wafer make this method of optical image transfer unacceptable for VLSI manufacturing.

In proximity printing, a gap is introduced between the mask and wafer surfaces. This reduces the contamination problems associated with contact printing but also degrades the resolution performance to a level unacceptable for VLSI device fabrication.

Projection printing systems utilize optical elements, lenses, and/or mirrors to project the mask image onto the wafer surface. Projection printing completely avoids mask and wafer contamination problems associated with contact exposures and significantly improves the resolution capabilities of proximity systems. Projection optical cameras, or *projection aligners,* are almost exclusively used in VLSI production.

13.2.1 Projection Aligners

Two types of projection aligners are employed in VLSI manufacturing: *scanning aligners,* based on reflective optics, and step-and-repeat printers, or *steppers,* based on refractive or refractive-reflective optics. Scanning aligners, exemplified by the Perkin-Elmer Micralign 600 HT Series Projection Mask Alignment Systems,[6] use spherical mirrors to project the arc-shaped illuminated region of the mask onto a resist-coated wafer, Fig. 13.5. The mask and wafer are scanned simultaneously on the same carriage so that an entire mask image is transferred to the wafer surface. The scanning exposure method allows one to minimize the distortions and aberrations of the optical system by keeping the mask and wafer within the optimum zone of the illuminator. The 1:1 magnification ratio of the telecentrically designed optical system is independent of focal position. The addition of refractive elements, known as the strong and weak shells, allows the correction of magnification and optical aberration errors over the entire slit area. The broadband illuminator of the Micralign condenser system, Fig. 13.6, offers the capability of selecting any exposure band within the 240 to 440-nm ultraviolet (UV) radiation spectrum, adjusting the illuminated slit width, and setting the aperture to accommodate a broad range of desirable exposure conditions. Today, scanning aligners can expose wafers up to 150 mm in diameter with a resolution approaching 1.0 μm and a registration accuracy of ± 0.30 μm (3σ). The basic limitation of full-field scanners lies in the ever increasing wafer size trend, requiring the dimensions of the optical system and the mask size to grow proportionally in order to cover an entire wafer area.

The alternative to full-field aligners are optical steppers designed primarily around refractive projection optics. In such a system, the wafer surface is exposed sequentially by projecting a reduced or 1:1 mask image in a step-and-repeat fashion. The major advantages of this technology are (1) unlimited wafer size, (2) reduced sensitivity to mask-related defects (reduction steppers), (3) local correction for process-induced wafer distortions at each exposure step, and (4) improved resolution and registration performance. The stepping exposure method significantly limits productivity or throughput of the aligner. For improving stepper throughput and increasing the single exposure field size, 5:1 reduction cameras are the instrument of choice, as opposed to the 10:1 steppers used primarily for advanced VLSI manufacturing. Nonreduction or 1:1 steppers, though offering

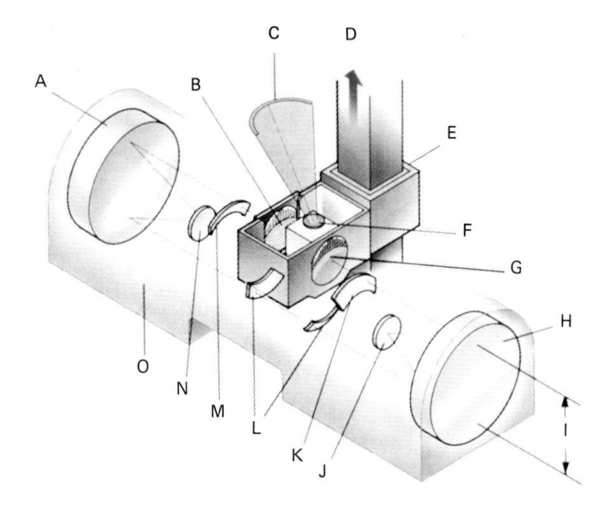

A. Primary Mirror
B. Illuminated Arc Passes Through Mask
C. Illuminated Slit Image/Condenser
D. Linear Scan Direction
E. Scanning Carriage Assembly on Air Bearing
F. Actinic Filter/Stationary Relay Package
G. Illuminated Arc Exposes Wafer
H. Primary Mirror
I. Telecentric Optical Path
J. Secondary Mirror
K. Weak Shell
L. Strong Shells
M. Weak Shell
N. Secondary Mirror
O. Helium-Filled Environmental Chamber

FIG. 13.5 Micralign 600 HT Series Projection Optics. (*Courtesy of Perkin-Elmer Corporation.*)

higher throughput, lack one of the major advantages of the reduction steppers—their relative insensitivity to mask-related defects. Regardless of the lens employed, almost all optical steppers feature a highly precise mechanical stage with a closed-loop positioning system controlled by a laser interferometer. Field-by-field registration and focusing, combined with wafer-plane-leveling corrections, allow steppers to match previously distorted patterns and compensate for the shallow depth of focus associated with the high-resolution refractive lenses.

The refractive optical lenses are the heart of the reduction step-and-repeat cameras. The resolution R and depth of focus F of a projection optical system are functions of the exposure wavelength λ and numerical aperture NA of the system:

$$R = \frac{k\lambda}{\mathrm{NA}} \tag{13.3}$$

$$F = \pm \frac{\lambda}{2(\mathrm{NA})^2} \tag{13.4}$$

where k is a constant, which is commonly taken as 0.5 for a theoretical diffraction-limited optical system, 0.6 in a laboratory-controlled environment, or

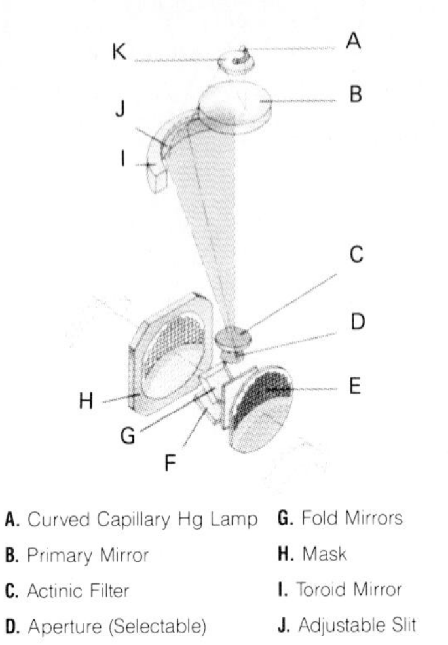

FIG. 13.6 Micralign 600 HT Series Condenser System. (*Courtesy of Perkin-Elmer Corporation.*)

0.8 in production manufacturing.[7] In practice, the constant k value reflects the integrated performance of the lithographic process, including such factors as exposure camera tuning, substrate surfaces and topography, and photoresist materials and processing. It is clear from the above expressions that reduction of the exposure wavelength and increase in the numerical aperture, while yielding higher-resolution performance, reduce the depth of focus of the exposure system. Nevertheless, by employing shorter-wavelength exposure radiation and larger numerical aperture optics, projection lithography is rapidly approaching the 0.5-μm feature size resolution limit, which just a few years ago was thought to be achievable only by e-beam or x-ray lithography.[8–10]

13.2.2 Photoresist Materials and Processes

Positive resist systems are predominantly used in optical VLSI lithography. A typical positive resist system consists of a low-molecular-weight *novolac resin,* which forms a resist film and acts as a *matrix* for the appropriate diazonaphthaquinone *sensitizer,* or *photoactive compound* (PAC). The novolac resin and sensitizer are dissolved in organic *solvent.* The resulting solution can be spin-coated on a wafer and then baked to remove the solvent and form a uniform novolac film coating. The PACs used in positive photoresists are *diazonaphthoquinones,* which inhibit the dissolution rate of the novolac resin in an aqueous base *developer* solution. The novolac film containing about 20 wt % of diazonaphthoquinone sensitizer typically dissolves several orders of magnitude slower than the novolac resin without PAC. Upon exposure to uv radiation, the diazonaphthoquinones photochemically decompose and undergo a chain of molecular reactions with a carboxylic acid as a final product.[11] The complete photodecomposition of PAC leads to a significantly increased dissolution rate of the novolac matrix in the exposed area during the resist development process. The unexposed resist remains several orders of magnitude less soluble in the developer solution, leaving a resist relief image on the wafer surface after development. That the unexposed resist region is not permeated by the development solution greatly improves resolution capability of positive resist systems (absence of postdevelopment swelling), compared with that observed for negative resists.

The variety of positive resist systems available commercially allows optimization of lithographic performance for a specific application. Among major considerations affecting selection of a photoresist system are lithographic sensitivity in the spectral range of exposure camera; high contrast and resolution capabilities with wide processing latitude; reduced sensitivity to variations in substrate reflectivity and resist thickness changes over substrate topography; good adhesion to the different thin films, such as poly-crystalline and single-crystal silicon, oxides, nitrides, and aluminum alloys; ability to withstand plasma and wet etch environment; high-temperature stability and tolerance to the ion implantation processing; low particle and metal ion content; and consistency of overall performance in high-volume manufacturing. A detailed description of the individual processing steps affecting resist performance can be found in Ref. 3. While the overall optimization of the photoresist process can be carried out to enhance certain resist system parameters, the compromise of other performance criteria is often unavoidable.

With the minimum feature size in advanced manufacturing rapidly approaching 1.0 μm, it is becoming more and more apparent that the three-dimensional resist image control (vertical profile) is as important as linewidth control (lateral resist dimensions). The requirements imposed on photoresist thickness are often determined by the plasma etching or ion implantation processing steps rather than by lithographic (resolution) considerations. In many instances, vertical dimensions of topographical features create significant local resist thickness variations, and bulk exposure effects may dominate linewidth control across such topographies.

Figure 13.7 illustrates exposure variations within a photoresist region due to local resist thickness changes, as well as differences in reflectivity over substrate materials, combined with interference effects. The photoresist thickness h on top of narrow topographical features can be substantially lower than the photoresist thickness H in the field area, since photoresist tends to planarize the substrate surface. These variations in the resist film thickness lead to an uneven exposure dose distribution across topographical features, and a corresponding linewidth variation. Similar linewidth distortions may result from variations in reflectivity of the substrate material within the exposed region (metals, polysilicon, oxide, etc.), with the reflected radiation contributing to the local exposure dose.

Incident rays may interact with the rays reflected from different optical interfaces and produce interference effects in the bulk of photoresist, known as *standing wave* effects. The magnitude of this effect is related to the substrate reflectivity, photoresist and dielectric film thicknesses, and their refractive indices at the exposure wavelength. Standing waves contribute to the periodic exposure energy distribution throughout the photoresist layer, and can substantially distort feature dimensions, especially over sharp steps where resist thickness changes abruptly. The incident light can also be scattered at the edges of highly reflective topographical features, causing stray photoresist exposures. The resulting local variations in linewidth are known as the *notching effect*.

A large literature exists describing the various methods designed to decrease the undesirable bulk photoresist exposure effects. An overview of these tech-

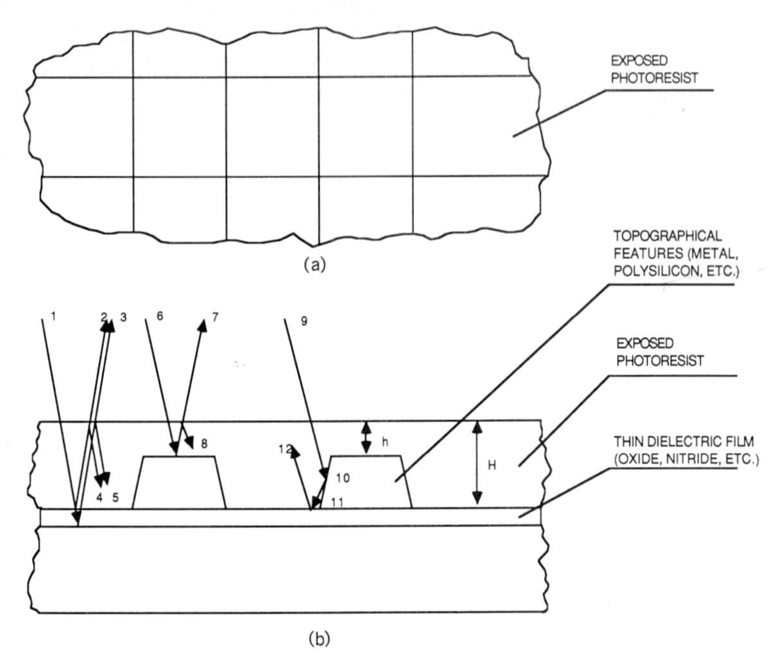

FIG. 13.7 Illustration of the bulk exposure effects in photoresist. (a) Top view; (b) cross-sectional view.

niques can be found in Ref. 12. While some of these methods complicate photoresist processing, they substantially improve the lithographic performance of optical cameras and extend their resolution limits into the submicrometer range. The use of contrast-enhancement materials (CEM) on top of photoresist films,[13] antireflective coatings (ARC) below the photoresist surface,[14] and the addition of unbleachable dyes in the bulk of photoresist[15] all relieve some of the resist exposure problems and do not require involvement of the additional type of processing equipment.

One of the most effective (and most complicated) methods of avoiding reflectivity and resist thickness nonuniformity problems involves anisotropic reactive ion etching (RIE) image transfer into the thick planarizing polymer layer.[16] The thick layer of polymer is spin-coated on top of the wafer and covered with the thin silicon dioxide film. The active photoresist layer is applied on top of this structure. The underlying polymer coat substantially smooths topographical irregularities and absorbs exposure radiation so that the imaging photoresist layer exhibits good film uniformity and is not affected by various substrate reflectivity effects. After exposure and development, the photoresist image is transferred by the RIE process into the thin silicon dioxide layer and then into the thick polymer layer, using the delineated SiO_2 as an RIE mask. The imaging photoresist layer can be coated very thin, since it has to withstand only the SiO_2 etch step.

The described RIE-based image transfer process does not impose any additional requirements on photoresist masking properties, so resist selection and processing can be optimized to enhance resolution performance. Planarized topography and a thin top photoresist layer also improve the resolution limit of optical cameras by relaxing depth-of-focus requirements. Very similar resist processing sequences offer many comparable advantages in electron beam and x-ray lithography by enhancing resist layer sensitivity or suppressing undesirable scattering effects in the imaging layer.

13.3 ELECTRON BEAM LITHOGRAPHY

The ability of a focused electron beam to expose radiation-sensitive polymers and form appropriate circuit patterns is the basis of electron beam lithography. An electron beam (e-beam) can be focused to a diameter of a few nanometers spot size and rapidly positioned by electrostatic or electromagnetic deflection systems. The capability of directly generating a pattern from CAD data has made electron beam lithography the dominant technology for optical mask generation in VLSI manufacturing. Direct e-beam wafer exposure is also used when rapid turnaround time or submicrometer pattern generation is required. The resolution of e-beam lithography is not limited by diffraction effects. However, electron scattering in the resist and underlying substrate material imposes limitations on the ability of an electron beam to resolve small features placed in close proximity to each other. The same electron scattering makes it possible to detect alignment marks on the substrate surface and perform precise pattern overlay. Pattern placement accuracy, resolution capabilities, and productivity of e-beam exposure systems very much depend upon the particular system design and writing strategy.

13.3.1 Electron Beam Lithography Systems

All electron beam lithography (EBL) systems employed in VLSI manufacturing use a scanned, focused, electron beam to expose resist material in a serial fashion, which means that every pattern element is addressed sequentially. For exposure time to be reduced, the current density in the focused e-beam must be high, and the beam must be shaped to a size consistent with the required resolution. In some EBL systems the beam shape can be dynamically varied to optimize writing time by matching the beam and feature size. *Variable-shaped* e-beam systems have much higher writing speed than *gaussian-shaped* round beam tools,[17] while fixed-beam-shape systems can deliver superior resolution performance.[18] Two basic strategies are used for addressing individual pattern elements: in the *raster scan* mode the total exposure field is scanned while the beam is turned on and off according to the pattern density; in the *vector scan* mode the e-beam is deflected in two dimensions, and only exposed features are addressed during the writing sequence. In general, the maximum deflection of the e-beam is much smaller than the required total exposure area, and the substrate is either continuously moved under the scanning e-beam or positioned in the step-and-repeat fashion by a laser-interferometer-controlled mechanical stage.

The design of EBL machines is very much tailored to the particular application—optical mask writing, high-resolution lithography, or direct wafer exposure for rapid turnaround manufacturing. In all instances, EBL systems consist of the *electron gun* (or electron source), the *electron-optical column* (various lenses, apertures, and deflection systems to focus, shape, position, and modulate the electron beam), the *mechanical stage* to move the substrate under the focused e-beam, the *substrate handling system,* and the *control computer.*[19] Two EBL systems manufactured by the Perkin-Elmer Corporation undoubtedly represent the state of the art for both applications in VLSI manufacturing[20,21]—mask writing (MEBES III) and direct wafer exposure (AEBLE 150).

The MEBES III machine was designed around a *gaussian-shaped* round beam and *raster scan* deflection philosophy. The raster scan approach using periodic beam movement has several advantages in a mask writing task where an extremely high pattern placement accuracy is required for multiple layer-to-layer overlays. A one-dimensional periodic scan can be accomplished with a low deflection bandwidth, and scan distortions are easy to compensate; the pattern data processing from the CAD to the e-beam machine format can be accomplished very efficiently, and the tone of the pattern can be changed from positive to negative by simply inverting the signal to the beam modulator. Low writing speed, associated with the raster scan approach, limits MEBES III output to about one mask level per hour, which is quite acceptable for optical mask production.

The AEBLE 150 system was designed to maximize productivity requirements for direct wafer exposure, where a *variable-shaped* e-beam and *vector scan* strategy are advantageous in order to achieve the high throughput goals. The beam shape in the AEBLE 150 can vary dynamically with a repertoire of 0.25- to 2.0-μm rectangles and right triangles and current densities between 10 and 200 A/cm^2. The deflection electronics positions the e-beam within the 2- by 2-mm major fields and executes pattern exposures in the 64- by 64-μm minor fields. High beam current makes it possible to use almost any electron-sensitive resist with the laser-controlled stage continuously scanning an entire 150-mm-diameter wafer area under the focused e-beam. Direct e-beam wafer exposure requires modification of the pattern placement in real time to accommodate substrate distortions due to the wafer processing steps or distortions introduced by the optical exposure tools at the previous levels. The ability of the AEBLE 150 to detect and measure the location of resist-covered registration marks makes it possible to perform global wafer registration, update beam drift, map distortions, and check registration accuracy.[21] Economic considerations make it very desirable to overlay electron beam exposures with optically defined features. In hybrid lithography the e-beam patterning can be done to achieve optimum resolution and placement requirements, while optical lithography can be used at less critical levels to minimize manufacturing cost.[22,23]

13.3.2 Resolution and Registration Limits

During electron beam exposure the electrons are scattered through inelastic and elastic collisions with the atomic nuclei of the resist and substrate materials. Inelastic collisions result in electron energy loss and are largely responsible for exposure effects in the resist material by breaking or linking the chemical bonds of the polymer. Elastic collisions change only the trajectories of the electrons and result in spreading out the impinging electron beam. Forward scattering of the electrons broadens the incident beam in the resist material, while electron backscattering from the substrate material causes scattered electrons to return to the resist-coated surface a great distance from the impinging point, as shown in Fig. 13.8. The resist material is exposed by the incident electron beam and by contributions from the numerous scattered electrons. The combined effect of this integration, known as the *proximity effect,* is that the exposure of any particular area is affected by the exposure of resist in neighboring areas. The result is that

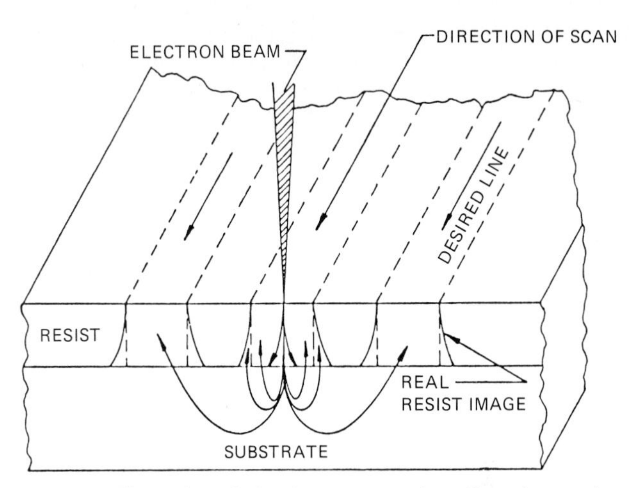

FIG. 13.8 Illustration of the electron scattering effects in a resist-coated substrate. (*Reprinted with permission from J. S. Greeneich, "Electron-Beam Process," in G. R. Brewer (ed.), Electron-Beam Technology in Microelectronic Fabrication, Academic Press, New York, 1980.*)

the feature size depends upon the proximity of adjacent exposure points within any particular figure (*intraproximity* effect) and between figures spaced in close proximity (*interproximity* effect). While intraproximity effects could be corrected by empirically adjusting the exposure dose, depending upon the feature size, interproximity corrections require detailed consideration of the exposure dose and the position of patterns in relation to each other.[24,25] For high-density and large-area VLSI patterns, the proximity correction task becomes a very time-consuming process, which requires extremely elaborate computer algorithms and results in much longer e-beam writing time. Proximity effects can be suppressed rather than corrected by adjusting the electron beam energy, writing strategy, and resist thickness, and by using multilayer resist structures to reduce exposure contributions from electron scattering. One of the proximity correction methods involves exposure of the field area at a lower dose with a defocused electron beam after the pattern exposure has been completed.[26,27] By equalizing energy profiles for all exposed patterns, this method greatly reduces proximity effects at the expense of lowering image contrast and requiring additional exposure time to address the field area. The particular proximity correction strategy employed very much depends on the application and pattern fidelity requirements, as well as upon the electron beam exposure system utilized for a given exposure task.[28] It is quite acceptable to assume a 0.5-μm resolution limit in the EBL systems for both mask making and direct wafer exposures in advanced VLSI fabrication.

In overlay performance the EBL systems have demonstrated their superiority to optical cameras. The same electron scattering effects that limit ultimate resolution of the focused e-beam allow topology and atomic number contrast imaging for registration mark acquisition. The registration capabilities of EBL systems are limited by absolute, field-butting, and registration errors.[29] The absolute, or machine-matching errors, are important for mask writing, if a good overlay between a set of masks produced on different machines is required. The absolute errors include unique uncompensated magnetic fields, electron optics, and deflection system distortions within EBL machines, as well as metrological errors of laser interferometry. The absolute errors are not as important for direct-write machines, where local registration is performed to a previous layer. The butting or stitching errors are local overlay errors that result in boundary displacement of adjacent deflection fields. The main sources of butting errors are related to the beam deflection errors (electronic noise, static charging, and eddy current magnetic field distortions), electron-optics distortions, and least laser count errors. The overlay errors are usually associated with the long-term stability of the EBL system and largely depend upon thermal effects, stage and substrate flatness, electron-induced charging, and registration mark measurement accuracy. The pattern placement accuracy of the mask-writing EBL can be better than ± 0.12 μm (3σ),[20] and approach ± 0.15 μm in electron beam to optical camera overlays[21,23] for a direct-write e-beam machine.

13.3.3 Electron Beam Resists

The required properties of electron beam resist materials are very similar to those of optical resists and include sufficient sensitivity to radiation as measured by incident electron charge per unit area (C/cm^2), high resolution and contrast, resistance to wet and dry etching, adequate thermal stability to withstand ion implantation and other heat-producing process steps, good adhesion, defect-free uniform application properties, and ease of stripping after the pattern transfer has been completed. Exposure of resist by an impinging electron beam occurs by cross-linking polymer chains (negative electron resists) or by breaking polymer bonds (positive electron resists). The typical electron energies in the exposure beam (10 to 20 keV) are far greater than the polymer chemical bond energies, and both bond formation and bond scission processes are possible, with the predominant one determining the type of resist action.

It is very difficult to make a general comparison between different electron resists, since the specific application largely determines resist selection. Electron resist sensitivity is one of the primary concerns for all applications because of the serial nature of EBL exposures and significant throughput penalties associated with the slower resist. The availability of positive and negative resists for identical applications also affects economy of the e-beam exposure, since low-density patterns can be written efficiently only when the desired pattern tone and resist polarity are the same. Various positive- and negative-acting electron resists are presented in Table 13.1; generally, negative electron resists demonstrate much higher sensitivity, while positive resists have better resolution characteristics.

Optical mask generation and direct wafer exposures are the major applications for e-beam lithography in VLSI manufacturing. The resist requirements for optical masks are significantly different from those for direct wafer writing, although with the increased demand for high resolution near micrometer and submicrometer optical masks, those needs are gradually converging. The absence of topology in photooptical mask fabrication sets limits on electron resist thickness to about 0.5 μm. Silicon wafers usually exhibit topographical features and etch requirements that increase the electron resist thickness range to 1.0 to 2.0 μm. Thick resists not only demand a larger exposure dose but also reduce the ultimate e-beam resolution limit, due to electron scattering and negative resist swelling in the developer solution. Multilayer processing schemes, similar to those used in optical lithography, can be successfully em-

TABLE 13.1 Performance of Electron Beam Resists

	Image type	Trade name	Manufacturer*	Sensitivity, μ Cl/cm^2	Resolution, μm
Poly(glycidyl methacrylate)	−	OEBR 100	TO	0.7	1.5
PGMA co poly(ethyl acrylate)	−	COP	MD	0.7	1.5
Poly(chloromethylstyrene)	−		IIP	0.5	1.0
Poly(iodostyrene)	−	RE4000N	II	1.8	1.0
Poly(methylmethacrylate)	+	PMMA	N	80	0.05
PMMA co poly(acrylonitrile)	+	OEBR 1030	TO	30	>0.5
Poly(trifluro-α-chloroacrylate)	+	EBR 9	T	15	>0.5
Poly(fluroalkylmethacrylate)	+	FBM	DK	??	>0.5
PMMA Xlinked	+	PM	M	40	>0.5
Poly(butene sulphone)	+	PBS	MD	2.0	0.5
Novolac-diazoquinione resist	+	AZ 1350	AZ	15/50	0.5
Novolac-poly(methylpentene sulphone)	+	RE 5000P	H	5.0	>0.5
Poly(styrene)-tetrathiofulvalene	±		IBM	8.0	>0.5
Inorganic GeSe/AgSe	−		N	500	>0.5
Poly(siloxane)	−		IBM	1.5	>0.5

*Company codes: TO = Tokyo Okha, MD = Mead, HP = Hewlett-Packard, TS = Toyo Soda, N = Numerous, DK = Daikin Kogyo, M = Microimage, AZ = AZ, H = Hitachi, IBM = IBM

Source: Adapted from M. P. S. Watts, "Electron Beam Resists Systems—A Critical Review of Recent Developments," *Solid State Technology,* Feb. 1984, p. 112.

ployed for electron resist exposures to optimize resist thickness-sensitivity-resolution requirements and greatly reduce proximity effects. The availability of high current density, variable-shape EBL systems for direct wafer writing, such as AEBLE 150, will undoubtedly stimulate new ideas in electron resist formulations, such as self-developing (ablative) and plasma-developable electron resists.[30]

13.4 X-RAY LITHOGRAPHY

Using x-rays for high-resolution pattern replication has potential for economical, high-throughput, submicrometer, VLSI production lithography. X-rays with wavelength of 4 to 50 Å essentially eliminate the diffraction limitations of optical lithography in proximity printing systems. The technology also offers immunity to organic particle defects and to the reflection and scattering problems associated with the optical and electron beam exposure methods.

13.4.1 Proximity Printing with X-Rays

Because there are no high-resolution x-ray optics available, x-ray lithography is currently limited to proximity printing. The lithographic process includes a mask (patterned x-ray absorbing material on a thin x-ray transparent membrane), an x-ray source, and an x-ray–sensitive resist material. For all practical considerations, the x-rays travel along a straight line, and resolution limitations of x-ray lithography are strongly associated with geometrical realities of proximity printing, Fig. 13.9. The sharpness of the feature edges defined in the resist material is determined mainly by resist and x-ray source characteristics and by geometrical factors, such as penumbra shadow, mask absorber thickness, and wall profile.[31]

A penumbral blur p occurs as a result of the finite extent S of the x-ray source, the distance D from the source to mask, and the gap g between the mask and the resist-coated substrate surface:

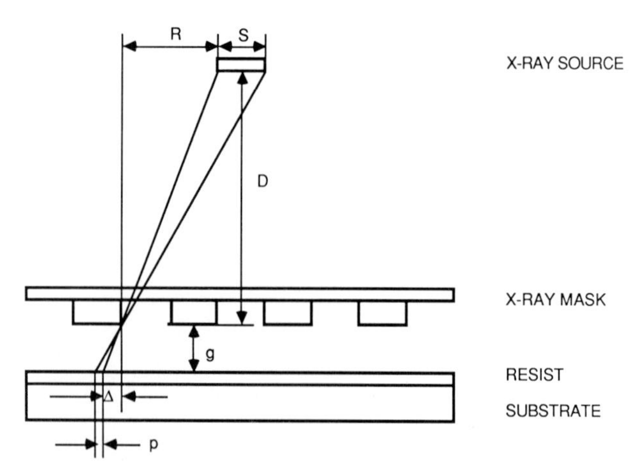

FIG. 13.9 X-ray proximity lithography.

$$p = \frac{gS}{D} \tag{13.5}$$

The penumbral blur alone can be as large as 0.2 μm for a typical x-ray exposure system (g = 20 μm, S = 3 mm, and D = 30 cm).

The same geometrical factors are responsible for the field magnification error Δ, which increases linearly from zero in the center to the maximum runout value at the edges of exposure field:

$$\Delta = \frac{gR}{D} \tag{13.6}$$

where R is the radius of the exposure field. For a field size of R = 100 mm this runout will be almost 7 μm at the edges of the wafer, assuming g = 20 μm and D = 30 cm. This magnification error must have the same value for each exposure level in the fabrication sequence, with the variations being less than the registration tolerance. Stringent mask-to-wafer gap control is required; for R = 100 mm and D = 30 cm, the 1.0-μm-gap variation will result in misregistration of 0.33 μm at the edges of the exposure field. Mask and substrate nonflatness aggravate this problem, since wafer warping in processing does not allow equal mask-to-wafer separation over the entire exposure field.

To reduce magnification error and technological problems associated with the large field x-ray masks, the equipment suppliers moved from full-field x-ray aligners[32] to the step-and-repeat machines.[33,34] However, the loss of productivity associated with the low sensitivity of available x-ray resists and limited brightness of conventional x-ray sources penalize the step-and-repeat approach.

13.4.2 X-Ray Resists and X-Ray Sources

Absorption of x-rays in the resist material produce photoelectrons, which are largely responsible for chemical changes in resist. That is why most electron-sensitive resists can also be used for x-ray exposures.[35] The high correlation between electron beam and x-ray resists sensitivity suggests that the radiation chemistry is basically the same for both radiations. The energy of x-ray–generated photoelectrons is in the 0.3- to 3.0-keV range, and their path lengths in resist increase from 50 to 650 Å as a function of the wavelength of x-ray radiation. Electron scattering proximity effects for x-ray exposures are negligible compared with those in EBL, and resolution well below 0.5 μm has been routinely demonstrated. For most polymeric resist materials only a small amount ($\not\approx$ 10 percent) of x-rays get absorbed in resist itself, resulting in excellent exposure uniformity across resist thickness at the expense of reduced effective resist sensitivity. This is why all major development efforts in x-ray resists are aimed at increasing the radiation sensitivity in order to compensate for the low intensity of conventional x-ray sources. Negative x-ray resists are generally more radiation-sensitive with lower resolution, compared with less sensitive, high-resolution, positive x-ray resists.[36]

Conventional electron impact x-ray sources generate a broad x-ray spectrum centered around the characteristic line of the target material. The high-energy electrons are focused onto a rotating or stationary liquid-cooled target and x-rays are generated with an efficiency of less than 1 percent. Laser and pulsed plasma sources are under development and can emit x-rays with energy densities about 50 times brighter than the electron impact x-ray sources.[37,38] Nevertheless, electron impact and laser plasma x-ray sources deliver very low x-ray fluxes on the order of 10 to 100 $\mu W/cm^2$ at the wafer plane and require extremely sensitive resist materials.

The synchrotron, or storage ring, x-ray sources generate a broad spectrum of radiation by electron energy loss in circular motion at relativistic speed. The radiation is highly collimated, and x-ray fluxes can exceed 10 mW/cm^2 at the wafer plane.[39] At these energy densities conventional electron beam and even optical resists can be employed. The collimated beam of the synchrotron x-ray sources eliminates most of the problems associated with the geometrical limitations of x-ray proximity printing, such as penumbral blur, mask absorber thickness shadowing, and runout magnification errors. However, the high construction and utilization cost of storage rings limits current application of synchrotron x-ray sources to universities and national laboratories.

13.4.3 X-Ray Masks

The x-ray mask fabrication presents the most difficult and complicated aspect of x-ray lithography. An x-ray mask consists of a transmissive membrane that supports patterned x-ray absorbent material. The thicknesses of the x-ray mask components are determined by the transmission of these materials for the exposure radiation. Gold has been widely used as an absorber, with boron nitride and polyimide as a supportive membrane structure. The x-ray mask patterning is done by EBL, and image transfer into the gold absorber is achieved either by additive (electroplating) or subtractive (ion milling) processing. The detailed description of the x-ray mask fabrication process can be found elsewhere.[40,41] The requirements for the x-ray masks capable of printing VLSI structures in the submicrometer region combine qualities of submicrometer optical patterning (image size control, placement accuracy, and low defect density) and spatial stability of three-dimensional mask structure (thermal and mechanical). Considering the 1:1 pattern transfer nature of x-ray lithography, the mask fabrication becomes very challenging in a technical sense even without taking into account the cost involved in producing such a complicated three-dimensional mask structure. That is why applications of x-ray lithography are currently limited to demonstrations of its resolution capabilities rather than to submicrometer VLSI manufacturing.

13.5 TRENDS

The continual advancement of semiconductor manufacturing toward higher density and greater operating speed of VLSI circuits places tremendous demand on lithographic technology. The most important requirements can be expressed in terms of smallest circuit geometries (resolution), pattern overlay (registration), largest chip size (die area), and substrate dimensions (wafer size).

EBL is firmly established as a mask generation method, and is making steady inroads into the small-volume custom and semicustom VLSI manufacturing market. The superior resolution and registration capabilities of the e-beam systems make them indispensable for advanced development work.

At present, the x-ray lithography applications are limited to pilot demonstrations. The major technical challenges of this technology are availability of high-intensity x-ray sources and x-ray masks. As soon as these issues are resolved, x-ray lithography may emerge as a strong contender to optical lithography in high-volume VLSI manufacturing.

Optical lithography is firmly asserting itself as being capable of resolving 0.5-μm features, not so long ago thought to be possible only with the e-beam or x-ray exposure tools.[7–10] The alignment accuracy required to match this resolution performance, about 0.1 μm ($|\text{mean}| + 3\sigma$), remains the most challenging task for the next several years. Increasing die area and wafer size eventually will impose resolution and throughput limitations on performance of the step-and-repeat cameras, and new generation of subfield scanning optical aligners may extend high-volume VLSI manufacturing down to a 0.5-μm limit.[42] Only a very small fraction of the VLSI market calls for submicrometer resolution; furthermore, optical lithography will undoubtedly continue to dominate IC manufacturing in the foreseeable future.

ACKNOWLEDGMENTS

The author would like to thank J. L. Saltich and J. E. Martin for encouraging this work, and B. S. Stallard, J. N. Helbert, and A. J. Gonzales for critically reading the manuscript.

REFERENCES

1. W. B. Glendinning, "A VHSIC Lithography Overview," *Solid State Technology,* March 1986, p. 97.
2. J. D. Buckley, "Expanding the Horizon of Optical Projection Lithography," *Solid State Technology,* May 1982, p. 77.
3. L. F. Thompson and M. J. Bowden, "Resist Processing," in *Introduction to Micriolithography,* L. F. Thompson, C. G. Willson, and M. J. Bowden (eds.), Advances in Chemistry Symposium Series, **219,** American Chemical Society, Washington, D.C., 1983.
4. M. C. King, "Principles of Optical Lithography," in *VLSI Electronics, Microstructure Science,* N. G. Einspruch (ed.), Academic Press, New York, 1981, chap. 2.
5. P. S. Gwozdz, "Positive Versus Negative: A Photoresist Analysis," *Proc. SPIE,* **275:**156 (1981).
6. J. J. Greed, Jr. and D. A. Markle, "Variable Magnification in a 1:1 Projection Lithography System," *Proc. SPIE,* **334:**2 (1982).
7. H. L. Stover, M. Nagler, I. Bol, and V. Miller, "Submicron Optical Lithography: I-Line Lens and Photoresist Technology," *Proc. SPIE,* **470:**22 (1984).
8. J. Austin, G. Keller, and G. F. Witting, "A Submicron Optical Lithography Process," preprints of papers presented at Interface '86 Kodak Microelectronic Seminar, November 20–21, 1986.
9. M. Tipton, V. Marriott, and G. Fuller, "Practical I-Line Lithography," *Solid State Technology,* January 1987, p. 55.
10. V. Pol, "High Resolution Optical Lithography: A Deep Ultraviolet Laser-Based Wafer Stepper," *Solid State Technology,* January 1987, p. 71.

11. J. Pacansky and J. R. Lyerla, "Photochemical Decomposition Mechanisms for AZ-Type Photoresists," *IBM J. Res. Develop.,* **23**(1):42 (1979).
12. B. J. Lin, "Multi-Layer Resist Systems," in *Introduction to Micriolithography,* L. F. Thompson, C. G. Willson, M. J. Bowden (eds.), Advances in Chemistry Symposium Series, **219,** American Chemical Society, Washington, D.C., 1983.
13. D. R. Storm, "Optical Lithography and Contrast Enhancement," *Semiconductor International,* May 1986, p. 162.
14. R. D. Coyne and T. Brewer, "Resist Processes on Highly Reflective Surfaces Using Anti-Reflective Coating," *Kodak Interface Seminar Proceedings,* 1983, p. 40.
15. I. I. Bol, "High Resolution Optical Lithography Using Dyed Single-Layer Resist," *Kodak Interface Microelectronics Seminar,* October 1984.
16. J. M. Moran and D. Maydan, "High Resolution, Steep Profile Resist Pattern," *J. Vac. Sci. Technol.,* **16**(6):1620 (1979).
17. E. V. Weber, "Shaped Beams for Integrated Circuit Fabrication," *Proc. SPIE,* **333:**94 (1982).
18. M. Isaacson and A. Murray, "In-situ Vaporization of Very Low Molecular Weight Resist Using ½ nm Diameter Electron Beams," *J. Vac. Sci. Technol.,* **19**(4):1117 (1981).
19. D. R. Herriot and G. R. Brewer, "Electron-Beam Lithography Machines," in *Electron-Beam Technology in Microelectronic Fabrication,* G. R. Brewer (ed.), Academic Press, New York, 1980.
20. J. Freyer, K. Standiford, and R. Sills, "Enhanced Pattern Accuracy with MEBES III," *Proc. SPIE,* **471:**8 (1984).
21. J. L. Freyer and R. M. Sills, "AEBLE 150 Performance in a Mix and Match Environment," *Proc. SPIE,* **632:**5 (1986).
22. J. N. Helbert, P. A. Seese, A. J. Gonzales, and C. C. Walker, "Intralevel Hybrid Resist Process for the Fabrication of the Metal Oxide Semiconductor Devices with Submicron Gate Length," *Optical Engineering,* **22**(2):185 (1983).
23. K. A. Barnett, R. A. Metzger, and O. W. Otto, "AEBLE 150 E-Beam/Optical Hybrid Lithography," *Proc. SPIE,* **632:**17 (1986).
24. M. Parikh, "Self-Consistent Proximity Effect Correction Technique For Resist Exposure," *J. Vac. Sci. Technol.,* **15**(3):931 (1978).
25. M. E. Haslam and J. F. McDonald, "Submicron Proximity Correction by the Fourier Precompensation Method," *Proc. SPIE,* **632:**40 (1986).
26. G. Owen and P. Rissman, "Proximity Effect Correction For Electron Beam Lithography By Equalization of Background Dose," *J. Appl. Phys.,* **54**(6):3573 (1983).
27. Hua-yu Liu and E. D. Liu, "Application of GHOST Proximity Effect Correction Method to Conventional and Nonswelling Negative E-Beam Resists," *Proc. SPIE,* **632:**244 (1986).
28. B. Carlson and D. Burbank, "Practical Proximity Correction," *Proc. SPIE,* **632:**2 (1986).
29. C. K. Chen, "Electron Beam Error Sources," *Proc. SPIE,* **471:**2 (1984).
30. E. D. Roberts, "Recent Developments in Electron Resists," *Solid State Technology,* June 1984, p. 135.
31. R. P. Jaeger and B. L. Heflinger, "Linewidth Control in X-Ray Lithography: The Influence of the Penumbral Shadow," *Proc. SPIE,* **471:**110 (1984).
32. B. Fay, W. T. Novak, and I. Carlsson, "Automatic X-Ray Alignment System for Submicron VLSI Printing," *Proc. SPIE,* **471:**90 (1984).
33. B. S. Fay and W. T. Novak, "Advanced X-Ray Alignment System," *Proc. SPIE,* **632:**146 (1986).

34. R. B. McIntosh, G. P. Huges, J. L. Kreuzer, and G. R. Conti, Jr., "X-Ray Step-and-Repeat Lithography System for Submicron VLSI," *Proc. SPIE,* **632:**156 (1986).

35. L. F. Thompson, E. D. Feit, M. J. Bowden, P. V. Lenzo, and E. G. Spencer, "Polymeric Resists for X-Ray Lithography," *J. Electrochem. Soc.,* **121**(11):1500 (1974).

36. C. G. Willson, "Organic Resist Materials—Theory and Chemistry" in *Introduction to Microlithography,* L. F. Thompson, C. G. Willson, and M. J. Bowden (eds.), Advances in Chemistry Symposium Series, **219,** American Chemical Society, Washington, D.C., 1983.

37. B. Kuyel, "Prospects of a Plasma Focus Device as an Intense X-Ray Source for Fine Line Lithography," *Proc. SPIE,* **275:**44 (1981).

38. S. M. Matthews, R. Stringfield, I. Roth, R. Cooper, N. P. Economou, and D. C. Flanders, "Pulsed Plasma Source for X-Ray Lithography," *Proc. SPIE,* **275:**52 (1981).

39. R. P. Haelbich, J. P. Silverman, W. D. Grobman, J. R. Maldonado, and J. M. Warlaumont, "Design and Performance of an X-Ray Lithography Beam Line at a Storage Ring," *J. Vac. Sci. Technol. B.,* **1**(4):1262 (1983).

40. G. E. Georgiou, C. A. Jankoski, and T. A. Palumbo, "DC Electroplating of Sub-Micron Gold Patterns on X-Ray Masks," *Proc. SPIE,* **471:**96 (1984).

41. A. R. Shimkunas, J. J. LaBrie, P. E. Mauger, and J. J. Yen, "Mask Technology for X-Ray Step-and-Repeat System," *Proc. SPIE,* **471:**106 (1986).

42. D. A. Markle, "The Future and Potential of Optical Scanning Systems," *Solid State Technology,* September 1984, p. 159.

P • A • R • T • 4

VLSI TECHNOLOGY SELECTION

CHAPTER 14
BASIC CIRCUIT TECHNOLOGIES

Bhadrik Dalal
Amdahl Corporation
Sunnyvale, California

14.1 DIGITAL LOGIC CIRCUIT TECHNOLOGIES

Circuit approaches for implementing logic functions are influenced by the choice of semiconductor material, device type, and IC product requirements. The choice of semiconductor material, whether elemental silicon or compound gallium arsenide, is governed by the key material properties discussed earlier (see Sec. 5.1.3), namely, the bandgap, carrier velocity, and carrier mobility. Silicon ICs represent more than 98 percent of all ICs used today. Gallium arsenide (GaAs) and related III-V compound digital ICs have been in the research stage for more than a decade, and are just now beginning to find some applications in the industry.[1]

The choice of device, whether field-effect transistor (FET) or bipolar junction transistor (BJT; see Sec. 5.2), depends largely on the desired key device properties (see Sec. 5.2.3), namely, the size, input capacitance, transit time, threshold voltage, and transconductance.

The interdependent key IC product requirements are cost, integration level, and performance. The cost of an IC depends primarily on the processing complexity, chip size, yield, test complexity, and package. The practical integration level is largely dependent on the digital logic IC product classification, namely, application specific integrated circuits (ASICs), standard logic circuits, custom circuits, and microprocessor circuits. The integration level is also limited by the maximum allowable power dissipation, which is a system requirement imposed by the application. The performance is affected by the choice of device, the integration level, and the maximum allowable power dissipation.

The two major classifications for circuit technologies are unipolar circuits and bipolar circuits. The unipolar circuits employ FET structures and are subdivided into MOSFET circuits and metal-semiconductor field-effect transistor (MESFET) circuits. The MESFET circuits are predominantly fabricated in GaAs material. However, the MOSFET circuits are fabricated only in silicon and constitute the dominant production technology. The bipolar circuits employ BJT structures and are subdivided into saturated circuits and nonsaturated circuits, based on whether the BJT is allowed to operate in the saturation region. If the BJT is allowed to saturate, minority carrier charge storage occurs in the base region, which degrades the transistor performance and, subsequently, the circuit performance. Consequently, only nonsaturating bipolar circuits are considered when the highest circuit performance is desired.

14.1.1 MOSFET Circuit Techniques

The design of MOSFET digital circuits falls into two major categories:[2] static circuits and dynamic circuits. Static circuits dissipate power in one or both of the logic states and do not suffer a lower limit on clock frequency for proper operation. Dynamic circuits employ charge storage on capacitors for memory functions and use clocked load elements and transmission gates. Periodic clock signals are necessary to ensure proper operation for dynamic circuits. Circuits in each of these categories can be implemented with the use of *n*-channel and/or *p*-channel MOSFETs. PMOS circuits, once popular, are now seldom used. NMOS circuits have in the past several years seen decreasing usage, while the usage of CMOS circuits has increased. The major reason for this shift is the nearly zero standby power of CMOS circuits, which makes it possible to have increased chip integration levels while still maintaining a low-chip-power dissipation budget.

NMOS:* n*-Channel MOS Logic. Static NMOS circuits can be implemented with a variety of different load structures.[3] Resistor loads, saturated enhancement transistor loads, linear enhancement transistor loads, and depletion transistor loads constitute the different load structures used with enhancement transistor drivers. The enhancement-driver–depletion-load (ED) structures are shown in Fig. 14.1 for NMOS ED circuits. The voltage transfer characteristic of the circuit differs for different load structures. The sharpest transfer characteristic is realized for a depletion-load structure. For the NMOS ED technology, the typical values of key circuit parameters are as follows:

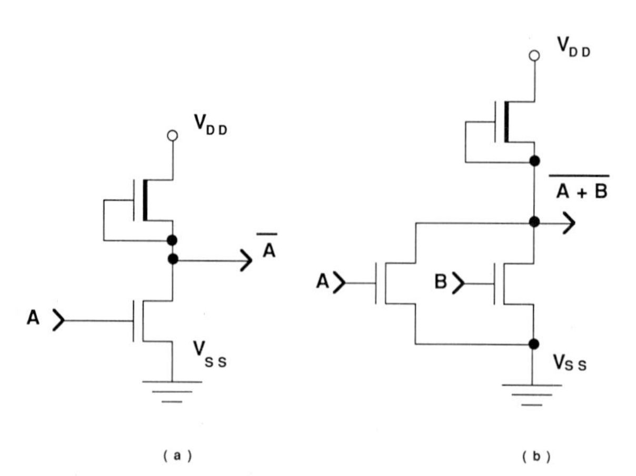

FIG. 14.1 NMOS ED circuits: (*a*) Inverter; (*b*) two-input NOR gate.

Supply voltage	$V_{DD} = 5$ V
Logic high voltage level	$V_{OH} = 5$ V
Logic low voltage level	$V_{OL} = 0.3$ V
Power dissipation per gate	$P_D = 118\ \mu$W

CMOS: Complementary MOS Logic. Complementary pairs of n-channel and p-channel transistors are used in implementing CMOS logic circuit functions[3] as shown in Fig. 14.2. Because only one transistor in each complementary pair is

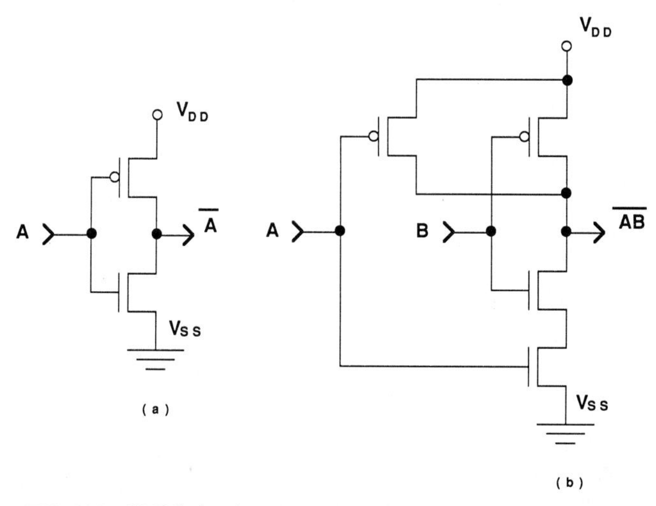

FIG. 14.2 CMOS circuits: (*a*) Inverter; (*b*) two-input NAND gate.

conducting during either of the logic states, the static power dissipation is essentially zero, determined only by the junction leakage current of the transistor that is cut off. This is the most attractive feature of CMOS circuits. Because of the need to fabricate both n-channel and p-channel transistors, the process complexity is increased over that of NMOS circuits. The increase in component count for multi-input circuit functions, and the generally larger size of transistors because of the need to isolate one polarity transistor from another, causes the circuit area to be somewhat larger than that of NMOS circuits. However, because of the complementary nature of the circuit, the voltage transfer characteristic can be designed to be almost ideal. The logic high level is equal to the positive supply voltage, and the logic low level is equal to the ground level. The supply voltage V_{DD} is generally 5 V. However, a major industry effort is underway to create a new second standard around 3.3 V to alleviate the problems of hot carrier injection found in scaled devices.

14.1.2 MESFET Circuit Techniques

MESFET circuit techniques fall into two major categories: depletion-mode (D-mode) circuits and enhancement-depletion-mode (ED-mode) circuits. n-Channel

GaAs depletion-mode MESFET technology was reported by Liechti in 1972.[4] In 1974, Van Tuyl and Liechti[5] reported a high-speed integrated logic gate implemented with GaAs MESFETs. This gate implementation has subsequently come to be known as buffered FET logic (BFL). Though BFL usually enables one to design the fastest MESFET integrated gate, it consumes a large amount of power. Source-coupled FET logic (SCFL) and direct-coupled FET logic (DCFL) are two other MESFET circuit techniques.

BFL: Buffered FET Logic. BFL uses only depletion-mode MESFETs.[5,6] The circuit comprises two sections: a logic switch-amplifier section, and a load-driver–level-shifter section. The number of components used per gate is large, and, as such, the circuit layout area tends to be rather large. Power dissipation per gate ranges from 1 to 10 mW, with a logic voltage signal swing ranging from 1 to 2.5 V, depending on the supply voltage and the number of level-shifting diodes used. This circuit approach uses two supply voltages and is very tolerant of process-related variation of device parameters.

DCFL: Direct Coupled FET Logic.[7] The simplest form of MESFET circuit generally requires both enhancement- and depletion-type devices, as shown in Fig. 14.3. However, the depletion-type device can be replaced by a resistor, if necessary. This circuit technique is called DCFL and is very similar to the bipolar direct-coupled transistor logic (DCTL) circuits of the early 1960s.

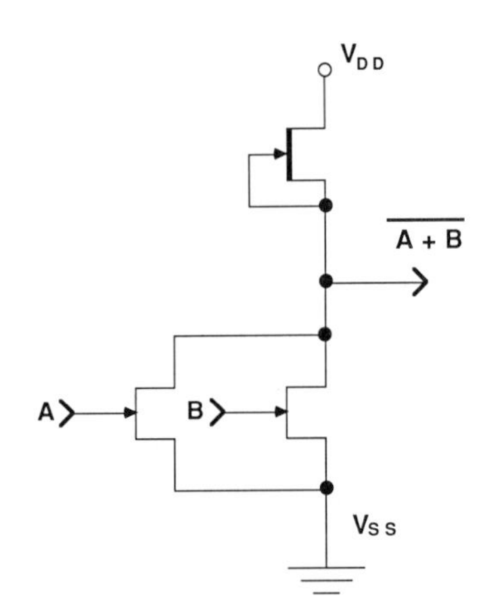

FIG 14.3 DCFL circuit: two-input NOR gate.

Because this circuit uses the least number of components per gate, the circuit layout area is very small. The logic voltage signal swing is 0.6 V, which is the difference between the turn-on voltage of a Schottky diode, about 0.8 V, and the logic low level of approximately 0.2 V. DCFL requires only one supply voltage, which ranges from 1.0 to 1.5 V. The operating current ranges from 100 to 200 μA, so the power dissipation per gate is much less than a milliwatt. The already small logic voltage signal swing is adversely affected by increasing fan-out, and the power dissipation per gate varies considerably with supply voltage variations. The circuit sensitivity to process-related variations in threshold voltage is the highest of all MESFET circuit techniques. Therefore, a very tight process control on threshold voltage is required.

SCFL: Source-Coupled FET Logic.[8] SCFL uses a source-coupled differential pair to provide the logic switch and amplifier function, and a source-follower to provide the level-shifting function, as shown in Fig. 14.4. The circuit is very tolerant of device threshold voltage variation and can operate with either enhancement- or depletion-mode transistors for the switching function. The logic voltage signal swing is about 1.2 V, and the circuit uses a single power supply. The number of components used per gate is very large. Consequently, the circuit layout area per gate is also very large. Power dissipation per gate ranges from 2 to 10 mW.

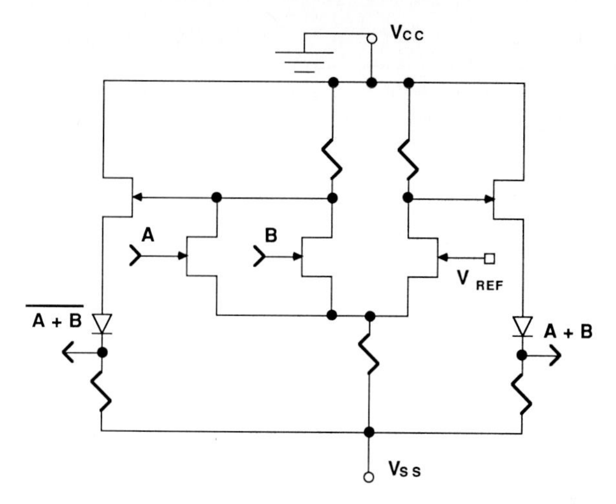

FIG. 14.4 SCFL circuit: two-input OR/NOR gate.

14.1.3 Bipolar Saturated Circuit Techniques

The earliest ICs were made using bipolar transistors. The simplest of the bipolar saturating logic circuits is the DCTL gate. The DCTL gate was improved and eventually evolved into a saturating circuit called transistor-transistor logic (TTL). TTL soon became the industry standard and dominated the IC industry for almost two decades. Innovations in bipolar circuit concepts such as integrated injection logic (I^2L) and Schottky transistor logic (STL) appeared in the early 1970s but did not make a significant impact on the industry. Most of the saturating bipolar circuits have been replaced by NMOS and CMOS circuits in new system applications.

DCTL: Direct-Coupled Transistor Logic.[9] The DCTL gate is the simplest bipolar saturating logic circuit, as shown in Fig. 14.5. It uses the least number of components to implement a logic gate. Early DCTL circuits did not have series input base resistors, but were added subsequently to reduce "current logging" effects caused by the process-related variation in base-emitter turn-on voltage V_{be}. The

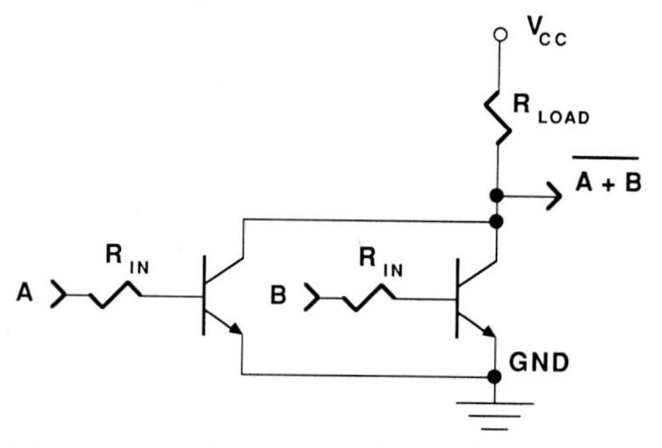

FIG. 14.5 DCTL circuit: two-input NOR gate.

logic voltage signal swing is about 0.6 V. The logic high voltage level is given by the V_{be} of a saturated bipolar transistor, which is approximately 0.8 V. The logic low voltage level is given by the V_{ce} of a saturated bipolar transistor, which is approximately 0.2 V. The logic voltage signal swing is dependent on the fan-out of the gate and reduces in value with increasing fan-out. Consequently, the fan-out is limited to four or five. The DCTL circuit uses a single power supply, which can range from 1.5 to 3.0 V. The operating current ranges from 0.5 to 2 mA, with power dissipation per gate ranging from 1 to 5 mW.

TTL: Transistor-Transistor Logic. The TTL family was introduced by Texas Instruments as a standard product line in 1964. The TTL circuits used in SSI and MSI circuits were complex structures that used many components.[10] The basic TTL circuit[11] used in LSI circuits is a much simpler structure, as shown in Fig. 14.6. The typical circuit uses a single power supply of 1.7 V, with operating currents from 30 to 270 μA, resulting in a power dissipation of 0.3 to 0.6 mW. The logic high voltage level is approximately 1.7 V, and the logic low voltage level is approximately 0.2 V, thus resulting in a logic signal voltage swing of about 1.5 V.

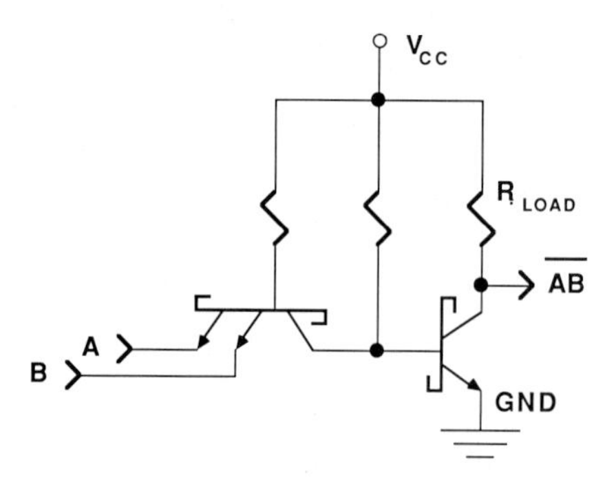

FIG. 14.6 TTL circuit: two-input NAND gate.

I²L: Integrated Injection Logic. In 1972, a new form of bipolar logic circuit technique was reported.[12,13] This circuit was named integrated injection logic and merged transistor logic (MTL) by the two teams who independently invented this concept. An example of this circuit technique is shown in Fig. 14.7.

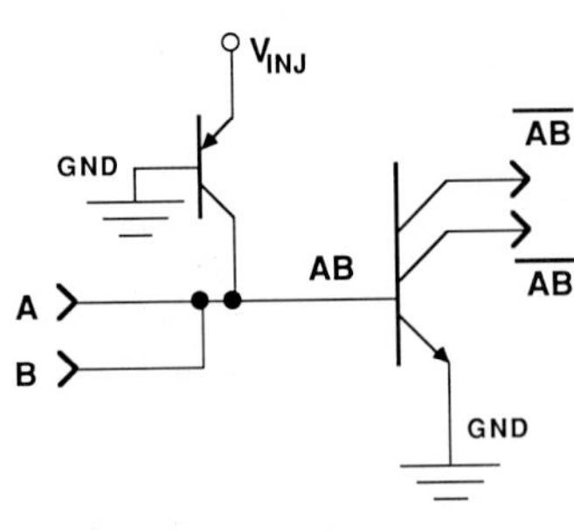

FIG. 14.7 IIL circuit.

In contrast with conventional circuits, which have multiple inputs and single outputs, the I²L circuit has a single input with multiple outputs. The multicollector *npn* transistors act as inverters with multiple outputs, whereas the logic function is realized by connecting collectors of different inverters in a wired-AND connection. This circuit concept utilizes merged complementary transistors, wherein lateral *pnp* transistors are used for current sources, and vertical *npn* transistors operated in the inverse region are used as switching transistors. Because of the merging of the transistors, a minimum number of components are needed to implement a logic function. The resulting circuit layout area is the smallest, even when compared with FET circuit techniques. Through control of the *pnp* current source, the power dissipation per gate can be varied from 1 nW to 1 mW.

The circuit actually requires a current supply rather than a voltage supply. However, a supply voltage of 5 V is generally used in conjunction with a current-limiting series resistor. The circuit has a logic voltage signal swing of about 0.7 V. The main disadvantage of this circuit is its limited high-speed performance because of the degraded gain-bandwidth product of the inversely operated *npn* transistor. However, some recent development efforts[14] have been successful in eliminating this performance limitation.

An evolution from I^2L circuit results in a circuit technique referred to as complementary constant-current logic (C^3L) or STL. This technique uses a Schottky clamped *npn* transistor with multiple Schottky diode outputs. The logic voltage signal swing is reduced to approximately 0.3 V, thereby speeding up the circuit performance. The processing complexity associated with the generation of two different threshold Schottky diodes and the system noise susceptibility due to the very small logic swing have been the key problem areas associated with this concept.

14.1.4 Bipolar Nonsaturated Circuit Techniques

As discussed earlier, to exploit the superior intrinsic performance of a bipolar transistor, one must operate the transistor out of the saturation region to prevent the charge storage of excess minority carriers in the base region. Therefore, bipolar nonsaturated circuit techniques provide the best intrinsic circuit performance. Current-mode logic (CML), emitter-coupled logic (ECL), and emitter-function logic (EFL) are such circuit techniques. Though they are often referred to as distinct and separate circuit techniques, they belong to the unique class of current-steered logic circuits. In the design of custom high-performance ECL ICs, the circuit designer often uses all of these circuit techniques in the design of a single IC to optimize performance. Because the logic interface levels are different in each of these circuit techniques, special care has to be exercised by the circuit designer to translate the logic levels appropriately. However, in ASICs only one circuit technique is generally used to keep the logic interface levels simple and consistent.

CML: Current-Mode Logic.[15,16] The basic function of a differential amplifier is to amplify the difference between two input signals. The differential amplifier, or the emitter-coupled pair as it is often called, represents the basic current switch used in all ECL circuits and its modifications. A CML circuit represents the simplest differential current switch, as shown in Fig. 14.8. The input to the switch can be either common mode or differential mode.

If the switching transistors are to be kept out of saturation, the input and output logic voltage levels of the basic switch need to be offset by V_{be}. However, to eliminate the need for level translation, we must allow the switch to go into soft saturation, such that the base-collector diode is never forward-biased by more than 500 mV. Consequently, to prevent the switching transistor from going into hard saturation, we must keep the input and output signal amplitudes to less than 500 mV in magnitude, centered around 250mV below ground level.

The basic attributes of a CML circuit are the small number of components per gate, the low power dissipation per gate, and the ability to provide complemen-

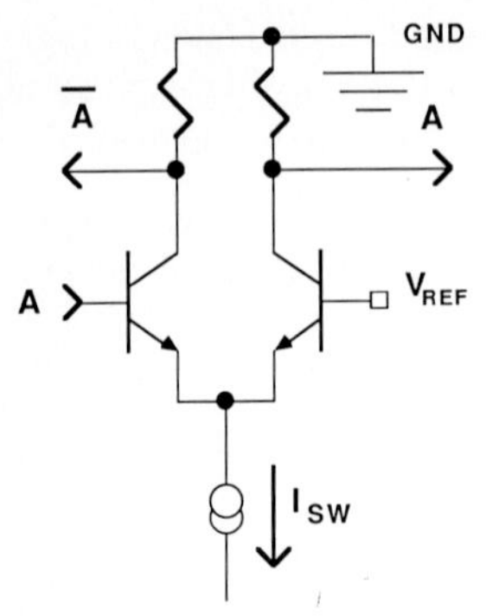

FIG. 14.8 CML circuit.

tary outputs. The disadvantages of a CML circuit are its slower speed, because of transistor operation in soft saturation, its low noise margin, because of the small signal swing, the lack of wired-OR capability at the outputs, and the lack of output drive capability. Nevertheless, in applications where a high degree of structured logic appears, CML circuits could be very effective. The typical logic voltage levels are 0 V and −0.5 V, with power dissipation per gate ranging from 160 μW to 3 mW.

ECL: Emitter-Coupled Logic.[16,17] Like the CML circuit, the ECL circuit uses the emitter-coupled pair as the basic switch. However, output emitter followers are added for output signal level translation. Because the differential pairs can be arranged in a series-parallel cascode arrangement, input-level translators are also used whenever multilevel ECL circuits are constructed. The ECL circuit concept preceded the invention of the transistor. In IC circuit form, it has been available since the mid-1960s. It has dominated the market for the highest-speed ICs for over two decades, and its reign may continue for possibly another decade.

The basic ECL circuit is comprised of a current switch (CS) and output emitter followers (EF), as shown in Fig. 14.9, and is often referred to as an ECL CS-EF circuit. The addition of inputs is achieved by adding input transistors to the emitter-coupled transistor group. The switching transistors in this ECL configuration do not saturate, and the EF's low output impedance enables it to drive large fan-out and signal interconnect loads. The ability to realize collector wired-

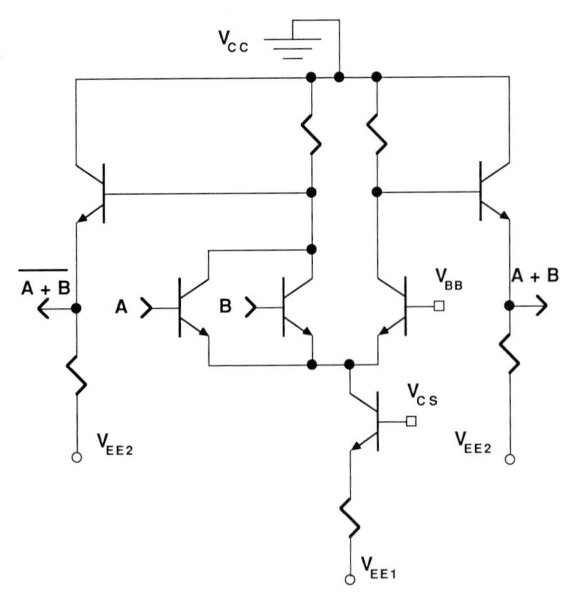

FIG. 14.9 ECL circuit: two-input OR/NOR gate.

AND and emitter wired-OR functions without incurring an additional gate delay penalty, the ability to have complementary output signals that have equal delays, and the ability to implement complex logic functions through multilevel series gating make ECL a circuit technology of choice for the highest-performance digital applications. The typical ECL logic voltage signal swing ranges from 500 to 800 mV, with logic voltage levels of −0.8 V and −1.6 V and power dissipation ranging from 0.5 to 10 mW. The price paid for this high performance is (1) high power dissipation, (2) large number of components per gate, and (3) limited integration level.

EFL: Emitter-Function Logic.[18,19] The basic structure of an EFL circuit is shown in Fig. 14.10. This circuit technique was proposed by Skokan in 1973.

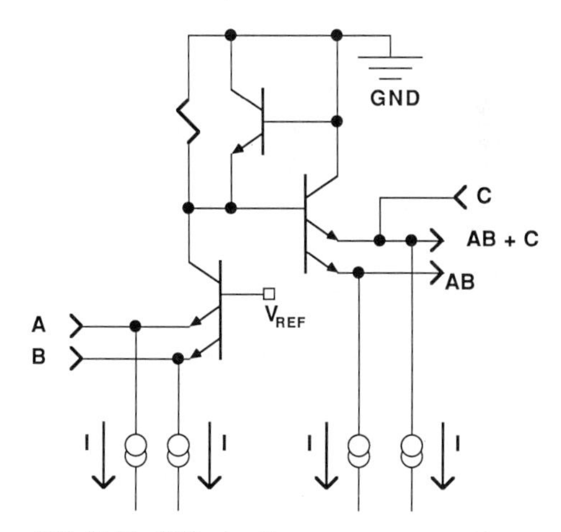

FIG. 14.10 EFL circuit.

EFL is based on a noninverting gate structure. However, inversions are realized through traditional ECL techniques. Because the circuit concept is based on multiemitter transistor structures for both inputs and outputs and uses a very small number of components, it is very area-efficient for noninverting functions. The multiemitter inputs perform the AND function, and the multiemitter outputs, which can be wire-ORed with outputs from other EFL gates, provide the OR function. This circuit concept, when used in conjunction with ECL circuits, provides a powerful combination for circuit optimization in terms of integration level, circuit performance, and power dissipation.

14.2 CIRCUIT CHARACTERISTICS

The characteristics of digital logic ICs can be described by circuit properties and circuit parameters. The properties of the circuit are its physical, functional, and electrical attributes. The parameters of the circuit are the values assigned to such attributes.

14.2.1 Circuit Properties

The major IC product requirements are cost, function, and performance. Circuit properties influence one or more of these requirements. Consequently, key circuit properties may be grouped into categories based on these product requirements.

Cost Considerations. The IC industry is driven by the desire to provide low-cost functions of ever increasing complexity. The major components of cost for an IC product are

Chip fabrication cost
Package and assembly cost
Product test cost

The package and assembly costs are influenced by the physical, thermal, and electrical considerations of the package. Among these considerations are the package size, number and type of chip connections, maximum allowable power dissipation, package thermal resistance, and the electrical high-frequency bandwidth of signal I/O connections.

The product test cost depends on the type of product, whether logic or memory, integration level, functional complexity, performance level, and the degree of built-in test designed into the product.

The chip fabrication cost is influenced by the wafer processing cost and yield. The wafer processing cost is related to the process complexity, number of mask levels, wafer size, cost of equipment, and other factors. The yield depends on chip area and defect density. The chip area depends on the product type and integration level.

Integration Level. Integration level is measured in number of components per IC. Other measures of integration level are the number of storage bits per IC for memory products and the number of logic gates per IC for logic products. The integration level of an IC depends on the circuit packing density for logic gates and memory bits. Circuit packing density is inversely proportional to the logic cell size and the memory cell size. The cell size is influenced by the number of devices required per function, device type, and device size. The size of a device depends on the minimum feature size.

Functional Considerations. The functional attributes of a circuit describe its ability to implement logic and storage functions, to tolerate random and self-generated noise, to swiftly and reliably communicate with other circuits within the same IC and other ICs, and to operate over a wide temperature range.

The value of a circuit in implementing logic functions depends on its ability and limitation in realizing multiple fan-in, multiple fan-out, logical OR, logical AND, logical invert, and "wired" functions. The ability to provide complementary outputs is also a valuable attribute.

The ability of a circuit to operate in a noisy environment depends on its

Logic voltage signal swing
Voltage transfer characteristic
Transition width

High- and low-level noise margins

Self-generated circuit switching noise

Coupled noise between signal interconnect lines

External noise sources

The ability of a circuit to swiftly communicate with other circuits in the same IC depends on the magnitude of its output current drive, the transconductance and intrinsic response time of the driving device, the fan-out load, and the on-chip interconnect net load. The ability of a circuit to swiftly communicate with other circuits in different ICs depends on its ability to drive the off-chip interconnect net load, its output current drive, its output impedance, and its intrinsic response time. The ability of the circuit to operate over a wide temperature range depends on the temperature dependence of the key device parameters that influence circuit characteristics.

Performance Considerations. The major factors influencing circuit performance are the internal gate propagation delay, the output driver propagation delay, and the worst-case delay distribution. The internal gate propagation delay is composed of three basic parts: the intrinsic internal gate delay, the fan-in and fan-out delay penalties, and the on-chip interconnect net delay penalty. The output driver propagation delay is composed of two basic parts: the intrinsic output driver delay, and the external reactance loading delay penalty. The worst-case delay distribution is given by the ratios of maximum to nominal and minimum to nominal propagation delays. This distribution is influenced by the variations in delay due to

Process variation

Temperature variation

Power supply variation

14.2.2 Circuit Parameters

A comparison of key circuit parameters for different circuit techniques is helpful in understanding their suitability in a given system application. Three representative production technologies are selected for a comparative analysis. The purpose is to merely compare technologies, not to do a competitive analysis. Silicon CMOS technology is chosen because it is the dominant low-cost production technology. Silicon ECL is chosen because it is the dominant high-performance production technology. GaAs DCFL is chosen because it is the most promising contender to ECL for future high-performance production technology. ASIC technology is chosen because of its emerging dominance in the industry. The three announced products chosen are as follows: LCA100K compacted array plus, VSC4500 gate array, MCA10000ECL macrocell array. These products are manufactured by LSI Logic Corporation, Vitesse Semiconductor Corporation, and Motorola Semiconductor Products, Inc., respectively. The three technologies are compared in Tables 14.1 and 14.2, based on published information on these products.[20–22] For a detailed analysis, the reader is advised to contact the manufacturer directly.

TABLE 14.1 Circuit Technology Functional Comparison

Parameter	Silicon MOSFET CMOS LCA 100K+	GaAs MESFET DCFL VSC4500	Silicon BJT ECL MCA10000ECL
Integration level, equiv. two-input gates	100,000	4500	10,000
Minimum feature size, μm	1.0	1.5	1.5
Internal gate cell area, μm^2	772	640	3325
Chip size (approx.), mm	15 × 15	4 × 7.1	9.8 × 9.8
Power supply voltages, V	0, +5	0, −2	0, −5.2, −3.3
Internal cell drive current, μA	2500 (switching)	120	400–1600
Logic voltage signal swing—internal gate, mV	5000	600	500
Internal gate power dissipation	12 μW/MHz	240 μW	826–1430 μW
Output driver type	4 mA	50 Ω	50 Ω
Output driver power dissipation	25 μW/MHz/pF	10 mW	10.4–15.6 mW

Source: Refs. 20, 21, and 22.

TABLE 14.2 Circuit Technology Performance Comparison

Parameter	Silicon MOSFET CMOS LCA 100K+	GaAs MESFET DCFL VSC4500	Silicon BJT ECL MCA10000ECL
Internal gate propagation delay			
Inverter (FI = 1, FO = 1), ps	215	150	100–200
Two-input gate, ps	NAND 420	NOR 155	NOR 130–230
Additional fan-in delay penalty, ps/FI	50	5	20–35
Additional fan-out delay penalty, ps/FO	50	75	15–30
Output driver prop. delay (no load), ps	1300	770	350
Additional fan-out delay penalty (fan-out load = 5 pF), ps	250	50	90

Source: Refs. 20, 21, and 22.

ACKNOWLEDGMENTS

The author would like to express his gratitude for the support and encouragement provided by Mr. Michael Clements and Mr. Bruce Beebe. The author would also like to thank Mrs. K. Bellew for her assistance in the preparation of the manuscript.

REFERENCES

1. Dave Kiefer et al., "CRAY-3: A GaAs Implemented Supercomputer System," *IEEE GaAs IC Symp. Tech. Dig.,* October 13–16, 1987, pp. 3–6.
2. David A. Hodges and Horace G. Jackson, *Analysis and Design of Digital Integrated Circuits,* McGraw-Hill, New York, 1983, p. 70.
3. David A. Hodges and Horace G. Jackson, *Analysis and Design of Digital Integrated Circuits,* McGraw-Hill, New York, 1983, chap. 3, pp. 68–124.
4. C. A. Liechti, et al., "GaAs Microwave Schottky-Gate FET," *IEEE ISSCC Dig. Tech. Papers,* February 16–18, 1972, pp. 158–159.
5. Rory L. van Tuyl, and Charles A. Liechti, "High-Speed Integrated Logic with GaAs MESFETs," *IEEE J. Solid-State Circuits,* **SC-9**(5):269–276 (October 1974).
6. Ryuichiro Yamamoto et al., "Depletion-Type GaAs MSI 32b Adder," *IEEE ISSCC Dig. Tech. Papers,* February 23–25, 1983, pp. 40, 41, 283.
7. Hajime Ishikawa et al., "Normally-Off Type GaAs MESFET for Low Power, High Speed Logic Circuits," *IEEE ISSCC Dig. Tech. Papers,* February 16–18, 1977, pp. 200, 201.
8. Akio Shimano et al., "A 4GHz 25mW GaAs IC Using Source Coupled FET Logic," *IEEE ISSCC Dig. Tech. Papers,* February 23–25, 1983, pp. 42, 43.
9. David K. Lynn, Charles S. Meyer, and Douglas J. Hamilton, *Analysis and Design of Integrated Circuits,* McGraw-Hill, New York, 1967, chap. 8, pp. 200–245.
10. Robert L. Morris and John R. Miller (eds.), *Designing with TTL Integrated Circuits,* Texas Instruments IC Application Staff, McGraw-Hill, New York, 1971, p. 18.
11. Stewart Brenner et al., "A 10,000 Gate Bipolar VLSI Masterslice Utilizing Four Levels of Metal," *IEEE ISSCC Dig. Tech. Papers,* February 23–25, 1983, pp. 152, 153, 298.
12. H. H. Berger and S. K. Wiedman, "Merged Transistor Logic—A Low Cost Bipolar Logic Concept," *IEEE ISSCC Dig. Tech. Papers,* February 16–18, 1972, pp. 90, 91, 219.
13. C. M. Hart and A. Slob, "Integrated Injection Logic—A New Approach to LSI," *IEEE ISSCC Dig. Tech. Papers,* February 16–18, 1972, pp. 92, 93, 219.
14. Tohru Nakamura et al., "High-Speed IIL Circuits Using a Sidewall Base Contact Structure," *IEEE J. Solid-State Circuits,* **SC-20**(1):168–172 February (1985).
15. Jacob Millman, *Microelectronics: Digital and Analog Circuits and Systems,* McGraw-Hill, New York, 1979, pp. 155–157.
16. David A. Hodges and Horace G. Jackson, *Analysis and Design of Digital Integrated Circuits,* McGraw-Hill, New York, 1983, pp. 271–283.
17. David K. Lynn, Charles S. Meyer, and Douglas J. Hamilton, *Analysis and Design of Integrated Circuits,* McGraw-Hill, New York, 1967, chap. 7, pp. 159–199.
18. Z. E. Skokan, "EFL Logic Family For LSI," *IEEE ISSCC Dig. Tech. Papers,* February 14–16, 1973, pp. 162, 163, 216.
19. Z. E. Skokan, "Emitter Function Logic—Logic Family for LSI," *IEEE J. Solid-State Circuits,* **SC-8**(5):356–361 (October 1973).

20. LSI Logic Corporation, *LCA 100K Compacted Array Plus—Published Product Specification Data,* 1551 McCarthy Blvd, Milpitas, Cal.
21. Vitesse Semiconductor Corporation, *VSC4500 Design Manual,* Revision 01, November 1987, 741 Calle Plano, Camarillo, Cal.
22. Motorola Semiconductor Products Inc., *MCA10000ECL Macrocell Array Design Manual,* Preliminary Version 1.41, 1300 N. Alma School Road, Chandler, Ariz.

CHAPTER 15
BIPOLAR GALLIUM ARSENIDE HETEROJUNCTION INTEGRATED INJECTION LOGIC

D. A. Roberts
N. A. Schmitz
J. D. Watkins
Texas Instruments
Dallas, Texas

15.1 INTRODUCTION

The GaAs heterojunction integrated injection logic (HI^2L) described in this chapter was developed by Texas Instruments during the 1970s and is currently being used to fabricate VLSI circuits with complexities of greater than 12,500 gates.[1]

Major HI^2L advantages are high speed-power product, small gate size, radiation tolerance, operation from -55 to $+150°$ C, ease of design, single-voltage operation, and ability to drive high current loads (50 Ω or ECL).

15.2 ADVANTAGES

15.2.1 Speed-Power Product

Inherent differences in the material properties of GaAs yield several advantages over silicon. When used in bipolar digital circuits, these material properties realize devices with better speed-power performance, packing density, radiation tolerance, and higher temperature operation.

The intrinsic speed-power product of GaAs provides the designer with the potential to build devices that switch many times faster than silicon. However, the state of GaAs lithography lags that used with silicon by about four years. Most comparisons between the two materials do not consider these differences in crit-

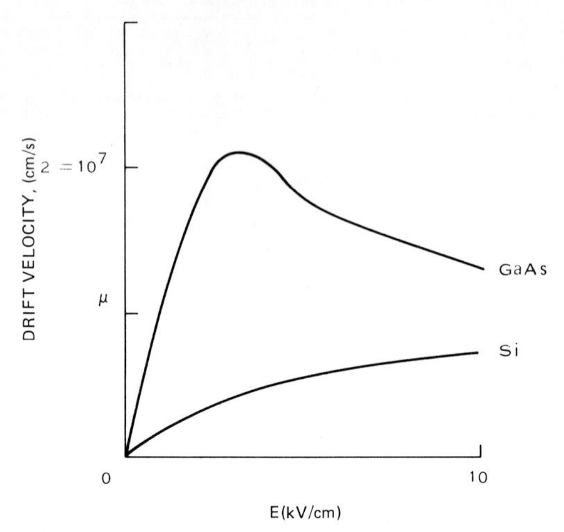

FIG. 15.1 Comparison of gallium arsenide and silicon drift velocities.

ical geometry sizes. Smaller, more advanced silicon structures with less capacitance approach speeds achievable with the current large GaAs transistors. For identically sized transistor structures built in both GaAs and silicon, the superior electron mobility of GaAs yields superior speed and power performance.

With GaAs electron mobilities of over six times that of silicon, as logic devices switch off and electric fields decrease, the last quantity of electrons present in a charged junction is rapidly reduced.[2] Figure 15.1 shows this advantage, which contributes to high device switching speeds. Fast npn minority carrier recombination limits carrier lifetimes, fostering high current operation, shifting devices into and out of saturation very quickly.

15.2.2 Gate Packing Density

GaAs is a semi-insulator, whereas silicon is semiconducting. The high resistance of the substrate facilitates simple isolation techniques, minimizing device spacing. This semi-insulating characteristic also reduces parasitic capacitances.

15.2.3 Radiation Tolerance

Radiation latch-up effects and single-event upsets affect GaAs much less than silicon. The HI^2L devices described in this chapter have been demonstrated to be particularly immune to single-event upset.[3] These devices are fabricated on a conductive $n+$ doped substrate, which precludes the formation of trapped charge. Further, the generation of radiation induced carriers is more difficult in GaAs than in silicon because of the much larger bandgap, as shown in Fig. 15.2. Shorter minority carrier lifetimes eliminate minority charges quickly.

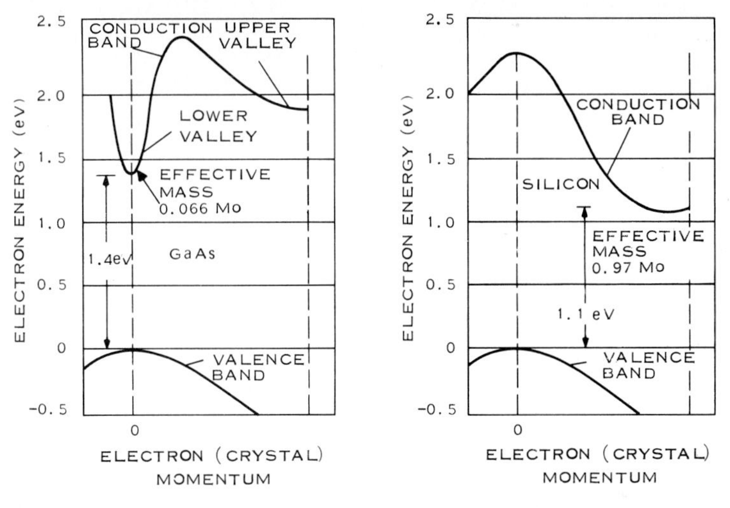

FIG. 15.2 Comparison of gallium arsenide and silicon bandgaps.

15.2.4 Temperature

The large bandgap affords higher temperature operation. GaAs circuits have been reported operational up to +150° C.[4] As temperatures climb, both silicon and GaAs become more *n*-type as thermally excited electrons jump from valence to conducting bands. A larger bandgap reduces the number of thermally induced conducting electrons. In contrast with GaAs, germanium, with a bandgap less than that of silicon, has a very limited temperature range of operation.

15.2.5 Bipolar Technology

Bipolar GaAs circuits require only one 2-V power supply voltage, operate over larger temperature ranges (−55 to +150°C), and are less susceptible to radiation upsets than are voltage-mode GaAs devices.

One of the simplest bipolar transistor structures is the HI^2L transistor shown in Fig. 15.3. As in silicon I^2L, this is the slowest bipolar device, yet in GaAs it will switch in less than 100 ps.

A few subtle differences separate GaAs HI^2L and silicon I^2L. Instead of the merged *pnp* injector transistor used in silicon, HI^2L uses a simple thin-film resistor pull-up. Omitting *pnp* transistors provides for more straightforward fabrication. Adding Schottky diodes to silicon I^2L outputs improves speed by reducing logic swing. However, in silicon the forward bias voltages of the Schottky diodes and the base-emitter junctions do not track over temperature, which results in degradation of noise margin over temperature. In GaAs the Schottky diodes and base-emitter junction forward-bias voltages track very well over temperature. Schottky diodes also eliminate the need for isolation wells surrounding individual collector outputs, allowing a more tightly packed, higher-density design.

A slow-technology HI^2L process in GaAs with 2.0 μm feature sizes still outpaces smaller, 1.0-μm silicon ECL. Ultimately, a GaAs heterojunction ECL

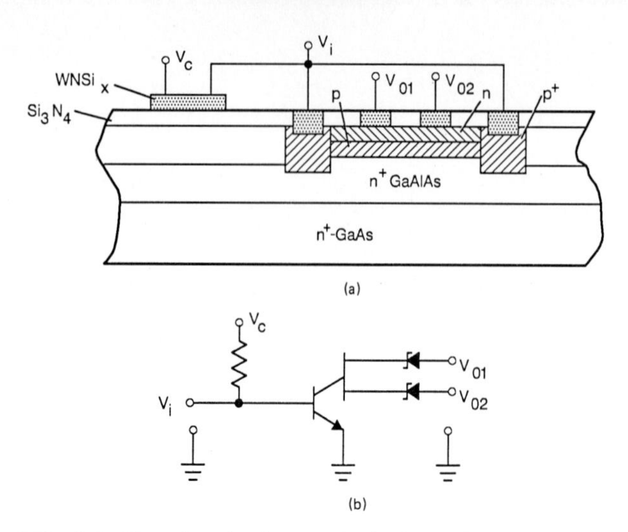

FIG. 15.3 Heterojunction integrated injection logic transistor. (*a*) Cross section; (*b*) circuit schematic.

(HECL) will replace HI^2L. Figure 15.4, shows the HECL basic transistor. As HI^2L matures, the development of more complicated transistor structures is progressing. Rather than using a substrate ground in HI^2L, HECL transistors allow access to each emitter. Besides high-speed logic applications, these transistors are useful for wide-bandwidth analog circuits.

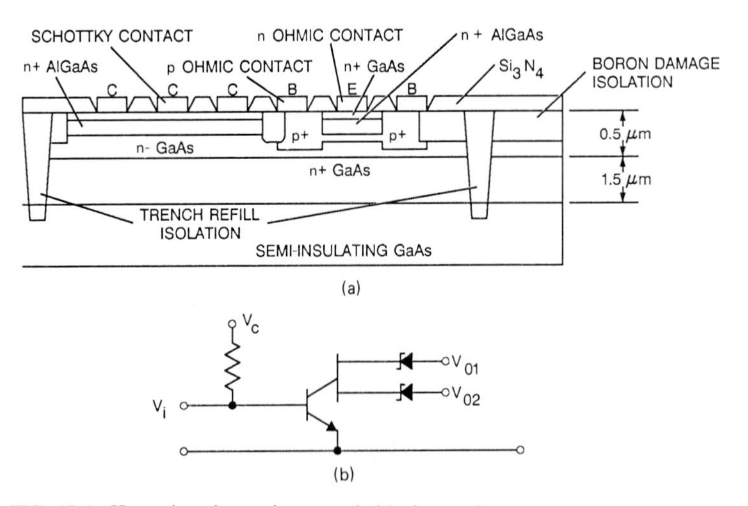

FIG. 15.4 Heterojunction emitter-coupled logic transistor.

15.3 DEVICES

15.3.1 Theory of Operation

GaAs HI^{2}L operates in much the same way as silicon I^{2}L.[5] The major difference in the two technologies is that the basic HI^{2}L gate uses a resistor connected to a power supply as the injector current source instead of a *pnp* transistor.

The basic logic gate is an inverted transistor structure that uses a GaAs-AlGaAs heterojunction as the emitter-base diode. An extrinsic base *pn* AlGaAs homojunction is also present, but this diode, with a turn-on voltage that is 200 mV higher than the transistor base-emitter diode, can be neglected to a first approximation. Since the transistor is inverted, all of the emitters are common. The collectors are made by making Schottky contacts to the top of the wafer, causing the internal logic swing to be from $V_{ce(\text{sat})}$ of the transistor to $V_{be(\text{on})}$ as defined in Fig. 15.5. Because a Schottky diode is in series with the collector, the $V_{ce(\text{sat})}$ of

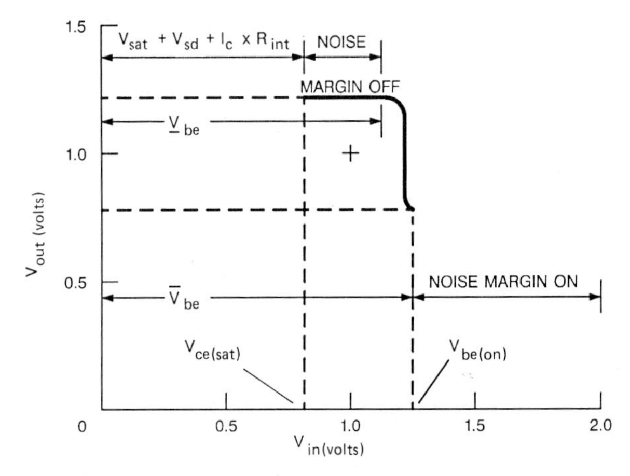

FIG. 15.5 Noise margin in HI^{2}L logic.

the transistor includes the Schottky diode drop of about 700 mV, raising the $V_{ce(\text{sat})}$ of the logic gate to about 800 mV. For a logic high, the $V_{be(\text{on})}$ for a GaAs heterojunction transistor is approximately 1.2 V, giving an internal logic swing of 400 mV. Gate speeds can be increased by lowering base resistors or by increasing the power supply, giving substantial flexibility to the designer. Signals are carried off the chip by means of open-collector output buffers that easily drive 50-Ω loads. Several important device characteristics have an impact on circuit design and performance.

15.3.2 Device Characteristics

The designer needs to be concerned with several important transistor parameters. The most important property of the logic gate is the current gain of the device. This current gain determines the maximum fan-out, and is determined by the size of the device and by the width of the transistor base region. For a gate fan-out of unity, the current gain must be greater than the ratio of the collector current to the base current. In the on state the base current is

$$I_b = \frac{V_{\text{inj}} - V_{be(\text{on})}}{R_b}$$

and the collector current is

$$I_c = \frac{V_{\text{inj}} - V_{ce(\text{on})}}{R_c}$$

where V_{inj} = logic power supply voltage
$V_{be(\text{on})}$ = transistor base clamp voltage (typically 1.2 V)
R_b = the base resistance
$V_{ce(\text{on})}$ = transistor saturation voltage (typically 0.8 V)
R_c = collector resistance

For a fan-out greater than unity, the total collector current must be summed and divided by the base current to find the current gain required of the logic gate.

The designer also needs to be aware of the Schottky diode voltage drop because of its impact on noise margin and internal logic swing. Figure 15.6*a* shows the typical HI^2L transistor characteristics. Figure 15.6*b* shows a similar transistor with merged Schottky collector contacts. The 0.7-V offset of V_{ce} represents the Schottky diode voltage drop. Lower diode voltage drops will increase logic swing, while higher voltage drops decrease the noise margin.

Another parameter of importance to the designer is the collector sheet resis-

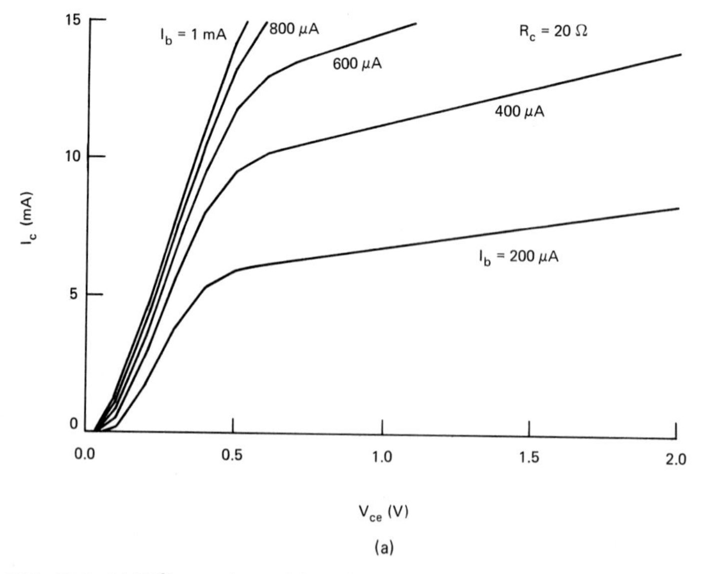

FIG. 15.6 (*a*) HI^2L transistor without Schottky collector contacts and (*b*) with merged Schottky collector contacts.

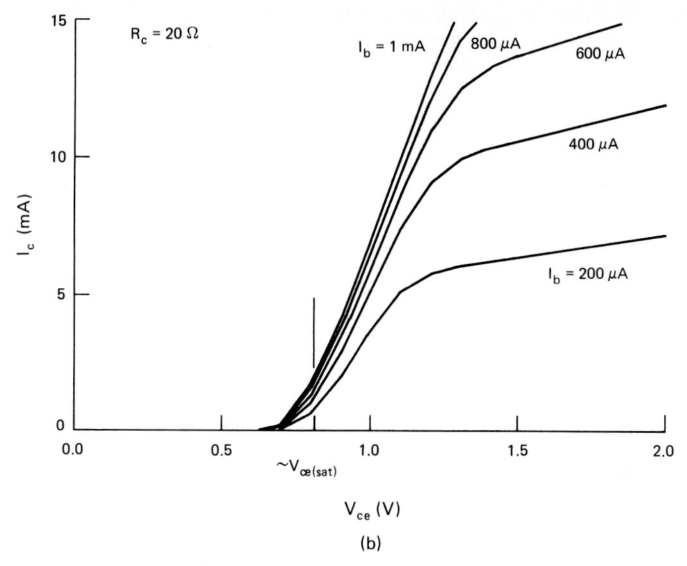

FIG. 15.6 (*Continued*)

tance. The sheet resistance is determined by the collector doping level and by the thickness of the collector layer. The collector sheet resistance determines the internal collector resistance that impacts the noise margin. Figure 15.7 shows the effects of high collector sheet resistance. The voltage drop across the internal collector resistance reduces the internal logic swing by increasing $V_{ce(\text{on})}$. Designing

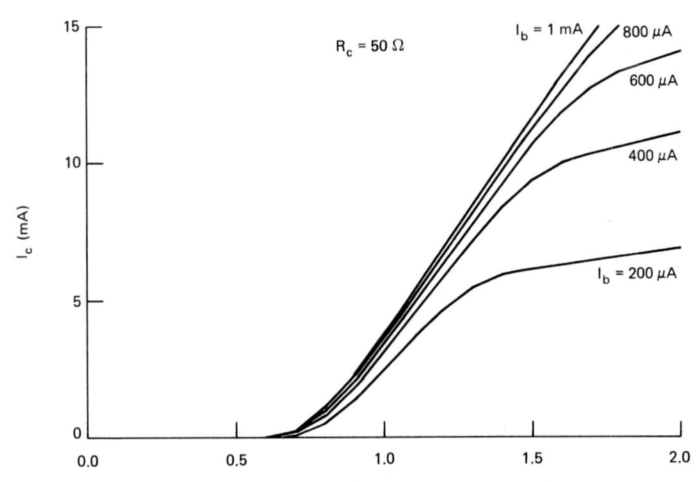

FIG. 15.7 Effect of high collector sheet resistance on transistor characteristics.

with lower currents reduces the circuit dependence on collector sheet resistance, but at the expense of reducing gate switching speed. A typical gate switching speed for a 1-mW four-collector gate is 400 ps.

15.4 HI²L CIRCUIT DESIGN

To implement a circuit design in GaAs, designers use a process that is similar to that employed for silicon circuits. The major steps are logic design, layout, and design verification.

15.4.1 Logic Design

The basic HI²L gate shown in Fig. 15.8*a* is logically equivalent to a TTL NAND gate. Wire-ANDing is performed at the base of a transistor, with the transistor executing the logic inversion. Thus the HI²L gate can be represented by an inverter with a single input and multiple outputs, as shown in Fig. 15.8*b*. Typical designs will use one, two, three, or four output gates. Any digital circuit can be implemented with the GaAs HI²L logic gate, but the circuit must comply with the logic design rules for fan-in, fan-out, current hogging, and noise margins.

Fan-in is limited in HI²L technology by the current leaking through the collectors wire-ANDed to the input of the logic gate. The worst-case condition, shown in Fig. 15.9, is when all the transistors in the previous stage are off and the total leakage current, the sum of all prior-stage leakage currents, is sufficient to force the transistor out of saturation. While an absolute fan-in limitation can be calculated, the leakage is variable with processing, operating temperature, and the number of outputs on the gates; therefore a conservative number is used.

Fan-out is limited by the current gain capabilities of the HI²L transistor and the geometric layout. When the transistor is ON, each collector output sinks current, as shown in Fig. 15.10. The worst-case condition will occur when the sum of the collector currents equals the base gain requirements. The gain is also variable; therefore conservative constraints are used.

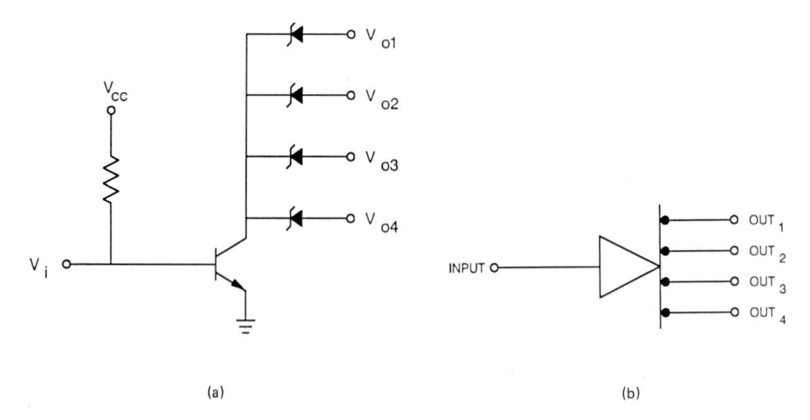

FIG. 15.8 GaAs HI²L. (*a*) Gate schematic; (*b*) logic gate (wired-AND logic).

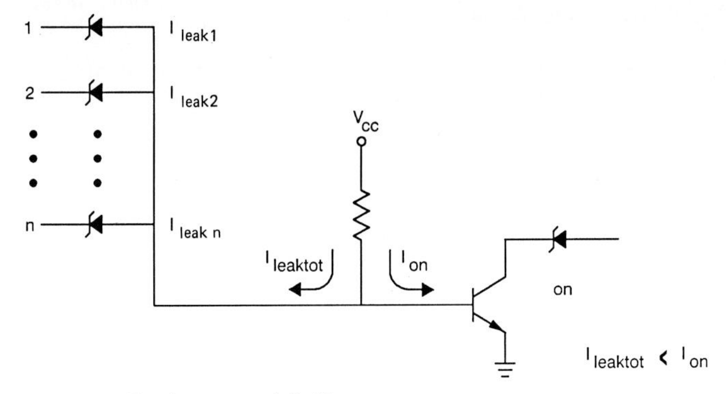

FIG. 15.9 Fan-in current definition.

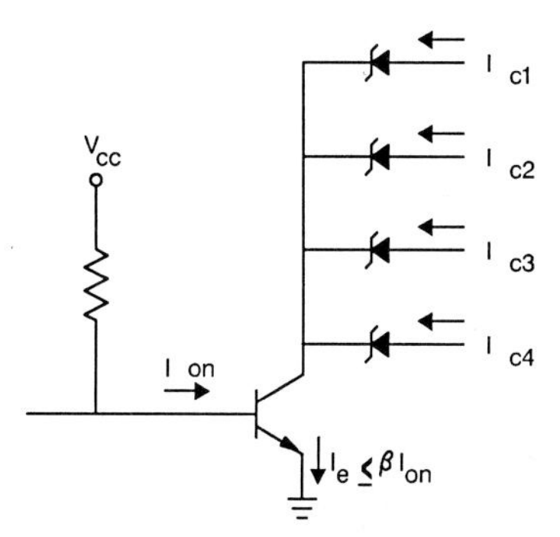

FIG. 15.10 Fan-out current definition.

Slight variations in V_{be} require that only one base be connected to a collector contact. If two or more bases are connected together, the one with the lowest V_{be} would turn on first and prevent others from turning on (current hogging). To prevent this, the designer limits the number of gates that are driven by a single output.

Noise margins are centered around V_{be} (approximately 1.2 V), as shown in Fig. 15.5. An advantage of HI^2L is its high noise margin with low material and processing sensitivity. The OFF noise margin is the primary concern for the designer, since it will vary with lead length. The saturation voltage (V_{sat}), Schottky diode voltage (V_{sd}), and intrinsic collector resistance voltage drop ($I_c R_{int}$) are the material and processing elements that affect the OFF noise margin.

The lead-length voltage drop is determined by the designer. A lead-length analysis of each layout is done to determine if the current through the lead will cause an excessive voltage drop. Careful planning of the layout will reduce noise margin problems; however, very large, high-power designs may have signal leads that must be widened to decrease the resistance and lessen the resistive voltage drop. A good design practice is to maintain at least a 300-mV OFF noise margin to ensure operation over a wide temperature range.

15.4.2 Selectable Gate Delays

A significant advantage of I^2L circuits is the ability to vary the speed of the circuit by increasing or decreasing the power. The GaAs HI^2L technology varies the speed of each transistor with a single base (injector) resistor by typically varying the injector current from 160 μA to 1 mA. Mixed power levels are used. A good example of where varied speeds are used is a register file. Fast, high-power transistors are used in the address decode and control sections, while the data storage cell uses slow, low-power transistors to moderate the power consumption. With the mix of high- and low-power transistors, caution must be used where the two interface. Low-power transistors have limits on their ability to drive high-power transistors. It is necessary to have two collectors from a low-power transistor drive the base of a high-power transistor so that current gain capabilities are not exceeded. Design verification is more difficult since an added check must be performed to verify that two collectors are driving a high-power transistor. Also, the additional current may cause excessive resistive voltage drop or current density conditions. Variable speeds are a powerful design advantage for high-performance custom HI^2L circuits and are frequently used. Changing the resistor value requires only a single layout level change since, unlike MESFET devices where the device size is altered and many mask changes are needed, only the injector resistor value is altered. In practice, the injector resistors are tapped, and the designer merely selects that resistor tap point that corresponds to the speed desired for each individual gate.

15.4.3 IC Layout

The transformation of the logic schematics into the many photolithographic processing levels required is heavily dependent on CAD tools, such as the Apollo or GE Calma systems. A top view of a typical HI^2L stick transistor with a base resistor is shown in Fig. 15.11. The resistor supplies current to the transistor from a power bus with space on either end of the stick for base to collector interconnection. Each collector has a merged Schottky diode for signal isolation. Remote Schottky diodes can be used, but are not discussed here.

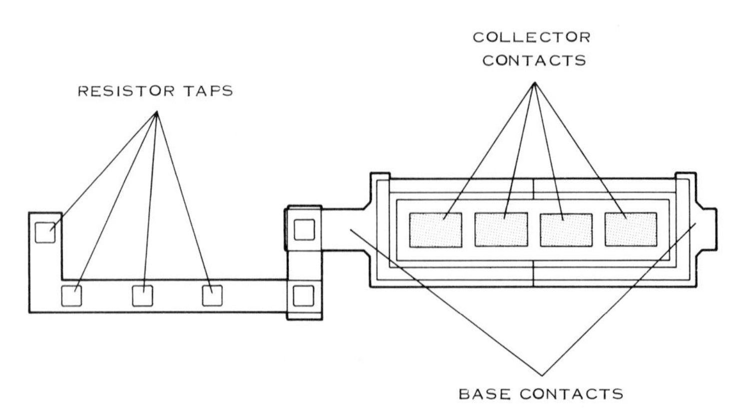

FIG. 15.11 HI^2L stick transistor layout (top view).

Transistor interconnect can be done manually for small circuits or, more usually, with computer-aided autorouters. In either case the goal is to locate the transistors as densely as possible. Lead length is critical for maximum performance. Long leads slow the circuit with capacitive delay and reduce the noise margin due to resistive voltage losses. Capacitance is not as crucial with HI^2L as with voltage-driven technologies, due to low impedance inherent in the current-mode bipolar devices.

The effect on noise margin of a 2500- by 2.5-μm lead with a sheet resistance of 0.1 Ω/square would be 100 mV if the transistor were operating at 1 mA. Such a combination is possible with a custom HI^2L VLSI design. Autorouting can be inefficient in lead-length selection, so manual intervention is often used to layout macros such as flip-flops. The macro inputs and outputs are used as the targets for interconnections rather than the input and outputs of the individual gate. This prevents the gates of the macro from being spread unnecessarily all over the chip. Minimum metal width and spacing are set by the processing limitations and are usually adequate for interconnection. Unavoidably long leads or leads with excessive current density are widened at the discretion of the designer as he or she trades off capacitive delay, functionality, and performance margin.

Voltage drop and current density within power buses are carefully assessed. The use of gold for interconnection with GaAs eases the situation since it is a good conductor with excellent resistance to metal migration. Power buses are usually much wider than the transistor interconnects and must be planned and placed to maximize the space available. From the voltage input pad, the voltage drop to any base resistor should not be allowed to exceed 100 mV. This allows for power supply variations and operation over temperature. Current density limits are determined by device reliability requirements. Because accelerated metal migration is caused by excessive current density, a 5.0×10^5 A/cm^2 limit on maximum current density is adhered to.

15.5 DESIGN VERIFICATION

Application of standard IC design tools to HI^2L is straightforward.

15.5.1 Simulation

Most design work in bipolar does not require extensive SPICE simulations. A complete HI^2L device library contains only a few basic transistor types (e.g., one-, two-, three-, and four-collector transistors). For a given base current drive, a simple, lumped delay model for each type suffices. Speed variations versus fan-in are negligible, and simulation accuracy suffers only slightly by ignoring interconnect delay. Back annotation, also known as postrouting simulation, combines lumped-gate and interconnect delay models to achieve an error of less than 10 percent for propagation delay and maximum frequency predictions.

15.5.2 Load Checking

Although certain design rule violations may pass simulation, real circuits will not function. For example, connecting two or more bases in parallel causes a slight difference in base-emitter voltages to rob current from the other devices. Similarly, operation over temperature degrades severely when fan-in design rules are violated, which allow transistor leakage currents to get out of specification. Load checking guarantees proper fan-in, fan-out, and device connections.

15.5.3 Schematic-Layout Comparison

Schematic verification ensures that the circuit's physical layout matches the logical description. Automated programs recognize both intended and parasitic devices and produce a schematic equivalent of the drawn layout. These programs calculate values of injector resistors and compare them against the design. If necessary, metal interconnect area, parasitic capacitance, and point-to-point delay computations further refine simulation accuracy (back annotation).

15.5.4 Geometric Verification

By checking for violations of physical manufacturing rules, device yields greatly improve. Specifically important in the bipolar manufacturing process are the contact metals on transistor base and collectors. Since a lift-off process patterns these metals, closed annular geometries are more difficult to fabricate and may result in base-to-collector shorts.

15.5.5 Power Bus

Programs similar to those used for schematic verification decompose a power bus into resistive components. For input to SPICE, the program converts bond pads into voltage sources and injector resistors into current sinks. Back-end processing after SPICE alerts the designer to excessive voltage drops and current density violations, recommending proper widths of leads for corrective action.

15.6 FABRICATION

15.6.1 Material

HI^2L circuits are fabricated on 3-in *n*-type GaAs wafers doped with silicon or tellurium to a concentration of 5.0×10^{17} to 3.0×10^{18} electrons/cm^3. Gold is deposited on the back side of the wafer to form the ohmic contact, and three epitaxial layers are grown on the top side, forming the active device regions on the slice. The first layer grown on the substrate is an *n*-type AlGaAs region about 1.35 μm thick and doped with silicon to a concentration of 1.5×10^{18} electrons/cm^3. The second layer is an undoped region of GaAs 0.2 μm thick, which will be doped *p*-type to form the base region of the HI^2L transistors. The third layer grown is *n*-type GaAs 0.3 μm thick and doped to a concentration of 5.0×10^{17} electrons/cm^3, which serves as the collector region. Several methods are used to grow these epitaxial layers on the substrate. Two of the more common methods are molecular beam epitaxy (MBE) and metal-organic chemical vapor deposition (MOCVD).[6] Each process has its advantages and its disadvantages.

With MBE, epitaxial (epi) doping and width can be very accurately controlled, thus yielding uniform device characteristics. However, the limited throughput of most available MBE reactors and their high cost limit their use in a production environment. Another problem with limitation with MBE is formation of particulates present inside the reactor. These particles result in excessively large numbers of defects on the wafer surface, which lower yield.

Presently, considerable success has been achieved with epilayers grown by the MOCVD process. With MOCVD, layers can be quickly grown on the substrate with fewer particles, but the tradeoff is that there is less control over the doping concentration and layer thickness.

15.6.2 Processing

The HI^2L circuits described here are fabricated with 13 mask levels.[7] Figure 15.3 shows a cross section of a logic inverter. The first step in the process is a $p+$ Be implant 0.7 μm deep, which makes contact with the p base epilayer. This also forms the pn extrinsic homojunction diode already discussed. Next, a thin Zn layer is deposited and diffused into the existing $p+$ implanted area, which reduces the resistance of the base ohmic contact. A high-energy boron implant is then performed to transform the epi GaAs layers into insulating GaAs to provide isolation between the devices. Next, the resistor metal is patterned, and the two ohmic contact metals and the Schottky metal are deposited. After a nitride layer is deposited on the entire wafer surface, vias are etched to allow contact of the four bottom metal layers to the first-level metal layer. The first-level metal is then deposited over the entire wafer and selectively etched to form the first-level interconnect layer. Second-level vias are then etched following a second nitride deposition, to allow contact from the second-level metal to the first-level metal. The final step forms the second-level interconnect system and a final protective nitride overcoat layer to provide passivation of the second-level metal. During these processing steps, several factors can impact the yield of the HI^2L circuit.

Data taken from early HI^2L 1K gate arrays showed that the most critical yield limiter is lateral zinc diffusion. This problem can occur due to misalignment of the zinc mask with the $p+$ implant mask. Should this misalignment occur, zinc may diffuse into the collector region, raising the collector sheet resistance to an unacceptable level or causing a short between the Schottky contact and the base layer. Wafers are closely monitored to detect this problem before further processing is attempted.

The second major yield limiter is interconnect leakage. This leakage is caused by cracks or defects in the dielectric layer between the first- and second-level metals. With these yield limiters under control, epi uniformity becomes the remaining yield inhibitor. Epi uniformity is a function of the epi growth process as described previously. A nonuniform base epilayer causes variation of the transistor gain as well as the breakdown voltage. These two characteristics are inversely related, so a well-regulated base epi growth process is required to keep these parameters within specification. The depth of the grown heterojunction also has an impact on circuit performance. If the heterojunction is too deep from the wafer surface, then the $p+$ implant will not be deep enough causing the GaAs-AlGaAs homojunction turn-on voltage to be too low, which also degrades the gain of the transistor.

Other yield factors include particle contamination, complexity or integration level, and total *n* contact metal area. Large circuits require many interconnects and give rise to a large chip. These factors increase the risk of encountering a killing defect.

ACKNOWLEDGMENTS

The authors wish to thank Mr. Philip Congdon for his support and guidance in preparing this chapter. They also wish to thank the staff responsible for making this technology a success.

REFERENCES

1. P. A. Congdon et al., *High Speed GaAs Microprocessor Program Report,* Texas Instruments and Control Data, October 1985.
2. S. M. Sze, *Physics of Semiconductor Devices,* Wiley, New York, 1981.
3. J. F. Salzman, P. J. McNulty, and A. R. Knudson, "Intrinsic SEU Reduction from Use of Heterojunction in GaAs Bipolar Circuits," *IEEE Trans. Nuc. Sci.*, vol. NS-34, Dec. 1987, pp. 1676–1679.
4. S. Evans et al., "GaAs HBT LSI/VLSI Fabrication Technology," *Proc. 1987 GaAs IC Symposium.*
5. S. C. Blackstone and R. P. Mertens, "Schottky Collector I^2L," *IEEE J. Solid State Circuits*, vol. SC-12, June 1977, pp. 270–275.
6. R. Dingle, M. D. Feuer, and C. W. Tu, "The Selectively Doped Heterostructure; Materials, Devices, and Circuits," *VLSI Electronics Microstructure Science*, vol 11, *GaAs Microelectronics*, 1985, pp. 215–259.
7. H. T. Yuan, *Interim Report for GaAs/AlGaAs HBT Gate Array Technology,* Texas Instruments, May 1986.

CHAPTER 16
SUPERCONDUCTOR DIGITAL ICs

T. Van Duzer
Department of Electrical Engineering and Computer Sciences and the Electronics Research Laboratory
University of California
Berkeley, California

16.1 INTRODUCTION

Computer projects based on thin-film switches in which superconductivity was quenched by a magnetic field were being phased out in the mid-1960s since it was recognized that the developing semiconductor field would produce faster devices. The situation changed with the introduction of the Josephson junction. It was soon appreciated that this device would switch very rapidly and would operate at power levels several orders of magnitude below those for semiconductor devices. Of particular importance was the work of Matisoo at IBM in 1966 on the use of the Josephson junction as a switch.[1] It also became clear that circulating currents in a superconducting loop could be used for information storage in a nondestructive-read out (NDRO) cell.[2] On these bases a project began to grow at IBM and became a major research effort by the mid-1970s.

Josephson junctions have the highest switching speed of any known device, and terminated, nearly dispersion-free transmission lines can be used to interconnect devices. Also the voltage is at millivolt level, so the power is about three orders of magnitude lower than in semiconductor circuits. The lower power allows much greater packing density, so the limitation on system speed caused by transmission time can be weakened. Furthermore, electromigration problems that plague semiconductor circuits, especially at high packing density, are eliminated.

There are also factors that tend to inhibit introduction of superconductive systems. Both the circuit principles and the fabrication technology are different from those for semiconductor circuits. The logic circuits are of latching type and do not invert, and memory is totally different from that in semiconductor computers. The fact that superconductive devices operate only close to the absolute zero of temperature means that they can only be tested at those temperatures. Furthermore, they must be thermally cycled many hundreds of times from room temperature to cryogenic temperatures and withstand the resulting stresses.

In 1983 IBM terminated the effort to build a mainframe computer with Josephson technology. The low gain of Josephson logic circuits resulted in tight design margins and strong requirements on the fabrication technology. The development of a 4-kbit memory, which was deemed necessary, appeared to be a difficult task, and projections of the semiconductor and superconductor technologies indicated a system speed advantage for superconductors of only about 2.5 over semiconductors.[3] Meanwhile, in about 1980, the Japanese began an intensive effort on Josephson digital applications. They were able to quickly adapt the IBM lead-alloy technology to start doing circuit work. At the same time, they began to develop the use of refractory materials, which had been introduced in a particularly useful form at the Sperry Research Center in 1981.[4] In 1985–1987 the organizations in Japan that were participating in the effort to evaluate Josephson technology for computer use made the transition from using lead alloys to using refractory materials. Organizations in the United States involved in superconductive IC work have also made the change from the use of a complete lead-alloy technology. Large logic and memory circuits are understood to require the use of wholly refractory materials to achieve sufficient yield.

The present status of the technology development is that wholly refractory IC processes give junction parameter variance of a few percent; this represents a considerable improvement over the lead-alloy technology. Logic circuits such as adders, multipliers, and analog-to-digital converters with remarkable operating speeds have been demonstrated. Memory development has been less successful, but the new circuit processes, with their improved control of parameters, are expected to enhance its development. The digital effort in the United States is directed toward signal processing applications presently including analog-to-digital converters, shift registers, and multipliers. Some important signal processing devices can be made with smaller memories than are needed for mainframe computers.

The emergence this year of the new copper oxide superconductors with transition temperatures of about 100 K will affect new directions for the work reported in this chapter. The higher temperature may make possible combinations of superconductor and semiconductor devices or new single devices that combine the phenomena of superconductivity and properties of semiconductors. Many other important issues have only begun to be considered at this writing. How high can the transition temperatures be made? How much current density can be carried in films? Can high-current-density tunnel junctions be made? Should other types of logic circuits be considered that do not require the tunnel-junction type of Josephson device? What is the optimum tradeoff between logic swing and power dissipation? How can interfaces between superconducting devices and semiconductor devices be worked out? This paper gives the state of the Josephson digital circuits as it exists at the onset of these new considerations.

16.2 SUPERCONDUCTORS

Many of the elements in the periodic table have been shown to exhibit superconductivity below a characteristic transition temperature T_c; there are also literally thousands of alloys and compounds that have been shown to be superconductors. Of these materials, a few have been prominent in superconductive electronics. Most IC work has used alloys of lead, which have T_c almost equal to that of lead (7.2 K). These have low melting points and are fairly easy to deposit, but they also are unstable in aging and thermal cycling. Niobium is a more stable refractory material and has the highest T_c of any element (9.2 K); it is quite reactive, and only in the past few years has it become possible to make high-quality Josephson junctions with it. Numerous alloys and compounds have transition temperatures well above 4.2 K. Most of them require formation at highly elevated temperatures. This is not a problem for deposition of the first electrode of a tunnel junction, but, when used for the other electrode, the junction barrier material must be stable at that temperature. At this time the most useful high-T_c material is NbN; films with a T_c as high as 17 K can be made on a substrate at room temperature.[5]

Below the transition temperature, some of the electrons in a superconductive material form pairs, as described by the BCS theory. These electron pairs have wave functions that are heavily overlapping. Their energy is reduced when the wave functions of the pairs lock phases; the result is a macroscopic wave function that extends throughout the superconductor with phase coherence. This wave function representing the paired electrons exists in the same space as the unpaired component of the conduction-electron sea.

When a dc current is imposed in a superconductor, it is carried only by the paired component, which is not scattered by collisions. The electric field inside the superconductor is zero, and the unpaired electrons have no drift component of velocity. The situation is different for rf currents, since it is necessary to provide acceleration of the electron pairs. An rf electric field then exists inside the superconductor, and the single electrons are accelerated and have collisions. The effective conductivity has both real and imaginary parts. The result is that superconductors are not absolutely lossless; however, they are far less lossy than any normal metal. As the frequency of the electromagnetic fields increases to about 1000 GHz, the photons have enough energy to break electron pairs, and the losses increase rapidly.

Another result of the phase coherence of the pair wave function is the so-called Meissner effect, which is that magnetic flux is excluded from the interior of bulk superconductors. The phase coherence also leads to magnetic flux quantization. The flux quantum has a value equal to $h/2e$, where h is Planck's constant and e is the electron charge. It is found that, for type II superconductors and sufficiently thin superconducting films, no flux penetrates for very weak perpendicular fields, but, as the field is raised, magnetic flux can penetrate as single flux quanta called *fluxons* (also called *vortices* in recognition of the circulating currents that accompany the flux quanta). The exclusion of flux from the interior of superconductors is never complete; the flux of a field tangential to the superconductor surface decays exponentially into the superconductor with a $1/e$ distance of about 0.1 μm.*

16.3 COMPONENTS

The unique components important for superconductive digital circuits include nearly lossless transmission lines, loops that carry lossless circulating currents,

*For background on superconductivity see, for example, Ref. 6.

and Josephson junction switches. The first provides for a minimum of signal transmission time and very little pulse dispersion. Lossless circulating currents either in fabricated loops or in self-defined circular patterns in thin films can provide nonvolatile memory as long as the temperature is maintained. The Josephson junction is a tunneling structure and switching involves only a change from the tunneling of electron pairs to the tunneling of single electrons; that change can be effected in less than a picosecond, but actual switching speeds are limited by parasitic circuit elements.

With signal rise times becoming shorter and approaching the picosecond range, dispersion in transmission lines is of increasing importance. Calculations comparing pulse propagation on superconducting and normal metal, parallel plane transmission lines have been made.[7] Figure 16.1 shows that 1.0-ps pulses are badly distorted or virtually vanish after propagation for 1 cm on normal metal lines, depending on the insulator thickness. Because of the very small rf loss for frequencies below the superconductor energy gap ($f = eV_g/h \approx 1000$ GHz for Nb at 4.2 K), the pulses are well preserved on the superconducting lines. The oscillatory tail on the pulse results from the sharp increase in loss at the energy gap frequency; it has been seen recently in a coplanar line using optical sampling techniques.[8]

Another factor of considerable importance regarding the use of superconductive signal lines is the avoidance of electromigration, a pervasive problem in semiconductor ICs that is exacerbated by miniaturization. The fields in the superconductors are too small to cause electromigration.

It is possible to establish circulating current in a superconductive loop. The current will persist without degradation as long as the material is kept in the superconductive state; experiments have shown that its half-life is at least 10,000

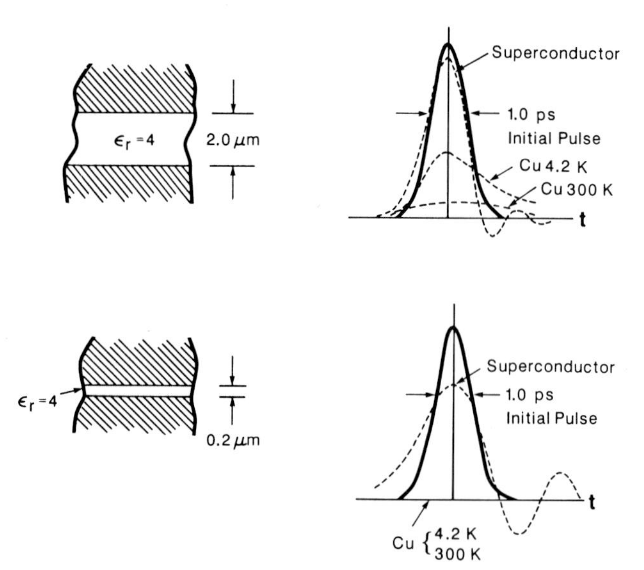

FIG. 16.1 Propagation of 1.0-ps pulses a distance of 1 cm on superconducting (Nb at 4.2 K) and normal metal (Cu at 4.2 and 300 K) parallel plane transmission lines for two different insulator thicknesses.

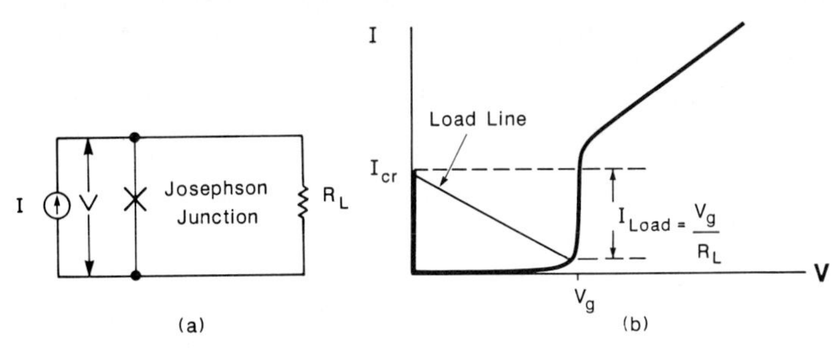

FIG. 16.2 (*a*) Model circuit for Josephson junction loaded by a matched transmission line ($R_L = Z_0$). (*b*) *I-V* characteristic of a tunnel junction with a resistive load R_L. When I is raised above I_{cr}, switching to the nonzero-voltage state occurs. The current transferred to the load is the ratio of the energy gap voltage to the load resistance.

years. If a quantum of magnetic flux is passed through a thin film, a circulating current is set up in the film in the same region. Again, the current will circulate indefinitely without degradation. Most memory cells have been based on circulating currents.

The active element for superconductive circuits is the Josephson junction. In the form used for digital circuits, it is a tunnel junction. (Although a number of other types of Josephson junction have been demonstrated, their *I-V* characteristics are not as well suited to the type of digital circuits that has been considered most extensively.) Typically, the tunneling barrier is an oxide of about 3-nm thickness. Consider a Josephson tunnel junction connected to a shunt load resistor R_L and a current source, as in Fig. 16.2*a*. The load is commonly a matched transmission line, and, therefore, the value of R_L is the characteristic impedance of the line. Notice in Fig. 16.2*b* that there are two branches of the *I-V* characteristic. As the current supply is raised from zero up to a critical value I_{cr}, all of the current is carried through the tunneling barrier by electron pairs, and there is no voltage drop (the oxide barrier behaves like a superconductor). If the current is further increased, switching to the other branch of the characteristic occurs, and the time-average current through the junction is carried almost exclusively by single-electron tunneling. In the stable state on the single-particle branch, most of the source current is transferred to the load if its value is selected as shown.

Another way to switch between states is to raise the source current to a value somewhat below I_{cr} and then reduce I_{cr} to a value below the source current by means of a magnetic field passing through the plane of the junction. The field can be produced by an overlaid insulated thin-film control line. The critical current is reduced because of quantum interference effects induced in the junction electrodes.

It is more advantageous to replace the single junction with two or more very small junctions connected in a superconducting loop or an assembly of loops. Such a circuit element is called an *interferometer* or *superconducting quantum interference device* (SQUID) and has an *I-V* characteristic like that of the single junction. In this case magnetic flux coupled into the loop(s) reduces the critical current of the SQUID. The dependence of SQUID critical current on the control current (called a *threshold characteristic*) for a two-loop, three-junction SQUID (Fig. 16.3*a*) is shown in Fig. 16.3*b*. Resistors are added in the loops of the SQUID to damp resonances that arise from interaction of the junction capacitances and

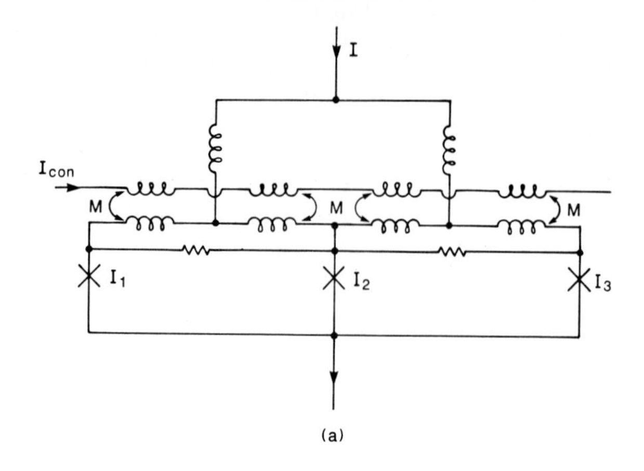

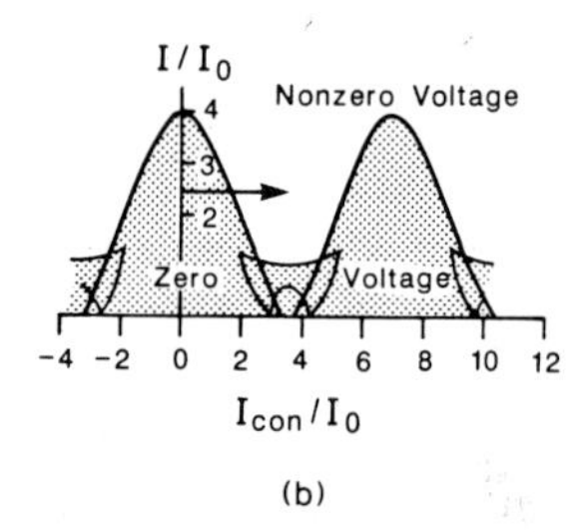

FIG. 16.3 (*a*) Schematic of two-loop SQUID with a control line. (*b*) Threshold characteristic of the SQUID in (*a*) with critical currents $I_2 = 2I_1 = 2I_3 = 2I_0$. The pattern repeats for I_{con}/I_0 outside the range shown.

the loop inductances. The SQUID switches from an initial zero-voltage state to a nonzero-voltage state when the currents are applied (as shown, for example, by the arrow in Fig. 16.3*b*). The repetitive nature of the threshold characteristics is useful for some applications, but is disadvantageous for others.

Capacitors and resistors are formed from thin films, and no especially remarkable properties are found in this technology. Inductors made of films thicker than about 0.1 μm (as is usual) behave essentially the same as ones made of normal metal except that transformer coupling from an overlaid insulated film line can occur even for direct currents.

16.4 LOGIC CIRCUITS

The discussion regarding Fig. 16.2 reveals the basic ideas of two different types of Josephson logic circuits. Switching between the zero- and nonzero-voltage states can be accomplished either (1) by overdriving or (2) by inductive reduction of the critical current of the junction or SQUID. The switching speed is limited by parasitic circuit effects; the record speed for a gate with unity fan-out is 2.5 ps.[9] Several different types of Josephson gates have been demonstrated to have nearly that speed.

Josephson logic circuits differ from their semiconductor counterparts in several important respects, in addition to being faster and using several orders of magnitude lower power. When a gate is switched from the zero-voltage state to the nonzero-voltage state, it stays in the latter until the power supply is reduced to near zero for a sufficient length of time. That is, the gates are *latching* gates. Also, they do not invert naturally. To provide inversion, one may employ a circuit provided with a timing pulse.[10] Inversion can also be obviated by using a dual-rail system wherein both TRUE and COMPLEMENT are carried throughout.[11] The principal disadvantage of Josephson logic circuits is low gain: this imposes limitations on design margins.

An effect that can set a limit on the speed of operation of Josephson logic circuits comes from the need to reset the gates to the "0" state at the end of each logic cycle. The symmetry of the Josephson circuits allows them to be operated with either positive or negative supply current; thus an ac power source can be used, and the circuits should reset to the "0" state while the supply is changing polarity. A suitable ac power supply for the logic circuits can be made, as shown in Fig. 16.4*a*. The string of junctions is driven by an ac current source; the voltage is shaped as shown in Fig. 16.4*b*. The dead zone around zero results from the Josephson current at $V = 0$ and serves to enhance resetting of the logic gates. There is a nonzero probability that the gate will switch from a "1" state of one polarity to the "1" state of the opposite polarity rather than to the "0" state. The probability is a function of the circuit parameters and the time during which the polarity of the supply changes. The phenomenon is called *punch-through*.[12] In-

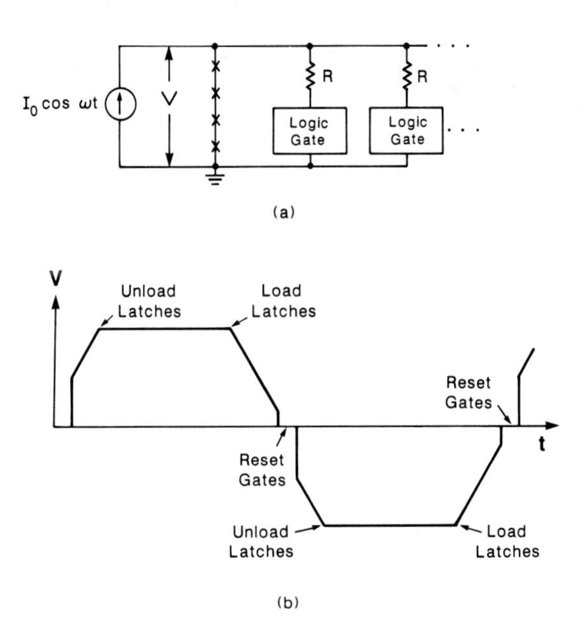

FIG. 16.4 (*a*) Power-supply-shaping circuit for Josephson logic circuits; (*b*) supply voltage waveform.

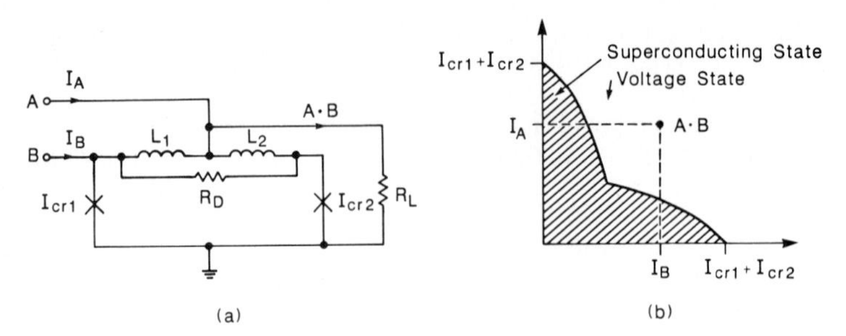

FIG. 16.5 (*a*) Current injection logic AND gate; (*b*) threshold characteristic for the CIL AND gate. (*After Gheewala.*[10])

creasing the time reduces the punch-through probability but slows down the logic.[13] A dual-powering system has been devised to circumvent the punch-through problem.[14] In this approach, the circuits are divided into two groups having different clock phases. The critical logic path usually contains about 10 to 15 logic stages, so the delay is approximately equal to the time required for gate reset. The two clocks must overlap by two to three gate delay times to assure data transfer. The result is that the processing speed is approximately doubled over a system with a single-phase clock.

The logic family last used in the IBM project is called *current injection logic* (CIL); its unique feature is an AND gate comprising three Josephson junctions in a pair of coupled superconducting loops, as shown in Fig. 16.5*a*[10]. The importance of the CIL AND gate is in its design margins, which are larger than those for a SQUID used as an AND gate. The control characteristic is shown in Fig. 16.5*b*. The input signals are on the axes. If the combination of the inputs corresponds to a position on the graph outside the crosshatched region, switching to the nonzero-voltage state occurs. Notice that if only I_A or I_B is present, its value must be about three times that required if both are present. Since currents are coupled directly into the CIL AND gate, there is no I/O isolation. The OR gate for the CIL family is a SQUID with magnetic control lines; this does provide I/O isolation. In this family, the AND gates are always driven by OR gates, as shown in Fig. 16.6.

The logic gate used at the Electrotechnical Laboratory (ETL) in Japan has four junctions in a loop arrangement, where the loop itself has negligible inductance. Its operating characteristics are determined mainly by the phases of the individual junctions. Elimination of the loop inductance leads to a reduced device

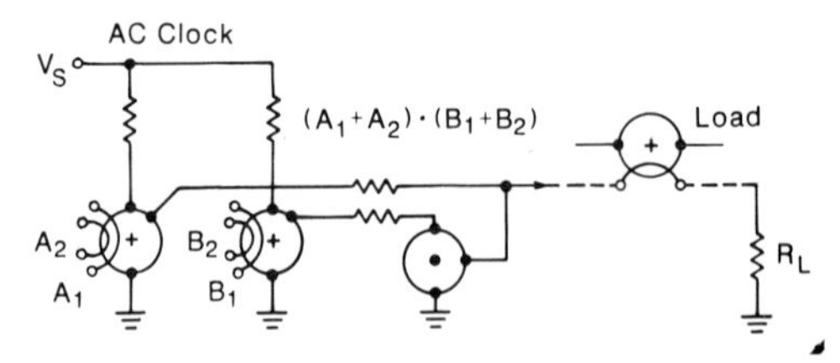

FIG. 16.6 Two-input OR-AND gate. The CIL circuit is fed by two isolation OR gates. The output of the circuit is supplied as control line current to the load circuits in a serial fan-out. (*After Gheewala*[10])

size. A logic family has been built around this gate, and logic systems of several hundred gates have been demonstrated.[15]

The first resistor-junction logic circuit, published in 1979, was called a *Josephson atto-weber switch* (JAWS).[16] These circuits have two advantages over those using interferometers. Since there are no totally closed loops, the possibility of accidentally trapping error-producing stray magnetic flux is minimized. Also, the area of a gate is smaller. The need for isolation in resistor-junction gates is supplied by switching a junction in series with the input. The logic used by Hitachi combines resistor-junction AND gates (having no isolation) with interferometer OR gates that provide isolation.[17]

A number of other resistor-junction gate families have appeared in the literature. The *direct-coupled logic* (DCL) was a simple bridge type of circuit developed at IBM.[10,18] A family of gates with wider margins called *Resistor-coupled Josephson logic* (RCJL) was devised at NEC,[19] has been tested extensively, and is used in some logic and memory.

The OR, AND, and 2/3 majority gates in the RCJL family are shown in Fig. 16.7. In the OR gate with both input signals and the gate current I_g flowing, critical currents of the junctions connected to ground are exceeded one at a time. The series input junction switches last and serves to isolate the input. In the AND gate, all three junctions switch only when both inputs are present; then the input currents are diverted to the output. In the 2/3 majority circuit, the output switches if two of the three inputs are high.

Several types of logic circuits have been developed at the Nippon Telegraph and Telephone Co. (NTT). One basic cell is called *resistor-coupled logic* (RCL) and is shown in Fig. 16.8*a*. Modifications of this cell include the *high-voltage resistor-coupled logic* (HRCL), in which series pairs of junctions are used to raise the output voltage level (Fig. 16.8*b*), and *integrated resistor-coupled logic* (IRCL), which connects amplifying stages to the RCL gate to raise the output current (Fig. 16.8*c*). These circuits have been reviewed recently.[14] Another cir-

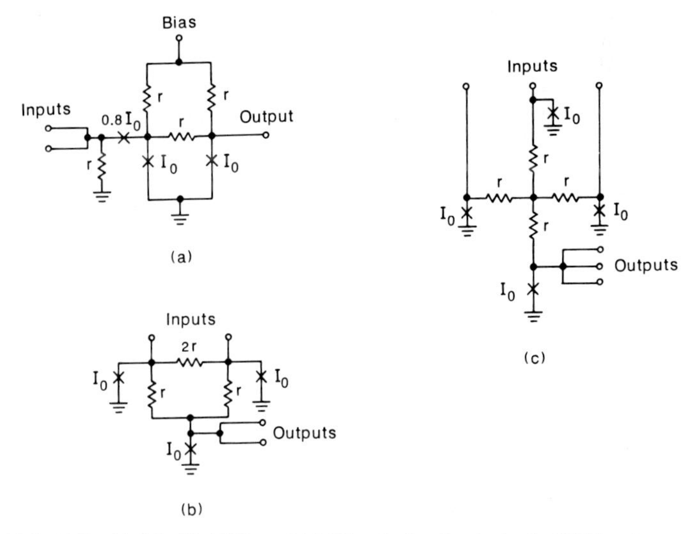

FIG. 16.7 (*a*) OR, (*b*) AND, and (*c*) 2/3 majority circuits in the NEC resistor coupled Josephson logic family.

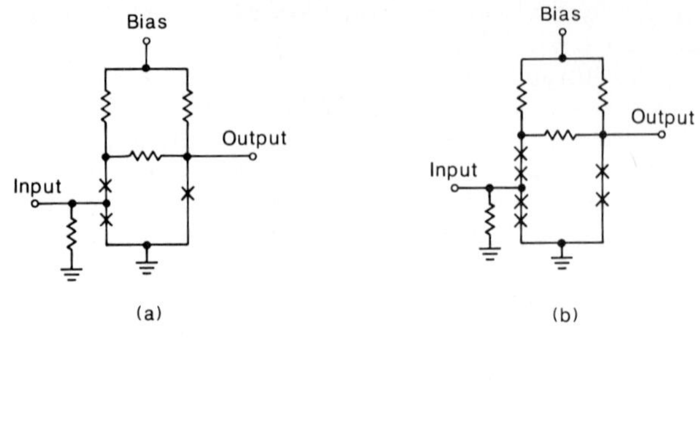

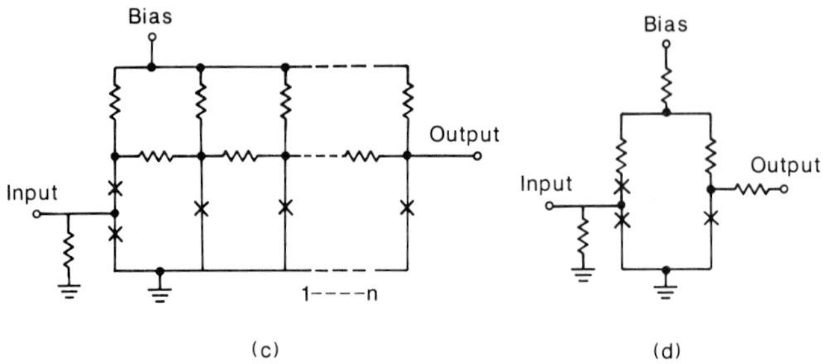

FIG. 16.8 Logic gates investigated by NTT. (*a*) Resistor-coupled logic gate; (*b*) high-voltage resistor-coupled logic; (*c*) integrated resistor-coupled logic; (*d*) high-gain direct-coupled Josephson logic.

cuit reported by NTT, the *high-gain direct-coupled Josephson logic* (HDCL), is shown in Fig. 16.8*d*. A chain of six OR gates of this type with $2 \times 2\ \mu m^2$ junctions ($J_c = 10^4\ A/cm^2$), made by employing a technique that reduces junction current spread, had an overall delay of 25 ps, or 4.2 ps/gate.

Several studies have been made of scaling of Josephson logic circuits. Projections of the gate delays for several families, taking account of realistic interconnections, lead to approximately the same result. For these studies the chip is divided into unit cells as shown in Fig. 16.9; each cell contains two OR gates and

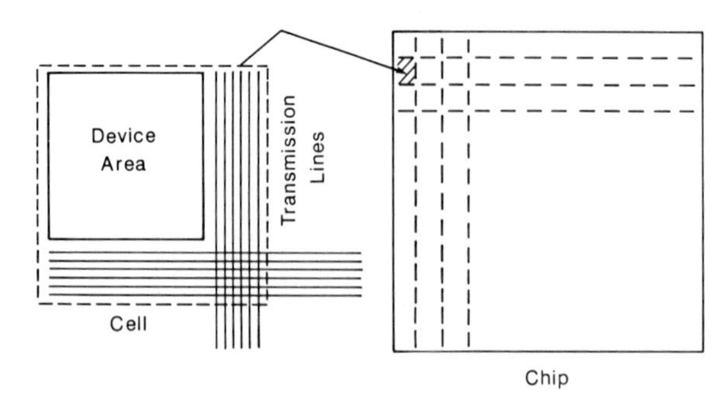

FIG. 16.9 Model for scaling study of delays in OR-AND gate cells.

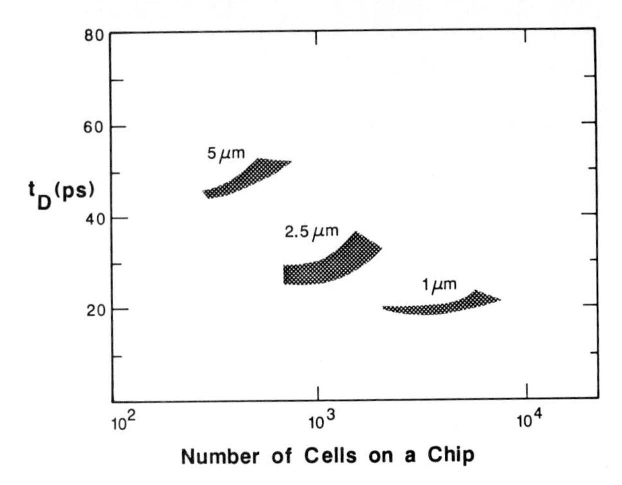

FIG. 16.10 Tradeoff of delay per cell against cell density, based on the model in Fig. 16.9, for the CIL family.

one AND gate.[20,21] Then parameters are varied while remaining consistent with a set of constraints imposed either by basic physics of the devices or by limitations of technology. The result of such a study[21] for CIL gates is shown as a tradeoff between speed and density for different minimum feature sizes in Fig. 16.10. Variations of insulator dielectric constant and thickness in the transmission lines lead to the range of results represented by the shaded regions. Speed is largely controlled by junction current density and the maximum practical current density of 10^4 A/cm^2 is used for the 1-μm^2 case, so further reduction in size affects mainly density. Still further incremental improvements in speed and density can be achieved by using an additional overlaid ground plane. Similar studies of the RCJL family project that cell delays are similar to the CIL for 1-μm features, but that circuit density is much higher. Miniaturization of the IRCL family developed at NTT has been analyzed.[14]

16.5 LOGIC SYSTEMS

The most complex Josephson logic circuits made to date are gate arrays. Workers at Hitachi have reported a gate array of 544 cells on a 9- by 10-mm^2 chip.[17] Each logic cell contains four SQUID-type OR gates and two direct-coupled AND gates, giving a total of 3264 gates. These were fabricated using Nb-NbO$_x$-Pb-alloy 1.5- by 1.5-μm^2 junctions. Total power dissipation for the array is about 12 mW. Testing was done on cascades of OR and AND gates to confirm performance.

A number of SSI and MSI logic systems have been demonstrated. These have included 4-bit and 8-bit adders, 4- by 4-bit multipliers, gate arrays, analog-to-digital converters, and counters. The critical path of a 16- by 16-bit multiplier has also been demonstrated. These circuits are considerably faster than any reported comparable semiconductor circuits, and the power dissipation is much lower. Table 16.1 shows a comparison of the speeds and powers of several of these circuits. Notice that the best result with Josephson circuits is 5 to 10 times

TABLE 16.1 Comparison of Small Superconductor and Semiconductor Logic Systems

	Superconductor				Semiconductor			
Type of circuit	Delay, ns	Power, mW	Junction materials, size	Ref. no.	Delay, ns	Power, mW	Material, device type	Ref. no.
4-bit full adder	0.173	0.85	Pb alloy 7-μm dia.	22	1.9 1.25	45 180‡	GaAs FET	23
4 × 4-bit multiplier	0.210	3.0*	Nb-AlO$_x$-Nb-1.5 × 1.5 μm^2	17	1.6	55	GaAs heterostructure, 1-μm gate	25
	0.280	1.0	Pb alloy 5-μm dia.	24				
16 × 16-bit multiplier (critical path)	1.10	250†	Nb-AlO$_x$-Nb 2.5 × 2.5 μm^2	26	6.0	2050	Si bipolar	27

*The authors have indicated in a private communication that this power could be decreased by a factor of 4 to 5 with improved testing conditions.
†Estimate for entire multiplier, based on an experiment on the critical path, in which power dissipation was about 10 times the design value.
‡Excluding output buffers.

faster than the best semiconductor result, and that the power dissipation is one to two orders of magnitude lower. If correction is made for the deviation from design noted in the footnote to the table for the 16- by 16-bit multiplier, all power-delay products for the superconductor circuits are two to three orders of magnitude better than those for the semiconductor circuits.

Figure 16.11 shows an example of the 4- by 4-bit multipliers; the unit shown is the one with the lowest critical-path delay (210 ps). Figure 16.12 shows a 4-bit analog-to-digital converter with a set of subharmonic latches added for testing; the converter was shown to have a 3-dB bandwidth of 500 MHz, even though it was in a rather imperfect experimental situation. Work is currently in progress on superconductive shift registers. The simulations indicate clocking speeds as high as 60 GHz.

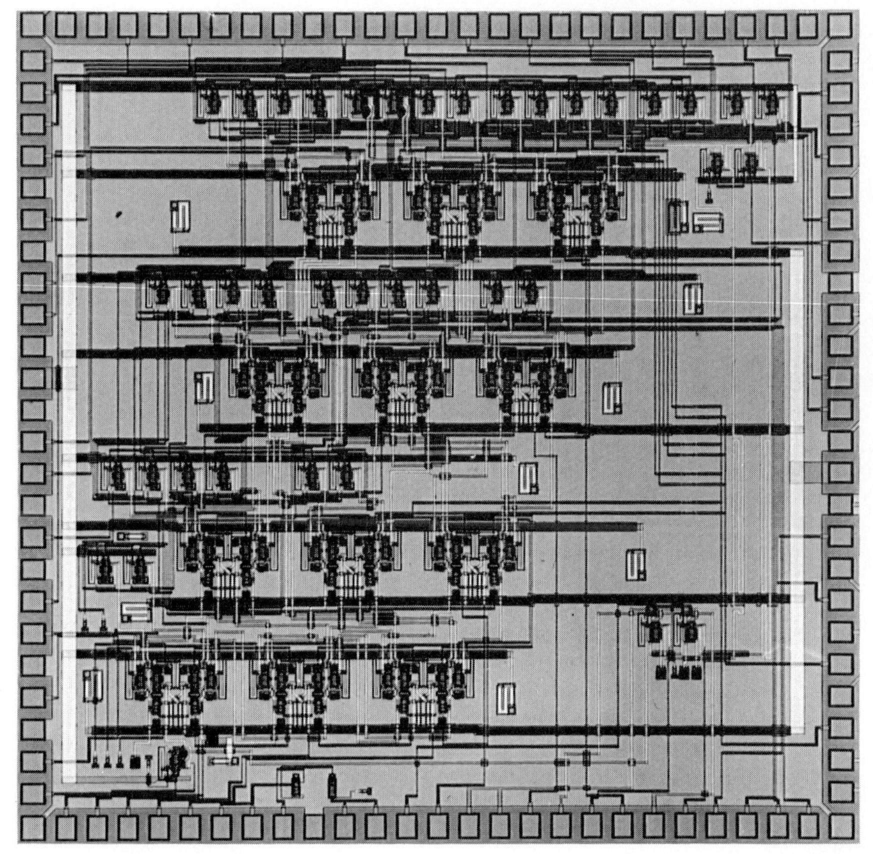

FIG. 16.11 Superconductive 4- by 4-bit multiplier with 210-ps overall delay in the critical path. (*With permission from Ref. 17.*)

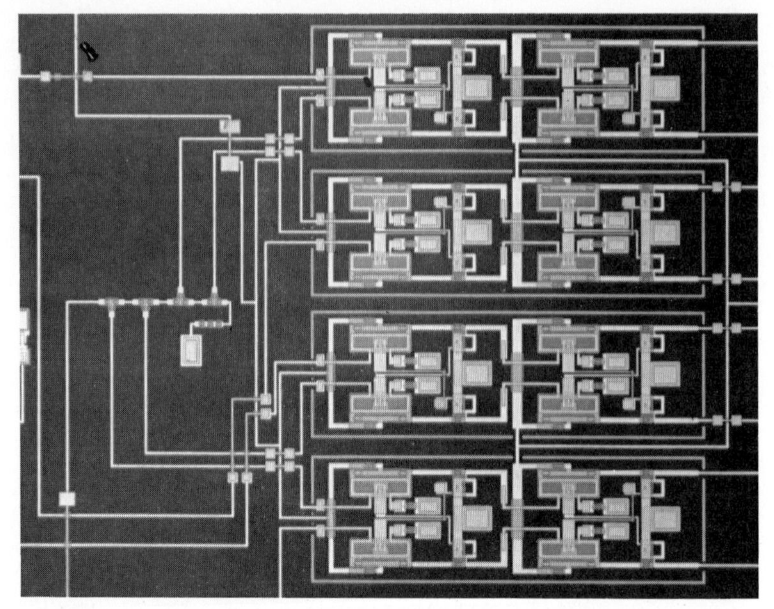

FIG. 16.12 Four-bit analog-to-digital converter that was clocked at 1.0 GHz and had analog 3-dB bandwidth of 500 MHz. The four circuits in the left column are comparators fed by a binary resistor divider. The four circuits in the second column are subharmonically clocked latches included to facilitate measurements.

16.6 MEMORY

The lossless property of superconductors and Josephson junctions for small currents provides an opportunity for memory. Most memory work has used circulating current in a superconducting ring, interrupted by a Josephson junction or a SQUID, to store a bit. The memory cell shown in Fig. 16.13 is the one that was under study by IBM for a cache memory at the time of termination of their mainframe computer project.[28] The two states of the cell are zero circulating current to represent a "0" and a clockwise circulating current to represent "1." A "1" is written in an empty cell by the following procedure. The supply current I_y is supplied first, and it splits up because of flux conservation into left- and right-hand branch currents I_L and I_R, respectively, as determined by $I_L L_L = I_R L_R$, where L_L and L_R are the inductances of the left and right branches of the loop. The currents I_x and $I_{y'}$ are applied to the cell after I_L is established in the write gate. This forces the gate to switch, and I_L transfers to the right branch. Upon removal of all currents, a clockwise circulating current is left in the loop because of flux conservation. A "0" is written in a cell with an initial "1" by dissipating the circulating current by coincident application of I_x and $I_{y'}$. The cell is read nondestructively by coincident application of I_y and I_s; the current in the right-hand branch is detected by the sense gate. If the cell contains a "1," the clockwise circulating current adds to the fraction of I_y that goes to the right-hand branch and is sufficient to cause switching of the sense gate. If there is no circulating current ("0" state), I_R is insufficient to switch the sense gate.

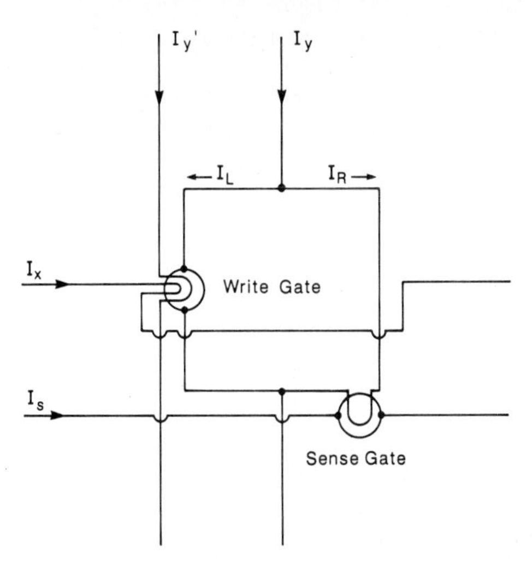

FIG. 16.13 A nondestructive-readout (NDRO) memory cell. (*After Henkels et al.*[27])

The IBM 1-kbit memory unit was first made with lead alloys and later with Nb edge junctions (Pb alloy counterelectrode) and was close to being operational in 1983. The large spreads in the junction critical currents made the design difficult. The entire project was terminated before any of the newer material technologies could be applied.

The project at NTT and the Japanese Ministry of International Trade and Industry (MITI) project also have produced 1-kbit memories, but none of them has worked completely. The required tolerances for writing and sensing gates seem to be smaller than is achievable in the technologies applied to date. There is hope that the all-refractory fabrication technologies presently coming on line will give sufficient parameter control. Otherwise, new memory concepts will be required. Some ideas have appeared, but need further evaluation. A new cell using superconducting loops and having a wide-margin sense gate is being studied by NEC at the time of this writing. In a new type of cell under study at NTT, the bit is stored as a group of vortices in a thin film; a SQUID in the cell is affected by the nearby vortices and acts as a readout for the cell. At ETL a memory with word organization, rather than bit organization, is being developed. With word organization, addressing can be done with two lines rather than three, and this simplifies the peripheral circuits and the cell, thus reducing the access time.

Evidence of the potential of fabrication using all-refractory materials (Nb-AlO_x-Nb) is given by the successful evaluation of an array of 8000 memory cells, which were wired but were without driver and decoders.[29] The initial-failure results were 100 times better than for similar arrays made with lead alloys and 10 times better than for Nb-NbO_x-PbBi. This is a test of the integrity of the metal films, insulation layers, and tunnel barriers; it is not yet a working memory.

The lack of a successful cache memory unit was a major reason for ending the IBM mainframe computer project and remains today a deterrent to the development of the technology. However, there is currently strong optimism in some quarters about making subnanosecond 4-kbit memories. This optimism is based on the existence of new technologies (especially Nb-AlO_x-Nb junctions).

In the course of the IBM project, a compact single-flux quantum memory cell was developed for a proposed main memory. Long before the termination of the mainframe project, work on the main memory was discontinued. Since then it has been understood that the main memory would be in semiconductor technology for a computer with superconductive logic and cache memory.

16.7 CIRCUIT PROCESS TECHNOLOGY

The goals of a circuit process technology for superconductive ICs include small spread of parameters of circuit elements, highly repeatable parameter values from run to run, stability with aging and thermal cycling, and ruggedness. In order to bring to bear the techniques developed extensively for semiconductor circuits, the various layers also should be capable of withstanding the highest possible processing temperature.

As stated earlier, most circuit studies have been done using the process based on alloys of lead that was developed by IBM and carried to a high degree of perfection by them and others. The process was adopted by groups in the United States and abroad, notably Japan. The junction-base electrode is a Pb-In-Au alloy and the counterelectrode is either Pb-Au or Pb-Bi. The barrier is an oxide of the base electrode which is formed by rf sputter oxidation; typically, it is mostly indium oxide. Resistors are made of a Au-In alloy. After the base electrode formation, an SiO junction-area mask is defined by photolithography and formed by lift-off. After the exposed base electrode is sputter-cleaned, the tunnel barrier is formed. These junctions change properties upon aging and have higher than allowable (for a mainframe computer application) failures with thermal cycling. The maximum allowed processing temperature is 70°C. Further improvements in Pb alloy junctions have been reported, and, at the time of this writing, some circuit studies still depend on this technology as the new all-refractory processes are being brought on line.

During 1981 to 1983, IBM changed to the use of a niobium base electrode, niobium oxide tunnel barrier, and a lead-alloy counterelectrode. To counteract the increase of capacitance caused by the appreciably higher dielectric constant of Nb_2O_5 compared with the oxide of the Pb-In-Au alloy, a much smaller junction area is used. This is done by forming the barrier on an oblique edge of a film so that one dimension of the junction area is on the order of the film thickness. These "edge junctions" have sufficiently low capacitance and are much more stable than the wholly lead-alloy junctions. With a highly disciplined production facility for circuits with edge junctions, IBM was able to meet their goals for logic demonstration.

The present approach is to use only refractory materials for circuit fabrication. The refractory materials are more stable with regard to aging and thermal cycling and allow higher processing temperatures. In most projects, the process starts with the formation of a Josephson tunnel junction over the entire wafer. Work of this type was first done at the Sperry Research Center and was reported in 1981; the procedure was called the selective niobium anodization process (SNAP).[4] Anodization is used to define the junction areas. The tunnel barrier is formed of partially hydrogenated amorphous silicon. The thickness can be rather large since the barrier height is low; the result is low junction capacitance.

Other circuit processes based on the initial formation of a whole-wafer junction have been devised. At ETL a process was developed in which the electrodes are composites of Nb and NbN layers such that NbN is at the interface with the Nb_2O_5 tunnel barrier.[15] The aims are to achieve a higher tunneling energy gap (advantageous particularly in memory circuits) and to appropriate the greater chemical stability of NbN (compared with Nb). The use of Nb is for the interconnecting transmission lines, since magnetic flux penetration is greater in NbN and that causes greater signal propagation delay. They achieved a very small spread (σ = 1.6 percent) in the critical currents for junctions of area 2.5 by 2.5 μm^2.[30] The process uses dry etching for patterning and has a self-aligned feature for the insulation surrounding the junction. It is called a *self-aligned niobium nitride insulation process* (SNIP). Processing temperatures as high as 300°C can be used without changing the junction characteristics. One disadvantage of this process is that Nb_2O_5 has a high dielectric constant. This leads to large junction capacitance, which slows the circuits (edge junctions are not possible with this approach). Quite complex circuits have been made; these include a functional 360-gate 8-bit adder, a 650-gate 4- by 4-bit multiplier, and a 1-kbit word-organized RAM.

A number of other insulators have been evaluated as barriers in the whole-wafer junction approach with Nb electrodes. These have included Ge and oxides of Al, Mg, and Y as well as nitrides and fluorides. In some cases, high-integrity oxides can be obtained by first depositing the element and then oxidizing. For other materials (e.g., MgO), the most successful junctions have been made by sputter-depositing the oxide. A great advantage of oxidizing after deposition is that ambient oxidation can give excellent uniformity over the entire wafer. The most highly developed process of this kind uses aluminum oxide; it was first employed in a process involving dry etching and was called the *selective niobium etching process* (SNEP). This aluminum oxide barrier process is currently the one most widely used.[31]

In the Nb-AlO_x-Nb process, a 5- to 10-nm film of Al is deposited onto the Nb base electrode. The oxide is then grown in an oxygen atmosphere within the vacuum system; its thickness depends on the temperature and pressure. The Nb counterelectrode is then deposited to complete the structure. Junctions have been reported with current densities up to 10^4 A/cm^2 and I-V characteristics that are quite suitable for digital circuits. Processing temperatures must be restricted to less than 150°C to avoid any changes of the junction properties. Changes of current density only, which can be calibrated, occur from 150 to 250°C. Also, difficulty has been experienced with failure of small junctions (less than 3×3 μm^2), apparently as a result of stresses in the Nb, but some techniques have been developed to circumvent this problem. Most of the current work with Nb-AlO_x-Nb uses the whole-wafer junction approach. Much of the spreads of critical currents can be ascribed to junction-area differences, though nonuniformities of the barrier also will make contributions that are area-dependent.

Zirconium appears to be a possibly advantageous alternative to aluminum for barrier information.[32] Like aluminum, its oxide can be formed by ambient oxidation. The principal advantage over aluminum oxide barriers evident at this point is that higher processing temperatures can be used (200°C compared with 150°C for Al).

Another important issue is run-to-run repeatability. So far, reports on Nb-AlO_x-Nb junctions indicate that ± 10 percent is the common experience. Since the whole-wafer junctions apparently have good shelf life (without changing critical current density), it may be possible to develop measurement techniques to accurately evaluate the current density of a wafer without invasion of the main body of the wafer; an alternative would be to have test wafers in the same batch.

The calibrated wafers could then be sorted and stored, as is done in resistor manufacture.

Another combination of materials for junctions in which there is current interest is NbN-MgO-NbN. Niobium nitride is less chemically active than niobium, and experience with the MgO barrier indicates the following advantages compared with Nb-AlO_x-Nb junctions: (1) it has a higher superconductor tunneling energy gap, which increases circuit speed; (2) NbN is more stable than Nb; (3) junctions can withstand a processing temperature of 300°C, thus allowing a greater range of processing flexibility, such as the use of spun-on glass and formation of plasma-enhanced chemical vapor deposition (PECVD) SiO_2 insulation. On the other hand, the voltage step at the gap is less abrupt than desired (especially for use in power supply circuits). The major drawback is that the MgO must be deposited and critical currents vary over a 2-in wafer by factors of 2 to 3. Ambient oxidized barriers may be the only way to get satisfactory wafer-scale uniformity; so far it has not been possible to get good junctions by oxidizing Mg.

The most common resistor material used in refractory superconductive ICs is molybdenum. Accuracy for resistor values is just as important as for critical currents in some cases. The determining factors are film thickness and structure, ratio of length to width, and resistance of the contact to the superconductors. Few data have appeared on spreads in resistors. In the IBM project (where the materials were different but the sources of variances were the same) the spreads (3σ) in resistor values were reported to be about $\pm$ 20 percent. Much better results have been obtained with refractory materials.

16.8 PACKAGING

As computer cycle times are decreased, the time for transmission of signals from place to place in the computer becomes predominant. The limitation imposed by the speed of light has led in today's computers to increased packaging density and heroic means for cooling. Superconducting circuits consume much less power and thus present an opportunity for greater packing density and shorter cycle time.

In superconductor technology, signals can be communicated by matched transmission lines on-chip and also for interchip communications. In the project at IBM, a packaging scheme was developed and tested. All structural components were made of silicon in order to avoid stress problems resulting from differential expansion. The chips were flip-mounted via solder balls to superconducting transmission lines on the card that carried the chips. The cards were provided with an array of 300 pins each of 75 mil diameter for connection to the main boards. The fragile pins were inserted into drops of mercury held in place by surface tension in cavities etched in the board. Interconnections between cards were made by transmission lines on wiring modules with pins inserted from the other side of the board. The inductance of the pins required special matching capacitors and driving circuits. The packaging concept was tested in a cross-sectional model experiment.[33] A data path that simulated a jump control sequence and a cache memory access in each machine cycle was operated with cycle times as low as 3.7 ns. All of the key components were evaluated in the test. A package very similar to IBM's was also tested by NTT.

The low power consumption of Josephson circuits and their independence of the substrate make them good candidates for multilayer, three-dimensional packing. Since they are not affected by defects in the substrate, whole-wafer integration is a possibility. This depends on yield and wafer-scale uniformity.

16.9 REFRIGERATION

A refrigerator requirement was described in connection with the IBM mainframe computer project. It was not considered to be one of the project's principal problems. At the time when the power estimates were made, it was assumed that the main memory, as well as the central processor unit (CPU) and cache memory, would be in Josephson technology. The estimated power dissipations were 1.2 W for the CPU (120,000 circuits), 0.8 W for the cache memory (32 kbytes), and 4.8 W for the main memory (16 Mbytes). The stray heat loss was estimated at 0.2 W. For this system, the refrigerator would be about the size of a small office desk.

As mentioned in Sec. 16.6, the expectation of having a superconductive main memory was dropped long before the termination of the project; in all quarters it has come to be understood that the main memory will be in semiconductor technology. Nothing has appeared in the literature on how this decision will affect the refrigerator. It may be possible to operate the semiconductor main memory at 77 K; in that case it could be physically quite close to the Josephson CPU, and cache memory and access time would be satisfactory. How this would affect the refrigeration requirement will require further analysis.

The refrigeration requirements for new systems can be expected to be profoundly affected by the new high-T_c copper oxide superconductors.

ACKNOWLEDGMENTS

I would like to express appreciation to the Naval Research Laboratory for hospitality during the writing of this chapter, with specific thanks to Martin Nisenoff and William L. Carter for the arrangements and to Ingham A. Mack for help with data. I am grateful for the encouragement of Edgar Edelsack and the support of the Office of Naval Research for our research in superconductor electronics over a number of years. Finally, many thanks are due to Bettye Fuller for her patience and skill in preparation of the manuscript.

REFERENCES

1. J. Matisoo, "Measurement of Current Transfer Time in a Tunneling Cryotron Flip-flop," *Proc. IEEE,* **55**:2052 (1967), "Subnanosecond Pair-Tunneling to Single-Particle Tunneling Transitions in Josephson Junctions," *Appl. Phys. Lett.*, **9**:167 (1966).
2. W. Anacker, "Potential of Superconductive Josephson Tunneling Technology for Ultrahigh Performance Memories and Processors," *IEEE Trans. Magn.*, **MAG-5**:968 (1969).
3. *Superconductive Electronics,* National Academy Press, Washington, D.C., May 1984, p. 5.

4. H. Kroger, L.N. Smith, and D.W. Jillie, "Selective Niobium Anodization Process for Fabricating Josephson Tunnel Junctions," *Appl. Phys. Lett.*, **39**:280 (1981).
5. J.C. Villegier, L. Vieux-Rochaz, M. Goniche, P. Renard, and M. Vabre, "NbN Tunnel Junctions," *IEEE Trans. Magn.*, **MAG-21**:498 (1985).
6. T. Van Duzer and C. W. Turner, *Principles of Superconductive Devices and Circuits,* Elsevier, New York, 1981.
7. R.L. Kautz, "Miniaturization of Normal-State and Superconducting Striplines," *NBS J. Res.*, **84**:247 (1979).
8. C.C. Chi, W.J. Gallagher, I.N. Duling III, D. Grischkowsky, N.J. Halas, M.B. Ketchen, and A.W. Kleinsasser, "Subpicosecond Optoelectronic Study of Superconducting Transmission Lines," *1986 Applied Superconductivity Conference,* 28 September–3 October 1986, Baltimore, Md.; *IEEE Trans. Magn.*, **MAG-23**: 1666 (1987).
9. S. Kotoni, T. Imamura, and S. Hasuo, "A 2.5 ps Josephson OR Gate," Technical Digest of the *International Electron Devices Meeting*. Washington, D.C., Dec. 6–9, 1987, p. 865.
10. T.R. Gheewala, "Josephson-Logic Devices and Circuits," *IEEE Trans. Electron Devices,* **ED-27**:1857 (1980).
11. H. Yamada, Y. Ichimiya, and A. Ishida, "Josephson Dual-Rail Two-Bit Adder Circuit Utilizing Magnetically Coupled OR-AND Gates," *IEEE Trans. Electron Devices,* **ED-31**:307 (1984).
12. T.A. Fulton, "Punchthrough and the Tunneling Cryotron," *Appl. Phys. Lett.,* **19**:311 (1971).
13. E.P. Harris and W.H. Chang, "Punchthrough Analysis of Josephson Logic Circuits," *IEEE Trans. Magn.,* **MAG-19**:1209 (1983); H. Yamada, T. Tanaka, and Y. Ichimiya, "Punchthrough Probability of Josephson Logic Gates," *Japanese J. Appl. Phys.,* **23**:1446 (1984).
14. H. Yamada, "Configuration of High-Speed Superconducting Logic Circuits," *SQUID'85—Superconducting Quantum Interference Devices and Their Applications,* Walter de Gruyter, Berlin, 1985, pp. 1077–1096.
15. S. Kosaka, A. Shoji, M. Aoyagi, F. Shinoki, S. Tahara, H. Ohigashi, H. Nakagawa, S. Takada, and H. Hayakawa, "An Integration of All Refractory Josephson Logic LSI Circuit," *IEEE Trans. Magn.,* **MAG-21**:102 (1985).
16. T.A. Fulton, S.S. Pei, and L.N. Dunkleberger, "A Simple High-Performance Current-Switched Josephson Gate," *Appl. Phys. Lett.,* **34**:709 (1979).
17. Y. Harada, Y. Hatano, K. Yamashita, M. Hirano, Y. Tarutani, and U. Kawabe, "Prototype Josephson Logic LSIs," *Extended abstracts, 1986 International Conference on Solid State Devices and Materials,* Tokyo, 1986, pp. 451–454.
18. T.R. Gheewala and A. Mukherjee, "Josephson Direct Coupled Logic," in *Proc. 1979 Electron Devices Meet.,* Washington, D.C., December 3–5, 1979.
19. J. Sone, T. Yoshida, and H. Abe, "Resistor-Coupled Josephson Logic," *Appl. Phys. Lett.,* **40**:741 (1982).
20. J. Sone, J-S Tsai, and H. Abe, "Picosecond Josephson Logic Gates for Digital LSIs," *Picosecond Electronics and Optoelectronics,* G.A. Mourou, D.M. Bloom, C.-H. Lee (eds.), Springer-Verlag, New York, 1985.
21. H. Ko and T. Van Duzer, "Miniaturization of Josephson Logic Circuits," *IEEE Trans. Magn.,* **MAG-21**:725 (1985). Ko, H., *Scaling of Josephson Logic Circuits,* Ph.D. Dissertation, University of California, Berkeley, 1987.
22. J. Sone, T. Yoshida, and H. Abe, "High-Speed Four-Bit Full Adder with Resistor Coupled Josephson Logic," *Technical Digest of International Electron Devices Meeting,* Washington, D.C., December 5–7, 1983, pp. 682–685.

23. E.H. Perea, F. Damay-Kavala, G. Nuzillat, and C. Arnodo, "A GaAs Low-Power Normally-On 4 Bit Ripple Carry Adder," *IEEE J. Solid-State Circuits,* **SC-18**:365 (1983).

24. J. Sone, J-S Tsai, S. Ema, and H. Abe, "A 280 ps Josephson 4b × 4b Parallel Multiplier," *Digest of Technical Papers, 1985 IEEE International Solid-State Circuits Conf.,* February 13–15, 1985, pp. 220–221.

25. A.R. Schlier, S.S. Pei, N.J. Shaw, C.W. Tu, and G.E. Mahoney, "A High-Speed 4 × 4 Bit Parallel Multiplier Using Selectively Doped Heterostructure Transistors," *Technical Digest of 1985 GaAs Symposium,* Monterey, Cal., 12–14 November 1985, pp. 91–93.

26. S. Kotani, N. Fujimaki, S. Morohashi, S. Ohara, and S. Hasuo, "High-Speed Unit Cell for Josephson Logic LSI Circuits Using Nb/AlO_x/Nb Junctions," *Applied Superconductivity Conference,* 28 September–3 October, 1986, Baltimore, Md. *IEEE Trans. Magn.,* **MAG-23**:869 (1987).

27. T. Sakai, S. Konaka, Y. Yamamoto, and M. Suzuki, "Prospects of SST Technology for High Speed LSI," *Technical Digest of International Electron Devices Meeting,* Washington, D.C., December 1–4, 1985, pp. 18–21.

28. H.H. Zappe, "Memory-Cell Design in Josephson Technology," *IEEE Trans. Electron Devices,* **ED-27**:1870 (1980). [The memory work at IBM was also described in a series of papers after termination of the project: W. H. Henkels, L. M. Geppert, J. Kadlec, P. W. Epperlein, H. Beha, W. H. Chang, and H. Jaeckel, "Josephson 4K-Bit Cache Memory Design for a Prototype Signal Processor." I, II, III. *J. Appl. Phys.,* **58**:2371–2399 (1985).]

29. H. Hoko, A. Yoshida, H. Tamura, T. Imamura, and S. Hasuo, "Material Dependence of Initial Failure Rates of Josephson Junctions," *Extended Abstracts of 18th (1986 International) Conference on Solid State Devices and Materials,* Tokyo, 1986.

30. H. Hayakawa, "Josephson Computer Technology," *Phys. Today,* March 1986, pp. 46–52.

31. M. Gurvitch, M.A. Washington, H.A. Huggins, and J.M. Rowell, "Preparation and Properties of Nb Josephson Junctions with Thin Al Layers," *IEEE Trans. Magn.,* **MAG-19**:791 (1983). [A review of the current status of junction formation is given by A. I. Braginski, J. R. Galaver, M. A. Janocko, and J. Talvacchio, "New Materials for Refractory Tunnel Junctions: Fundamental Aspects," *SQUID-85-Superconducting Quantum Interference Devices and Their Applications,* Walter de Gruyter, Berlin, 1985, pp. 591–629.]

32. H. Asano, K. Tanabe, Y. Katoh, and O. Michikami, "Fabrication of All-Nb Josephson Junctions Using Oxidized Zr Overlayers," *Japanese J. Appl. Phys.,* **25**:L261 1986.

33. M.B. Ketchen, D.J. Herrell, and C.J. Anderson, "Josephson Cross-Sectional Model Experiment," *J. Appl. Phys.,* **57**:2550 (1985).

P • A • R • T • 5

VLSI DESIGN METHODOLOGY

CHAPTER 17
HIERARCHICAL DESIGN

Edward L. Hepler
Villanova University
Villanova, Pennsylvania

17.1 HIERARCHICAL DESIGN

The concept of generating a hierarchical VLSI design is one that has been interpreted differently by different people. As with most things, the perception of hierarchical design is often influenced by the inclination of the perceiver. Two common points of view concerning hierarchical VLSI design are those represented by physical designers (those who lay out circuits on silicon) and system designers (those who determine how functions are to be partitioned within a system). The first group considers hierarchical design the process of building larger functions out of smaller ones; the second group considers it the process of breaking down large functions into smaller ones. The difference is subtle yet distinct. Both groups have the common goal of easing the burden of designing complex systems and consider hierarchical design as a necessary and welcome tool.

System level designers like to generate top-down designs and view hierarchical design as a methodology that permits complex system-level problems to be analyzed and synthesized. This is accomplished by recursively decomposing functions into smaller, more manageable, and more understandable functions. This layered approach is supported by modern simulation systems and tools and permits a more holistic design approach, where more global system concerns may be analyzed. With this approach, each function is initially modeled based on how it behaves (algorithmically), with the focus on how each function interacts with others within the system. This permits more of the important initial time of a project to be spent on architectural concepts, global coordination of system functions, and function partitioning, using models that are easily generated and modified, rather than spending time on trying to determine how the functions will be implemented. Once an overall architecture has been formulated, details of particular functions may be pursued. As these details become available, they may be introduced into the analysis and simulated along with higher-level models of functions yet to be detailed.

Layout designers, while not necessarily advocating a bottom-up design approach are nevertheless very often concerned with the minute details of design rules—problems that compound themselves as a design grows. For them, hierarchical design also provides a methodology that permits parts of a system to be analyzed and checked at a reasonable level. For example, "correct by construction" is a hierarchical methodology that purports that a macro function, one which is composed of a collection of wired-together primitive functions, will be physically correct if its component primitive functions have been previously proven physically correct. This form of hierarchical design stresses physical (layout) functions.

In reality, both methods of hierarchical design must be considered. For example, the knowledge of the functions that must be performed (systems level) within a chip and how the data flows through the system impact the floor plan (how the major subsystems are arranged on the chip) of the chip and the shape (physical aspect ratio) of the underlying cells. Ideally, the concerns of both the systems designer and the layout designer can be balanced by the architect(s) of the system.

The methodology used for the design implementation will depend not only on the preferences of the designer (or the designer's boss or company) but also on the level of customization to be exploited. Additional degrees of design freedom require that more tools, audits, and so on, be used to ensure the success of a design. For example, in designs with gate arrays, normally only a predefined set of functions, whether they are primitive functions or macros, are available for use. These functions have been pretested, and their layout possibilities are well known or defined. This limits the amount of analysis required to ensure a working design. As the level of customization increases through the use of standard cell and custom design, the additional freedoms afforded the designer require that more extensive checks be employed. The methodology employed for gate arrays cannot be used for custom design!

17.2 A SAMPLE METHODOLOGY

Figure 17.1 shows a possible methodology for custom VLSI circuits. It is a combination of top-down and bottom-up design approaches. It is top-down since it considers the overall functions to be performed and how those functions interact with one another and the system. At the same time, low-level details (concerning capabilities, etc.) are considered to enable the designer to make informed, intelligent tradeoffs. It also attempts to consider both the systems designer's and physical designer's concerns.

17.2.1 Requirements

The time spent defining what is to be designed and how it fits into its environs is very important. The success of a system design effort is often a reflection of the level of effort placed on the definition of its requirements. While complete requirements are definitely a necessity, efforts must be tempered by a common problem of overspecifying a system. Care must be taken to define only what a system or subsystem must do and the performance requirements of the functions being specified. It is very important that the "hows" of the implementation *not* be included in the requirements. The implementation of requirements must be reserved for the tradeoff analysis that will occur during the architecture of the system. Implementation suggestions and examples of how the system will be used

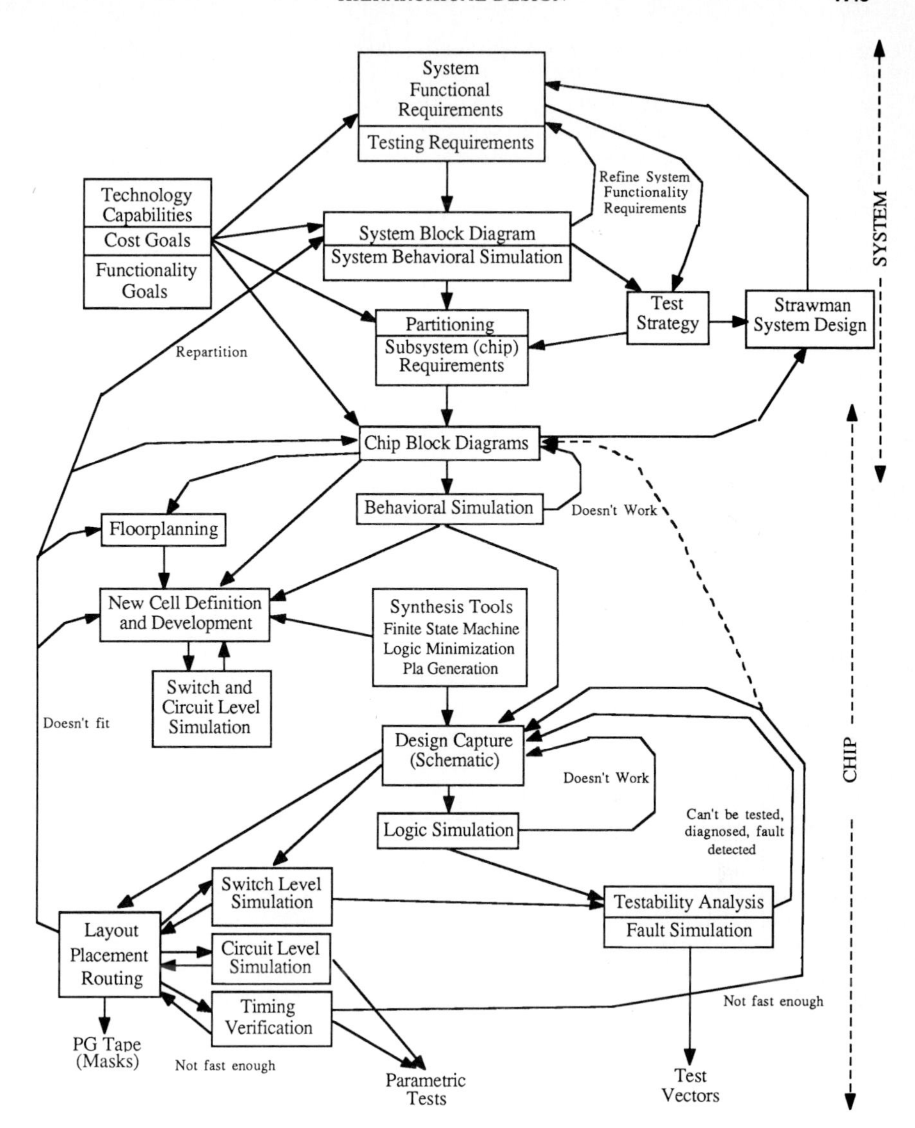

FIG. 17.1 Methodology.

can be of use during the design and should accompany the requirements when possible. Requirements are important because they not only define what is to be designed, but they also define a metric by which the designer knows when the design is complete. Requirements must be considered at many levels.

1. *System level:* The function of the system must be considered. After all, it is the need for this function that has caused the design effort to be initiated. However, system designers too often consider only the function that the system must perform and not how the system is to perform it (or whether it can be implemented) or how it is to be tested. System design for VLSI is often different than system design for older technologies. The overall system must be considered with the virtues and limitations of VLSI exploited and understood, respectively. The density of VLSI permits implementations of functions that were only dreamed about 5 to 10 years ago. Each subfunction should be considered and analyzed for the best possible method of implementation in the technology being used.

2. *Partitioning:* The partitioning of the system into chips is very critical. Pin limitations and the cost (delay time) of transferring a signal from one chip to another must be considered. Tradeoffs between data path implementations and the control logic necessary to control each of those data paths should be analyzed. Sometimes, a more complex data path may be easier to control, thereby requiring less control logic than a simple data path, which must be heavily controlled to perform a complex function. The number of transistors available on a chip versus the expected manufacturing cost of the chip will also influence the partitioning of the functions of a system across chips. Partitioning ends up being a delicate balance between function, pin count, power dissipation, and fabrication costs.

3. *Testing:* As the level of functionality being placed on a single chip has increased, so has the problem of testing the chip. The number of bonding pads available around the periphery of the chip has not kept pace with the number of transistors that may be placed on the chip. This has caused the level of controllability and observability of the circuit functions to be much lower than needed to perform acceptable tests. Various strategies for overcoming these problems, including scan path and LSSD, have been discussed in the literature. Important to the discussion here is that the testing strategy for the system must be defined hierarchically. The method of testing required for the system should influence and be influenced by the method of testing imposed on the modules (boards) that make up the system. These decisions in turn should be influenced by and influence the chip level testing. The methods used to test the chip after fabrication should be useful for testing those chips after they have been installed in the boards and the boards installed in the system. The testing needs of the system must be defined, including under what conditions the testing must take place and the level of fault detection that must be achieved. Related topics that should be considered include whether the system will have to be diagnosed (a slightly different form of testing) during its use and whether real-time fault detection (detecting both hard and soft faults as they occur) will be necessary.

17.2.2 Behavioral Simulation

Behavioral simulation is sometimes used to help define system requirements. Whether it has or not, once the requirements have been interpreted and a preliminary architecture has been determined, behavioral simulation should be used to verify that the selected architecture performs the required functions within the specified performance range. It also provides a convenient method for trying various architectures and performing tradeoff analysis. At every step of the methodology, it is very important to verify as much as possible before moving on to the next (and more detailed) level. Later in the development, as more details are learned about circuit functions, the behavioral models should be updated to incorporate any changes deemed necessary. The simulations may then be rerun to determine if the additional details, timing information, and so on, have affected the simulation results.

17.2.3 Floorplanning and New Cell Development

The chosen architecture should have considered the needs and limitations of VLSI technology. Part of the analysis that led to the architecture should have been how the data and control signals flow and how the functions to be placed on the chip interact with each other. These factors contribute to the planning of the physical layout of the functions to be placed on the chip—better known as *floorplanning*. The shape and space allocations for each of the functions to be placed on the chip will also influence the definition and development of the cells that will be used to generate the functions. The aspect ratio of the various functional blocks and the relative positions of their inputs and outputs will directly influence how the individual cells making up the blocks should be constructed. Ideally, the entire data path can be designed such that the individual parts within the path butt together, thereby completely eliminating the silicon area that might have been expended in interconnect material to route them together. The layout of the data path will likewise influence the size, shape, and positioning of the required control logic. Efforts should be made to effect a floorplan that "cleanly" moves data (signals) through the chip.

As the floorplan and functionality of the individual blocks become more defined, the actual cells used to implement the blocks must be determined. Often, cells from earlier designs or libraries may be used directly or modified to suit the needs of the function. Sometimes completely new cells must be designed. In custom design, much emphasis is placed on the layout of individual cells and how the cell will be used with neighboring cells. Interactive graphics editors and various design rule checkers and analysis tools aid in the layout process. Characteristics of cells may be predicted by using switch and analog simulators.

Any tools that may be used to eliminate the errors injected by human intervention while designing a system are welcome. Tools exist to minimize logic, automatically lay out programmable logic arrays (PLAs), generate and verify state machines, and stretch or compact layouts.

17.2.4 Schematic Capture

When designing with gate arrays or standard cells, the designer constructs the desired function from a set of predefined primitive functions. These primitive functions may be combined in a hierarchical manner to form macrofunctions, which may then be used to generate yet higher-level functions. This procedure is analogous to using TTL to create board designs.

Designers use schematic capture CAD tools to interactively (and graphically) enter the interconnection detail of primitive or macrofunctions into a design database. This permits the schematic to be modified interactively and the various levels of hierarchy to be viewed.

Once the database has been populated, other CAD tools may access the data in it. This database contains the interconnection and cell-type data necessary to drive placement-and-routing tools and create netlists for logic-level simulators. Often the database generated by the schematic capture step may contain "unfilled data items" for other CAD tools to populate. For example, a placement program may populate the topology area of the database while a timing analyzer may fill in delay information.

17.2.5 More Simulation

At each level of design, simulation plays a key role. Logic simulation verifies functions at the gate (logic) level, determining whether the appropriate function has been synthesized. Switch-level simulation verifies that the function has been implemented correctly at the transistor level. Circuit-level simulation provides insight into the actual analog operation of the circuits. Fault simulation indicates whether a set of tests adequately exercises a circuit in order to flush out any hidden manufacturing or fabrication problems. Modern simulators provide the capability of simulating various parts of a circuit using different modes (behavioral, logic, switch) simultaneously. This allows the designer to simulate with more detail the known portions of the circuit while driving those portions with higher-level models. It also allows blocks of circuitry to be replaced by behavioral models that act exactly like the underlying transistors once the function of those transistors has been verified. This decreases simulation time dramatically.

The necessity of simulation cannot be minimized. Every effort must be taken to verify as much of a design as possible before proceeding to the next step. As information from layout, routing, etc., becomes available, the derived delays (from parasitic resistances and capacitances, and distances) must be fed into the simulation models.

Simulation is an ongoing task and must be maintained throughout the life of the project. All models, from early architectural models to individual cell models, must be maintained and kept up to date, and regression tests must be run. Any change to the system should be verified for impact to other parts of the system through simulation.

17.2.6 Analysis-Audits

Various tools exist that predict power consumption and power density, and perform worst-case timing analysis. Design rule checkers and electrical rules checkers verify that physical parameters of the technology have been adhered to. Circuit extractors "reverse engineer" a completed layout into a circuit description (netlist) including resistances and capacitances. This netlist may then be used to simulate the extracted circuit for timing analysis and comparison with earlier simulations. In gate-array or standard cell design, the extracted netlist may be compared with the netlist created by the schematic capture step for consistency.

17.2.7 Feedback

Each time an analysis or simulation is performed, a determination must be made as to whether the circuit under test has met the requirements placed upon it. If a problem is uncovered, something must be done to correct it. The object of the methodology is to expose all problems as early as possible, thus ensuring a high degree of confidence in the design before a commitment is made to the expensive and time-consuming task of fabrication.

17.2.8 Documentation

Overshadowing the entire design process is the need for keeping records of design decisions, design alternatives, notes, and so on. Ideally, an on-line documentation system that has the capability of incorporating sketches, circuit drawings, and block diagrams, and works in concert with the rest of the CAD environment will be available.

17.3 IMPLEMENTING THE METHODOLOGY

Ideally, a single integrated CAD system will be available to support and enforce all of the ideas and tools mentioned here. Indeed, a number of commercially available CAD systems have recently appeared that integrate most of the steps outlined in the methodology. Readers are encouraged to test a number of systems in an actual design environment before committing themselves to a particular system.

When a single system is not available, efforts must be increased to ensure the integrity of data as it passes between the individual tools. Often a set of converter programs will have to be generated to provide the format conversion necessary among nonintegrated tools. This is currently the case at Villanova University, where the University of California at Berkeley VLSI design tools are used.

Whether an integrated CAD system or a set of loosely coupled tools exist, simply having the methodology will not ensure a correct design. The methodology must itself be a tool that is used as needed for each program. Some cases may choose to eliminate certain steps, while others may find it necessary to add additional tests or checks.

CHAPTER 18
SIMULATION

Edward L. Hepler
Villanova University
Villanova, Pennsylvania

18.1 NEED FOR SIMULATION

Simulation has become a requirement for successful system design. It has proven its usefulness in the following:

1. *Its ability to allow the exploration of alternative designs:* During the system architecture phase of a design, the functionality of a system may not always be known in its entirety. By performing behavioral simulation, one may try different functions, functional implementations, or design alternatives, usually by simply changing a few simulation statements. At a lower level, the implementation of a primitive function may be analyzed by using a switch- or circuit-level simulation to predict its characteristics before going through the expensive and time-consuming task of fabrication.
2. *Its ability to aid in the verification of the correctness of complex designs:* As designs become more complex, verifying that the functions being implemented are correct has become increasingly more difficult. Many designs would undoubtedly go through many fabrication iterations if it were not for the simulations that have flushed out design problems. Not simulating would be prohibitive from the viewpoint of cost and time. The practice of using simulation to completely verify a function before committing to the fabrication step has become a standard practice.
3. *Permitting the time allocated for design to be reduced:* If it were not for simulation, much of the analysis required to verify a design for correctness or timing would have to be done manually. This is prohibitive not only in terms of time and cost, but it might be nearly impossible for many designs.
4. *Decreasing the immense cost of waiting to find problems:* The longer a problem remains hidden, the more costly it will be to correct once it is found. Simulation at the various levels described here permits concepts and implementations to be tried and verified throughout the design process. By utilizing simulation to its fullest extent at each level, one can flush out problems early in the design life. Ignoring simulation results or waiting for the next level of detail before using simulation may simply allow problems to fester, thus ensuring that project costs will be high and schedules missed.
5. *Reducing the difficulty in debugging hardware:* Waiting for hardware to begin

debugging is not the most advantageous way of ensuring system success. The models used to verify the correctness of a VLSI chip may be very useful in verifying the system(s) that will ultimately use them. However, in order to be useful at the system simulation level (due to time constraints and computer resource constraints), we *must* usually use behavioral models. If the simulation models are written at a high enough level, software may often be debugged before the hardware exists. This step is sometimes useful in detecting deficiencies and may lead to modifications of the system architecture.

6. *Generating tests:* Various types of tests must be generated in the course of a system design. Simulation plays a major role in generating the stimulus and predicting the results for many of them. The types include the following:
   a. *Chip fabrication test:* This test determines whether the chip has been fabricated correctly. It exercises the functionality of the part and provides metrics for determining whether the part functions within the operating parameters specified in the requirements.
   b. *Factory system test:* This test is performed on the complete system and must determine whether it has been manufactured correctly and meets shipment standards. The test is performed on each system as it leaves the assembly line and must be executed in a reasonable amount of time.
   c. *System functionality:* This is a set of tests that are designed to verify that all requirements specified in the requirements document have been met. For example, in a processor all modes-options of all commands-instructions must be tested for correctness. Usually this test is performed only once to verify the correctness of the overall design. It is often a very lengthy test and is generated by a combination of simulation and actual usage.

18.2 HISTORY OF SIMULATION

Digital logic simulation has been used since the early 1960s. Early simulators were used mostly for logic verification or fault analysis and were one of three types:

1. *Compiled simulators:* Compiled simulators converted a model definition consisting of an interconnection of elementary gates into a series of instructions that were executed on the host machine. When executed, these instructions determined the logical function the model represented. Figures 18.1 through 18.3 depict an example of this process. The circuit to be simulated is shown in Fig. 18.1, while the instructions produced during the compilation process are listed in Fig. 18.2. The process of converting a circuit description into the host code is

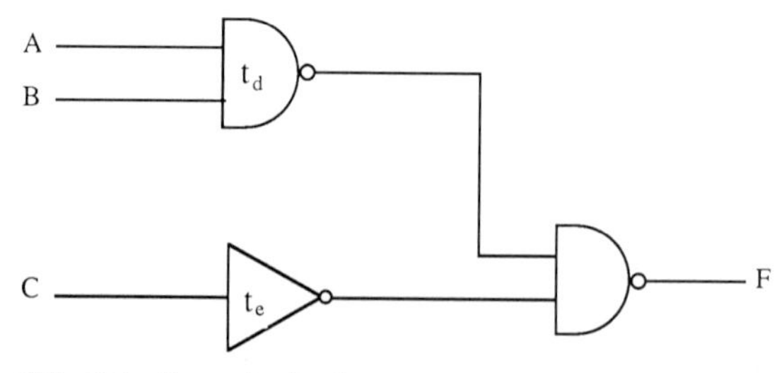

FIG. 18.1 Example circuit.

called leveling or levelizing. Compiled simulation has disadvantages and is not used in modern simulation systems. Consider the case where inputs A, B, and C are represented by the first three timing diagrams in Fig. 18.3. The actual circuit output will contain a glitch due to propagation delay differences in gates D and E, as depicted by the timing diagram in Fig. 18.3. (Gate delay is the time necessary to propagate the result of a logic element from its inputs to its outputs.) However, a compiled simulation output will present a timing diagram as shown in line G. Due to the compilation process, a circuit is evaluated by logic levels. A logic element is not evaluated until all the logic elements in the preceding level have been evaluated and until values for those elements have been determined. With this process, timing information such as signal skew due to variations in propagation delays is lost. Glitches are difficult, if not impossible, to detect by using this type of evaluation. During the evaluation of a circuit using a compiled simulator, each gate is evaluated each time a new set of inputs is applied to the model. This is often a waste of simulation time, when sections of a circuit remain inactive for some time. Compiled simulators generally can accommodate large circuits and they execute relatively fast.

```
LOAD     A
AND B
COMPLEMENT
STORE    D
LOAD     C
COMPLEMENT
STORE    E
LOAD     D
AND E
COMPLEMENT
STORE    F
```

FIG. 18.2 Compiled instructions.

The addition of new elements to the library of a compiled simulator is a difficult task. Each element to be simulated must be converted into a sequence of instructions that simulate the function the element performs. Since the compilation process produces code that is directly executed by the host machine, a working knowledge of the host machine is mandatory. Intimate knowledge of the data structures and subroutine linkage for the simulator is also necessary. This is not practical for the design engineer. The fact that host code is produced also seriously limits the portability of the simulation system.

2. *Table-driven simulators:* Modern simulation systems are usually table-driven. Table-driven simulation systems follow lists of logic elements stored in tables (queues) that have been produced by an interconnection evaluation algorithm.

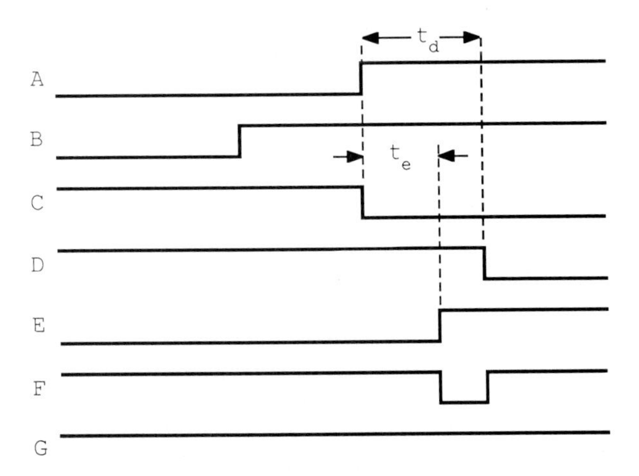

FIG. 18.3 Example timing diagrams.

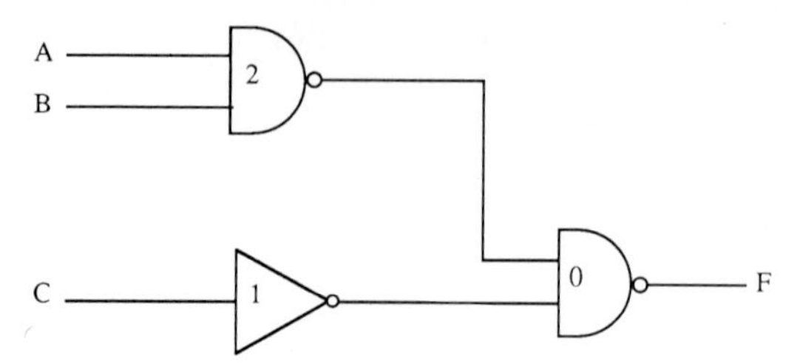

FIG. 18.4 Example circuit with delays.

A model definition of the circuit to be simulated is converted into a set of tables that describes the topology of the circuit, including the fan-in and fan-out of each element. When an element is being evaluated, its fan-in at the evaluation time is sampled and its output is determined according to the function the element performs. This is usually accomplished by table lookup or evaluation of a single copy of a model of the element. The fan-out of the element (elements whose inputs include the element being evaluated) is then scheduled for evaluation at a time determined by the delay of the element.

The time necessary to perform the simulation can be minimized by comparing an element's evaluated output with its previous output. If the values are the same, no further evaluation is scheduled; otherwise, the element's fan-out is scheduled for evaluation at some future time. This technique is called *selective trace* or *event-driven* simulation. Selective trace is valid, since no element can change if its inputs do not change. Therefore if an element's output does not change, its fan-out will not change, and the fan-out elements need not be evaluated.

Evaluation continues with the next element scheduled in the element queue. This method of logic simulation preserves timing information within a circuit. Since elements are scheduled for evaluation as a function of their topology within the circuit and their propagation delays, glitches such as the one lost during the previous example can be flagged and corrected.

Figures 18.4 through 18.8 consider the same example as was presented for the compiled simulator. In the table-driven example numerical values are assigned to gate delays to facilitate easy explanation of the simulation process. Thus, gates *D, E,* and *F* are assigned delays of 2, 1, and 0 units, respectively. Figure 18.4 shows the circuit with the delays in the gates.

TIME	CHANGING SIGNAL
3	D
5	D
5	E
⋮	⋮

FIG. 18.5 Queue at initialization.

At initialization of simulation, the element evaluation queue is filled with those elements affected by changes on input lines. In this case, element *D* is scheduled for evaluation at times 3 and 5 (relative to 0), corresponding to changes on input lines *B* and *A*. Element *E* is scheduled for evaluation at time 5 due to the change on input line *C*.

When simulation begins, the queue contains only those entries corresponding to changes on the input lines of the circuit. This is depicted in Fig. 18.5. For sake of example, it is assumed that signals *D, E,* and *F* have been predefined. The first entry indicates that element *D* should be evaluated. This is done and the results compared with the previous value of element *D*.

Since no change has occurred, no elements are scheduled due to this evaluation.

The next entry in the queue indicates that element D should again be evaluated. This time, however, the output has changed due to the changes of the input lines. The fan-out of element D is then scheduled for evaluation at the current time plus the number of units delay assigned to element D. This results in the scheduling of element F for evaluation at time 7. Since a new element has been added to the queue, it must be sorted. The queue then appears as in Fig. 18.6. Element E is next evaluated, again resulting in the queuing of element F. After sorting, the queue appears as in Fig. 18.7. Element F is now evaluated twice, with results being reported immediately due to a delay of zero. No elements are scheduled because element F represents a circuit output and therefore has no fan-out elements. The queue is now empty and simulation ceases.

TIME	CHANGING SIGNAL
5	E
7	F
⋮	⋮

FIG. 18.6 Updated queue.

TIME	CHANGING SIGNAL
6	F
7	F
⋮	⋮

FIG. 18.7 Updated queue.

As can be seen from the resulting timing diagram (Fig. 18.8), the glitch due to element propagation delay differences has been detected and correctly reported.

3. *Register transfer simulators:* Register transfer simulators are useful in computer architecture studies where circuit detail is less important than data flow analysis. Arbitrary registers may be defined and information moved from register to register based on conditions, values, etc. Using this type of simulator, one can study questions such as word length, number of registers, and instructions to be implemented without committing to a detailed design effort. Relative timing, in terms of cycles needed to accomplish a function,

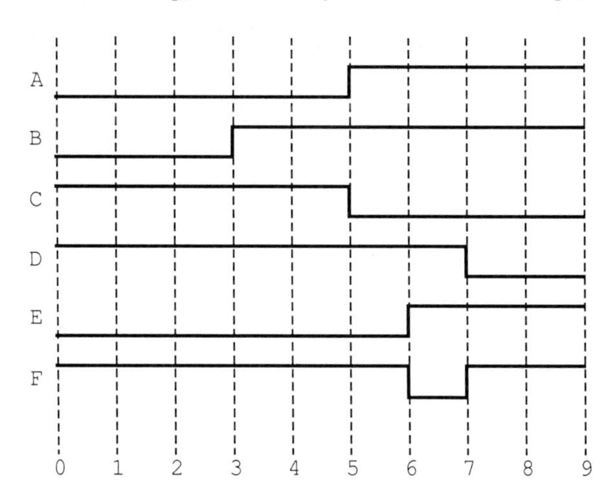

FIG. 18.8 Timing diagram.

can be found by using this type of analysis. Detailed timing, such as propagation delay, hazard analysis, and determination of cycle time are usually not determined with a register transfer simulator.

18.3 MAJOR USES OF SIMULATION

Simulation is the act of developing and exercising a software model of a hardware circuit to predict the response of the circuit to a set of known inputs; it is useful during the entire course of a design.

18.3.1 Architectural Studies

During the architectural definition phase of a project, simulation provides a tool for testing ideas and predicting performance of straw-man system configurations. It is obviously much easier to code and exercise a model of a system (especially one where all the details may not be known), to gain insight into its characteristics, than to attempt to define all the details necessary to generate a hardware prototype and to construct, debug, and instrument a system for testing purposes. An architect will be much more willing to try new ideas when those ideas may be tested in an interactive way with little cost.

18.3.2 Design Verification

Once an architecture has been selected and a detailed implementation of that architecture generated, simulation may be used to verify that the implementation is logically correct and that certain design rules have been met.

Logical-Functional. Logic simulation provides a means of verifying that the logic function that has been designed provides the functionality called for in the requirements. It also may be used to verify that "extra functionality" has not inadvertently been placed into the logic. Often, all conditions have not been specified in a requirements document, and if these have not been taken into consideration during logic design, these conditions may cause unexpected results should they occur. Simulation may help to flush out these surprises. As with other levels of simulation, logic simulation gives the designer the opportunity to try new approaches without the cost of building and debugging a circuit.

Timing. Simulation provides the designer with a tool for performing timing analysis. Most modern simulators provide the ability to detect hazards, race conditions, and oscillations. Some are capable of performing worst-case analysis, where, for example, a worst-case data path is "raced" against a best-case clock path for a register. The data path must "beat" the clock path with sufficient margin to satisfy the setup time of the register.

Electrical. Some simulation systems provide audits for signal load analysis and can predict the power consumed over a set of vectors. Circuit (analog) simulation predicts the electrical characteristics of the circuit.

18.3.3 Fault Coverage

Fault simulators monitor all nodes within the circuit model and determine which nodes have been exercised with observability for a set of vectors. By ensuring that all nodes have been exercised (100 percent fault coverage), one can be confident that the vector set will find most (if not all) fabrication faults of the "stuck-at," open, or shorted variety. Many fabrication facilities require a fault coverage of more than 95 percent before beginning fabrication.

18.4 TYPES OF SIMULATION

Different forms of simulation exist to serve the differing needs of system and logic designers.

18.4.1 Behavioral

Behavioral simulation is used to model systems at the highest level. In this type of simulation, functions are represented algorithmically. The architect captures the essence of a function by describing the behavior of the function, usually using a hardware descriptive language. This permits the designer to model the function without having to implement it with gates or transistors. The function's requirements can therefore be defined and refined and simulated with other functions before detailed implementations of any of them are known. This provides a forum to test ideas and concepts and predict rough system performance.

Figure 18.9 shows a behavioral description of a *D* flip-flop. While this is a simple example, it will be used to contrast against gate-level and switch-level models in later sections.

Register transfer simulation is a subset of behavioral simulation. Here, the architect normally has a better idea of the structure of the system and may represent the behavior of a function as a flow of information between registers. In

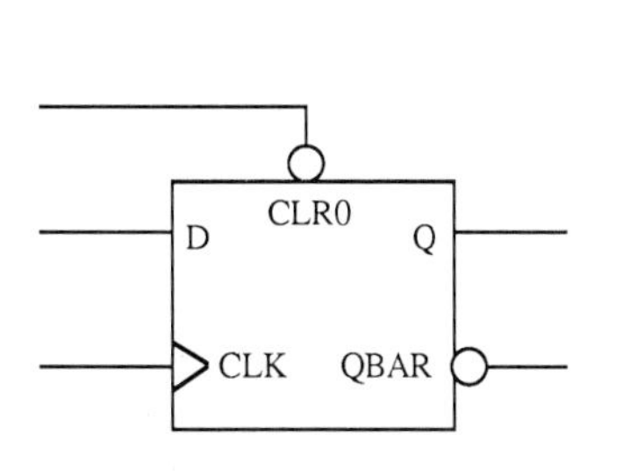


```
dff(d,clk,clr0,q,qbar)
        input d,clk,clr0;
        output q,qbar;
{
register bit;

if(clr0 == 0)
        {
        bit = 0;
        q = bit after Q_DELAY;
        qbar = ~ bit after QBAR_DELAY;
        }
else if(clk == rising)
        {
        bit = d;
        q = bit after Q_DELAY;
        qbar = ~ bit after QBAR_DELAY;
        }
}
```

FIG. 18.9 Behavioral *D* flip-flop.

practice, most models of systems at this level are a combination of behavioral and register transfer simulation.

18.4.2 Gate

In gate-level simulation, all functions are represented by their gate-level implementation. Larger elements are often hierarchically defined by using smaller elements that are defined at the gate level. Before simulation begins, the simulator usually "flattens" the model to generate a model that consists of only primitive gates. The number of levels or signal types represented within the simulator varies from one to another. All model at least three levels (0.1. *X*—unknown), and many model more (e.g., *Z*—high impedance). Many also provide multiple strengths per logic-level type. Figure 18.10 shows a gate-level representation of a *D* flip-flop and the simulation description of it.

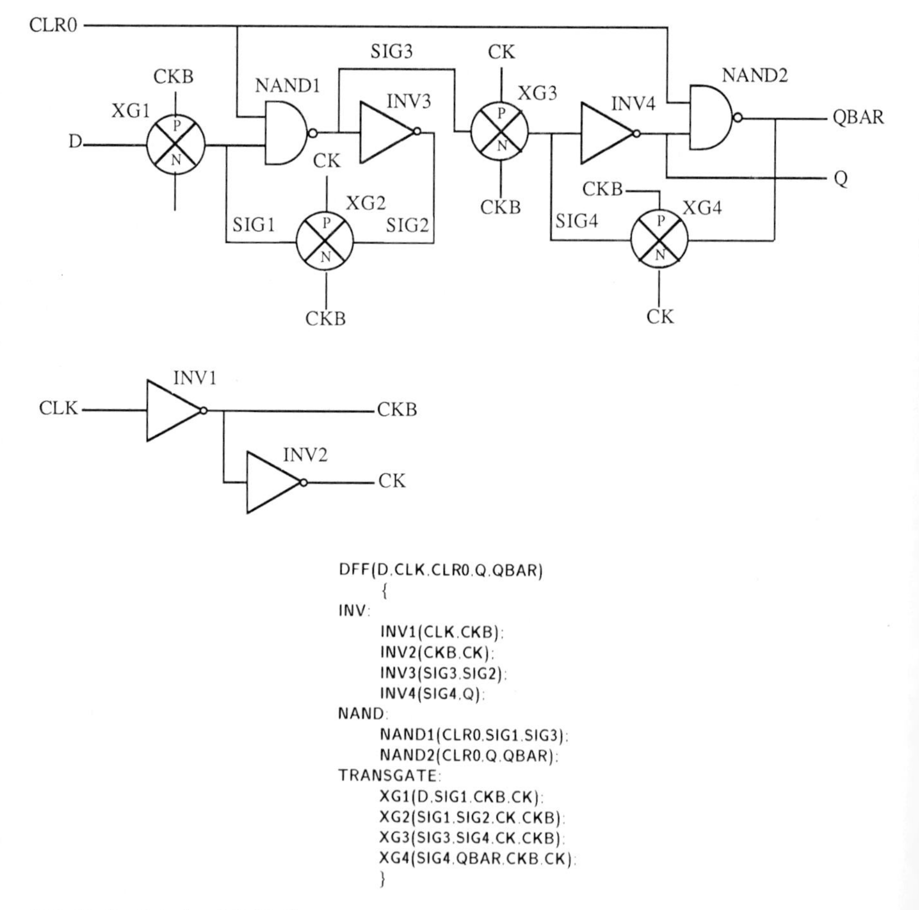
CLR0
SIG3
CK
CKB
NAND1
XG3
NAND2
INV3
INV4
XG1
QBAR
D
CK
Q
CKB
XG2
CKB
XG4
SIG1
SIG2
SIG4
CKB
CK
INV1
CLK
CKB
INV2
CK
DFF(D.CLK.CLR0.Q.QBAR)
{
INV:
INV1(CLK.CKB):
INV2(CKB.CK):
INV3(SIG3.SIG2):
INV4(SIG4.Q):
NAND:
NAND1(CLR0.SIG1.SIG3):
NAND2(CLR0.Q.QBAR):
TRANSGATE:
XG1(D.SIG1.CKB.CK):
XG2(SIG1.SIG2.CK.CKB):
XG3(SIG3.SIG4.CK.CKB):
XG4(SIG4.QBAR.CKB.CK):
}

FIG. 18.10 Gate-level *D* flip-flop.

18.4.3 Switch

Switch-level simulation represents functions at the transistor level, but the transistors are evaluated as perfect or near-perfect switches. Often the switch-level models are generated automatically by extracting a VLSI layout. In these cases, parasitic resistances and capacitances are often included in the extracted data, and timing may be generated by the simulator using crude *RC* models. Since the models are accurate at the transistor level, the simulator may be capable of handling charge sharing, multiple signal strengths, and other effects that may only be seen when MOS transistors are considered. Figure 18.11 shows a possible transistor-level representation of a CMOS *D* flip-flop and the switch-level simulator input for it.

18.4.4 Circuit

Analog-level simulation represents functions at the transistor level but treats the transistors as analog devices. Figure 18.12 shows the SPICE input required to model the *D* flip-flop.

18.5 MIXED-MODE SIMULATION

Obviously, as the level of detail in the simulation process is increased, the time required to evaluate the models also increases. There are times, however, when the designer may want to see detailed simulation of part of a system as it is driven by another part. Simulators that provide mixed-mode simulation permit the various modules that make up the system to be represented at different levels. This would permit one part of the system to be modeled at the switch level while other parts are modeled at the gate or behavioral level. This allows larger simulations to be performed at lower costs than if the entire system had to be modeled at the same level.

18.6 FEATURES TO LOOK FOR

Various commercial simulators are available.[1] The following features may be useful to the designer.

1. *Mixed mode:* Permits different levels of detail to be simulated simultaneously. Most simulators provide behavioral and gate simulation: many include switch-level simulation, and some include circuit-level simulation.
2. *Hierarchical:* Permits model definitions to be composed of other model definitions thereby allowing better control of models.
3. *Concurrency:* The system "knows" the correct order of evaluation to preserve correct values during simulation. This is necessary to simulate pipelined systems.
4. *Event-driven:* Also known as selective trace, the event-driven method schedules model evaluation to speed simulation. When a module is evaluated, only

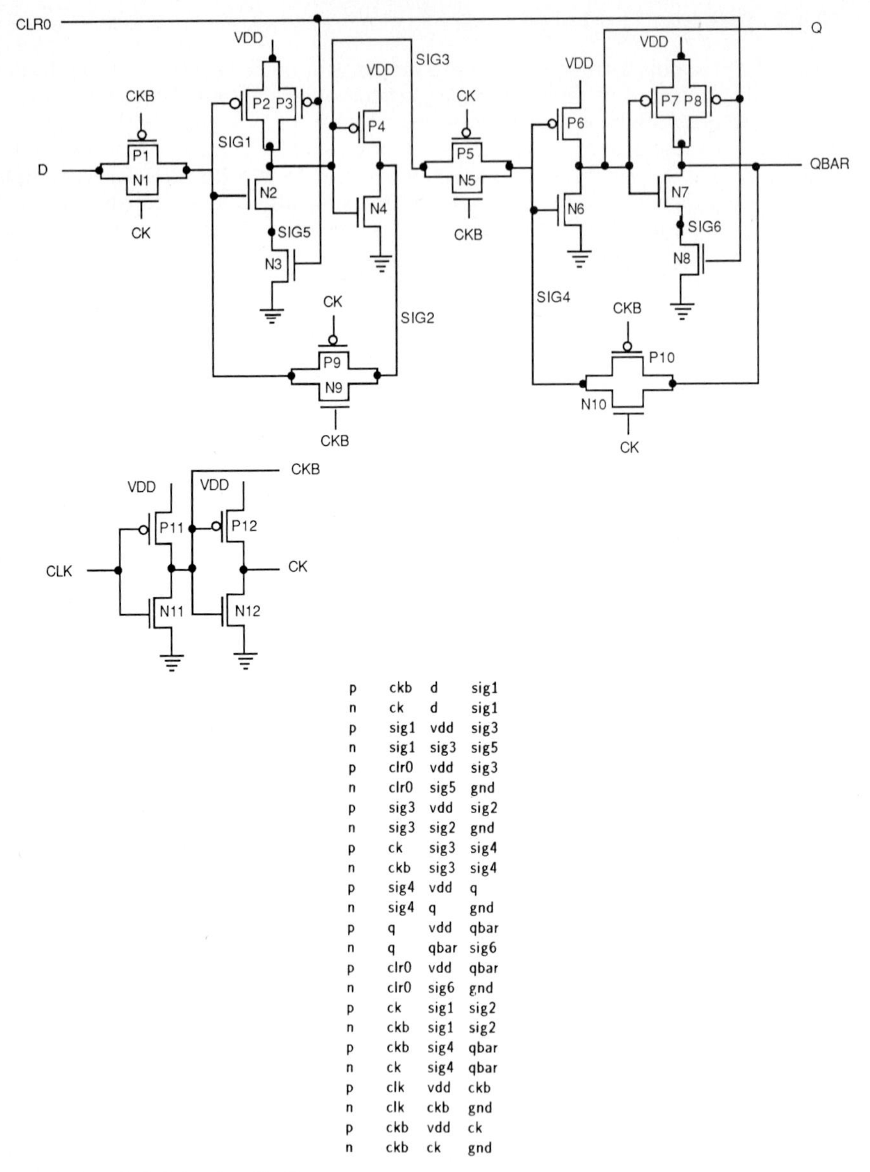
CLR0
Q
QBAR
D
CK
CKB
VDD
SIG1
SIG2
SIG3
SIG4
SIG5
SIG6
P1
N1
P2 P3
N2
N3
P4
N4
P5
N5
P6
N6
P7 P8
N7
N8
P9
N9
P10
N10
CLK
P11
P12
N11
N12
p ckb d sig1
n ck d sig1
p sig1 vdd sig3
n sig1 sig3 sig5
p clr0 vdd sig3
n clr0 sig5 gnd
p sig3 vdd sig2
n sig3 sig2 gnd
p ck sig3 sig4
n ckb sig3 sig4
p sig4 vdd q
n sig4 q gnd
p q vdd qbar
n q qbar sig6
p clr0 vdd qbar
n clr0 sig6 gnd
p ck sig1 sig2
n ckb sig1 sig2
p ckb sig4 qbar
n ck sig4 qbar
p clk vdd ckb
n clk ckb gnd
p ckb vdd ck
n ckb ck gnd

FIG. 18.11 Switch-level *D* flip-flop.

those outputs that change cause their fan-outs to be evaluated. This causes only modules whose inputs have changed to be evaluated, thereby reducing the time required for simulation.

5. *Interactive:* The simulator should provide interactive features such as single-step, symbolic examine-set, breakpoints, and trace. Providing an interactive environment increases the productivity of the designer.

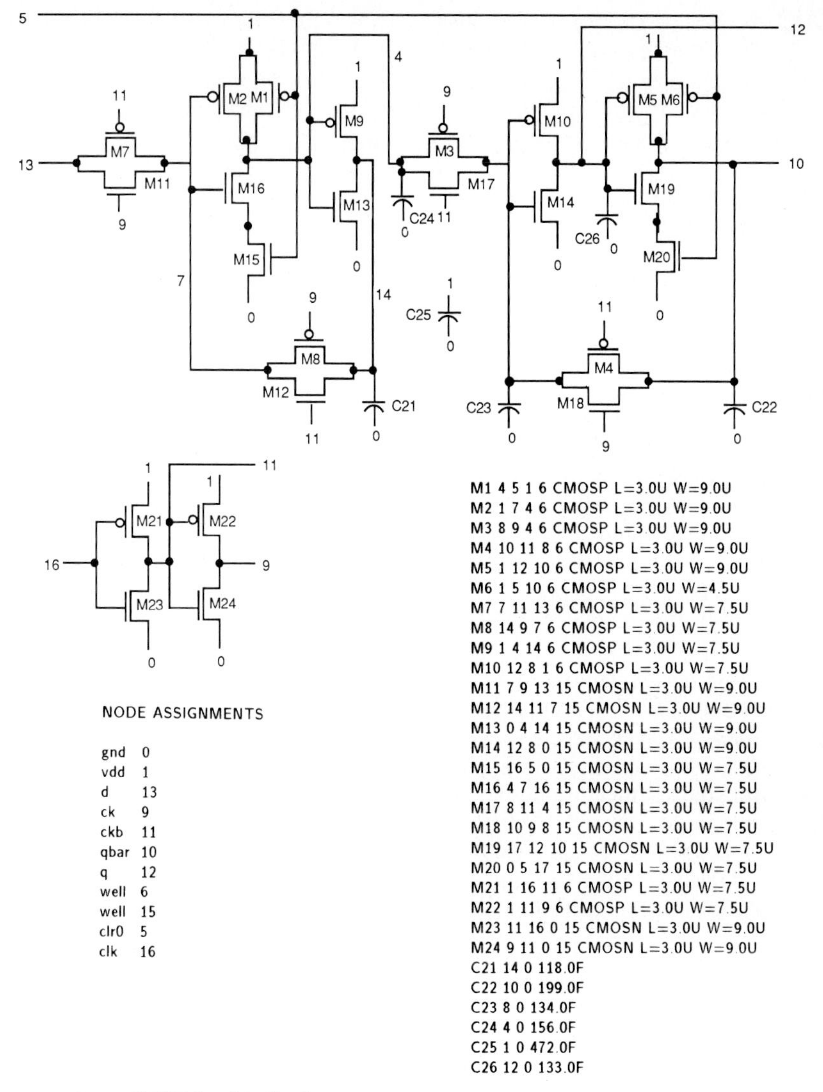
5
12
13
10
M2 M1
M5 M6
M7
M11
M9
M3
M10
M16
M13
M17
M14
M19
C24 11
C26
M15
M20
C25
M8
M4
M12
C21
C23
M18
C22
M21
M22
M23
M24
16
NODE ASSIGNMENTS
gnd 0
vdd 1
d 13
ck 9
ckb 11
qbar 10
q 12
well 6
well 15
clr0 5
clk 16
M1 4 5 1 6 CMOSP L=3.0U W=9.0U
M2 1 7 4 6 CMOSP L=3.0U W=9.0U
M3 8 9 4 6 CMOSP L=3.0U W=9.0U
M4 10 11 8 6 CMOSP L=3.0U W=9.0U
M5 1 12 10 6 CMOSP L=3.0U W=9.0U
M6 1 5 10 6 CMOSP L=3.0U W=4.5U
M7 7 11 13 6 CMOSP L=3.0U W=7.5U
M8 14 9 7 6 CMOSP L=3.0U W=7.5U
M9 1 4 14 6 CMOSP L=3.0U W=7.5U
M10 12 8 1 6 CMOSP L=3.0U W=7.5U
M11 7 9 13 15 CMOSN L=3.0U W=9.0U
M12 14 11 7 15 CMOSN L=3.0U W=9.0U
M13 0 4 14 15 CMOSN L=3.0U W=9.0U
M14 12 8 0 15 CMOSN L=3.0U W=9.0U
M15 16 5 0 15 CMOSN L=3.0U W=7.5U
M16 4 7 16 15 CMOSN L=3.0U W=7.5U
M17 8 11 4 15 CMOSN L=3.0U W=7.5U
M18 10 9 8 15 CMOSN L=3.0U W=7.5U
M19 17 12 10 15 CMOSN L=3.0U W=7.5U
M20 0 5 17 15 CMOSN L=3.0U W=7.5U
M21 1 16 11 6 CMOSP L=3.0U W=7.5U
M22 1 11 9 6 CMOSP L=3.0U W=7.5U
M23 11 16 0 15 CMOSN L=3.0U W=9.0U
M24 9 11 0 15 CMOSN L=3.0U W=9.0U
C21 14 0 118.0F
C22 10 0 199.0F
C23 8 0 134.0F
C24 4 0 156.0F
C25 1 0 472.0F
C26 12 0 133.0F

FIG. 18.12 SPICE-level *D* flip-flop.

6. *Desensitization:* The speed of simulation can be increased (or the time required to perform the simulation decreased) by using simulators to provide desensitization of signals. This is a technique that provides a conditional method of disabling certain events from scheduling a model evaluation. For example, the designer may know that the 32 bits of input to a register may not affect the evaluation of the register unless the clock signal is active. The designer may choose to desensitize the input bits when the clock is inactive, thereby inhibiting the evaluation of the register.

7. *Detection of hazards, races, spikes, oscillations:* Many simulators can detect these conditions and report them to the designer.
8. *Generation of tester input:* Allows test vectors used during design phase to be formatted in a way applicable for use by a test machine.
9. *Min-Max timing verification:* Automatically performs worst-case timing analysis. Performs race analysis between worst-case data and best-case clock for setup and best-case data and worst-case clock for hold and reports violations to the designer.
10. *Visual feedback (interaction with schematic-layout editor):* Many simulation systems provide timing diagram output. Many also interact with a schematic or layout editor to highlight points of timing discrepancies or other monitor points. The selection of signals to monitor may also be made by interaction via the graphics interface.
11. *Ease of use (user interface, number of files needed, etc.):* Ease of use will ultimately determine how much the designer uses the system. Does the interface allow the user to easily keep track of what is happening in the circuit? interact with the system? change inputs? trace a signal? How many steps are required to perform a simulation? How many files must be maintained for the system simulation?
12. *Source level debugging (behavioral):* This feature provides capability of "watching" the behavioral description being executed. The source statements currently being executed are displayed (many times with their results) as the simulation system evaluates them.
13. *Passing parameters into models (delay, etc.):* This allows parameters to be passed into modules and may permit a module to be more generic.
14. *Length of time required to cycle through a complete simulation-debug-fix-simulation procedure:* This often determines how useful the system will be during the architecture-design phase of a project. Does the simulation system promote a try-and-see attitude?
15. *Ability to link in high-level language procedures to model functions algorithmically:* This permits functions that may not be easily modeled using a hardware descriptive language to be modeled by a programming language (i.e., C, Pascal, etc.).

REFERENCE

1. 1987 Survey of Logic Simulators, *VLSI Systems Design*, **8**(2): (1987).

CHAPTER 19
LAYOUT VERIFICATION

William J. Haydamack
Valid Logic Systems, Inc
San Jose, California

This chapter describes methods for verifying IC layout to ensure it will meet the fabrication process requirements and will provide satisfactory manufacturing yield. It addresses not whether the circuit will perform the correct function or have the correct performance, but the question "can it be fabricated?" Verification of functionality is covered in Chap. 20.

19.1 INTRODUCTION

The purpose of layout verification is to determine whether the structures that have been designed can be fabricated. The process is one of identifying those structures that are too small, too large, too close together, or violate some other requirement with respect to neighboring geometric features. With modern wafer fabrication it is possible to separate the design process and the fabrication process by the specification of a set of topology or geometric design rules. The design rules are usually a set of permissible geometries that are within the resolution of the processing equipment. They also take into account device physics, allowing transistors to be constructed and interconnect to be formed. In most cases they are specified as minimal or maximal values, as opposed to absolute values. The designer has some freedom to adjust geometries to establish the electrical performance desired. In general, the design rules are a compromise between achieving acceptable yield in the fabrication process and the desire to increase performance and reduce die size.

19.1.1 Process Independence

Currently numerous IC processes are available,[1] including ECL, TTL, I^2L, NMOS, PMOS, CMOS bulk, CMOS sapphire, GaAs, and VMOS. Each of these processes has variations in the number of layers used (interconnect, for example) and in the way structures are implemented (*p* well, *n* well, and twin tub in CMOS). Finally, because of the variety of processing equipment used, the design rules must cover a wide range of limits. Currently CMOS is fabricated with processes labeled as 5, 3, 2, 1.5, 1.2, and 0.85 μm.

Tools designed to check the geometrical correctness of a layout must accommodate the intended process requirement range. Most design-rule checking (DRC) programs can be customized to handle any process. This customization is usually done in a text file that contains a set of operations and tests that must be performed on the geometrical data.

19.1.2 Data Formats

Although there are many systems from different manufacturers that generate and process mask layout data, a few standards have emerged. The GDSII format from Calma is the most popular interchange format. All design verification systems support it. The CIF format,[2] which originated at the California Institute of Technology, is popular in the university community and, to a lesser extent, in the industrial community. The EDIF committee[3] is also working on standards for both the transfer of design data and the command files for performing design verification operations.

19.1.3 Error Indications

Design verification tools are intended to detect errors in the topological geometry of a layout. Usually these errors will involve edges, or enclosed areas, in the layout database. The first DRC systems relied on plotters to report error conditions. Modern systems incorporate error reporting with graphical layout editors. These systems allow the user to view the errors superimposed on the layout, thereby utilizing the viewing and manipulation powers of the layout editor. These systems also provide the capability of managing the errors so that the designer can "check off" those errors that have been corrected.

19.1.4 False Errors

It is very difficult to customize a DRC program so that it reports every error that exists but does not report false errors. Users tend to bias their checking programs to reporting every error, including false errors, rather than risking an undetected error. Over the years DRC programs have become more sophisticated, and the false error problem has diminished; nevertheless it still exists.

19.1.5 Algorithms

The alogorithms for doing DRC have been structured to deal efficiently with large quantities of data. These data have been doubling every year, and the trend continues. Original algorithms had a computer time requirement that grew as $O(N^2)$, where N represents the number of features on a chip. Algorithms have improved, with $O(N^{1.5})$ common today and research has demonstrated that $O(N \log N)$ algorithms are possible.

DRC programs separate the rectangles and polygons that comprise IC layout into a set of edges, each with a layer property and a direction that determines which side the material is on. They make liberal use of sophisticated sorting capabilities to arrange data for efficient processing. An example would be a file that contained only those line segments that had a projection on the Y axis. It would be sorted by the lower vertex of each line segment, with the Y value the key sort element and the X value the secondary sort. The data would then be processed in a band starting at the lowest Y value and moving upward. The bandwidth would be determined by the operation being performed. A spacing test would require a bandwidth of at least the spacing distance. Only data contained in the active band area would be held in computer memory: the remainder would remain on disk.

Two significant changes have occurred since the first design verification systems. First systems could only handle geometries of orthogonal and 45° orientations. Modern systems handle geometries of all angles. The second change has been in the quantity of operations and tests that can be performed on the data. Probably the most significant change has been a methodology to handle design data in a hierarchical manner so that DRC programs do not operate on the complete database at any one time.

19.2 FABRICATION REQUIREMENTS

A number of factors are used to establish a set of geometries or design rules for a process. Detailed coverage is beyond the scope of this chapter; however, some insight will be useful. Design rules are a function of the process, device physics, mechanical tolerances, lithography, etching, cleanliness, regularly of materials, voltage potentials, economic tradeoffs, and many other things.

In the most advanced processes, the mask data are used to create reticles 5 or 10 times the size of the circuitry on the wafer. These reticles contain layout patterns for one or several dice, depending on the size of the die. A typical reticle can cover a maximum of about 1 cm^2 of wafer area. The reticle contains a frame that serves the dual purpose of being the scribe line for the die and for containing numerous verniers used to align the reticle. During the exposure process the wafer is moved to a position, an exposure is made, and the wafer is then moved to a new position. At each position the wafer is carefully aligned by using the verniers in the frame area. This method of lithography is termed *step-and-repeat*. Although the process is slow, it is the only way to obtain accuracy. During processing, especially at high temperatures, the wafer becomes distorted. The step-and-repeat process compensates for X-, Y-, rotational-, and Z-axis distortion.

Many of the processes in fabrication involve the etching of material. In some cases the material is photoresist; in others it is metal, oxide, or polysilicon. The etching process has been one of the most troublesome over the years. Wet etch processes had a limit of a few micrometers.[4] Modern processes use dry etch techniques such as reactive ion etching.

One of the interesting facets of the etching process is the reflection problem that occurs when the surface of the wafer is uneven. One example of this is when a metal trace runs parallel to a polysilicon trace. Since the polysilicon is elevated, it provides an edge that slopes toward the metal. Etchant is reflected from this surface and undermines the metal. Under these circumstances the design rules for metal width are usually increased to compensate for the undercutting.

Design rules vary with the material and structure being created. Diffusion spacing depends on the types of devices being formed and the possibility of current flow between adjacent diffusion regions. Diffusion width partly depends on the voltage potentials in the region. Contacts must be completely surrounded by the materials they are connecting to ensure that unwanted connections do not occur. Ion implantation must overlap the gate area of transistors to ensure that no part of the transistor remains unimplanted, or in the enhancement mode, when a depletion-mode transistor is required. Metal is required to traverse much rougher terrain than other materials. Therefore, the width and spacing design rules for metal are usually larger than for other materials.

A statistical viewpoint of yield and design-rule requirements is also of interest. During wafer processing defects will occur because of the dust particles in the air, silicon defects, insufficient etching, and/or improper alignment of masks. Defects are similar to grains of sand on the beach. There are a few big grains, some medium-sized grains, and many small grains. As geometries become smaller, vulnerability to the larger number of small defects increases. Design rules specify the size of features and the tolerance to these defects. Larger design rules increase the tolerance to defects. Economics and performance dictate smaller design rules.

19.3 GEOMETRICAL OPERATIONS

It is often necessary for the user to create new layers from combinations of other areas in order to properly perform a test. These operations may be used to isolate certain regions for testing, or they may be used to enhance a feature for test purposes. Often, complex sets of logical operations are performed on the original layers before tests are performed. It is sometimes useful to modify single layers as well.

19.3.1 AND, OR, XOR, and AND NOT

The logical operations AND, OR, XOR, and AND NOT produce a new layer that is a logical/geometric combination of two other layers. The AND operation gives the area of intersection of the two layers. The AND NOT operation gives those areas contained in one layer and not overlapped by the other layer. The AND NOT operation is different from the other operations because the result depends on the choice of original layer and overlapping layer. In all other operations the layers are interchangeable. The OR operation gives the logical combination of the two layers. Any regions in either of the two layers will be contained in the final layer. Finally, the XOR operation contains all regions in which there is no overlap. These operations are illustrated in Fig. 19.1.

19.3.2 Scaling

Scaling is a process by which every dimension in a layout is multiplied by a fixed value. The result is a circuit that is usually smaller than the original. As a process evolves and yields increase, a shrink of the artwork can be performed. The shrink gives smaller die size and increased performance. Most features can be scaled in this manner, with the exception of bonding pads. Often a new process is planned with the intent of performing one or two shrinks.

With this in mind, a decision must be made as to the original size and location

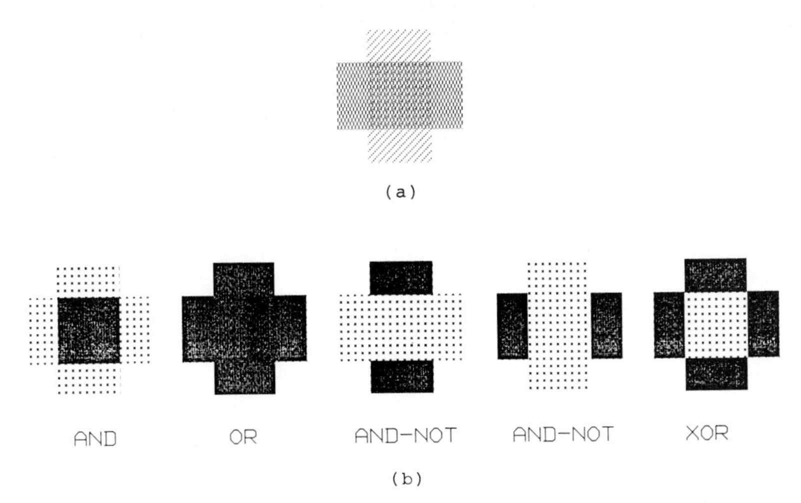

FIG. 19.1 Geometrical operations. (*a*) Original layers; (*b*) results of geometrical operations.

of bonding pads. If the design is not pad-limited, it may be possible to perform the shrink and then replace the shrunk pads with pads of normal size.

19.3.3 Sizing

The sizing commands work on individual polygons to increase or decrease their size by a specified distance. For example, a 5-μm square of polysilicon that was sized by plus 1 μm would become a 7-μm square of polysilicon. Although the operation appears to be very simple, tool developers have had problems dealing with vertices. Three algorithms have been used to place the new vertex. One algorithm simply states that the new sides should be extended until they intersect. For acute angles this can move a vertex a distance much greater than the size amount, as illustrated in Fig. 19.2.

Two other approaches have been explored. Researchers at the California Institute of Technology explored the use of arcs in the geometry database. An acute angle sized with this approach will not cause any of the region to extend by more than the sized amount. Unfortunately, arcs are not well-suited to most lithography equipment, which is designed to produce rectangles or trapezoids. Arcs are also relatively difficult to deal with in algorithms that perform logical operations and testing. The use of arcs in a sizing operation is illustrated in Fig. 19.3.

Another approach to sizing is to approximate the arcs with line segments. This approach is easier to deal with algorithmically and more closely represents the final lithography equipment. Expansion with line segments is illustrated in Fig. 19.4.

Although sizing operations are performed on single polygons, the result may cause adjacent polygons to touch and thereby merge together, or they may cause a single polygon to fragment into smaller pieces. Sizing operations may also be used to remove features from an object. An expansion by an amount X followed by a contraction of X will not necessarily yield the original object, as illustrated in Fig. 19.5.

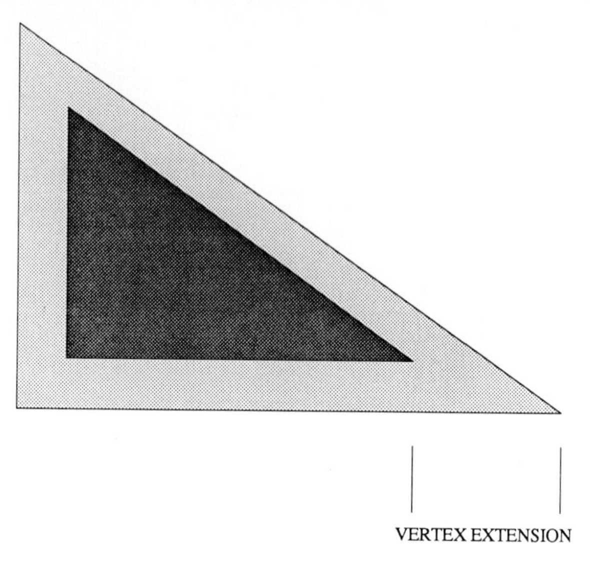

FIG. 19.2 Movement of vertex in sizing.

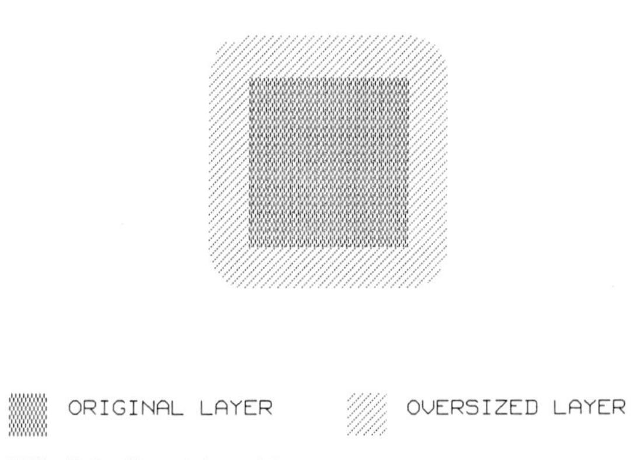

FIG. 19.3 Oversizing with arcs.

19.3.4 Relocation

Another operation used to create a new layer is relocation. Each feature of a layer is transformed in X and Y by a specified amount. A magnification factor can also be used to expand or contract the size of the features that are translated.

19.3.5 Area Sieves

Area sieves are used to isolate regions that have areas between certain limits. Each region is tested to see if it meets the area criterion. If it does, the region is

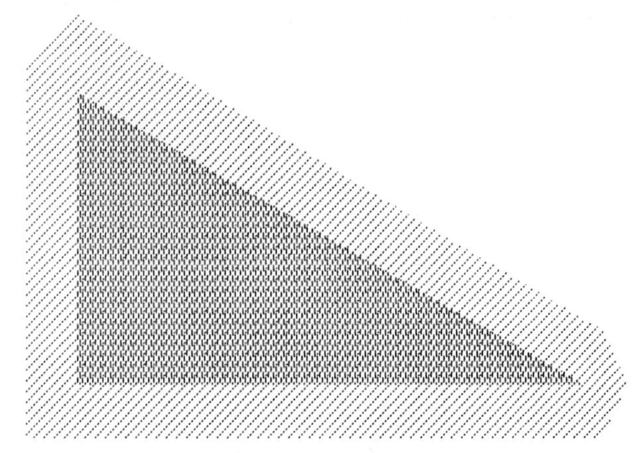

FIG. 19.4 Approximating arcs.

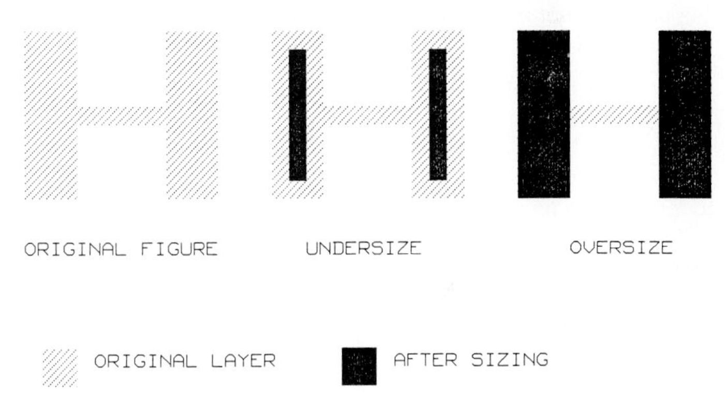

FIG. 19.5 Sizing operations. Original figure does not return to its original shape.

copied to the new layer. If it does not, the region is discarded. A variation of this command is used to select polygons that contain holes. In this case the criterion is specified on the area of the hole rather than the region.

19.3.6 Special Polygon Selection Commands

A set of commands has been devised to select polygons based on the number and lengths of edges. These terms can be specified as minimum, maximum, inclusive, or exclusive ranges.

Another type of command will select polygons that touch, overlap, enclose, or are outside of polygons on another layer. This command is different than the other logical operations, since the whole unmodified polygon, rather than just a portion, is included.

19.3.7 Manual Selection of Polygons

Although the operations just described can be used to isolate many of the features of a design, modern systems also provide a manual means of selecting features. The designer will attach labels to certain regions during the design process. The design verification system can then isolate those regions by recognizing the label.

19.4 TESTS

Tests are operations performed on one or two layers of the design. The tests may be on the original layers, but most often they involve layers generated by the geometrical operations. Tests involve the measurement of some distance or area. Distances are measured between line segments or edges of polygons. Area measurements are made on complete polygons and take into account the holes inside of polygons.

Four distance measurements are possible between two edges. Measurements from the starting edge can be made in an inward or outward direction. The measurement continues until it encounters another edge. It may approach the second edge from the outside or inside of that polygon. These measurements are illustrated in Fig. 19.6.

19.4.1 Width

The width test detects an error if any structure on the specified layer is less than the minimum width. A common use of the width test is to determine whether conducting layers will not have gaps after processing. Most IC processes have at least one width test for each layer.

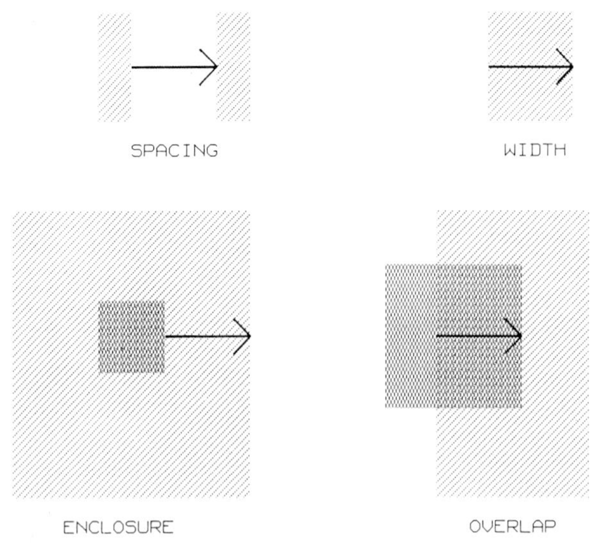

FIG. 19.6 Distance measurements.

19.4.2 Spacing

Spacing tests determine whether structures are closer together than a specified distance. The structures may be on the same layer or on one of two different layers. The test is specified with a minimum distance between features. Most IC processes have a spacing rule for each layer and several rules for interlayer spacing.

19.4.3 Enclosure

Enclosure tests are used to ensure that a region of one layer is entirely surrounded by a region of another layer. A common example of an enclosure requirement is the need for metal to surround a contact cut. During test specification, a minimal distance is used. Figure 19.7 illustrates the enclosure test.

19.4.4 Overlap

In some ways the overlap test can be considered a width test of two different layers rather than a single layer. When the overlap test is performed, the measurements are made from an edge, looking in an inward direction, until an edge from the second layer, which is approached from inside, is encountered. Some manufacturers refer to the overlap test as an overhang test.

19.4.5 Area

The area command checks for polygons that have less than the specified area on a given layer. When such a polygon is discovered, an error is generated. A typical use of the area command is in checking the area of contacts.

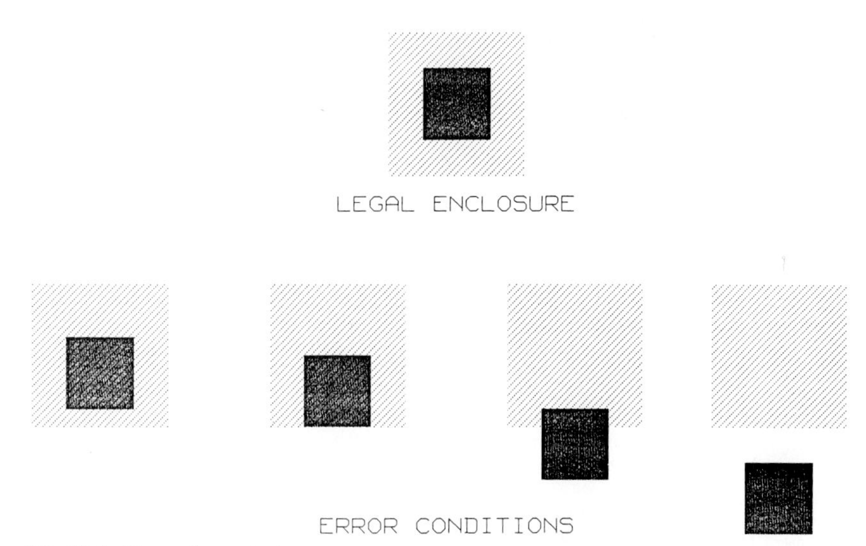

FIG. 19.7 The enclosure test.

A special case of the area command is the existence command which is used to detect the existence of anything on a layer. In any NMOS process it is illegal to place *n*-plus, poly, and cut at the same location. Logical AND operations can be used to create a layer that contains simultaneous occurrences of all three layers. The existence command would then report these as error conditions.

19.4.6 Edge Selection and Use

Often the errors reported by the simple operations we have discussed will yield false reports. It is not uncommon for geometries to touch or overlap in an acceptable manner, as illustrated in Fig. 19.8.

Earlier systems included options in spacing and overlap tests to handle touching and overlapping conditions. Modern systems have the ability to collect the edges involved in one error report for further processing before final error reports are generated. The processing may involve selecting edges of certain characteristics. Examples include selection of edges by length (between a minimum and a maximum value) or by orientation (vertical, horizontal, or angled). Edges generated from a test violation (i.e., spacing test) may be treated as pairs. Edges may be turned into polygons for further processing. A common method is to oversize the edge into a polygon perhaps 1 μm wide. Often, processing will involve sets of geometrical operations as well as other tests.

Once a set of edges has been isolated, edge-oriented tests may then be applied. As an example, Valid Logic[5] uses nine options in its edge-spacing command. These options are IGNORE-OVERLAPS, JUST-PARALLEL, JUST-OPPOSITE, NOT-PARALLEL, NOT-OPPOSITE, SAME-NODE, DIFFERENT-NODE, SELECT1, and SELECT2. The IGNORE-OVERLAPS option causes overlap violations in the spacing check to be ignored. The JUST-PARALLEL option causes only spacing violations between parallel edges to be reported. The JUST-OPPOSITE option causes only spacing violations between two edges that have an overlapping section to be reported. When edges have been identified as belonging to an electrical node (Chap. 20), SAME-NODE and DIFFERENT-NODE options can be used. The SELECT options are used when spacing tests involve two layers. The SELECT command selects error edges from one of the two original layers.

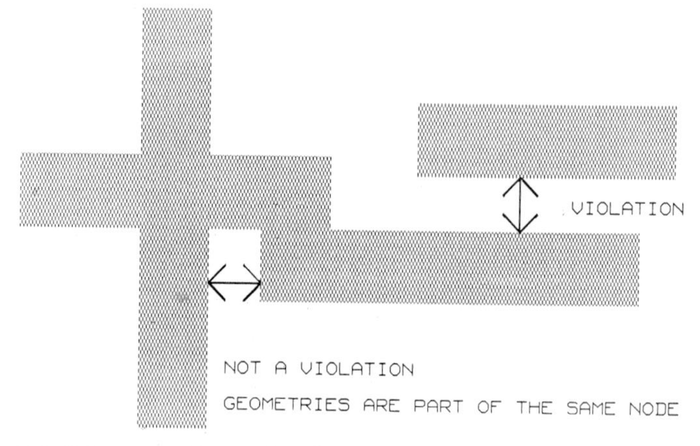

FIG. 19.8 Acceptable touching of geometries.

19.5 SAMPLE RULES CODING

The following is an example of a command file for the rule set published in *Introduction to VLSI Systems* by Carver Mead and Lynn Conway.[2] It is for an NMOS process. The text enclosed in braces {...} is for explanation.

LAYERS {definition of layers}

{The following layers are the original layers that were designed using a layout editor.}

diffusion {diffusion layer}
polysilicon {polysilicon layer}
metal {metal layer}
cut {contact cut layer}
implant {implant layer}
overglass {overglass layer}

{The following layers will be generated as a result of the geometrical operations that will be performed on the original layout data.}

transistor {transistors}
etran {enhancement transistors}
dtran {depletion transistors}
nplus {nplus regions}
bcontact {butting contacts}
nbcontact {nonbutting contacts}
bcdiff {butt-contact diffusion}
bcpoly {butt-contact polysilicon}
bcarea {area of poly and diffusion in butting contact region}
dcontact {diffusion contact}
pcontact {poly contact}
rtran1
rtran2
testpoly
temp1
temp2
temp3
temp4
ctran
ctranex
cdtran
cetran

{The following operations are used to identify transistors and the nplus region. They use commands of the syntax OPERATION, INPUT-LAYER, INPUT-LAYER, and OUTPUT-LAYER}

and diffusion polysilicon transistor	{identify transistors}
and transistor implant dtran	{identify depletion transistor}
andnot transistor implant etran	{identify enhancement transistors}
andnot diffusion polysilicon nplus	{identify nplus region}

{The following operation defines layers that touch other layers.}

touch cut metal polysilicon nplus

{The following section performs the design verification tests.}

width diffusion 2.0 "dif < 2.0"	{width test of diffusion and error message}
spacing diffusion diffusion 2.0	{spacing of diffusion}
width polysilicon 2.0	
spacing polysilicon polysilicon 2.0	{poly to poly spacing}
width metal 3.0	{error metal to thin}
spacing metal metal 3.0	

{The following operations are designed to test the various requirements for contact cuts. They include both geometrical operations and tests. The first section deals with butting contacts, and the remainder deals with the rest.}

and diffusion polysilicon ctran	{isolate butting contacts}
and ctran cut temp1	
expand temp1 2.0 temp2	{do an expand on geometry in temp1 to form temp2}
and cut temp2 bcontact	{butting contacts}
width bcontact 2.0	{width = 2.0 ?}
area bcontact 8.0	{area < 8.0 ?}
width temp1 1.0	{poly diffusion overlap width < 1 ?}
area temp1 2.0	{area poly-diff overlap < 2}
andnot bcontact polysilicon bcdiff	{isolate butting contact layers}
enclosure bcdiff diffusion 1.0	{check enclosure by diffusion}
width bcdiff 2.0	{contact diffusion}
area bcdiff 4.0	{contact diffusion}
and bcontact polysilicon bcpoly	{poly in butting contact}
width bcpoly 2.0	
area bcpoly 4.0	
or diffusion polysilicon bcarea	
enclosure bcontact bcarea 1.0	{butting contact enclosed by < 1}

{The next section deals with contacts that are not butting contacts.}

```
andnot cut temp2 nbcontact             {nonbutting contact}
and nbcontact diffusion dcontact       {diffusion contacts}
and nbcontact polysilicon pcontact     {poly contacts}
enclosure pcontact polysilicon 1.0     {poly surrounds cut}
enclosure dcontact diffusion 1.0       {diffusion surrounds cut}
enclosure cut metal 1.0                {metal surrounds cut}
width nbcontact 2.0                    {contact < 2 wide}
area nbcontact 4.0                     {contact < 4 area}
spacing nbcontact nbcontact 2.0        {contact to contact spacing}
```

{The following section performs test on the transistors.}

```
andnot diffusion polysilicon temp3    {diffusion extension past gate}
width temp3 2.0
```

{This operation and the next are used to remove small ears at the contact end of the depletion load.}

```
expand ctran -0.5 rtran1
expand rtran1 0.5 rtran2
andnot polysilicon rtran2 testpoly
width testpoly 2.0                    {poly overhang}
and rtran2 implant cdtran             {depletion transistor}
enclosure cdtran implant 1.5          {implant enclosure < 1.5}
expand ctran 1.0 ctranex              {poly to unrelated diff spacing}
andnot diffusion ctranex temp4
spacing temp4 polysilicon 1.0         {poly diffusion spacing < 1}
andnot rtran2 implant cetran          {enhancement transistor}
spacing cetran implant 1.5            {implant  < 1.5 from transistor}
spacing cetran dcontract 2.0          {contact < 2 from transistor}
end                                   {signifies end of rules file}
```

19.6 ALGORITHMS

Disk and memory utilization is quite different for flat verification systems and hierarchical verification systems. In flat verification systems the results of each operation are stored on a disk in a temporary file. These files are maintained until they are no longer needed, and then they are purged. It is not uncommon for designs to take 100 to 200 Mbytes of storage in order to complete processing. In hierarchical verification, a much smaller quantity of data is taken through the verification process during each pass. A cell is brought into memory, all processing is done, and all results all stored. There is no intermediate storage on disk.

Compute time has been improving with algorithm development. Current programs run between $N \log N$ and $N^{1.5}$. Hierarchical methodology yields programs whose execution increases at less than a linear rate with respect to data.

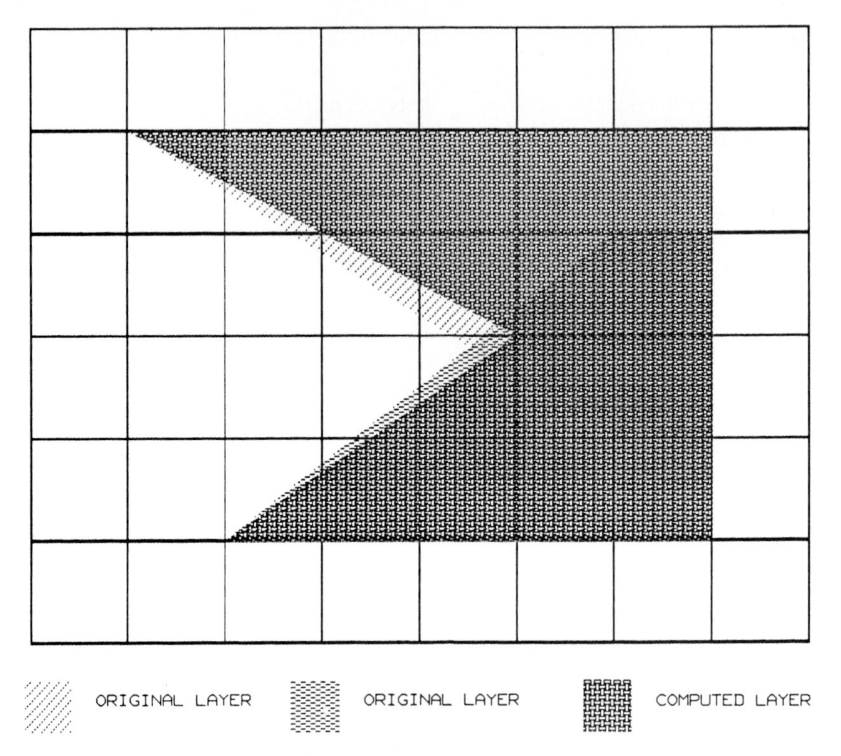

FIG. 19.9 Computed intersection point.

Early design verification systems only handled orthogonal geometries and features oriented at 45° angles. Pressure from the design community has now forced toolmakers to support geometries of all angles. Reluctance to add such geometries was founded in the added complexity of the algorithms, the data structures, and the special cases that required handling. For example, consider the case where two lines are intersecting and it is desirable to perform a logical OR operation. If the geometry is orthogonal, the intersection point will always fall on a grid point and the operation is relatively easy. If the lines are both oriented at 45° increments, the intersection point will fall off grid in half of the cases. Since the solution must involve line segments oriented in increments of 45° and must terminate on grid points, it is necessary to modify the intersection point and add a new line segment. Finally, in the case of geometries of all angles, it is highly unlikely that the intersection point will fall on a grid point. The computed intersection point is rounded to the nearest grid point, and the line segments are slightly altered. These examples are illustrated in Fig. 19.9.

19.7 HIERARCHICAL ANALYSIS

All of the original work in design verification programs required that the data be flattened before processing began. The primary reason was the simplification of the algorithms and data structures. The major problems with flat verification include the computational time, main memory requirements, disk requirements, and difficulty of reporting errors in the context of the cell in which they occurred. If an error was discovered in a cell, it was reported for each instance of the cell. Since the cell had been flattened, the identity of the cell was also lost. Hierarchical verification will make one report of the error in the context of the cell in which the error occurred.

There have been two basic approaches to hierarchical analysis. The first assumes that some constraints can be placed on the designers to prepare artwork in a format that is amenable to hierarchical analysis.[6,7] The second makes no assumptions about the design style. Strict hierarchical design (where cells have no overlap) is uncommon except in some semicustom methodologies, such as standard cell. Major overlap where one cell alters the functionality of another cell is also uncommon (ROMs and PLAs are exceptions). In most design styles used today there is some sharing of, or interaction between, geometries at the edges of cells.

Much work has gone into hierarchical analysis. One of the earliest attempts[8] used a filter to identify each unique cell-cell interaction. Only one of each case would be tested. Results were very dependent on how well structured the design data was for the algorithms. Ill-structured design data offered little savings in compute time. The technique was applicable to any hierarchy, which was one of its major advantages.

A constrained design methodology was proposed by Lou Scheffer.[9,6] In this form each cell has a boundary that primitives may not overlap. The advantages were linear time analysis, easy extension to the extraction process (Chap. 20), and error reports that match the design. Although it can be proved that this design style could be used for any design, the design community is reluctant to accept the constraints.

A less-constrained methodology introduced the concept of a separate boundary per layer.[10] This allowed more flexibility in permitting overlap, but made it difficult to handle cases where there was a strong interaction between layers.

Finally, the concept of using different abstracts for a cell dependent on the type of analysis to be performed was introduced by Scheffer.[6] In principle, a cell is analyzed without knowing the context in which the cell will be placed and then reanalyzed if the original assumptions are violated. The user must specify how the abstract is to be created and the circumstances that would cause the cell to be reanalyzed. Two exmples of abstracts follow: if only DRC is to be performed, then an abstract that contains the geometry within a maximum design-rule region of the boundary would be created. If connectivity analysis is to be performed, the abstract may only contain the occupied area and the pins used to connect the cell. In the first case any extension of geometry within the boundary of the cell would be cause for reanalysis.

Although there has been much research on hierarchical methods, the application of those methods is still in its infancy. The major obstacle is in the application of the technology, rather than the technology itself. It is certainly a part of standard-cell and gate-array methodologies. In the case of a standard cell, each cell is fully characterized so that it can be placed and connected to its immediate neighbors by abutment and interconnected to remote cells by selected routing layers. Gate arrays present another form of hierarchy. All layers except routing and via layers may be checked once. For each customized gate array, only the routing layers are checked. In custom design the problem is more difficult. Implementation requires cooperation of both the designers and the CAD people responsible for coding the rules. There is reluctance to apply new and more complicated technology to a process that is so critical to the success of a design. However, the desire to run CAD programs on cheaper and more accessible workstations should provide the impetus.

19.8 MASK CREATION AND MODIFICATION

Although early DRC systems were used for verification purposes, the value of the logical operations to create new mask layers soon became apparent.[11] The fabrication process is one of continually evolving technology. Yields from each run are analyzed, and minor process changes are made. Often it becomes desirable to bias the artwork to achieve some performance or yield enhancement. The active area of transistors may be reduced to give slight increases in performance. Metal width may be decreased to reduce the chance of bridging problems or to reduce capacitance. Complete designs may be shrunk as a process matures and the yields increase.

Another benefit of logical operations is the ability to create complete mask layers from the data contained on other masks. Often it is standard practice to design a subset of the mask set and to create the other masks by using logical operations and sizing. In some CMOS processes it is desirable to slope the edges of contact cuts. A second mask is used in this operation, which has slightly larger contacts than in the original drawn layout. Other CMOS examples include the *n*-field mask, which is an oversize of the *n*-well mask, and the *n*-plus clear field, which is a reversal of the *p*-plus implant mask.

19.9 TRENDS

Two strong forces are at work in the area of design verification. One force is toward "correct by construction methodologies," which should eliminate the need for design verification. The other, more conservative, approach is to continue to use design verification. With modern algorithms and computer performance the cost of running design verification software is small compared with the cost in turnaround of a bad IC design. Costs of a bad design include mask and material costs, market window loss, dissatisfied customers, and unscheduled use of extra resources, which could have been doing other design work. It is likely that the design verification process will remain a vital part of IC design in the foreseeable future.

The major changes will be in methodologies. The benefits of hierarchical verification have not been fully realized to date. Most designs are still done in a relatively flat methodology. We feel there will be a slow but steady movement toward hierarchical design in the future. The persuasive nature of the workstation and its ability to do hierarchical verification will contribute.

As the power of workstations increase, the ability to do on-line verification as part of the graphical editing process will increase. Currently, only simple rules such as width and spacing can be done interactively; however, more powerful systems will be able to handle the complex rule sets.

Perhaps the most significant development in the future will be to combine the technology that has been developed for verification purposes with those that are used in the construction of the layout. Compaction is a good example. A user may have prepared an elaborate technology file to direct the verification of artwork for that process. The user would like to utilize this knowledge base to establish the constraint equations for the compaction process. Similarly, the technology file could be used for directing technology requirements in automatic routing and placement systems and in silicon compiler systems. Much effort is now being placed in the transportability of design verification system files between various CAD systems. The EDIF committee is addressing this topic.

REFERENCES

1. S. Muroga, *VLSI System Design,* Wiley, New York, 1982.
2. C. Mead and L. Conway, *Introduction to VLSI System,* Addison-Wesley, Reading, Mass., 1980.
3. Electronic Design Interchange Format Steering Committee, "EDIF Electronic Design Interchange Format Version 1 0 0," The EDIF Users' Group, Design Automation Department, Texas Instruments, Dallas, Tex.
4. P. J. Hicks, *Semi Custom IC Design and VLSI,* Peter Peregrinus, London, 1983.
5. *Scaldstar Reference Manual,* Valid Logic Systems, San Jose, Cal., 1986.
6. L. Scheffer, "The Use of Strict Hierarchy for Verification of Integrated Circuits," Ph.D. Thesis, Stanford University, 1984.
7. L. Scheffer and R. Soetarman, "Hierarchical Analysis of IC Artwork with User Defined Abstraction Rules," *22d Design Automation Conference,* 1985.
8. T. Whitney, "Description of the Hierarchical Design Rule Filter," SSP file 4027, Silicon Structures Project, California Institute of Technology, Pasadena, Cal., October 1980.
9. L. Scheffer, "A Methodology for Improved Verification of VLSI Designs without Loss of Area," *Caltech Conference on Very Large Scale Integration,* 1981.
10. K. H. Keller, A. R. Newton, and S. Ellis, "A Symbolic Design System for Integrated Circuits," *Proc. 19th Design Automation Conference,* Las Vegas, Nev., June 1982, pp. 84–92.
11. M. Tucker and W. Haydamack, "A System for Modifying Integrated Circuit Artwork through Geometrical Operations," *Twelfth Asilomar Conference on Circuits, Systems, and Computers,* 1978.

SUGGESTIONS FOR FURTHER READING

Alexander, D., "A Technology Independent Design Rule Checker," *Proc 3rd U.S.A.–Japan Computer Conference,* San Francisco, Cal., October 1978, pp. 412–416.

Colclaser, R., *Micro Electronics Processing and Device Design,* Wiley, New York, 1980.

Johnson, S., "Hierarchical Design Verification Based on Rectangles," *Proc. Conference on Advanced Research in VLSI,* MIT, 1982, pp. 97–100.

Rowson, J., "Understanding Hierarchical Design," Ph.D. Thesis, California Institute of Technology, 1980.

CHAPTER 20
TOPOLOGY VERIFICATION

William J. Haydamack
Valid Logic Systems, Inc.
San Jose, California

20.1 INTRODUCTION

Chapter 19 dealt with the problems of determining whether the circuit, as expressed in mask layers, was manufacturable. Topology verification determines whether the layout represents the circuit the designer intended. It involves circuit elements, such as transistors, resistors, and capacitors, along with interconnect and parasitic elements.

To be useful, topology analysis must be tolerant of conditions where unexpected elements are present. For example, was a transistor accidentally formed by a poly layer crossing a diffusion? Labeling of nodes may help the verification algorithms, but they must also be tolerant of incorrect labeling. Topology analysis should answer the questions: "Is the interconnect right?" "Will the circuit perform as expected?"

Topology verification is a two-phase process. The first phase is the extraction of the netlist and associated parameters from the artwork. The second phase is the analysis of this data. Unlike design verification, which is a maturing technology, many problems remain to be solved in topology verification. There are still no good means of extracting parameters for bipolar transistors, and the problem of resistance extraction lacks an efficient and accurate solution. There is also no proven method of analyzing the extracted data from custom designs. A variety of tools and methodologies are used. In spite of these shortfalls, topology verification is a vital component of today's IC design methodologies.

20.2 PROCESS REQUIREMENTS

The extraction process involves deducing from the mask layer data the electrical components that will exist after fabrication.

Some masks such as the metal mask have a direct equivalence to electrical components. However, it is often necessary to use the geometrical operations described in Chap. 19 to identify distinct features. These features are usually the same features needed for design verification, and often the extraction process and design-rule checking (DRC) process are combined.

20.3 INTERCONNECT

Several different mask layers (for instance metal and polysilicon) are used for interconnect. Diffusion is used only for very short lengths of interconnect because its resistivity is too high for long interconnect. Two electrical nodes on separate interconnect layers can be joined by a contact cut.

If two features on the same interconnect mask touch, a connection is made. Often the geometrical OR operation is used to merge features. Some interconnect masks usually touch or overlap with no electrical connection being made. Metal and polysilicon are two such layers. A third case occurs when polysilicon overlaps diffusion to form a transistor. In this case of self-aligning gates, the diffusion is split at the overlap region and two electrical nodes are formed. The geometrical operations are used to identify these nodes.

Verification programs usually implement a touch command to identify which layers will be interconnected if they touch. If metal touches a contact, both are electrically equivalent. If polysilicon touches a contact, both are electrically equivalent. If metal and poly both touch the same contact, then the metal and polysilicon are electrically equivalent. Figure 20.1 illustrates the results of geometrical operations to isolate the electrical nodes in an NMOS circuit.

Once the nodes of a circuit are identified, the port connections must be determined. When devices are identified, various regions become port regions with identifiers. Interconnect that touches these regions becomes connected to that port.

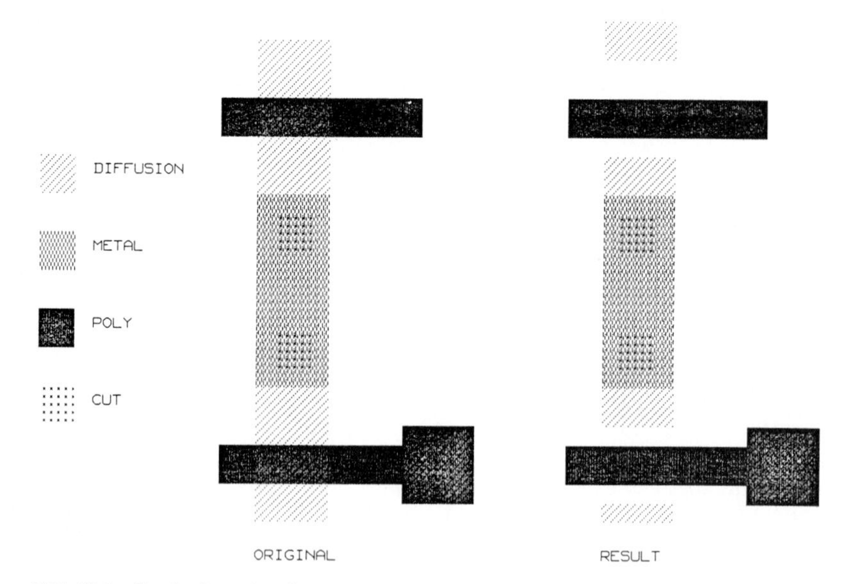

FIG. 20.1 Continuity extraction.

Although interconnect is relatively easy to identify with geometrical operations, the parasitics require substantially more work. For metal nodes the capacitance is important. For polysilicon interconnect the resistance is also very important.

20.4 CAPACITANCE EXTRACTION

Three major components of capacitance affect the performance of a circuit: (1) the capacitance from a node to ac ground, (2) the internodal capacitance between nodes that cross one another, and (3) the parallel plate capacitance of nodes that are routed next to each other.

The extraction of capacitances is much more complicated than the extraction of resistances. The capacitance of each polygon on a node is a function of the polygon's area, perimeter, and the material thickness.

When a new process is developed, test circuits are fabricated to characterize the devices and interconnect. Part of this procedure is the determination of the parasitic capacitance and resistance of various structures. These parameters are then used in the design and extraction process to predict the capacitance and resistance of the circuit being designed.

20.5 RESISTANCE EXTRACTION

Exact resistance extraction of every feature in a layout is a difficult and computational intensive task. Most systems employ some approximation scheme to estimate resistance. The need for resistance extraction varies with the process being used. For bipolar analog circuit design, it is very important to predict resistance values for interconnect and for the various contacts used. In digital MOS design only a few paths (and nodes) are involved in critical delays. Techniques that identify candidates for further analysis are often adequate, provided that tools are available for detailed analysis. Some of the techniques used for resistance extraction are detailed in the following sections.

20.5.1 Maximum Value Estimation

Perhaps the simplest method of estimating resistance is one in which only the area of each node is used to determine the value. If it is assumed that the region is of minimum width, then the area divided by the minimum width will yield an approximation of worst-case length. This value multiplied by a user-supplied constant (resistance per unit of minimum width) will yield a maximum value for the nodal resistance. This approximation is good for nodes with two ports that are constructed with minimum-width material. It breaks down for nodes with more than two ports and for nodes of greater than minimal width.

20.5.2 Interactive Measurement

Another method of resistance extraction involves separating the path into a set of resistors of uniform geometry and two ports. The area of each region is calculated, and the distance between ports is measured. The width is then calculated by dividing the area by the length, and standard resistance calculations, using the resistivity of the material, are made. Often this capability is implemented with a graphical editor. Although the method involves user intervention, it can yield fairly accurate results. For practical reasons it can be applied only to a small number of nodes.

20.5.3 Shapes Library

A variation of the above method was explored by Horowitz and Dutton.[1] They realized that it was possible to calculate accurately the resistance of a simple polygon by solving Laplace's equation. They also realized the difficulty of solving Laplace's equation for the quantity and complexity of polygons found in modern ICs. Why not break the complex polygons into a series of simpler polygons for which the solution is known? The resistances of a number of complex but common shapes are put in a library. The interconnect is then broken into recognizable structures. The resistance of each structure is obtained from library data, and programs sum the resistances. The researchers claim results within 10 percent of exact values for this method.

20.5.4 Finite Element

Other research[2] in resistance extraction involves applying finite element techniques to the problem. The work addresses the problem of bipolar design where resistors are designed and are inherent through process parasitics. The program addresses problems such as different materials, arbitrary shapes, multiple-connected mask regions (''holes''), multiterminal structures, multiple regions of different resistivity, and contact resistance.

20.5.5 Min-Max Estimation

Another interesting approach to the delays in circuits is not an attempt at exact calculation, but one that yields maximum and minimum delay values. Original work in this area was conducted by Penfield and Rubinstein.[3] This work takes into account capacitance, resistance, and the voltage threshold of devices. Variations of this technology have been used in standard-cell routing packages.[4]

20.6 DEVICE EXTRACTION

Device extraction involves identifying a set of geometries that constitute a device (transistor, diode, resistor, etc.). Once the device is identified, its ports must be located and its parameters extracted. Usually these parameters are those required to satisfy the device models of a simulator such as SPICE.[5]

MOS devices are relatively easy to extract. There is a uniqueness in MOS devices that makes them easy to identify (overlap of poly and diffusion). The variety of bipolar devices makes them difficult to identify. Often hierarchical methods are used with bipolar devices. An equivalence is established between a bipolar device, its circuit model, and a cell in the layout. The extraction process proceeds only until the cell is identified. An appropriate SPICE model is then substituted.

20.6.1 MOS Device Identification

MOS transistors are defined by four layers. The first defines the location, the second the source and drain, the third the gate, and the fourth the substrate.

A correspondence is established between the extracted transistors and circuit simulation models. In CMOS circuits there are different models for the *n* and *p* transistors. In NMOS the same transistor model is used for both enhancement and depletion node devices, although the extraction process will yield different threshold voltages. The layers that are defined as being source, gate, or drain must also be specified in the electrical connection definition.

20.6.2 MOS Parameter Extraction

Parameters extracted for MOS transistors include the width *W* and length *L* of the gate region and the areas of the source and drain regions. Width is usually calculated as half of the total linear distance that the transistor has in common with the source and the drain. The length *L* is computed as the area of the gate divided by the width *W*. This definition will handle bent transistors as well as regular transistors. Calculations of *W* and *L* are illustrated in Fig. 20.2.

MOS SPICE models also use the parameters area of source, area of drain, periphery of source, and periphery of drain as well as *W* and *L*. Once these areas

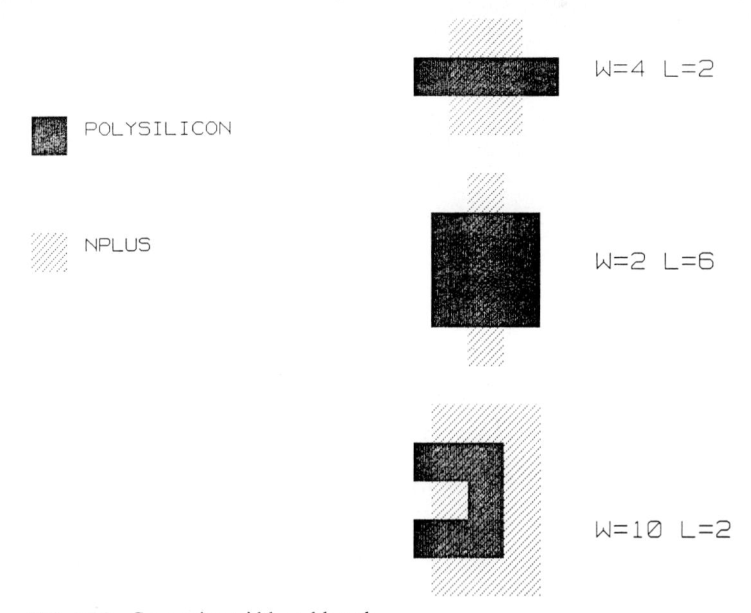

FIG. 20.2 Computing width and length.

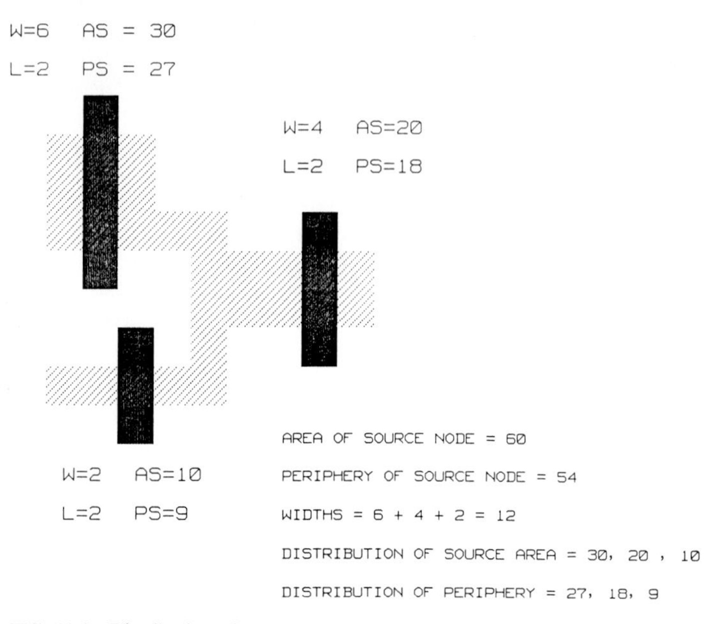

FIG. 20.3 Distribution of A_s and P_s.

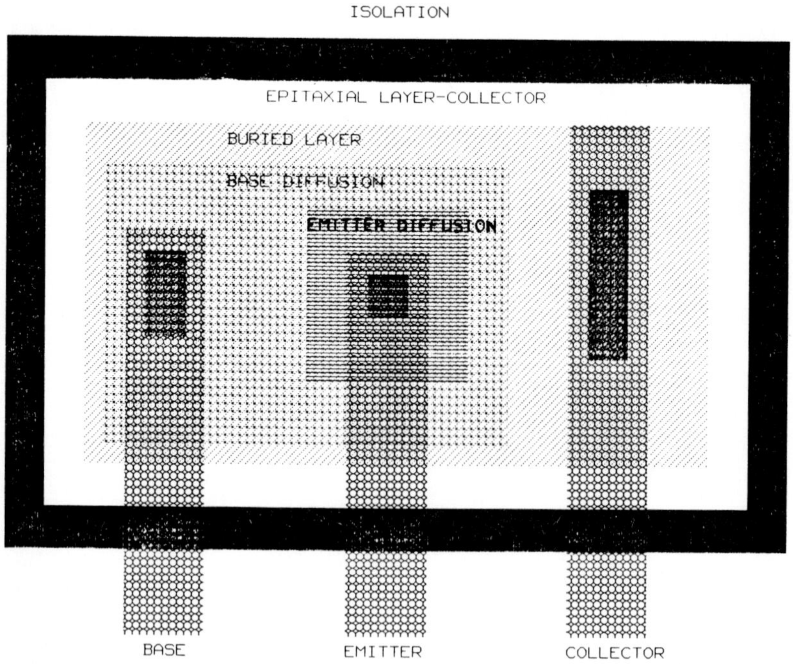

ISOLATION

EPITAXIAL LAYER-COLLECTOR

BURIED LAYER

BASE DIFFUSION

EMITTER DIFFUSION

BASE EMITTER COLLECTOR

FIG. 20.4 Bipolar transistor.

are isolated with geometrical operations, the measurements can be made. It is important for extract programs to be able to partition the area and periphery of the source according to the width of the transistors, as illustrated in Fig. 20.3.

20.6.3 Bipolar Device Identification

Identification of MOS transistors is relatively easy with geometrical operators because the transistor region is uniquely defined by the overlap of a diffusion region and polysilicon. This uniqueness does not exist for bipolar devices. At least 80[6] different structures are used in bipolar analog design. The geometrical operations described in Chap. 19 are not adequate to separate these structures. Figure 20.4 illustrates the layout of a typical bipolar transistor.

In one approach[7] contour lines are extracted from the geometrical data. A polygon with a hole in it would have an inner contour bounding the hole and an outer contour bounding the complete polygon. Each of these contours would be associated with a mask layer. The contours are organized into a directed graph in a hierarchical fashion. If two contours intersect, the nodes on the graph are doubly connected. The graphs can be used to identify devices, cells, blocks, and complete circuits.

The user of the system is required to uniquely identify each structure. Identification is by a directed graph of arcs and nodes. Each node is identified by a mask layer, a contour type (inside or outside), and a contour level. Devices may be described with the minimal set of nodes and arcs or with more complete descriptions. The latter will make it easier to identify faulty devices but will also increase the run time.

After the user has described a library of devices, the graphs developed in an extraction process will be compared with the user-specified library of graphs to determine matches. Since the user manages the device graphs, it is possible to extend the technology to nearly any process.

20.6.4 Bipolar Parameter Extraction

A more difficult problem for analog devices is the extraction of simulation parameters. Whereas it is relatively easy to determine the MOS device parameters (W, L, A_s, A_d, P_s, P_d), it is much more difficult to determine bipolar parameters. The Gummel-Poon bipolar junction model,[8] as implemented in SPICE, requires approximately 40 parameters.

Because of the difficulty in extracting parameters and characterizing different transistors, most bipolar designers limit themselves to a library of transistors. These transistors have been designed, their parts have been fabricated, and a complete set of characterization tests has been performed. When the extraction program recognizes a transistor cell, it simply substitutes a circuit model for that cell and halts further decomposition of the artwork.

20.7 DIRECT COMPARISON AGAINST THE SCHEMATIC

In most design methodologies, the designer creates a schematic and stimulates the circuit before beginning the layout process. An assumption can be made that if the schematic is properly implemented in layout, the circuit will perform as expected.

Comparison programs have been developed that will look at the netlist generated by the schematic and that which has been extracted from the layout and determine if they match. Modern programs work well and have facilities for displaying errors in both the schematic and layout editors. They can also account for simple differences, such as input pin swapping on a NAND gate. The quality of these programs is determined by execution time and by the isolation of the errors.

Comparison programs also have facilities for determining if the parameters associated with devices and the parasitics associated with interconnect are within tolerance. Usually in MOS design only W and L of the gate region are specified. It is unusual to have all the parasitic elements specified for a circuit. CAD systems have the capability of back-annotating this information into the schematic.

20.7.1 Compare Algorithms

The problem for a comparison program is to match nodes in one network with nodes from another network. Since all of the nodes are usually not labeled, other properties must be used to distinguish the nodes and to determine equivalence. The comparison problem is part of a larger problem, termed *graph isomorphism,* in which a program must determine whether two graphs of interconnected objects are connected in the same way. Although this is still an open problem in the general sense, adequate solutions are available for the IC problem.[9]

Compare algorithms should rapidly identify devices and nodes that are equivalent and remove them from further consideration. It becomes increasingly difficult to distinguish nodes as processing progresses, and more and more information is needed. An example of a comparison algorithm follows.

For each node in the two netlists, determine the number of ports. If there is a unique number of ports on a node in one netlist, then it must match a single node in the other netlist. If this is the case, the nodes are labeled equivalent and removed. Whereas there may only be one node in each netlist with 27 ports, there may be hundreds with three ports. In this case additional information will be added to the nodes. For example, of the three ports, two may connect to n-type transistors and one to a p-type transistor. Matches of this type of node are then made.

The algorithms also look for device matches, again using the characteristics of the devices and the signals on the ports to ensure a match.

This slow progressive adding of information to the nets and ports ensures that actual error conditions are isolated as tightly as possible. If all the information was used in the first comparison pass, it is unlikely that many nodes or devices would match, and the error condition would involve major pieces of circuitry, a condition of little value to the designer.

Hashing schemes are used to identify unique nodes and devices. Each netlist is hashed with identical algorithms. If a node or device has a unique hash number in the netlist, then a match is assumed. Additional information is added to either the devices or nodes, the netlists rehashed, and the process repeated until all nodes and devices that can be matched have been matched.

The remainder can fall into two categories. If a node or device has a unique hash number in one netlist but none in the other, an error condition exists. If two nodes have the same hash number and match an equivalent set in the other netlist, redundant circuitry has been identified that may not be in error.

20.8 INDIRECT COMPARISON BY SIMULATION

Although direct comparison of the schematic and artwork gives the designer a degree of confidence, it is often not enough. There is a large degree of abstraction between the 10 to 20 mask layers a designer views on plots and the electrical measurements that will be made on a final circuit. The desire to "simulate the artwork" is real.

The extraction process also yields a lot of parasitic information that is normally not accounted for in the original design. The effects of this information on the performance is significant.

Most extraction programs format the data in a netlist form suitable for entry into a circuit simulator. The problem is usually the size of circuit that can be simulated. For practical purposes this is usually limited to several hundred or a few thousand transistors.

Other transistor-level simulators[10,11] have been developed to speed up the simulation process. These simulators use much simpler models for the transistors. The results are less accurate, but the speed is better and they can handle larger circuits. Motis-C, which uses table models for the transistors, operates from 10 to 50 times the speed of SPICE. MOSSIM,[12] which uses a very simple switch model for a transistor, has a reported speed of about one-third of an equivalent gate-level logic simulator. Hierarchical methods can be used in the extraction process to accommodate logic simulators. These are described later in this chapter.

20.9 OTHER ANALYSIS TOOLS

Besides simulating the extracted artwork, designers often use electrical rules checking (ERC) programs and timing verification programs[13] to verify that the design will perform as expected.

20.9.1 Electrical Rules Checking

One of the analysis tools that can be applied to a complete design is the ERC program. This program examines the topology and parameters extracted from the layout to determine if they violate any of a set of rules. In some cases the rules may have tolerances that are user-programmable.

One of the most common checks is for opens and shorts. Several test configurations are illustrated in Fig. 20.5. Another form of rules checking that is particularly useful for MOS circuitry is β ratio testing. This is used to ensure that the appropriate voltage levels will be output. Normally two sets of checks are done: one for circuitry driven by pass transistors, and one when no pass transistor is involved. β ratios are normally specified as a nominal value and a tolerance, for example a ratio of 8.0 and a tolerance of 0.5. Any β ratio less than 7.5 or greater than 8.5 would be reported as an error. Besides the ratio of channel length and width, minimum values are also important, but it is normally assumed that these will be checked with DRC programs (Chap. 19). Figure 20.6 illustrates an NMOS

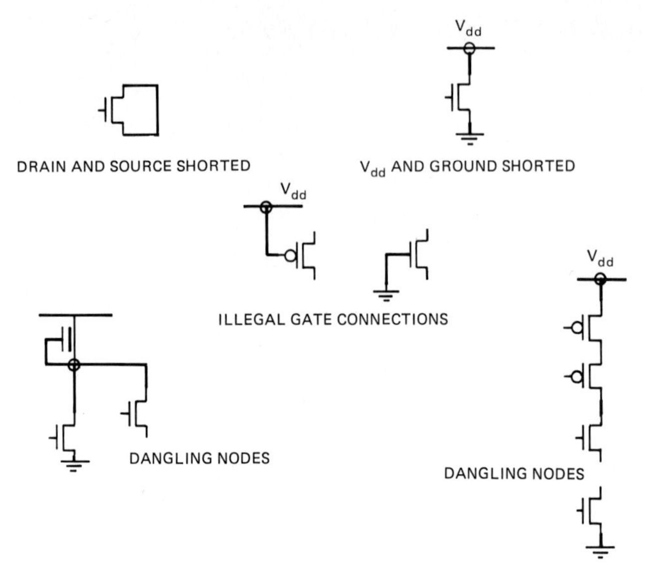

FIG. 20.5 Opens and shorts.

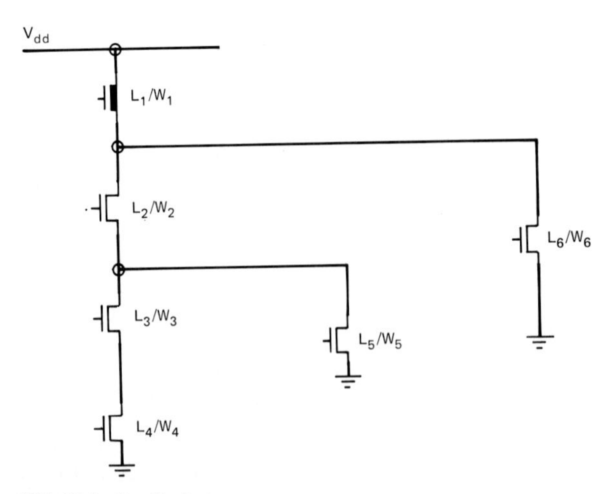

FIG. 20.6 β ratio tests.

circuit with three dc paths to ground. The following β ratio checks will be made on this circuit:

$$(\beta \text{ path})_1 = \frac{L_1/W_1}{(L_2/W_2 + L_3/W_3 + L_4/W_4)}$$

$$(\beta \text{ path})_2 = \frac{L_1/W_1}{(L_2/W_2)/(L_5/W_5)}$$

$$(\beta \text{ path})_3 = \frac{L_1/W_1}{L_6/W_6}$$

Another simplistic test is the dc net depth test. The number of transistors in each path from V_{dd} to ground are counted and checked against some maximal value the user specifies. Paths that would be checked for a CMOS circuit are illustrated in Fig. 20.7.

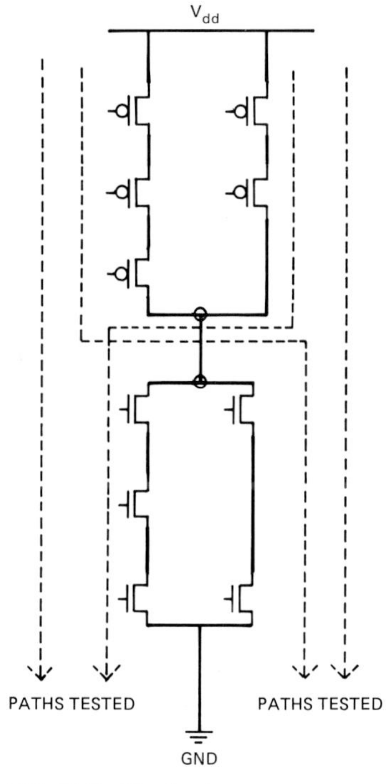

FIG. 20.7 Net depth test.

If the number of V_{th} drops in a path exceed a maximum, the circuit will cease functioning. This check is illustrated in Fig. 20.8. In the example, three V_{th} drops are shown. This may or may not exceed the design requirements that the designer will specify.

A check similar to the V_{th} check is the pass-level check in which the number of pass transistors connected in series is checked against a user-specified maximum value. This check is illustrated in Fig. 20.9.

Although ERC by itself will not ensure a working circuit, it is a useful capability. Many logic-level simulators do not model V_{th} drops or pass transistors with sufficient accuracy to detect problems.

20.9.2 Timing Verification

There are many types of timing verification programs. Some are interactive and allow the designer to measure the delays between various ports on the circuit. Others are designed for clocked synchronous circuits. They are aimed at checking propagation delays between various memory elements and ensuring that these delays are consistent with the clock frequency and the setup and hold times of the memory elements.

The extraction process is incapable of determining many of the parameters and attributes needed by these analysis tools. For these techniques to work, it is necessary to link cells in the artwork to other

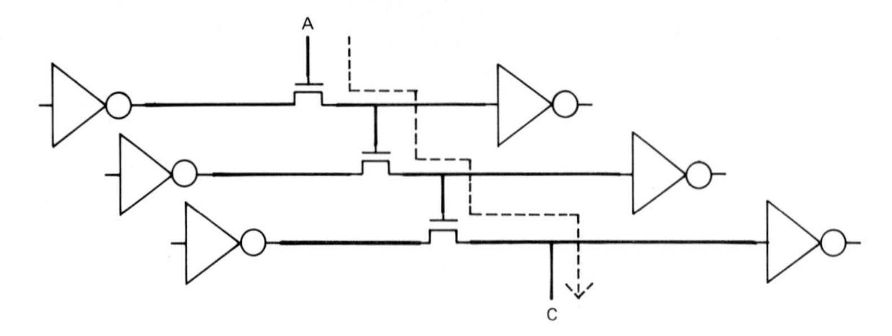

FIG. 20.8 V_{th} drops.

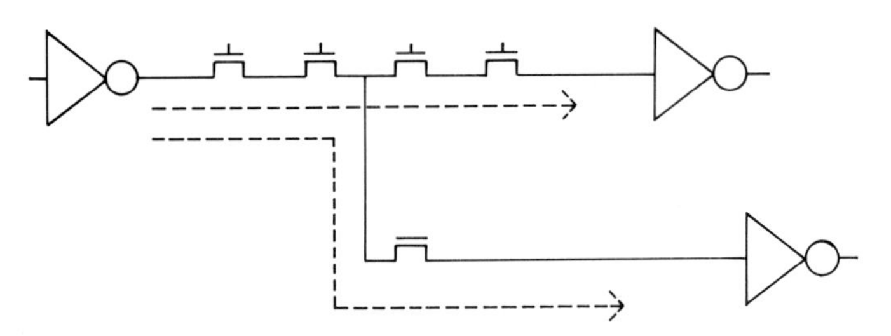

FIG. 20.9 Pass gate checking.

models that may have port attributes, timing assertions, and other information. This information is usually entered in the schematic. The extraction process may simply back-annotate its information into the schematic.

20.10 HIERARCHY AND HIERARCHY CREATION

Hierarchical design has many benefits. Designers are unable to deal with the quantity of information in an IC design unless it is organized into smaller pieces. Analysis tools also have difficulty with a large quantity of data. Most analysis tools are incapable of handling all the circuitry in an entire chip at a transistor level. SPICE simulation runs usually include a few hundred transistors, and rarely exceed a few thousand. Timing verification tools need higher-level models that have attributes that cannot be extracted. Although many logic simulators can handle transistor-level models, they are not capable of simulating complete circuits at a transistor level. Comparison and ERC programs are perhaps the only programs that can handle a complete circuit at a fully instantiated level.

Hierarchy can be used in two ways. If the design is done hierarchically, the hierarchy can be used in the verification process. If the design is not done in a hierarchical manner, or if the design must be flattened for extraction purposes, it may be possible to re-create a hierarchy suitable for analysis tools. Both methodologies are common in industry.

20.10.1 Hierarchical Design Verification

There are basically two forms of hierarchical design. One form is application specific integrated circuit (ASIC) methodology, in which a set of functions has been very well characterized from use in other designs and circuits. The verification process is performed on the interconnect between cells and on the interconnect between the cells and the pads. Often the extraction process is set up so that logic-simulation-level netlists are extracted directly. The other form of hierarchical design is for custom circuits in which most or all of the cells are new and verification procedures must deal with design in a hierarchical manner.

Usually designers will break the verification into two steps. They will select lower-level cells and verify their performance with circuit-level simulation. They will then rely on comparison programs to ensure that the design is correct with respect to a hierarchical schematic they have created. In some cases, they may use logic simulation on the hierarchical schematic. In these cases, the schematic would be expanded to a level that matched the simulator primitive elements.

20.10.2 Hierarchy Creation

Two approaches have been used to add hierarchy to a flat-transistor-level network. One is a batch-oriented algorithmically driven process[14,15,16] that yields gate-level structures, Boolean equations, transfer gates, and transmission gates. The second is an interactive pattern-recognition scheme where the user tries to find higher-level structures in a flat description. Each has its merits.

The algorithmic scheme attempts to identify generalized structures, such as AOI (AND-OR-invert) or OAI (OR-AND-invert) in NMOS. In CMOS the task becomes one of identifying connections in which *n*-channel and *p*-channel transistors are connected and are driven by the same signal.

Transistors that are not connected into either of the generalized schemes are examined to determine whether they are transmission gates or transfer gates. Transistors that do not perform a discernible logical function are removed in the early stages of the program.

The algorithmic extraction program has the ability to use the *W*/*L* values of transistors and nodal capacitance to add delay equations to the logic-level simulator.

Problems in the algorithmic approach have been in obtaining a rich-enough set of patterns to distinguish all circuitry. There is a need for a mechanism that will allow users to define new topologies that can be searched for.

The interactive pattern-recognition scheme is one with which the author is familiar, but on which no technical literature has been published. It was a simplistic approach in which the user was able to define the topology of a structure, and the program would then search for each occurrence of the structure in a netlist.

If a matching structure were found, the program would substitute the macro that had been defined for the topology. The algorithm worked in a hierarchical manner in which lower-level structures were first identified, and then higher-level structures were built from the intermediate levels. The program had the capability of handling equivalent sets of pins (such as inputs to a NOR gate) to facilitate matching topologies.

The algorithmic approach has been demonstrated for both NMOS and CMOS circuits. The pattern-recognition scheme has been demonstrated for NMOS circuits, which are simpler because of the implanted device that provides a good topological starting point.

20.11 TRENDS

It is difficult to describe trends in IC technology, especially as they pertain in CAD tools, unless one identifies the design methodology being used. Utilization of tools in a custom-design environment is much different than that of an ASIC environment. Even today many custom designs are initiated without a complete schematic. Often the schematic is generated to satisfy topology verification requirements. There is a convergence of the simulation of schematic netlists and the extracted netlists until the designer is satisfied.

In custom methodologies, topology extraction will continue to play an extremely important role. It is the only tie between the layout and the eventual performance of the circuit. The cost of proceeding through fabrication without some form of topology verification is too expensive and too error-prone to be a viable alternative. Hierarchy in custom-design methodologies is becoming much more common. Designers realize the benefits of leveraging previous work in cell design.

ASIC methodology has forced at least a two-level hierarchy. Cells at the lower level are fully characterized as to function, electrical characteristics, and timing characteristics. The higher level consists of interconnect between these cells. Although cells are compact and can be accurately characterized, the methodology often forces rather long interconnect paths between the cells.

20.11.1 Timing-Driven Layout

Most methodologies involve the design and layout of the circuit, followed by the analysis of the layout for design-rule correctness, topological correctness, and performance correctness. Gate-array and standard-cell methodologies create design-rule-correct circuits, assuming that each cell has been correctly verified prior to being placed in the library. These circuit layouts are usually driven from a schematic and therefore are topologically correct. The problem is one of performance verification.

Tangent Systems of Santa Clara, California,[4] has taken the "correct by construction" methodology a step further. Layouts from their system are driven by timing assertions as well as design-rule and topological requirements. Timing assertions include clock characteristics, I/O timing requirements, and timing requirements for selected paths within the design. Clock characteristics include the master clock period and the relationships among active edges of additional clocks in the system. I/O timing assertions include external delay characteristics and setup and hold-time requirements at the interface to the chip.

The Tangent systems receives the input netlist and timing constraints and then performs a series of steps involving two stages of global placement, global routing, detailed placement, and final routing. Between each stage, timing analysis is performed on the available data. In the early stages there is a degree of uncertainty in the analysis. As the layout proceeds, the data are refined and the analysis is more detailed.

The timing analysis performed in the Tangent system utilizes an enhanced version of the algebraic *RC* tree-modeling algorithms first developed by Paul Penfield.[3] Although these algorithms do not yield exact delays, they bound the delay time. If timing analysis satisfies the bound limits, then the actual design should work. The speed of analysis offered by the algorithms is important when many delay calculations are made.

The Tangent system is an example of a hierarchical approach in which the cells are carefully characterized; the system then uses these data at a higher level to complete the design. The system assumes that each cell has been properly characterized by conventional topology verification tools and that the data are correct. Compiler companies offer another approach to cells that is not only layout correct but also offers accurate prediction of cell performance.

20.11.2 Compiler Technology

Silicon compilers have the capability of generating layout for several functions and processes. Compiler companies[17] have put much effort into qualifying each of their compilers with each of the processes they support. Qualification involves ensuring that the cell works and performs in a predictable fashion. The performance is quantified in a SPICE simulation model for each member of the compiler set. Often, data are also available in logic simulation models and timing verification models.

Because compilers offer very fast execution times (often measured in seconds or a few minutes), it is possible to use the data in the early stages of design. This ability to do detailed analysis is a major attraction of compiler technology. Designers can explore many alternatives before committing themselves to a layout.

20.11.3 Simulation Technology

The major problem in verification today is in the analysis of extracted data for custom designs. ASIC methodologies have been developed with a two-level hierarchy that permits the use of logic-level simulators. A transistor-level simulator that can handle complete designs is needed. It must support reasonable transistor models, precharged circuits, and parasitic delay elements. It should also present a friendly interface to a designer who is dealing with over 100,000 elements. Accelerator technology, which is now common with logic-level simulators, needs to be applied to transistor-level simulators.

REFERENCES

1. M. Horowitz and R. W. Dutton, "Resistance Extraction from Mask Layout Data," *IEEE Transactions on Computer-Aided Design,* July 1983.
2. E. Barke, "Resistance Calculation from Mask Artwork by Finite Element Method," *22nd Design Automation Conference,* 1985.
3. Paul Penfield, Jr. and Jorge Rubenstein, "Signal Delay in *RC* Tree Networks," *18th Design Automation Conference,* Nashville, Tenn.
4. S. Teig, R. Smith, and J. Seaton, "Timing-Driven Layout of Cell-Based ICs," *VLSI Systems Design,* May 1986.
5. L. Nagel, "SPICE a Computer Program to Simulate Semiconductor Circuits," Memorandum No. ERL-M520, UCB, May 1985.
6. E. Barke, "A Technology Independent Approach for Device Recognition from IC Mask Artwork Data," *J. Digital Systems,* Winter 1982.
7. E. Barke, Dierker Werner, and John Werner, "Device Recognition from IC Mask Artwork Data for Bipolar Analog Circuits," *J. Digital Systems,* Winter 1982.
8. H. K. Gummel and H. C. Poon, "An Integral Charge Control Model of Bipolar Transistors," *Bell Systems Tech. J.,* May 1970.

9. R. Ellickson and J. D. Tygar, "Hierarchical Logic Comparison," *Proc. MIDCOM,* 1984.
10. B. R. Chawla, H. K. Gummel, and P. Kozak, "Motis a MOS timing simulator," *IEEE Trans. CAS,* **CAS-22** (1975).
11. R. E. Bryant, "An Algorithm for MOS Logic Simulation," *Lambda Mag.* fourth quarter, 1980.
12. R. E. Bryant, "MOSSIM: a Switch-Level Simulator for MOS LSI," *18th Design Automation Conference,* 1981.
13. T. M. McWilliams, "Verification of Timing Constraints on Large Digital Systems," *Proc. 17th Design Automation Conf.,* 1980.
14. R. Apte, "Logic Extraction for NMOS Circuits," *Proc ICCC Conference,* 1982.
15. S. Greenberg and Mahmud Buazza, "Logic Recognition in the Savvy Timing Verification System," *ICCAD,* 1984.
16. A. Kishimoto et al., "Logic Function Extraction Algorithm for MOS VLSI," *ICCAD,* 1983.
17. *Concorde Users Manual and Data Book,* Seattle Silicon Technology Incorporated, 1986.

SUGGESTIONS FOR FURTHER READING

Colclaser, R., *Micro Electronics Processing and Device Design,* Wiley, New York, 1980.

Ebeling, C. and Ofer Zajicek, "Validating VLSI Layout by Wirelist Comparison," *ICCAD,* 1983.

Hicks, P. J., "Semi Custom IC Design and VLSI," Peter Peregrinus, London, 1983.

Mead, C. and L. Conway, *Introduction to VLSI System,* Addison-Wesley, Reading, Mass., 1980.

Mori, S., I. Suwa, and J. Wilmore, "Hierarchical Capacitance Extraction in an IC Artwork Verification System," *ICCAD,* 1984.

Reed, J. and A. J. deGeus, "*RC* Delay Extraction for Gate Array and Standard Cell Design," *ICCAD,* 1983.

Rowson, J., "Understanding Hierarchical Design," Ph.D. Thesis, California Institute of Technology, 1980.

Rubinstein, J., Paul Penfield, Jr., and Mark Horowitz, "Signal Delay in *RC* Tree Networks," *IEEE Trans. Computer Aided Design,* July 1983.

Saporito, A. and M. Vanzi, "An Optimal Bipolar DC and AC Parameter Extraction for Circuit Simulation," *ICCAD* 1984.

Scheffer, L., "The Use of Strict Hierarchy for Verification of Integrated Circuits," Ph.D. Thesis, Stanford University, 1984.

Scheffer, L. and Apte, R., "LSI Design Verification Using Topology Verification," *Twelfth Asilomar Conference on Circuits, Systems, and Computers,* 1978.

Scheffer, L. and R. Soetarman, "Hierarchical Analysis of IC Artwork with User Defined Abstraction Rules," *22d Design Automation Conference,* 1985.

Tarolli, G. and Herman, W., "Hierarchical Circuit Extraction with Detailed Parasitic Capacitance," *20th Design Automation Conference,* 1983.

Trick, M., A. J. Strojwas, and S. W. Director, "Fast Simulator for VLSI Interconnect," *ICCAD,* 1983.

Tygar, J. D. and Ron Ellickson, "Efficient Netlist Comparison Using Hierarchy and Randomization," *22d Design Automation Conference,* Las Vegas, Nev., 1985.

Whitney, T., "Description of the Hierarchical Design Rule Filter," SSP file #4027, Silicon Structures Project, California Institute of Technology, Pasadena, Cal., October 1980.

P • A • R • T • 6

VLSI CAD TOOLS

CHAPTER 21
THE WORKSTATION AS A TIME MACHINE

Mark T. Fuccio

Herbert L. Hinstorff
Daisy Systems Corporation
Mountain View, California

The process of developing a VLSI chip demands a great deal of interaction between the designer and the design. Workstations provide computer-aided engineering (CAE) tools that organize and expedite this interaction. From schematic creation to simulation and physical layout, workstation tools help manage all the interrelated design tasks associated with VLSI development and allow easy updates to schematics and layouts. Workstations thus encourage accurate, complete designs and documentation.

The fast interaction possible with a CAE workstation makes designers more productive in many ways. It can be as simple as the designer's ability to "play" with a circuit without waiting for the tool to catch up. Or it can be as pervasive as the overall organization of the design process, so design changes become easy because the data are organized in an accessible way.

Thus, the workstation can be viewed as a time machine—a "what if" tool that collapses time as an engineer passes through several iterations of refining a design, and a tool that saves time by organizing the overall design process. This chapter goes into some detail about how a workstation does these things. Many hardware and software factors are explored here to help designers evaluate the benefits a workstation brings to VLSI design and to help designers decide which specific benefits are appropriate for a given task. The main subjects include

- Evolution of CAE
- CAE productivity
- Meeting the needs of VLSI designers
- Design database
- Graphics
- Hardware and software for VLSI design
- Trends

21.1 WHAT IS A WORKSTATION?

The first aspect of workstation CAE that must be dealt with is defining the nature of the workstation. This is important because the concept of an engineering workstation has become somewhat confused since the first one was introduced in 1982. The important aspects that distinguish a workstation from other types of computers are its accessibility (compared with mainframes), independent control (again as compared with mainframes), and interactivity (in terms of high-quality graphics and concurrent activities), as shown in Table 21.1.

TABLE 21.1 Workstation versus Mainframe Usage

Parameter	Workstation	Mainframe
Number of users	One at a time	Many
Control	User or system manager	System manager
Work environment	Office	Computer room with special power & cooling
Graphics display	Raster	Alphanumeric standard—raster optional

For the purposes of this discussion, a *workstation* is a stand-alone computer dedicated to a small number of users (usually only one). A workstation is also specialized to handle a certain type of task, such as VLSI design, in contrast to a general-purpose computer that deals with any sort of data users care to force into it—although a general-purpose computer can function as a workstation with the right bundled software.

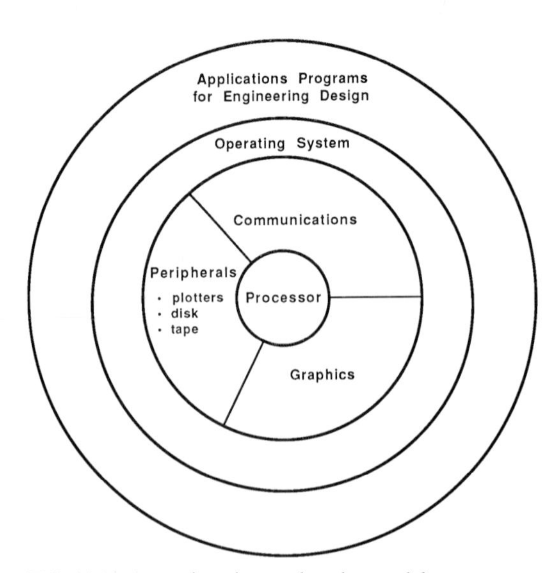

FIG. 21.1 An engineering workstation model.

A workstation also implies a certain architecture, shown in Fig. 21.1, that includes networking, an operating system, application software, database, and graphics facilities. Each of these elements is crucial to the workstation's success in VLSI design. For example, networking allows the workstation to act as a stand-alone resource while maintaining a link to other computers. This is important in VLSI design because there is usually more than one designer involved in a project, and everyone needs to share data. The people sharing the data must be able to work on different aspects of the design concurrently. It is also sometimes useful to archive design data on a company mainframe computer and distribute that data to workstations via networking.

An operating system provides the basic functions that application programs need to take advantage of the workstation's hardware. A standard operating system, such as Unix, also furnishes a straightforward way to adapt new application programs to a specific workstation. The software developer can quickly port new programs to a variety of workstations—if they all use the same standard operating system—or users may add their own special-purpose software. When putting programs together to configure a system, however, it is crucial that all the programs work together rather than as separate tools that all happen to reside on the same machine.

21.2 EVOLUTION OF CAE TOOLS INTO WORKSTATION TOOLS

The first CAE tools were developed on general-purpose mainframe computers through military, commercial, and university efforts in the 1960s. These tools were primarily concerned with simulation and provided little, if any, graphics capabilities. (Figure 21.2 shows the evolution of graphics technology and usage by CAE tools.) For the most part, graphics were limited to simulated voltage and current waveforms that were generated on printers using a standard alphanumeric character set—low-resolution graphics, to say the least. Probably the most characteristic aspect of this phase of CAE development was that all programs had to be run as batch jobs: a designer submitted a job and waited, usually for many

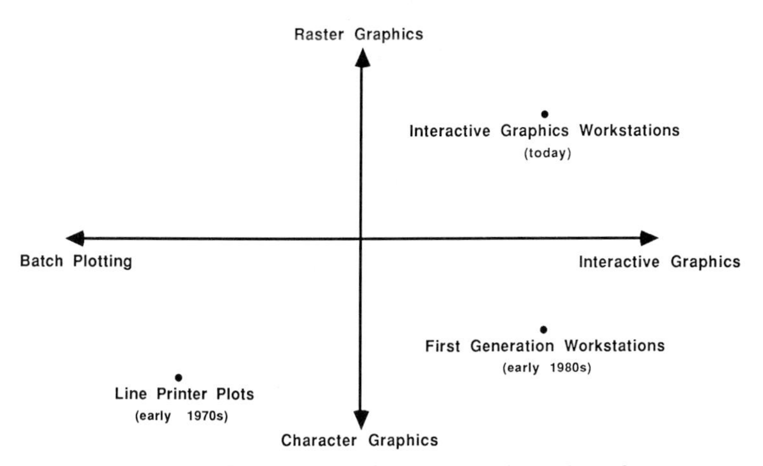

FIG. 21.2 Evolution of computer graphics for engineering design software.

hours, until the job came up in the queue and the mainframe was through crunching on the data.

CAE/CAD systems eventually began solving the graphics limitations, although the first graphics-based design tools actually provided only computer-aided drafting functions. These were CAD/CAM tools that started with printed circuit board layouts and later dealt with ICs. Engineers continued to create schematics on paper, which were then entered into the drafting systems by draftsmen, much as the job had always been done; the main difference was that the drawing could be managed more easily on the computer. A shared central processor unit (CPU) was still viable for the early automated drafting systems because drafting did not tax a shared CPU to its limits.

Another influence on CAE/CAD tools was the increasing complexity of integrated circuits. As this came about in the early 1970s, additional tools were developed to check masks for IC layouts. The design-rule checker (DRC) evaluated geometries to make sure the designer had created a layout that was manufacturable. A layout-versus-schematic comparison tool helped ensure consistency between these two design representations (this is especially important in today's CAE systems). Even before the layout stage was reached, designers could use an electrical rule checker (ERC) to evaluate the validity of conditions such as electrical interconnects, loading, and unconnected nodes. Even though

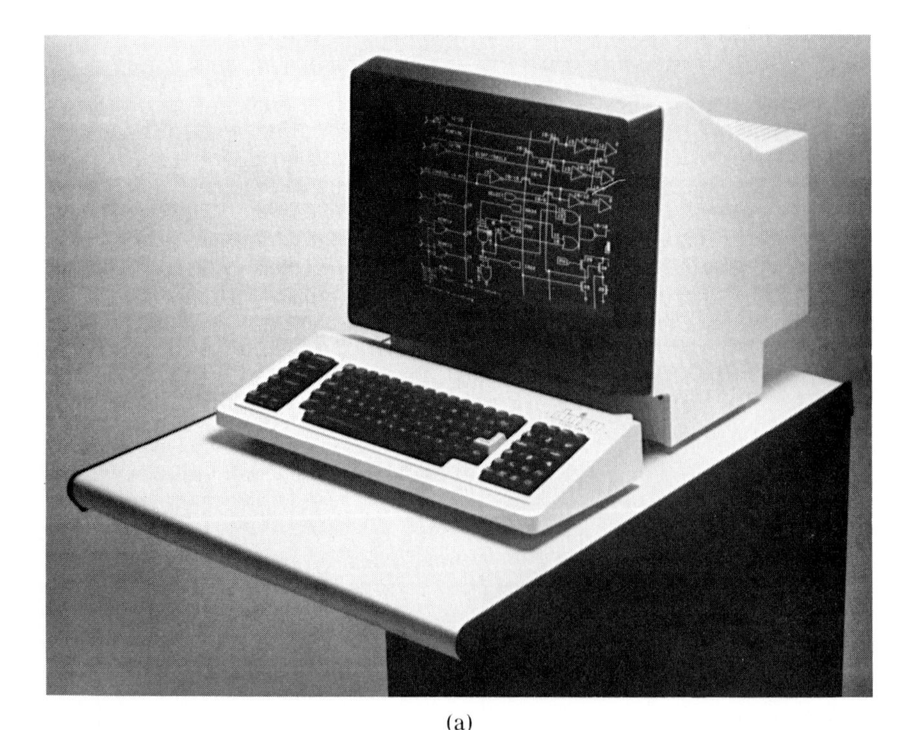

(a)

FIG. 21.3 CAE workstation evolution. (*a*) An early CAE workstation provided schematic capture and logic simulation. (*b*) Today's powerful but inexpensive personal computers are the dominant workstations for schematic entry and simulation. (*c*) General-purpose computers with high-performance graphics run a variety of CAE and layout software.

(b)

(c)

FIG. 21.3 (*Continued*)

these tools often dealt with a visual aspect of IC production, interactive graphics were not used.

With the growing popularity of semicustom ICs in the early 1980s, the widespread need for automated design tools became obvious. With the introduction of the workstation concept, CAE finally found the vehicle for putting design tools in the hands of many engineers. Now, instead of waiting for the slow response of a shared CPU, designers could quickly perform the necessary tasks with a dedicated, but versatile, resource.

As workstations have taken the CAE market by storm, the hardware has evolved mainly by delivering greater compute power and enhanced features for less money (Fig. 21.3). Starting from the foundation of a general-purpose

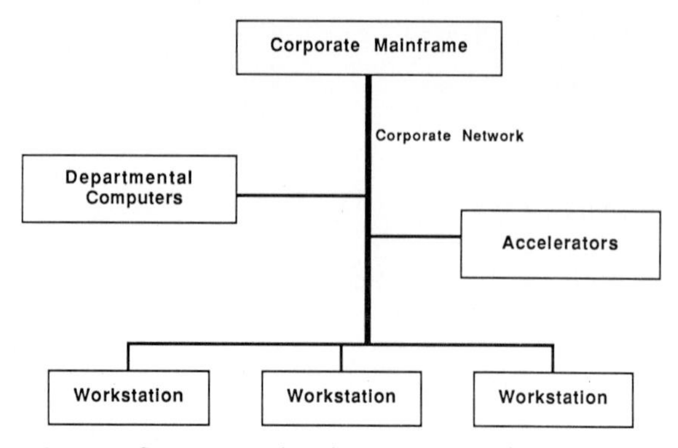

FIG. 21.4 Corporate engineering computer environment.

microprocessor-based computer, workstations have been enhanced with dedicated accelerator hardware that lets engineers handle the most demanding design tasks. As standard personal computers became powerful enough, they were brought into service as platforms for many design functions. Figure 21.4 illustrates the computing environment many companies are creating by "mixing" different computers into a single networked environment.

CAE software has progressed in terms of both the variety of available functions and the refinement of the functions. All aspects of chip design, from schematic capture to layout, are now available on workstations. Later, we explore the available functions.

21.3 PRODUCTIVITY: THE HOLY GRAIL OF WORKSTATION CAE

CAE productivity has proven difficult to quantify objectively, although it serves as the primary justification for the use of CAE tools. Despite the difficulties in measuring actual productivity improvements, however, users' subjective evaluations indicate that CAE tools provide a significant productivity gain even when the users do not consider all of the sources of potential gains. (See Fig. 21.5.) Even though quantitative data are scarce, the qualitative results of CAE use are informative.

One of the few aspects of productivity that has been evaluated statistically is transaction time, which consists of system and user response times. It turns out that reducing the system response time also reduces the user response time. This link has been demonstrated by Arvind J. Thadhani of IBM,[2] following the theoretical work of Walter J. Doherty, also of IBM.

In the typical programming and design environments Thadhani studied, the average transaction time was 20 s. System response accounted for 3 of the 20 s, and user response accounted for the other 17 s. When system response time was improved in increments from 3 to 0.3 s, user response time decreased from 17 to 9.4 s. Thus, reducing the system response time by 2.7 s produced a total time savings of 10.3 s—a 100 percent performance improvement over the original 20-s transaction time. These results were averaged over full workdays.

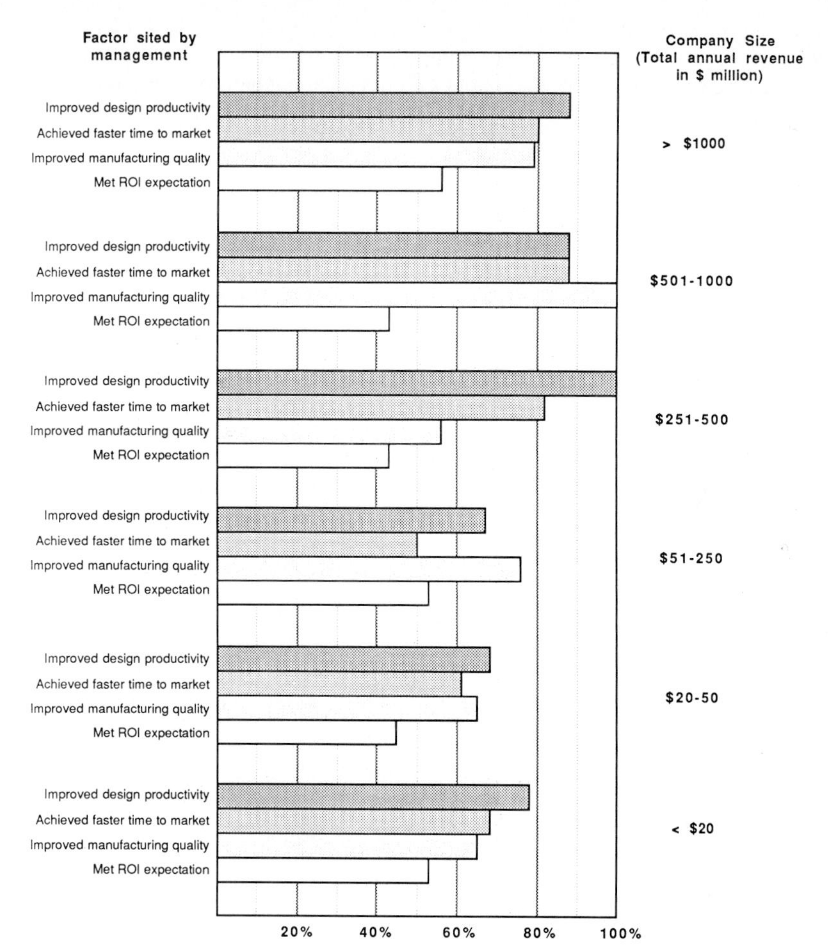

FIG. 21.5 Workstation layout accomplishments. (*From Coopers and Lybrand.*[1] *©1987 Cahners Publishing Company.*)

Thadhani and others have shown that similar results apply to a variety of user activities, including design engineers using graphics terminals. Studies at IBM revealed significant improvements in overall task times when system response was decreased by even small amounts. In one laboratory, for example, every 0.1-s reduction in system response time resulted in a 3.6-min improvement in task time. Similar results were borne out over a mix of applications.

As an added benefit, the quality of work improved in all cases. Such studies illustrate why graphics accelerators and other high-performance hardware are so highly valued in many CAE systems (see Sec. 22.4).

From a more intuitive point of view, the most obvious result of an increase in productivity is a shorter design cycle. The biggest reduction in design time comes from the elimination or reduction of iterative design passes to obtain a working product. Specifically, if it is not necessary to enter many netlists, the number of errors is reduced and engineering change orders are easier to process. CAE users rarely, if ever, take advantage of such productivity gains to reduce the number of engineers working on a project (although CAE does allow junior engineers to do sophisticated design work that would otherwise be handled by more experienced engineers). Instead, users cite their ability to design more complex devices in the same amount of time.

Despite increased IC complexity, users report that CAE gives them control over design schedules. Project development plans become more predictable and manageable.

Some firms take advantage of CAE to build gate arrays or even standard-cell chips instead of printed circuit boards. Costs are reduced by using fewer components on a board and designing more circuitry into the ICs, and the cost of gate arrays is kept to a minimum by using CAE/CAD layout tools in house, thus lowering charges for engineering services from gate-array vendors. The resulting cost savings are as evident as the productivity gain.

The most important productivity benefit to users is the reduction or elimination of design flaws. Many workstation users routinely turn out flawless chips with first silicon. This completely eliminates iterative design passes, which allows a tremendous reduction in the design cycle.

Several factors contribute to the ability to perfect a design in one pass. For example, schematic entry, compilation of graphic data, and automatic database extraction for netlisting improve data accuracy and consistency over that of manual efforts. But the biggest factor is simulation. Users who take the extra time for the use of verification tools in front-end design report a dramatic payoff in the form of error-free chips. Finding unexpected bugs and going through major engineering changes at the last stage is very expensive and causes tremendous delays in schedules. Pressure to get the product out of the door is especially high at the last stage of the project. Engineering managers can eliminate much of this headache by encouraging engineers to simulate more at the beginning. CAE workstations themselves encourage designers to simulate because of the relative ease of running simulation software in the same environment as schematic capture.

It is critical to bear in mind that extensive verification is usually necessary to significantly boost productivity. If only limited simulation is used so that the number of design passes is not reduced, there is little gain from detecting a few more errors in advance. Unless the number of passes decreases, the overhead of prototype fabrication, debugging, and rework still exists even though there are fewer errors to work on. Fortunately, diligent simulator use can locate all errors in most designs.

Users seldom consider the productivity benefits of CAE beyond the design phase. To most users, shorter design cycle means reducing the time from the start of the design to the first working prototype. (Figure 21.6 shows the engineering design steps involved in bringing a product to market.) From the viewpoint of higher-level management, however, there are also productivity gains to be had from better manufacturability and testability. These benefits are increasingly seen as one of CAE's richest sources of productivity improvements. As it becomes standard practice to simulate and verify entire systems before prototyping, more high-level CAE benefits will become apparent.

Productivity gains should eventually manifest in higher profits, and there are two ways that this can occur. A McKinsey & Co. study revealed that competitive pressures result in penalties for bringing products to market late (Fig. 21.7). Some observers feel that even greater rewards come from CAE's ability to extend a product's life: the longer a product's life, the greater its profitability. Further, with CAE tools, a company can make greater strides in product evolution on the

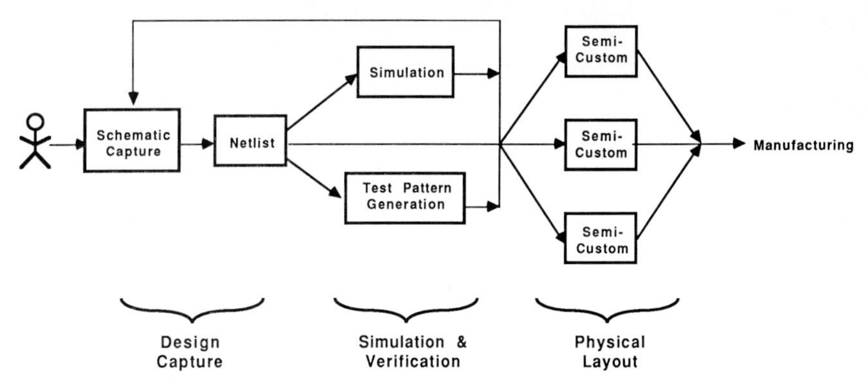

FIG. 21.6 VLSI product development flow. CAE tools help in all stages of VLSI product development.

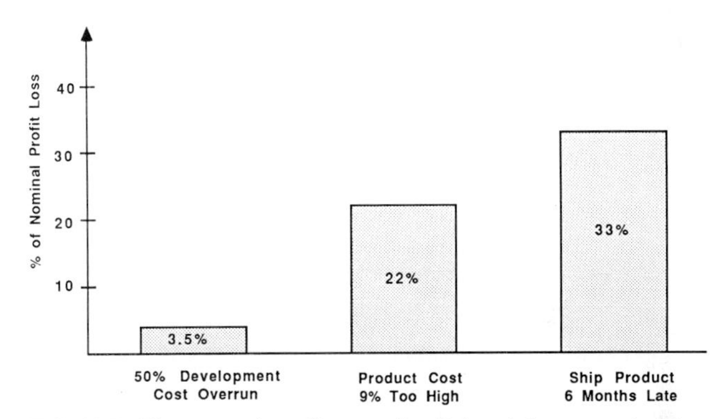

FIG. 21.7 Time to market affects profits. Delayed time to market dramatically reduces profits over the product life cycle. (Assumes 20% growth rate market, 12% annual price erosion, 5 year product life.) (*From Reinerstein.*[3] *© 1983 Cahners Publishing Company.*)

first design pass. By using CAE to design a better product (which will have a longer life expectancy) in addition to shortening time to market, users can realize higher profitability.

21.4 TWO TYPES OF IC DESIGN ON WORKSTATIONS

Having considered some factors that apply to all types of design, the next step is to focus on the needs of specific types of IC design. Workstation-based tools cover both of the main areas of IC design: full-custom and semicustom ICs. The latter category, also known as application specific IC (ASIC), includes gate arrays and cell-based designs. (See Fig. 21.8.)

In terms of software tools, full-custom design involves three main categories:

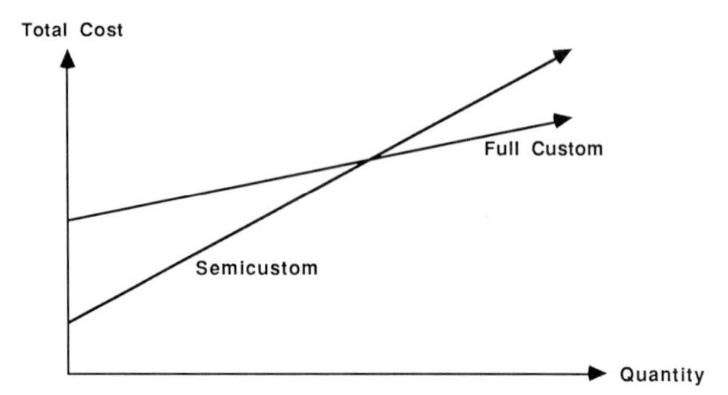

FIG. 21.8 Cost comparison between semicustom and full-custom ICs. Semicustom methodologies like gate array and cell-based are cost effective for low volume IC designs. High nonrecurring engineering costs of full custom chips are offset by their lower manufacturing cost.

layout, analysis, and postprocessing. Layout tools include polygon editors for creating and changing mask features and place-and-route functions for connecting various blocks to form a full-custom VLSI chip.

The other full-custom tools will be considered in detail later, but it is useful here to note the difference between the mask editor and the graphics editors used for other types of IC designs. Due to the nature of full-custom design, it is important that the mask editor be able to produce any complex shape quickly. Although an editor for CMOS designs can be restricted to 90° and 45° angles without limiting the designer's options, a mask editor must be more versatile for full-custom bipolar or GaAs work, where rings and irregular angles are required.

More advanced tools in this category include block generators and symbolic editors. The latter show components, such as transistors, as generic symbols rather than as actual geometric forms; this allows designers to lay out a circuit quickly and make changes easily. After a layout has been done symbolically, another program, called a *compactor,* must be used to convert the generic symbols into process-specific geometries. To get tight layouts, today's compactors require significant manual interaction. Despite this limitation and the slow speed of many compactors, the symbolic layout process shows great promise.

Another promising tool, the block generator, is a knowledge-based facility that "compiles" functional descriptions of a chip and information about process technology (two-metal CMOS, etc.) to generate a mask-level layout. There are currently three types of block generators: those that handle regular structures of predefined objects, those that employ preprogrammed blocks (for data path blocks such as ALUs), and language-based generators that require IC knowledge to create layout geometries.

The second category, design analysis, includes electrical rules checkers, layout verifiers, and layout parameter extraction tools. The latter extract actual resistances and capacitances of transistors and interconnects for accurate logic and circuit simulation. Electrical rules checkers and layout verifiers examine schematics and mask data, respectively, for consistency.

Postprocessing includes plotting utilities and tools that "fracture" data for mask making. The fracturing procedure converts the complex polygons in layouts into a data format suitable for mask making. Analysis tools check the mask-level data to ensure the accuracy of the fracturing step.

Another tool that proves useful is a procedural interface to give users access to the design database. This is an especially important item for full-custom design because IC houses often have their own software that has been developed over the years. The procedural interface provides a straightforward way to connect in-house developed tools into the system as well as directly modify information in the database. The interface also makes it easy to create new in-house functions and bring in a mix of other software. (The section on databases will present more information on procedural interfaces and their associated programming languages.)

Full-custom design can also make good use of networking. If both the engineering and layout departments in a company use the same design tools, they can pass information back and forth in a compatible format, without the delays and miscommunications inherent in manually transferring data via tapes or floppy disks, as shown in Fig. 21.9. Unlike the traditional approach, everyone involved in a network-based project always has immediate access to all the data associated with the project, from schematics and notes to simulation models. CAE can thus boost team productivity as well as the productivity of individual designers.

Networked access also makes it easier for design engineers to handle some crucial layout tasks. This becomes increasingly important with leading-edge technologies, where shrinking geometries entail greater parasitic effects on circuit performance.

For this occasional layout work, there is usually little need to give each designer a complete set of layout tools on a workstation that is optimized for layout tasks. A low-cost workstation (generally based on a personal computer architecture) can serve design engineers' mask editing needs, so long as the tools on the company's various design and layout systems are compatible. The low-cost workstation might not be best to deal with large designs all at once or provide very fast response times, but some basic layout viewing and editing capabilities usually suffice for occasional users.

The set of capabilities required for semicustom IC design generally resembles those needed for full-custom work. The main difference lies in the fact that semicustom design does not usually demand that designers deal with chip geometries. The design process focuses on a logical level for schematic design and a physical level only for connecting and/or arranging preconfigured logic elements. Despite this difference, there is a large area of interplay between cell-based semicustom designs and full-custom designs. Cell-based blocks of random logic can be inserted in full-custom designs, for example, and custom blocks can clearly be used in cell-based designs. In fact, in many cases, the only difference between full-custom and cell-based designs is that most of the cells for a full-custom design are not preconfigured and precharacterized. Block generators also blur this border between semi- and full-custom designs. Whatever type of design is involved, compatibility between full-custom and semicustom design tools can prove helpful.

The most versatile area of semicustom design is the cell-based IC, which can involve cell-based blocks, handcrafted blocks, programmable logic arrays (PLAs), and memory blocks. The cells might be designed especially for a specific IC or imported from another library or an IC that includes the necessary logical function. To deal with all these types of blocks, a cell-based design tool must avoid rigid constraints on block size, block pin location, or pin layers.

Gate-array designs present a simpler case because they involve a limited set of logical elements in fixed locations, because most of the masks for a gate array are predefined. (Users only customize interconnections of transistors.) In fact, many

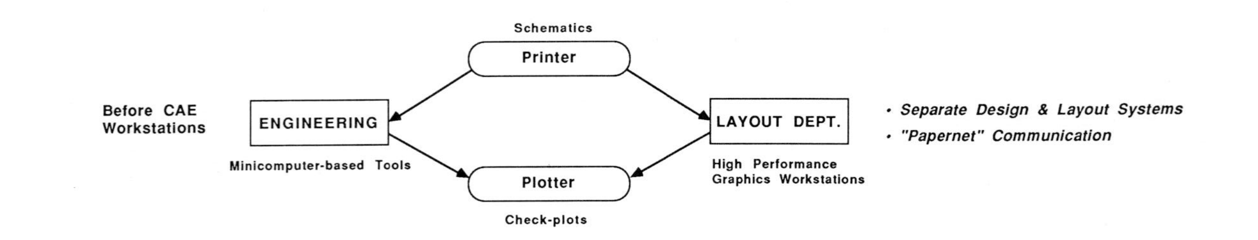

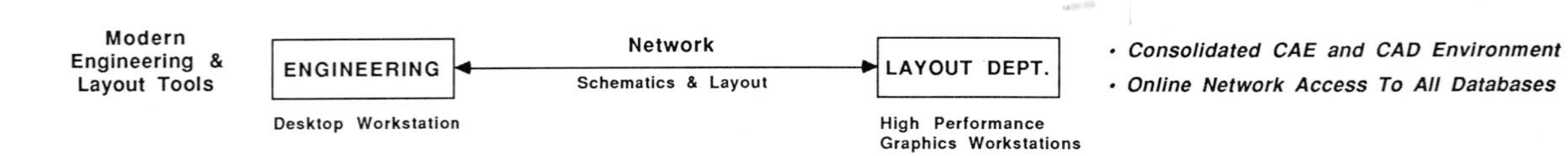

FIG. 21.9 VLSI design communication bottlenecks. Networked CAE improves communications between members of the design team.

gate-array design systems include only design entry and simulation tools; layout and final simulation are often left to the gate-array vendor. With this approach, the designer loses control and must wait for prototypes; the method does not allow the designer to run simulations with parasitics included. By including gate-array layout in the in-house design system, users gain greater control over critical parts of the design and increase the likelihood of producing a working part on the first pass.

21.5 AN IC DESIGN SEQUENCE

It is possible to complete the entire design sequence for full-custom and semicustom chips on a workstation, using a five-step procedure. For simplicity, the description that follows deals only with gate arrays. But note that, with some variations, this sequence applies to full-custom and semicustom cell-based ICs as well as to gate arrays. However, the sequence described here does not include the steps involved in the creation of cells or gate-array macros. (When these steps are necessary, cell and macro creation can begin with TTL equivalents that are then optimized; a library of components supports the design, characterization, and simulation functions. Some array cells, such as RAM, are based on partial schematics that are then fully expanded at the layout level.)

First, the designer creates a schematic using the gate-array vendor's component library. These components usually consist of familiar SSI-MSI logic functions. When the schematic is complete, two types of verification programs are typically used: one checks for flaws, such as duplicate names and connectivity errors, and the other evaluates technology and vendor-specific design considerations.

The second step of the design procedure involves more extensive types of design verification: logic simulation and timing verification. For maximum accuracy, the tools used in this step can employ vendor-specific delay models that account for variables such as fan-out and wire length. Voltage, temperature, and process variation can also be adjusted to enable worst-case simulation.

In the third step, the simulation stimulus patterns are converted into a test vector format suitable for the vendor's automatic test equipment. Workstation utilities can not only translate one pattern format into the other, but they can also check to make sure the test vectors will accurately test the IC. For example, a fault simulator can use toggle testing, gate collapsing, and statistical or exhaustive fault analysis to grade test patterns.

Layout is performed in step 4. This process requires the availability of the vendor's base array data (footprint), which specifies the size of the array, any obstructed areas, available cell or transistor locations, and layer information. The latter includes metal and polysilicon layers. To initiate the layout, a program merges the user's database with the vendor's base array data (the predefined mask layer information). During this merger, the vendor macro library (which defines the transistor interconnects for various logic functions) is referenced to ensure that the macros meet the pin and naming requirements of the design's schematic components.

If place-and-route options are available on the workstation, the user can usually perform placement (the assignment of logical components to the physically available components in the array) manually for critical components and/or allow the workstation to execute automatic placement. Once placement is complete, routing can generally be performed using a similar combination of manual and/or automatic techniques. The ability to predefine placement and routing in critical areas often gives the designer a great deal of control over an IC's performance.

Step 5 involves postlayout verification. Unlike the earlier simulation and timing verification step, this procedure takes actual interconnect parasitic effects into account. The workstation program that performs the verification usually extracts the wiring capacitance from the layout database automatically; this capacitance is then added to the load capacitance of logic elements to calculate delay times for a final accurate logic simulation.

An additional simulation step is available on some workstations to see how the gate array operates within its intended system design. A hardware modeler (which inserts the responses of actual chips into the simulation) can be used to verify a gate array's performance within the context of an entire system. This context check reveals any problems with either the gate array, the system specifications, or the rest of the circuitry. One of the biggest problems with gate arrays is incorrect specifications; 40 percent of the chips work according to their specifications, but not in the system. When the gate array has been checked out thoroughly, the final step is to transfer the design data to the vendor in the appropriate format for manufacturing.

REFERENCES

1. Coopers and Lybrand, "Design and Engineering: Where the Dollars are Headed," *Electron. Business,* April 15, 1987.
2. A. J. Thadhani, *The Economic Value of Rapid Response Time,* IBM Corporation.
3. D. G. Reinerstein, "Whodunit? The Search For New-Product Killers," *Electron. Business,* July 1983.

CHAPTER 22
THE COMPONENTS OF CAE SUCCESS

Mark T. Fuccio

Herbert L. Hinstorff
Daisy Systems Corporation
Mountain View, California

Having gone through an example design procedure in the previous chapter, it is useful to focus on the main aspects of the CAE system involved. One of the most important elements of such a system is the database, which contains all the information about the IC's design and layout. Because a company stores all its designs in database files, over time the database becomes a highly valuable resource, one that must be protected and kept accessible no matter how CAE technology evolves.

Another element in a CAE system that has a global impact on how the system works is graphics, which is at the heart of the user interface. An overview of a CAE system reveals that graphics act as the top layer in a pyramid of functions, while the database furnishes the foundation on which all the other functions rest.

22.1 THE DATABASE FORMS THE FOUNDATION BENEATH CAE FUNCTIONS

The capabilities and efficiency of a CAE system depend to a large extent on the structure of the database. Data structures are designed to support specific operations; various structures make some operations inefficient or impossible. If the database is structured a certain way, for example, the user of a schematic editing program can move a right-angled interconnection by dragging its corner. Without that certain database structure, however, this convenient operation becomes slow or awkward, if not impossible. It would be necessary to deal with the wire only as a whole or as line segments; this would restrict the ways in which wires could be moved around in a schematic.

There are many more examples of such operations. The reason they are so difficult with some database structures is that correlating the necessary information is too inefficient; these databases would have to look at every piece of a schematic or layout to move something, for example. Thus, the database must have some way of quickly filtering information. This simplifies many activities, such as inserting a block of logic quickly within an existing block.

Fast access to certain types of data must be possible even when the volume of data is quite large. In general, databases must be designed to accommodate very large schematics and layouts, in keeping with today's trend toward more complex ICs. For this reason, almost all such databases employ a compact binary format. Some databases also employ a technique for limiting the number of times an object (such as a logic device) must be defined in the database. When an object is defined once and all instances of that object are referred to the definition, the database need not store redundant information. Thus, database sizes are minimized and easier to handle.

A more fundamental influence on a database's performance lies in the technology used to implement the database structure. There are several database structures appropriate for CAE work, including relational, network, and hierarchical models. The relational model provides a great deal of flexibility in how data will be organized. Specifically, relational databases establish no predefined relationships among data elements, giving CAE system designers and/or users the ability to change the way data is accessed at any time. New data elements can also be added easily at any time to accommodate new functions.

Relational databases are not perfect, however. Because of this model's loose structure, a great deal of sorting and bookkeeping are required to relate all the appropriate data elements for any given task. Relational databases thus tend to yield their information more slowly than other types of databases.

This speed problem can be corrected by predefining the relationships among data elements, as network and hierarchical databases do. Conceptually, these models group together data that relates to an object such as a logic gate; each element in the description of the gate is given a specific amount of space, and only the predefined elements can be included. When it is necessary to obtain some information from the database, that information can be found in rapidly calculated locations. There is no need to sort all the data elements, as would be necessary with a relational database, to find all the related elements that should be included in the body of information.

The drawback with the network and hierarchical models is that making changes to the information stored in the database is difficult. For example, a database might be designed without including values for a gate's temperature response. If at some future time it becomes desirable to analyze circuits over a range of temperatures, the database will have to be restructured to support this function, because the original structure offers no way to attach temperature-related data. The network and hierarchical models thus make it difficult to expand a CAE system's functions beyond what is currently desired or at least envisioned.

No one can foresee every option that designers might need in future years. With so much time and energy invested in design databases, companies need a straightforward way to upgrade their databases when technologies and priorities change.

The solution arrived at for advanced CAE databases has been to combine aspects of the relational, network, and hierarchical models. Data that are clearly necessary for a CAE database and that must be available quickly can be stored in network and/or hierarchical formats. An associated relational structure allows the addition of data elements at any time. The database's relational part can contain information for user-defined and third-party functions, as well as for new tools introduced by the CAE firm at a later date.

22.2 GAINING DIRECT ACCESS TO DATABASE CONTENTS

Another requirement for an advanced database is that it allow easy access for manipulating the stored information. In essence, the database needs a general-purpose database manager so that users can add, delete, and change design information. CAE databases have evolved from the simple collection of data files used by early tools to the sophisticated entities of today. Formerly, each data file in a CAE system was specific to one tool or set of tools. Now, the database serves a variety of functions—schematic creation, netlisting, simulation, test—and the entire CAE system revolves around the central database.

As CAE databases have become more sophisticated, it has become increasingly important to provide methods for users to gain access to data. Many companies have their own technology-specific CAE/CAD tools that have been developed in house. Users want these tools to operate on data from the main CAE database.

A procedural interface offers a means to gain this access in a straightforward way, insulating the user from delving into the binary format in which data is coded. This method of accessing the database supports uniform handling of all objects to make it easier for users to manipulate data, at the same time shielding the application from any future changes to the structure of the database itself, as shown in Fig. 22.1.

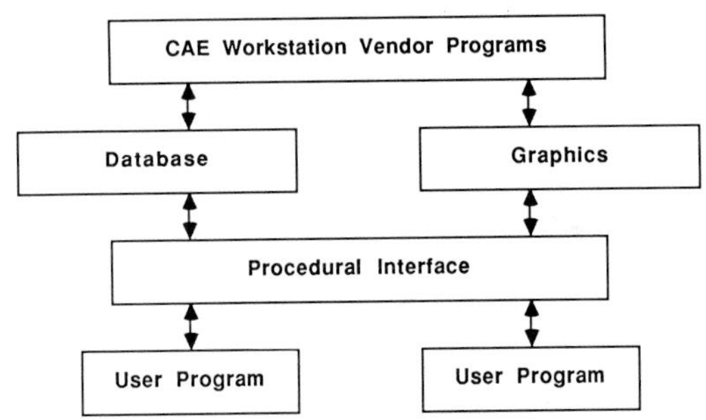

FIG. 22.1 A procedural interface insulates users from changes to the database or graphics, and at the same time simplifies creating programs to extend system capabilities.

Working with the procedural interface, a standard programming language (such as C, Pascal, or FORTRAN) can be used to access the database. CAE/CAD departments in companies often use the standard languages to funnel selected data into in-house programs in specific formats. The procedural interface language allows a CAE/CAD department to customize a CAE system to the company's existing design environment.

22.3 STANDARD INTERCHANGE FORMATS ALLOW DATA PASSING BETWEEN TOOLS

An alternative to a procedural interface in some cases is the use of a standard interchange format. Such a format would be able to contain all the data required in a CAE design. The CAE system would not use the interchange format as a database format, but the information in the database could be completely represented in terms of the interchange standard. Then other tools that needed to access the data would simply have to be able to read the standard interchange format.

Several interchange formats are in use to some extent in the CAE industry. Most of them cover only segments of the design process, usually either electrical or layout data. The one exception to this segmentation is EDIF (the electrical data interchange format). This public-domain format can include data describing libraries, schematics, behavior, functional and logical structures, tests, circuits (for simulation), and geometry (Table 22.1). Even processing technologies can be specified. The treelike structures used in EDIF files are similar to those used in Lisp, but EDIF is not itself a programming language.

EDIF's popularity has been growing as its benefits become more obvious. However, the scope of the format (which necessitated many proprietary extensions) and the use of previously accepted formats for specialized purposes has prevented instant implementation across the CAE industry. There is also the limitation that an EDIF file representing a complex IC layout is enormous. Despite these drawbacks, EDIF shows promise for at least some aspects of design interchange, especially schematics, netlists, and logic models.

Three other interchange formats—GDS II Stream, CIF, and simulator languages, which cover segments of the IC design process—are widely supported in the CAE industry. Two of these, GDS II and CIF, are designed to contain data relating to physical layout. GDS II Stream is a de facto standard created by GE-Calma Corporation for archiving data and transferring it between systems. Unlike

TABLE 22.1 EDIF Data Representations

EDIF supports a range of design representations for data interchange between different workstations

Data Type	Design Data Described	Tools Using This Data	Stage of Design Process Supported
DOCUMENT	Textual documentation	Text editors	All stages
GRAPHIC	"Dumb" graphics & text representation of design information	Text editors Drawing systems	All stages
SCHEMATIC	Graphical representation of components and connectivity	Schematic editor	Logic design
NETLIST	Circuit components and interconnections	Routers Logic & circuit simulators Timing verifiers Automatic test pattern generators Fault analyzers	All stages
BEHAVIOR	Functional description	Functional & logic simulators	Architectural design
LOGICMODEL	Logic simulation model	Logic simulators Timing verifiers Automatic test pattern generators Fault analyzers	Logic design Test program generation
SYMBOLIC	Symbolic integrated circuit layout	Polygon editors IC placers & routers	Integrated circuit layout
MASKLAYOUT	Integrated circuit layout	Polygon editors IC placers & routers Design rule verifiers PG & E-beam mask makers IC test equipment	Integrated circuit layout Layout verification Production mask fabrication
PCBLAYOUT	Printed circuit board layout	Printed circuit board placers Printed circuit board routers	Printed circuit board design
STRANGER	Undefined data	Future tools	All stages

EDIF, a GDS II Stream file is viewed as a sequence of binary records that describe the database.

CIF (Caltech Intermediate Form) was developed at the California Institute of Technology. This format is ASCII-character based and includes data describing mask features. It is intended to serve as a generic format for transferring data between tools and converting to the specific formats used to drive various plotters, video displays, pattern-generation machines, and other output devices. As with EDIF, the ASCII nature of CIF files makes them very large compared to the size of binary-coded files. (A binary scheme can reduce the size of a file by a third or more.)

Simulator netlist formats are another sort of "standard" in wide use. These include the formats for TEGAS, SPICE, and SILOS, which were originally intended for organizing inputs to the simulator program. However, the formats have taken on a broader role as a way of structuring netlists. As a typical example, the TEGAS language can include descriptions of components, interconnects, and some timing relationships. Although the TEGAS language is useful, there are several versions and extensions that must be dealt with simultaneously.

22.4 GRAPHICS SHOW OFF CAE CAPABILITIES

Probably no other aspect of CAE indicates the power of the design technology so well as graphics. In fact, as described earlier in this chapter, the advent of high-resolution graphics defined the industry's transition into what we think of as CAE today. A signature of today's graphics systems is the multiwindow environment that gives designers easy access to several tools at once.

On single-user, multitasking systems such as engineering workstations, each window on the CRT screen shows a process that is running concurrently with all the others. A designer might run the schematic capture and simulation tools simultaneously, for example, so that simulated waveforms can be viewed in the context of the schematic being simulated as in Fig. 22.2. With each process working from the same database (or linked databases), the designer can point to the schematic and automatically see the corresponding simulated signals in the simulator window. The simulator might even be running on a different system from the designer's workstation; this illustrates how a window can give access to processes running elsewhere on a network.

The multiwindow environment is therefore a source of productivity gain for two reasons: It eliminates the need for the designer to go visit another computer (or another user) to tap other resources, and it provides a quick way to monitor concurrent tasks. The designer can now perform operations across multiple applications.

Some design tasks, such as full-custom IC layout, demand extensive graphics capabilities. Thus, for the greatest productivity, a workstation for this purpose could benefit from the use of a dedicated graphics accelerator. This accelerator offloads the graphics-generation chores from the workstation's CPU so that the system can respond quickly to users' graphics manipulations and handle CAE computations at the same time.

As the section on productivity in the previous chapter showed, the faster the workstation responds, the higher is the user's productivity level. An accelerator furnishes the fastest response possible by handling some graphics-intensive functions in hardware. For instance, the accelerator might provide capabilities such

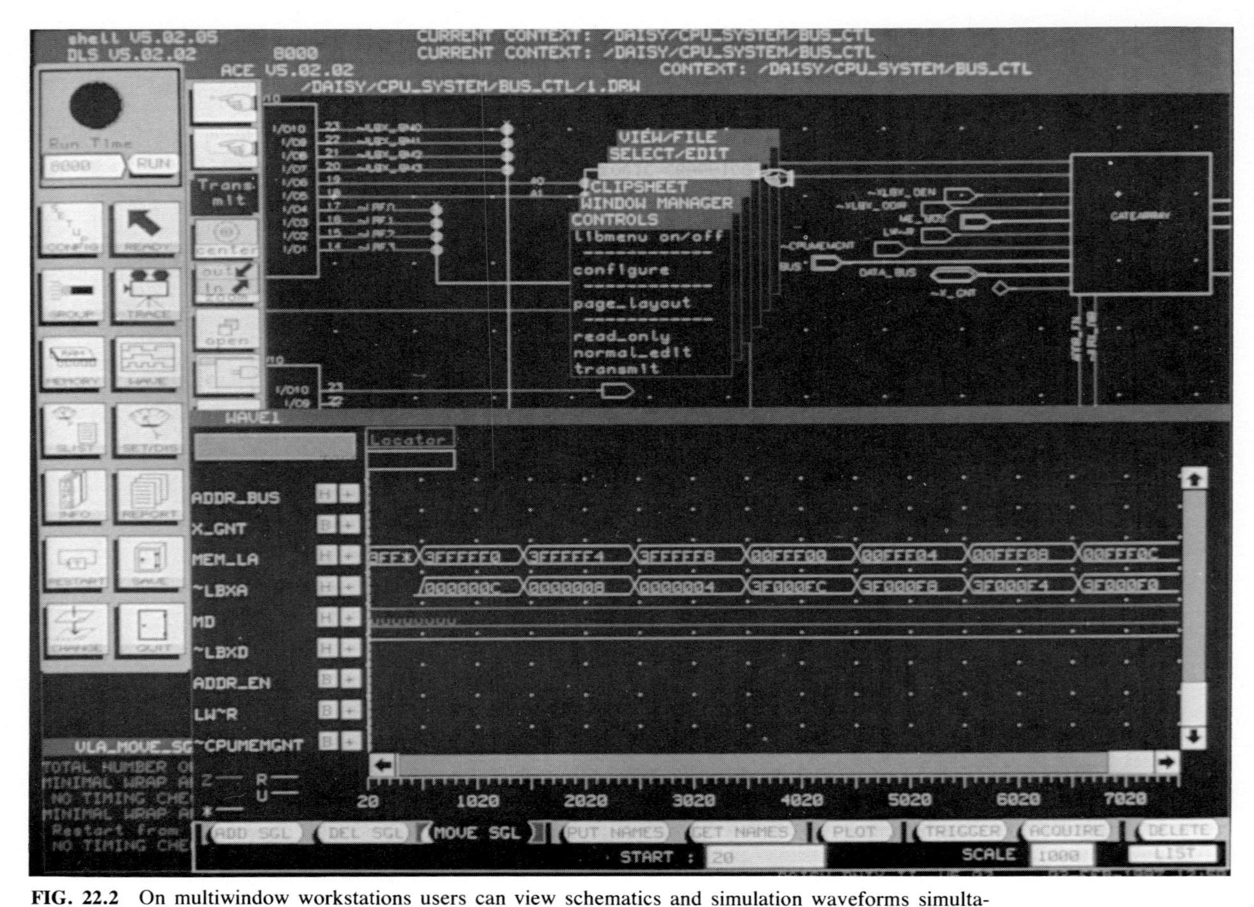

FIG. 22.2 On multiwindow workstations users can view schematics and simulation waveforms simultaneously, eliminating the need to ever refer to a netlist.

as intelligent repaint of only the areas in a design that have changed and clipping of an image to show only the parts of an object that should be visible from the user's point of view. The accelerator might also provide sophisticated window management functions.

Schematics do not generally demand this same level of graphics power because the image complexity is not usually as great and resolution is not as critical as for IC layout. Thus, the various CAE tasks differ in the graphics sophistication they require. Because high-resolution, high-speed graphics workstations cost more than less powerful systems, designers often want more than one workstation configuration to perform the steps in the design process. Each user wants a workstation that optimizes the graphics for the most frequently performed tasks. While systems with ever greater power are needed for increasingly complex VLSI layout chores, personal-computer-based systems have become popular for interactive tasks, such as schematic capture.

22.5 A TOOLBOX FOR VLSI DESIGN

In the environment provided by the database and a user interface implemented in a graphical format, a CAE system contains a variety of tools to perform the different steps in the design process (Fig. 22.3). The main tools handle functions such as

- Schematic capture
- Electrical rules checking
- Simulation
- Polygon editing
- Automated layout
- Design-rules checking

The schematic capture tool allows a designer to choose parts from a semicustom library and assemble them into a circuit. The library is a crucial element in this process because it contains all the raw data for schematic capture, simulation, and possibly layout. The electrical data can include schematic symbols, component-specific logic models, timing models, and electrical design rules. Physical (layout) data typically include base array or die information, physical design rules, placement-and-routing obstructions and the macro library (the phys-

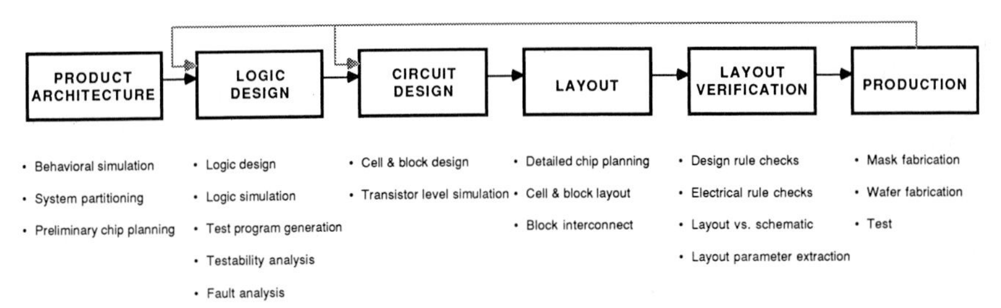

FIG. 22.3 The IC design process.

ical representation of the schematic symbol library). Libraries are produced through the cooperative efforts of both CAE firms and semicustom IC vendors. Established CAE companies typically make available long lists of IC libraries supported on the company's systems.

In schematic capture, a CAE system can do more than just automate the drafting process. Through hierarchical methods, the system can organize the efforts of a team of engineers. This is essential for creating complex VLSI designs, which are broken down into manageable blocks, with each block handled by a specialist. The blocks are then laced together into a single chip.

Hierarchical design systems implement this approach by giving views of the overall design at several levels. The highest level (behavioral) is a block diagram of the chip's major functions, and the lowest level contains the most detailed transistor-level circuit schematic. Blocks at the highest level can be divided into smaller blocks at lower levels, depending on the IC's complexity. A block at any level of the hierarchy can be separated and filled in independently, then dropped back into the total design; the high-level representation in the hierarchy defines how that block will connect to other blocks via the appropriate interface signals and buses.

In the actual process of creating or editing a schematic, the designer usually employs a mouse to move components and interconnects into position. Large schematics are usually divided into pages to limit the amount of data the user must deal with at once.

As part of the design entry process, the schematic's graphic form is compiled into an electrically meaningful netlist for use by other tools. Compilation links multiple pages together and usually includes a check of electrical design rules. For example, the electrical rules checker might provide a listing of intrapage electrical errors and ensure that all signals crossing schematic boundaries are used consistently according to their specifications.

After schematic capture, all or part of the design can be simulated, using logic and circuit (analog) simulators. There are several levels at which logic simulation can be performed: gate, functional, behavioral, and physical. (The latter applies primarily to board-level designs, in which the responses of an actual physical component can be substituted for a software model. This is useful in cases where the software model is not available or is too complex.)

A gate-level simulation maps each gate into one simulation element, while at the more advanced simulation levels (functional, behavioral, physical) the ratio between simulation primitives and gate equivalent models can easily extend to 10 to 1. The higher simulation levels thus provide a more efficient simulation approach—one that is compatible with the way many VLSI designs are specified. Completed blocks in an IC lend themselves to detailed gate-level simulation; blocks that have not been completed can be described using a functional or behavioral language. A mixed-mode simulator can simulate all these types of descriptions together to characterize an entire chip design at any stage of the design process.

Whatever form the input takes, the simulator output usually consists of waveforms that a user would see if he or she were using a logic analyzer to look at signals in a real chip. The simulated waveforms contain more data than the logic analyzer would provide, however, because of the large number of states monitored by the simulator. In addition to the usual high, low, and unknown states, a simulator must deal with how strong a logic level is to fully describe a chip's activity, especially for MOS designs. Strengths are generally classified in categories such as supply, driving, resistive, and high impedance. By combining the various strengths with the basic logic states, a logic simulator provides a good level of accuracy.

For evaluation of analog designs, as well as greater precision in simulating logic, CAE workstation users can turn to a circuit simulator. One of the most popular circuit simulators is SPICE, developed at the University of California at Berkeley. Many versions of this simulator have been produced commercially, some with a graphical user interface (Fig. 22.4), integration with other design tools, and improved algorithms that converge on a solution reliably.

SPICE is a powerful simulation package that allows users to employ several simulation approaches. These approaches fall into three main categories: dc analysis, frequency response, and transient analysis. Within these categories, SPICE can provide dc transfer curves, sensitivity analysis, noise analysis, distortion analysis, and Fourier analysis.

In addition to simulating analog and digital circuits, CAE systems often include related tools that perform testability analysis, fault analysis, and timing verification. Testability analysis examines the structure of a design to determine whether all the components can be controlled and observed from the I/O pins; this analysis is done without test vectors. In fault analysis, on the other hand, test vectors are checked against the circuit to determine if they are adequate for manufacturing test. During operation, the fault analyzer inserts errors in the circuit and then runs the flawed design using the test pattern as a stimulus to locate any defects. A report is generated showing any faults that could not be detected with the existing test vectors.

The timing verifier tests a design's timing constraints and analyzes the design over a range of operating and manufacturing variations. The worst-case check performed by this tool ensures that a chip will work reliably under a variety of conditions.

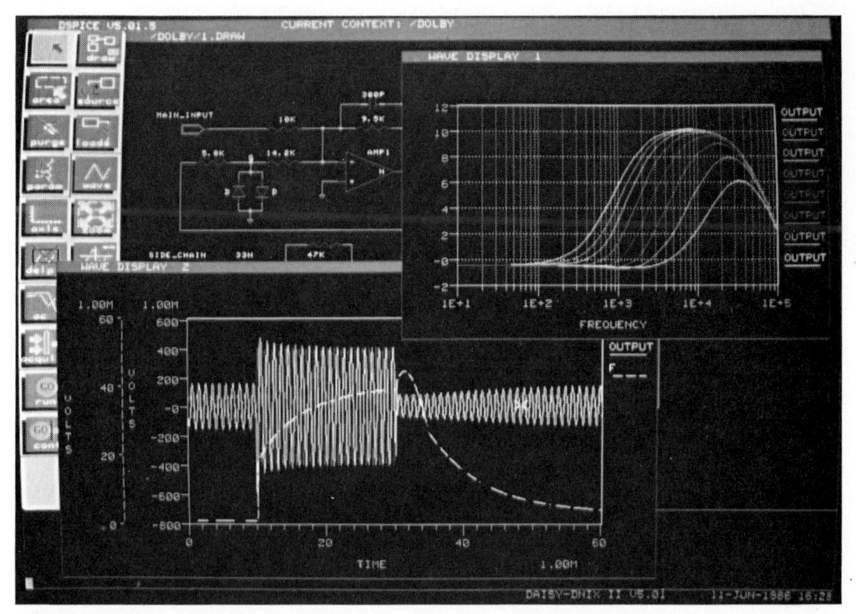

FIG. 22.4 Modern workstation-based circuit simulator interfaces simplify circuit debugging and running "what if" scenarios.

As mentioned in the section on CAE productivity, simulation is a crucial element in shortening the design cycle. By using simulation tools to verify a design, engineers can usually produce flawless, manufacturable, testable chips on the first pass. This eliminates all the work that usually goes into debugging prototypes and reworking layouts.

Simulation is important, but it can take a great deal of time on a general-purpose computer, even a powerful workstation or a minicomputer. The solution in many cases is to add a simulation accelerator. This is a special-purpose unit that uses dedicated hardware to perform the evaluations that would otherwise be done more slowly in software. When the hardware accelerator is compatible with the workstation's software simulator, users run simulations just as they would in software, only at much higher speeds. The accelerator thus reduces or eliminates dependence on mainframe computer simulation, an especially cost-effective solution for engineers who run many simulations. Another benefit is faster turnaround than is generally possible using mainframe-based tools.

22.6 AUTOMATION SPEEDS LAYOUT TASKS

Once a VLSI design has been created in schematic form and verified with simulation tools, it is time to convert the schematic into a layout (at least for those portions of a chip that begin in schematic form). While automated layout tools are getting better, most layouts are still created manually to obtain the best use of chip real estate in terms of flexibility and density. With today's polygon editors, however, the use of the term *manually* can be misleading because of the automated aids that support the layout process.

One of the layout tasks that a polygon editor can greatly simplify is the creation of complex geometries. Using functions such as Boolean operators, designers can transform simple shapes into any geometries imaginable. For example, Boolean operators can be used to subtract one shape from another and then add the result to other shapes to build a nonorthogonal polygon (Fig. 22.5).

Some aspects of the layout task, such as forming devices, can be automated. If the layout system knows how a device is formed, it can draw the appropriate polygon structure. The designer simply specifies the device's location, width, and length, and the layout system generates the details automatically.

Another aspect of layout involves array structures (RAMs and PLAs, for example) that do not usually originate fully formed in schematics; a schematic might cover only one cell in the array. Polygon editors that support the construction of stepped, random, and lattice arrays allow designers to build complex blocks.

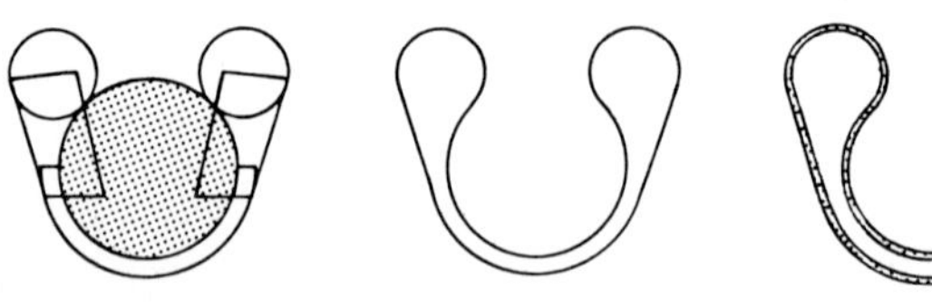

FIG. 22.5 Modern polygon editors provide sophisticated functions to accelerate creation of complex shapes.

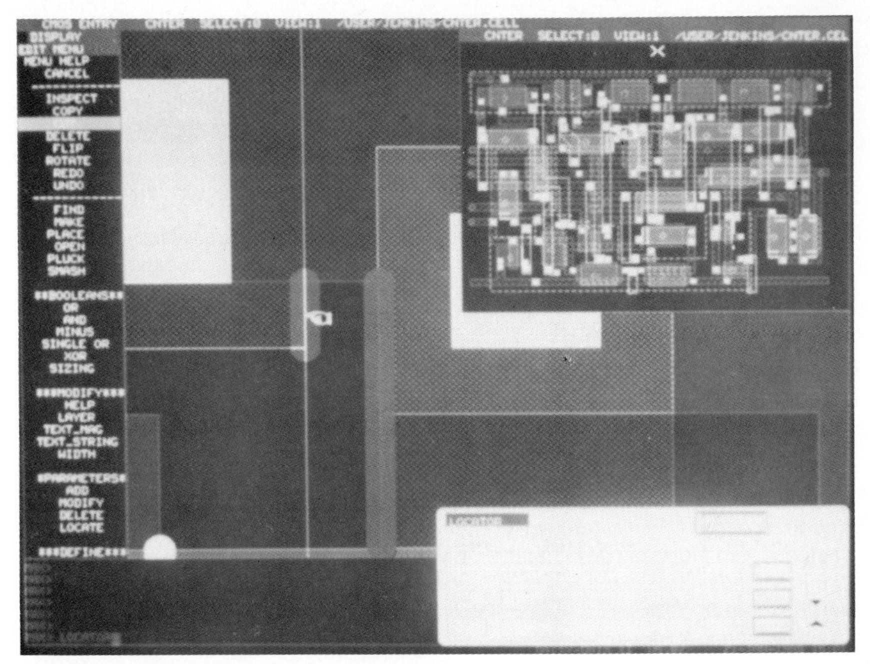

FIG. 22.6 Layout verifiers such as design-rule checkers and electrical rule checkers identify layout problems before masks are made. These typically run on-line on workstations to verify each cell as it is built, and batch on mainframes to verify complete chips.

Functions such as the ability to extract one cell from an array allow designers to easily create many circuit options.

After creating the layout, the designer can use the verification tools described earlier to make sure the layout adheres to appropriate design rules and is consistent with the schematic. Some layout systems automatically identify and display design-rule violations for users, replacing the traditional method of having engineers manually locate and decipher the cause of a design-rule violation (Fig. 22.6). Without the automatic location feature, it can be surprisingly difficult to determine precisely why a specific polygon is improperly formed and to keep track of which errors have been corrected. In simplifying this and other layout tasks, CAE/CAD tools have improved the speed and accuracy of chip design.

22.7 CAE/CAD MOVES INTO THE FUTURE

Several trends promise to shape CAE/CAD tools in years to come. Although it is difficult to predict exactly how these trends will affect the industry or what the time frame of the effects will be, it will be worthwhile for CAE/CAD users to note how these influences play out.

A trend that should prove significant in coming years is the move toward closer working relationships between workstations and more powerful computers. This is now happening with minicomputers and mainframes to some extent, and the entire CAE environment will probably become more heterogeneous with regard to computing platforms.

The bigger machines can be useful for CPU-intensive work (such as simulation, verification, and automated mask data generation), central file management, and high-volume disk and tape storage. Workstations, on the other hand, excel at handling interactive tasks. One of the goals in integrating workstations with larger computers should be to make the resources of the latter transparently available to users as needed.

Standardized networking is a must for this trend to continue. Network protocols are required in two areas: for improved data communication capabilities and for network-based program execution environments.

Ethernet has become a strong industry standard for data communication, but high-level protocols are also necessary. The Ethernet specifications describe how to move data bytes from one machine to another, but higher-level protocols describing network-based file systems (directories) are also needed. Transmission control protocol–internet protocol (TCP–IP) has been widely used in this regard.

To advance heterogeneous computer environments, we need standards that allow parts of CAE programs to execute on different computers. For example, an interactive workstation might be used to view and probe circuit simulation results that are generated on a number-crunching mainframe computer. The X-window standard, developed at MIT, and Sun Microsystems NEWS (network extensible windowing system) provide for device-independent graphics. Using these standards, different types of display devices can present graphics objects according to the devices' own capabilities.

The need to tie together many users with many types of resources is creating an increasing dependence on networking. This, in turn, will drive improvements in managing databases across a variety of platforms. More software will become available to track an engineer's work over time, creating an "audit trail" for a design. Although such capabilities will further increase the need for more disk space, the added functionality will be well worth the cost of higher-capacity disks (which continue to decline in price anyway).

Supercomputers and massively parallel processors offer another avenue of expansion for CAE. As computer scientists develop algorithms for the parallel machines that are appropriate for CAE work, CAE will probably make use of the power available, especially for tasks such as simulation, fault analysis, and IC layout verification. Just as CAE is moving to tap the power of conventional minicomputers and mainframes today, the ever expanding need for additional computer power will undoubtedly drive the industry toward the best available source of that power in the future.

Another trend that applies across all CAE functions is the application of artificial intelligence (AI) research. In developing and modifying the CAE tools themselves, AI-based software design techniques should allow CAE companies to more quickly respond to changing needs.

On the application side, expert systems should be especially useful in helping with layout tasks such as schematic partitioning, IC floor planning, and advanced routing. Whether layout is directly automated or an expert postprocessor helps pack layouts tighter, the AI influence here should be profound. This is due to the many choices involved in the place-and-route function. When routing power signals, for example, an expert layout system could automatically give these interconnects the special attention they require. The speed with which such techniques become available in CAE will depend to a large extent on the rate of progress in converting AI research into practical results.

Several trends are also influencing the development of simulation and design analysis. For instance, a greater use of mixed analog and digital simulation and better simulation models will make it easier to refine designs. In addition to determining whether a design works, tomorrow's simulators will allow designers to maximize performance. Improved analysis tools will also assist in this refinement process by giving engineers the ability to generate a variety of "what if" reports.

Testability will also improve, and there will be more self-testing devices—perhaps even devices that employ redundant circuitry to perform self-repair. These benefits will depend on the development of improved test methodologies and design analysis.

These and other trends will affect CAE tool development by CAE vendors as well as by in-house CAE/CAD departments. The latters' efforts will get more focused as broader and more versatile capabilities become available from CAE firms. However CAE techniques progress from here, there will be a great deal of attention on the latest developments. Companies engaged in VLSI design have come to recognize clearly that advanced CAE tools provide a profound advantage over competitors.

P • A • R • T • 7

VLSI PACKAGING

CHAPTER 23
ELECTRONIC PACKAGING AND IC PACKAGING PROCESSES

Daniel I. Amey
Du Pont Electronics
Wilmington, Delaware

23.1 ELECTRONIC PACKAGING

What is electronic packaging? Charles Harper[1] defines the field as "the conversion of electronic or electrical functions into optimized, produceable, electromechanical assemblies or packages." The key words are *optimized* and *produceable,* two of the most difficult tasks facing a packaging engineer. Harper goes on to define electronic packaging as an area that "requires an overlapping of disciplines and breadth of knowledge." The knowledge required is in the area of materials, components, processes, and interconnection techniques and devices; it requires the exercising of the mechanical, electrical, chemical, and thermal engineering disciplines to properly package an electronic circuit. Electronic packaging has suffered (although some may say advanced) from the rapid advancement in semiconductor technology. Semiconductor technology has far outpaced interconnection technology to the extent that interconnections have become the primary limit to circuit and system performance. To overcome these performance limitations, packaging density has continued to increase, and interconnection lengths have been reduced to minimize the effective length of the electrical propagation of signals. More and more interconnections are being placed on silicon to increase performance, reduce costs, increase packaging density, and increase reliability.

23.2 LEVELS OF INTERCONNECTION

The levels of interconnection have been defined to quantify the types of interconnections. Defined about 15 years ago,[2] level 1 is described as the connections within the component case: the wire bonds, thick-film interconnections, the metalized connections on a package. Level 2 connections are from the component to a printed circuit board: solder joints, socket terminals, printed circuit paths on the board. Level 3 connections are between printed circuit boards, from a printed circuit board to a wire, or to another printed circuit board. Interconnections such as wire-wrap, printed wiring, and other discrete wiring techniques are typical level 3 interconnections. Cabling is also used in level 3. Level 4 connections are from one internal chassis to another within an electronic equipment cabinet. Cabling and optical interconnections are typically used for level 4. Level 5 connections are from one piece of equipment to another; these are external connections typically using the same technologies as level 4 but with the means to protect them from those stresses that affect the signal interconnections as a result of the external connections (the environment, mechanical abuse, temperature, humidity, and electromagnetic radiation). These generally accepted definitions of the levels of interconnection are described in Fig. 23.1.

The technology has changed such that some are now referring to level 0 as the aluminum connections on silicon (as in wafer scale integration) and level 6 as the connections beyond equipment in local area network applications or large system interconnects. These definitions have yet to be generally adopted, but will most likely be added to the basic definitions.

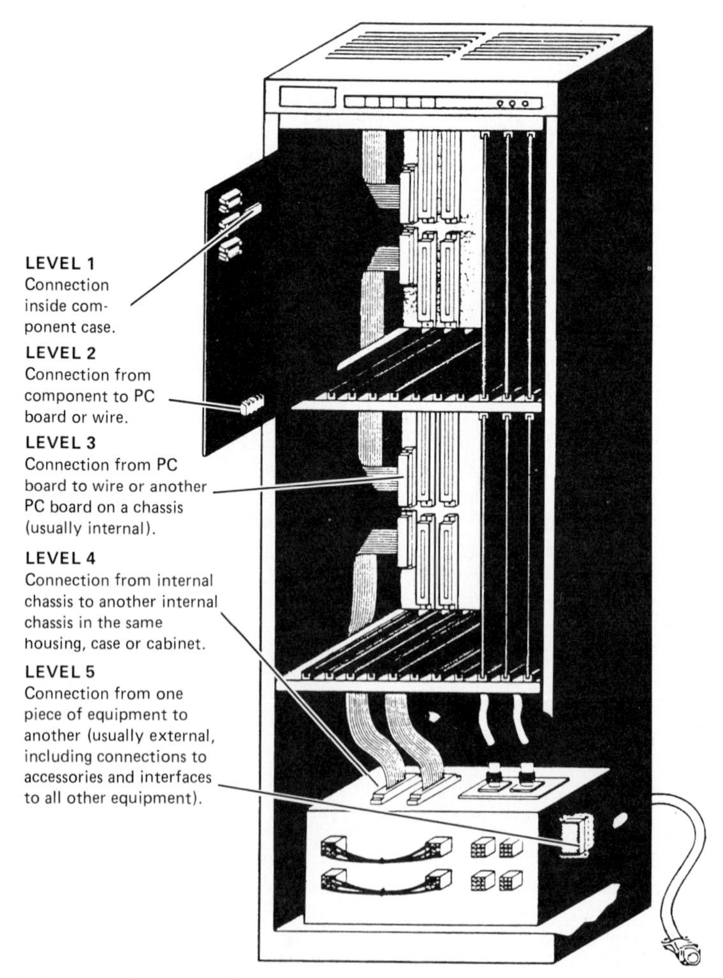
LEVEL 1
Connection inside component case.
LEVEL 2
Connection from component to PC board or wire.
LEVEL 3
Connection from PC board to wire or another PC board on a chassis (usually internal).
LEVEL 4
Connection from internal chassis to another internal chassis in the same housing, case or cabinet.
LEVEL 5
Connection from one piece of equipment to another (usually external, including connections to accessories and interfaces to all other equipment).

FIG. 23.1 The levels of interconnection.[3]

23.3 SEMICONDUCTOR PACKAGE TRANSITIONS

This chapter will primarily address level 1 and level 2 interconnections and the effect of the microelectronic connection on packaging systems. Interconnection and package technology have been driven by semiconductor technology. Figure 23.2 shows some of the component types that have been used in electronic systems.

In the 1950s and early 1960s discrete semiconductor devices and axial lead components such as transistors and resistors and capacitors with axial and radial lead terminals were predominant. As semiconductor technology improved, and more and more components could be put on a silicon chip, the number of pins on the TO-type semiconductor packages were not satisfactory. Circular TO5 packages with 10 and 12 leads in a circular pattern were used, but that was the pin limit for the package and the printed circuit interconnect technology at the time (about 1963). This resulted in the need for, and introduction of, the dual in-line package (DIP). The dual in-line package has served as the workhorse package for many years and will continue to be the primary semiconductor package for many, many years.

However, in the mid-1970s, again as the semiconductor technology advanced into MSI and LSI, more and more components and capabilities were on the silicon chip, but there were not enough pins or terminals to support the logic that could be placed on a silicon chip. This led to the need for different package styles: the large pin (or terminal) count packages for LSI and VLSI, such as chip carriers and pin-grid arrays, two package types that will be discussed in some depth. These styles are becoming more and more popular as terminal counts increase. Figure 23.3 shows the package transition, the types of interconnect tech-

FIG. 23.2 Typical electronic package components.

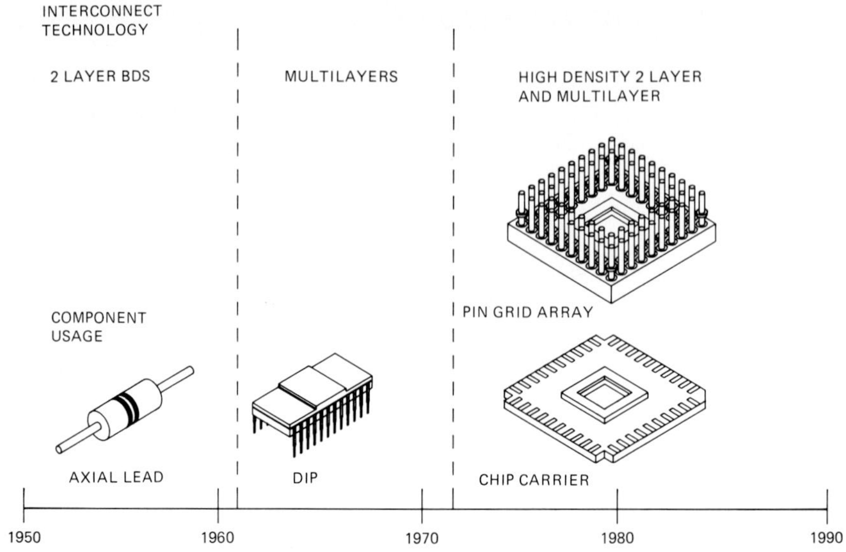

FIG. 23.3 Electronic component and interconnection technology transition with time.[5]

nology needed to support larger pin counts, and the fact that printed circuits and interconnections of these packages require greater density driven by the semiconductor package.

23.4 RENT'S RULE

Microelectronic packaging is also used to describe the level 1 and level 2 connections from the silicon chip to the outside world of packages and printed circuit boards. One of the driving forces of packaging technology can be seen in Fig. 23.4, where the increasing IC density (in both memory technology and microprocessors) has moved at a rapid rate.

Increased IC component density, as measured by the number components (logic gates) per chip, will continue in the foreseeable future. As a result, the number of package input and output terminals has significantly increased to where microprocessor terminals counts are between 200 and 300 and moving into the 300- to 400-terminal range. Projections for general-purpose logic show VLSI pin count continuing to increase in a linear fashion. Why is this? There is a relationship between the number of terminals and the number of logic gates in an electronic function. In the mid-1960s IBM studied their computer logic packages on printed circuit boards and developed a relationship called *Rent's rule*,[6] a relationship between the number of pins and the number of logic gates in the assembly. The relationship has a proportionality factor α multiplied by the number of gates raised to a power β (less than 1).

$$P = \alpha g^{\beta}$$

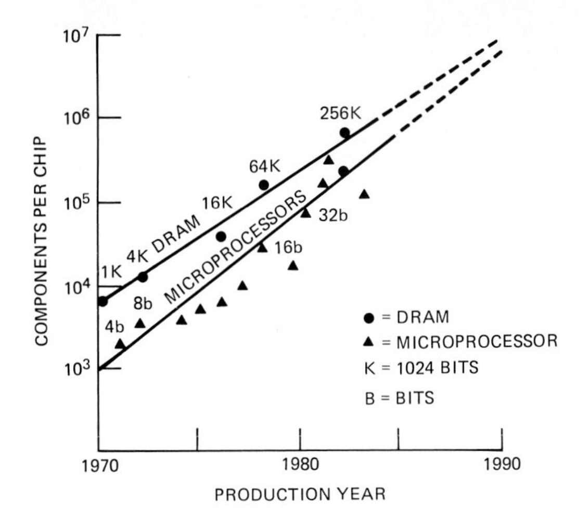

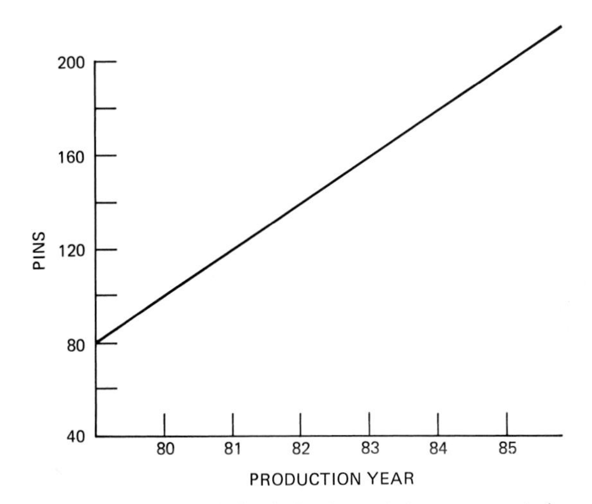

FIG. 23.4 Integrated circuit density and pin count trends.[4]

Rent's rule was empirically derived for random logic on PC boards but has been shown to apply to individual semiconductor circuits.[7] Bell Labs studies[8] have shown the Rent's relationship that best fits their system designs to be

$$P = 4.5g^{0.5}$$

This is the number of signal pins needed to support the logic gate count. Sperry Univac (now Unisys) studied some of their gate-array logic functions[7] and found the relationship that best fits their designs to be

$$P = 2.2g^{0.6}$$

The power pin relationship will vary with circuit technology. For the Bell Labs model the power pins required were 25 percent of the number of signal pins. Another rule of thumb is to make the power pins a function of the number of active outputs; one ground pin for every three switching outputs has been used for ECL circuitry. Other considerations are needed to determine the number of power pins. One must consider the current required for the logic assembly. It is also important with high-speed logic to consider the number of signal outputs. More power pins are needed for a larger number of outputs that may potentially switch simultaneously. The increased number of power pins reduce overall lead inductance, an important parameter for simultaneous switching to minimize noise and ensure signal integrity, reduce lead inductance, and increase performance. Additional power pins may also be used for impedance control, where the pins serve as an impedance reference for the signal pins to maintain a uniform impedance across the package or connector interface. Lastly, system noise immunity must be considered, and a sufficient number of power and ground pins must be provided to minimize power distribution losses and keep the noise levels within circuit system limits. All of these considerations result in an increasing number of pins for logic devices.

23.5 INTEGRATED CIRCUIT PROCESSING

Figure 23.5 depicts the integrated circuit manufacturing process from the raw silicon wafer through the mask and wafer processing steps, sometimes referred to as *front-end* processing. This chapter examines the *back-end* processing or packaging: wafer dicing (the cutting up of the wafer to create an individual die or silicon chip), die attachment or die bonding (the physical attachment of the die to the package for mechanical attachment and thermal integrity), wire or tape bonding (the interconnection from the die to the package), and sealing (the protection of the die from the environment).

Note the various points where test is performed. Wafers are tested in probe test, where each individual die in the wafer is tested to determine if it functions. Typically, only key dc characteristics are tested. Faulty circuits are dabbed with a spot of ink to indicate to the assembly equipment in subsequent steps that they are not functional. The next stage of testing is final test, after all of the packaging and assembly operations are complete. A great deal of value is added between these test steps so that packaging represents a significant portion of the overall cost of a semiconductor circuit. Figure 23.6 graphically depicts the back-end processes: the probe test, the dicing or the sawing of the wafer into individual dice or chips, the die bond, the wire bonding, and the sealing processes.

23.6 DIE ATTACHMENT

There are three primary ways to connect the die to the package: alloy or eutectic bonding, solder attachment, and adhesive bonding. Dice that have been separated from the wafer may be in waffle pack form (individual cases) for manual bonding to a package or hybrid circuit. Dice may also be mounted on a releasable plastic substrate that keeps the dice in the uniform position (as they are in wafer

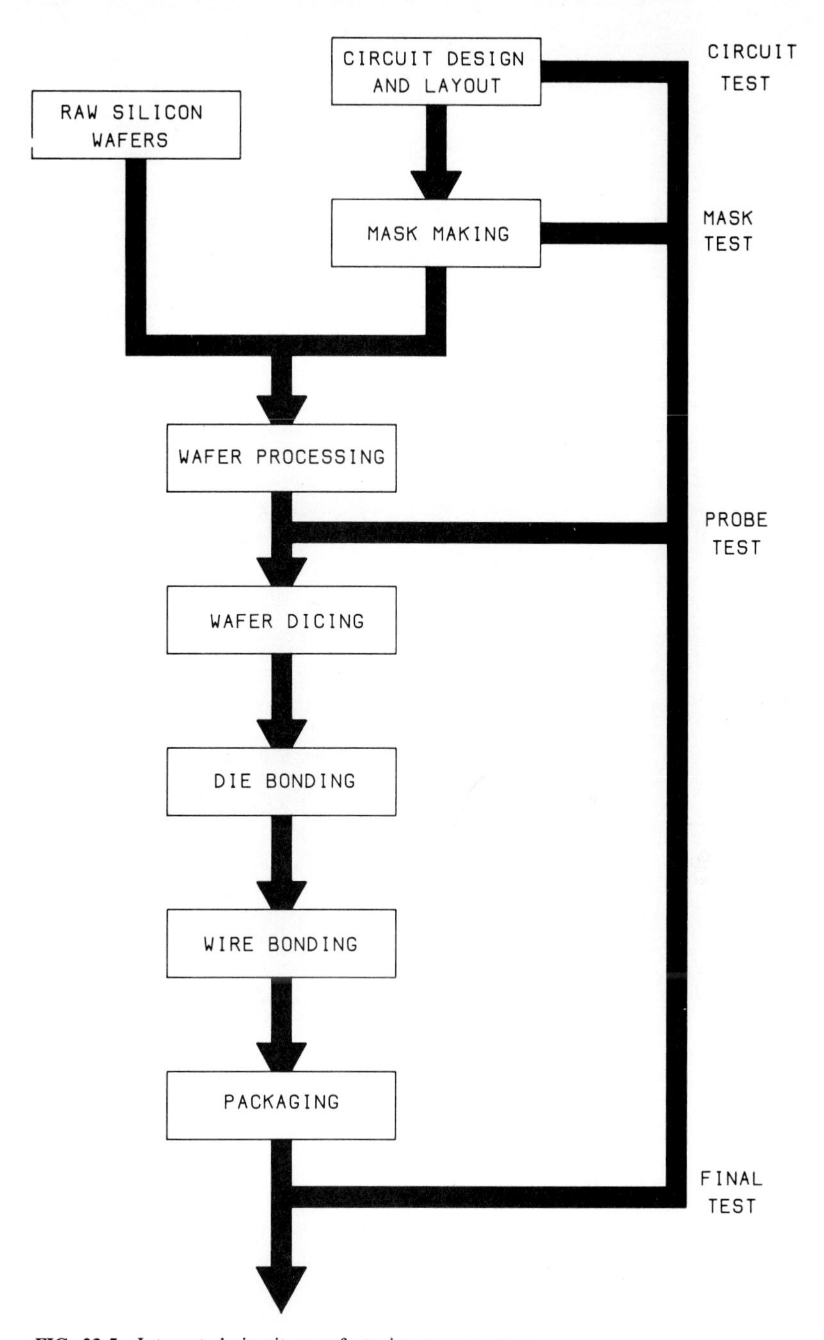

FIG. 23.5 Integrated circuit manufacturing process steps.

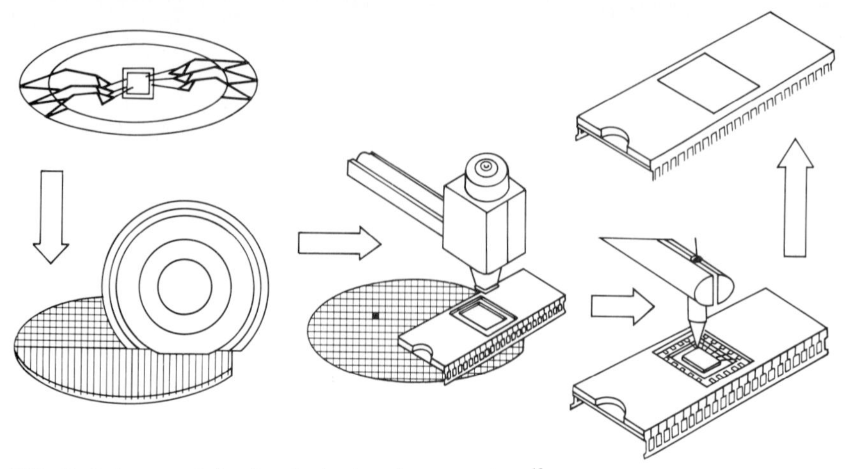

FIG. 23.6 Integrated circuit wafer back-end process steps.[10]

FIG. 23.7 Integrated circuit die attach equipment.[9]

form) after the dicing operations for attachment with automatic equipment. Figure 23.7 shows the die attach head for a typical pick-and-place operation. A heated vacuum tip picks up the die and moves it to the package, which is on a heated platen.

In alloy attachment the solder or alloy material is a preform wafer about 1 mil thick, placed between the package and the die. The backside of the die is typically metalized with gold, and gold is on the package cavity base. The assembly is heated to about 300°C range, where the solder flows and attaches the die to the package. Epoxies are also used and have been widely used in hybrid circuits and multichip modules. In the use of epoxies there are organics present in the package that may be of concern for some applications where contamination may result. The epoxies are typically one-part materials and may be filled with metallic particles for good thermal and electrical conductivities.

Criteria have been established for the amount of material around the perimeter of the die. Typically, military packaging requires a good fillet between three sides of the die to the package base. This does not, however, give an indication of the overall integrity of die attach, which is important when one has to consider high thermal dissipation. Figure 23.8 shows an epoxy-bonded die on a hybrid substrate. Note on the perimeter of the die the circular area of epoxy that has flowed out around the sides of the die to attach the die to the package.

X-rays have been used as one means to determine the integrity of the die attach. Figure 23.9*a* is an x-ray that shows a die attach that has about three quar-

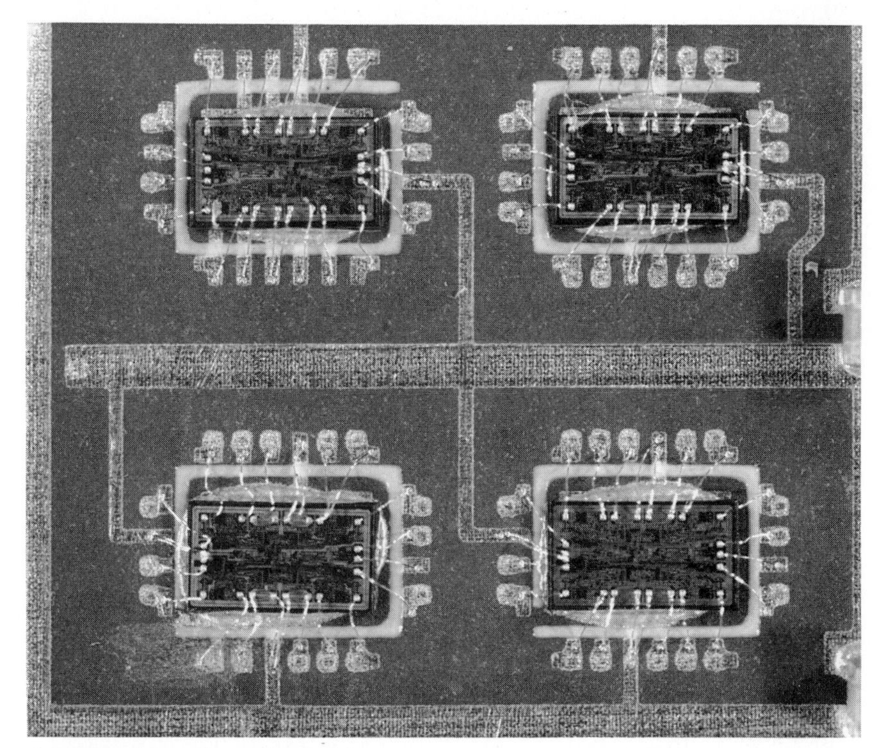

FIG. 23.8 Epoxy die attach material.

(a)

(b)

FIG. 23.9 X-ray photographs of (*a*) partially attached die and (*b*) fully attached die.

ters of the area attached to the package base. Note the voids in that area. Figure 23.9*b* shows a higher percentage of die attach, but there are still voids. This can affect the thermal properties of an integrated circuit and a parameter known as the thermal resistance.

23.7 THERMAL RESISTANCE

Thermal resistance is defined as

$$\theta_{JA} = \frac{T_1 - T_2}{P}$$

where T_1 = average die junction temperature
T_2 = ambient air temperature
P = dissipated power
θ_{JA} = junction-to-air thermal resistance, °C/W

Junction-to-air thermal resistance comprises the individual thermal resistances of the elements that are in the thermal path of the package:

$$\theta_{JA} = \theta_D + \theta_{J1} + \theta_P + \theta_{J2} + \theta_H + \theta_{J3}$$

where θ_{JA} = junction-to-ambient thermal resistance of the test assembly
θ_D = thermal resistance of the die
θ_{J1} = thermal resistance of the die-to-package interface
θ_P = thermal resistance of the package material
θ_{J2} = thermal resistance of the package-to-heat-sink interface
θ_H = thermal resistance of the heat sink
θ_{J3} = thermal resistance of the heat sink to ambient air (film resistance)

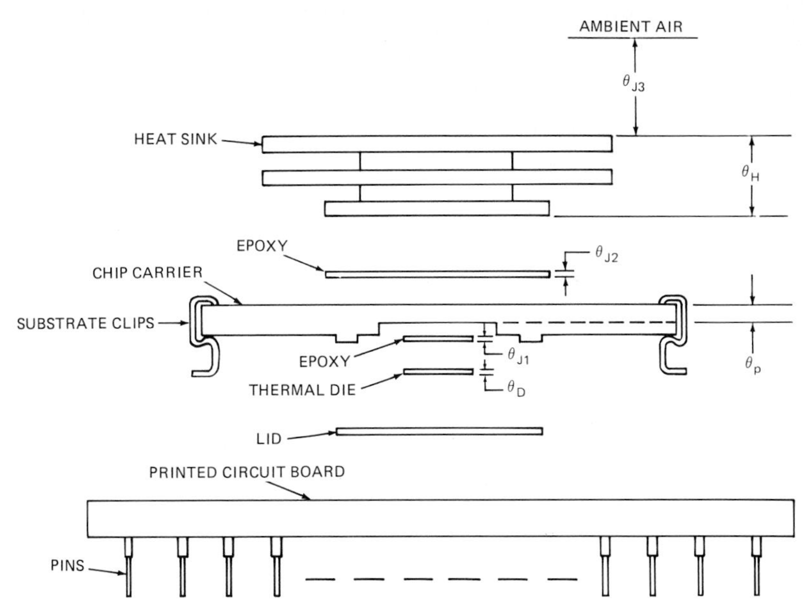

FIG. 23.10 IC package cross section and component thermal resistances.[11]

θ_{JA} is the summation of the thermal resistances of the package, the die, and the materials used to attach the die to the package. If a heat sink is present, the material used to attach the heat sink to the package, the thermal resistance of the heat sink, and the effect of airflow over the package are the factors that ultimately remove the heat to the airstream. This is shown in Fig. 23.10, a cross-sectional exploded view of a board-mounted chip carrier package and its thermal resistances.

The thermal resistance from the junction to the case or exterior of the package is called θ_{JC}. This is a common reference point for an integrated circuit manufacturer to specify thermal performance. Heat sinks and the means to remove heat external to the package are functions of individual designs and are difficult for an IC manufacturer to specify. θ_{JC} is typically specified by semiconductor manufacturers for their package design. The thermal resistance of DIP packages ranges from 50 to 200°C/W, depending on materials and construction. Each package type has a unique thermal resistance. Figure 23.11 shows the thermal resistance of a 68 I/O chip carrier. Note that the effective thermal resistance θ_{JA} varies with the airflow (air velocity is a measure of the amount of air over the package measured in linear feet per minute, LFPM). For these chip carriers θ_{JA} is about 25°C/W in still air. At 1000 LFPM the thermal resistance is about 12°C/W.

To determine thermal performance, one must consider the package, not the system environment. A piece of equipment may have a maximum operating temperature of 40°C, as shown in the examples in Fig. 23.12. The outlet temperature of the equipment may also be limited to 60°C for personnel safety and to control the maximum temperature of the devices within the equipment. Those ICs at the top of the equipment will be subject to the heat from devices beneath them and experience a 60°C maximum ambient temperature. If a package is dissipating 5 W (not unusual with today's VLSI circuits) with θ_{JA} = 12°C/W, the temperature difference from the junction to the airstream will be 60°C (5 W × 12°C/W). The 60°C

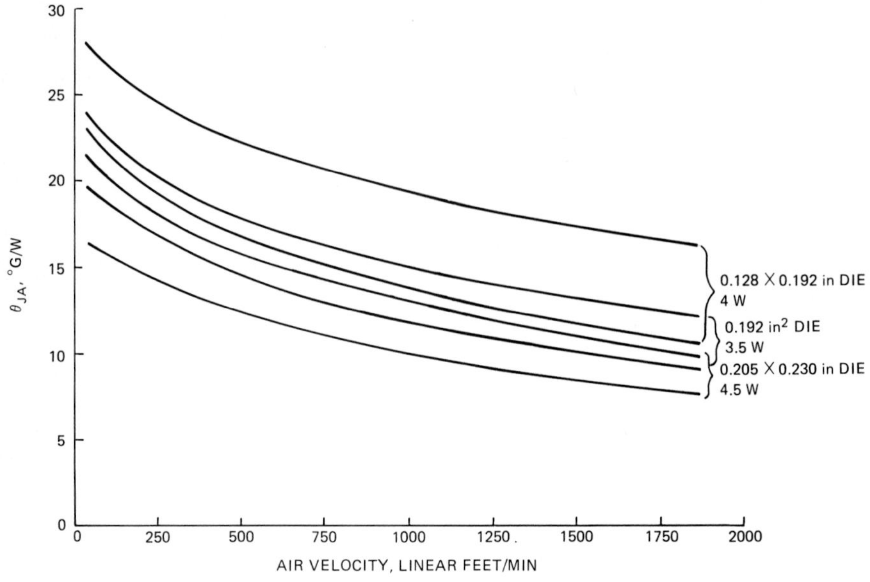

FIG. 23.11 Typical thermal resistance of a 68-terminal leadless chip carrier.[12]

air cooling the package is added to that of the junction temperature rise, resulting in a junction temperature of 120°C.

It is best to minimize the die operating temperatures for improved reliability and performance. Electrical parameters will degrade, and nonuniformity of the thermal environment will cause wide differences in the electrical characteristics (i.e., threshold level). It is best to have temperature differences minimized. Thermal performance of packages is a major concern with new high-speed circuits, and the die attach has a primary effect on the thermal resistance of an integrated circuit package.

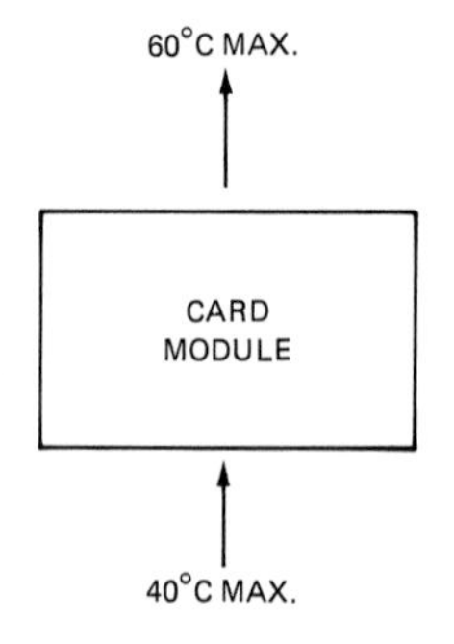

PACKAGE $\Delta T = P_D \times \theta_{JA}$ = 5 WATT × 12°C/WATT = 60°C
PACKAGE AMBIENT + 60°C
JUNCTION TEMPERATURE 120°C

FIG. 23.12 Maximum junction temperature example.

23.8 CAVITY-UP

Another important definition relating to thermal performance is the terminology *cavity-up* and *cavity-down,* which refer to the orientation of the package cavity. Most DIP packages are in a cavity-up configuration. Chip carriers offer the opportunity for mounting in two orientations because they are leadless and can be metalized to allow them to be mounted with the die cavity adjacent to or away from the mounting surface. For air-cooled systems the cavity-down configuration, where the die cavity is down and adjacent to the mounting surface, was developed. This arrangement allows the primary heat-dissipating surface, the back side of the die and the back of the package, to have direct conduction of heat from the die surface through the die through the back of the package to the heat sink, as shown in Fig. 23.10.

The cavity-down chip carrier construction has been used extensively in air-cooled systems. The cavity-up construction is still quite popular for indirect cooling, such as cold bars or surface pads, to remove the heat through the base of the package into the mounting substrate or printed circuit board. Packages now are becoming more and more complex to where they have been designed for 10-W dissipation for a single integrated circuit die. The package shown in Fig. 23.13 uses beryllia ceramic materials to dissipate the heat, multilayer alumina ceramic for signals, and special heat sinks. Thermal performance has always been important in electronic packaging, but it is now essential to consider thermal performance in the earliest stages of systems design for successful electronic packaging.

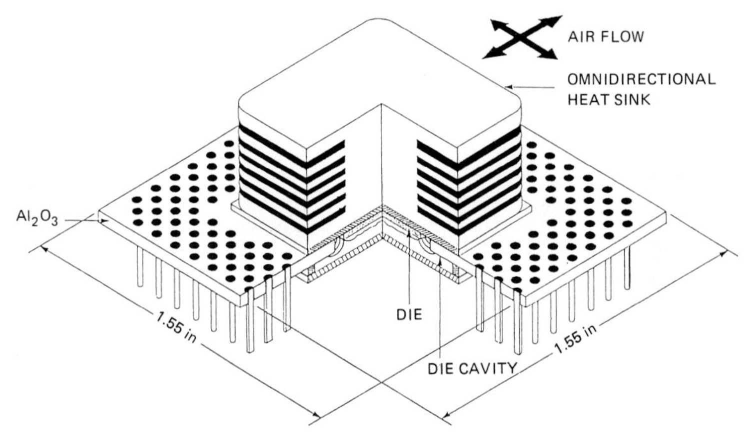

FIG. 23.13 A high-performance thermally enhanced pin-grid-array package.[13]

23.9 BONDING

Earlier we examined the increase in logic components, increasing pin counts, and the number of inputs and outputs. The number of inputs and outputs affects the overall die size. As the number of terminals or pads increases, the spacing between them must decrease if one is to maintain die sizes that are practical for economic production. Reducing the pitch or spacing of terminals to minimize die sizes for large-terminal-count devices has an effect on the bonding process.

Figure 23.14 shows four types of bonding: wire bonding, flip-chip (also called solder bump or the C4 process—controlled collapse chip connection), beam lead, and tape automated bonding (TAB). The left of the figure shows the orientation of the die. In wire bonding the die is in a "face-up" orientation. The input-output terminals (and active surface of the die) are up and away from the package mounting surface. The solder bump technology has a "face-down" orientation, where the active surface with its plated bumps is down and adjacent to the mounting surface. A solder-bumped die is attached by reflow soldering to interconnect the die to the package. IBM developed the technology for automated assembly, where it is used in high volume throughout their product line.

The beam lead technology was developed by Bell Labs as their means of automated interconnection. It uses a relatively thick gold-plated lead that extends over the surface of the die for thermocompression bonding to a substrate. This is also a face-down technology. It is inefficient in the use of silicon area and was not widely used but was quite suitable for AT&T when developed. Tape automated bonding technology is primarily a face-up technology with bumps plated on the die terminal pads. The bumps will be thermocompression-bonded to copper leads supported by a film carrier.

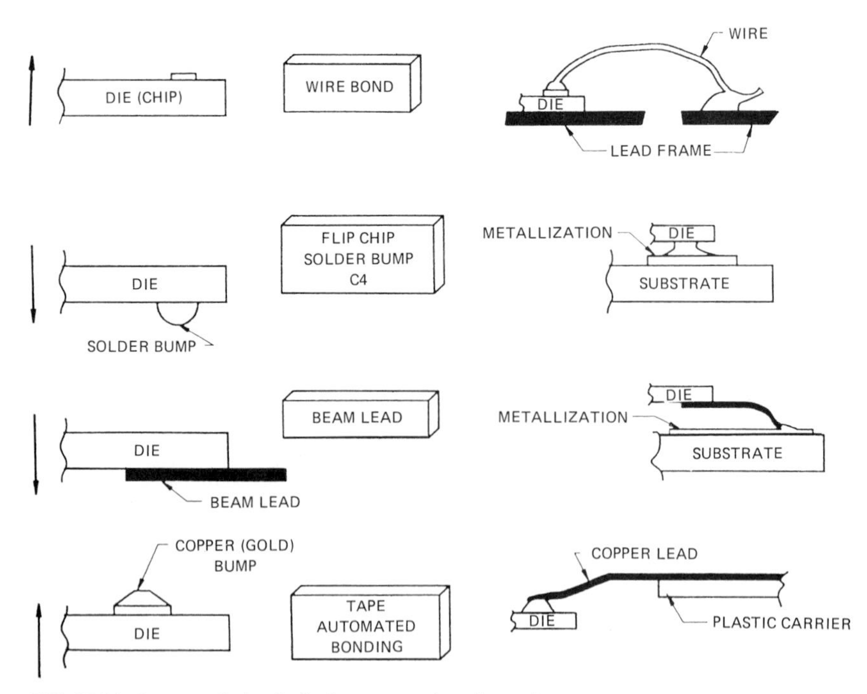

FIG. 23.14 Integrated circuit die interconnection alternatives.

23.9.1 Wire Bonding

Figure 23.15 shows the basic steps of wire bonding. Thin gold or aluminum wires about 1 mil in diameter are bonded to connect from the die input-output terminals to the package, lead frame, or substrate metallization. Gold ball bonding is used in hybrid electronics and high-reliability circuit interconnections. A gold ball is formed on the end of the 1-mil-diameter gold wire. It is then placed on the die surface and, with the application of heat and pressure, is attached to the input-output terminal pads on the die; the capillary or tip holding the wire moves above the package lead frame (or the package terminal) and with the application of heat and pressure forms a bond. The tip moves, the wire is broken from the bond, a

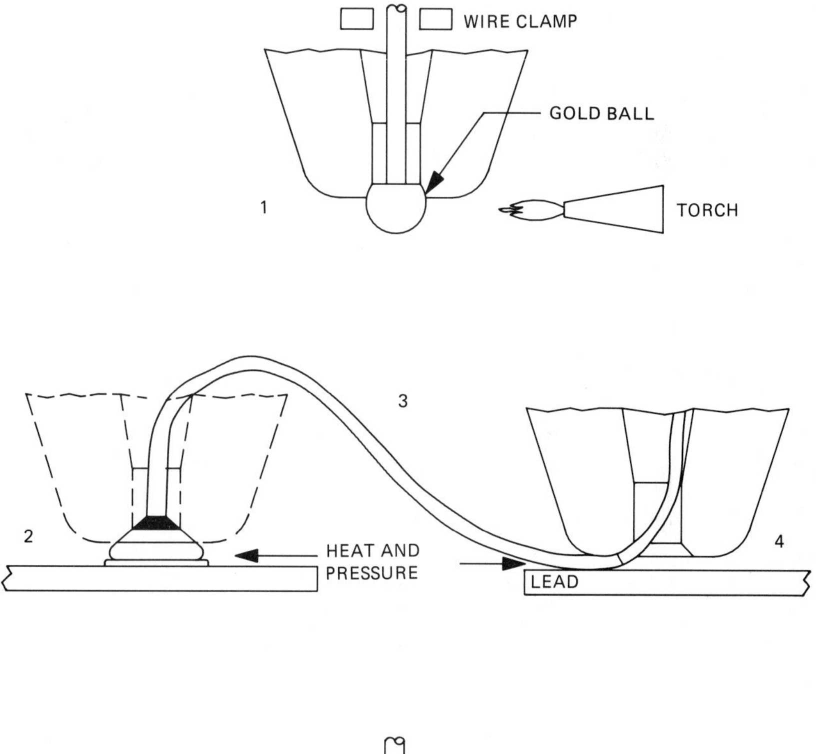

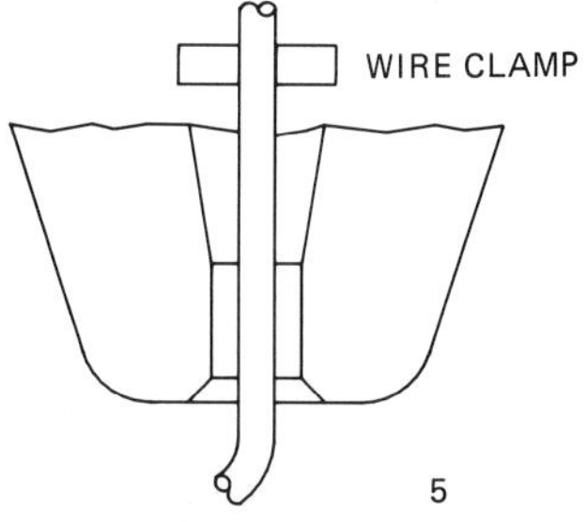

FIG. 23.15 Typical gold ball wire-bonding process steps.

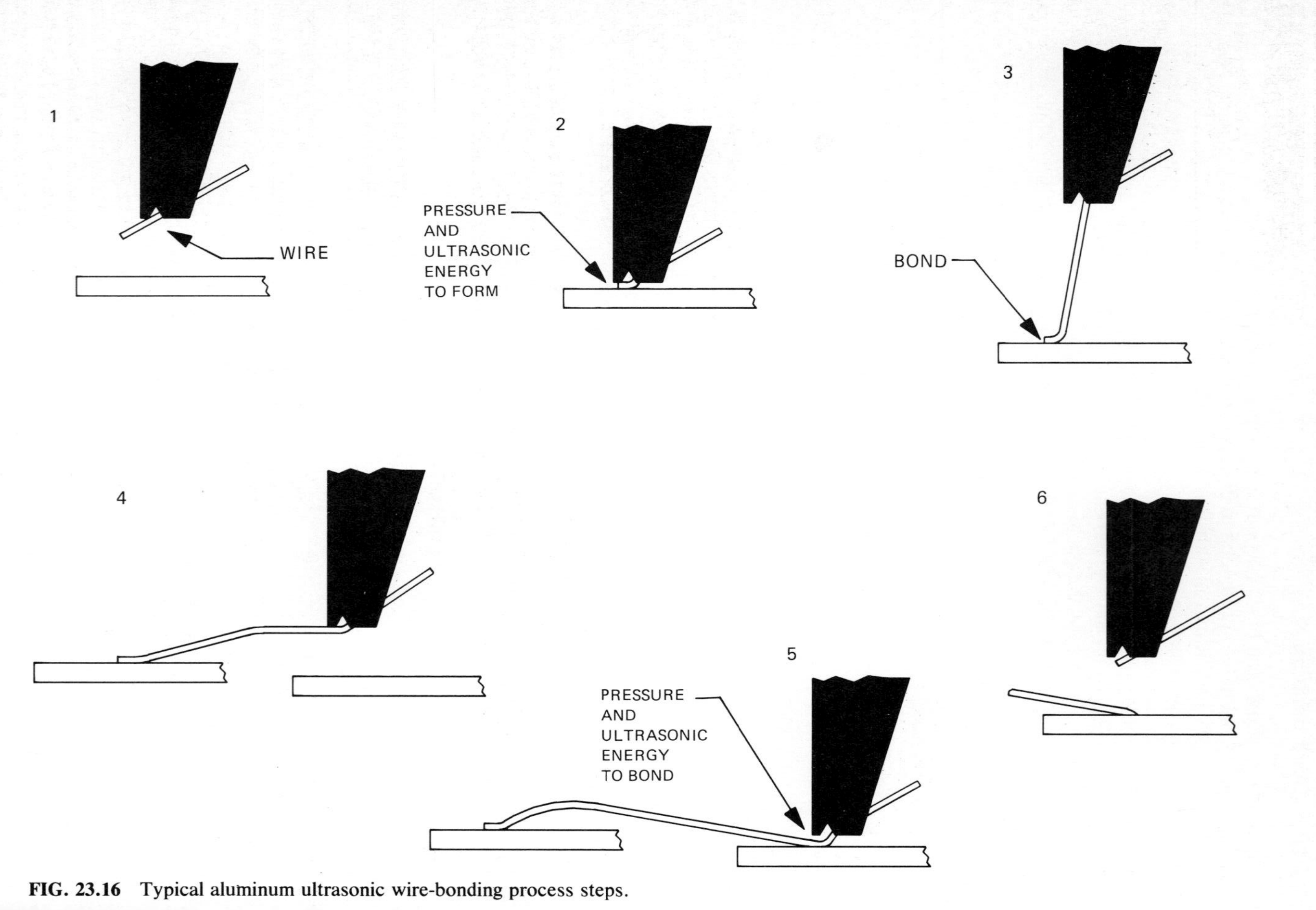

FIG. 23.16 Typical aluminum ultrasonic wire-bonding process steps.

ball is formed and moves to the next die pad. This is a sequential operation forming one bond at a time. Automatic bonding machines bond at a rate of five to seven bonds per second, a significant increase over manual operations.

Ultrasonic bonding is typically used with aluminum wire. Figure 23.16 shows the process steps. The wire is placed on the die, heat and pressure are used to form the bond, and, in this case, pressure and ultrasonic energy bond the wire to the die. The head raises and pulls the wire over and above the terminal pad. Pressure and ultrasonic energy bond the wire to the terminal pad on the package. The head raises, the wire breaks, and the head moves on to the next bond pad. Wire bonding is the most popular interconnect technology used in the merchant market. Figure 23.17 shows the typical geometry of package terminals on about 20-mil centerline spacing and the terminals on the die on about 8-mil spacing. The package bond shelf or platform for bond termination is typically 30 to 35 mils wide for tool access and ease of viewing in manual bonding.

Table 23.1 shows some of the wire sizes. The most popular use 1.0- and 1.3-mil-diameter gold and aluminum wire. Note the length restrictions. The maximum wire lengths are a function of the ability of the wire to support itself, to maintain its integrity under shock and vibration, and to not short to adjacent wires. Also note the resistance values. Significant resistance (0.5 to 0.8 Ω are not unusual) can exist in the signal wire bonds. This can affect electrical performance for both current-carrying capabilities and signal propagation.

The maximum length restriction causes the package cavity size to vary based on die size. A variety of integrated circuit packages exist, which, from all exter-

FIG. 23.17 Integrated circuit package terminal spacing.

TABLE 23.1 Wire Bond Mechanical and Performance Parameters

Material	Wire diameter	Min. wire length	Max. wire length	Resistance per foot, Ω	Max. current, A
Au	0.0010	0.040	0.080	13.5–14.4	0.200
Au	0.0013	0.040	0.110	8.03–8.53	0.500
Al/Si	0.0010	0.020	0.120	17.7–19.5	0.125
Al/Si	0.0013	0.020	0.150	1.3–12.5	0.250

Parameter	Wire bond*		TAB lead†
	Alum.	Gold	Copper
Electrical properties			
Lead resistance, Ω	0.142	0.122	0.017
Lead to lead capacitance, (pF 0.008″ spacing)	0.025	0.025	0.006
Lead inductance, nH	2.621	2.621	2.10
Thermal resistance			
Lead conduction, °C/mW	79.6	51.6	8.3
Lead convection (free), °C/mW	336.5	336.5	149.5

* 0.001-in dia. by 0.10-in long wire bond.
† 0.001-in by 0.004-in by 0.100-in long tab lead.

nal appearances, are the same; however, the internal details to accommodate different die sizes make them unique components that the semiconductor manufacturers must stock. Figure 23.18 shows a typical package cavity. Note that some of the interconnect metallization is wider than others to obtain lower resistance. For power connections multiple wire bonds are used to minimize resistance and increase the current-carrying capacity. If these exceed acceptable limits, a smaller cavity size must be used.

As terminal counts increase and die sizes become larger, one must be concerned with package terminal spacing. One way in which wire bond length is minimized is the use of a double-bond shelf. Figure 23.19 shows a double-bond shelf or two-level bonding. In the earlier example we saw a single bond shelf for one level of wire bonding. Large-terminal-count double-bond-shelf packages have the advantages of wider spacing on the wire-bond interconnections and a smaller cavity size, which minimizes wire-bond lengths. More and more large-terminal-count packages will have this feature because of the terminal counts and the wire-bond limitations. Smaller chip carrier packages have been designed with a dual bond shelf not for density or length, but to allow one package type to be used in high volume to accommodate a variety of die types where different wire-bond connections establish the pinout for each type.

23.9.2 Solder Bumps

Of the total number of silicon interconnects used worldwide there are probably more solder bump terminations than any other type. Figure 23.20 shows the IBM

FIG. 23.18 Typical ceramic integrated circuit package cavity.

solder bump or C4 process. IBM developed the technology in the early 1960s and has perfected this automated mass interconnection process. It results in short length and is more efficient because the full area of the chip can be used for input-output. This is not practical with wire bonding because the pressures that are involved with the bonding process can change device characteristics if placed beneath input-output pads. With wire bonding there are no active circuits beneath the interconnect area. Area solder bumps are also more efficient, for all bonds are made in one process step as opposed to serial one-at-a-time termination with wire bonds. The process has not been widely used outside of IBM, for the process controls are quite stringent.

23.9.3 Beam Leads

Beam lead technology was widely used within AT&T. Developed in the 1960s by Bell Labs, it was another approach to automated die interconnection. The process was also used in some military and high-reliability programs. Beam leads are

FIG. 23.19 A two-level wire bond ceramic integrated circuit package cavity.

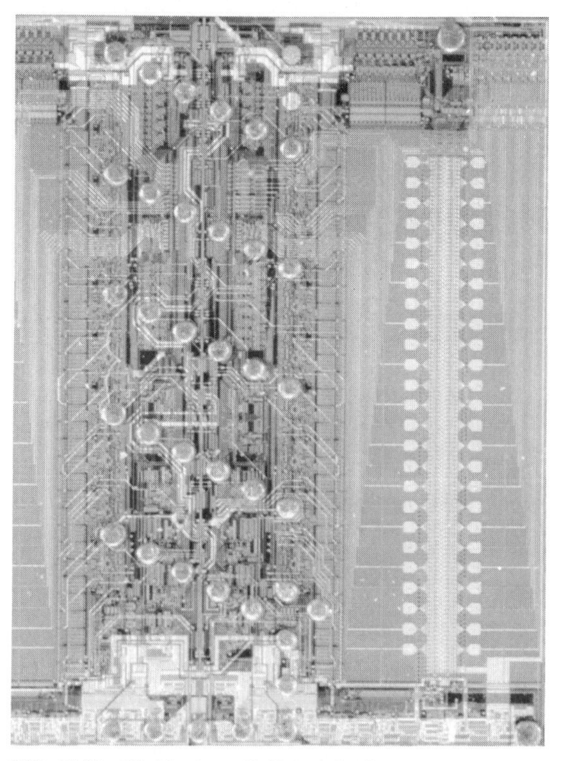

FIG. 23.20 "Solder bump" die terminations.

not used in the marketplace nor at this time within AT&T, but in its time it was an effective mass bonding process that avoided the serial one-at-a-time wire-bonding steps.

23.9.4 Tape Automated Bonding

In the early 1970s TAB was developed for automatic interconnection. Figure 23.21 shows the overall construction and basic process steps.[15] Initially developed as a high-volume mass termination and production technique for small-pin-count "jelly bean" (high-volume) circuits with 14 to 16 terminals, it is a technology that holds great promise for use with large-terminal-count dies and for high-performance circuits. There are active developments within the industry applying TAB technology to solve VLSI interconnect problems.

The TAB process places a die on a tape or plastic film with an etched copper pattern, automatically taking it from the wafer and placing it on the tape. Automated equipment makes the connections from the die to the tape, a process called *inner-lead bonding*. An advantage of this process is that the integrated circuits may be tested or burned-in (subjected to temperature extremes that will cause the circuits that are not as reliable or prone to failure to fail) prior to the expensive packaging and assembly steps. The outer-lead bonding process applies the tape-carried integrated circuits from the inner-lead bonding process to the package or a multichip substrate, as would be done with hybrid technology.

Kapton (polyimide) base material is typically used for the tape to withstand the process temperatures. The Kapton tapes are prepunched and laminated with copper foils 1.0 to 1.4 mils thick. (Copper foils may also have the plastic material cast on them to make the tape.) The copper-clad tape is etched to form the lead features. Note the cinematic format of the tape (initially 35-mm film format was used), where the sprockets on the film provide the means for the indexing and alignment. To use TAB an additional wafer plating step is necessary which is not required for wire bonding. This plating or bumping of the wafer in Fig. 23.22 shows bumps plated on the terminal pads. This system was widely used in high-volume production.[16]

Some dice are fabricated with copper and gold bumps. The plated bumps are about 4 mils square and about 1 mil high. Bumped wafers are mounted onto a plastic carrier with a heat-releasable adhesive. The wafer is diced or sawn so that the individual dice are separated but still precisely oriented as fabricated in wafer form. Thus the dice can be individually removed automatically and attached with the proper process and equipment.

Inner-Lead Bonding. The inner-lead bonding process uses tape unique to the die to be mounted and an adhesive-mounted diced wafer and semiautomatically attaches the die to the tape, resulting in a precisely mounted die in tape form. To keep the die from being damaged, a plastic tape with scalloped edges separates the devices. The process steps are shown in Fig. 23.23.

A heated thermode is used on the equipment, and the mounted wafer is on an *xy* table. It is positioned so that the thermode is aligned above the tape. The bad dice are marked from prior testing so that only the good dice are assembled. The thermode comes down and applies heat and pressure to connect the copper lead to the bump on the die. The thermode is raised, the table goes down, and the die releases from the adhesive and is attached to the tape. The tape is fed, placing the next open area on the tape beneath the thermode, the next good die is positioned

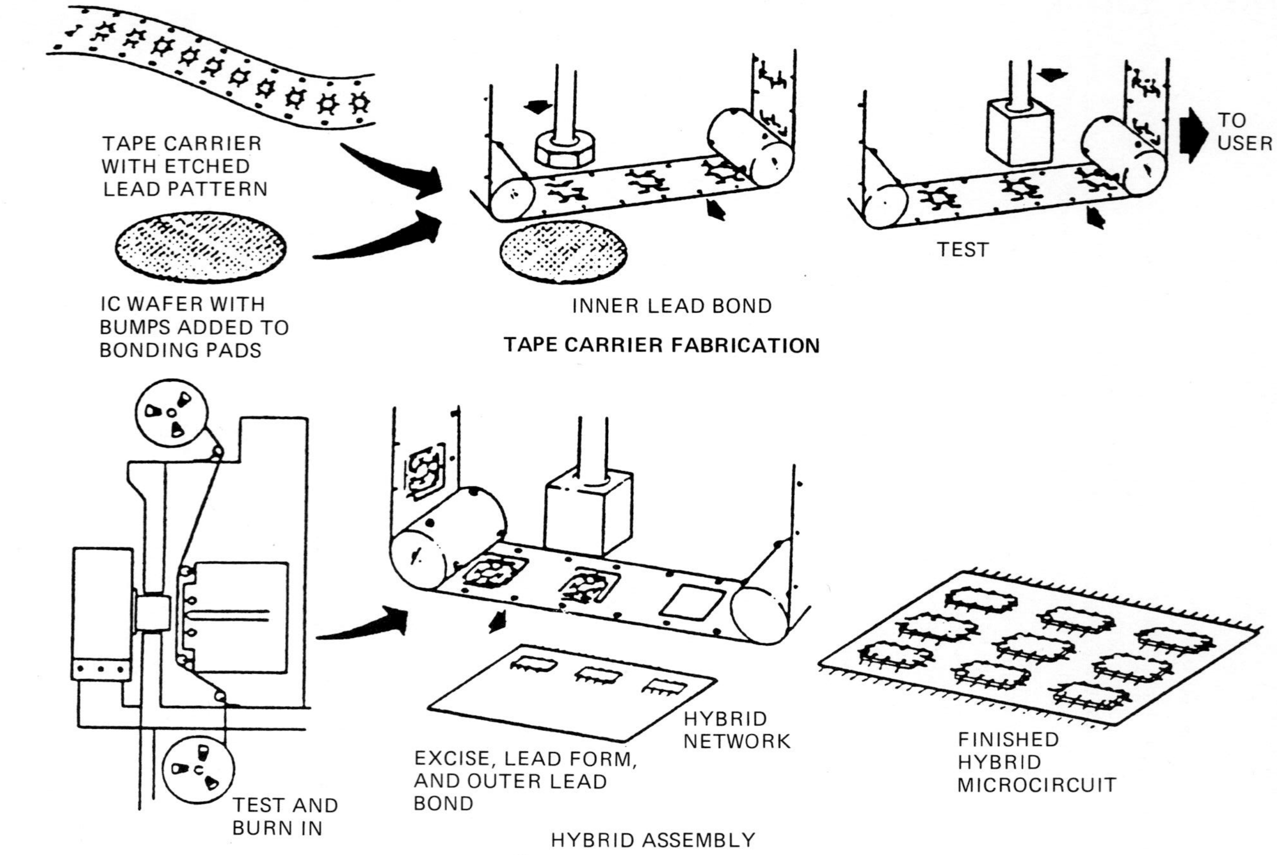

FIG. 23.21 Tape automated bonding process steps.[14]

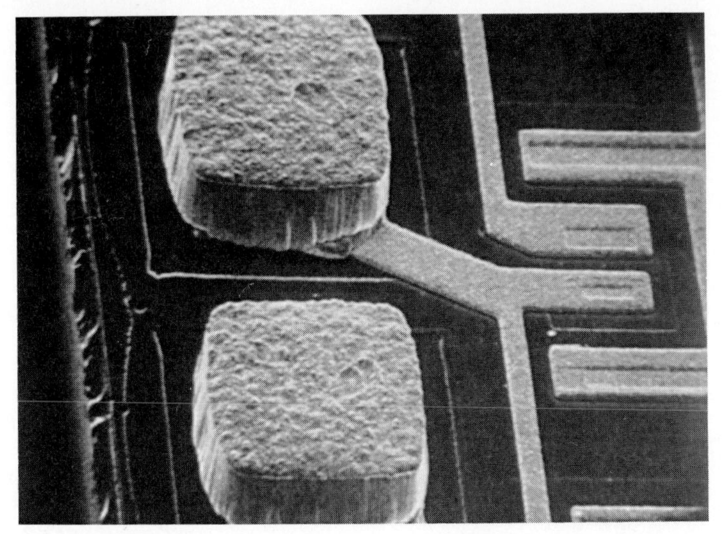

FIG. 23.22 A plated bump for tape automated bonding. (*Courtesy of IMI.*)

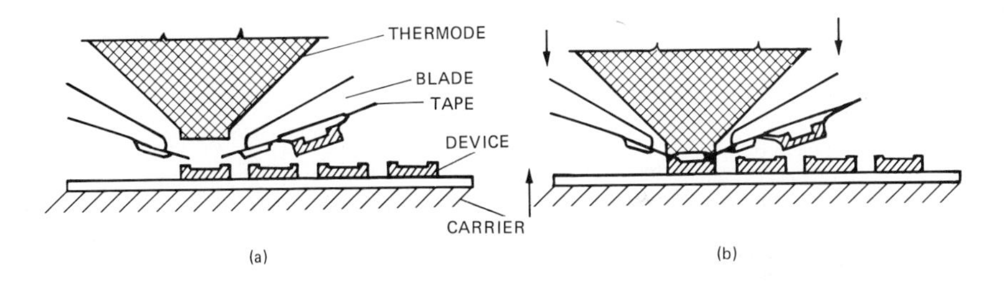

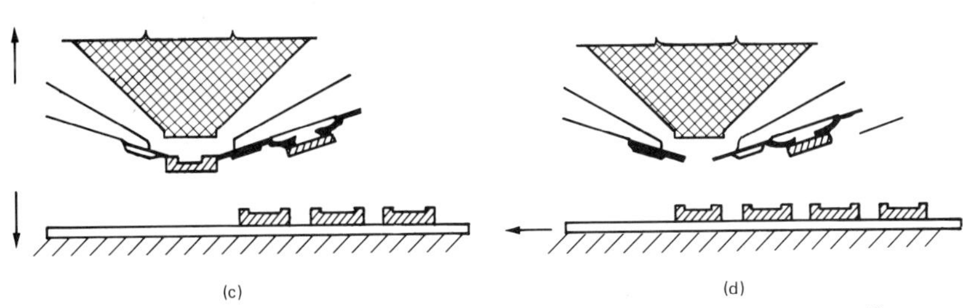

FIG. 23.23 Inner lead bonding process steps. (*a*) Align; (*b*) bond; (*c*) pick; (*d*) feed.[16]

beneath the thermode, and the process is repeated. The result is shown in Fig. 23.24.

This completes the inner-lead bonding process. The circuits can then be tested. They may be electrically tested to ensure that they are operating, and they may also be subjected to burn-in or other environmental stresses that will weed out the weak circuits.

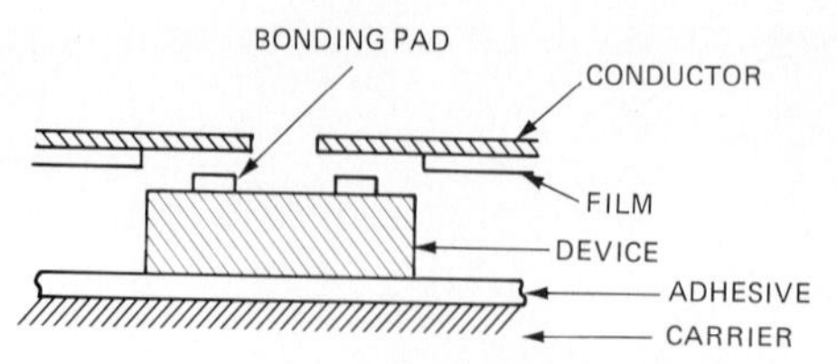

FIG. 23.24 Inner lead bonding cross section.

Outer-Lead Bonding. The next process step is outer-lead bonding. Outer-lead bonding attaches the tape-mounted circuits (that have been bonded with the inner-lead bonder) to a substrate, package, or dual in-line package lead frame. Figure 23.25 shows the basic process steps. In this example DIP lead frames are sheared into strips of 10 lead frames and loaded in a magazine that holds 50 strips. The magazine is used in the subsequent transfer molding process encapsulating the die in plastic and creating a finished DIP package. Operations similar to inner-lead bonding are performed in this process. First there is a feed step where an open area of the lead frame, which is in continuous reel form, is positioned above the thermode. The tape-mounted die is positioned above a punch that excises the die from the tape. A vacuum is used to hold the die in position after excising, thus maintaining the precise orientation of the conductors. The die is then presented to the lead frame, and the thermode applies heat and pressure (a

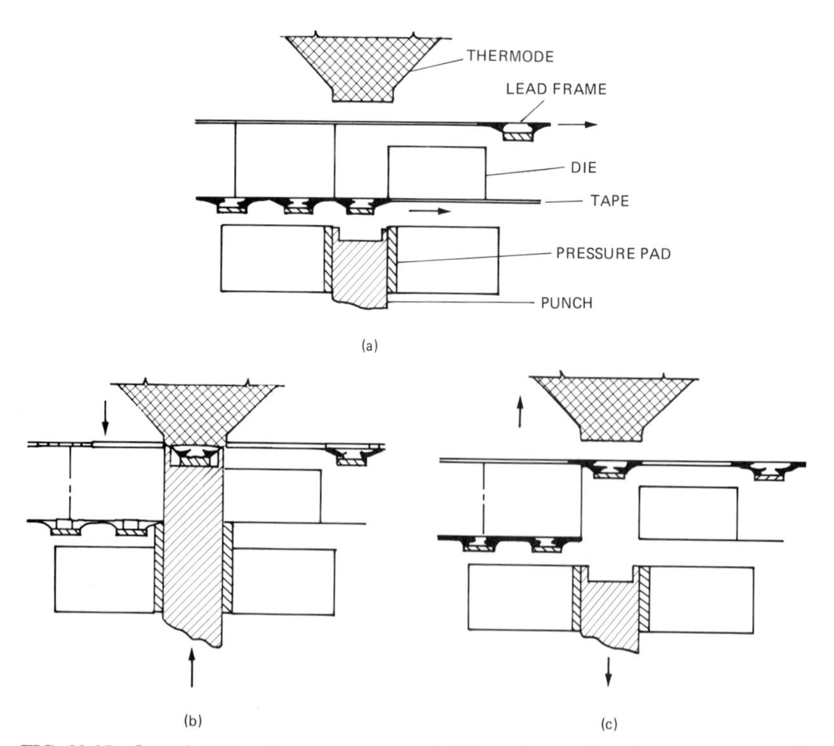

FIG. 23.25 Outer lead bonding process steps. (*a*) Feed; (*b*) punch and bond; (*c*) retract.[16]

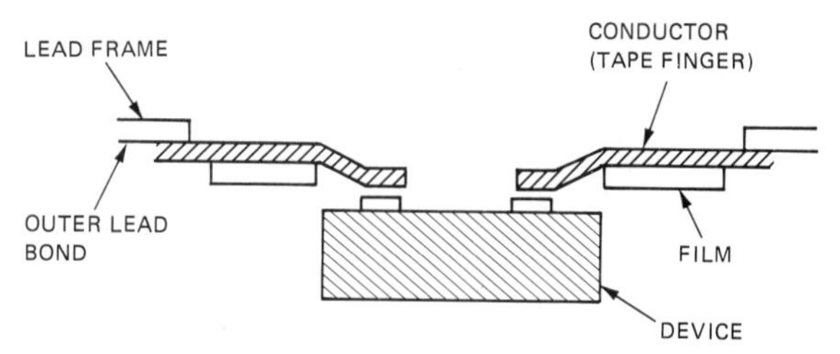

FIG. 23.26 Outer lead bonding cross section.

thermocompression bond) to bond the copper leads on the tape to the lead frame. A retraction step moves the thermode away, the punch moves down, and the process repeats. The end result is shown in Fig. 23.26.

Other types of lead formation can be used. Sometimes the term *spider bonding* is used to describe this technology, since the leads are formed away from the circuit in a spider-leg fashion where they can be attached by either solder or thermal compression bonding to a substrate. This lead form is also referred to as *gull-wing*-shaped leads. This formation is typically used in high-performance multichip circuits for the advantages of high packaging density and the elimination of the individual IC package and its electrical limitations. Figure 23.27 shows a multichip TAB circuit.

TAB Applications. TAB bonding is not limited to large-terminal-count ICs. It is also a high-volume automated technique suitable for three- and four-terminal devices. Another emerging TAB application is its use in *smart cards*. Because of the very low profile of TAB compared with the height required for reliable wire

FIG. 23.27 A multichip hybrid using tape automated bonding.[17]

bonding, TAB-bonded integrated circuits can be mounted within the 30-mil thickness of a plastic credit card. Smart cards will be used for financial transactions, personnel identification, and a variety of other uses. TAB has also been widely used in consumer applications, such as calculators.

A related technology is bumped tape.[18] Instead of bumping the wafer, bumps are formed on the tape. The bumps are needed for the bonding process and to ensure that the lead is above the surface of the die to minimize any possibility of edge shorting. Tape bumping eliminates the additional wafer plating steps for TAB and the potential high-cost fallout should damage occur at that stage. It uses a lower-cost wafer and less expensive tooling. Bumped tape has been successfully developed by some manufacturers, but it is not a widely used technique.

Another new technique based on TAB technology is *tape-pack*[19] for high-density, fine-pitch interconnections. Initial development has been demonstrated with 20-mil centerline terminal spacing, but the technology has the capability for finer pitches, to 10 mils or less. Tape-pack uses TAB circuit fabrication, which is then overmolded with plastic to create a finished package. Leads may be formed in a number of configurations. This approach is an excellent example of high-density, fine-pitch packaging using tape automated bonding.

TAB Advantages. TAB offers the advantages of IC assembly automation and mass interconnection technique, improved power distribution through the use of copper conductors, and cost reduction through the use of an all-copper system (gold is used in some applications for bump and tape plating, but an all-copper system has been shown to be reliable and produceable.) The advantages sparking renewed interest beyond the ''jelly bean'' applications are that TAB allows the use of smaller bond-pad geometry. This results in a smaller die, if die size is determined by the number of terminals. If size depends on the active circuit area, then more terminals can be placed on the same size die. The smaller bond-pad capability will become more and more of an advantage as terminal counts increase.

TAB technology with 4-mil-pitch bond pads, one half of that current wire-bonding technology, has been demonstrated for 328-terminal circuits.[20] TAB has also been shown to have improved reliability over wire bonding.[21] This is obtained primarily through the 100 percent pretest and burn-in, which is practical with TAB but not with loose dice. TAB has a more rugged bond (higher mechanical strength) than wire bond, which will be significant as bond geometry gets smaller. TAB has been used as a high-volume DIP IC assembly technique and as a basic packaging technique for high-performance applications. There is more and more interest in chip-on-board technology with large-terminal-count dice using TAB, which eliminates the IC package and allows circuits to be placed much more closely together. Reliable, high-performance chip-on-board technology becomes practical with tape automated bonding.

23.10 PACKAGE SEALING

The last packaging process step is package sealing. Ceramic packages use two basic processes: glass and solder sealing. Figure 23.28 shows a ceramic package using glass to attach a ceramic base and cap to one another and encapsulate the leads. Glass may also be used to attach the package lid to the seal ring. Solder attachment of metal lids to a metalized seal ring is widely used in military appli-

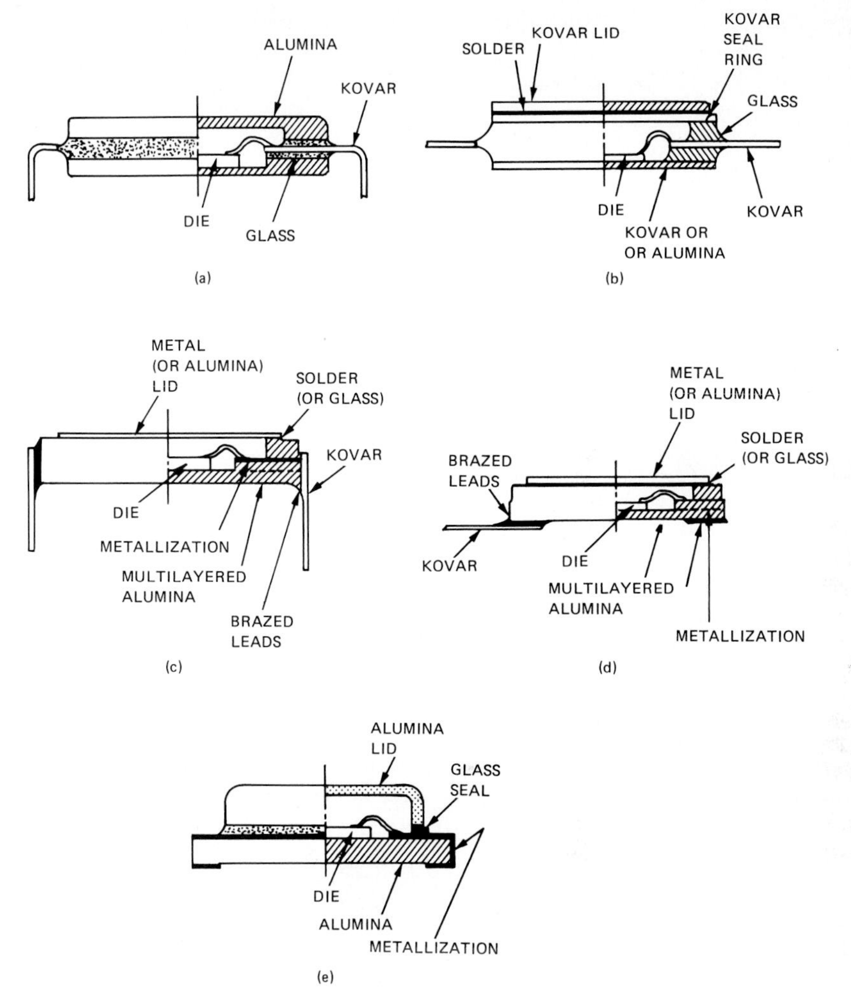

FIG. 23.28 Ceramic integrated circuit package construction alternatives. (*a*) CERDIP; (*b*) loaded glass–hard glass; (*c*) side braze; (*d*) chip carrier; (*e*) chip carrier (slam).

cations. Welding is used in very high reliability applications. The metal or glass sealing results in a hermetic package providing protection of the IC and its internal connections from the corrosion-inducing moisture of the environment. In plastic packages, the die is attached to a lead frame, and plastic is molded so that it surrounds the die. Figure 23.29 shows a cutaway view of a plastic package with a die wire bonded to a lead frame.

Some manufacturers use secondary protection materials to protect the surface of the die from moisture. Plastic packages are nonhermetic, since the plastic materials allow the ingress of moisture. Although plastic packages have been continually improved and are very reliable due to improved plastic and die passivation (where the surface of the die is protected by glass or oxides) or sec-

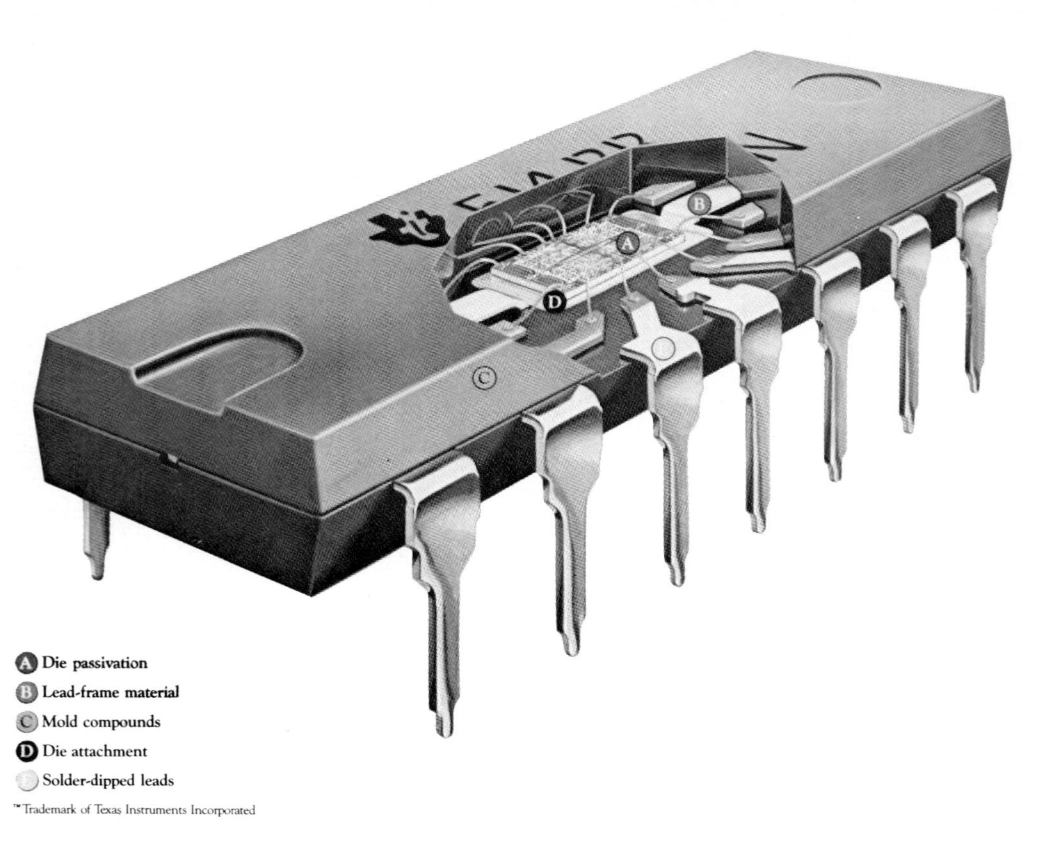

FIG. 23.29 Plastic integrated circuit package construction.[22]

ondary protection (where gels, room temperature vulcanizing compounds, or other materials are used), plastic packages do not provide the high-reliability of ceramic packages.

REFERENCES

1. C. A. Harper, *Handbook of Electronic Packaging* McGraw-Hill, New York, 1969.
2. M. Lazar, "Connector Selection Roadmap," *Electronic Packaging and Production*, May 1969.
3. Brochure, Kierulff Electronics, 1983.
4. E. Winkler, "High Performance Packaging", *Semiconductor International*, May 1985, pp. 350–355.
5. R. Wright, "Metal Core Interconnect Substrate Technology," IEPS Workshop, June 21, 1982 and "Polymer/Metal Substrates for Surface Mounted Devices," *Proceedings IEPS*, 1982, pp. 445–451.
6. B. S. Landman and R. L. Russo "On A Pin Versus Block Relationship for Partitions of Logic Graphs," *IEEE Trans. Computers*, **C-20**:1469–1479 (1971).
7. T. Steele, "Terminal and Cooling Requirements for LSI Packages," *IEEE Trans. Components Hybrids and Manufacturing Tech.*, **CHMT-4**:192, (1981).
8. D. Schmidt, "Circuit Pack Parameter Estimation Using Rents Rule," *IEEE Trans. Computer-Aided Design Integrated Circuits Systems*, **CAD-1**(4):(1982); "A Model of the Impact of Integrated Circuits on Printed Wire Routing," *Proc. IEPS*, 1981, pp. 143–147.
9. "Model 6300 Automatic Die Bonder," Kulicke and Soffa Inc.
10. "IC Equipment Market", *Electronic Business*, May 15, 1985, pp. 113.
11. M. J. Hutfless and F. W. Short, "Thermal Performance of The 68 Terminal JEDEC Leadless Type A Beryllia Ceramic Chip Carrier," *Proc. Int. Microelectronics Conf.*, 1981, pp. 125–132.
12. D. Amey, "The JEDEC Chip Carrier and LSI Standard: A Summary" *Semiconductor International*, June 1981.
13. L. M. Mahalingham, J. Andrews, and J. Drye, "Thermal Studies on Pin Grid Array Packages for High Density LSI and VLSI Logic Circuits" *Proc. IEPS*, 1982, pp. 77–96.
14. P. W. Rima, "The Basics of Tape Automated Bonding," *Hybrid Circuit Technology*, November 1985, pp. 15–21.
15. P. W. Rima, "The Basics of Tape Automated Bonding," *Hybrid Circuit Technology*, November 1985, pp. 15–21.
16. "Technical Bulletin—Massbond 4810," The Jade Corporation, 1977.
17. H. Miller, "Hybrid Approaches to Chip Interconnection Offer Many Alternatives," *Semiconductor International*, July 1984, pp. 82–85.
18. "Data Sheet—All Metal Automated Bonding Tapes," Mesa Technology.
19. Product Announcement, *Computer Design*, September 15, 1986.
20. D. Brown, M. Freedman, "Is There a Future for TAB," *Solid State Tech.*, September 1985, pp. 173–175.
21. W. Chaffin, "Reliability of TAB Products," *Solid State Tech.*, September (1981) and J. Lyman "Study Applauds Chips on Tape," *Electronics*, April 7, 1981, pp. 42–44.
22. "Managers Guide to Linear Q & R," Texas Instruments, 1986.

CHAPTER 24
SEMICONDUCTOR PACKAGE TYPES AND PACKAGE SELECTION

Daniel I. Amey
Du Pont Electronics
Wilmington, Delaware

24.1 PACKAGE TYPES

A wide variety of package types exist. Figure 24.1 shows some of the popular package types: the dual in-line package (DIP), pin-grid array packages, plastic chip carriers, and small outline (SO) packages. The SO package is suitable for low terminal counts (up to 64). The plastic-leaded chip carrier package with leads on four sides, which are folded under the package in a "J" form, is suitable for surface mounting with lead counts into the low hundreds. Flat packs are typically used for military applications, such as the four-sided "quad" flat pack with gull-wing-shaped leads. Pin-grid arrays are used for very large terminal count packages. The costs shown are relative costs compared with the SO package. Note that chip-on-board technology is projected for very large terminal count applications. No one package type will satisfy all application needs or be as broadly applied as the DIP. In the future there will be more and varying package types based on terminal counts and application. Figure 24.2 is a projection of the terminal pitch and maximum pin count for each package type. Note there is still improvement in DIP technology.

The 0.1-in (100-mil) lead-to-lead spacing is the most widely used, but 70-mil and even 50-mil DIPs are available; however, the DIP is still limited to relatively low lead counts. Flat packs are now on 20- and 25-mil centers, but they will be limited in package area and lead counts in the low hundreds primarily because of lead-frame stamping limitations. The pin-grid-array package, now a very popular package, has leads on 100-mil centers. Finer-pitch packages with 50-mil centers are being developed for very high speed integrated circuit (VHSIC) military applications. Pin-grid packaging has the potential for terminal counts in the thousands, as has been applied by IBM and others.[1] Chip carriers in their leaded or leadless form are now widely used with 50-mil terminal spacing. Fine-pitch 20-mil-terminal-spacing ceramic carriers with more than 200 lead counts have been developed, and leadless ceramic chip carriers with 10-mil spacing (spacings that are approaching silicon bond pad pitches) have been demonstrated.

FIG. 24.1 Integrated circuit package types. (*Courtesy of Texas Instruments.*)

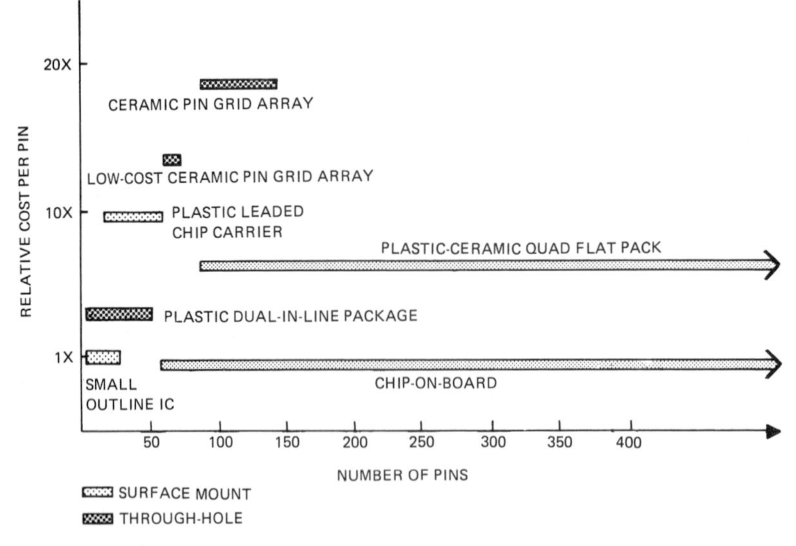

FIG. 24.2 Integrated circuit package cost and pin count projections.

As discussed earlier, tape automated bonding (TAB) currently achieves 4-mil pitch, with 2-mil pads on 2-mil centers, and terminal counts over 300. A popular DIP package construction is called CERDIP. The cross section is shown in Fig. 23.28*a* with glass-sealed ceramic bases and covers. They are available in component parts or as a finished assembly. Standardization of many package piece parts has been achieved. This construction offers a relatively low cost hermetically sealed package.

The flat pack has been used since the early 1960s in military applications, mainly with low (14 and 16) lead counts. The construction has been applied in larger terminal counts, for example, 84 leads on 50-mil centers. A 20-mil center flat-pack construction package was used in production as early as 1975 in the first Amdahl computer.[2] A consideration with flat packages is the lead forming and the "footprint" (or the interconnect pattern) on the mounting substrate.

With constructions such as the CERDIP package with glass-to-metal seals, the leads must be formed beyond the edge of the glass seal to avoid fracturing the glass seal. Typically 30 to 50 mils extension of the lead forming beyond the package is needed. This impacts density. A package with gull-wing leads with a body size of 0.4 in square would typically have a footprint of 0.750 in square. Leadless packages and other lead formations, such as "J" leads, do not require as much additional area beyond the package body size because of sealing and lead form differences.

Other types of packages are cofired ceramic and metal packages for high-reliability and high-dissipation circuits. These are typically used for hybrid circuits in military and aerospace applications. The metal packages use glass-to-metal seals for the leads, and package sealing can be welded, soldered, or sealed with glass.

24.2 THE SO PACKAGE

The SO package is a DIP-type package with smaller body size and smaller lead spacing. Table 24.1 compares the SO package with comparable package types.[3] The comparisons are relative to the DIP. The SO package has been standardized and used in high volumes in consumer and commercial products. The surface-mount, gull-wing-shaped leads are on 50-mil spacing. SO packages are typically used for memory circuits and small SSI-MSI logic circuits.

24.3 PLASTIC-LEADED CHIP CARRIER

The plastic-leaded chip carrier (PCC, PLCC, or LPCC) is predicted to be the second-highest-volume integrated circuit package next to the DIP. The PCC uses the same proven plastic packaging methods now used for high-volume DIP circuits and has the advantages of the chip carrier's density and surface mounting. Figure 24.3 shows the standard PCC mechanical outline with "J" formed leads, which minimizes the overall package footprint. The smaller-terminal-count plastic chip carriers are used primarily for SSI logic circuits and memory circuits. Terminal counts for PCC packages will extend above 84 up into the low hundreds.

24.4 CERAMIC CHIP CARRIERS

What is a chip carrier? Formally defined, it is "a low profile rectangular component package, usually square, whose semiconductor chip cavity or mounting area is a

TABLE 24.1 Low Terminal Count Package Comparison*

	16 terminal				20 terminal			
	DIP	SOIC	0.040″ ℄ LCC	0.050″ ℄ LCC	DIP	SOIC	0.040″ ℄ LCC	0.050″ ℄ LCC
Body (lid) size	0.785 × 0.260 (1)	0.394 × 0.158 (0.31)	0.186 × 0.186 (0.17)	0.253 × 0.253 (0.31)	1.02 × 0.280 (1)	0.511 × 0.299 (0.54)	0.217 × 0.270 (0.21)	0.305 × 0.305 (0.33)
Board area	0.785 × 0.325 (1)	0.244 × 0.394 (0.38)	0.190 × 0.190 (0.14)	0.308 × 0.308 (0.37)	1.02 × 0.325 (1)	0.511 × 0.419 (0.85)	0.290 × 0.290 (0.25)	0.360 × 0.360 (0.39)
Body thickness	0.150 (1)	0.061 (0.41)	0.070 (0.47)	0.072 (0.48)	0.195 (1)	0.096 (0.49)	0.070 (0.36)	0.072 (0.37)
Max. height	0.200 (1)	0.069 (0.35)	0.090 (0.45)	0.100 (0.50)	0.210 (1)	0.104 (0.50)	0.103 (0.49)	0.100 (0.48)
Board area/body size	1.25	1.54	1.04	1.48	1.16	1.40	1.44	1.39

*All dimensions in inches. Maximum dimensions used for comparisons. Quantity in ( ) is ratio of the package dimension to the corresponding DIP dimension.

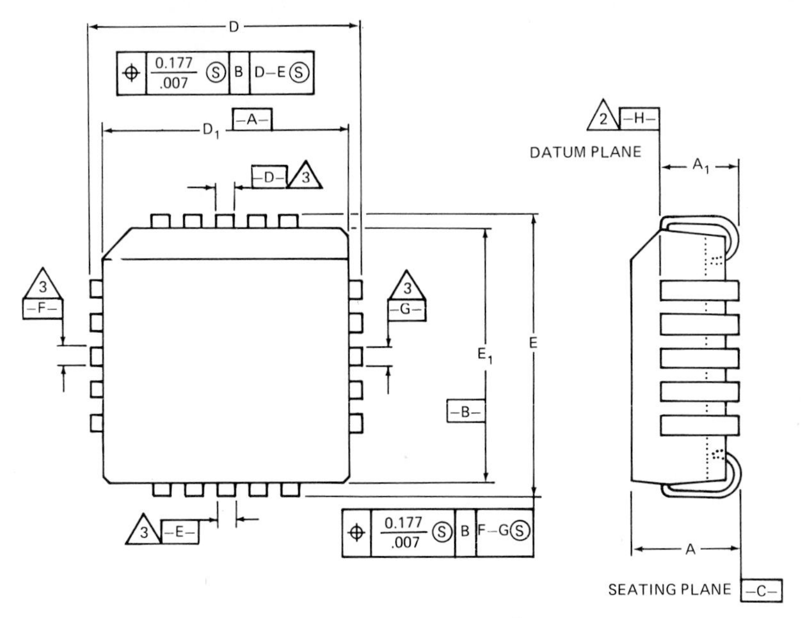

FIG. 24.3 Plastic leaded chip carrier outline dimensions.[3]

large fraction of the package size and whose external connections are usually on all four sides of the package."[4] It is really just the functional cavity area of the DIP that houses the die. Figure 24.4 shows the ceramic multilayer cofired construction of a chip carrier and the corresponding construction of the DIP, the seal ring, the die bond-pad metallization layer, the layer that creates the die cavity with wire bond pads, and the base. This construction is typical of many ceramic packages.

The chip carrier is a hermetic package and has the advantage that there may be interconnections between layers. Layers may have unique connections for the interconnection of multiple dice or for power distribution where large ground planes are needed for low inductance, or parallel interconnects for low resistance and alternative ground paths. Figure 24.5 shows the via configurations used to connect metalized planes and exterior features for a typical ceramic chip carrier. The semicircular features are called *castellations*. These areas are metallized to connect from the internal bond-pad shelf to the base of the package where the input-output terminals are located.

Chip carriers offer advantages over dual in-line packages in size, electrical performance, and cost. Chip carriers have a footprint or area advantage over DIPs anywhere from 2.7:1 to 5.6:1 (depending on terminal count) for packaging the same size die. The chip carrier uses a much smaller area primarily because of its reduced 50-mil terminal spacing with terminals on all four sides instead of two sides for the DIP. This area advantage can also be viewed as packaging efficiency or the ratio of the cavity area containing the semiconductor die to the overall package area. Chip carriers have shorter leads than DIPs, resulting in lower resistance and smaller propagation delay. Chip carrier interconnects are more uniform, as shown in Fig. 24.6, which compares the ratios of the longest traces on the packages. The longest to

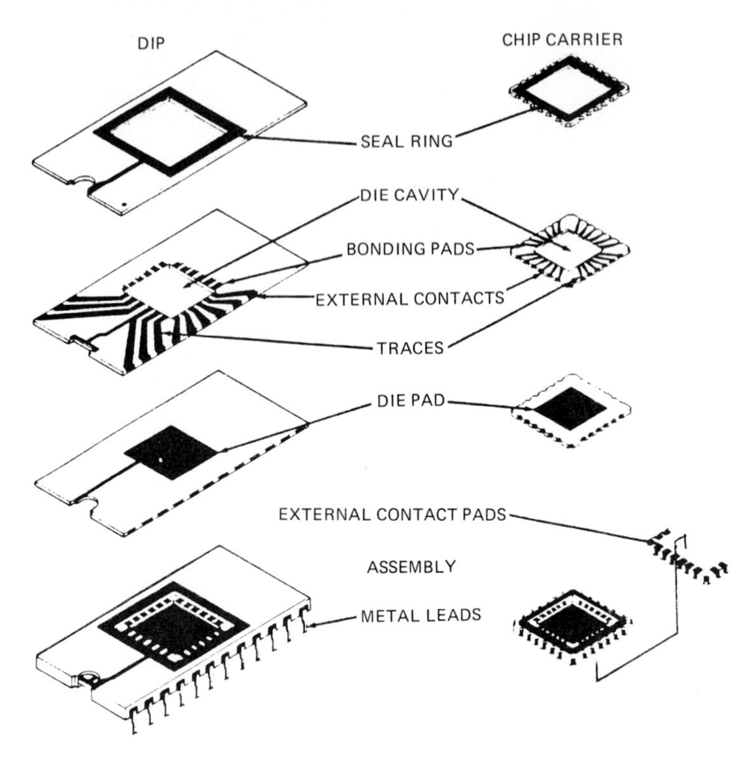

FIG. 24.4 Construction comparison of a dual in-line package and a chip carrier.[5]

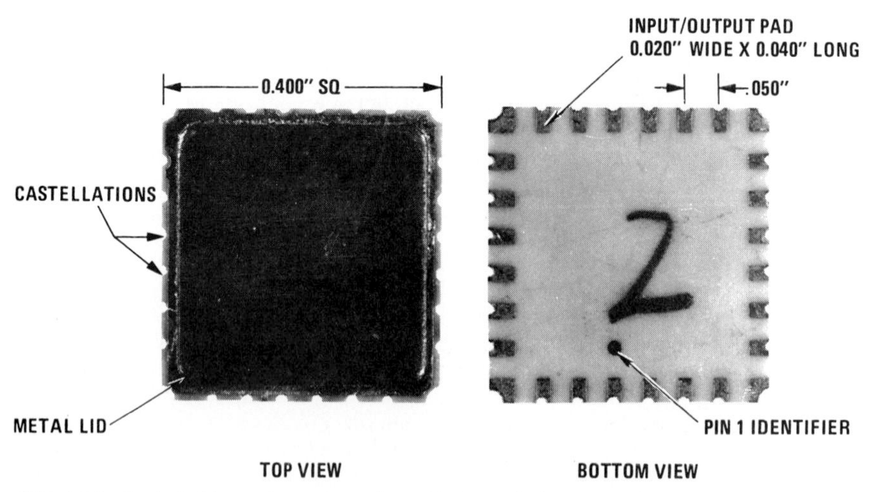

FIG. 24.5 Typical chip carrier construction with "castellations."

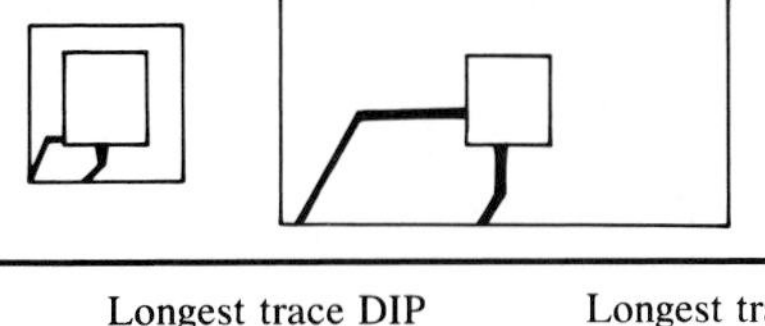

Lead count	Longest trace DIP / Longest trace CC	Longest trace / shortest trace	
18	2:1	1.5:1	6:1
24	4:1	1.5:1	3:1
40	5:1	1.5:1	5:1
64	6:1	1.5:1	7:1

FIG. 24.6 Lead length comparison of a dual in-line package and a chip carrier.[5]

shortest trace ratio is uniform with chip carriers because of the square form factor. The wide variation in DIP length is due to its two-sided I/O and rectangular form. For large packages this can result in a significant timing differential in very high performance applications.

In hybrid circuit applications chip carriers offer cost advantages because the package makes pretesting of individual circuits practical. Leadless chip carriers also reduce costs by eliminating the losses that occur because of lead damage and chip carriers use less material with much less gold content than the corresponding DIP package. Chip carriers are available in strip or array form, thus minimizing the number of pieces that are handled. In this form they are attached to one another by a fine web of ceramic, which is broken after the final assembly step so that the die attachment, wire-bonding, and sealing processes may use mass assembly methods.

Chip carriers can be fabricated in a number of configurations. Figure 24.7 shows typical single-layer, two-layer, and three-layer constructions. The number of layers refers to the layers of ceramic, not metallization. A flat lid or cup-shaped lid is sealed with glass or epoxy; with the proper metallization, solder sealing may be used. A metalized seal ring is not needed with an epoxy or a glass seal.

Single-layer and three-layer packages have been developed and designed with compatible footprints, and there are cost advantages with these different constructions. Figure 24.8 shows relative costs of different chip carrier constructions. The three-layer chip carrier with a gold-plated lid and solder preform (the most reliable construction) as specified in Mil-M-38510 is the reference. With a preglassed flat ceramic lid the high-cost gold-plated lid, preform, and metalized package seal ring is eliminated, saving about 25 percent. With a single-layer construction or SLAM package (single layer alumina-metalized) about a 50 percent cost savings over the three-layer construction could result.

Figure 24.9 shows a multichip "dipstrate" application of chip carriers, a popular approach for memory packaging where individual chip carriers mounted on a dual in-line substrate are internally connected in the multilayer cofired substrate. A major advantage of chip carriers is their ability to mount on both sides of the mounting substrate. This has been used for memory and very high density military packaging in this form.

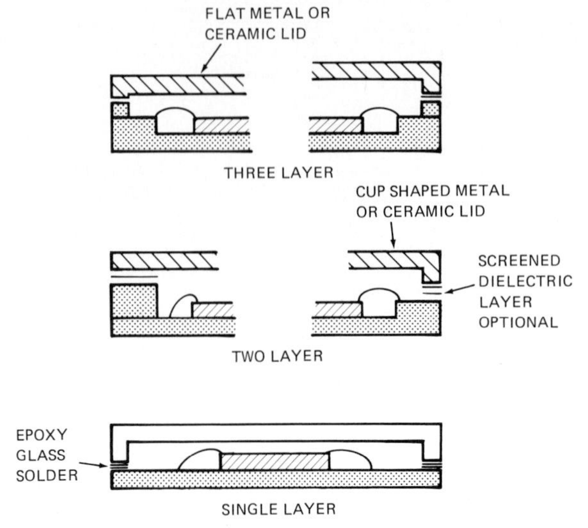

FIG. 24.7 Ceramic chip carrier construction alternatives.[5]

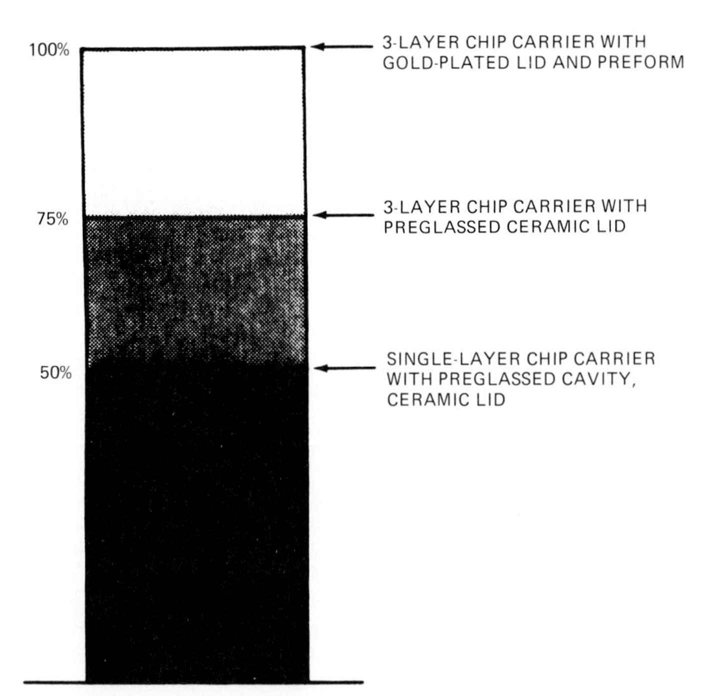

FIG. 24.8 Relative cost comparison of ceramic chip carrier construction alternatives.

FIG. 24.9 A ceramic substrate or "dipstrate" with chip carriers soldered on both sides.

24.5 PACKAGE STANDARDS

JEDEC, formerly the Joint Electron Devices Engineering Council, is a function of the Electronics Industry Association (EIA) and defines microelectronic standards. The EIA/JEDEC JC-11 committee on mechanical standardization is responsible for microelectronic package outlines and has developed standards for all package types. This committee is the recognized U.S. standards organization for microelectronic packaging and represents the United States in the International Electrotechnical Commission (IEC).

JEDEC also has committees that are involved in other aspects of microelectronics and semiconductor technology specializing in digital, bipolar, MOS, linear memory, gate arrays, and other areas. The committee has an ongoing standardization effort for packages in both ceramic and plastic construction. Ceramic chip carrier package styles have been standardized with leadless packages referred to as Leadless Types A, B, C, D, E, and F. Leadless Types A through D are square packages, and types E and F are rectangular packages specifically for memory circuits. JEDEC package standards allow for a variety of construction methods and techniques and do not specify internal details. Figure 24.10 describes the leadless package features. The different ceramic types exist because of the need for different performance factors for various applications.

The Leadless Type A was optimized for air-cooled systems with a cavity-down configuration. However, the configuration was not suitable for the small-terminal-count (14-, 16-, and 20-terminal) cavity sizes desired for military applications. This resulted in the Leadless Type B and low-cost, single-layer Leadless Type C packages, which were designed for cavity-up mounting. With the cavity-up construction the input-output surface is on the opposite side of the package sealing and bonding surface, and therefore Leadless Types B and C can have a larger cavity size than Leadless Type A for the same overall package size.

Leadless Type D resulted from the need for a cavity-down package that could be directly attached to the mounting surface (which is not possible with Leadless Type A because the seal ring prevents direct attachment). An additional layer of ceramic was added to create Leadless Type D and provide for direct attachment. However, this results in the same cavity size restrictions as Leadless Type A. The tradeoffs in the use of Leadless Types A, B, C, and D involve thermal orientation (cavity-up or cavity-down) and cavity size for small terminal counts to accommodate existing dies. For 44 terminals and above the cavity size is not an issue, so the primary tradeoff in the use of the different package types becomes one of thermal performance.

The JEDEC standard defines other forms, such as the Leaded Type A, defined for a premolded plastic "J" leaded package (the first to use the "J" lead configuration,[7] and the Leaded Type B, which uses a substrate clip, a means to attach a leadless ceramic package to a mounting substrate. The Leaded Type B package has been used with printed wiring boards where the substrate clip provides compliance for the difference in coefficient of thermal expansion between the printed wiring board and the ceramic package. The clips come in a variety of forms and can be applied with automatic machinery. The clips have the advantage of minimal protrusion of the lead beyond the edge of the package, resulting

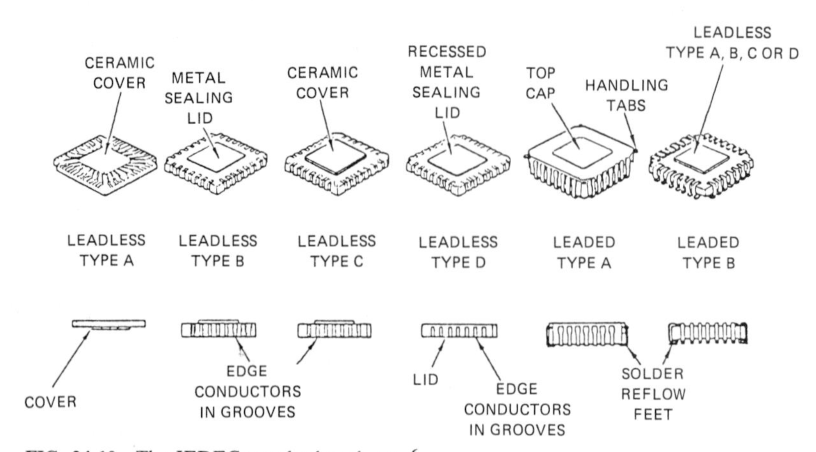

FIG. 24.10 The JEDEC standard packages.[6]

in a minimal board footprint. The substrate clips offer a "leads-last" assembly approach. A leadless package can be used in semiconductor fabrication, test, and handling, thus eliminating losses due to lead damage. The leads are put on the package by the user at the last stage of manufacturing, minimizing any potential lead damage. In addition to thermal compliance, substrate clips allow leads to be repaired, which is not practical with other leaded packages.

Chip carriers with 40-mil centerline terminals have also been standardized by JEDEC. This seemingly small difference in terminal pitch from 50-mil centerline packages was important to those military and aerospace applications that require the smallest size and weight. For similar reasons there are both 20-mil and 25-mil centerline terminal spacing for fine-pitch package terminal counts extending to over 300 terminals.

24.6 PIN-GRID ARRAYS

Pin-grid-array packages are typically ceramic packages with a matrix of pins brazed to the package. Standards have been defined for pin-grid-array packages with up to 400 pins. Pin-grid-array packages with 1800 and more pins have been produced.[8] A wide variety of pin-grid-array packages exist; it is a highly efficient, widely used packaging method for large-terminal-count packages. They are constructed in either a cavity-up or cavity-down configuration; however, the cavity-down construction significantly reduces pin count.

Pin-grid arrays are readily available and offer density improvements over standard chip carriers as well as the dual in-line packages. One major advantage of pin-grid arrays is that they can be wave soldered and assembled just like a DIP. Leadless grid arrays (grid-array packages that have pads for termination for surface mounting instead of pins) have been proposed, but they are not readily available. Most pin-grid-array packages have 0.1-in terminal spacing, which is compatible with available test fixtures for ease of board assembly testing. Military programs, however, are developing large-terminal-count (400+), 50-mil-terminal-spacing pin-grid packages.

A consideration in the use of a pin-grid-array package is that the package dictates the features of the next level of interconnection (the PC board). The board hole sizes and printed circuit land features are fixed by the package pin diameter. This can limit interconnection density. In addition to the many other performance and economic advantages of surface mounting, an advantage of surface mounting such as with chip carriers or SO packages is that the package does not limit the hole sizes. The hole size (limited only by plating capability) can typically be much smaller than with through-hole package types, allowing for more interconnection paths between holes and higher-density interconnection. The board mechanical features and interconnects are not restricted by the package leads.

Which package does one use? No single style can satisfy the many varied applications of semiconductor or VLSI circuitry. A variety of package types will continue to be used. Surface-mount technology will be widely used as well as the DIP and pin-grid-array package. Figure 24.11 shows terminal density in terminals per square inch versus terminal counts for the packages discussed.

A cavity-up 100-mil pin-grid array has a uniform density of 100 terminals/in^2. Pin-grid arrays have been and will be used for 200- to 300-terminal-count packages and beyond. With a cavity-down orientation pin-grid-array terminal counts are reduced because the pins are on the same surface as the package cavity. Chip

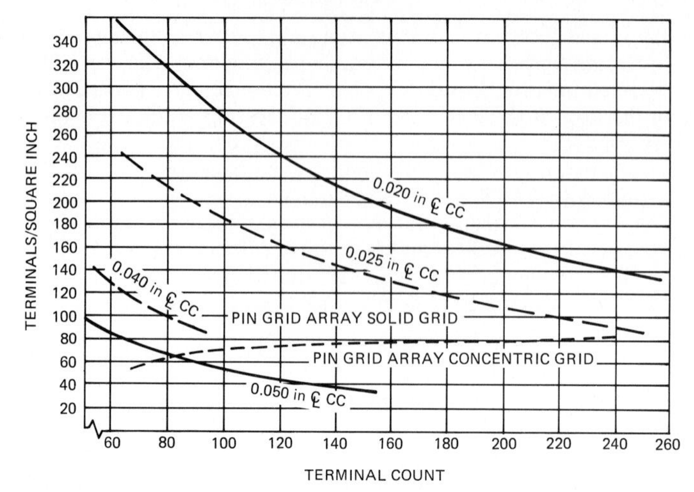

FIG. 24.11 Integrated circuit package terminal densities versus terminal count.

carriers can be mounted cavity-up or cavity-down without sacrificing terminal density because they are a perimeter terminal package. The pin-grid array in a cavity-down configuration has a reduced terminal count. The 20-mil and 25-mil centerline chip carriers offer density advantages up to about 300 terminals in the case of the 20-mil centerline package. Beyond that the pin-grid array is a better choice for density. A 50-mil centerline pin-grid array has a density of 400 terminals/in^2. Its broad use is limited by board interconnection capability.

24.7 PACKAGE MATERIALS

A key consideration in package choice is the package material. For high-performance applications the dielectric constant of the package material can significantly affect signal propagation delay. Table 24.2 shows the dielectric constant of some common materials used for interconnections. Note that ceramic, which is widely used for IC packages, has a relatively high dielectric constant

TABLE 24.2 Dielectric Constants of Typical Interconnection Materials

Material	Dielectric constant
Teflon	2.1
Kapton	3.5–3.7
Valox	3.7
Ryton	4.0
Epoxy-glass	5.0
Berylla ceramic	6.5–6.9
Alumina ceramic	8.8–10.1

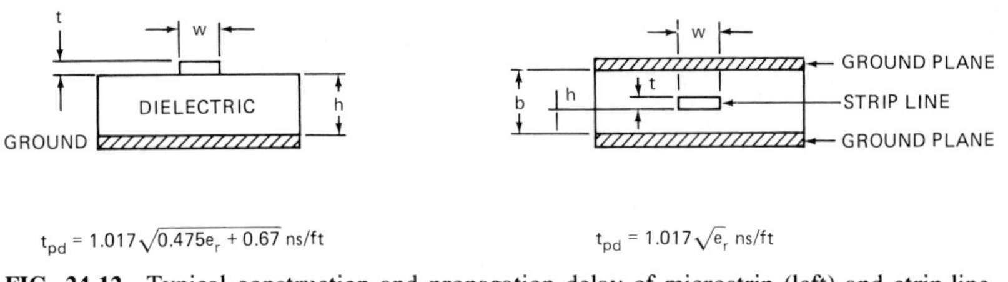

$t_{pd} = 1.017\sqrt{0.475e_r + 0.67}$ ns/ft

$t_{pd} = 1.017\sqrt{e_r}$ ns/ft

FIG. 24.12 Typical construction and propagation delay of microstrip (left) and strip-line (right) interconnections.

compared with printed circuitry and typical engineering plastics. This difference will affect the propagation of signals in either the microstrip form, typical of signals routed on external layers, or in strip-line form, which is typical of innerlayer signals. See Fig. 24.12. Signal propagation delay in the equation in Fig. 24.12 is a direct function of the dielectric constant, so a lower dielectric constant will result in faster signal transmission (shorter delay). Figure 24.13 plots the delay difference for the microstrip and strip line with varying dielectric constants.[9]

Figure 24.14 is a comparison of a 132-terminal, 100-mil pin-grid-array package and a smaller 132-terminal, 25-mil centerline, fine-pitch chip carrier. This relatively small size difference will affect the propagation delay within the package not only as a result of the length in high-dielectric-constant media (typically signals travel through multiple packages), but the shorter length also reduces the resistance, capacitance, and inductance, all important electrical parameters to minimize in interconnections.

Smaller package size improves lead length uniformity, and lead resistance will be lower when there are external plated circuit paths. Overall, smaller is better when one considers very high performance applications. In the choice of a package, yield loss is an important consideration. Losses can occur in die attachment, bonding, sealing, and the test operations. The choice of a leaded or leadless package has a significant effect on losses in each area.

24.8 PACKAGING TRADEOFFS

No one packaging technique can satisfy all applications. Individual applications and their package requirements will ultimately determine the best choice. The final technique must be determined by a system analysis (which is usually complex) that takes into account the application requirements, the manufacturing considerations, the risk in the use of new technology, maintenance considerations, and overall costs in the application of a package type. The semiconductor package cannot be considered independently from other interconnections and packaging hardware. The effect of the package choice on the different levels of interconnection must be considered not only from a user perspective but also from the perspectives of the package manufacturer, the semiconductor device manufacturer, and the equipment customer. Figure 24.15 shows those areas where a package choice can have an impact.

The semiconductor package manufacturer must consider the effect the package has on the IC test equipment, the availability of test sockets (a difficult problem with high-density devices), and handling and assembly equipment for the

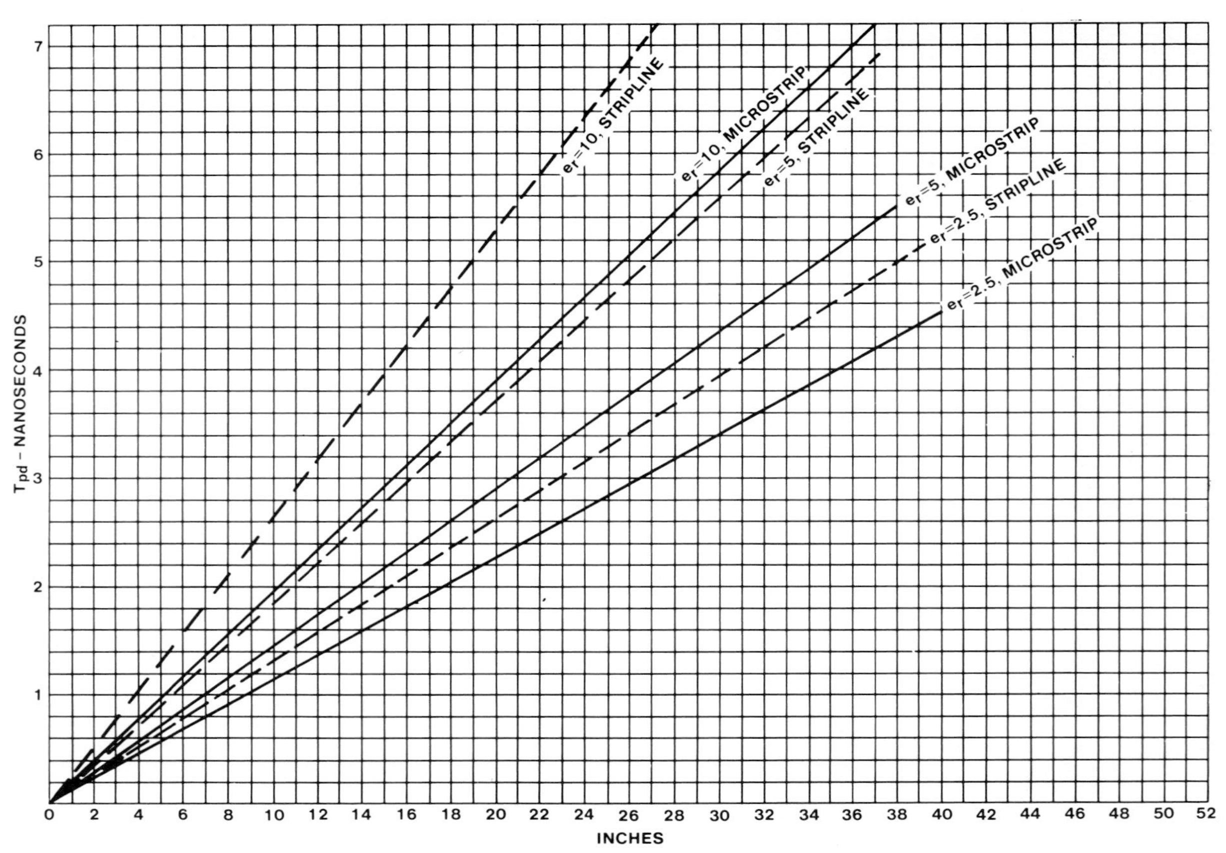

FIG. 24.13 Propagation delay versus length for microstrip and strip-line interconnections with varying dielectric constant materials.

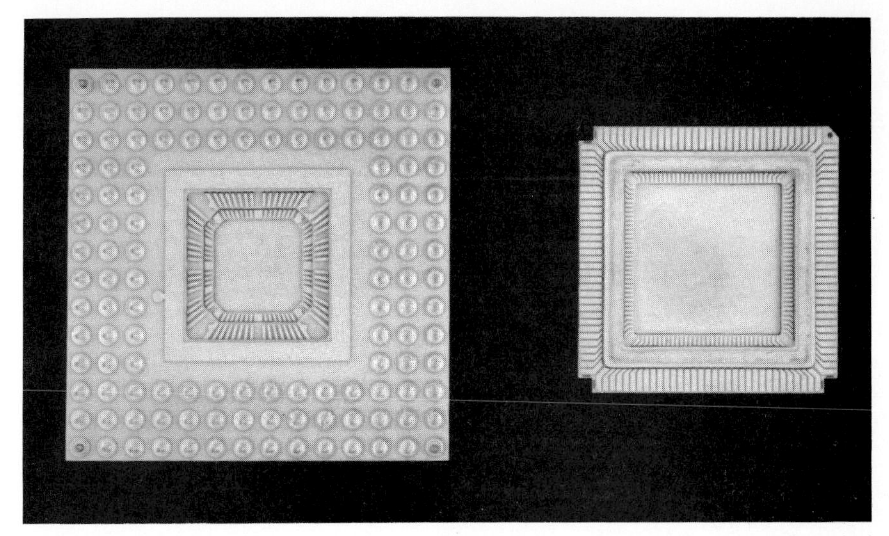

FIG. 24.14 Comparison of a 132-terminal 0.1-in pin-grid-array package (left) and a 132-terminal 0.025-in chip carrier (right).

new components. New package types can cost a semiconductor manufacturer on the order of a million dollars to tool for high-volume assembly and test. The choice and definition of a new package is not a trivial matter. Semiconductor device users must be concerned with the effect of the package on insertion equipment, the assembly operations, cleaning, component test and assembly, the interface with connectors or sockets, and compatibility with other components used on an electronic assembly, in addition to many other performance and application factors. Not only will the IC device influence package choice, but connectors and other electronic or electromechanical devices used on the printed circuit board must be considered.

Also of concern in the package choice are the thermal implications and package compatibility with heat sinks (if required), the printed circuit board or substrate, and system cooling. The tradeoffs are many and difficult. Specifying a very high density integrated circuit package that cannot be effectively interconnected may lower overall system performance or result in higher costs than another combination of package and board interconnection.

The package choice has an impact on the end user in its ease of test, repair and maintenance, the cost of board or package spares, and the ease of replacement of failed components. For example, maintenance by board replacement can be very expensive for multiple-package boards, far offsetting the costs of socketing high-cost individually replaceable semiconductors. It may also be worthwhile to add cost to a package, making it easier to repair. In the choice of packages and the evaluation of a packaging system, one has to look at the package choice from a variety of perspectives. One may not be able to influence the choice in all areas or at all levels of interconnection. Typically there are many "givens" and practical constraints. However, in viewing the packaging problem from each level of interconnection and from the above perspectives, considering all that is involved in a package choice, one will have a better understanding of the increased costs

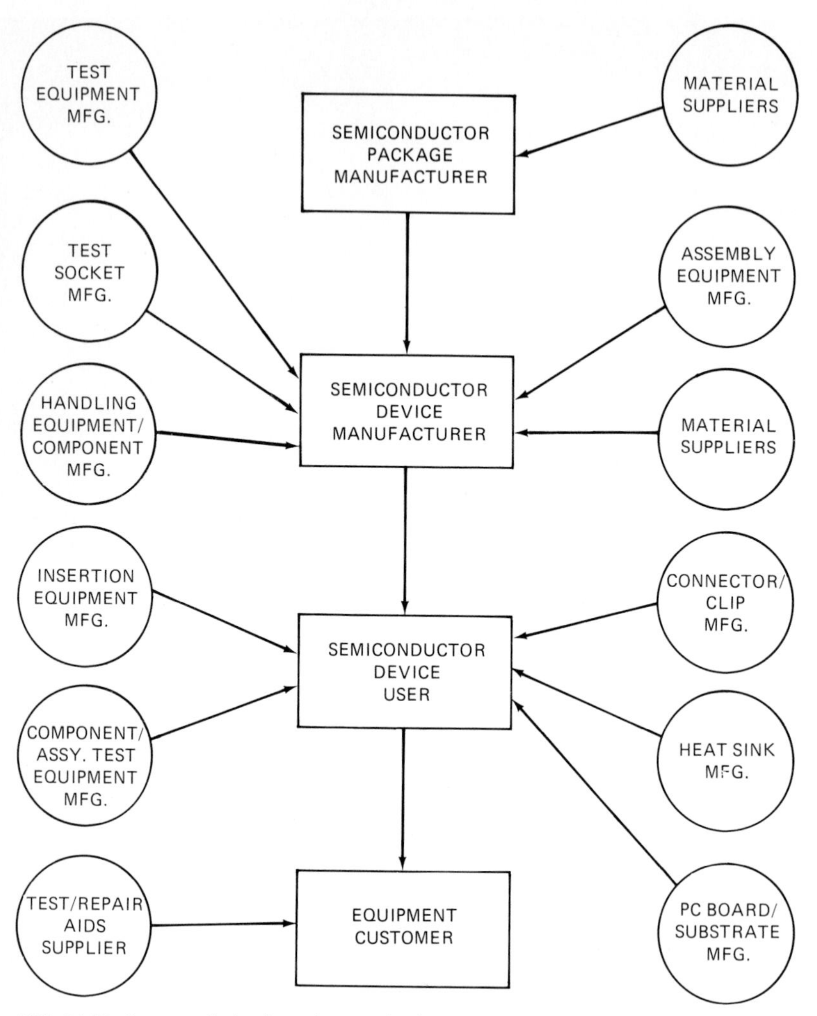

FIG. 24.15 Integrated circuit package technology considerations.

or performance degradation or design implications of a particular package choice. In this way one can get the most from what can be controlled and understand the tradeoffs.

24.9 SYSTEM CONSIDERATIONS

The rapid advancement of semiconductor technology has far outpaced package and interconnection technology. This has been evidenced in a number of ways where the packaging is limiting system performance, has become a major hard-

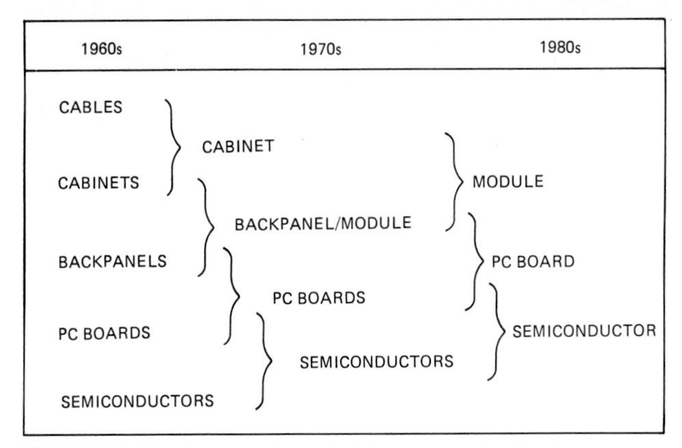

FIG. 24.16 The downward trend of interconnection levels.

ware cost factor, and is an increasingly larger percentage of the overall costs of the electronic system (a larger percentage than the IC costs). Figure 24.16 illustrates how the levels of interconnection have changed over time with a continued downward trend in the movement of the levels of interconnection.

For example, computers of the 1960s had several cabinets, each containing multiple backpanels and printed circuit boards carrying the integrated circuit devices. The cabinets were all interconnected with cables. In systems of the 1970s, as semiconductor technology improved and higher-density packaging evolved, the same electronics could all be packaged into one cabinet. Where level 4 and level 5 interconnections were required, they were now all within a single cabinet with only level 4 interconnections. The systems of the 1970s that took numerous backpanels and PC boards can now be packaged in a single module, further reducing the number of interconnections. For systems of the 1980s, where it previously required cabinets, backpanels, and multiple PC boards, a single module (or PC board) replaced a large number of printed circuit boards. Those functions previously packaged using levels 4, 3, 2, and 1 interconnections were all packaged with levels 3, 2, and 1 interconnections.

This downward trend in interconnection levels has been occurring for many years, with more and more interconnections moving toward level 1, and this trend will continue. This packaging trend is driven by lower cost, higher performance, and the reliability advantage that semiconductor technology offers. Figure 24.17 shows areas of relative interconnection density and relative cost for silicon, printed wiring, and hybrid technology showing orders of magnitude improvement in cost and density of integrated circuit interconnections over hybrid interconnects (which plays an important intermediate role in density and cost improvements) over printed wiring.[10] Cost and density are the primary forces moving more and more interconnections into silicon.

A good example of what can be achieved is Digital Equipment Corporation's J11 processor,[11] (Fig. 24.18), which consists of ceramic chip carriers on a ceramic dipstrate mounted on a single printed circuit board. The ceramic substrate is about 1.3 by 3 in, packaging two ceramic chip carriers. One chip carrier contains about 46,000 semiconductor devices, the other about 92,000. It dissipates 1 W. This is the same logic used in the PDP 1170 computer system, which required 20 printed circuit boards with 800 W of power dissipation. The cost saving of this

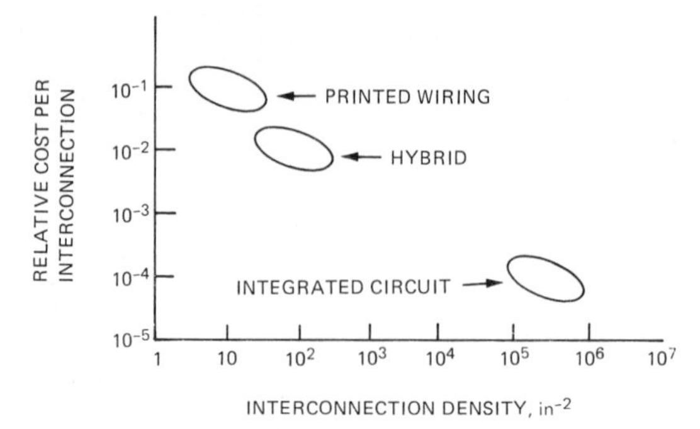

FIG. 24.17 Interconnection technology cost and density comparison.

FIG. 24.18 The Digital Equipment Corporation J11 processor circuit.

assembly, through the use of the chip carriers and the ceramic substrate, reduced the overall system price by a factor of 20 to 1. That is the payoff in semiconductor technology improvements and high-density packaging.

Other examples as dramatic as this demonstrate the advantages of high-density semiconductor packaging. The future will bring even larger terminal count packages, much higher density interconnection, both within the package and off the package, and the continued movement of interconnections into silicon. The impact on the semiconductor package and the next interconnection level substrate or printed circuit board is significant and will require the use of new materials, packaging techniques, and processes to successfully respond to the forces that are driving electronic packaging.

REFERENCES

1. I. Feinberg and D. Seraphim, "Electronic Packaging Evolution in IBM," *IBM J. Res. Develop.*, **25**(5) (1981).

2. R. Beall, "Packaging for a Supercomputer," *IEEE Intercon.*, **74** (1974).
3. D. I. Amey, "Surface Mounted Components," *Surface Mount Technology*, ISHM, Silver Springs, Mo., 1984, pp. 11–44.
4. Electronic Industries Association, JEDEC Standard 100 "Terms, Definitions, and Letter Symbols for Microcomputers and Memory Integrated Circuits,"
5. M. L. Burch and W. M. Hargis, "Ceramic Chip Carrier—The New Standard in Packaging?" *Proc., NEPCON 1977.*
6. D. I. Amey and R. P. Moore, "An LSI Package Standard and Its Interconnection Variations," *Proc. NEPCON 1977.*
7. D. G. Grabbe, "A Premolded Chip Carrier with Compliant Leads," *Proc. NEPCON 1977.*
8. D. R. Barbour, A. J. Blodgett, "Thermal Conduction Module: A High-Performance Multilayer Ceramic Package," *IBM J. Res. Develop.*, **26**(1) (1982).
9. D. I. Amey, "Integrated Circuit Package Selection: Pin Grid Array vs Chip Carriers," *Proc. IEPS 1981.*
10. J. S. Mayo "Trends in Solid State Technology," *Circuits Manufacturing,* April 1982, pp. 29.
11. P. Rubinfeld, "Two-Chip Supermicroprocessor Outperforms PDP-11 Minicomputers," *Electronics,* December 15, 1982, pp. 131–133.

P • A • R • T • 8

VLSI ECONOMICS

CHAPTER 25

ECONOMIC ASPECTS OF TECHNOLOGY SELECTION: LEVEL OF INTEGRATION, DESIGN PRODUCTIVITY, AND DEVELOPMENT SCHEDULES*

Curt F. Fey
Demetris E. Paraskevopoulos
Xerox Microelectronics Center, Xerox Corporation
El Segundo, California

25.1 INTRODUCTION

This chapter and the next provide a guideline for design methodology selection for systems companies rather than for semiconductor manufacturers. They address both the tactical question of how a designer should select an IC for a required function and the strategic problem of how an engineering organization should select design methodologies that will enable it to be or to remain competitive in product development. They provide insight into the fundamental underlying factors that drive the current economics and that determine future trends, such as schedule, functionality, cost, and the risk of these factors. Some basic references are provided. More detailed references are contained in Refs. 1–4. Periodic updates of VLSI economics are provided by Refs. 5 and 6.

The assessment is based on estimates of representative U.S. costs. The assessment focuses on fundamental relationships, not on numbers, and on relative, rather than absolute costs. These relative costs, reflecting the competitive advantage of the various design methodologies, are not very sensitive to our assumptions. We will show that varying our data within the range of the estimating error will affect the values but not the conclusions.

This assessment confines itself to the economics of application specific integrated circuits (ASIC). For an analysis of the U.S. semiconductor industry, see

*Adapted with permission, © 1985, 1986, 1987, IEEE. From *IEEE J. Solid State Circuits,* **SC-20**(2):555–561; **SC-21**(2):297–303; **SC-22**(2):223–229.

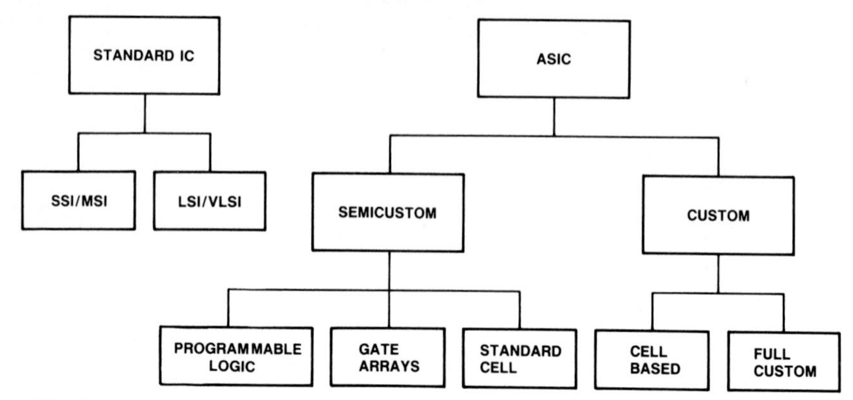

FIG. 25.1 Logic IC device design methodologies. The classification of logic IC alternatives used in this chapter. There is no generally accepted classification.

Hazewindus[7] and Linvill et al.;[8] for a review of the industry competition see Ernst,[9] Malerba,[10] Okimoto et al.,[11] and Paul;[12] for future trends see Petrocelli.[13]

There is no universally accepted method of classifying logic ICs. We classify them in the manner shown in Fig. 25.1. Standard ICs refer to those devices commercially available from merchant vendors, such as an Intel 80386 microprocessor or a 74LS00 quad two-input NAND gate. Standard cells are based on well-tested libraries that are developed prior to the device design. Cell-based designs are hybrids of full-custom and standard cells, with cells being developed concurrently with the device design. This chapter does not specifically discuss cell-based designs. Their economics lie between the economics of full-custom and of standard cells. Some authors combine standard cells and cell-based designs into one category called cell-based ICs.

This chapter provides some introductory material and discusses two of the four methodology selection criteria, namely schedule (productivity) and functionality. Chapter 26 deals with the remaining two selection criteria, namely cost and risk.

25.2 ALTERNATIVE DESIGN METHODOLOGIES

The market share of the various technologies is shown in Fig. 25.2. Market share is a function of economics; therefore, ASICs as a group will gain market share due to their favorable economics. Programmable logic devices (PLDs) will have a small niche. In 1985 CMOS gate arrays accounted for 60 percent of the gate-array dollar value, and bipolar accounted for the remaining 40 percent. Full-custom will maintain a constant dollar volume but a declining market share. The rate at which the standard-cell market will develop is difficult to predict.

We will examine the economic factors that influence the selection of one of the following design methodologies:

1. Standard SSI-MSI logic (bipolar and CMOS)
2. PLDs (bipolar and CMOS)
3. Bipolar gate arrays

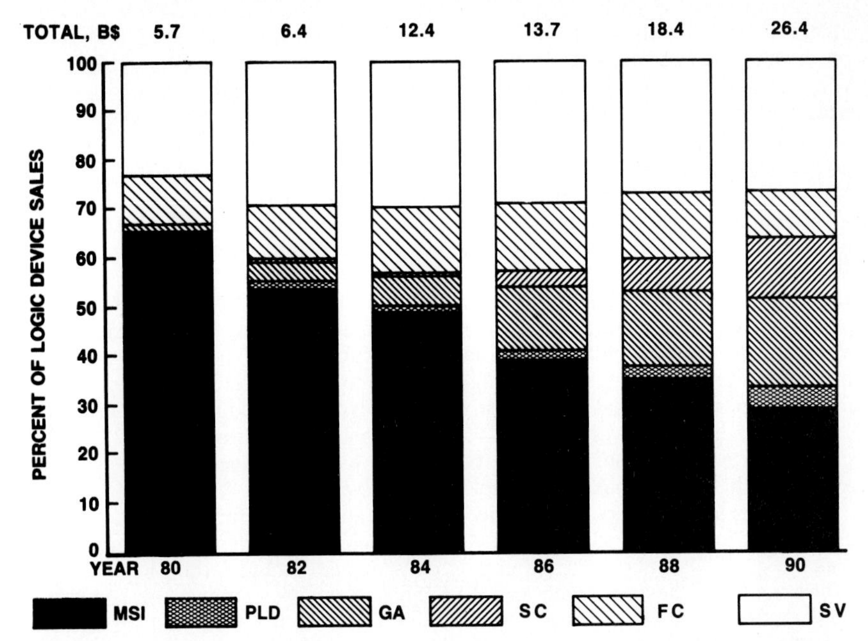

FIG. 25.2 Worldwide merchant and captive logic IC device sales by type, 1980–1990. Market share is a function of economics, therefore ASIC will increase its market share. PLD will have a small niche. Full-custom will have constant dollar volume, but declining share. The split between gate arrays and standard cells is difficult to predict. (Legend: MSI = SSI-MSI, PLD = programmable logic, GA = gate arrays, SC = standard cells, FC = full-custom, SV = standard VLSI.)

4. 2-μm CMOS gate arrays
5. 2-μm CMOS standard cells
6. 2-μm MOS full-custom
7. Standard MOS LSI-VLSI (mixture of mature and newer devices)

For brevity we will sometimes refer to MSI-SSI as MSI.

25.3 DESIGN METHODOLOGY SELECTION CRITERIA

We recommend four main selection criteria:

1. Development schedule
2. IC functionality (discussed in Sec. 25.4)
3. Total cost per gate
4. Risk

Development schedule is the elapsed time between the completion of specification and verification that prototypes meet full functional specifications. It excludes fabrication times. This definition applies both at the IC level and at the product level. Product development schedule is the most important economic variable, and therefore it weighs heavily in the choice of design methodology. For example, the Japanese introduced a conservatively designed 64K RAM on a large die early, set the standard, and eventually captured most of the market. This example illustrates that an engineering organization with a schedule advantage will eventually be a generation ahead in lower cost per function.

Schedule is difficult to convert directly into costs, since its effect depends on market factors, such as the rate of product price erosion and market penetration. We have estimated the cost effect of a 1-, 3-, 6- or 12-month schedule slip to be equivalent to a 0.2 to 0.4 percent, 1.5 to 3 percent, 6 to 12 percent, or 25 to 50 percent increase, respectively, in the unit manufacturing cost (UMC) of a product.[1] Reinertsen of McKinsey & Co.[14] estimates a 33 percent decrease in profit for a product that is six months late, assuming a 20 percent market growth rate, a 12 percent annual price erosion, and a five-year product life. If profit decreases quadratically with time, this implies a profit reduction of 1, 8, 33, or 100 percent for a slip of one, three, six or ten months, respectively. In other words, a ten-month slip can make a product unprofitable.

IC functionality refers to the logic, memory, and special functions available on an IC device, such as designer-defined logic functions, pin assignments, megacells, PLDs, memory, and analog cells. It also includes the physical characteristics of IC devices, such as speed and power. The emphasis on speed and power often prevents engineers from examining what must be paid for them in terms of schedule and total cost. Functionality is discussed in Sec. 25.7. For trade-off of functionality against economics, see Muroga.[15] The reliability of semicustom ICs is reviewed by Beasely.[16]

We distinguish between IC devices and systems. A logic IC device consists of a die and its package. To function in a product, devices must be mounted on a printed circuit board (PCB) and supplied with power. The assembled boards (PCBAs) must be held in the product and interconnected. The product must be maintained in the field. All these parts and services, which are required for the device to function in a product, are called *systems*.

One fundamental characteristic of a device is its complexity. Here, we measure device *complexity* by the number of logic gates or the number of equivalent gates when memory or programmable logic arrays (PLAs) are present.

Systems cost includes development, manufacturing, and interest expense, as well as maintenance costs of the end user of the product. The PCBA cost excludes the cost of logic IC devices but includes the cost of passive devices and connectors. The systems cost was determined by two detailed studies in 1983 and 1986 on five different types of products. *All costs are measured in cents per gate in 1986 dollars.*

Total cost refers to the sum of the costs of IC devices and their supporting systems. The cost elements are shown in Table 25.1. Device design costs include nonrecurring engineering (NRE) expenses, i.e., charges to vendors for prototyping, engineering, tooling, and computer time. Device design ends with verification that prototypes meet specifications. Sometimes engineering continues to improve manufacturability and to reduce cost. This effort is called *productizing*.

Strictly speaking, our first three selection criteria are stochastic variables with varying distributions. The measures of dispersion of these distributions are called *risks* of schedule, functionality, and cost. Fey[17] has described a method of evaluating stochastic decision criteria. We prefer to simplify the analysis by treating

TABLE 25.1 Elements of Total Cost

Device cost	Design Productizing Unit manufacturing cost (UMC)
Systems cost	Systems design cost PWBA UMC Service Power Harnessing Frames Cooling Systems test Back panel
Interest cost	

the three measures as deterministic and by adding a fourth one that expresses the risk of not meeting specification of schedule, functionality, or cost.

Additional selection criteria are commonly used. There is an increasing preference of PCBA design organizations to do their own device design. Thus, design methodologies available to the PCBA group are preferred. This works to the disadvantage of full-custom designs, which are usually done by specialists in other organizations.

25.4 ASIC VERSUS STANDARD IC DEVICES: LEVEL OF INTEGRATION

All microelectronic design methodologies compete with one another. We will examine this competition in two steps. In this section we compare designing with ASICs to designing with standard IC devices. Later, we compare the various ASIC methodologies with each other.

We created a measure for comparing ASICs to standard ICs, called *level of integration (LOI)*. For an IC device, LOI is defined by

$$\text{LOI} = \frac{(\text{no. of gates}) \times (\text{degree of usage})}{\text{no. of pins}} \tag{25.1a}$$

For a system (PCBA), LOI is defined by

$$\text{LOI} = \frac{\Sigma(\text{no. of gates}) \times (\text{degree of usage})}{\Sigma(\text{no. of pins})} \tag{25.1b}$$

where gates are the equivalent gates of a logic IC device; pins are the pins of a logic IC device; the summation is over each logic IC device in the system; and degree of usage is the estimated proportion of logic IC device gates used.

Our analysis is based on an empirical model that assumes that devices have certain characteristics. These characteristics are a function of complexity and design methodology. They include the number of pins of the package, power consumption, device cost, and design productivity (gates designed per day). The device LOI can be estimated by the following empirically derived relationships for 1986:

$$\text{LOI} = \frac{\text{no. of gates}}{16 + 0.7(\text{no. of gates})^{0.5}} \quad \text{for full-custom and standard VLSI}$$

$$\text{LOI} = \frac{\text{no. of gates}}{16 + 0.9(\text{no. of gates})^{0.5}} \quad \text{for standard cells} \tag{25.2}$$

$$\text{LOI} = \frac{\text{no. of gates}}{16 + 1.3(\text{no. of gates})^{0.5}} \quad \text{for gate arrays}$$

These relationships are obtained by dividing the number of gates by the estimated number of pins. The estimated number of pins has a functional form similar to Rent's rule, although the coefficients differ.[18,19] Note that for a given number of gates, LOI increases from gate arrays to standard cells to full-custom devices. The sensitivity of the relative cost of various ASIC methodologies to these LOI assumptions will be examined later.

The relationship between LOI and total cost per gate for IC devices at a 10,000 units per year production volume is shown in Fig. 25.3. Please note that the figure deals with *IC devices,* not with systems. The ASIC curve represents the cost of the lowest-cost ASIC for a given LOI. The cost curves have the typical convex downward shape; costs at first decrease rapidly, then more slowly, and finally start increasing, first slowly and then rapidly. The minimum cost point varies for different design methodologies. At 10,000 units per year, the minimum cost for ASIC is at an LOI of 90, whereas for standard VLSI devices it is about 250 and, therefore, is not on the figure. The rapid cost increase at very high LOI is not seen in actual IC designs, since such devices are not economically viable. At 1000 units per year, the ASIC minimum is $0.025 per gate at an LOI of 60; at 100,000 units per year it is $0.005 per gate at an LOI of 150. Average LOI changes over time, but the change in cost per gate proceeds along a cost-versus-LOI curve that moves slowly to lower costs.

The key points of Fig. 25.3 are

- SSI-MSI have a low LOI of 1–3.
- PLDs also have low LOI of about 3–5.
- At 10,000 units per year, costs decrease rapidly as LOI increases to about 30; *high LOI significantly decreases device-related cost.*
- Cost curves diverge as LOI increases.
- At the same LOI, ASICs have a higher cost than standard IC devices do.
- To be cost-effective, ASIC must achieve an LOI of about twice that of a standard LSI-VLSI device at 10,000 units per year.

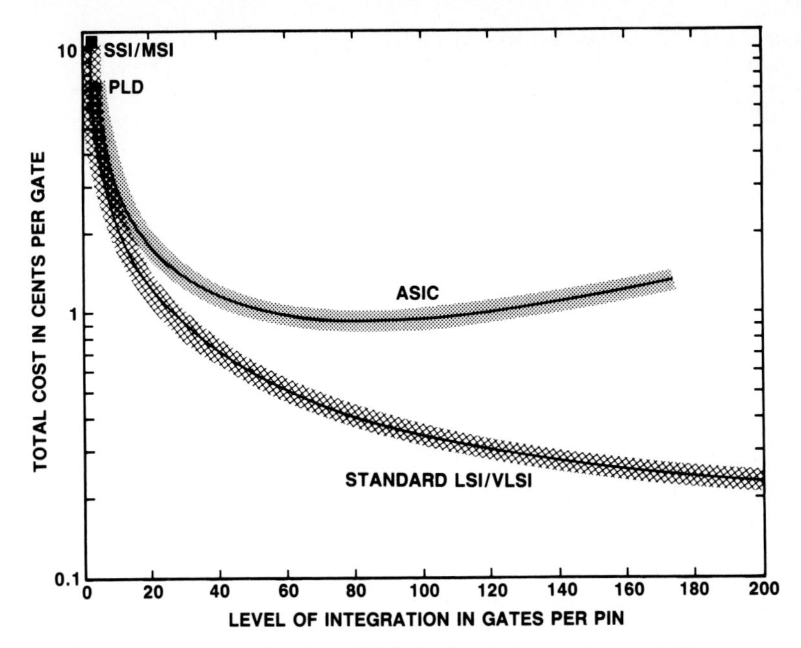

FIG. 25.3 Total cost as a function of IC device level of integration at 10,000 units per year for four years. At a given level of integration standard IC devices have lower cost than ASIC. To compete, ASICs must increase the level of integration. The ASIC curve is the envelope of the lowest-cost ASIC for a given level of integration.

However, the total costs of a product are determined not by the LOI of a device, but by the LOI of a system. The LOI of systems, containing "off-the-shelf" standard VLSI with high LOI, decreases rapidly as usage falls below 100 percent and as a few low-LOI peripheral devices are added. These phenomena are illustrated in Table 25.2. The microprocessor, at 100 percent usage, has an LOI of 125. However, the system has an LOI of 25. If the number of SSI-MSI is decreased from 6 to 4, LOI increases by 25 percent to 31. In 1986, even though 90 percent of the gates used in PCBA manufacturing were in devices with LOIs between 40 and 400, most PCBAs had LOIs between 10 and 100. However, 90 percent of the cost is in ICs with an average LOI of 17.

The LOI of an ASIC design is sometimes greater than expected due to a tendency of designers to increase its logic functions over those of an SSI-MSI

TABLE 25.2 Example of System Level of Integration

	No. gates	No. pins	Degree of usage, %	LOI
1 microprocessor	5000	40	50	62.5
1 LSI peripheral	1500	28	100	53.6
6 SSI-MSI (at 30 gates, 16 pins each)	180	96	100	1.9
Total	4180	164		25.5

design. This is a classic demand effect. As cost per gate decreases, the quantity of gates used increases. This not only adds functions, features, or performance to the product but also increases cost in terms of labor, schedule, LOI, and UMC. Managers should guard against undesirable additions of logic functions.

Discussion of the comparative costs of systems containing ASIC with systems containing standard VLSI devices is beyond the scope of this chapter. As a first-order approximation, one can extend the IC device relationships to systems. But the reader should note that this is a linear approximation to a nonlinear relationship.

To summarize, ASICs compete well against SSI-MSI, since they have a higher LOI. The LOI of systems of standard VLSI degrades rapidly with the addition of low-LOI peripheral devices, often required for specific applications. Replacement of these high-priced low-LOI VLSI peripheral devices allows ASICs to compete at the systems level.

25.5 FULL-CUSTOM DESIGN PRODUCTIVITY

The traditional design methodology involves the design of individual transistors. Productivity of this method is discussed first. Productivity for more recent design methodologies is summarized later. For a detailed method for estimating gate-array design productivity, see Fey and Paraskevopoulos.[19a]

Various types of transistors (e.g., unique random logic, RAM) require different amounts of design time. A measure called *equivalent transistor* (design time) is introduced to normalize these values. A unique random logic transistor is taken as unity. For any other transistor type, the number of equivalent transistors is equal to the relative design time.

Equivalent Transistor (Design Time). An equivalent transistor is a measure of the time required to design a transistor. Two transistors are equivalent if the times for designing each transistor are equal. The time to design a unique random logic transistor is taken as unity. A physical transistor that takes a fraction K of that time has a value of K equivalent transistors.

For example, a transistor that can be designed in 50 percent of the time required for a unique random logic transistor has a value of 0.5 equivalent transistor. The model expresses design manpower as a function of equivalent transistors.

By our definition, the design effort starts after the requirements for the chip have been specified. It consists of design and layout. The effort ends after prototype chips have been successfully tested to verify that they accomplish the design intent. Productization efforts for yield or producibility improvements are not included. Table 25.3 specifies the relations that transform physical transistor counts to equivalent transistors. One unique random logic transistor is, by definition, one equivalent transistor. The quantities C, E, F, and G are parameters of the model. Memory exhibits economies of scale (the larger the memory, the fewer man-days per transistor), hence, the fractional exponent of RAM and ROM. Note that transistors are measured in units of 1000. Only active PLA transistors are counted. Memory is also measured in transistors, not in bits.

We distinguish between unique and repeated random logic transistors. Suppose one designs 16-bit parallel processing random logic. The first transistor designed is unique random logic. The 15 repetitions of this transistor are repeated random logic.

For logic, VLSI design manpower per device increases with complexity (number of transistors); for memory, it decreases with complexity. Thus, pure

TABLE 25.3 Transistor Equivalence

Type of transistor	*K* equivalent transistors
Unique random logic (UNQ)	1(UNQ transistors)
Repeated random logic (RPT)	*C*(RPT transistors)
Programmed logic array (PLA)	*E*(PLA transistors)
RAM	$F(\text{RAM transistors})^{-0.5}$
ROM	$G(\text{ROM transistors})^{-0.5}$

logic without memory shows diseconomies, while pure memory without logic shows economies of scale. For example, 1000 unique random logic transistors can be designed at the rate of four transistors per day. This productivity drops to two transistors per day for a 60,000-transistor design. The additional amount of interconnection makes the latter task more difficult. This decrease in productivity is shown in Fig. 25.4. Such diseconomies have been observed in other similar activities, such as software engineering.[20]

Pure memory, on the other hand, exhibits economies of scale. As Fig. 25.5 indicates, a 1000-transistor RAM can be produced at six transistors per day, while a 300,000-transistor RAM can be designed at 10 times that rate. Much effort is expended in developing the first few memory cells. Adding additional cells requires a decreasing amount of effort.

Most manpower is a function of the number of transistors designed. However, there may also be a fixed amount of manpower *A* required for each design, irrespective of its size. The model allows for such a fixed start-up cost.

Both productivity and requirements (e.g., frequency) change over time. The model approximates these irregular changes by a constant percentage change in productivity per year. This parameter reflects the combined effect of an increase in productivity and an increase in requirements, which decreases productivity. The increase in productivity is a function of technology, training, and design learning (i.e., the number of designs produced by the firm and the industry). An

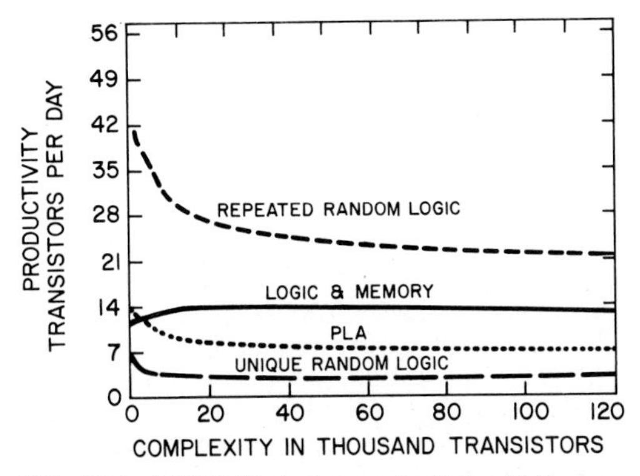

FIG. 25.4 LSI-VLSI design productivity, U.S. random logic, 1984.

organization learns most from producing its own designs, but it learns also, with some time delay, from the designs produced by others.

Putting all of these considerations together, one obtains the following model for design manpower M in man-months:

$$M = (1 + D)^{\mathrm{YR}}[A + B(K \text{ equivalent transistors})^{H}] \tag{25.3}$$

where YR is 1984 minus the year of the bulk of the design effort, and A, B, D, and H are parameters of the model. Note that the model adjusts productivity from year to year. D is the average annual improvement factor; A is the start-up manpower; B is a measure of the productivity, the time required to design one equivalent transistor (i.e., one unique random logic transistor); and H is a measure of economies-diseconomies of scale due to changes in complexity.

Substituting Table 25.3 in (25.3), one obtains the following equation for the number of design man-months M:

$$M = (1 + D)^{\mathrm{YR}}[A + B(\mathrm{UNQ} + C\cdot\mathrm{RPT} + E\cdot\mathrm{PLA} + F\cdot\mathrm{RAM}^{0.5} + G\cdot\mathrm{ROM}^{0.5})^{H}] \tag{25.4}$$

To summarize, the model states that design manpower is as a first-approximation B man-month per equivalent transistor. This first approximation is improved by three refinements:

1. *Complexity:* Complex designs (many equivalent transistors) may have lower or higher productivity, as indicated by the exponent H. These changes in productivity are called diseconomies and economies of scale, respectively.
2. *Start-up:* Some manpower A may be independent of the number of equivalent transistors.
3. *Learning:* Each year productivity increases by $100D$ percent.

The foregoing model is a compromise between two conflicting objectives: to have but a few parameters and, at the same time, to include all important factors. The relatively small number of parameters requires changes of parameter values, where design requirements, designer experience, technology, or other circumstances differ from the sample average. Hence, we provide not only an estimate of the parameters but also low and high values. The extreme values may apply for design conditions that differ from the average of our sample. The parameter values in Table 25.4 are preliminary estimates from a limited amount of data.

The model has some important implications. Figures 25.4 through 25.7 display the model relations graphically. Each figure shows productivity as a function of complexity. Productivity is measured by the number of transistors designed per day, complexity by the number of transistors on the chip (1 year = 252 working days). The figures represent the results for the average of the designs of a particular class (sample mix). Note that the scales of the figures vary.

Over most of its range, productivity is not a strong function of complexity (Figs. 25.4 and 25.5). Only productivity of pure memory is (Fig. 25.5). Since pure ROM memory design takes a relatively small fraction of the total design time for both memory and logic chips, this dependence has little effect on the overall productivity complexity relationship. The largest diseconomies of scale are seen for repeated random logic (Fig. 25.4). Memory productivity, on the other hand, increases with complexity (Fig. 25.5). The effect of complexity is largest for small

TABLE 25.4 Manpower Model Parameters Values

Parameters	Low value	Estimate	High value
A. Constant	0	0	3
B. Productivity	6	12	20
C. Repeated logic	0.05	0.13	0.25
D. Improvement	−0.05	0.02	0.10
E. PLA	−0.1	0.37	0.7
F. RAM	0.1	0.65	1.3
G. ROM	0.05	0.08	0.15
H. Complexity	1.05	1.13	1.40

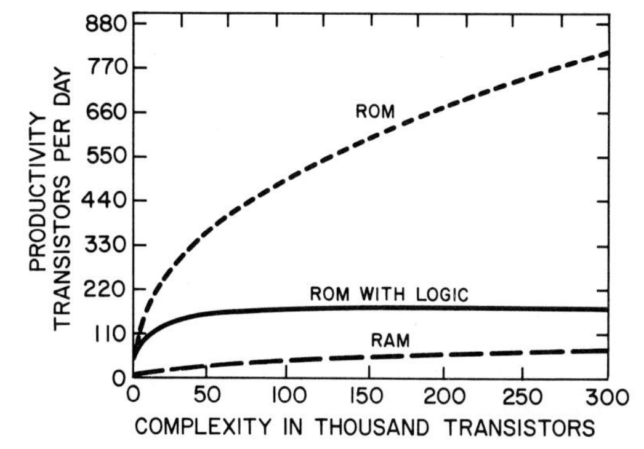

FIG. 25.5 LSI-VLSI design productivity, U.S. memory, 1984.

designs, due to the exponential relationship. Every time the number of equivalent transistors doubles, productivity changes by a fixed percentage.

Most designs contain several different types of transistors. Even for logic chips, our sample mix has a productivity of about 13 transistors per day in most of the range of interest (Fig. 25.4). This is considerably higher than the proverbial eight transistors per day. For ROM designs, including the associated logic, productivity is still higher: 180 transistors per day (Fig. 25.5).

The model parameters differ in their effect on productivity. Table 25.5 lists the parameters in decreasing order of importance. Here, importance is measured at 120,000 transistors for logic and at 300,000 transistors for ROM.

The sensitivity of productivity to changes in the model parameters associated with a given transistor type (parameters *C* through *G*) depends on the relative amount of design time spent on that particular transistor type in our sample. This is a function of both the design time for one transistor and the proportion of all transistors that are of that type. For example, the repeated random logic parameter is an important determinant for productivity of ROMs but not for logic designs. In ROMs there is a relatively large amount of repeated random logic, and the time spent on it is large relative to the time spent on memory design. In logic designs, on the other hand, the time spent on the repeated random logic is small compared with the time spent on unique random logic.

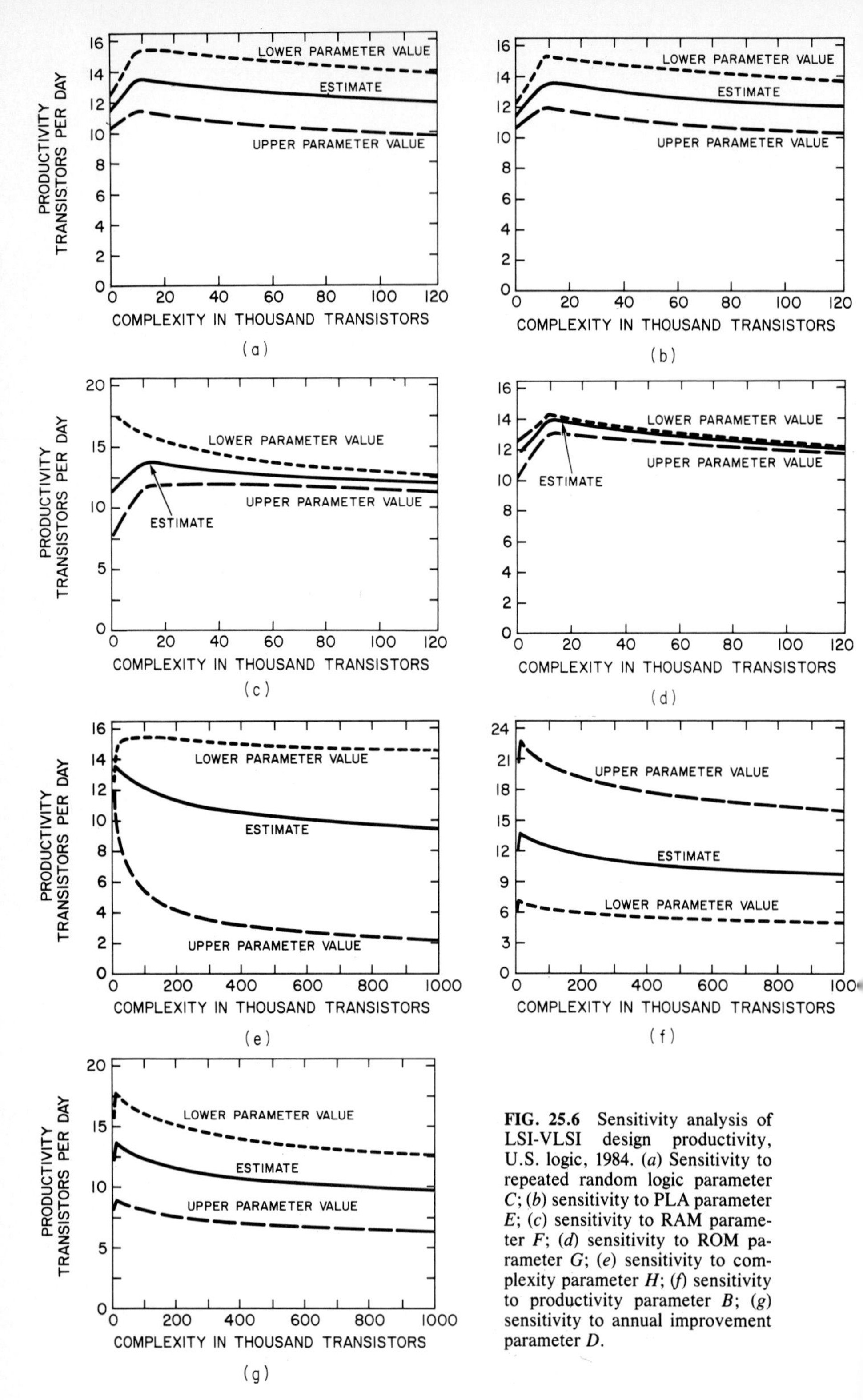

FIG. 25.6 Sensitivity analysis of LSI-VLSI design productivity, U.S. logic, 1984. (*a*) Sensitivity to repeated random logic parameter *C*; (*b*) sensitivity to PLA parameter *E*; (*c*) sensitivity to RAM parameter *F*; (*d*) sensitivity to ROM parameter *G*; (*e*) sensitivity to complexity parameter *H*; (*f*) sensitivity to productivity parameter *B*; (*g*) sensitivity to annual improvement parameter *D*.

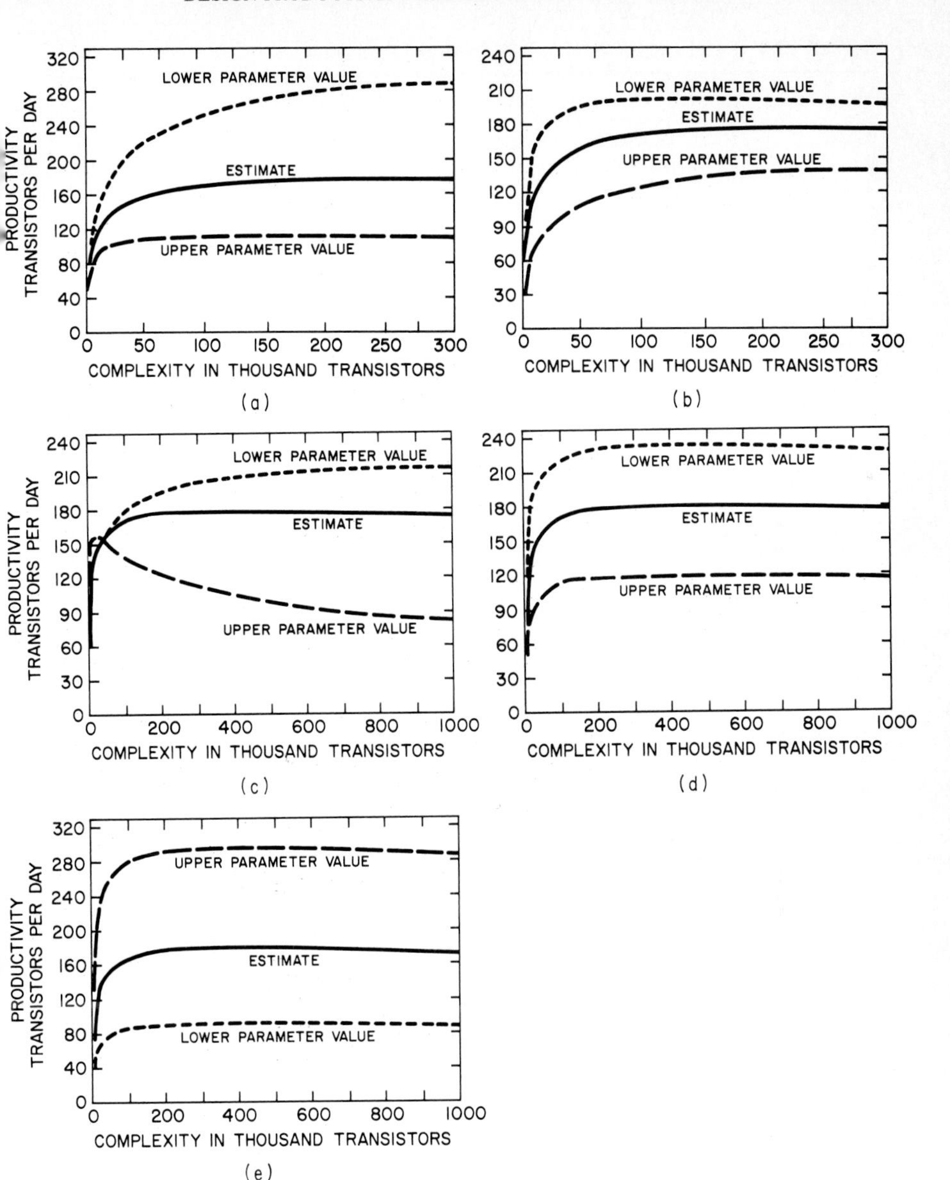

FIG. 25.7 Sensitivity analysis of LSI-VLSI design productivity, U.S. ROM memory, 1984. (*a*) sensitivity to repeated random logic parameter *C*; (*b*) sensitivity to ROM parameter *G*; (*c*) sensitivity to complexity parameter *H*; (*d*) sensitivity to annual improvement parameter *D*; (*e*) sensitivity to productivity parameter *B*.

TABLE 25.5 Sensitivity of Productivity to Model Parameters

Design	Parameter	Sensitivity	Figure
Logic	*H*. Diseconomies of scale	High	25.6*e*
Logic	*B*. Productivity	High	25.6*f*
Logic	*D*. Annual improvement	High	25.6*g*
Logic	*C*. Repeated random logic	Low	25.6*a*
Logic	*E*. PLA	Low	25.6*b*
Logic	*F*. RAM	Low	25.6*c*
Logic	*G*. ROM	Low	25.6*d*
ROM	*B*. Productivity	High	25.7*e*
ROM	*C*. Repeated random logic	High	25.7*a*
ROM	*D*. Annual improvement	High	25.7*d*
ROM	*H*. Complexity	High	25.7*c*
ROM	*G*. ROM	Low	25.7*b*

Note: Sensitivity for sample mix of transistors is measured at 120,000 transistors for logic and at 300,000 for MOS. Parameters are listed in decreasing order of sensitivity.

The productivity parameter *B* (Figs. 25.6*f* and 25.7*e*), the complexity parameter *H* (Figs. 25.6*e* and 25.7*c*), and the annual improvement parameter *D* (Figs. 25.6*g* and 25.7*d*) largely determine the productivity for logic and ROM. The productivity parameter *B* has a large effect both for logic and for ROM designs. This factor is a composite of many factors that would be included in a more detailed model. The detailed factors would account for differences in requirements, experience, tools, technology, etc., that vary for different organizations. Such wide variations in productivity factors are not unusual. Similarly, complexity effects of the magnitude, shown in Table 25.4 and Figs. 25.6*e* and 25.7*c*, are not unusual for engineering activities.

The value of *H* depends on the nature of the design problem and usually varies between 1 and 1.2. The more complex the interconnection of various parts of the design, the greater are the diseconomies of scale.

The low rate of annual improvements in productivity is surprising. The conundrum of a small change in productivity in the face of heavy investment in technology, such as CAD tools, may be due to increases in product requirements, such as speed, frequent changes in technology, rapid growth in number of designers and teams (low experience), and absence of standardization (not-invented-here factor). Together they may offset the productivity gains. This parameter could vary in the future as the relative emphasis on productivity and requirements changes, as technology stabilizes, or as standards develop.

Possibly the low rate of improvements over time is a function of our method of measurement. Technological progress is not a continuous function of time. Rather it is a function of the number of models designed with discontinuities at irregular intervals, whenever an innovation occurs.

Note that the repeated logic parameter is an important determinant for productivity of ROMs but not for logic design (Figs. 25.7*a* and 25.6*a*). The manpower is not sensitive to the ROM parameter (Figs. 25.6*d* and 25.7*b*), the RAM parameter (Fig. 25.6*c*), or the PLA parameter (Fig. 25.6*b*).

The model can aid in making predictions about the future of VLSI design, although extrapolation is risky. Under our assumptions, a 1,000,000-transistor logic chip in 1989 would take about 400 man-years to design. However, if the complexity of the interconnection requirements increases at the same time, the task design time could increase by a factor of 10. Such a design manpower (and time) is unacceptably large; hence this method cannot be used to design large ICs. It is for this reason that chips such as the Intel 80386 were designed first as cell-based designs and then shrunk.

Under our assumptions, ROM chips of the same size require only 20 man-years to design. Even under unfavorable diseconomies of scale, the size of the effort would only double. However, more favorable diseconomies of scale assumptions would not decrease the size of the effort significantly. Because of the relatively small size of the ROM design effort, it is not surprising that we already see 1,000,000-transistor ROM chips.

25.6 DEVELOPMENT SCHEDULE

25.6.1 Design Methodologies and Phases

Gates are selected as the basic unit of VLSI functionality. One gate is defined as three transistors for NMOS and as four transistors for CMOS. VLSI design manpower and schedule included the following phases:

1. Logic design and verification, including chip architecture
2. Circuit design and verification, including timing and simulation
3. Layout and verification
4. Test design, testing, and debugging

Design manpower, like scheduling, includes all activities up to confirming that prototypes meet full functional specifications; it excludes generation of specifications and redesign due to specification changes. Mask generation and fabrication were excluded, since they are not primarily a function of design methodology. These activities usually require an insignificant amount of design engineering manpower. Their elapsed time is primarily a function of the vendor, the vendor's backlog, and the buyer's schedule priority. In 1987 the mask generation and fabrication for gate arrays took about two to six weeks.

25.6.2 Schedule Model

We have studied IC device schedules of 81 designs from 21 firms. The results indicate that schedule is a function of a single parameter, namely the amount of effort or manpower (Fig. 25.8). This is true for all VLSI devices, such as gate arrays, standard cells, cell-based, and full-custom. The function consists of two segments. The initial segment, with a slope of 1, represents one-person projects. As projects increase in size beyond approximately 30 man-weeks, schedules become proportional to the cube root of manpower. Most of our full-custom schedules are longer than 30 weeks; thus their relationship shows only the previously reported exponential segment, called the *schedule line* (Figs. 25.8 and 25.9). The following schedule model summarizes the relationship between manpower and schedule:

$$T = M \qquad M < 29 \tag{25.5a}$$

$$T = 9.1M^{0.34} \qquad M \geqq 29 \tag{25.5b}$$

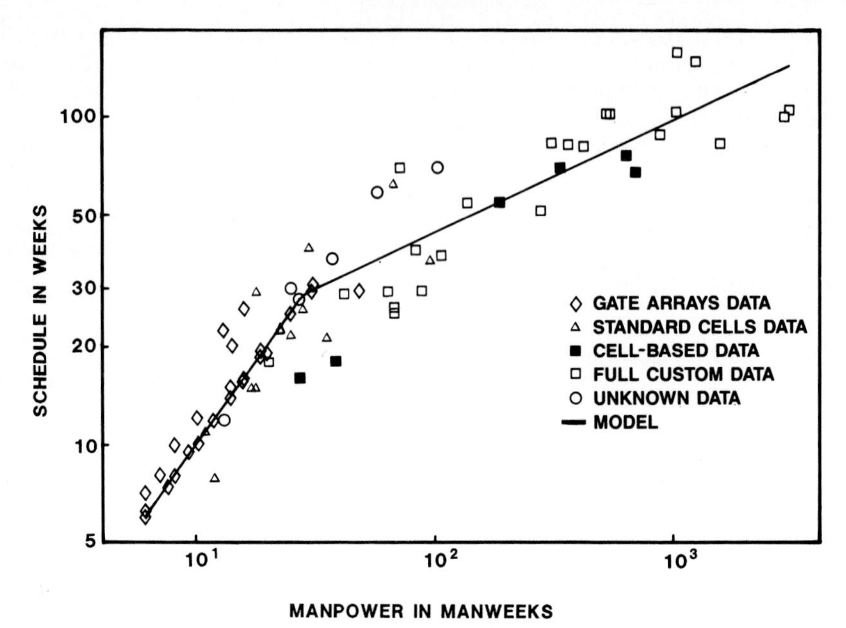

FIG. 25.8 VLSI schedules as a function of manpower. Function is piecewise log linear with a breakpoint at 29 weeks. Break occurs where one-person projects change to multiperson projects.

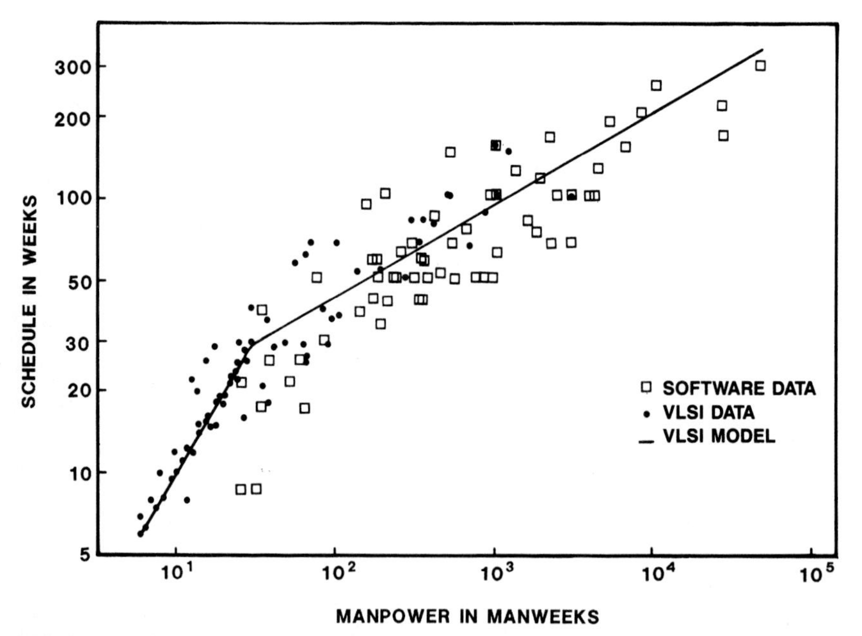

FIG. 25.9 VLSI and software schedules as a function of manpower. VLSI schedules are similar to software schedules but are longer for the same manpower and less variable.

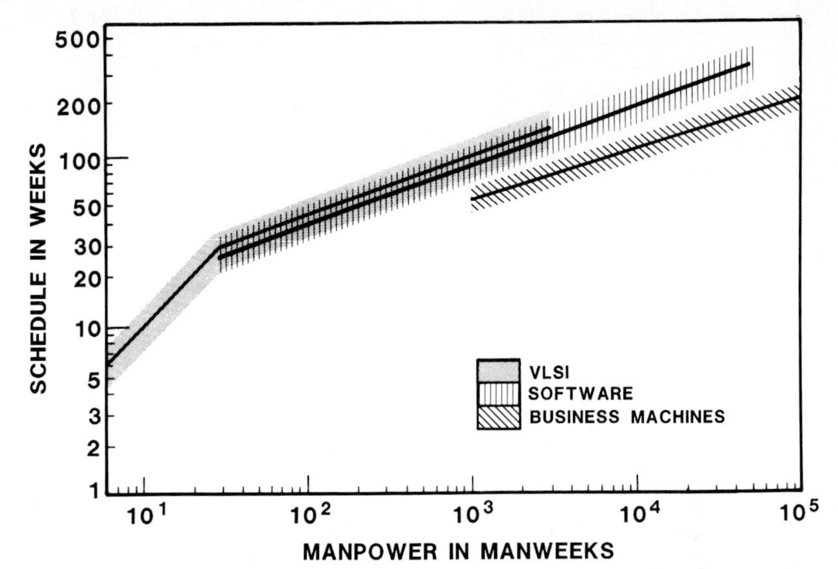

FIG. 25.10 VLSI, software, and business machine schedules as a function of manpower. Slopes of hardware and software schedules are the same. Business machine schedules are shorter for same manpower due to partitioning into subsystems.

where T is the schedule in weeks and M is manpower in man-weeks. Taken as a whole, the model fits the data reasonably well (Fig. 25.8). The median error is 13 percent of observed value; 75 percent of the errors were less than 29 percent. For details and for comparison with other models, see Paraskevopoulos and Fey.[3]

Figure 25.8 shows that the model fits the data for gate arrays, standard cells, cell-based, and full-custom VLSI. Figure 25.9 indicates that the VLSI model is very general; it fits both the software[20] and VLSI data. The slope of the VLSI schedule line (exponent) is similar to those found in other hardware and software projects (Fig. 25.10). However, the intercepts (multipliers) differ.

25.6.3 Effect of Partitioning

If a design project can be partitioned into independent subsystems, we recommend the following relation. The schedule T of a project with manpower M and n subsystems is given by

$$T = 3.5\left(\frac{kM}{n}\right)^{0.34} \tag{25.6}$$

where T is in months, M is in man-months, and k is between 1 and 2. For details see Paraskevopoulos and Fey.[3]

25.6.4 Distribution of Effort by Phases

Semicustom designs have higher productivities than do full-custom designs, as shown in Table 25.6 and Fig. 25.11. Therefore, they require less manpower and

TABLE 25.6 Average Productivity in Gates per Day

Methodology	Average productivity gates per day
Gate arrays	25
Standard cells	13
Cell based	4
Full custom	2.7

Source: R. C. Anderson.[21]

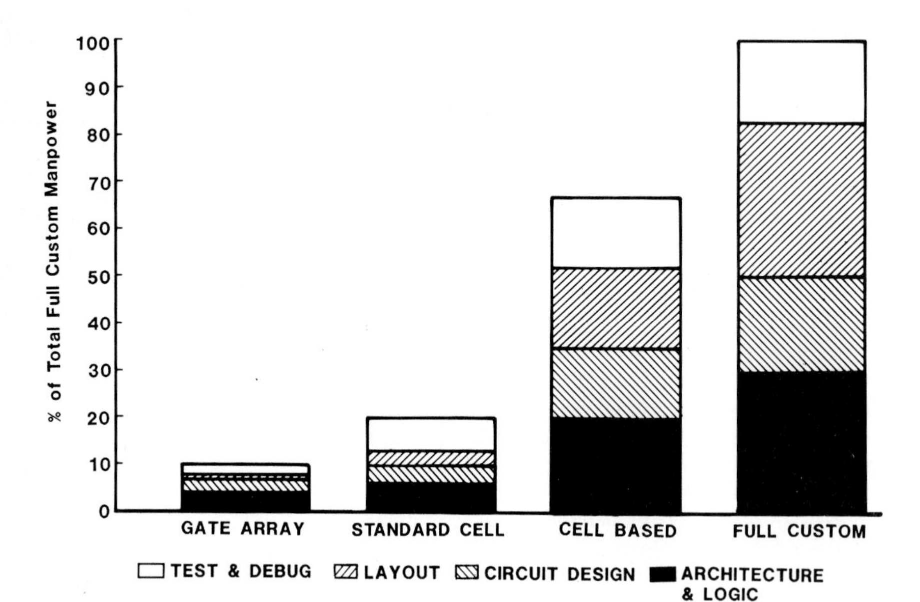

FIG. 25.11 Average relative amount of manpower by phase. Gate arrays (standard cells) have 10 times (5 times) the productivity of full-custom designs.

have shorter schedules. This higher productivity pervades all phases of the design (Table 25.7 and Fig. 25.11). The factors that promote this higher productivity affect all phases uniformly, since the percentage distribution is nearly the same for all methodologies (Table 25.7 and Fig. 25.12). The layout phase is the only exception.

25.6.5 Schedule as a Function of Complexity

Manpower is a function of productivity and complexity. Schedule is a function of manpower. Hence, schedule is a function of complexity for a given productivity. The latter relationship can be quantified by using our model and productivities, as shown in Fig. 25.13. Here we assume that productivities for gate arrays and standard cells are independent of complexity and that full-custom devices have a pro-

TABLE 25.7 Distribution of Effort by Phases

Mean ± one standard deviation, both in percentage of total effort.

Methodology	Arch. and logic	Circuit design*	Layout	Test and debug	Total	No. of designs
Gate arrays	40 ± 16	28 ± 11	8 ± 5	24 ± 10	100	23
Standard cells	30 ± 13	18 ± 13	16 ± 10	36 ± 14	100	15
Cell based	30 ± 8	22 ± 7	26 ± 6	22 ± 6	100	9
Full custom	30 ± 9	20 ± 8	33 ± 15	17 ± 5	100	16

*Including timing and simulation.

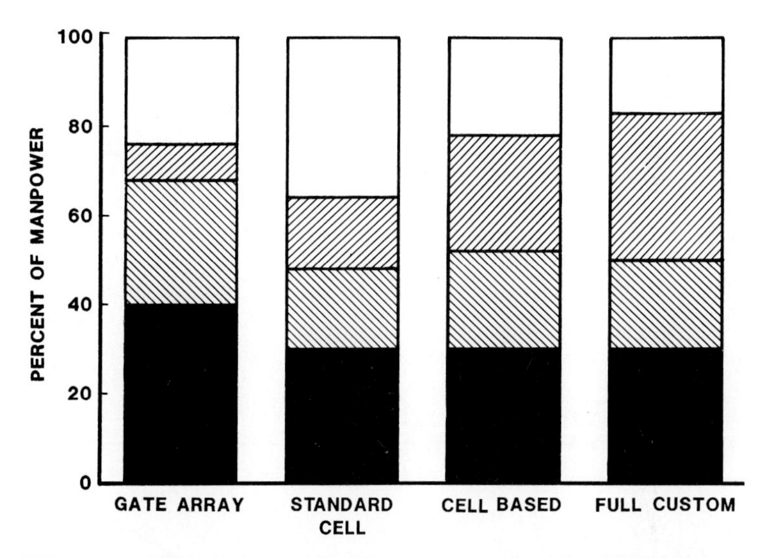

FIG. 25.12 Distribution of manpower by phase. Distribution is similar for various methodologies, except for layout.

ductivity of 2.7 gates per day at 4000 gates; at different complexities we scale their productivities by the relationship shown earlier.

25.6.6 Team Size

Average team size S can be determined from manpower M and schedule T as follows:

$$S = \frac{M}{T} = a^{-1}M^{1-b} \tag{25.7}$$

where $a = 9.1$ and $b = 0.34$ are the parameters in Eq. (25.5).

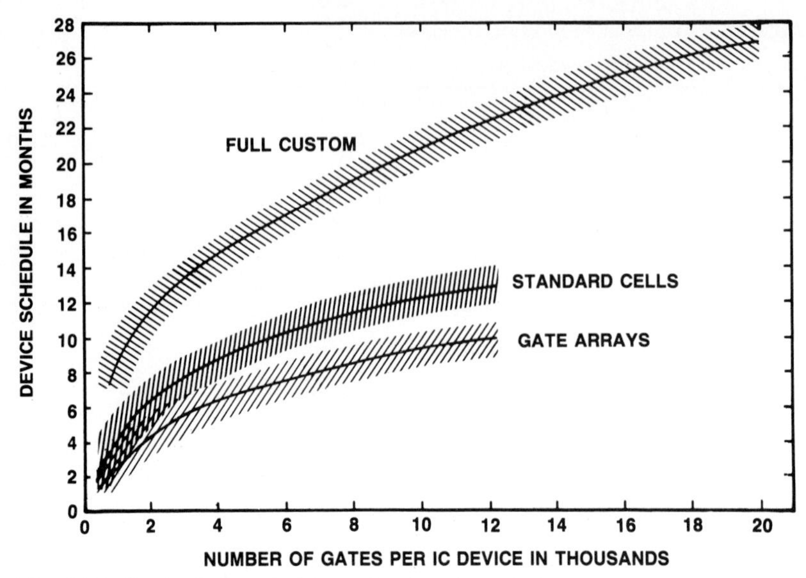

FIG. 25.13 Device design schedules as function of complexity. Gate arrays and standard cells can be designed in approximately half the time of full-custom devices of same complexity.

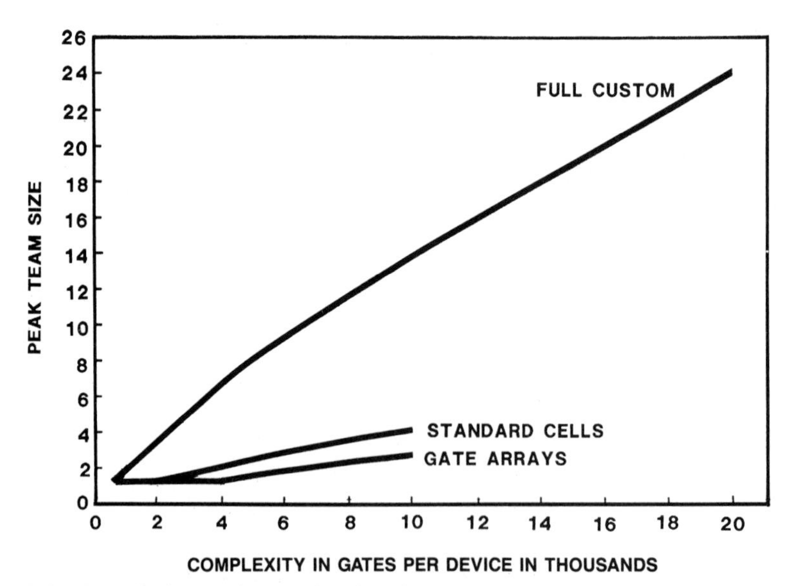

FIG. 25.14 Peak team size as a function of complexity. Semicustom devices require about one third the peak team size of full-custom designs.

More important than average team size is peak manpower. It can be developed with results from Norden,[22] Starr,[23] and Putnam.[24] Figure 25.14 shows that full-custom requires larger peak team sizes than do gate arrays and standard cells. For example, an 8000-gate design requires up to 11 to 12 people in full-custom, but only 2 for gate arrays and 3 to 4 for standard cells. The low productivity of full-custom designs is partially due to this larger peak team size.

25.6.7 Methods for Compressing Schedules

Increasing Productivity. Increasing productivity is the best and most practical way to reduce schedule. This can be accomplished through high-productivity design methodologies and through high productivity for a given design. Gate arrays and standard cells offer a 5- to 10-fold improvement in productivity over full-custom designs (Table 25.6, Fig. 25.11). This translates to a 50 percent reduction of design schedules for devices of the same complexity (Fig. 25.13).

The increase in productivity extends nearly evenly to all phases of design, but it is particularly pronounced in layout, which is highly automated in gate arrays and standard cells (Fig. 25.12). This uniform increase indicates that some fundamental changes have taken place in the design environment, and they affect *all* phases: the systems designer is close to device designs, team size is reduced, the design process is standardized, only proven technology is used, and rapid feedback of device performance is obtained.

Reduction in team size offers communication savings, particularly if the systems designer can design the device. This is generally the case with semicustom ICs. Standardization eliminates choices, introduces discipline, and facilitates learning (improvements in productivity). Standardization also enables the use of proven technology and prevents the use of ad hoc design rules or pushing the available technology beyond its reasonable limits. The net effect is a high rate of first-pass success. The short schedule itself enables rapid improvements in productivity, since good practices are reinforced early and bad practices are discovered soon. We believe this phenomenon, which we call *design learning,* is one of the key advantages of semicustom designs.

In conclusion, semicustom devices promote high productivity. Improving full-custom design productivity requires similar design practices. Such changes are gradually taking place. Pure full custom designs are now rarely done. There is a shift toward increasing standardization (cell-based design) in order to improve productivity, to using proven design rules, and to avoiding unproven technology. The latter is accomplished through increasing separation of cell and process technology development from device design and volume production.

Decreasing the Design Task (Chip Partitioning). Schedule is a function of the size of the design task (e.g., complexity measured in gates per IC, Fig. 25.13). Partitioning a design into several *independent* parts of lower complexity will reduce schedule, provided that no part affects another. This can be achieved only through writing *formal specifications* for each part at the start of the design, which is difficult to accomplish. For shortest schedules, each part should require about the same amount of effort as discussed previously. High-complexity designs carry a schedule penalty. A 20,000-gate full-custom device takes an average of 2.7 years to complete, compared to four 5000-gate gate arrays, which average seven months. Another way to reduce the design task is through reusing previous designs wholly or partially (design portability).

TABLE 25.8 Penalties for Short and Long Design Schedules

Schedule	−25%	−20%	−10%	Normal	+10%	+20%	+30%
Manpower, Tausworthe	+19%	+14%	+7%	0	−6%	−10%	−15%
Manpower, Boehm	+23%	+12%	+5%	0	+1%	+3%	+4%

Source: Boehm[20] and Tausworthe.[25]

Increasing Manpower. When schedule problems arise, project managers tend to ask for more manpower. Higher-level management should deny such requests and insist on finding productivity-increasing solutions that will reduce manpower and consequently shorten schedule. The effect of increased manpower is shown in Table 25.8.

25.6.8 The Value of Short Schedules

The value of short schedules depends on market growth rates, product life, and the customer's switching cost. In Sec. 25.3 we saw that a schedule slip could make a product unprofitable. However, many manufacturers still underestimate the importance of schedule. For example, in a study of seven European electronics firms, none gave schedule a higher priority than cost or performance. This is in marked contrast to the priorities of Pacific rim firms.

The important economic quantity is not IC device design schedule but *systems development schedule*. The systems schedule advantage of ASICs derives from their high LOI. SSI-MSI have an LOI of one to three gates per pin. One hundred 16-pin dual in-line packages (DIP) require about a 100-in^2 board that must be designed, placed into a machine, powered, cooled, interconnected, and serviced. This 100-in^2 PCBA can often be replaced by a single ASIC device (LOI = 100 gates/pin) requiring 4 to 7 in^2 of PCBA. Therefore, the product containing ASIC will generally take less time to develop, since the amount of PCBA design is reduced by a factor of 10 to 20. PCBA design is often less automated than chip design. Hence, converting board design to chip design increases the level of automation and therefore productivity. SSI-MSI will typically take more time, since they require more iterations, more physical volume, more mechanical design, more physical interconnection, and more labor. Thus, we see that semicustom designs increase the level of automation and productivity, thus reducing manpower and, consequently, development schedule at the PCBA level. The schedule effect can be estimated by applying Eq. (25.5), since the relationship is also a first-order approximation to systems schedules.

Systems design schedules can be compressed by reducing the required manpower through system-level high-productivity design methodologies (gate arrays, standard cells) or through partitioning of IC devices into devices of lower nominal complexity. Increasing team size offers only limited opportunity for schedule reduction.

TABLE 25.9 Functionality*

	SSI/MSI	PLD	Standard VLSI	Gate array	Standard cell	Full-custom
Logic functions (designer defined)	N	L	L	Y	Y	Y
Pin assignments	N	N	N	L	Y	Y
Mega cells	N	N	L	L	Y	Y
PLA	N	N	N	N	Y	Y
Memory	N	N	Y	L	Y	Y
Analog cells	N	N	N	N	L	Y

*Y = yes, typically available; L = limited, or through software programming; N = not readily available.

25.7 IC FUNCTIONALITY

We will briefly discuss functionality. An increasingly important aspect of systems functionality is the tradeoff between software and hardware, discussed by Hartman.[26] For a basic understanding of gate arrays, see Read;[27] for details of ASIC functionality see Hurst[28] or Walsh;[29] for systems implications of functionality see Muroga.[15] Systems case studies can be found in Weste and Eshraghian.[30] For subjects not covered in the above, see the 12-volume treatise edited by Einspruch.[31] The reliability of semicustom ICs is reviewed by Beasely.[16] Functionality of various IC design methodologies is summarized in Table 25.9.

High functionality, or complexity, typically demands a price in terms of schedule, cost, or risk. When technology is pushed to the limit for a fast-paced program, one often inadvertently oversteps that bound. To avoid this difficulty, one should carefully separate technology development (extending the limits) from product design. We should also be willing to examine alternatives and make tradeoffs and compromises. In particular, we recommend careful examination of the effect of high functionality on schedule and its risk. A common trap is replacing a known technology with too high a cost, or known limitations with a new technology having unknown cost and limitations.

25.8 SUMMARY AND CONCLUSIONS

1. ASIC CAD tools, libraries, and design support organizations enable high productivity, reusability commonality, and standardization among diverse design groups.
2. ASICs have a systems development schedule advantage over SSI-MSI. The schedule advantage derives from high productivity, reusability, and standard-

ization enabled by CAD tools. The high productivity results in low systems development manpower and therefore short schedule. Gate arrays and standard cells have a significant IC device development schedule advantage over full-custom designs.

3. The key economic challenge to ASIC is achieving improved design productivity (schedule), particularly for designs of ever increasing complexity.

ACKNOWLEDGMENTS

The authors would like to thank R. Markle, Professor S. Muroga, and S. Shohara for their critical comments. The authors also thank R. C. Anderson, Inc., for permitting the use of data from their research reports and for providing additional data. Also, they thank those merchant semiconductor companies and electronics manufacturers who contributed their data anonymously to the original studies, and the Xerox Microelectronics Center and the Xerox Technology Strategy Office for their support.

REFERENCES

1. C. F. Fey and D. E. Paraskevopoulos, "A Technoeconomic Assessment of Application Specific Integrated Circuits, Current Status and Future Trends," *Proc. IEEE,* **75**:829 (1987).
2. C. F. Fey and D. E. Paraskevopoulos, "Studies in LSI Technology Economics II: A Comparison of Product Cost Using MSI, Gate Arrays, Standard Cells, and Full Custom VLSI," *IEEE J. Solid State Circuits,* **SC-21**:297 (1986).
3. D. E. Paraskevopoulos and C. F. Fey, "Studies in LSI Technology Economics III: Design Schedules for Application Specific Integrated Circuits," *IEEE J. Solid State Circuits,* **SC-22**:223 (1987).
4. C. F. Fey, "Custom LSI/VLSI Chip Design Productivity," *IEEE J. Solid State Circuits,* **SC-20**:555 (1985).
5. Integrated Circuits Engineering Corp., *Status 1987: A Report on the Integrated Circuit Industry,* Scottsdale, Ariz., 1986.
6. Semiconductor Users Information Service, San Jose, Cal., Dataquest, August 26, 1985.
7. N. Hazewindus, *The U.S. Microelectronics Industry: Technical Change, Industry Growth & Social Impact,* Pergamon Press, New York, 1982.
8. J. G. Linvill et al., *The Competitive Status of the U.S. Electronics Industry,* National Academy Press, Washington, D.C., 1984.
9. D. Ernst, *The Global Race in Microelectronics,* Campus Verlag, Frankfurt, W. Germany, 1983.
10. F. Malerba, *The Semiconductor Business: The Economics of Rapid Growth and Decline,* Univ. Wisconsin Press, London, 1985.
11. D. I. Okimoto et al. (eds.), *Competitive Edge: The Semiconductor Industry in the US and Japan,* Stanford University Press, Stanford, Cal., 1984.
12. J. K. Paul, *High Technology, International Trade and Competition,* Noyes, Park Ridge, N.J., 1984.
13. Petrocelli, *The Future of the Semiconductors Computer, Robotics, and Telecommunications: A Source Book,* Petrocelli Books, Princeton, N.J., 1984.

14. G. Reinertsen, "Whodunit? The Search for the New-Product Killers," *Electronic Business,* **9**:62, 64, 66, (1983).

15. S. Muroga, *VLSI Systems Design: When and How to Design Very-Large-Scale Integrated Circuits,* Wiley, New York, 1982.

16. K. Beasely, "Semi-custom I.C.'s—Reliable and Productive," *IEEE, Proc. Annual Reliability and Maintainability Symp.,* Orlando, Fl., January 1983, pp. 469–474.

17. C. F. Fey, "Putting Numbers Where Your Hunches Used to Be," *IEEE Spectrum,* **11**:34 (1974).

18. B. S. Landman and R. L. Russo, "On a Pin Versus Block Relationship for Partitions of Logic Graphs," *IEEE Trans. Computers,* **C-20**:1469 (1971).

19. D. K. Ferry, "Interconnection Lengths and VLSI," *IEEE Circuits and Devices Mag.* **1**:39 (1985).

19a. C. F. Fey, and D. E. Paraskevopoulos, "Gate Array Design Productivity: An Empirical Investigation," *Proc. 1988 Custom Integrated Circuits Conf.,* Rochester, N.Y., May 1988.

20. B. W. Boehm, *Software Engineering Economics,* Prentice-Hall, Englewood Cliffs, N.J., 1981.

21. R. C. Anderson, E. J. Kuuttila, and P. Matlock, *IC Design Productivity Report,* R. C. Anderson, Inc., Los Altos, Cal., 1985.

22. P. V. Norden, "Curve Fitting for a Model of Applied Research and Development Scheduling," *IBM J. Rsch. Dev.,* **2**:232 (1958).

23. M. K. Starr (ed.), "Useful Tools for Project Management," in *Management of Production,* Penguin Books, Baltimore, Md., 1970.

24. L. H. Putnam, "A General Empirical Solution to the Macro Software Sizing and Estimating Problem," *IEEE Trans. Software Eng.,* **SE-4**:345 (1978).

25. R. C. Tausworthe, *Deep Space Network Software Cost Estimation Model,* Jet Propulsion Lab, Pasadena, Cal., 1981.

26. A. C. Hartman, "Software or Silicon? the Designer's Option," *Proc. IEEE,* **74**:861 (1986).

27. J. W. Read (ed.), *Gate Arrays Design Techniques and Applications,* McGraw-Hill, New York, 1985.

28. S. L. Hurst, *Custom-Specific Integrated Circuits,* Dekker, New York, 1985.

29. M. J. Walsh, *Choosing & Using CMOS,* McGraw-Hill, New York, 1986.

30. N. Weste and K. Eshraghian, *Principles of CMOS VLSI Design: A Systems Perspective,* Addison-Wesley, Reading, Mass., 1985.

31. N. G. Einspruch (ed.), *VLSI Electronics,* 12 vols., Academic Press, New York, 1981–1985.

CHAPTER 26
ECONOMIC ASPECTS OF TECHNOLOGY SELECTION: COSTS AND RISKS*

Curt F. Fey
Demetris E. Paraskevopoulos
Xerox Microelectronics Center, Xerox Corporation
El Segundo, California

26.1 INTRODUCTION

The previous chapter provided some introductory material and discussed design methodology criteria selection in general. Two of the four selection criteria were discussed, namely schedule (productivity) and functionality. This chapter discusses the remaining two criteria, cost and risk.

A key concept defined in the previous chapter will also be used here. It is the level of integration (LOI), the number of gates used per IC pin. We recommend that the discussion of LOI in the previous chapter be read before reading this chapter.

26.2 COSTS

26.2.1 Methodology

Costs are shown in a series of figures as cost per gate (cost per function) versus number of gates (complexity) for various design methodologies. For simplicity, we will generally use cost to mean cost per gate. Costs are discounted at 20 percent to adjust for differences in timing of expenditures. For details see Fey and Paraskevopoulos.[1]

Standard-cell and gate-array costs are generally shown for 500 to 12,000 gates per device, currently the dominant range of complexity. LSI Logic, Inc., and Toshiba offer arrays of up to 50,000 gates in up to 256-pin packages and up to

*Adapted with permission, © 1986, 1987, IEEE. From *IEEE J. Solid State Circuits,* **SC-21**(2):297–303; *Proc. of the IEEE,* **75**(6):829–841.

129,000 gates in up to 362-pin packages, respectively. Beresford reports, for July 1985, a median complexity of new designs for both gate arrays and standard cells of 3000 gates.[2] Mackintosh Consultants reports for 1988 a median of 5500 gates for both gate arrays and standard cells.

We will first present data at an annual production of 10,000 units (approximately the median volume for standard cells in Beresford's study). These data include development cost, IC device cost, and total cost. Nomograms of total costs for production volumes ranging from 1000 to 100,000 units per year for a four-year period are presented later.

26.2.2 Device and Systems Development Cost

We will consider the total development cost of devices and systems (PCBA, etc.) together, as shown in Fig. 26.1. The breakdown of the total development cost into its two components will be discussed later. The importance of development cost per gate varies with production volume. However, since volumes are not accurately known at the time of development, a high development cost usually carries a considerable risk of a high development cost per IC. Low fixed costs limit the maximum loss if production volume is low; hence they reduce risk.

Development cost consists of two components: expenses for the system design team and *charges by the device manufacturers.* This distinction can be important. The desire to maintain a stable design work force may hinder or facilitate the buying of outside services. Thus, management of a systems company needs

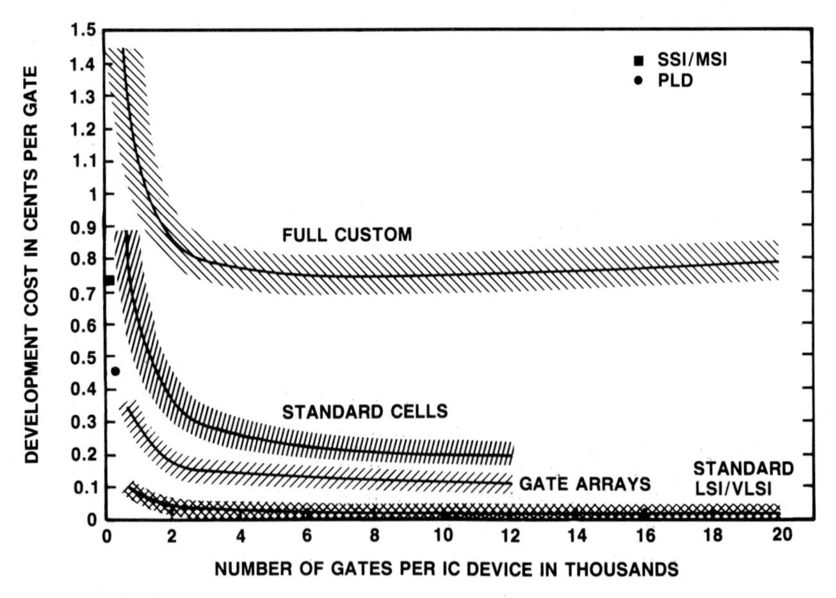

FIG. 26.1 IC device and systems development cost at 10,000 units per year for a four-year production. Products containing logic devices require design of PCB and other systems. In the case of ASIC they also require design of IC devices. The latter is highly automated, the former typically is not. Therefore, development cost per gate is large for low-integration methodologies such as SSI-MSI, and small for ASIC devices (PLDs excepted).

to install special administrative procedures to encourage cost-effective application specific integrated circuit (ASIC) designs, especially in lean times, when they are most needed. The investment in vendor charges, which enables the cost-saving ASIC design, has a very high return.

Development cost is a function of design productivity, which itself is a function of complexity and regularity. Complexity is generally measured in terms of transistors, or gates, since other measures do not yet readily exist. Regularity may be measured by the proportion of the logic gates that are repetitions of previously generated logic blocks on the IC device (e.g., user-defined macros). For a detailed study of the factors that influence gate array design productivity, see Fey and Paraskevopoulos.[3a]

Productivity is typically stated in transistors, or gates, per day at the IC device level. However, it should be measured at the printed circuit board assembly (PCBA) level. Here, productivity is the number of gates on the PCBA divided by the manpower required to design devices and PCBAs. PCBA complexity is measured by the number of equivalent integrated circuits (EICs) that take the same amount of PCB area as a 16-pin DIP. Manpower to design PCBA by leading European firms is given in Table 26.1.

TABLE 26.1 PCBA Development Manpower

Man-month per equivalent integrated circuit.

EIC/PCBA	60	100	200
Range	0.2–0.6	0.2–0.5	0.2–0.4
Median	0.27	0.26	0.21

Source: Seven leading manufacturers in six different European countries, 1986.

The various methodologies differ in all phases of design productivities. Productivity is important primarily because it affects schedules, as discussed earlier. But, there are conflicting influences: higher functionality and complexity requirements tend to decrease productivity, while better tools, availability of larger functional blocks, and better management techniques will increase it.

We discussed earlier the factors that determine current productivity. We have shown that as design manpower increases beyond six man-months (3500 to 7000 gates for semicustom in 1986), the design team size increases to more than one person, reducing overall productivity.

Development costs per gate are shown in Fig. 26.1. They include both IC device and PCB design plus prototyping. Costs are amortized over the assumed total production of 10,000 units per year for four years, i.e., 40,000 units. For different production volumes, readers can easily calculate the appropriate development cost by multiplying our cost per gate by 40,000 divided by the total number of units produced. Note that *at the systems level,* SSI-MSI and programmable logic device (PLDs) have a higher development cost per gate than do gate arrays and standard cells.

26.2.3 IC Device Cost

To the system house, device unit manufacturing cost (UMC) or price is a function of complexity. It does not include development cost. The cost relation-

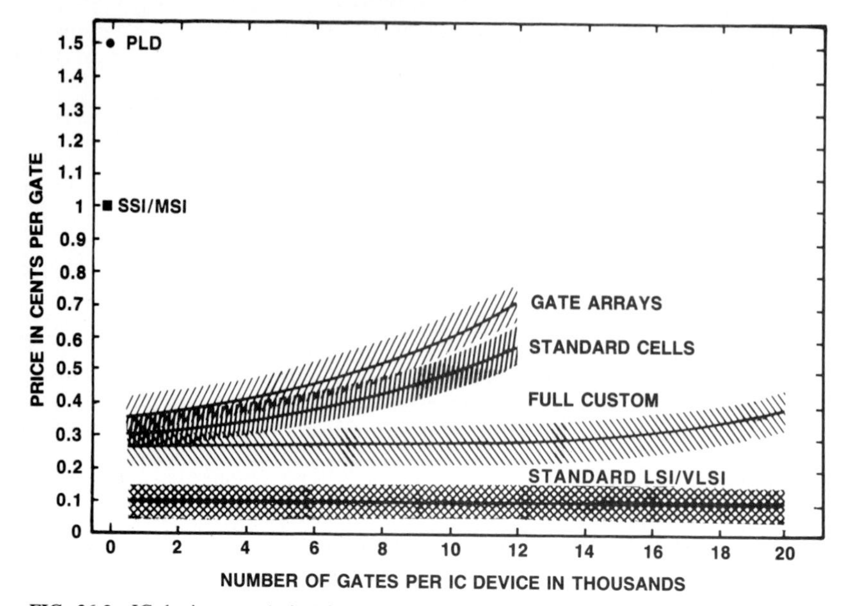

FIG. 26.2 IC device cost (price) in cents per gate at 10,000 units per year for a four-year production. SSI-MSI have an IC device cost that is much higher than that of ASICs (PLDs excepted).

ships in Fig. 26.2 are, as expected, increasing in the following order: standard LSI-VLSI, full-custom, standard cells, gate arrays, SSI-MSI, PLDs. The cost of standard LSI-VLSI varies. Standard ICs with hundreds of gates can cost up to 10¢ per gate; ICs with several thousand gates can cost in excess of 1¢ per gate for high-demand chips produced by a single vendor. The cost per gate of PLDs is about two to four times that of gate arrays; SSI-MSI is about one and one-half to three times.

Figure 26.2 shows that full-custom devices exhibit almost a flat cost per gate at the complexities shown, but semicustom devices have costs that increase due to their larger die size and lower LOI. Bipolar gate arrays cost 2 to 5¢ per gate, ECL arrays 3 to 6¢ per gate. ASIC costs vary not only with complexity but also with the amount purchased (annual production). However, cost per gate of standard devices does not depend on the quantity purchased by a firm for a particular application but is a function of a much broader demand and supply involving a multitude of purchasing firms, applications, and suppliers.

26.2.4 Total Cost

In Fig. 26.3 the total costs per gate are L-shaped functions. Above 3000 gates per device, the total cost for each ASIC device changes little and maintains its rank order. In the same range, the total cost of standard LSI-VLSI devices decreases. The difference is due to a very low, constant device cost per gate of standard devices and a systems cost that is decreasing as the LOI increases. ASIC devices, on the other hand, have a development cost per gate that is roughly con-

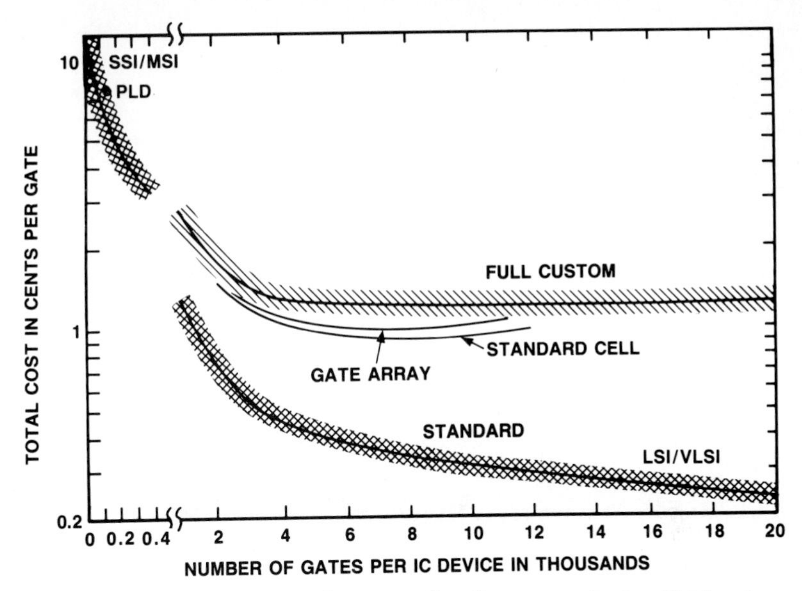

FIG. 26.3 Total cost at 10,000 units per year for a four-year production. Total cost per gate of ASICs is lower than that of SSI-MSI. Standard VLSI has the lowest cost.

stant, a UMC that is increasing with complexity, and a systems cost that is decreasing with complexity, resulting in little overall change in cost per gate. (LOI was defined in the previous chapter as the gates used per pin.)

Below 3000 gates, total costs increase rapidly due to the rapid decrease in LOI and the increase in fixed (development) costs per gate. Cost curves coincide for low gate counts, diverging as gate count increases. The low level of integration results in systems costs of 3 to 7¢ per gate for PLDs and of over 9¢ for SSI-MSI. Note that the total cost per gate differences in Fig. 26.3 are far larger than the IC device cost differences in Fig. 26.2 and do not always have the same sign.

At 10,000 units per year, gate arrays and standard cells have similar total costs. Full-custom devices cost slightly more. Since cost, schedule, and risk of full-custom are higher than those of semicustom, the choice is usually between standard cells and gate arrays. The selection in any specific case will depend on differences in design productivity and LOI for the particular situation.

The production volume of 10,000 was discussed as an example. Figures 26.4, 26.5, and 26.6 show components of total costs at production of 1000, 10,000, and 100,000 per year, respectively. The rank ordering of total costs is standard VLSI (lowest), gate arrays, standard cells, full-custom at 1000 per year production (Fig. 26.4). It changes to standard VLSI (lowest), full-custom, standard cell, gate arrays at 100,000 per year (Fig. 26.6). In between, at 10,000 per year, there are only small total cost differences between the various ASIC methodologies (Fig. 26.5). However, note the change in relative importance of various cost components for different complexities and production volumes.

Figures 26.4, 26.5, and 26.6 indicate the source of difference in total cost between standard LSI-VLSI and full-custom IC devices. We assume that both have the same LOI, systems UMC, field service, and systems development cost. However, a full-custom design can sometimes achieve a higher LOI and there-

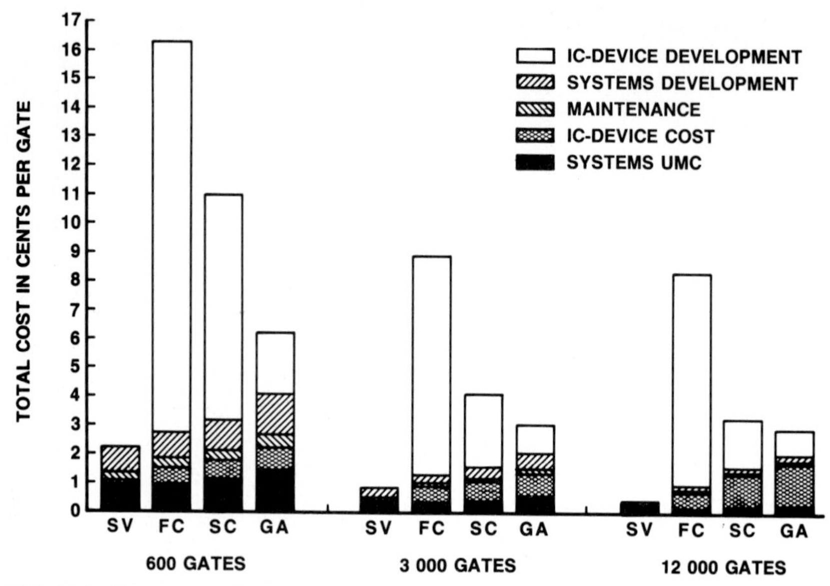

FIG. 26.4 Components of total cost at 1000 units per year for a four-year production. Cost per gate decreases for IC devices of any number of gates from full-custom to standard cells, to gate arrays, to standard devices. This is due to decreasing device development costs, which dominate IC device and system UMC differences. Cost differences between standard cells and gate arrays decrease as the number of gates per device increase. (Legend: GA = gate arrays, SC = standard cells, FC = full-custom, SV = standard VLSI.)

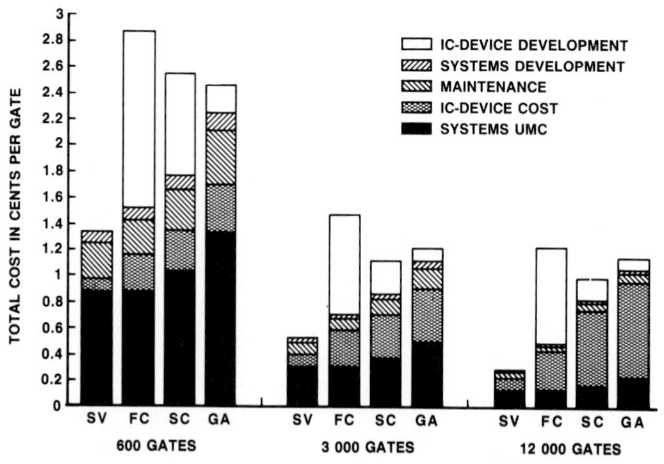

FIG. 26.5 Components of total cost at 10,000 units per year for a four-year production. Cost per gate differences between the various ASIC methodologies are small. High development costs are balanced by low UMC. (Legend: GA = gate arrays, SC = standard cells, FC = full-custom, SV = standard VLSI.)

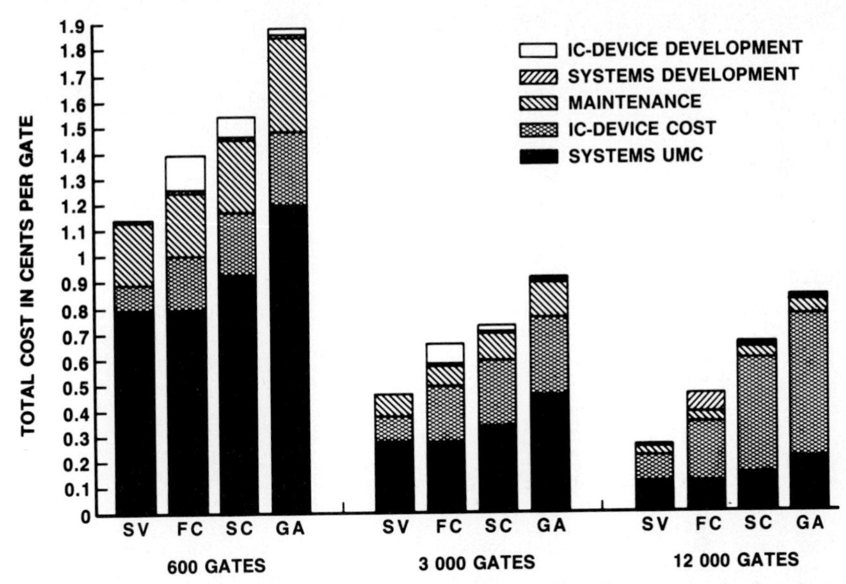

FIG. 26.6 Components of total cost at 100,000 units per year for a four-year production. Cost per gate for IC devices of any number of gates increases from standard VLSI devices to full-custom to standard cells, to gate arrays, as development costs become negligible. Cost differences between standard cells and gate arrays are largest for low complexity devices. (Legend: GA = gate arrays, SC = standard cells, FC = full-custom, SV = standard VLSI.)

fore a lower systems cost. Standard IC devices have no device development cost for the systems house and a lower device cost, since they are produced and purchased in high volumes.

A key component of the total cost is systems cost (Figs. 26.4, 26.5, and 26.6). Table 26.2 compares our cost estimates with those of others. Dataquest estimates the systems UMC to be 5.7 times the SSI-MSI device UMC. The principal differences between various cost estimates stem from the cost components considered. Given this difference in cost components, the agreement is surprisingly close.

TABLE 26.2 Comparison of System Cost Estimates

Dollars per equivalent integrated circuit.

	This paper (10,000 per year)	Sabo[4]	EDN[5]	Dataquest[6] (10,000 per year)
Systems UMC	2.10	1.44	2.05	2.83
Field service	0.64			
System development	0.22	0.002		
Total	2.96	1.44	2.05	2.83

26.3 COMMONALITY AND STANDARDIZATION

Many organizations have various electronic design groups developing a great variety of products. It is desirable for schedule, cost, design learning, and compatibility to follow design standards and to reuse technology to achieve commonality. To follow proven design practices and reuse large parts of previous designs will reduce risk, development cost (manpower) and, consequently, shorten schedule.

It is very difficult to achieve commonality and enforce standards among creative professionals working in dispersed groups. The CAD tools and libraries of ASIC, particularly of semicustom designs, enforce common design practices and promote commonality. These benefits can be amplified by company-wide design support organizations. The key benefit of standardization and commonality is the improved schedule.

26.4 BREAK-EVEN ANALYSIS

Gate arrays are available both as bipolar and as MOS devices. Bipolar devices have certain performance advantages at a substantial cost premium. The previous sections dealt with CMOS gate arrays; this section deals with bipolar (ALS) arrays.

26.4.1 Bipolar Gate Arrays versus MSI-SSI

Figure 26.7 compares 1300 ALS gate arrays to commercial MSI-SSI. The figure shows the number of equivalent integrated circuits (EICs) that must be replaced by the gate array to break even as a function of total (not annual) build. Each type of gate array is represented by a band. The band is the area between the envelopes obtained by varying the cost assumptions in accordance with the sensitivity analysis discussed in the following paragraphs.

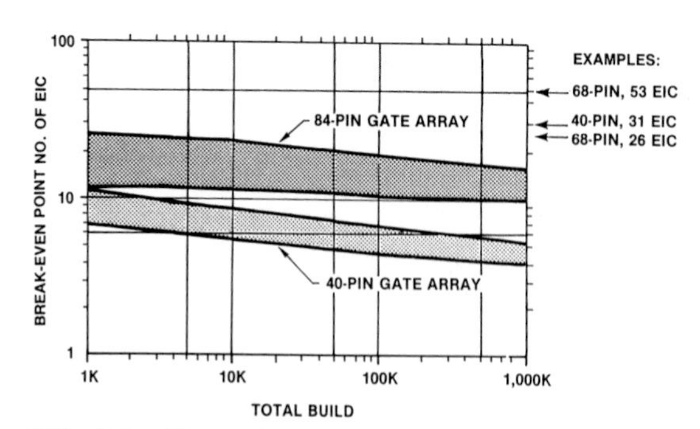

FIG. 26.7 1300 ALS gate array versus commercial MSI-SSI. Number of MSI-SSI circuits that must be replaced by ALS gate arrays to achieve break-even cost. The band represents variations in cost assumptions. Gate arrays are cost effective at most production volumes.

The bands are remarkably flat. The 40-pin gate-array band is steeper than the 84-pin band because the development costs represent a larger proportion of the total cost. The band for 68-pin gate arrays would lie between the two bands shown. The figure indicates that gate arrays have a lower cost than MSI, withbuild volumes as low as 1000 devices if they replace at least 5 to 10 EICs. These gate arrays save approximately the amounts shown in Table 26.3. The 1300

TABLE 26.3 Savings due to Replacing MSI by Bipolar Gate Arrays

Savings	Measure
\$3.00–5.00	Per machine per EIC replaced, above 10 EIC
\$0.08–0.15	Per machine per gate replaced, above 400 gates

ALS gate arrays typically replace between 20 and 60 EICs. Thus they are cost-effective. CMOS gate arrays, discussed earlier, have even larger cost advantages.

How sensitive is this conclusion to our assumptions? Two sensitivity analyses are shown in Fig. 26.8: one for 84-pin arrays and another for 40-pin arrays. They

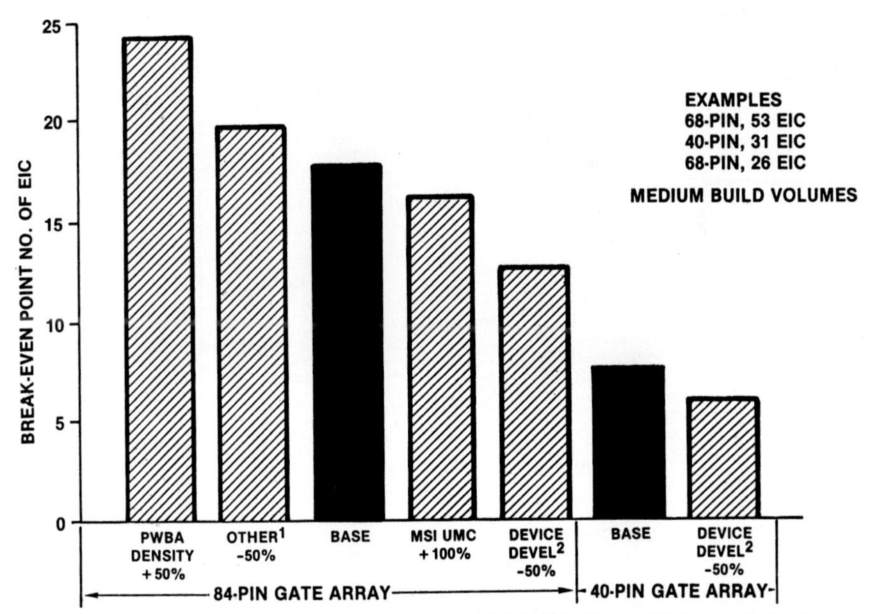

FIG. 26.8 1300 ALS gate arrays versus commercial MSI-SSI, sensitivity analysis. Large changes in cost assumptions do not change the conclusion that gate arrays are cost-effective. The break-even point is sensitive to LOI. (1) Frames, cooling, system test, back panel. (2) Development and productizing.

indicate that for relatively large variations in assumptions, break-even points vary relatively little (less than linearly), except for changes in LOI.

26.4.2 Bipolar Gate Arrays versus Custom Devices

Applying the above method to the comparison of custom devices and gate arrays yields the data in Fig. 26.9. The dependent variable is the number of 1300 ALS gate arrays that must be replaced by a custom device for break-even; the independent variable is the total number of devices built. Three types of tradeoffs are shown:

1. Small custom device, 40-pin versus 1300 ALS gate array, 84-pin
2. Small custom device, 84-pin versus 1300 ALS gate array, 84-pin
3. Large custom device, 84-pin versus 1300 ALS gate array, 84-pin

The small custom device is assumed to perform the same function as the gate array; the large custom device is in the same package and has about four times the complexity of the arrays.

The shape of the trade-off bands differs from those shown in Fig. 26.7. The bands are L-shaped, falling fairly steeply up to about 100,000 units total build. The trade-offs are sensitive to gates per pin. If the custom design can be implemented in a smaller package, or if it can achieve a higher level of integration, then break-even occurs at a modest volume of less than 10,000 devices. However, if gates per pin (LOI) cannot be increased, then volumes of 100,000 to 1,000,000 devices may be required for break-even.

There is another important aspect of this comparison that was discussed earlier. A custom design will take far longer schedule to complete than a gate array. Figure 26.10 shows the results of a sensitivity analysis. The sensitivity, except for

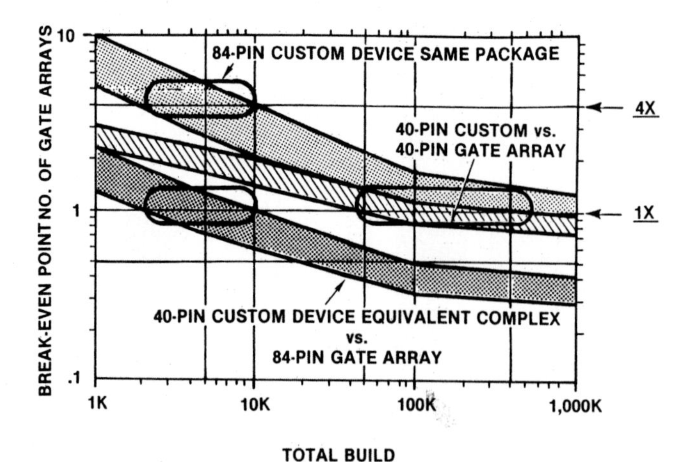

FIG. 26.9 1300 ALS gate arrays versus custom devices. Number of gate arrays that must be replaced by custom devices to achieve break-even cost. The bands represent variations in cost assumptions. Custom devices that allow reduction in pin count or increase the level of integration are cost-effective at low-build volumes.

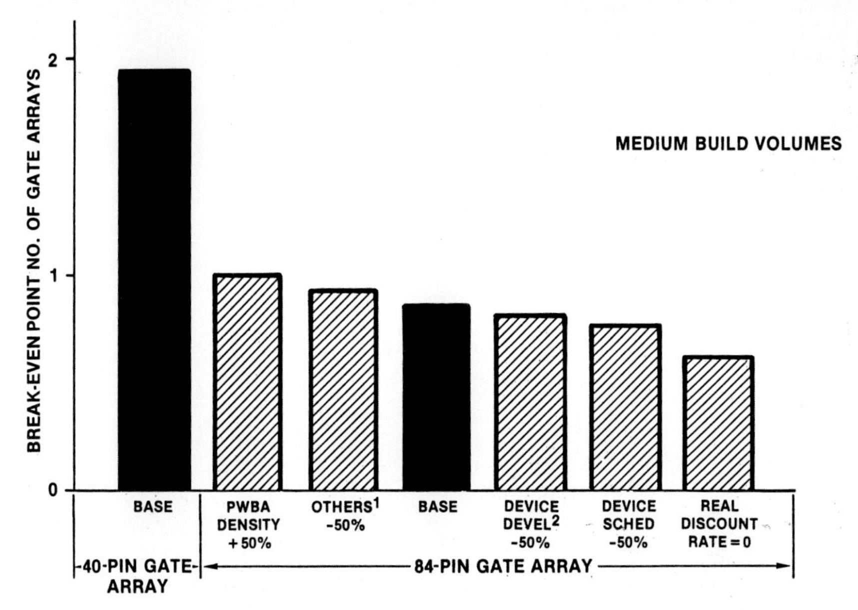

FIG. 26.10 1300 ALS gate array versus custom devices, sensitivity analysis. Large changes in cost assumptions do not change the conclusion that gate arrays are cost-effective if they decrease gates per pin. However, custom devices do have longer schedules. The break-even point is sensitive to the number of gates per pin. (1) Frames, cooling, system test, back panel. (2) Development and productizing.

LOI, is relatively small. This indicates that even if our assumptions contain a large error, our conclusions are still correct.

26.4.3 CMOS Gate Arrays

Small CMOS gate arrays have a cost that is lower than ALS, making gate arrays even more attractive. The break-even point of CMOS gate arrays versus full-custom devices can be determined from Figs. 26.4 to 26.6.

26.5 COMPARISON OF RISKS

The principal perceived risk for ASIC is the development schedule uncertainty due to a higher-than-planned number of design iterations. This is a subjective judgment that depends on the experience of the designer or manager. Indeed, the novice without support can have a harrowing experience. However, with proper training, support, and design tools, even a novice designer can expect a high degree of first-pass success for semicustom designs.

The perceived low cost of making design changes in SSI-MSI and PLDs is both an advantage and a handicap because it encourages an iterative design approach. Software engineering experience has shown that when change is inex-

TABLE 26.4 Comparison of Risks

Risk	Nature of risk		SSI-MSI	PLD	Standard VLSI	Gate array	Standard cell	Full-custom
Schedules slip	No. of iterations	New technology	....	Low	High	Medium	Medium	High
		Mature technology	Low	Low	Low	Low	Low	High
High cost, low production volume	Amortization of R&D		Low	Low	Low	Low	Medium	High
First-pass failure, (system)	Management, support, tools	Novice	Medium	Medium	Medium	Low	Medium	Very high
		Experienced	Low	Low	Low	Low	Low	High
Required skills unavailable	....	....	Low	Low	Low	Low	Medium	High
Chip size underestimated	Design rules	....	....	....	....	Low	Low	High

pensive (software on disks), change is frequent. Conversely, when change is expensive (software in ROM), change is rare. In both cases, the cost of changes is usually far greater than the estimate.

One could start a design with a methodology of low complexity or perceived risk for prototypes and early production. Later the IC device cost can be reduced with a pin-compatible redesign, once the design has stabilized, thus saving schedule. At present, two-step full-custom designs are primarily used when the original design results in chip sizes that create serious yield problems. It is rarely done for semicustom. For example, both the Apple Macintosh and IBM AT personal computers still contain several PLDs. Furthermore, some gate arrays are sold as standard products and are never redesigned as full-custom. Certain gate arrays can be easily converted to standard cells. Because LOI remains unchanged (same number of gates and pins), the board area and other systems costs do not decrease. Therefore, the total cost savings to the systems house are limited to the small IC device cost differences. Note, however, that the economics for the IC device manufacturer can be more favorable for the conversion approach.

Our assessment of risks is shown in Table 26.4. There is currently a high probability of failure to achieve QCD (quality or functionality, cost, and delivery on schedule) for pure full-custom designs. For this reason there is a trend away from that methodology toward cell-based designs.

ASIC risks differ from those of SSI-MSI. Design changes can be easily made after prototyping; such changes are still costly for ASIC in terms of money and perhaps in time. On the other hand, CAD tools reduce the need for change, since they produce simulation results that can be counted upon to predict circuit behavior, enabling a high probability of first-pass success.

26.6 SENSITIVITY ANALYSIS

We will now discuss the key factors that determine the economics of MOS ASICs, present a sensitivity analysis of these underlying factors, and examine their effect on our conclusions. LOI is seen by users as the most important ASIC characteristic, and for a good reason. LOI determines total cost. In the range of devices considered here, the LOI varies from 1 to 3 for SSI-MSI to 200 gates per pin for a 20,000-gate, 100-pin full-custom device. The relative cost of various ASIC methodologies changes significantly as LOI varies (Fig. 26.3). Figure 26.11 shows the difference in trade-offs between design methodologies for two different LOI assumptions. In the first case, LOI is derived from Eq. (25.2). In the second case, gate arrays are redesigned as pin-compatible standard cells, and standard cells as pin-compatible full-custom devices. Therefore standard cells have the LOI of gate arrays at the lower boundary of the standard-cell region, and full-custom has the LOI of standard cells at the upper boundary. As shown, a gate-array conversion into a pin-compatible standard cell requires a higher production volume to be justifiable. A similar effect is also shown for standard cells and full-custom.

The effect of production *volume* is illustrated by nomograms for volumes ranging from 1000 to 100,000 units per year. Many gate arrays and standard cells are produced at volumes of less than 1000 per year. This indicates that semicustom is economically viable at low production volumes. In 1986, less than 10 percent

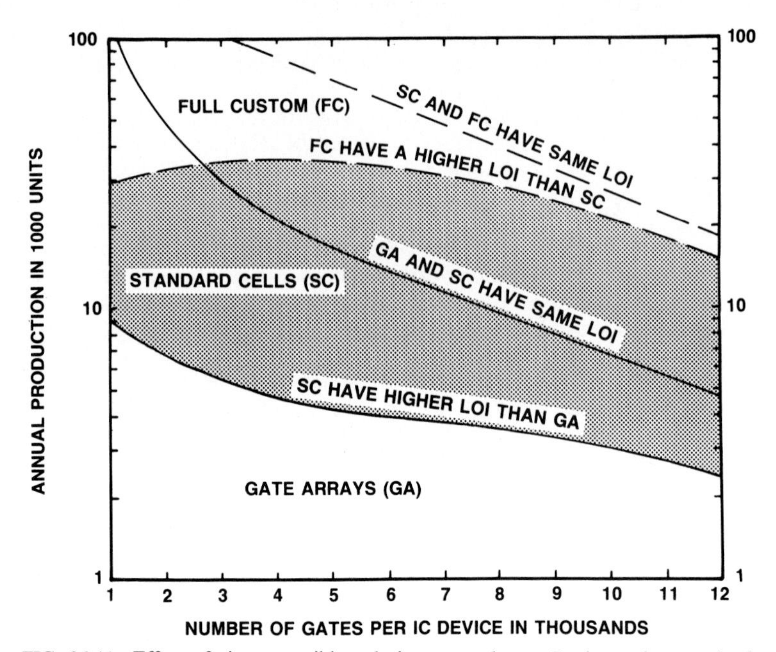

FIG. 26.11 Effect of pin-compatible redesign on total cost. Regions where each of the three ASIC methodologies has the lowest cost are shown under the assumption that the level of integration increases from gate arrays to standard cells to full-custom devices. The region boundary migrates upward to higher production volumes when standard cells have the same level of integration as gate arrays (i.e., pin-compatible redesign). A similar migration occurs when full-custom has the same level of integration as standard cells.

were produced at more than 100,000 units per year. Figure 26.12 shows the regions where each design methodology dominates—i.e., has the lowest total cost per gate, however small the cost difference. Here, regions are defined in terms of complexity and annual production volume. There are large regions where cost differences between the various methodologies are small.

Let us illustrate the use of the nomogram. Let point *X* represent a design with a 4000-gate device, produced at 10,000 units per year for four years. Equal cost-per-gate curves are shown for the various methodologies. For example, point *X* lies between the 1¢ and 1.5¢ per gate curves of both gate arrays and standard cells. Interpolating, we can estimate the total cost for gate arrays and standard cells at 1.3 and 1.1¢ per gate, respectively. The cost difference between the two methodologies is small. Thus, the nomogram indicates the total cost per gate for each of the three design methodologies as a function of complexity and production volume.

Examining Fig. 26.12 as a whole, we see that below 3000 to 10,000 units per year, gate arrays have the lowest cost. Above 15,000 to 30,000 units per year, full-custom devices dominate. In between, standard cells have the lowest cost. The volume effect on total cost is the sum of the effects of volume on each cost component, as shown in Figs. 26.4, 26.5, and 26.6.

There are large regions, or market segments, of *small total cost differences* between gate arrays and standard cells, on the one hand, and standard cells and

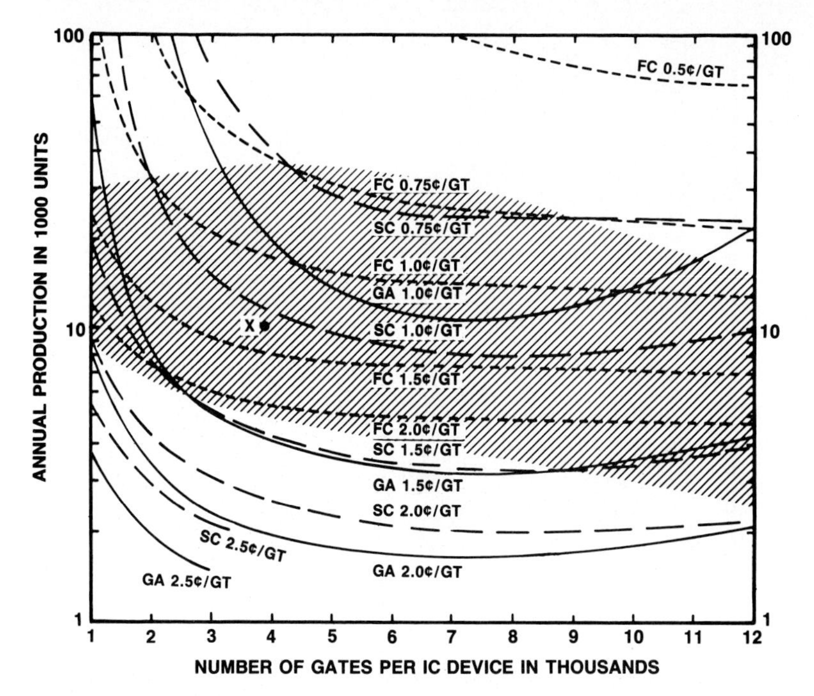

FIG. 26.12 Total cost in cents per gate (¢/GT) of various design methodologies as a function of complexity and production volume. Each design methodology has the lowest cost in a region, i.e., gate arrays (bottom), standard cells (middle), full-custom (top). (Legend: GA = gate arrays, SC = standard cells, FC = full-custom.)

full-custom devices, on the other. Figure 26.13 highlights regions where each of the three methodologies dominates—i.e., has the lowest total cost per gate. These regions are separated by bands of small cost differences, defined as *equal cost ± 10 percent*. In each of the bands, actual cost differences for a particular design will depend on LOI, application, design practices, and cost structure of the manufacturing and servicing organization.

These regions of small total cost differences are expected to remain large because the differences between the various methodologies are expected to decrease. Currently, they are predominantly the domain of the old methodologies, namely gate arrays and full custom. However, they provide a future opportunity for standard cells. Standard cells could capture these market segments if they gain competitive advantages—e.g., if they increase LOI and improve design productivity. Contrary to prevailing opinion, the competition among the design methodologies will generally be on the basis of schedule and LOI (total cost), not on the basis of device cost.

Full-custom has a small niche. However, even there it incurs a schedule penalty. The future of full-custom depends primarily on design productivity and on its ability to increase systems LOI over standard cells or over a mixture of various standard devices. Device cost is also important since the market for full-custom is mostly in the high-volume, high-complexity region.

Systems cost dominates the total cost of electronics. This has been true for the

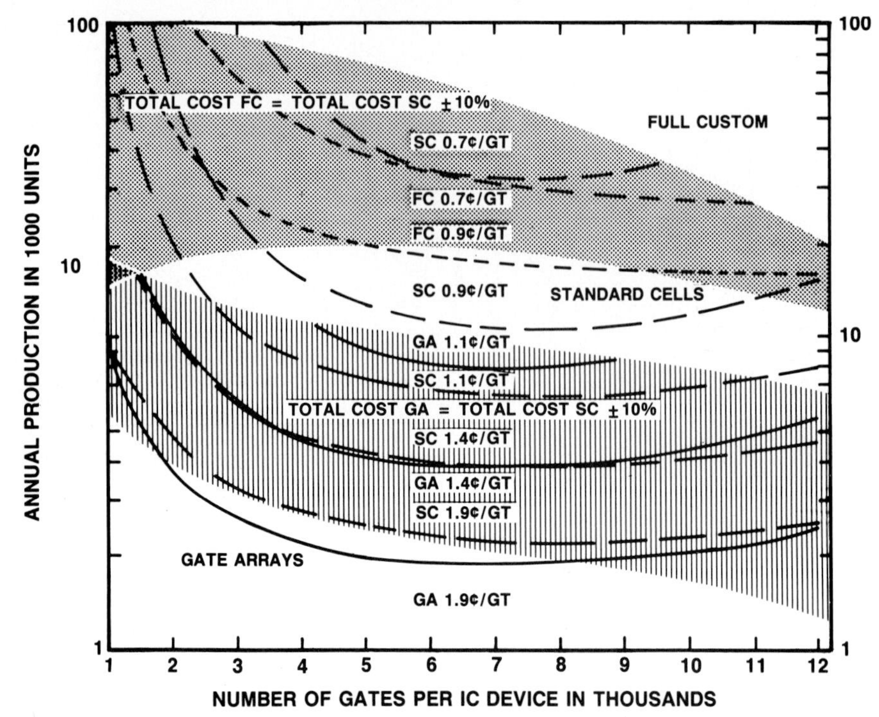

FIG. 26.13 Regions of small and of large total cost differences of various design methodologies. Each of three design methodologies has the lowest cost in a different region. These regions are separated by highlighted bands of small cost differences (equal cost ± 10 percent). There is a wide range of complexity and volume where two design methodologies have approximately the same overall cost. Currently, these regions are predominantly captured by the old methodologies. They provide future opportunities for standard cells. (Legend: GA = gate arrays, SC = standard cells, FC = full-custom.)

last 30 years, and it is expected to remain so, as shown in Fig. 26.14. At present, systems cost accounts for about 80 percent of total cost.

Systems cost is driven by LOI. Even device testing is a function of LOI. The historical data based on Phister[7] in Fig. 26.14 support this assertion. The ENIAC with an LOI of less than 0.1 gate per pin weighed 30 tons. This weight translates into costs, since computers used to sell for about $200 per pound.

Thus, the greatest opportunity for both cost and schedule reduction lies in increasing the average LOI of a product. The relationship between LOI and total cost is the cornerstone of our analysis. Flaws in the details of this relationship would be the largest source of error.

26.7 TIME TRENDS

Based on our work to date and the physical limits reviewed by Rideout,[8] we project the following future trends:

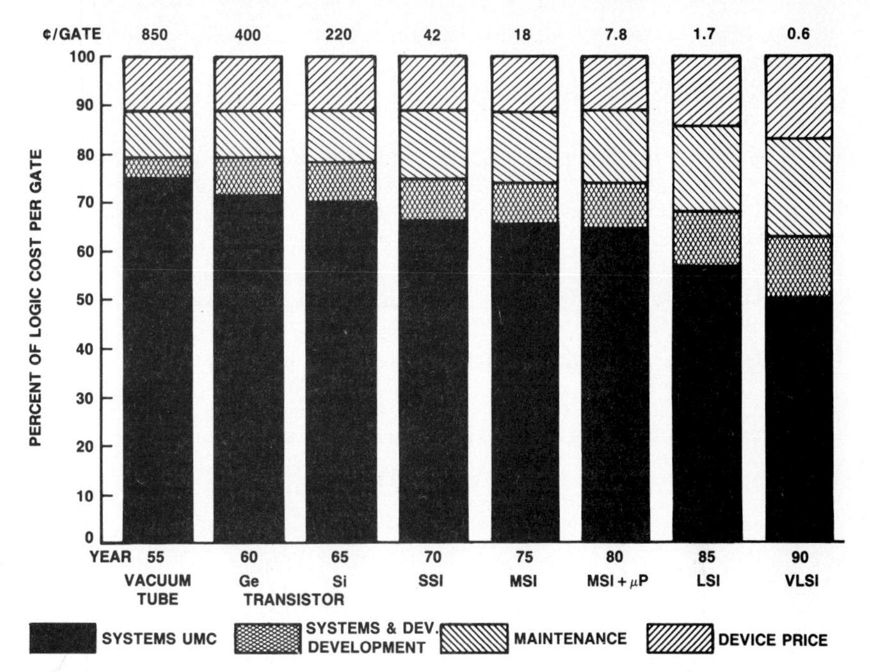

FIG. 26.14 Components of logic cost, 1955–1990. Averages of the technologies used at various dates are shown. Higher levels of integration decrease systems cost. The result is a sharp decrease in total cost per gate. The percentages of maintenance, systems, and IC device development cost increase due to their large labor content. The percentage of silicon cost increases with each change in device complexity (technology).

1. Since the 1960s, increasing LOI has been a key enabler of decreasing total cost. LOI will continue to increase, since it leads to lower total cost and shorter schedule. However, the rate of decrease in total cost will slow down.
2. Density (gates per square inch silicon) will increase to enable higher LOI cost effectiveness.
3. Differences between the various technologies will decrease, since the economics of the various ASIC devices are similar.
4. Increase in LOI will be limited by design productivity and testability. The market for ASICs with design times of more than one year will be limited.
5. Designers will demand high probability of first-pass success and high productivity to achieve schedule goals.
6. Systems cost will continue to be the overwhelming cost component, but device cost will gain in importance, as will maintenance and systems development cost (Fig. 26.14).
7. Gate arrays and standard cells will become the dominant ASIC methodologies (Fig. 25.2). The rate at which the standard-cell market develops will depend on the level of integration, design productivity, and first-pass success. The exact split in the market share is difficult to predict due to the high overlap of total costs (Fig. 26.13).

8. LOI of gate arrays will increase, enabling a better competive position relative to standard cells.
9. Design productivity and first-pass success of standard cells will approach that of gate arrays.
10. Productivity and first-pass success of full-custom IC devices will improve, while maintaining a lower UMC than standard cells. Full-custom ICs will achieve an LOI that is higher than that of both standard cells and standard VLSI devices.
11. Pin-compatible redesigns of gate arrays into standard cells, and of standard cells into full-custom devices, without change in LOI will occur only in a limited number of cases.
12. PLDs will be confined to a niche market since they have a high total cost (Fig. 25.3).

The rewards for increasing average LOI will be substantial, as shown in the next section.

TABLE 26.5 Size of Worldwide Cost-Avoidance Opportunity

	Function		Device cost			Total system cost*		
Logic type	billion gates	%	$B	%	¢/gate	$B	%	¢/gate
			1984					
Vendor LSI	1,346	70	3.5	36	0.20–0.30	5.8	13	0.4
Full custom	173	9	0.9	9	0.40–0.70	1.5	3	0.8
Standard cells	5	0	0.1	1	1.60–2.20	1.0	2	3.0
Gate arrays	26	1	0.5	5	1.60–2.20	1.0	2	3.8
MSI	362	19	4.7	48	1.10–1.50	35.1	81	9.7
Total	1,912	100	9.7	100		43.5	100	
			1987					
Vendor LSI	4,615	77	6.0	36	0.10–0.20	10.9	19	0.2
Full custom	577	10	1.5	9	0.20–0.30	2.8	5	0.5
Standard cells	155	3	1.7	10	0.80–1.40	2.7	5	1.8
Gate arrays	142	2	1.7	10	0.90–1.50	3.2	6	2.3
MSI	500	8	6.0	36	0.90–1.50	37.6	66	7.5
Total	5,989	100	16.9	100		57.2	100	
			1990					
Vendor LSI	17,667	83	10.6	37	0.04–0.10	23.6	28	0.1
Full custom	1,923	9	2.5	9	0.10–0.20	5.5	7	0.3
Standard cells	571	3	4.0	14	0.50–0.90	6.6	8	1.2
Gate arrays	526	2	4.0	14	0.50–1.00	7.9	9	1.5
MSI	718	3	7.9	27	0.80–1.40	40.3	48	5.6
Total	21,406	100	29.0	100		84.0	100	

*Device driven.

26.8 SIZE OF COST-AVOIDANCE OPPORTUNITY

High integration provides substantial opportunities for cost-avoidance. Table 26.5 lists some of the assumptions for the three time periods by device type. Vendor LSI refers to off-the-shelf devices such as microprocessors. The table shows the distribution of function (number of gates), device cost, and total cost. Device cost refers to the cost (price) of the IC device to the user. The total cost refers to all the cost elements identified previously. Thus, it includes ASIC development, UMC, service, and systems costs.

MSI accounts for the bulk of the total systems costs (device cost and systems costs) but only for a small part of the electronic functions (Fig. 26.15). Even in 1990, the 3 percent of the gates implemented in MSI will account for 30 percent of the device cost and for 50 percent of the total cost.

In other words, very small amounts of MSI can account for a large portion of the device-related cost. That means semicustom and full-custom devices will

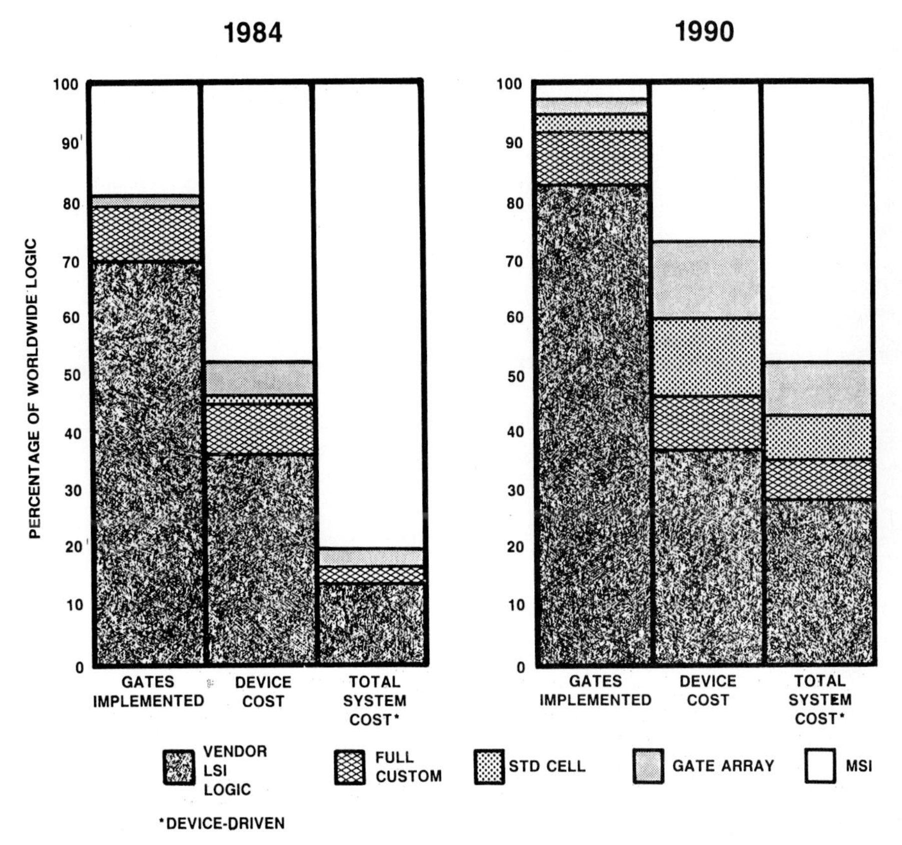

FIG. 26.15 Size of logic cost-avoidance opportunity. MSI logic accounted in 1984 for 19 percent of the logic gates but 81 percent of the total systems cost. By 1990, the 3 percent of MSI gates will still account for 48 percent of the total systems cost.

offer significant cost-avoidance opportunities for years to come. To gain full advantage of the cost-avoidance opportunity of custom and semicustom devices, a systems manufacturer must virtually eliminate all MSI.

26.9 SUMMARY AND CONCLUSIONS

1. Level of integration largely determines total cost.
2. SSI-MSI and PLDs have a low level of integration compared with ASICs. Therefore, they have cost and system development schedule disadvantages at nearly all levels of production. ASICs can also compete successfully against combinations of SSI-MSI and LSI-VLSI if they improve level of integration.
3. Large market segments of small total cost differences exist between gate arrays and standard cell, on one hand, and standard cells and full-custom devices, on the other.
4. Systems cost largely determines total costs. It did so in the past and will do so in the future. It currently accounts for about 80 percent of the total cost. In the future, IC device cost (price) will gain in importance, as will maintenance, and device and systems development cost.
5. Redesign of gate arrays as standard cells or standard cells as full-custom devices will produce only small device cost advantages to the systems house, unless the level of integration is significantly increased. The economics of redesign differ for systems houses and for IC device manufacturers.

ACKNOWLEDGMENTS

The authors would like to thank R. Markle, Professor S. Muroga, and S. Shohara for their critical comments. The authors also thank the Xerox Microelectronics Center and the Xerox Technology Strategy Office for their support.

REFERENCES

1. C. F. Fey and D. E. Paraskevopoulos, "A Technoeconomic Assessment of Application Specific Integrated Circuits, Current Status and Future Trends," *Proc. IEEE,* **75**:829 (1987).
2. R. Beresford, "A Profile of Current Applications of Gate Arrays and Standard Cell ICs," *VLSI Systems Design,* **6**:62 (1985).
3. Mackintosh Consultants, Inc., *The Industry Brief,* vol. 2. Mackintosh Consultants, Saratoga, Cal., August 1985.

3a. C. F. Fey and D. E. Paraskevopoulos, "Gate Array Design Productivity: An Empirical Investigation," *Proc. IEEE 1988 Custom Integrated Circuits Conf.,* Rochester, N.Y., May 1988.

4. D. Sabo, "Kostenanalyse für Gerate mit Gäte Arrays," *Feinwerktechnik & Messtechnik,* **93**:199 (1985).

5. EDN Editorial Staff & Cahners Research, *The Semicustom-IC Revolution: An EDN Research Report,* Cahners, Newton, Mass., 1984.
6. "Application Specific IC's," in *Semiconductor Industry Service,* Dataquest, San Jose, Cal., February 1986.
7. M. Phister, Jr., *Data Processing Technology and Economics,* 2d ed., Digital Press, Bedford, Mass., 1979.
8. V. L. Rideout, "Limits to Improvement of Silicon Integrated Circuits," in *VLSI Electronics,* vol. 7, N. G. Einspruch (ed.), Academic Press, New York, 1983.

P • A • R • T • 9

VLSI RELIABILITY AND YIELD ANALYSIS

CHAPTER 27
RELIABILITY

L. J. Gallace
GE Ceramics, Inc.
Chattanooga, Tennessee

27.1 INTRODUCTION TO RELIABILITY

The physics of reliability is concerned with the type and rate of change in material properties and the relationship between these changes and past history. Failure patterns for VLSI ICs are determined uniquely by the process used to create them and by their basic physical structure. Therefore, faults in VLSI circuits are manifestations of physical defects caused by process instabilities and contamination during the fabrication process. IC processes can be considered to be the layering of surfaces formed through various chemical processes. A given layer can be either insulating, semiconducting, or conducting. Figure 27.1 shows a cross section of such a process for a 2-μm CMOS VLSI IC.

The reliability of a VLSI circuit or, indeed, of any system is a statistical parameter, derived by the methods of mathematical probability theory, that evaluates success or failure; it determines the likelihood that components or equipment will continue to operate satisfactorily. The reliability figure indicates the probability of failure of a device during its normal operating life. Thus, the reliability of equipment is directly related to the reliability of its components.

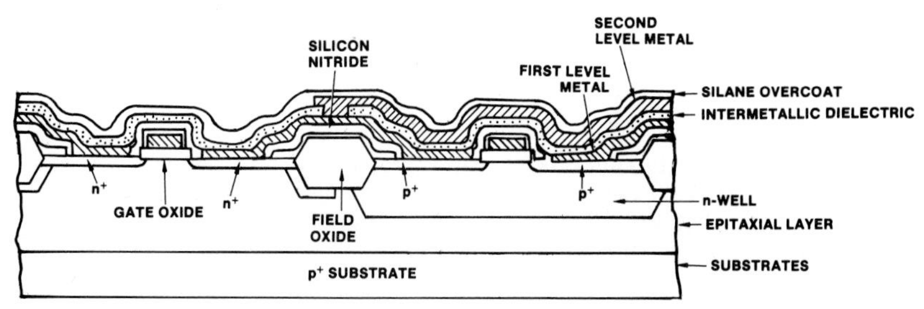

FIG. 27.1 Cross section of 2-μm CMOS integrated circuit.

Modern reliability principles are based heavily upon statistics and probability. Therefore, some elementary concepts, which are often the basis of published reliability data, are reviewed.

27.1.1 Probability

The two main definitions of probability are the classical definition and the relative frequency definition. In the classical definition, if an event can occur in N mutually exclusive and equally likely ways, and if n of these outcomes are of one kind A, then the probability of A is n/N. For example, the probability of a head or tail in the toss of a coin is 1/2. The classical definition is not widely used in real applications.

In the relative frequency definition of probability, a random experiment is repeated n times under uniform conditions and a particular event E is observed to occur in f of the n trials. The ratio f/n is called the *relative frequency* of E for the first n trials. If the experiment is repeated sufficiently many times, the ratio of f/n for the event E approaches some value P, the probability of event E. This definition indicates that the probability is a number between 0 and 1:

$$0 \leq P \leq 1$$

The probability of E plus the probability of not E, $\overline{E}$, equals 1:

$$P(E) + P(\overline{E}) = 1$$

Table 27.1 summarizes basic probability laws used in reliability mathematics.

27.1.2 Probability Distributions

A probability distribution describes how often different values of a given characteristic are expected to occur. These distributions may be discrete or continuous. Discrete random variables assume distinct values, such as the integers, while continuous random variables assume any value within a defined range.

Discrete Distributions. If $f(x)$ generates probabilities that a random variable X will take on certain discrete values, it is called a probability function.

The binomial probability function is a discrete distribution associated with repeated trials of the same event. For an event (for example, success, or no failure) where the probability of its occurrence on any trial is p, the probability of nonoccurrence is $1 - p$. Then $f(x)$ is defined by

$$f(x) = \frac{n!}{x!(n-x)!} p^x (1-p)^{n-x}$$

where $x = 0, 1, 2, \ldots, n$. This function describes the number of successful trials expected in a series of n independent trials.

The Poisson probability function, in addition to being an approximation of the binomial probability function when n is large and p is very small, is a useful probability function in its own right to describe the occurrence of isolated or rare events. The Poisson probability function is expressed as

TABLE 27.1 Basic Laws of Probability

Note: This table assumes an understanding of set theory.

Law of complementation
If the probability of event A not occurring is represented by $P(\bar{A})$, then $P(A) + P(\bar{A}) = 1$ and $P(A) = 1 - P(\bar{A})$
Laws of addition
The probability that either event A or event B (or both) occurs is written $P(A \cup B)$. Then $P(A \cup B) = P(A) + P(B) - P(A \cap b)$ If A and B are mutually exclusive (i.e., if A occurs, B cannot, and if B occurs, A cannot), then $P(A \cap b) = 0$.
Laws of multiplication
The probability that both event A and event B will occur is written $P(A \cap b)$. Then $P(A \cap b) = P(A)P(B \mid A) = P(B)P(A \mid B)$ $P(A \mid B)$ is the probability that event A will occur given that event B occurs. ($P(B \mid A)$ is the same statement for B.) If events A and B are statistically independent, then $P(A \mid B) = P(A)$ and $P(B \mid A) = P(B)$ and the law of multiplication reduces to $P(A \cap b) = P(A)P(B)$ Statistical independence implies that the occurrence or nonoccurrence of an event A in no way influences the occurrence or nonoccurrence of the other event B.

$$f(x) = \frac{\lambda^x \exp(-\lambda)}{x!}$$

where $x = 0, 1, 2, \ldots, n$ (number of times rare event occurs), and λ = the average number of times the event occurs. When $x = 0$, the Poisson function reduces to the reliability formula or negative exponential:

$$R = \exp(-\lambda t)$$

Continuous Distributions. When a chance variable X is free to take on any value within an interval, the resulting probability distribution is continuous. Figure 27.2 shows the relationship between the probability density function $f(x)$ and the cumulative distribution function $F(x)$.

$$F(x) = P(X \leq x)$$

For a continuous distribution

$$f(x) = \frac{dF(x)}{dx}$$

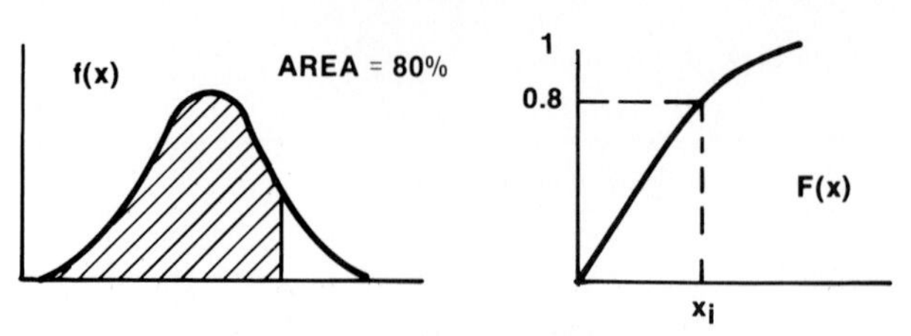

FIG. 27.2 Relationship between the probability density function $f(x)$ and the cumulative distribution function $F(x)$.

It should be clear from these last statements that, for $f(x) \geq 0$,

$$\int_{-\infty}^{+\infty} f(x)\, dx = 1$$

The cumulative distribution function never decreases as the variable increases:

$$F(x_2) \geq F(x_1) \qquad \text{if } x_2 \geq x_1$$

and

$$\int_{-\infty}^{x_1} f(x)\, dx = F(x_1) \qquad F(-\infty) = 0 \quad F(\infty) = 1$$

as shown in Fig. 27.2.

In general, engineers are familiar with the "normal" distribution, which is the basis for many statistical techniques. The fact that this distribution is called normal does not imply that other distributions are nonnormal. The probability density function for the normal distribution is

$$f(x) = \frac{1}{\sigma\sqrt{2\,\pi}} \exp - \left[\frac{(x - \mu)^2}{2\sigma^2}\right]$$

where x ranges from $-\infty$ to $+\infty$, μ is the mean, and σ is the standard deviation. The cumulative distribution function $F(x)$ for the normal distribution cannot be integrated as an algebraic equation, but it has been evaluated by numerical integration techniques and is extensively tabulated in statistics books.

In the lognormal distribution, ln X is normally distributed. The density function is

$$f(x) = \frac{1}{\sigma x\sqrt{2\pi}} \exp - \left[\frac{(\ln x - \mu)^2}{2\sigma}\right]$$

An interesting fact about the lognormal distribution is that the failure rate function $Z(t)$ increases at first and then decreases. The log-normal density is the only density for which this occurs.

The Weibull distribution was developed for the study of fatigue of materials. The density function for the Weibull distribution is

$$f(x) = \beta(x - \gamma)^{\beta-1} \exp\left[\frac{-(x-\gamma)^\beta}{\eta}\right]$$

and

$$F(x) = 1 - \exp\left[\frac{-(x-\gamma)^\beta}{\eta}\right]$$

where η = scale parameter
β = shape parameter
γ = location parameter

If the failures can start as soon as the devices are operated, then $\gamma = 0$. The β parameter of the Weibull distribution is important in determining the failure rate:

For $\beta < 1$, the failure rate is decreasing.

For $\beta = 1$, the failure rate is constant.

For $\beta > 1$, the failure rate is increasing.

Therefore, the Weibull distribution can be used to characterize components that are subject to infant mortality, chance failures, or wearout failure.

27.1.3 Reliability Distribution Theory

Nearly any discussion of reliability begins and ends with the statement of failure rates for either components or systems. Some very basic and interesting reliability equations can be developed. For example, if the probability of a successful event is represented by $R(t)$ and the probability of an unsuccessful event, a failure, is represented by $F(t)$, then

$$R(t) \cup F(t) = 1 \qquad R(t) \cap F(t) = 0$$

The failure probability is defined as

$$F(t) = \int_0^t f(t)dt$$

the probability of success as

$$R(t) = 1 - \int_0^t f(t)\, dt$$

where $F(t)$ is the probability-of-failure distribution function—the probability that a device will fail by time t—and $R(t)$ is the probability-of-success distribution function, the probability that a device will not fail by time t.

The probability that failures will occur between any times t_1 and t_2 can be calculated from the probability function

$$P = \int_{t_1}^{t_2} f(t)\, dt$$

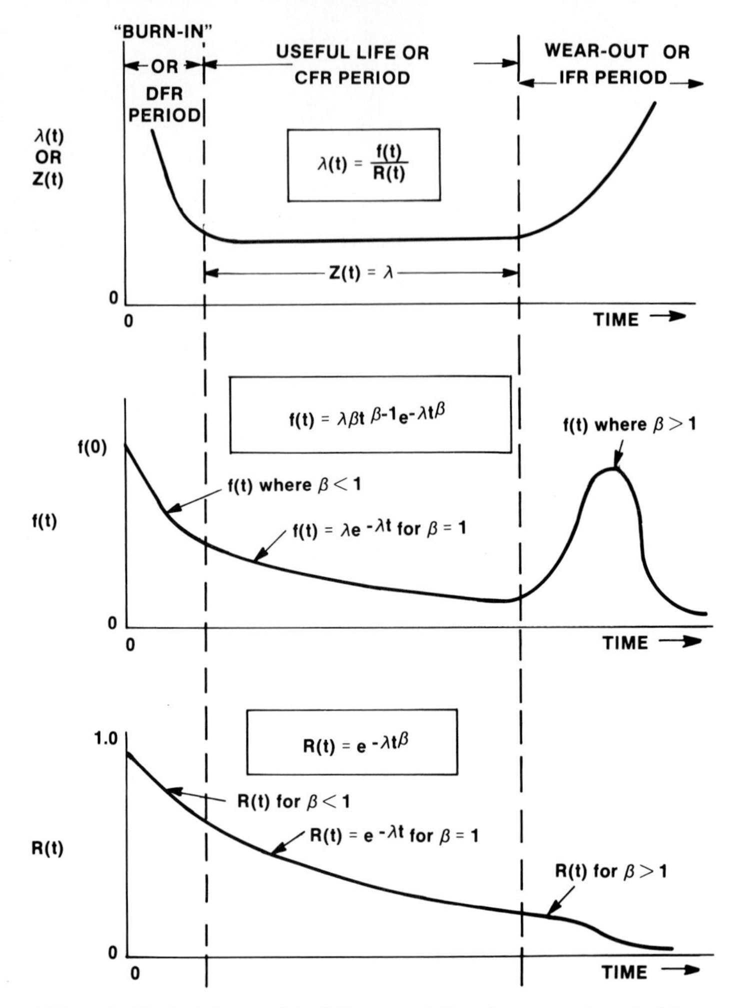

FIG. 27.3 Typical shapes of the failure rate, failure density, and survival functions. The DFR (decreasing failure rate), CFR (constant failure rate), and IFR (increasing failure rate) periods are indicated.

Since all devices and systems have a finite lifetime,

$$P = \int_0^t f(t)\, dt = 1$$

The density function $f(t)$ may be derived from

$$R(t) = 1 - \int_0^t f(t)\, dt$$

By differentiating this expression, we obtain

$$f(t) = \frac{dR(t)}{dt} = R'(t)$$

One more expression that is always part of every reliability discussion is mean time to failure (MTTF). By use of the mathematical expectation theorem, MTTF can be expressed as

$$\text{MTTF} = \int_0^\infty tf(t)dt$$

The failure rate $Z(t)$ can be determined for any probability distribution by taking the ratio of the failure density function $f(t)$ and the reliability function $R(t)$:

$$Z(t) = \frac{f(t)}{R(t)}$$

Figure 27.3 shows the relationship of the shapes of the failure rate, failure density, and survival functions.

One of the best well-known distributions is the negative exponential distribution. The reliability formula

$$R = \exp(-\lambda t)$$

where λ is the failure rate and t is time can be derived from the Poisson distribution by using the first term of this distribution, which is the probability of no failure. The probability density function of the exponential distribution is

$$f(t) = \lambda \exp(-\lambda t) \qquad t \geq 0$$

The MTTF can be calculated as

$$\text{MTTF} = \int_0^\infty tf(t)\, dt$$

Substituting for $f(t)$ gives

$$\begin{aligned} \text{MTTF} &= \int_0^\infty t\lambda \exp(-\lambda t)\, dt \\ &= \lambda \int_0^\infty t \exp(-\lambda t)\, dt \\ &= \frac{1}{\lambda} \end{aligned}$$

The MTTF of the negative exponential is equal to the reciprocal of the failure rate; this relationship holds only for the negative exponential. The failure rate of the negative exponential is

TABLE 27.2 Methods of Presentation of Data When Failure Rate Is Constant

Failure rate equivalent statements
0.00001 failure/h
10 failures/10^6 h
0.1%/100 h
1%/1000 h
MTTF = 100,000 h
When failure rates are in the 0.001%/1000-h range, the term FIT is often used
1 FIT = 1 failure/10^9 h
= 0.0001%/1000 h

$$Z(t) = \frac{f(t)}{R(t)} = \frac{\lambda \exp(-\lambda t)}{\exp(-\lambda t)} = \lambda$$

Very often, as a first approximation, it is assumed that electronic component failures follow an exponential distribution. One of the properties of the exponential distribution is that failure is independent of time. This allows one to vary the combination of devices and hours in unit hours of reliability testing. For example, if 100,000 unit hours are required for test, 100 units can be tested for 1000 h or 10 units can be tested for 10,000 h to demonstrate a given reliability level. Failure rates are then usually expressed as percent per 1000 h. If the exponential distribution does not apply, which means the failure rate is not constant with time, then reliability cannot be expressed by percent per 1000 h. When the failure rate is constant, there are a number of ways of presenting data. Table 27.2 shows the most common forms found in reliability papers.

27.1.4 Estimation of Failure Rate and MTTF

Assume that an estimate of the failure rate λ and MTTF is required on the basis of the system life test data given in Table 27.3. The mean time before failure (MTBF) is

$$\text{MTBF} = \frac{70{,}900}{3} = 23{,}633$$

(MTBF is used in system work rather than MTTF because systems can be repaired.) The failure rate FR is

$$\text{FR} = \frac{3}{70{,}900} = 0.000042/\text{h}$$

TABLE 27.3 Sample System Live Test Data

Unit	Operating time	Number of failures
1	7,000	0
2	6,000	0
3	5,500	0
4	6,800	1
5	7,200	0
6	9,000	0
7	7,800	1
8	6,100	0
9	9,100	1
10	6,400	0
	70,900	3

In a component failure rate calculation assuming exponential distribution, we have that *if 1000 transistors are life tested for 5000 h with two failures, the number of device hours is equal to 1000 × 5000, or 5,000,000 device hours. Then*

$$\text{MTTF} = \frac{5{,}000{,}000}{2} = 2{,}500{,}000$$

$$\text{FR} = \frac{2}{5{,}000{,}000} = 0.0000004/\text{h}$$

$$= 0.04\%/1000 \text{ h}$$

These estimates are known as *point estimates* because the estimate is a point and not an interval. Conceivably, if the life test were conducted again, different values would be obtained for the MTTF and failure rate.

By using a statistical distribution known as χ^2 (chi squared), one can calculate a confidence interval whose range of values includes the MTTF a given percentage of the time, for a confidence level.

$$\text{Confidence limit} = \frac{2cM}{\chi^2}$$

where M = MTTF based upon c observations
c = number of observations
χ^2 = statistical distribution, selected for the confidence level or probability desired

The χ^2 distribution tables have associated with them degrees of freedom (DF); in this case, DF = $2c$.

With this information the 90 percent confidence interval for the first example above would be

χ^2 distribution, tabled values	Confidence limit (true MTTF) $2cM/\chi^2$, h
12.592 (lower)	$\frac{2(3)26{,}633}{12{,}592} = 11{,}260$
1.635 (upper)	$\frac{2(3)26{,}633}{1.635} = 86{,}726$

Therefore, the true MTTF estimates from this data can be anywhere in the interval of 11,260 to 86,726 h.

27.2 SOME CONCEPTS OF MODELS

No discussion of reliability models is complete without a definition of randomness, or random failure. The best definition states that random failures are a class of failures that are statistically independent of past history. Certain processes, such as the Poisson process, have this unique property.

In practice, the "bathtub" failure rate model, Fig. 27.4, with its constant or

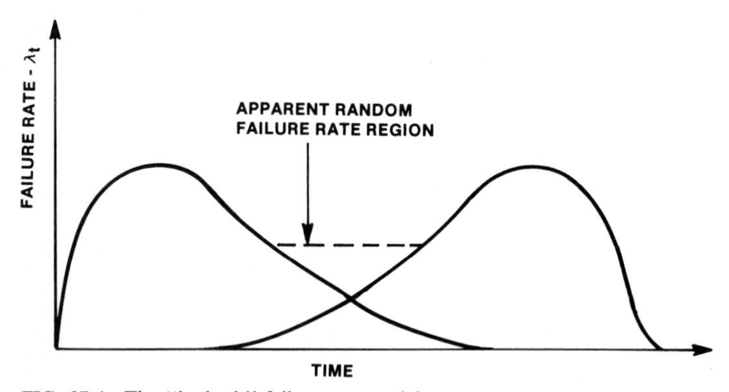

FIG. 27.4 The "bathtub" failure rate model.

random failure rate region, rarely exists. In this model, the failures generally considered to be in the "random" failure rate region are found more often to be late, early wearout or early, late wearout failures. They do not fit the definition of random; i.e., they are not independent of past history.

Failures in semiconductors follow the laws of cause and effect. The physics of failures and failure mechanisms are either already understood or are capable of being studied, and we know that failures and failure mechanisms are predictable. The statistical concept of randomness is a valid one in semiconductor studies, but must be coupled with knowledge of the physics of failure before any definite commitment to particular models can be made. Figure 27.5 is a more comprehensive view of modeling based upon the analysis of failures.

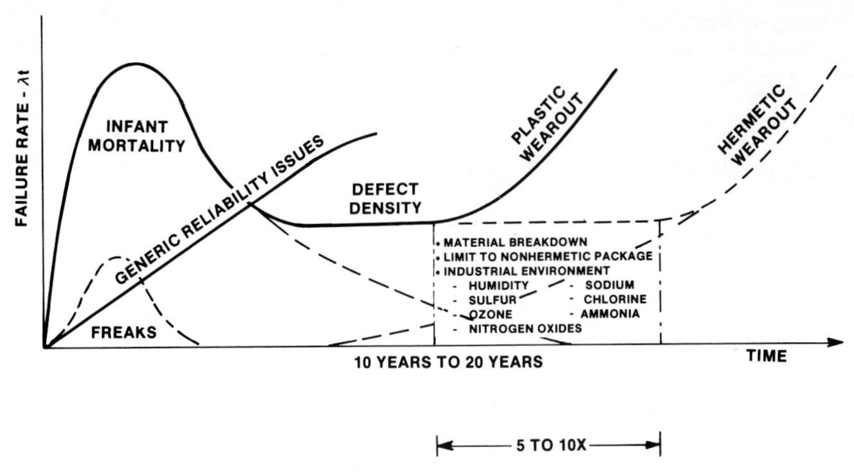

FIG. 27.5 View of modeling based on the analysis of failures.

27.2.1 Acceleration of Process Rate Reaction

The reliability of solid-state devices and processes is very high, especially when excessive temperatures are avoided. Normally degradation occurs very slowly by means of diffusion and ionic conductivity. Environmental stress, however, can sometimes alter the failure rate of devices and processes dramatically.

27.2.2 Arrhenius Reaction Rate Equation

Before we discuss some of the key areas of semiconductor reliability, we should review the basis of the Arrhenius equation. We will then use the Arrhenius formula for reaction rate kinetics to introduce the concept of activation energy. The Arrhenius reaction rate equation is

$$R = A \exp\left(\frac{-E}{KT}\right)$$

where R = reaction rate
E = activation energy, eV
K = Boltzmann's constant, 8.63×10^{-5} eV/K
T = absolute temperature, K (°C + 273)
A = frequency factor constant

Activation Energy. The *activation energy* is the minimum kinetic energy a molecule in the initial state of a process must acquire before it can take part in a reaction, whether it be physical or chemical (Fig. 27.6). Therefore, a failure can be caused by changing the activation energy at a given temperature or by increasing the temperature. The technique of accelerating failure (by increasing temperature) to predict years of useful life and to screen semiconductors for failures that occur early in system life is a common practice in semiconductor reliability

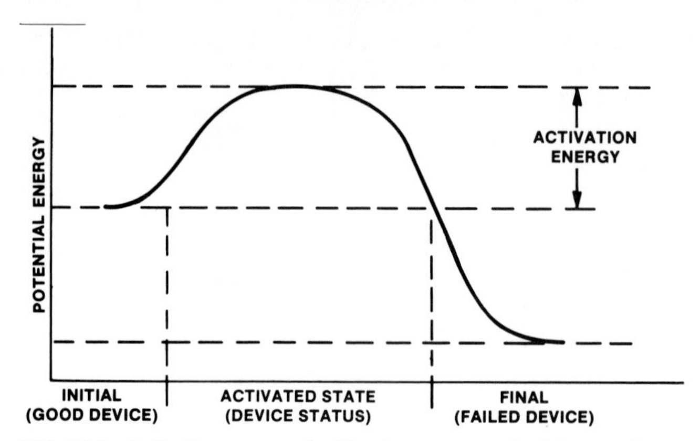

FIG. 27.6 Activation energy, the kinetic energy required for reaction.

studies. Some assumptions implicit in this technique are a single process, a single energy, and small deviations from the initial state.

Determination of the Constants *A* and *E*

For an electronic component:

1. Establish stress test failure distributions (≥ 50 percent failures) at three or more elevated temperatures.
2. Verify that the failure mechanism is the same at all stress levels.
3. Plot failure distributions (cumulative percent failure versus time) on probability paper (lognormal, Weibull, etc.).

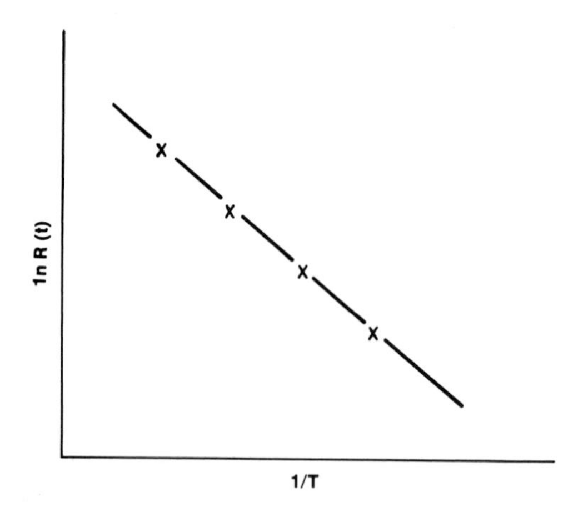

FIG. 27.7 Plot of stress test distribution, 50 percentile points, for obtaining the slope and intercept.

4. Choose a common point on each failure distribution (usually the median time to failure) and plot on Arrhenius paper, ln $R(t)$ versus $1/T$. The constants A and E can be obtained graphically.
5. By performing a linear transformation of the Arrhenius equation, we can obtain the values of A and E by linear regression:

$$R = A \exp\left(\frac{-E}{KT}\right)$$

$$\ln R = \ln A - \frac{E}{K}\left(\frac{1}{T}\right)$$

where E/K is the slope and ln A is the intercept. See Fig. 27.7.

27.3 VLSI PROCESSING-RELATED FAILURE MECHANISMS

Because of the complexity of VLSI circuits and their large die sizes, the most common defects are the result of shorts or breaks in active devices or interconnections. These types of defects lead to *hard* failures. Statistical variations in processes that lead to parametric variations, but not inoperable ICs, are called *soft* failures.

Table 27.4 details some of the typical failure mechanisms associated with IC fabrication process steps and the associated activation energy. Figures 27.8 through 27.20 depict actual failures. Physics-of-failure studies and experiments are helpful in identifying these failure mechanisms. As we have already stated, many chemical and physical processes that lead to failure are accelerated by an increase in temperature. The higher the activation energy, the easier it is to detect a mechanism at higher temperature.

The objective in the die manufacturing process for VLSI circuits is the elimination of defects from the active regions of the die for both yield and reliability considerations. Defects can be present in the initial starting wafer, or they can occur during subsequent processing. Failures can be broadly characterized as occurring either in the bulk material or in the surface of a device. MOS devices are particularly sensitive to surface-related contamination since the current flows on the surface of these structures. This sensitivity mandates the need for clean processes, especially those involving oxide layers.

Mobile positive charges, such as those found on sodium ions, are a constant presence in semiconductor processing and manufacturing. (Sodium and potassium are usually found in the manufacturing process; lithium, in packaging materials.) The most common MOS failure mode, caused by the presence of ions under gate structures, is the threshold drift phenomenon. Because positive ions are more likely than negative ones, most of these failures are caused by sodium ions. These charges are especially mobile in silicon dioxide and present a major problem in that they alter the characteristics and limit the high-temperature operation of a VLSI circuit. Fortunately, many process techniques, such as extensive cleaning of wafers with hydrogen chloride (HCl), purging of diffusion and oxidation tubes, and the use of phosphorous-boron-doped glass oxide, are effective in minimizing the effects of these ions.

TABLE 27.4 Typical Failure Mechanisms and Related Process Steps and Activation Energies

IC process step	Failure mechanism	Activation energy (eV)
Silicon, oxides, and respective interfaces	Surface charge accumulation (ionic drift, contamination)	1.0–1.4
	Slow trapping (charge buildup causing V_T shifts)	1.0–1.3
	Dielectric breakdown	0.3
	Metal interlevel dielectric breakdown	
	Silicon defects	0.3–0.5
	Hot carrier injection	
Metallization	Electromigration	0.5–1.2
	Corrosion (chemical, galvanic, electrolytic)	0.3–1.0
	Metal deformation	Temperature cycle induced
	Metallization defects (microcracks, etc.)	
Bonding interfaces	Intermetallic degradation (Al/Au)	1.0–1.05
	Metal fatigue (bond wires)	Temperature cycle induced

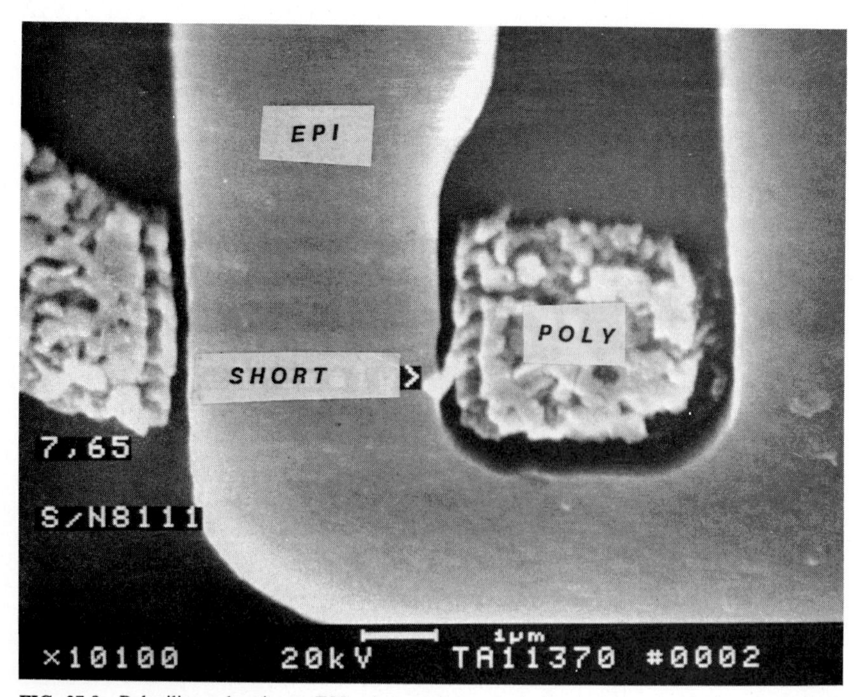

FIG. 27.8 Polysilicon shorting to EPI substrate is exposed after etching away poly gate and gate oxide.

FIG. 27.9 Cross section shows effects of hillocks in first level of metal. (3-μm double-layer metal.)

FIG. 27.10 Residual poly bridging (arrow) between two poly runs, causing a leakage current path. (3-μm ROM.)

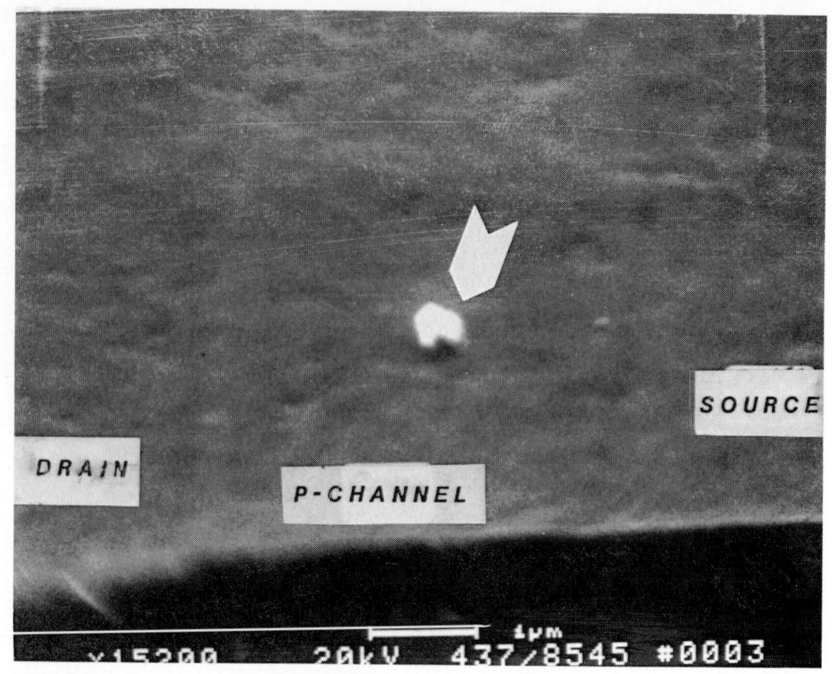

FIG. 27.11 A poly nodule is observed at gate oxide breakdown site after removal of the poly and oxide. (3-μm SOS gate oxide defect.)

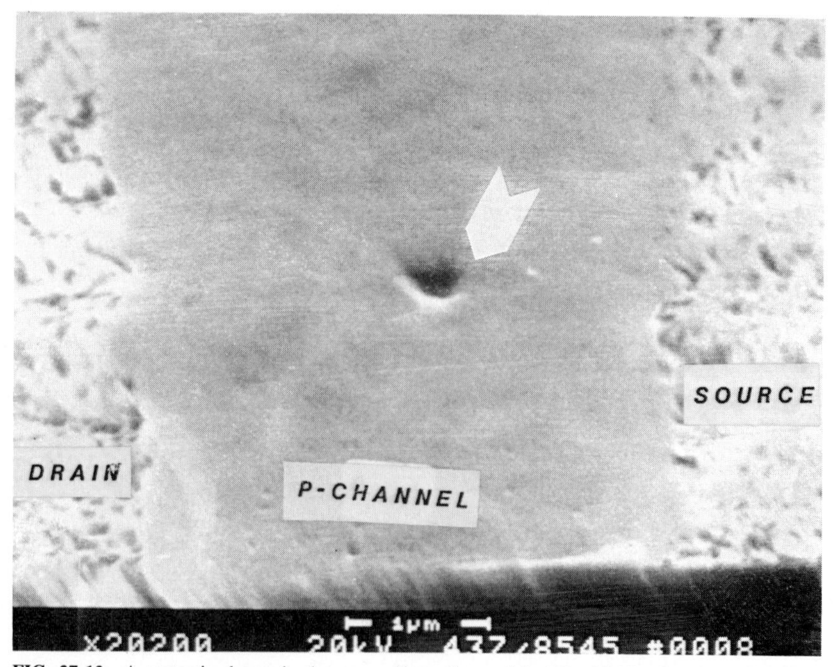

FIG. 27.12 A crater is shown in the same site as the nodule (Fig. 27.11) after the application of 5-s, Wright's crystal etch.

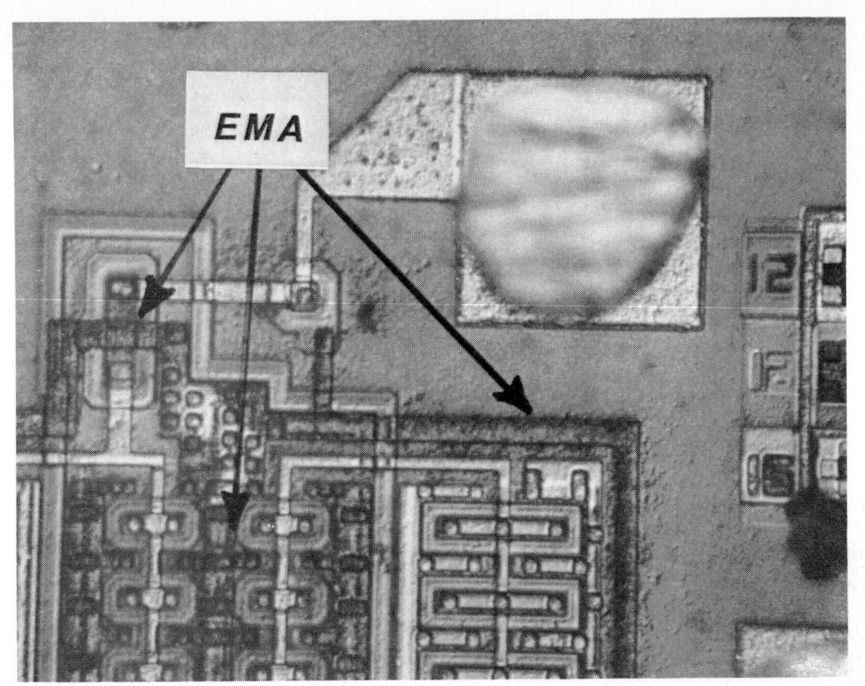

FIG. 27.13 Metal lines showing evidence of moisture-induced electrolytic metal attack (EMA).

FIG. 27.14 Inboard sites of electrolytic metal attack.

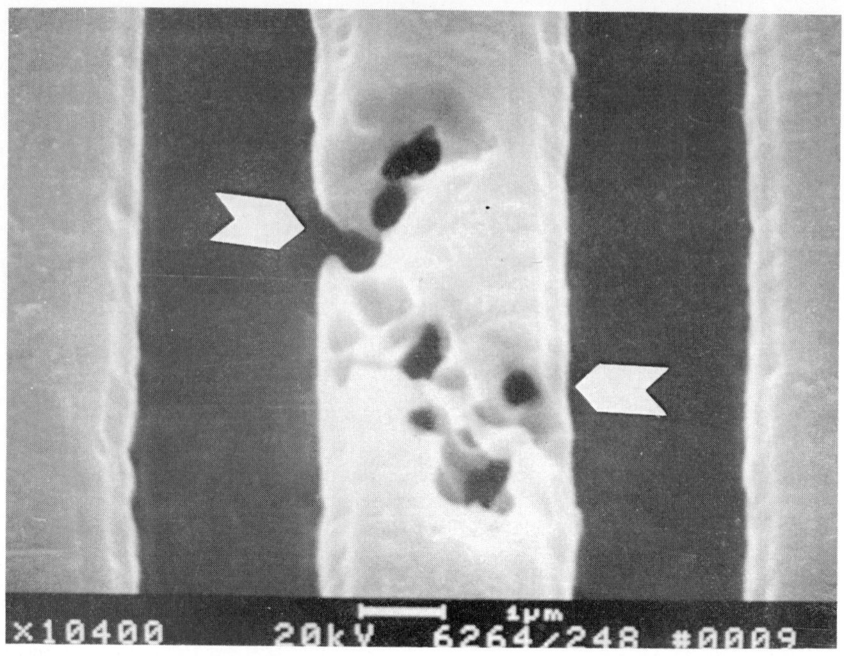

FIG. 27.15 Higher magnification of an EMA site in Fig. 27.14.

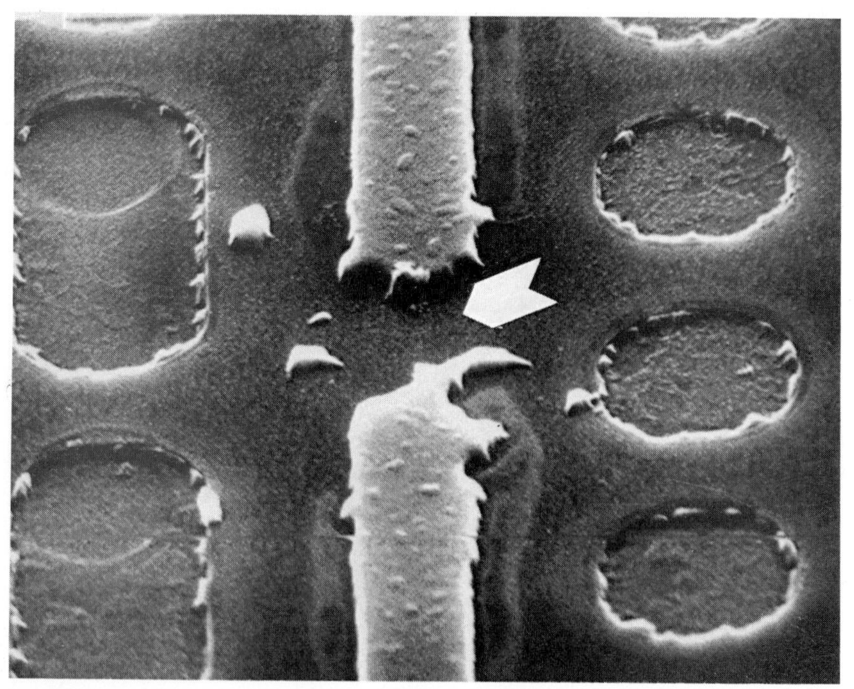

FIG. 27.16 A wafer processing defect resulting in a discontinuous 3-μm poly gate.

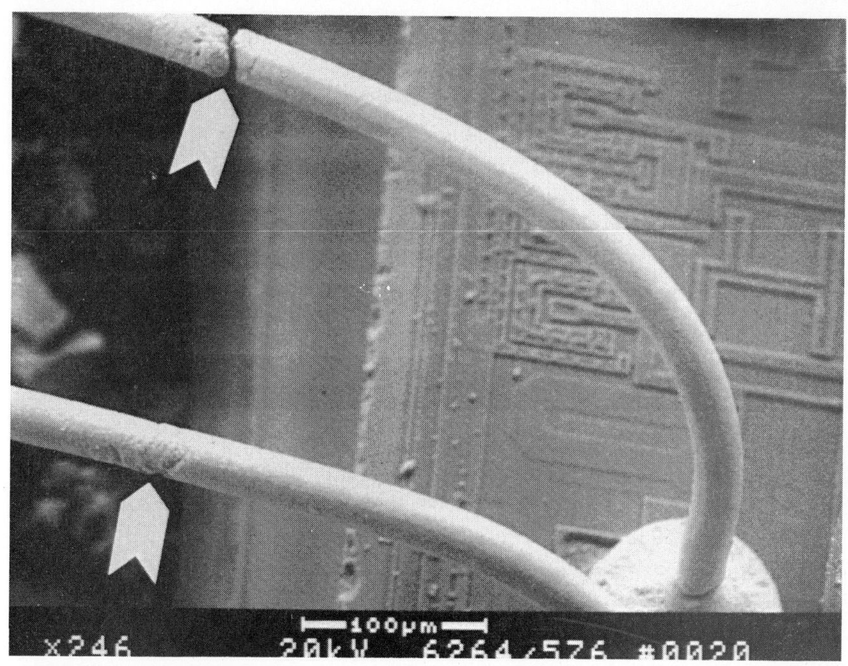

FIG. 27.17 Area of bond wire that opened due to fatigue under excessive thermal cycle stress.

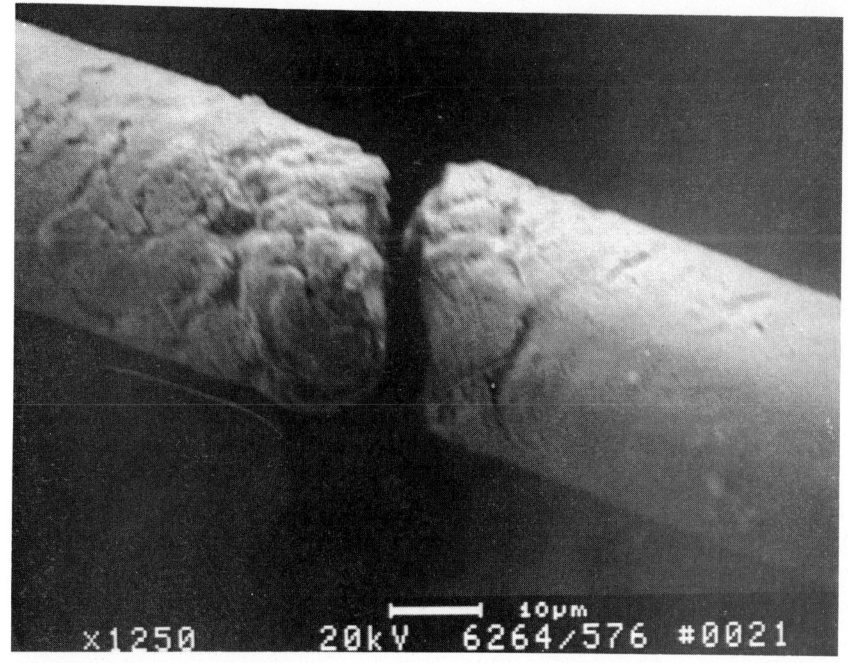

FIG. 27.18 Higher magnification view of open in wire showing fatigue type of metal failure.

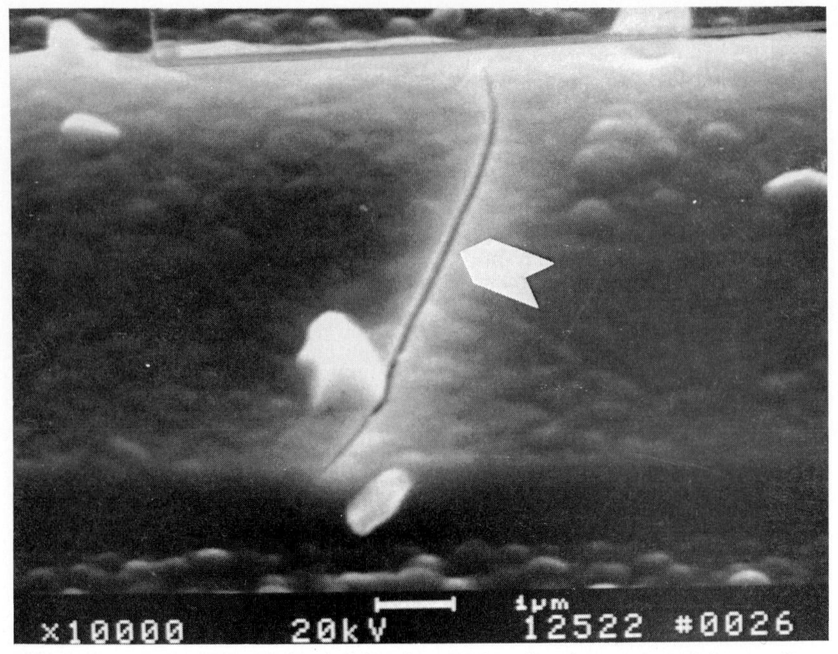

FIG. 27.19 Crack in the interlevel dielectric caused a short between metal layers. (3-μm double-layer metal.)

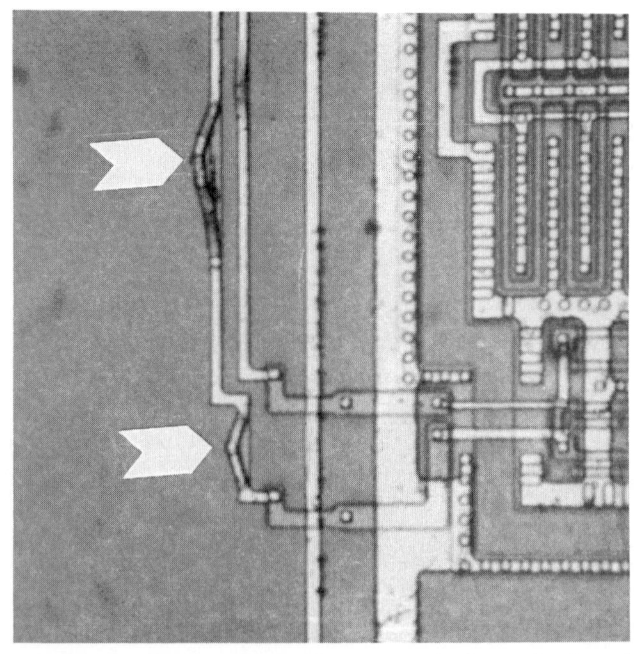

FIG. 27.20 Deformation of metal lines as a result of excessive thermal cycle stress. (3-μm standard cell.)

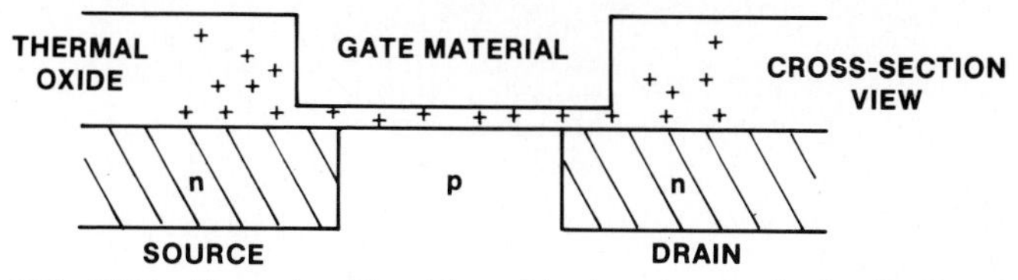

FIG. 27.21 Thermal oxide with positive-ion sites randomly distributed throughout.

Another problem related to mobile ionic charge is surface contamination, which may cause leakage between device source and drain. This leakage can lower the threshold voltage and seriously reduce the noise immunity of the device. Mobile ionic charges in the thermal oxide layer can also reduce or even increase threshold voltage, depending on charge polarity and the type of device (*n*-channel or *p*-channel) involved. Sodium ions (with a positive charge) in the thermal oxide layer of *n*-channel devices are the most difficult to control. Although very clean gate oxides are grown, sodium ions may still move in from the field oxide.

Figure 27.21 illustrates a thermal oxide with positive-ion sites randomly distributed throughout as a result of the natural diffusion of those ions at room temperature (290 K). Although these sites have some effect on the initial threshold, their influence can be compensated for, if the concentration of ions is known, by varying the oxide thickness in the manufacturing process.

As soon as bias is applied to the gate metal, however, ions begin to move as a result of electric field forces. The charges finally relocate, as shown in Fig. 27.22,

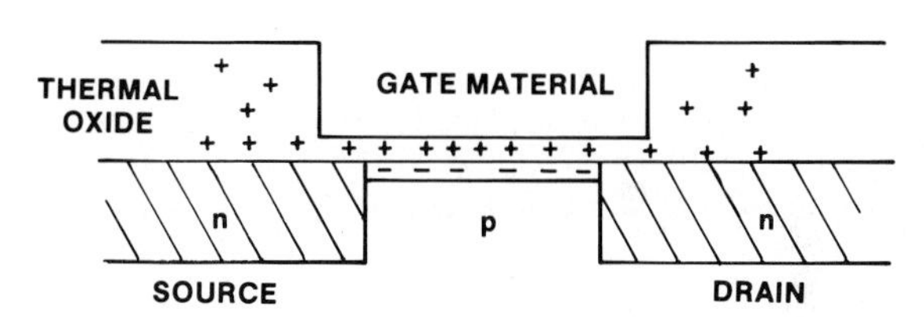

FIG. 27.22 Relocation of charges after application of bias to gate metal.

where the positively charged gate has pushed the majority of positive ions under the metal toward the silicon–thermal oxide interface. This concentration is sufficient to reduce the threshold voltage and may be enough to cause the new *n*-channel to remain, forming a permanent channel from *n* to *n* (source to drain) after the gate bias is withdrawn. The rate at which ions are driven into this condition is proportional to the temperature and the gate voltage. When gate bias is removed, ions begin to diffuse and revert to initial conditions.

If a negative polarity is applied to the gate, a transverse field is established that sweeps the positive ions into the gate oxide, where accumulation takes place. This mode of operation can cause an increase in leakage failure eight times that when there is positive gate polarity.

When, in MOS units, positive-ion concentrations are high, or high voltages are used, field inversion (*p*-layer inversion under the field oxide) can occur, causing the formation of parasitic MOS devices and resulting in the unwanted coupling of two or more circuits, as shown in Fig. 27.23. The inversion and its undesirable current paths

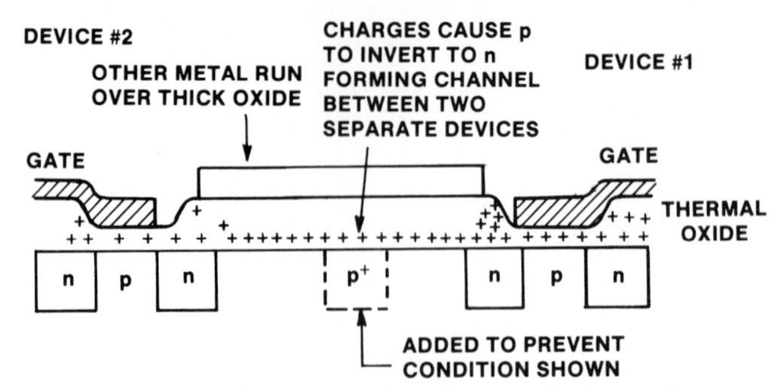

FIG. 27.23 Guard rings used as channel stoppers in *n*-type substrates.

can be prevented by installing *channel stoppers,* in the form of extra diffusions, between adjacent devices (Fig. 27.23). When *n*-type substrates are used, separate *p*-type wells or guard rings can be employed as channel stoppers for each device.

CMOS units include *n*-channel and *p*-channel devices on the same substrate. In an *n*-channel device, positive ions cause permanent inversion of the *p*-type well. When the gate is negative with respect to the substrate, positive ions are driven in a direction that tends to slightly increase threshold voltage above that for normally dispersed ions. Comments that apply to positive-ion contamination in thermal oxides apply equally to negative-ion contamination, except that the results for *p*-channel and *n*-channel units are reversed. However, negative-ion content is easier to control in processing and is not a major factor in leakage failure.

In the CMOS structure, only the NMOS device appears to be affected by sodium contamination. The NMOS transistor in the inverter will not change under positive gate potential, and the PMOS transistor will not change under either gate potential. The NMOS devices will eventually go into a depletion mode under gate ground potential in the presence of sodium ions. Experiments have shown that, in a CMOS inverter, the combination of both gate and source-to-ground potential is necessary to observe sodium instability. Only the NMOS devices in the CMOS inverter can fulfill this condition.

Several life tests are used to check for the presence of contamination. The most effective is the high-temperature, reverse-bias test (HTRB), where gate voltages are dc (100 percent duty cycle) and temperatures are elevated to accelerate ion movement. Gates are normally biased with their "on" polarities; the resulting changes in threshold voltage indicate the type of ion contamination present. HTRB tests are preferred to ac operating life tests for detecting ionic problems. In ac testing, gates that are operated at, for example, 50 percent duty cycle, allow time for diffusion of ions between pulses at rates equal to that for recovery at the life test temperature. As a result, devices having low levels of contamination may not drift enough to be failures. HTRB tests, however, would cause these devices to fail. Worst-case conditions should be used in these tests because the device could be subjected to dc bias in field use.

27.4 VLSI SCALING AND THE IMPACT ON RELIABILITY

Dennard's MOS transistor scaling principles served the industry well as it moved from the 5- to 10-μm range to the 1- to 3-μm range. These principles do not apply

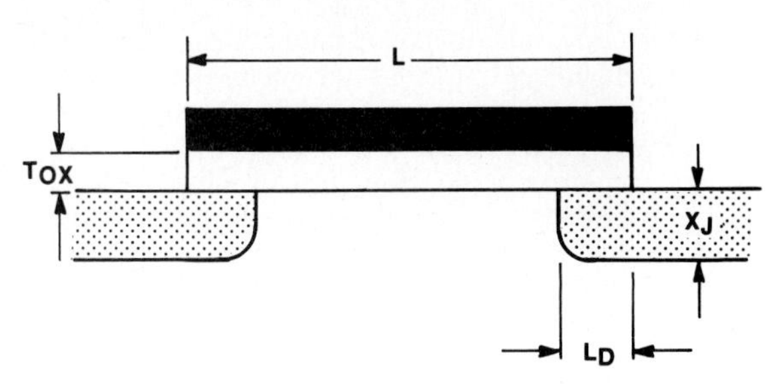

FIG. 27.24 Cross section of a silicon-gate n-channel device.

to the bipolar transistor. While MOS is planar, bipolar involves three dimensions. Also, the bipolar operation relies on more complex phenomena, such as minority carrier injection and collection.

The scaling principles state that if dimensions such as oxide thickness, channel length, channel width, and junction depth are reduced by the same factor, $1/k$ (k 1), and the electric field is constant, then device current, depletion-region width, and threshold voltage are reduced by the same factor.

Calculations show that as all lengths and thicknesses are reduced by a factor of $1/k$, all areas are reduced by $1/k^2$ and that complexity increases by k^2. Capacitance scales down by a factor of $1/k$, as does delay time, and the power dissipation per device scales down as $1/k^2$.

A key element in the scaling laws is that the electric field strength be constant. This means that as length is scaled by a factor of $1/k$, voltage is reduced by that same factor.

Figure 27.24 shows the cross section of a silicon-gate n-channel device, where L is the channel length, T_{OX} is the gate-oxide thickness, X_J is the junction depth, L_D is the lateral diffusion, and C_a is the substrate doping level. Table 27.5 shows circuit parameters and scaling factors for such a circuit. First-order scaling theory states that the characteristics of an MOS device can be maintained and the desired operation ensured if the parameters of the device are properly scaled.

TABLE 27.5 Circuit Parameters and Scaling Factors for a Silicon-Gate n-Channel Device

Device/circuit parameter	Scaling factor*
Device dimension, T_{OX}, L, L_p, W, X_J	$1/S$
Substrate doping, C_a	S
Supply voltage, V	$1/S$
Supply current, I	$1/S$
Parasitic capacitance, WL/T_{OX}	$1/S$
Gate delay, $VC/I(r)$	$1/S$
Power dissipation, VI	$1/S^2$
Power delay product	$1/S^3$

*S = scaling factor.

27.4.1 Second-Order Effects on Reliability

As devices are reduced in size, scaling theory states that, ideally, they should maintain the same qualitative characteristics. But, in reality, second-order phenomena become significant. Some of these phenomena affect the circuit design, while others relate to reliability. In general, all second-order effects arise for two reasons. First, as dimensions are reduced while a constant supply voltage is maintained, the average electric field is increased; this field activates many second-order effects. Second, the edges of a small device are so close together that the nonideal electric fields at these edges significantly affect its performance.

Small devices are vulnerable to the two second-order effects illustrated in Fig. 27.25. The second-gate effect occurs when the electric field lines emanating from the drain junction end up on the oxide-silicon interface. Punch-through can be relieved by careful control of the substrate impurity profile through ion implanation and the choice of a thin gate oxide.

Although not a second-order effect, impact ionization represents a third source of leakage in VLSI circuits. The effects of impact ionization are illustrated in Fig. 27.26. At a very large drain voltage, approximately 20 V, the junction avalanches for all channel lengths greater than about 4 μm. Impact ionization occurs when, activated by high electric fields, a population of electrons and holes with energies much higher than the normal channel electrons is created. The holes flow into the substrate and place a small load on the back-bias supply. Some of the electrons have enough energy to be injected into the gate oxide, as shown in Fig. 27.26*a*, where they can cause a gate current or be trapped. These trapped electrons cause a shift in the threshold voltage, Fig. 27.26*b*.

Because of the mechanisms found in VLSI devices as feature sizes are reduced, higher failure rates are predicted than would be expected. In actual practice, however, with fully developed volume-produced VLSI circuits, the predictions do not seem to be valid. There appear to be two reasons for this fortunate discrepancy. First, as feature size is being reduced, improved processes are being developed to minimize the mechanisms described as well as many other mechanisms that lead to surface instability in devices. Second, as transistor counts on a

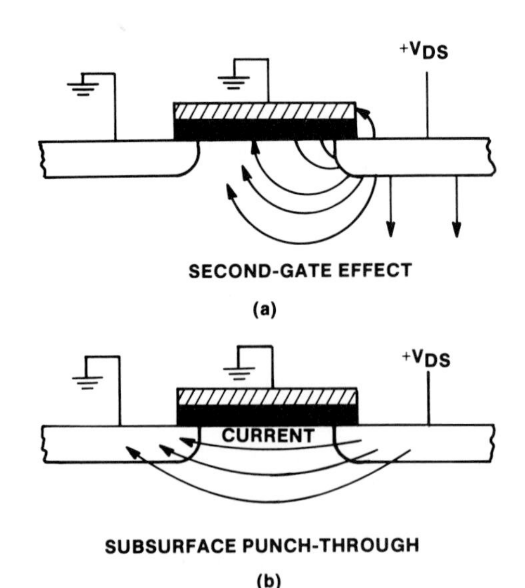

FIG. 27.25 Second-order effects on reliability. (*a*) Second-gate effect; (*b*) subsurface punch-through.

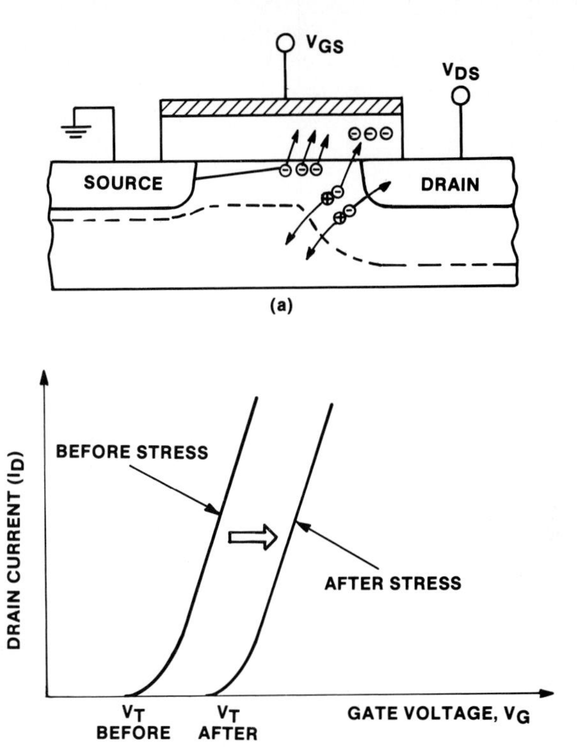

FIG. 27.26 Representation and effect of trapped electrons or impact ionization on VLSI circuit.

die increase, more and more CMOS designs are being used, making the use of NMOS devices (as noted above, those most susceptible to failure) unnecessary and impractical.

The CMOS technology enhances the reliability of VLSI devices in a number of areas. First, the complexity of the CMOS process has not changed much over the years, even as devices become more complex. Second, PMOS devices in the CMOS inverter do not suffer from hot carrier injection. Third, CMOS devices do not draw static current, a property that minimizes electromigration and degradation due to dc stressing. Figure 27.27 shows how failure rates have been reduced on CMOS circuits as feature size has been reduced. Figure 27.28 demonstrates that as complexity (the number of transistors per chip) increases steadily in the product designed each year, quality and reliability continue to improve.

27.5 VLSI PACKAGE RELIABILITY

The packaging of a VLSI device includes connection of the die to the package leads; the package itself provides protection for the die. A packaged die is con-

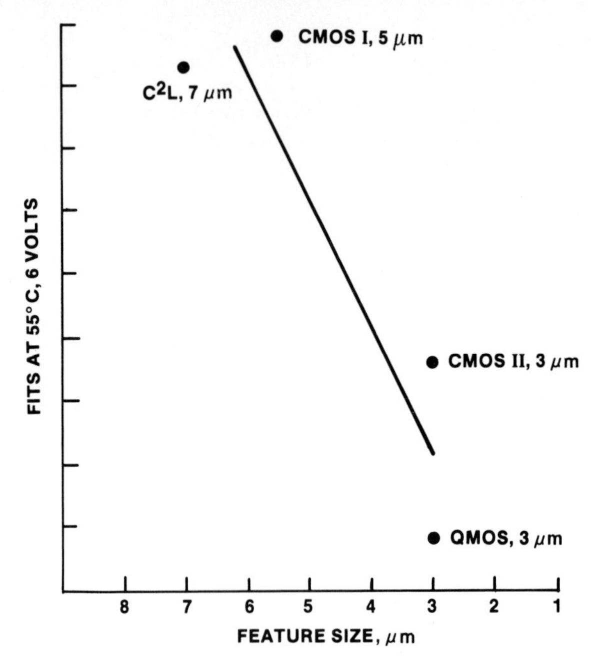

FIG. 27.27 Failure rate versus feature size.

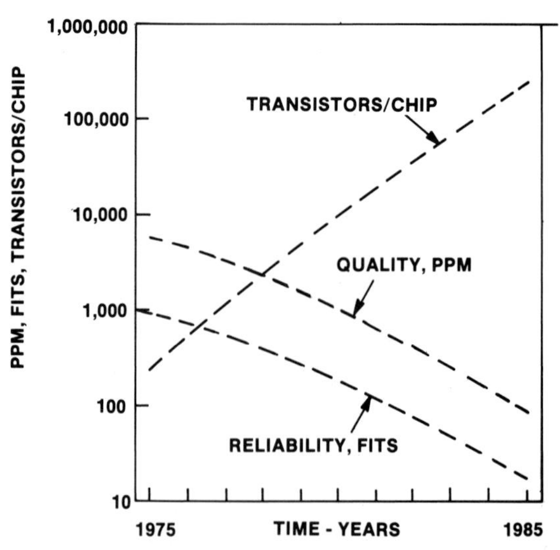

FIG. 27.28 Integrated circuit quality and reliability versus chip complexity for the past decade.

sidered a *system*. The protection afforded by the package prevents mechanical, chemical, electrical, and environmental degeneration—the causes of failures. Determination of the correct package for the silicon die, completion of the final system, is a procedure requiring careful consideration. A number of questions must be answered concerning package technology, size, performance, long-term reliability, and cost. The attributes required are available in a variety of packaging media, including both hermetic and plastic encapsulation (Table 27.6). Plastic encapsulation is generally selected for high-volume circuits because of its ruggedness and lower cost.

TABLE 27.6 Available Plastic and Hermetic Package Configurations

Plastic
Dual in-line (DIL)
Small outline package (SOP)
Plastic leaded chip carrier (PLCC)
Plastic pin-grid array (PPGA)
Hermetic
Frit seal [CERDIP (DIL)]
Ceramic (DIL)
Ceramic (flat packs)
TO-5
Pin-grid array (PGA)
Leadless ceramic chip carrier

Reliability requirements for hermetic packages mandate control of the internal sealing environment (low moisture and contaminant levels) and maintenance of hermeticity over the life of a system. Plastic-encapsulated packages are nonhermetic by nature, a fact that complicates the reliability issue substantially because of the package–silicon die interaction. Ultimately, package selection for VLSI devices involves a set of trade-offs among many package types, and must take account of their size and high lead count.

27.6 PLASTIC PACKAGE RELIABILITY

Plastic-encapsulated ICs are the predominant systems used by electronic system manufacturers today, again, primarily because of their significantly lower cost and good mechanical strength. However, the use of plastic materials brings with it certain reliability issues that must be resolved. The most important of these issues is the ability of a plastic package to resist various temperature and humidity conditions or, simply, its moisture resistance capability. This discussion describes several advances in wafer and assembly processes that have resulted in increased reliability of the plastic package.

27.6.1 Plastic Package System

In the design of any plastic package system for a semiconductor device, no single design factor is predominant. There is a tendency to believe that the entire issue

TABLE 27.7 Plastic Package System

Chip design, process, junction seal	Silicon dioxide Silicon nitride
Metallization	Aluminum Aluminum-silicon Gold
Passivation (over metal)	CVD PSG (chemical vapor deposited phosphosilicate glass) Plasma-deposited nitride Silicon resin
Chip mounting	Epoxy Eutectic Polyimide
Lead-frame or header material connections	Wire
Plastic	Epoxy (low chloride, low stress) Plastic mechanical strength Plastic thermal stability Plastic to lead-frame adherence
Plastic-molding system requirements	Time Temperature Pressure Postmold cure

is determined by the selection of plastic materials—that is, that a single plastic material will solve all problems in plastic-encapsulated-device reliability—but this is not true. Table 27.7 indicates the many factors involved in the design of packages for a plastic system, any one of which can be a complete study in itself. When a plastic package is designed, all of these factors must be considered as part of an experimental design because of the possible interaction between them. Attempts to improve the plastic package by simply changing materials and formulations can cause more problems than they solve.

Figure 27.29 illustrates the present plastic system used for most narrow-body (0.250-in) packages. The basic components of the package are a lead frame on which the chip is mounted and wire bonded and the plastic case material that provides mechanical protection. The plastic package is nonhermetic, but, practically, owing to the potential for chemical reaction of the circuit metallization, a degree of hermeticity is necessary to protect the chip.

Cost is the primary reason for use of the plastic package; a hermetic package may cost 3 to 10 times more than a plastic package. Moreover, the plastic package components lend themselves to assembly by mechanized techniques. These techniques result in improved yields and quality and lower assembly costs by eliminating manual handling and operator error.

Each package lead count can pose different problems in designing for reliability. For example, the eight-lead package, because of its size, has a unique lead-frame design, and the amount of material around the chip is less than that around the chip in packages of larger lead count. Thus, when silicon devices fabricated by the same chip technology are put in packages with different lead counts, the reliability results for each can be different.

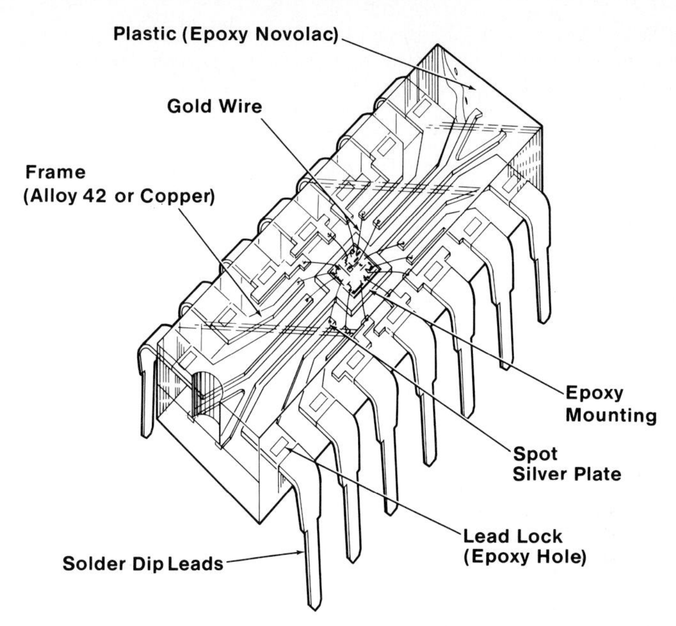

FIG. 27.29 Plastic package system [dual in-line plastic (DIP) 14-lead] used today for most narrow-body package needs.

In the design of the plastic package, the chip size and its layout play an important role. If the chip is too large for the package, the walls surrounding the chip can be very thin, and the amount of plastic material available to protect the chip from the external environment can be inadequate. Obviously, this condition in itself can result in a potential moisture problem, since the plastic material is the first line of defense.

In addition to the chip size and layout, consideration must also be given to the lead-frame material, the chip-mounting method, the chip-to-frame interconnection, the encapsulation materials, and the lead finish. Because plastic-packaging assembly is done in very high volume, automation of the assembly is a necessity. A mechanized system is needed not only to produce the volume required but to ensure consistent quality. Again, this mechanization reduces operator error and handling of the product.

27.6.2 Industry Status of Plastic Packaging

The plastic package is the dominant form of packaging for ICs in use today worldwide because it is mechanically stronger than its hermetic counterpart and more cost-effective, both in manufacturing and application.

Each year there is an improvement in the reliability of plastic-packaged ICs because of the ever increasing demand by the customer for plastic in more diverse, more hostile applications, and because of the introduction and use by the manufacturer of improved materials, techniques, and processes. The information provided in the following sections indicates the kind of reliability improvement that can be achieved with identification of proper models, thorough engineering programs, and statistical experimentation, and will provide the basis for the transition to the next level of plastic-package reliability.

27.7 HERMETIC VERSUS PLASTIC PACKAGING

There are two major factors to consider when VLSI circuits are packaged in hermetic packages. The sealing environment should be free of contaminants and moisture, which can be harmful to the silicon die, and package hermeticity should be maintainable over the system life. History indicates that 1×10^{-8} atm cc/sec leak rate has been acceptable in hermetic packages over the past 25 years.

As indicated, there are also some factors to be considered when plastic-encapsulated devices are used, especially when long life (20 years or greater) is required. The atmospheric environment of the application is critical because harsh chemical environments can, in the presence of moisture, cause severe corrosion of the metallization on the silicon die. The thickness of this metallization is most often about 10,000 Å. Table 27.8 lists some of the most harmful chemicals.

Because of the interactive effects between die and plastic package, there are usually more potential mechanisms of failure present in these packages than in hermetic packages. These potential mechanisms exist because of the intrinsic nonhermeticity of plastic packages and the severe mechanical stresses present in the application during large delta excursions in temperature. Any use of plastic-encapsulated VLSI circuits must consider the factors listed in Table 27.9 as part of the reliability picture.

Figure 27.30 illustrates the reliability engineer's challenge as it relates to plastic-encapsulated and hermetic packages: that of matching the failure mechanism curves of plastic and hermetic packages until each package is of equal reliability. As the figure shows, failure rate reduction in plastic can only be accomplished by mechanism reduction. In the plastic package, a total reduction of

TABLE 27.8 Chemicals Causing Atmospheric Corrosion

Chemical	Reaction
Hydrogen sulfide	Corrodes copper, silver, and other metals in the presence of water vapor
Reactive sulfur	Corrodes like hydrogen sulfide
Sulfur oxides	Corrode metals in the presence of water vapor. Sulfur dioxide deteriorates polymers
Chlorine	Corrodes metals in the presence of water
Nitrogen oxides	Deteriorate polymers
Ozone	Deteriorates plastics and is potentially harmful to any organic material
Ammonia	Decreases insulation resistance, corrodes aluminum
Water (humidity)	Relative humidity above 40% increases the corrosion rate of many gases on metals

TABLE 27.9 Plastic-Encapsulated-Package Reliability Considerations

Mechanical defects
Thermomechanical failures
Die metallization corrosion
Die surface long-term stability
Lead-frame corrosion
External leakage currents
Plastic long-term stability

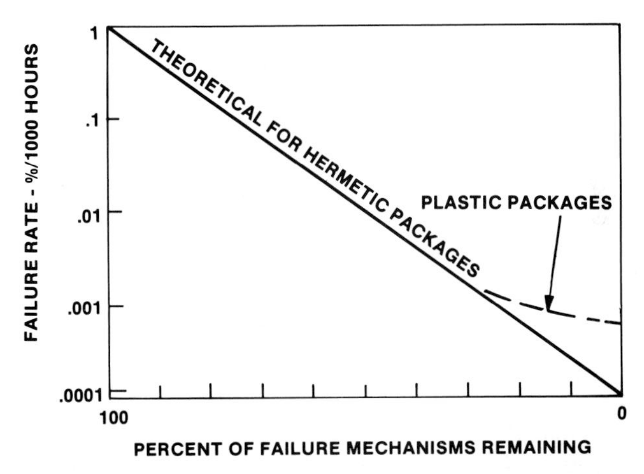

FIG. 27.30 Plastic-encapsulated versus hermetic-package failure reduction.

mechanisms may not even be theoretically possible; with hermetic packages, it may.

But even taking into account the issues discussed so far, plastic-encapsulated VLSI circuits are still the most reliable for the price. No other semiconductor product can be produced in such high volume with such a high comparative reliability to price ratio. Figure 27.31 shows some of the reliability results for plastic packages subjected to severe testing. These results make clear the reasons why plastic-encapsulated circuits are used in so many applications.

27.8 DIE PROTECTION IN PLASTIC PACKAGES

One of the most significant factors impacting the life of a plastic-packaged VLSI circuit is the method of deposition of the low-temperature, chemically vapor-deposited phosphosilicate glass (CVD PSG) and plasma-enhanced silicon nitride (PEN) protection layer used to cover the die before encapsulation. The purposes of this protect layer are as follows:

1. *Scratch Protection:* The soft-metal interconnections on the die, typically aluminum, can be easily abraded and shorted to each other. If pure SiO_2 is used

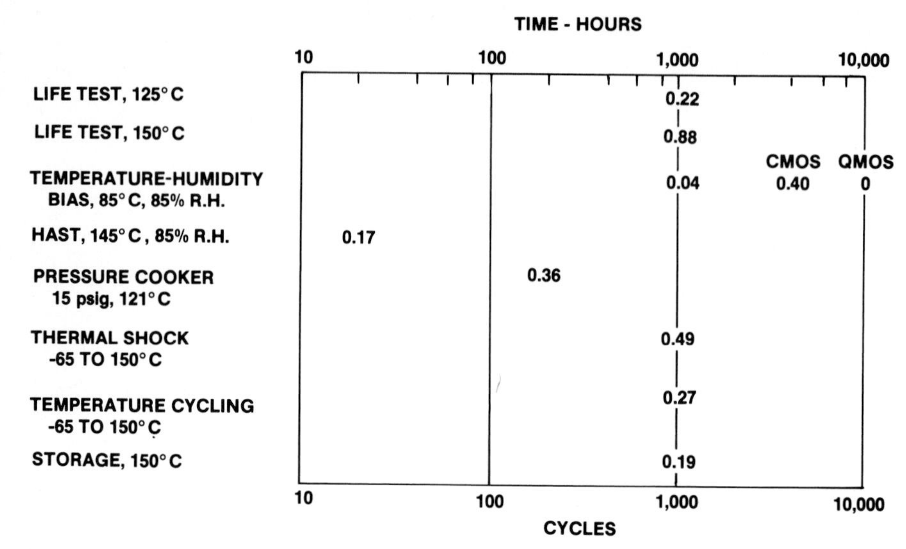

FIG. 27.31 Reliability data for plastic packaged integrated circuits.

as an overcoat, it *crazes* (forms a cracked, patchwork pattern) because of tensile stress and mismatch of the silicon, aluminum, and silicon dioxide. Crazing is prevented in the CVD PSG layer by the addition of the phosphorus.

2. *Sodium Contamination:* MOS device thresholds are adversely affected by the presence of sodium near critical Si-SiO_2 interfaces. Phosphorus helps getter, or collect and neutralize, sodium ions. The plasma-enhanced silicon nitride forms an even more effective barrier to sodium and is in compression on the die. The property of compression helps create a layer with physical integrity. PEN is usually used with the CVD PSG to form a *dual-dielectric* protect layer.
3. *Corrosion Resistance:* Water vapor can penetrate plastic packages to the die. If left uncoated, the aluminum parts of the die would be corroded. The amount of phosphorus in the CVD PSG layer must be tightly controlled, since this layer becomes more permeable by water with excessively high phosphorus concentration.
4. *Loose Particle Protection:* The CVD PSG layer isolates the conductive elements on the die and prevents foreign material from shorting them.

27.9 ALUMINUM CORROSION

A number of types of corrosion, or electrolytic metal attack (EMA), can take place in a plastic package. The amount of EMA that occurs is a function of the many factors listed in Table 27.10; several of these factors must be present for EMA to occur. The abundance of these factors gives a good idea of why plastic package reliability is so difficult to attain. From the time a plastic package is assembled to its final application, its lack of hermeticity leaves it open to exposure to a variety of adverse conditions.

TABLE 27.10 Factors Determining Amount of Corrosion of IC Metallization

System pH
Metal
Encapsulation material
Passivation glass
Ionic contamination
Temperature
Relative humidity
Applied voltage
Moisture resistance of package

One of the most severe of these conditions, and one of the main causes of corrosion, is the interaction of moisture with chlorides. The package materials and processes in use today minimize the amount of chloride content, as explained later, but chlorides may still be prevalent in many processes and applications to which the package may be subjected after manufacture.

The problem of corrosion begins with a tendency, during lead forming, for a separation to occur between the lead frame and plastic at the point where the leads enter the package body. This separation allows contaminants (e.g., chloride) to migrate along the wire onto the die and then onto the exposed aluminum bond pads where the gold ball bonds are made (Fig. 27.32). The interface of dissimilar metals, gold to aluminum, sets up a potential of approximately 3 V, and in the presence of water and chloride an EMA reaction can take place. The equations for the EMA reactions involving chlorides are described shortly.

Metallic aluminum and its alloys are normally protected from atmospheric conditions by a layer of passivating oxide. The oxide is dissolved in a moist environment by the absorption of Cl^- on the surface according to the reaction

$$Al(OH)_3 + Cl^- \rightarrow Al(OH)_2Cl + OH^-$$

After the surface oxide is dissolved, the exposed metallic aluminum then reacts with the chloride ion as follows:

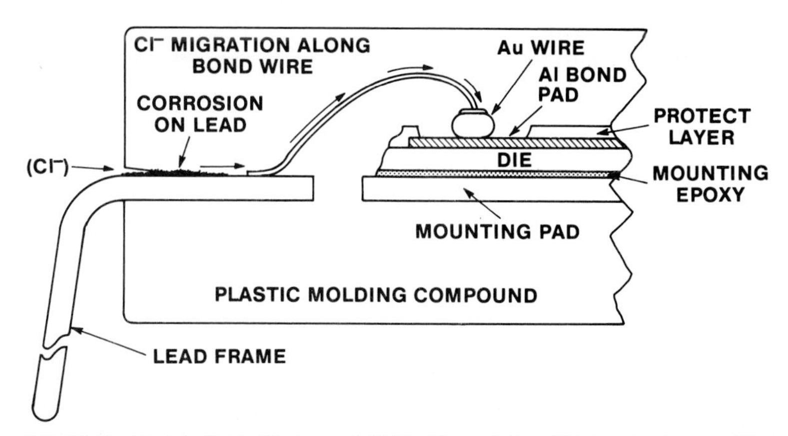

FIG. 27.32 Model of chloride-induced EMA. The path for Cl^- ions is shown, with the primary area affected being Al bond pads.

$$Al + 4Cl^- \rightarrow [AlCl_4]^- + 3e^-$$

The resulting complex aluminochloride anion $[AlCl_4]^-$ then reacts with water:

$$2[AlCl_4]^- + 6H_2O \rightarrow \tfrac{1}{4}Al(OH)_3 + 6H^+ + 8Cl^-$$

The chemical reaction described between the aluminum and the chloride is one that allows chloride to be continuously released into the system. This chloride is free to react as long as water is present.

Since 100 percent screening of all product for moisture resistance is impractical, the approach preferred is to eliminate the chlorides—more specifically, the sources of the chlorides and/or the means by which they enter a plastic package. The areas of investigation that have proved most fruitful and that have led to actual improvements in package fabrication techniques and reliability, are assembly-area soldering, lead-frame design, and plastic-molding processes.

27.10 SCREENING FOR RELIABILITY

There is always the question of the amount of testing necessary to ensure optimum reliability in an IC along with optimum profit for the manufacturer and optimum cost for the user. This question can only be answered after some very complex issues have been considered, since there is no cookbook method for determining correct testing procedures. Techniques that work with some devices are useless with others. The advisability of the use of burn-in of ICs to reduce infant mortality (or early failures) is part of the larger question of which total screening methodology is best for solid-state components.

Depending on the reliability level desired, the appropriate screening procedures can include as many as five or six different stress tests, each designed to reduce the occurrence of failure. No screening or burn-in test should be applied to any device unless sufficient reliability engineering work has been performed on the part and related failure mechanisms are understood. The indiscriminate application of screening test techniques may not reduce infant mortality and may actually increase failure rates. Burn-in is generally considered one of the most expensive methods of reducing infant mortality.

Figure 27.33 illustrates the basic theory of screening. As IC component counts increase, screening must be given more consideration, not only at the component level but at the system burn-in level as well. Most semiconductor users do not clearly specify the reliability levels they want and very often buy off-the-shelf components that have been final tested but have received little other preconditioning. In most cases, these devices are satisfactory if adequate system debugging has taken place. System debugging is crucial; even if the semiconductor component is burned-in, component failures can still occur in equipment in systems for which the design has not been optimized. Generally, however, the system burn-in will identify any failures that escape even the most rigorous IC quality control inspection system.

Component burn-in is effective in screening out temperature- and time-dependent mechanisms that normally escape detection under a 100 percent final electrical test. Improvements in reliability resulting from burn-in to avoid these types of mechanisms may range from an order to several orders of magnitude. Products with initially high failure rates generally show more improvement in re-

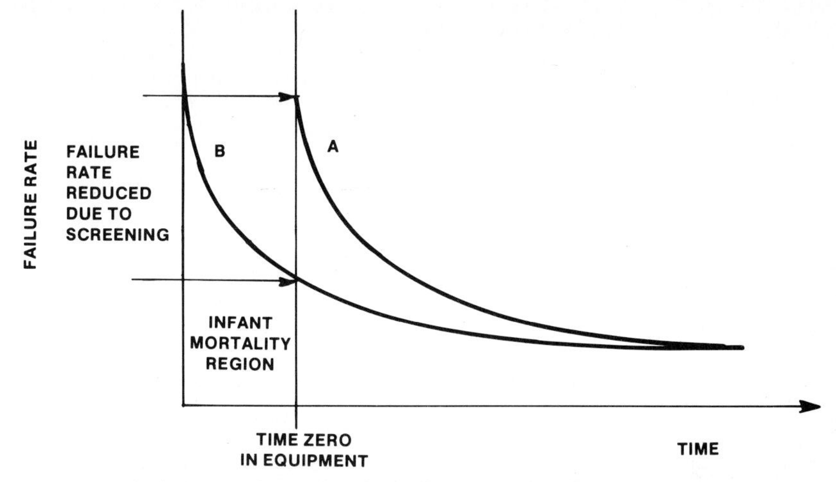

FIG. 27.33 Basic theory of screening. Screening, or burn-in, eliminates a major percentage of the infant mortality. Component life in equipment is translated from curve *A* to curve *B* as a result of burn-in.

liability when burned in. Other mechanisms, those mechanically oriented, for example, are better culled out by tests such as temperature cycling. Even simple storage life tests can be used to accelerate chemical degradation caused by surface contaminants. It is important, therefore, to understand the mechanisms of failure and the tests most effective in removing components that have a high probability of failure from a specific mechanism. The use of screening to increase MTBF levels is only justified when there is a resultant substantial decrease in the probability of failure. It is a fact that increases in MTBF will sometimes produce very little decrease in probability of failure.

The basic reliability equation illustrates this point:

$$R = \exp\left(\frac{-t}{m}\right)$$

where t = operating time
m = MTBF
R = probability of no failure

For values of R above 0.90, the MTBF must increase very rapidly to produce a small improvement in reliability.

In summary then, the question of burn-in cannot be answered simply. However, with the proper choice of circuit for application, a known reliability goal, and reliability engineering help, any user of integrated circuits should be able to obtain the reliability levels he or she requires.

27.11 SEMICONDUCTOR BURN-IN

Semiconductor burn-in is normally performed as an operating test under power or voltage conditions at high temperature, usually at the maximum rated temperature for the device. The duration of the test can vary from one day to a few weeks. Burn-in is intended to do the following things:

1. Weed out early failures; i.e., those devices that have a high probability of failing in system use
2. Stabilize devices
3. Monitor final performance of the entire lot subjected to burn-in.

Burn-in is also intended to remove the "freak" proportion of the distribution without affecting the central population strength, as shown in Fig. 27.34.

A burn-in test is effective when the product has a decreasing failure rate. If it does not, then the purpose of the burn-in will not be realized. Most semiconductors exhibit a decreasing failure rate for the initial period of the life cycle. However, there are still problems in specifying the correct amount of burn-in for a specific device. The faster the decrease in failure rate, the more the burn-in will improve reliability. Conversely, if the failure rate decreases slowly with time, then the burn-in will have only a small effect.

Because life is related to log time, some lot-to-lot variation may exist when

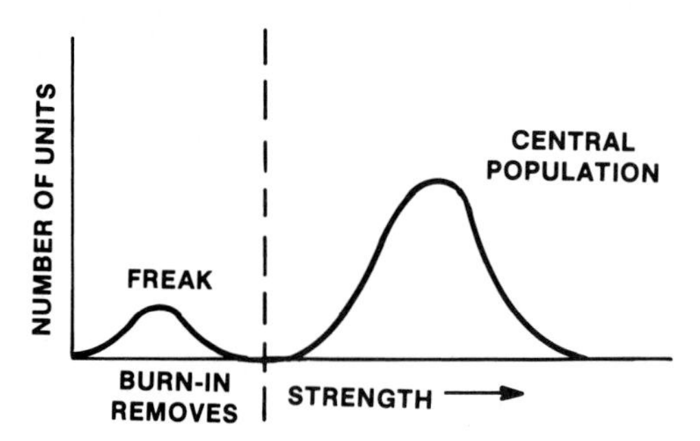

FIG. 27.34 Burn-in removes the "freak" proportion of the distribution without affecting the central population strength.

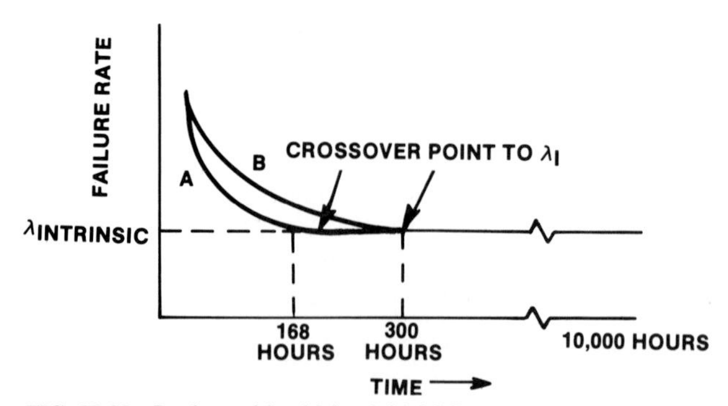

FIG. 27.35 Product with a higher initial failure rate can have a lower failure rate (λ) than product with a low initial dropout rate.

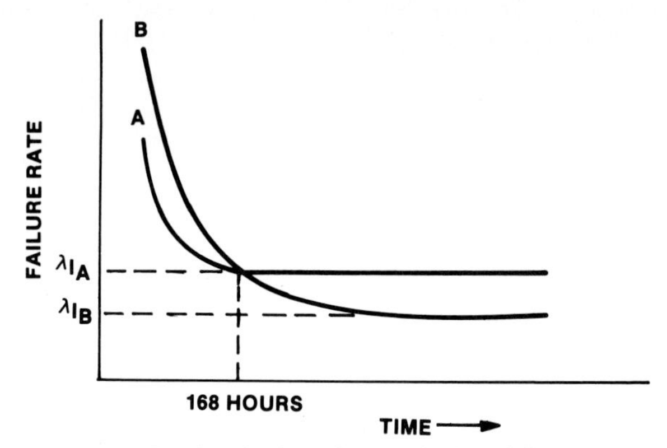

FIG. 27.36 Two lots that have the same intrinsic failure rate but that achieve the rate at slightly different times during the life distribution.

burn-in conditions are set at higher temperatures (125°C). Figure 27.35 shows this relationship for two lots that have the same intrinsic failure rate λ_I but achieve the rate at slightly different times during the life distribution.

If the crossover point to the intrinsic failure rate region could be accurately predicted for any predetermined temperature, devices could be provided with an intrinsic failure rate (usually constant or decreasing) as long as cost was not a major factor. In Fig. 27.36 λ_I is shown for the preset burn-in temperature; at lower application temperatures λ_I will be lower (i.e., below the λ_I shown in Fig. 27.36). High-temperature burn-in creates a certain amount of overkill (i.e., more devices than necessary are destroyed in setting the failure rate).

Products with a higher initial dropout on burn-in are not necessarily inferior, as has already been stated. Figure 27.35 shows that products with a higher initial failure rate can have lower λ_I than products with low initial dropout rates.

REFERENCES

1. A. M. Mood and F. A. Graybill, *Introduction to the Theory of Statistics,* McGraw-Hill, New York, 1963.
2. P. G. Hoel, *Introduction to Mathematical Statistics,* Wiley, New York, 1962.
3. W. Feller, *An Introduction to Probability Theory and Its Applications,* Wiley, New York, 1968.
4. Igor Bazobsky, *Reliability Theory and Practice,* Prentice-Hall, Englewood Cliffs, N.J., 1961.
5. Mrs. L. R. Goldthwaite, "Failure Rate Study for the Log-Normal Life Time Model," *Proceedings, Seventh National Symposium on Reliability and Quality Control in Electronics,* Philadelphia, Pa., January 1961.
6. R. H. Dennard, "Design of Ion-Implanted MOSFETs with Very Small Physical Dimensions," *IEEE J. Solid State Circuits,* **SC-9**:256–268.
7. L. J. Gallace, "Reliability of Plastic Package CMOS Devices," *Solid State Tech.* **23**(9):102, 108 (1980).

8. L. J. Gallace and M. Rosenfield, "Reliability of Plastic Encapsulated Integrated Circuits in Moisture Environments," *RCA Review*, **45**:249–277 (1984).
9. S. Gottesfeld and L. Gibbons, "Reliability Characterization of High-Speed CMOS Logic ICs," *RCA Review*, 45:179, 193 (1984).
10. L. J. Gallace and C. D. Whelan, "Accelerated Testing of COS/MOS Integrated Circuits," RCA Solid State Technical Paper ST-6379.
11. G. L. Schnable, "Reliability of MOS Devices in Plastic Packages," *Proc. Tech. Program Int. Microelectronics Conf.*, 1976, pp. 82–91.
12. K. C. Kampur and L. K. Lamberson, *Reliability in Engineering Design*, Wiley, New York, 1977.
13. S. Glasstone, K. J. Laidler, and H. Eyring, *Theory of Rate Processes*, McGraw-Hill, New York, 1941.
14. H. Semat, *Fundamentals of Physics*, Rinehart & Co., New York, 1958.
15. K. J. Pascoe, *An Introduction to the Properties of Engineering Materials*, Van Nostrand Reinhold, New York, 1972.

CHAPTER 28
YIELD ANALYSIS

Emory B. Michel
GE Solid State
Findlay, Ohio

28.1 INTRODUCTION

Simply put, yield is a number that describes the ratio of output to input. As related to the production of integrated circuits, this ratio is the percentage of good dice on a wafer relative to the total number of dice on the wafer factored by the percentage of good wafers left from the total wafers started. Yield analysis is simply the evaluation of the factors producing the yield.

In today's technology, yield analysis is not simple. Through advances in nearly every phase of semiconductor design and manufacture, we have reached the era of very large scale integration. We have come from one junction to hundreds of thousands of junctions per die, from one gate component per chip to thousands of gates contained in the same active area, and from the handling of one device at a time to the handling of batches of wafers, each containing hundreds of LSI circuits. Yield analysis in the manufacture of VLSI devices has become so complex and detailed that it demands not only the best equipment and techniques in analysis, but also the best techniques in communicating the results of those analyses to the manufacturing personnel.

In today's VLSI manufacturing facility, yield analysis becomes the evaluation or comparison of a complete cycle of manufactured output or yield against the maximum capability of the product cycle, and the investigation, understanding, control, and necessary adjustments, of all those process factors contributing to the observed yield. All this must be done cost-effectively and quickly enough so that any process changes needed can be made rapidly to improve yield and to keep the manufacturer competitive. In this environment, the need to effectively communicate analytical results to decision-making personnel is imperative.

The rewards of an effective yield analysis system are improved process control, reduced defect densities, increased product yield, reliability, and quality, lower inventory and manufacturing costs, an accelerated manufacturing learning curve through reduced cycle times, and a database providing data graphics and plots for the identification of long-term historical data trends. The costs of a poor yield analysis system can be the credibility and reputation of the manufacturer.

28.2 OVERVIEW

Two critical data points exist in yield analysis: where one is, and where one ultimately "wants to be"—in other words, the actual and theoretical yields. The comparison of these two data points produces an evaluation number or percentage of attainment. Such comparisons are made at each process step, and for sequential groups of processes and products as well as for whole product lines.

The first data point, the actual yield, is provided by the direct gathering of information on the process or product in a given manufacturing cycle. The second point, the theoretical yield, depends on the ultimate calculated capability of the cycle. In practice, the theoretical yield is rarely attainable, although it always remains as a goal. An attainable yield level becomes an important data point for cost control and price considerations. This attainable capability begins as a "norm" for a cycle and is then continually updated as experience with the cycle dictates needed improvements to it. Since such experiences are often sequential, the term *learning cycles* has been coined to describe them.

28.2.1 Learning Cycles

Experience has shown that the unit cost of fabricating an integrated circuit product through a VLSI process sequence declines exponentially with the accumulated number of units produced.[1] Studies have shown that the IC industry has a 70 percent learning curve (Fig. 28.1), and there is every reason to expect the 70 percent learning curve to continue to represent the IC technology progress. The significance of the 70 percent is that every time cumulative output doubles, cost

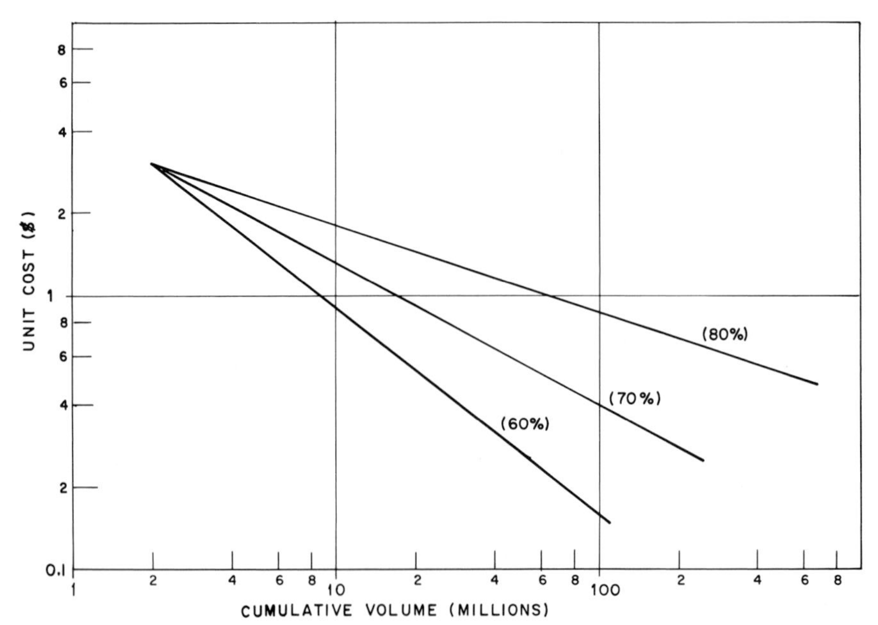

FIG. 28.1 IC learning curves, percent cost reduction for doubled volume.[3]

is reduced to 70 percent of the starting cost. However, the cost reduction is not "automatic"; to move down a learning curve (experience higher volume at less cost), a manufacturer must combine manufacturing skill and careful engineering analysis to effectively apply the information obtained in each cycle of learning.

This cycle is the basis for the experience curve slopes of Fig. 28.1, and seems to be independent of time and place and manufacturing skill to some degree, although only skillful analysis and meaningful feedback can optimize it. Every production lot of wafers that passes through the manufacturing cycle is a vehicle for a cycle of learning. In fact, each of the many individual process steps along the sequence of fabrication produces its own cycle of learning, which goes into making up the total learning cycle; any effective yield analysis technique must include methods capable of identifying critical factors at every process cycle step.

Critical factor analysis becomes even more complex in large-volume production where parallel process stations or types of equipment are involved. However, the comparison of such critical factors or variations in different locations within a factory or even with remote laboratory facilities enhances the learning curve process by providing correlating information. Nevertheless, the primary or overall learning cycle remains the key to gaining experience by showing how each complete process yields.

As all yields in the process are interrelated, all are critical. In fact, there are dynamic process interrelationships that constantly challenge effective yield analysis. For example, if a process produces an excellent yield, but later units from that yield fail to be reliable in a final package (due, perhaps, to contaminant residues from the process), there is *no* practical yield.

The results of every product cycle must be communicated to process personnel in order to move the system efficiently down the learning curve. When yields are not satisfactory, process adjustments must be made before excessive losses occur and costs increase.

28.2.2 Costs[3]

Although process cycle time may not appear to be directly involved in the learning process, the speed of cycle time is a critical determinant of how quickly problems can be found and corrected and, ultimately, of how much the part being manufactured will cost. Therefore, cycle time becomes the key to timely analysis and correction of yield variations. The relative importance of this cycle time effect on the learning curve is shown in Fig. 28.1 by the 80 percent curve for slow cycle time and by the 60 percent curve for fast cycle time. In a highly competitive market, rapid cycle times and detailed yield analysis are the marks of a leader.

In competitive comparisons, the cyclic experience becomes the dominant factor in economic growth. When mature products are no longer competitive, new designs using new technologies become economic necessities. Yield analysis plays an important role in evaluating such decisions.

The learning curve is useful for determining cost projections for pricing and margin considerations. In fact, the determination and reduction of VLSI manufacturing costs are directly dependent on the application of prior experience. A slow learning curve on mature product lines may occur when attention and capital investment are diverted to newer product lines and designs where rapid progress is critical to meet market needs.

Each time a process or product is completed and analyzed, a new and better perspective is obtained with respect to the next cycle of learning. The inputs to the yield analysis system that produce learning growth are the determining factors in effective yield analysis. As rapidly as one can complete a cyclic yield analysis, one can improve cost and functional reliability through process control.

28.2.3 Process Control[1,4]

Yield analysis has made wafer fabrication process controls extremely important. Ultimately, the best yield performance involves the constant improvement of these controls, an activity that ensures consistent design parameters and that optimizes defect density (the number of defects in a given area on a chip) and unit cost.

Specifically, statistical process control is the method used to keep the manufacturing cycle under control; it distinguishes random, expected variation in a process from variation caused by a systematic source (i.e., a process problem). The statistical process control method cannot cure process problems, but it can alert the manufacturer to processing situations where something has shifted or departed from the norm. The method is a tool to efficiently direct engineering and production effort to the correction of the real problems, and ensures that time is not wasted in controlling a process that has not really changed.

28.3 THE YIELD ANALYSIS PROCESS

Perhaps the best way to understand the yield analysis process is to follow a new circuit through a manufacturing cycle, such as the wafer fabrication process, the process by which a wafer of silicon becomes a matrix of operational VLSI circuits. Throughout this cycle, a key link between the manufacturing personnel and the process is the interactive computer system. The number of possible process factors, up to 400, that can contribute to a defect capable of destroying any one of thousands of active components on a VLSI circuit system has made mandatory the use of computerized reporting of statistical control and analysis procedures. This system monitors the process throughout the cycle, stores historical data, is the means of communication of process data to manufacturing personnel, and represents the vehicle for evaluation and control of the key parameters that influence yield.

28.3.1 Computer System

Statistical correlation, graphical display, and report generation among process, equipment, and facility is desirable in the computer system. All data inputs throughout the manufacturing process must be available to the user community in real time. There must be room for simultaneous users, and the system should be interactive and on line so that it allows the process-product engineering groups to extract and analyze production data. The computer system should have one integrated database that allows users to perform logical retrievals of data of any system parameter through a simple command. Any analysis program should be capable of producing trends, scatter plots, histograms, statistical analysis, and rapid printout reports. The system should also permit the interfacing of each measurement process monitor and intelligent fabrication equipment directly to a central computer for automatic data collection and correlation at specific process steps.

28.3.2 Data Collection Points

In semiconductor IC manufacturing, there are typically four key cost-determining data-collection points:

Wafer acceptance test (WAT) (electrical parameter testing)
Circuit probe (CP) (circuit functionality testing)
Assembly
Final test (FT)

Of this array of data collection points, those providing the most comprehensive data for yield analysis purposes are the wafer acceptance test and the circuit probe test. Both tests are computerized; the WAT measures design process parameters (the adherence of the circuit in manufacture to its design parameters); the CP, circuit characteristics. The yield figure at final test will be almost wholly dependent on the results of these two tests. Reporting of data at CP, assembly, and FT are necessarily type oriented while WAT uses the same test key on all types. Type-oriented data feedback should indicate primary and secondary (or more) failure tests on a by-lot basis (wafer-fab lot number) on every lot below standard cost or expected-norm levels. Thus, cross-discipline characteristics can be investigated.

An important factor that limits the WAT and CP yields is the density of electrically active defects in a given area. The whole system of defect prevention and control at every wafer fabrication process step, as embodied in the statistical process control system, is driven by this defect density figure. Naturally, as the process matures and experience increases, the process moves down the learning curve, and defect density decreases.

After WAT and CP testing, FT yield should be 100 percent except for any problems from assembly, or reliability characteristics, or dynamic electrical testing for characteristics that cannot be measured directly by probes. The main concern in the final test stage is the relatively high cost of yield losses at that stage as compared with losses in the fabrication stages. As a wafer moves through each process step, the cumulative yield decreases and the product cost increases as value is added for labor and materials. Some packages may eventually cost much more than the circuit unit they contain. Table 28.1 demonstrates the cost implications of yield losses at each step of the manufacture of a molded plastic unit. Again, the importance of this information is the relatively low cost of yield losses in wafer fabrication as opposed to the cumulative cost of losses in assembly or at final test. These considerations may be important to decisions concerning parameter limits at WAT that affect FT yields.

Wafer Acceptance Testing. The primary purpose of the wafer acceptance test is to measure the electrical parameters associated with the circuits on the wafer to ensure that design conditions set by the customer are being met. Table 28.2 lists some typical WAT tests and the process function used to control the limit associated with that test. Figure 28.2 shows a parameter histogram of daily data and

TABLE 28.1 Effect of Yield on VLSI Unit Cost for Molded Plastic Package with Original and Tightened Controls

a. Unit cost with original controls

Yield Point	Expected yield per operation, %	Cumulative yield, %	Add on $ cost/unit	$ Cost/unit
Wafer fab	90	90	0.192	0.213
Wafer test (WAT)	92	81.8	0.001	0.232
Circuit probe (CP)	40	32.72	0.033	0.663
Assembly	86	28.14	0.232	1.041
Final test (FT)	90	25.33	0.009	1.167
Burn-in test	98	24.82	0.063	1.255
Net effect	25	25	0.530	1.255

b. Unit cost with tighter controls on V_{TNO}

		Initial $ cost/unit		Add on $ cost/unit		Yield, %		Final $ cost/unit
WAT	=	(0.213	+	0.001)	/	0.87	=	0.246
CP	=	(0.246	+	0.033)	/	0.40	=	0.698
ASBLY	=	(0.698	+	0.232)	/	0.86	=	1.081
FT	=	(1.081	+	0.009)	/	0.95	=	1.147
BI	=	(1.147	+	0.063)	/	0.98	=	1.235

a 100-day chart used for trend comparisons. Limits and relationships are evaluated and corrected as a result of the WAT test, all with a view toward preventing costly yield losses during reliability screening and dynamic operation at final test.

The actual WAT is made by probing special test keys fabricated on the wafer along with the circuit die. These test keys, consisting of separate, actual-sized, operational devices similar to those being fabricated in the circuit chips on the wafer, allow manufacturing personnel to monitor and control the characteristics of the product circuits. The WAT database becomes a consistent and reliable composite picture of the process sequence. Through it, each process step can be set up in advance to yield the manufacturing and product characteristics required at WAT. That is, each WAT parameter can be characterized for both the range of data typical of the process and the design limits required for the product.

Where extremely narrow parameter limits are required, and on new processes, the learning curve progresses until the required parameters are in control. In that sense, wafer acceptance testing is the prime discipline in yield analysis. Good WAT is essential for maximizing yields at all subsequent test levels of a production line. WAT sampling should mature to only a few sample wafers per lot, but it should never be eliminated or its learning curve effect would be lost. The experience or learning curve must continue as long as yield analysis controls production. Further details on WAT are given in Sec. 28.3.3.

Circuit Probe Test and Setup. Since the wafers supplied to circuit probe testing have passed the wafer acceptance test and have been found to meet design parameters, the circuit probe test becomes a comprehensive test of circuit function-

TABLE 28.2 Wafer Acceptance Testing Parameters and Related Controls[6,7]

Test	Limits	Control conditions
$p+$ diff	35–100 Ω/□	Implant dose
$n+$ diff	5–30 Ω/□	Implant dose
$p+$ poly	25–75 Ω/□	Implant dose in poly
$n+$ poly	10–50 Ω/□	Implant dose in poly
p-Well w/oP.	3–7 kΩ/□	Implant dose in substrate
p-Well w/P.	1.5–3.5 kΩ/□	Implant dose on well
P_{LD} 10 V	10–100 GΩ/□	Poly purity
$n+$ Poly/MC	2–50 Ω/contact	Poly $n+$, poly oxygen
$p+$ Poly/MC	2–40 Ω/contact	Poly $p+$ vs poly
$p+$ AA/MC	2–40 Ω/contact	Implant and etch
$n+$ AA/MC	2–50 Ω/contact	Implant and etch
$p+$ Poly BC	0–150 Ω/contact	Pre poly etch and furnace seal
$n+$ Poly BC	0–100 Ω/contact	Pre poly etch and furnace seal
V_{TNO}	0.4–0.7 V	Well resistivity, gate oxide
BV_{DSSN}	15–35 V	Junction integrity
IDN_{SAT}	15–1500 μA	Gate function
$L_{EFF}N$	2–4.5 μm	Side diffusion
$K_{N(SAT)}$	1–7 μA/V²	Surface integrity
V_{TFN}	15–35 V	Field oxide
V_{TPO}	0.4–0.8 V	Substrate resistivity, gate oxide
BV_{DSSP}	15–35 V	Junction integrity
$I_{DP(SAT)}$	15–900 μA	Gate function
$L_{EFF}P$	2–4.5 μm	Side diffusion
$K_{P(SAT)}$	0.5–0.5 μA/V²	Surface integrity
V_{TFP}	15–35 V	Field oxide

ality. Any losses of yield in CP are determined by defects in the circuits. These defects may be caused by photomask, environmental contamination, handling, photodefinition and alignment, process, and/or wafer defects. When stated as a general formula, the circuit probe yield is made up of the product of the limits of the various steps in the process leading up to circuit probe:

$$\text{Yield}_{CP} = Y^n_{(\text{photo limited})} \times Y^n_{(\text{align limited})} \times Y^n_{(\text{handling limited})} \times Y^n_{(\text{process limited})} \times Y^n_{(\text{wafer limited})}$$

where Y is the fractional yield and n is the number of process steps.

In a process sequence where $n = 10$, the probability that trace defects will ruin the yield at CP is very large because, even though the individual yield at each process may be acceptable, the cumulative effect of the exponential factor can take its toll. Since the defects affecting the yield at CP are usually very small (compared with the size of a circuit) and often randomly distributed, the probability of encountering a killer defect increases as the area of each circuit on a wafer increases.

Yield analysis at CP usually begins with a routine checkup. Every new probe test setup must include a probe tip inspection for placement, probe tip wear, and contact resistivity (pressure). A previously tested and recorded correlation wafer for each design type must be provided to establish that the new software and

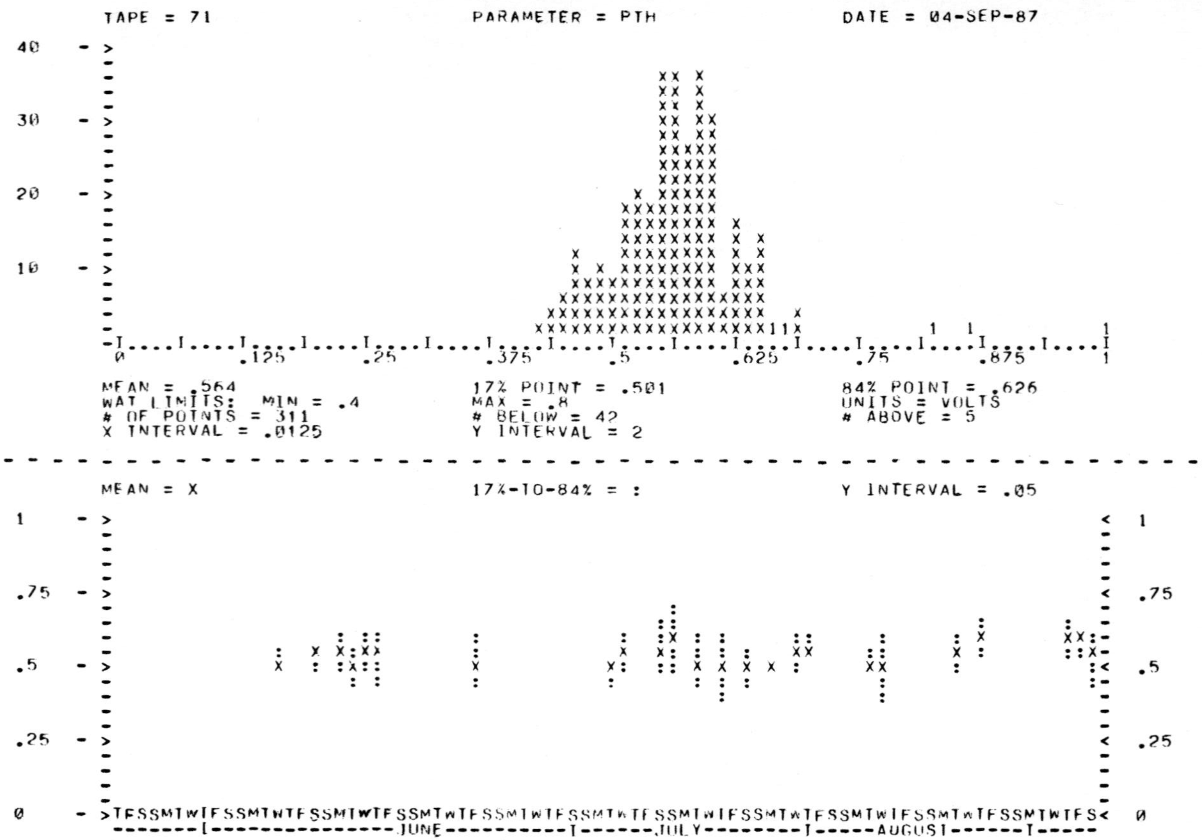

```
          TAPE = 71                    PARAMETER = PTH                DATE = 04-SEP-87

40    - >
        -
        -                                            XX X
        -                                            XX X
        -                                            XX X
30    - >                                            XX XX
        -                                            XX XX
        -                                            XXXXX
        -                                            XXXXX
        -                                            XXXXX
20    - >                                          X XXXXX
        -                                         XXXXXXXX
        -                                         XXXXXXXX X
        -                                         XXXXXXXX X X
        -                                     X   XXXXXXXX X X
10    - >                                     X X XXXXXXXX XXX
        -                                     XXXXXXXXXXXX XXX
        -                                    XXXXXXXXXXXXXXXXX
        -                                   XXXXXXXXXXXXXXXXXX  X
        -                                  XXXXXXXXXXXXXXXXXXX11X          1  1          1
        -I....I....I....I....I....I....I....I....I....I....I....I....I....I....I....I....I
         0        .125      .25       .375      .5        .625      .75       .875      1

          MEAN = .564                  17% POINT = .501               84% POINT = .626
          WAT LIMITS:  MIN = .4        MAX = .8                       UNITS = VOLTS
          # OF POINTS = 311            # BELOW = 42                   # ABOVE = 5
          X INTERVAL = .0125           Y INTERVAL = 2
- - - - - - - - - - - - - - - - - - - - - - - - - - - - - - - - - - - - - - - - - - - -
          MEAN = X                     17%-TO-84% = :                 Y INTERVAL = .05

1     - >                                                                          <  1

.75   - >                                                                          <  .75

.5    - >                                                                          <  .5

.25   - >                                                                          <  .25

0     - >TFSSMTWTFSSMTWTFSSMTWTFSSMTWTFSSMTWTFSSMTWTFSSMTWTFSSMTWTFSSMTWTFSSMTWTFSSMTWTFS<  0
         -------I--------------JUNE-----------I-------JULY---------I-----AUGUST-------I-----
```

FIG. 28.2 WAT parameter plot and 100-day history.[6]

hardware setup are functioning satisfactorily. The first new wafer probed must then be inspected for probe placement accuracy and alignment on fresh metal pads, which show the probe impression clearly. Excessively deep probe marks may indicate possible crystal damage under contact pads. A good design technique places diodes under contact pads to neutralize possible probe damage. Light or misplaced probe marks mean that a new probe holder setup is required.

These inspections should be repeated at regular intervals on each probe setup, as well as whenever a tested wafer shows a significant change from normal yield. Further details on CP testing are given in Sec. 28.3.3.

Final Test. If all circuits have passed the WAT and CP tests, then, as explained above, the FT yield should be 100 percent.

The dynamic operating tests run at FT cannot be accomplished using probe contacts alone. This kind of test is sensitive to conduction doping levels and design patterns that no prior test in the yield analysis process can evaluate accurately. Problems discovered as a result of this test can subsequently be dealt with at lowest cost by exercising tighter control over wafer fabrication process parameters.

The tight centering of WAT parameters is one means of preventing yield losses at FT. The addition of controls at the WAT stage and the absorption of losses (the sacrifice of yield) there usually provides the lower final cost. The best way to tighten the controls is through yield analysis techniques, which operate by refining parameter controls, as explained below. Very sensitive parameter controls at WAT can maximize FT yield without significantly affecting WAT yield. In fact, the improvement of materials and controls made in wafer fabrication to improve yields in assembly and FT frequently also result in higher yields at WAT.

For example, as shown in Table 28.1*b*, the control of the V_{TNO} parameter was tightened in a wafer lot that demonstrated a WAT yield of 92 percent. The loss at WAT due to the tightening was 5 percent, which reduced the WAT yield to 87 percent. But the FT yield increased 5 to 95 percent. The actual cost at each process step, recalculated using the new percent yields, is also shown in the table. The result of this exchange of 5 percent yield is a net saving of 2¢ per unit (a total actual cost of $1.235 per unit after tightening versus $1.255 before).

A similar situation prevails in the final testing and inspection of high-reliability products. The high cost of low yields in the FT stage can be improved radically by very tight photomask and alignment controls and wafer inspection in the wafer fabrication process and by careful attention to the interrelationship of assembly-bonding and metal-coating techniques.

An additional concern at final testing is the reliability of products that undergo extended periods of stress testing after FT, and the effect of that stress testing on final yield. Handling conditions that do not take static discharge seriously enough and assembly procedures that allow contamination can cause circuit losses that affect reliability. Most circuit designs incorporate static protection and protective glass films to prevent static-discharge damage and contamination effects, but careful inspection and control of the processes are necessary, even though electrical test yields may not be affected directly.

28.3.3 Analysis

Yield analysis is a dynamically balanced system. Each photo, etch, coating, and furnace operation is adjusted as required by operators recording statistical process control (SPC) data. They analyze the data, and may call for technical assistance, on a run-by-run basis. At the WAT, data are published for engineering evaluation daily by product type and lot as well as through a combined histogram of each key parameter. WAT parameter data verify that the SPC system is working. When variations are evident at WAT, the SPC process involved is investigated and corrective action is taken. Equipment may be shut down for cleaning, repairs, and adjustments. Operators may require retraining and instruction. Extensive baseline data must be taken under controlled conditions at all operations to ensure a firm knowledge of normal, good, operating characteristics. For example, in a vacuum metallizing system, the pumpdown time and the residual gas analysis (RGA) after pumpdown must be known so that trace leaks that affect electrical-contact resistance, protective-glass integrity, and bonding characteristics in assembly will not go undetected.

The ultimate goal of yield analysis is to minimize all special limits and special process variations. Special limits may be specified for custom products and special design requirements; however, such practices may create problems by disturbing the balance of control for the best process sequence. One process with one specified set of limits, though they may be the tightest limits, provides the best capability for maximum yields. Where process variations are considered necessary, separate, parallel processing lines and limits should be defined, and testing should be done separately. For example, a high-reliability product should be processed and tested separately from a standard commercial product.

On a mature wafer fabrication line, manufacturing personnel use WAT data for design-parameter target-centering by analyzing composite data histograms daily (Fig. 28.2). Daily data taken over several months are then plotted and observed for key parameter trends. When trends appear, a detailed analysis is made on a wafer-by-wafer basis. The detailed data analysis may expose a lack of lot integrity caused when a few wafers or an entire wafer lot is mixed in such a way that an entire process step is missed.

When a lot integrity error is found, the detailed process lot follower (Fig. 28.3), which shows time, equipment, and operator assigned at each process step, becomes the necessary final reference in the yield analysis. Now the error can be assigned and investigated, and the necessary corrective action taken. The administrator involved circulates a description of the cause of the lot integrity error and the correction taken to prevent repetition of the error. The WAT yield will approach 100 percent in any standard mature process, except when there are process integrity errors.

A practical formula has been set up that reduces the many factors involved in random defect analysis to a fairly simple model:

$$Y_{CP} = \frac{1}{1 + D_t A}$$

where Y_{CP} = percent yield at circuit probe/100
D_t = total killer defects per square inch
A = area per circuit, in^2

This simple formula bypasses the complexity of dealing with the many factored details of a more rigorous equation and provides a working summary yield relationship of killer defects to the size of the circuit in the process. The formula provides a way to isolate product anomalies and mask errors quickly and decisively by the comparison of a D_t already established for a given wafer fabrication process to any one of the product types using that process. The effect of mask

LOT

PROCESS CMOS 1 PLD

TYPE ALIS

	QTY	OP	DAT	SH	EQ
START					X
VENDOR LOT#	X	X	X	X	X
STD CLEAN					
SUB IMP (PHOS)					
DOSE =					
STD CLEAN					
WELL OXIDE					
THICK =					X
REPORT AT 0290					X
WELL PHOTO					
COAT-NEG.					
ALIGN WELL					
MASK NO	X	X	X	X	X
DEVELOP					
POST DEV INSP					
REPORT AT 0380					X
PRE-ETCH BAKE					X
OXIDE ETCH					X
CHEM RMVL					X
POST ETCH INSP					
REPORT AT 0470					X
WELL IMPL & DIFF					
STD CLEAN					
WELL IMPL (BORON)					
DOSE =					X
REPORT AT 0500					X
BOE ETCH 3 MIN					X
STD CLEAN					
QMOS WELL DIFF					
SH RES =					X
BOE ETCH 9 MIN					X
REPORT AT 0660					X
PAD OX/THICK NIT DEP					
STD CLEAN					
PAD OX					
THICK =					X
THICK NITRIDE DEP					
THICK =					X
NITRIDE INSP					X
REPORT AT 0720					X
ACTIVE AREA PHOTO					
COAT-NEG.					
PROJ. ALIGNER AA					
DEVELOP					
POST DEV INSP					
CD MEAS =					X
REPORT AT 0860					X
PRE ETCH BAKE					X
ACT AREA ETCH 801					
PLASMA STRIP					
CHEM RMVL					X
POST ETCH INSPECT					
CD MEAS =					X
REPORT AT 0960					X

LOT

PROCESS CMOS 1 PLD

TYPE ALIS

	QTY	OP	DAT	SH	EQ
N. IMPLANT PHOTO					
VAC HMDS					X
COAT. NEG.					
ALIGN N –					
MASK NO	X	X	X	X	X
DEVELOP					
POST DEV INSP					
REPORT AT 2150					X
N– IMPLANT					
PRE IMP BAKE					X
N – IMPLANT					
DOSE =					X
REPORT AT 2160					X
PLASMA STRIP					
CHEM RMVL					
FIELD OXIDATION					
STD CLEAN					
FIELD OXIDATION					
THICK =					
REPORT AT 2200					X
NITRIDE STRIP					
BOE OX. ETCH					
30 SEC					X
NITRIDE STRIP 'WET'					
POST STRIP INSP					X
BOE OX. ETCH					
40 SEC					X
REPORT AT 2300					X
GATE I					
STD CLEAN					
GATE OXIDATION 1					
THICK =					
REPORT AT 2400					X
GATE II					
OXIDE ETCH 45 SEC					
STD CLEAN					
Cmos 1 Gate ox/on					
RUN NO					X
THICK =					X
REPORT AT 2800					X
BURIED CONTACT PHOTO					
COAT-NEG.					
PROJ. ALIGN BC					
DEVELOP					
POST DEV INSP					
REPORT AT 2850					X
PRE ETCH BAKE					X
OXIDE ETCH					X
CHEM RMVL					X
POST ETCH INSP					X
REPORT AT 2860					X

FIG. 28.3 Lot-follower process ticket.

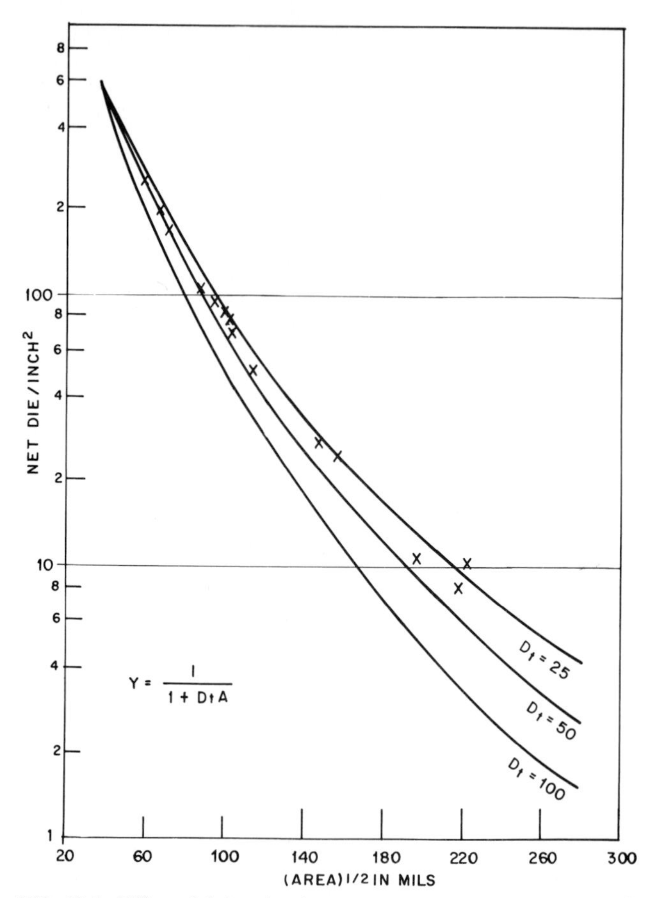

FIG. 28.4 Effect of defect density D_t on units-per-wafer yield at CP.[5]

changes, photo controls, handling procedure modifications, and process refinements on yield can be evaluated on a comparative basis with established results obtained before these modifications were instituted, and variations in yield due to circuit size can be estimated.

The curves in Fig. 28.4 show examples of the relationship between area, yield, and defect density. The use of this simple yield analysis tool provides a means of comprehending killer defect factors in a VLSI wafer fabrication process and in each product using that process. When a product type or design is not in line with other similar designs, the product can be given special attention in the form of a more detailed analysis, such as a HI-LO analysis.

The sample SPC chart in Fig. 28.5 shows how comprehensive this tool is for yield analysis on batch processes, where it is used to determine parametric control. The system displays specific wafer lots and sample data averages and readings for rapid reference analysis when some or all of the lots show abnormal characteristics at testing. The range identifies the possibility that both reject and acceptable lots could exist from a single batch process run. The run average can be compared to previous and subsequent run averages for batch process consistency.

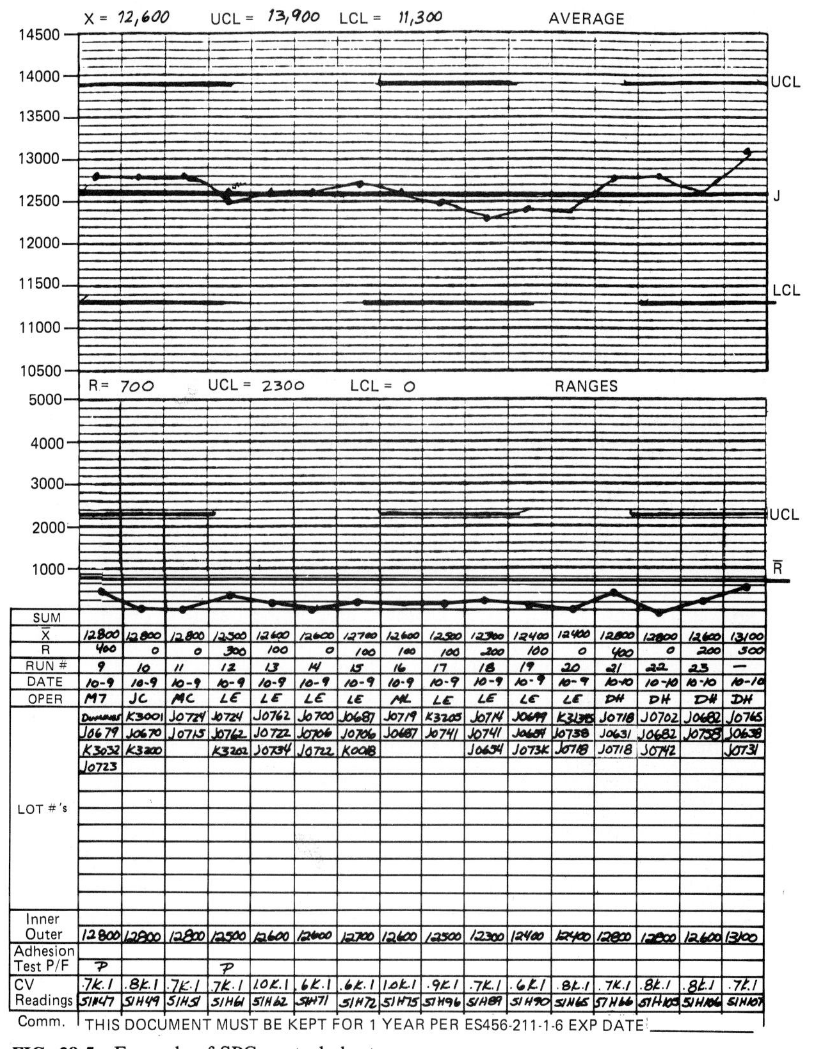

OPERATION/STEP METAL EVAP EQUIPMENT NO. AIRCO REJECT LIMITS 11,000–14,000 A

SUM																
X̄	12800	12800	12800	12500	12600	12600	12700	12600	12500	12300	12400	12400	12800	12800	12600	13100
R	400	0	0	300	100	0	100	100	100	200	100	0	400	0	200	300
RUN #	9	10	11	12	13	14	15	16	17	18	19	20	21	22	23	—
DATE	10-9	10-9	10-9	10-9	10-9	10-9	10-9	10-9	10-9	10-9	10-9	10-9	10-10	10-10	10-10	10-10
OPER	M7	JC	MC	LE	LE	LE	LE	ML	LE	LE	LE	LE	DH	DH	DH	DH
LOT #'s	Dummies	K3001	J0724	J0724	J0762	J0700	J0687	J0719	K3205	J0714	J0699	K3135	J0718	J0702	J0682	J0765
	J0679	J0670	J0715	J0762	J0722	J0706	J0706	J0687	J0741	J0741	J0654	J0738	J0631	J0682	J0758	J0638
	K3032	K3200		K3202	J0734	J0722	K0018			J0634	J073K	J0718	J0718	J0742		J0731
	J0723															
Inner																
Outer	12800	12800	12800	12500	12600	12600	12700	12600	12500	12300	12400	12400	12800	12800	12600	13100
Adhesion Test P/F	P			P												
CV	.7K.1	.8K.1	.7K.1	.7K.1	1.0K.1	.6K.1	.6K.1	1.0K.1	.9K.1	.7K.1	.6K.1	.8K.1	.7K.1	.8K.1	.8K.1	.7K.1
Readings	51H47	51H49	51H51	51H61	51H62	51H71	51H72	51H75	51H96	51H89	51H90	51H65	57H66	51H105	51H106	51H107

Comm. THIS DOCUMENT MUST BE KEPT FOR 1 YEAR PER ES456-211-1-6 EXP DATE ________

FIG. 28.5 Example of SPC control chart.

Screening Tests. Screening tests that make use of marked wafers to identify variations in process techniques at any stage of the wafer fabrication sequence continue to be a useful technique for analyzing yields. The use of split lot screening tests is the first stage of any innovation in an established production process. These screening tests permit the fast gathering of feasibility data that are invaluable for yield analysis when minor changes and comparison evaluations are required quickly to maintain production flow. Such data should be evaluated using Student t for statistical accuracy.

WAT Wafer Check. A difficult WAT yield loss analysis result is one in which several parameters shift in a manner that indicates either an error in the starting substrate resistivity or a contaminating "skin" effect across all or part of a wafer surface. A spreading resistance profile of an angle-lapped wafer section can demonstrate whether an abnormal skin effect is present and to what depth. A wrong starting wafer resistivity may occur either at the first process step or at the lot formation.

A simple WAT for determining the approximate resistivity of the starting wafer is called *back biasing*. It requires an MOS unit in a test key with a floating substrate contact. By dividing an appropriate threshold voltage change by the square root of a bias voltage applied to the substrate contact, the exponential relationship to the wafer substrate resistivity can be plotted and used for identification of the starting wafer resistivity:

$$\text{Wafer resistivity} = \text{function}\left(\frac{V_{\text{th,bias}} - V_{\text{th}}}{V_{\text{bias}}^{1/2}}\right)$$

A range of possible wafer substrate resistivities should be run through the wafer fabrication process and used to set up a reference curve.

WAT Skin Contaminants. Assignment of a wafer surface contamination cause, in the case of a skin effect, requires that all process and environmental possibilities be isolated and investigated. First, a furnace treatment must have taken place to diffuse and activate the contaminant in the silicon crystal. Second, the contaminant must have been present when all the contaminated areas were exposed and capable of being contaminated. Phosphorous contamination (n-skin) is a good example of the contamination process.

When vaporized $POCl_3$ is used as an n-dopant source in a furnace, that $POCl_3$ may transfer as vapor from a furnace opening onto nearby wafers. Similarly, traces of phosphates in water rinses attach to wafer surfaces. Silicon wafers with phosphorous contaminants require special cleaning, or they will cause an n-skin contamination in a subsequent furnace treatment.

A sodium ion contamination causes a CVBT (capacitance voltage bias test) shift. Sodium ion contamination of oxides threatens the reliability of silicon devices because these ions often migrate after extended use of the device and cause unstable bias conditions. Process chemicals must be adequately specified and controlled to prevent sodium traces. The most likely source of sodium ion contamination is salt from human spittle or perspiration. Extreme caution must be emphasized in training personnel in handling techniques. Both clean gloves and spittle shields should be used wherever handling contamination could occur. Control of every furnace process step for possible sodium contamination of oxides can be done regularly by exposing known clean oxide coated wafers to the furnace conditions, then performing CVBT. Most process sequences use a standard cleaning process just ahead of every furnace process to ensure sodium ion removal.

CVBT is a necessary and accurate method for determining the presence of mobile ion contaminants on and in oxide dielectric coatings by their effect on capacitance stability. Extended time under voltage bias conditions can cause mobile ions such as sodium to move through oxide lattice structures and cause small shifts in electronic switching speeds. These shifts in speed create circuit inaccuracies that cannot easily be diagnosed except by reliability testing. Therefore, this test is a simple screening test in reliability analysis.

CP HI-LO Analysis—Testing. Any yield loss at CP testing is a defect density alarm for the wafer fabrication area and must be checked out by the probe test operator. The analysis of each wafer yield data summary for a typical distribution of specific test failures is the normal procedure for recognizing the causes for probe yield changes. Occasionally, CP data histograms for samples of current production wafers should be stored for reference along with repetitive data recordings on reference units held for this purpose. A reference data file must be kept on normal, good product, so that when problems arise there can be a comparison of characteristics.

CP data comparisons for leakage test levels can be an accurate alert to subtle wafer fabrication variations that affect junction integrity. Reference to the wafer fabrication lot follower and to process files may show some notation that would explain a unique probe yield variance. Probe data analysis techniques must be applied routinely to establish a reference database.

When CP yields are poor, an analysis technique that provides useful data for wafer fabrication yield improvement is HI-LO wafer sampling. This technique should be exercised routinely for the fast informative feedback needed to complete the basic learning experience cycle. The HI-LO analysis data may be statistically summarized or simply compiled as annotated comments on a CP data summary form, but it must be rapidly distributed along with copies of the wafer fabrication lot follower to all interested wafer fabrication engineers and administrators. This rapid communication is a key part of the learning experience system, as it closes the production cycle with a yield analysis that ensures that the requirements for subsequent production starts will be correct, and it builds a database.

The initial objective of HI-LO wafer sampling is a simple visual verification of the wafer testing. Ink used to mark reject circuits may show telltale patterns, such as every other row all reject or matching areas on all wafers are all reject. A quick microscopic inspection will verify probe alignment. Since the HI-LO evaluation is a comparison of the best and worst wafers from the same wafer fabrication process lot of wafers, the two wafers will generally have been tested on the same probe setups. If different probe setups were used, it is necessary to rerun the bad wafer on the same setup as that used for the good wafer for this comparative HI-LO yield analysis to work.

HI-LO wafer sampling is not meant to hold up production lots or even to sample wafers unless severe yield or reliability problems are apparent. Wafer production lots sampled in this manner will generally be those lots yielding lower than standard or projected yield levels.

Occasionally, the HI-LO wafer sampling technique should be applied to good wafer lots. Each wafer test data summary and wafer fabrication lot follower should be retained after the HI-LO wafers are sampled at CP. Next, WAT parameter test data should be obtained from both wafers. A parameter data comparison analysis can then be made to assign possible cause for the variance in yield between the high and low wafers. Any significant parametric difference between wafers or areas on the HI-LO wafer samples should be reported to ensure uniformity of database. When the electrical test data comparison defines the failure mechanism sufficiently to give it a name or category well known in the process or product involved, there may be little value in continuing the HI-LO analysis. However, it is generally wise to continue the investigation through a visual comparison to be sure that other problems have not been overlooked.

CP HI-LO Analysis—Visual. A visual inspection of HI-LO wafer samples can be time consuming, but it provides extremely valuable information. Each circuit is tested and identified as acceptable or reject. Where good and reject units are adjacent, there may be visual evidence such as damaged or discontinuous patterns. When a damage or pattern defect is identified on the LO yield wafer circuit, the circuit in the same location on the HI yield wafer is inspected for the same defect. When a duplicate defect identification is made, a photomask defect is usually the problem. The identity of the offending photomask is recorded on the wafer lot follower card (Fig. 28.3).

There are many varied visual defects to become familiar with in every VLSI process. The advantage of a HI-LO wafer inspection comparison at CP is that only killer defects need be a prime concern. It is valuable experience to observe the general appearance characteristics of both HI and LO wafers even though there may be no apparent yield loss or gain involved. This observation provides a valuable visual database. A microscopic inspection may find narrow conductor lines, narrow spaces, wavy definition lines, marginal alignments, metal notching, overetch, underetch, contact opening sizes, polysilicon stringers, glass flow, uneven topography, steep oxide slopes that affect conductor continuity, roughness or particles that disrupt pattern continuity, and holes in oxide levels. Measurements are taken to evaluate any concern for meeting design specifications.

CP Defect Analysis Techniques. Killer defects are a prime concern in the detailed HI-LO wafer yield analysis. Various electrical diagnostic techniques have been developed to quickly isolate killer defects in the vast array of failure possibilities in VLSI systems. These techniques range from taking scanning electron microscope (SEM) pictures of a dynamically functioning circuit to watching for a circuit zener diode to light up in a dark microscope field. A most useful failure analysis technique involves the coating of a known bad circuit with a liquid crystal solution and observing the failure hot spot that quickly appears on the electrically biased circuit. Once the killer defect location has been established, the circuit involved can be visually inspected and the defect may be identified. This last step is not always easy, since hidden defects are often involved.

Considerable patience and skill in the use of destructive analysis procedures is necessary to carefully strip the circuit layer by layer at a defect location. But the destructive analysis technique is quick and useful for establishing any silicon crystal damage in active circuit junction areas. When electrical breakdown of junctions warrants analysis, the oxides and metal structures are removed by acid stripping. The exposed silicon surface is etched in specially formulated acids (e.g., Sirtl etch, Wright's etch)[2] that cause defect locations to produce identifiable patterns that establish where crystal damage has occurred. When damage has been identified, a careful search must be made in the wafer fabrication process for sources of heavy-metal contaminants and other process variations that can produce the damage condition found. Use of the surface etch for crystal damage identification after each furnace step can usually identify a problem.

28.4 STATISTICAL PROCESS CONTROL TECHNIQUES[1,4]

28.4.1 General

In the past, yield analysis in wafer fabrication process areas had been given over to constant attention to process adjustments that only flirted with the extremely tight parameters required to assure the proper functioning of VLSI circuit designs. As these inadequacies of technique were then compounded through many interdependent process steps, the yields at WAT and CP became erratic, and analysis was so complex that much of its effect was lost. The concept that evolved from the need to make yield analysis workable again, statistical process control, relies on defect prevention instead of defect detection.

At periodic intervals in the manufacturing process, samples of some critical parameter are measured and the average and range (or standard deviation) are calculated. Control charts of the average, $\overline{X}$ (with $\pm$ 3 standard deviations control limits), and the range, R (with $\pm$ standard deviations control limits), are updated every time a sample is taken. A major criterion for being in process control is that 99 percent of all observations lie within the $\pm$ 3 standard deviations control limits. Another major criterion for being in process control is that the average show no trends. Eight consecutive averages above or below the overall average signal are considered a statistically significant shift in the average.

In SPC, statistical control charts (Fig. 28.5), while permitting normal fluctuations, ensure quick response to abnormal process variations. The statistical use of averages and data distribution stabilizes the perception of each process. Subsequently, each process step may be characterized for its capacity to meet the necessary parameter requirements. If a process step is not capable of meeting WAT parameter requirements, a better process must be developed.

SPC has established yield analysis on a real-time basis as part of the wafer fabrication process; manufacturing personnel no longer have to wait for formal test results to make a correction. SPC also creates a demanding emphasis on quality in processing that is often mistaken as unnecessary and expensive, until the alternatives of erratic yields in subsequent operations are considered. VLSI processes cannot be competitive and survive without extensive use of the SPC concept, which also involves specifying that vendors of materials used in processing must consistently meet statistically controlled standards.

When SPC is used, tight communication between vendors and users is needed, especially when multiple vendors are involved, a situation that could excessively complicate the process and product learning curves. With SPC, the internal communication of a variant result can prevent minor process discrepancies in one process from causing unnecessary delays in a subsequent process, but the main goal of the system is consistency at each process step. Rework is expensive and generally deteriorates the quality of the product. Scrap may be preferable to dependence on special handling.

Whenever possible, processes and products should be kept moving until defined control limits are exceeded, then the product should be scrapped. The dynamics of production must be understood and exercised for the full impact of statistical controls and cyclic learning systems to be realized.

It should be kept in mind that it may be difficult to bring a process into statistical control through the use of SPC alone. SPC is an effective method for determining whether a process is in control, but it cannot be easily used to quantify the effects of the process variables or to determine whether the process can be controlled. For that reason, and to be totally effective, SPC may be complemented by other areas of statistics, such as experimental design, which is discussed later.

28.4.2 Choosing Controls

The choice of the control condition to be checked and plotted can be challenging in that it must correlate to a WAT parameter directly controllable. For example, the rate of diffusion can be accurately measured and related to the oxide growth to provide a quickly measured means of control as compared to the more laborious removal of oxides for wafer resistivity readings or the much slower cross-sectioning technique. However, rapid correlation controls do need the regular backup control validation provided by the more laborious techniques.

To prevent production delays, there should be backup techniques, correlated to appropriate WAT parameters, for every primary reference control. For example, the phosphorous content of protective glass can be measured by x-ray, infrared transmission, or, where such equipment is not available, by diffusing a sample into a semiconductor surface and reading the resistivity change using a four-point probe. Electrical breakdown characteristics of sensitive junctions and characteristics of wafer patterns can be used for backup reference when the need arises.

Photoprocess Control. Photoprocess control techniques are the most challenging part of VLSI processing. The number of steps and the intricate relationship between each subsequent step relative to alignment and definition accuracy make this control the major component of the defect density equation. The masks, the equipment, and the materials used in photoprocessing must be carefully controlled to maintain consistent production quality and flow. Persistence in maintaining adequate controls on automatic aligning equipment will directly improve defect density results at CP.

Trace residues of photoresist contribute to adhesion problems between process layers. Photoresist flow consistency, though each batch is accurately formulated, can change with age, room temperature, and application technique. Application spin speeds, exhaust airflow, and cleanliness affect the uniformity of the material on wafer surfaces and light exposure definition consistency. Exposure lamp age and intensity adjustments will ensure even light exposure over all areas of each wafer.

Photo techniques can be adapted to provide special etch techniques by varying curing temperatures and surface adhesion treatments. Special resists and bake-and-etch techniques provide sloped oxide steps for smooth metalized pattern continuity.

The interaction between the photoprocesses and each oxidation, diffusion, and conduction process requires extreme accuracy and integrity at each of the microscopic inspections associated with each photoprocess step. The inspectors require experience in recognizing process variations and in determining which variations are significant and which are harmless. Since process variations are rarely defined adequately, technical experts must be available for referral at all times. The technical experts are the front line of yield analysis and yield control. They must be kept accurately informed concerning process variations and changes so that they can perform effective yield analysis judgments without causing unnecessary delays in production flow. The learning curve is very important to inspection positions and there is no substitute for experience in these jobs.

Wafer Crystal Control. There are many processes other than photoprocesses that must be carefully controlled for VLSI yields to be acceptable. Controls on purchases involving wafer crystal composition and characteristics have become increasingly important. The "flatness" of a wafer surface, the degree to which it is a flat plane, is critical to photo exposure accuracy, so wafer thickness and warpage are controlled attributes, both in the wafer as received and after the normal heat treatments to which it must be exposed. Single-crystal integrity and reference alignment systems, such as one or more flat-edge reference planes, must be dependable for automatic alignment purposes and for the cleavage characteristics needed in separating chips on a wafer for assembly.

The retention of oxide concentrations in the bulk substrate crystal while the surface of the wafer is denuded of oxygen has been found to getter impurity levels that can then harmlessly concentrate in the bulk center. The denuded zone must be deep enough to isolate the active components on the surface from the impurity zone in the bulk crystal. This bulk impurity zone is also very useful for preventing crystal lattice slippage during normal process heat treatments.

A low-oxide crystal is susceptible to surface crystal damage after heat reprocessing. Excessive leakage across diffused junctions is caused by such crystal dislocations. Extreme crystal stability is attained by growing high-purity epitaxial silicon surface films for the active component zone on the surface while the bulk zone is characterized by heavily doped, high-oxide crystal.

Yield analysis sets the degree of control of crystal characteristics through vendor certification and occasional verification at incoming inspection. Screening techniques can be used to compare vendors by mixing marked wafers in process lots so that vendor characteristics may be established through duplicate processing. Then yield analyses can be compared easily at any point in wafer fab, at CP, and at FT.

28.5 EXPERIMENTAL DESIGNS

Statistically designed experiments are the most efficient tool for making accurate studies of complex interacting process combinations. Such studies bring range limits into focus at key process steps and lead to a comprehensive yield analysis. Certain processes, such as the chemical vapor deposition (CVD) of complex glass formulations onto a wafer surface at various temperature and pressure conditions, must be studied through the use of designed experiments, or the complexity of interacting variables will confuse and deteriorate the process controls. Statistically designed experiments should be utilized when new process combinations are proposed and when causes for yield variations are elusive.

Control charts are commonly used as the statistical analysis method for controlling a process. However, when a process cannot be brought into statistical control after known assignable causes have been corrected, experimentation is necessary to determine the unknown assignable causes and whether the process is, indeed, controllable. Experimentation can define not only which variables must be controlled but whether two-factor effects, or interactions, exist. Experimental design defines the causes for a process being out of control, quantifies the factors affecting the process, and demonstrates the existent inherent process variability. If tighter control of both the process average and dispersion, for example, are necessary, only through the use of experimental design can the optimum conditions be determined, and the capability of the process quantified, to conclude whether the process can be used.

Experimental designs fit nicely into the philosophy of defect prevention by providing a comprehensive understanding of process interactions. When unexplained variations in data distribution require this technique, the objective should be clearly defined. All independently controllable variables that appear to contribute to the objective must be determined and prioritized. Then the dependent response variables used to measure the process and product performance can be listed.

When control analysis experience is limited, the number of independent variables may be large and require some screening tests to ensure realistic priorities and limit ranges. Simple experimental design then requires the use of a block matrix so that each variable and range extreme can be run in combination with every other variable and range extreme of the independent process variations. The objective is attained by measuring the product for the dependent effects.

28.6 COMMUNICATION

All aspects of the manufacture of VLSI products rely on the processes involved working well together. The role of yield analysis is to creatively balance the picture by controlling each process so that it can contribute its function properly. The variants we call problems are deviations from the normal cyclic process systems and should be obvious to trained personnel. The first, and most important, function, then, for optimizing yields is to make sure that those who contribute to yields are aware of the factors they control that will affect yield. To help this happen, good communication is of paramount importance.

Close communication between wafer fabrication process engineers and product engineers doing yield analysis at WAT and CP must be maintained. Daily and weekly meetings to evaluate data and emphasize requirements are a must. The methods used for the communication of yield results contribute directly to the yields themselves, as the yield analysis process is applied to correct and adjust subsequent process and product cycles based on the experience gained with each successive cycle.

The use of computer compiled reporting permits rapid, accurate feedback. Daily production staff and process staff meetings provide forums for evaluating the cyclic experience information so that improvements can be implemented quickly.

28.7 VARIABLES THAT AFFECT YIELD

Many variants, other than those mentioned above, exist in the VLSI production process sequence that are important to its success but difficult to assess. Pinhole defects in oxide coatings and in photoresist coatings, quality of cleaning steps, handling techniques, and the presence of ambient atmospheric particles and airborne chemical contaminants are important examples. Every contributing factor in the VLSI process must have at least one feature under control for the experience curve concept to properly mature. A good database and a comprehensive understanding of a process requires regular attention to all important variants.

28.7.1 Coating Integrity

Pinhole defect studies of oxides can be made with vacuum evaporated aluminum coatings over the oxide. After annealing to drive the metal into possible oxide defect holes, electrical tests are performed in defined areas on the test wafer surface. Pinhole defect tests must be run regularly on all critical oxidation processes. Wafers prepared in this manner can be run on circuit probe test systems at sequential voltage test levels to determine the characteristic breakdown capability of the oxide under test. Realize that pinhole defects may be only partial penetration points in the oxide, which may produce variable breakdown levels in the product.

The presence of pinhole defects in photoresist coatings can be studied by trying to acid-etch oxides through a photoresist coating after normal wafer processing through exposure, development, and cure. After removing the photoresist, the oxide is metallized and patterned areas are annealed and tested electrically for oxide breakdown voltage effects. This pinhole process is an effective control on photoresist coating cleanliness under normal line handling and use conditions. When pinhole tests are evaluated and recorded as defect density readings, they can be regarded as a limiting defect density factor in the process.

Process control personnel must be alert to keep this form of defect density as near zero as possible, since random pinhole defects are almost impossible to analyze by means other than that described above. Severe pinhole defects may surface as random electrical failures at CP, but partial deterioration of oxide integrity can produce a delayed failure susceptibility that endangers the reliability of VLSI products under stressed operating conditions.

28.7.2 Wafer Surface Cleaning

The cleaning process required prior to any furnace heat treatment is an important process control variant. The resistivity of water rinses must be known as an indication of when rinsing is adequate. Particle counts are taken on the water to ensure minimum bacteria and particle contamination levels. Processing through cascade rinses permits water conservation and a quick quench finish to any etchant reaction. Final rinse sprays and spinoff drying in a filtered atmosphere minimize particle residues on wafers. Particles can be material from the product itself, from handling equipment, and from materials such as gases and water. Regular control checks must be made for residue particle counts at each cleaning facility.

Oxides heat grown on cleaned wafers make an exceptionally sensitive test surface for detecting clean and rinse residues. These test oxides (500–1000 Å) are inspected for visible spots of uneven oxide growth caused by residue particles. Defect density is used as a statistical control procedure flag in determining oxide integrity. The particle residue contributed by the cleaning processes can be an extremely difficult and erratic variable to measure. When furnace processing permits cleaning gases such as HCl to chemically react with and evaporate alkali and heavy metals from wafer surfaces, the effect of contaminants on circuits is minimized. However, where such furnace cleaning is done at temperatures below 950°C, some undesirable metallic contaminants will likely remain; the lower the temperature, the worse the contamination possibilities become.

Wherever the wafer surface is to be used for growing a critical integrity oxide coating, precleaning control tests must be exercised to ensure good process control. A simple test can be used in the inspection and control of cleaning operations for possible organic residues. A cooling plate set in a clean laminar flow hood will produce a fine haze of water condensation on a test wafer. Any condensation pattern may be considered indicative of undesirable surface residues. Furnace oxidation treatments will display any inorganic residue by the spot effect on oxide growth just described, while organic contaminants may burn off in the oxidizing furnace atmosphere. Any surface residue found by the cooling plate condensation test must be considered potentially harmful.

Total organic carbon (TOC) can be accurately analyzed on wafer surfaces. This TOC process can be used for wafer surface database information on wafers after cleaning, as well as for the control of rinse waters. Effects associated with organic residues, especially bacteria, are not acceptable in VLSI processing.

The use of chlorine and peroxide to control bacteria growth in super-purity water systems is a costly necessity. The use of cool water or hot water supply systems limits the bacteria growth conditions. Ultrafine filters are used to remove the bacteria debris at convenient locations in the distribution systems. High-flow water lines are carefully designed to shear away any bacteria growth, and pipe filters are changed regularly to prevent bacteria build up. High-purity water systems are an expensive necessity for the controlled cleanliness of wafer surfaces during the many processing steps in VLSI fabrication.

28.7.3 Handling Techniques

Handling techniques have received increasingly more attention as various particulate and chemical contaminates have been shown to destroy yields in both subtle and spectacular ways. Stainless steel tweezers for handling individual wafers through the numerous loading and unloading sequences leave iron and nickel particles that migrate and damage single-crystal surface areas. The elimination of bare-metal tweezers has been a necessary wafer fabrication handling improvement. In some cases, ceramic or plastic tweezers are used, but the majority of transfer handling steps are done by plastic belt systems that pick up wafers on the back side from slotted plastic cassettes and convey them to work stations where movement can be remotely controlled, as under a microscope.

The wafers are conveyed onto other cassettes for either rework or for transfer in a sealed box to the next process step. Where individual wafer handling has not been eliminated, the use of flexible-head, plastic, vacuum pickup tools, touching wafer backs only, makes a satisfactory replacement for tweezers. Top-to-top cassette-to-carrier transfer systems are usually clean and reliable. Although some of these techniques for handling may seem awkward, clean handling is a necessity. CP yields are directly dependent on clean handling. Yield losses for causes referred to as *microplasms,* which represent tiny defects near junctions and which can be observed as current-leakage hot spots when a circuit system fails to function, usually correlate to metallic handling contaminants. All sources of metallic contamination should be a concern when considering the transfer, processing, and handling of wafers and materials that come into contact with wafers. The extreme difficulty in determining the sources of microplasm defects is sufficient reason for extremely tight control of the metallic contaminant possibilities.

Another handling variant is surface abrasion from wafer-to-wafer contact. The superficial scratches on the active surface areas will frequently transfer to the silicon crystal surface through the layers of oxide coatings. What appears as original wafer crystal damage may be caused by a handling scratch on the oxide film surface. Plastic handling tools will often leave trace films on active surface areas of wafers. These conductive carbon leakage films can prevent adhesion of subsequent coatings. Silicon and silica dust particles generated in processes and handling equipment may be incorporated into subsequent overlay films, which develop large bumps. Though these particles are not contaminants in the normal sense, the bumps cause photoreflection errors and may severely reduce yields in dense, active circuit systems.

Regular inspections for the presence of particles are necessary for determining the frequency of cleaning maintenance and the methods necessary for cleaning the process and handling equipment. A test for particle count is run on clean control wafers, which are then passed through the handling aspects of a process, including entry and exit from furnaces, evaporators, inspection equipment, and handling transfer mechanisms. After such handling, the control wafers are again tested for particle counts, and those counts are compared with the starting particle test counts.

28.7.4 Particle Control

The source of particles and contaminants that enter clean-room process areas on an operator's clothing, hair, and skin is a costly but controllable aspect of both wafer fabrication and assembly operations. Protective clothing is used to cover the operator's street attire at entry points to processing clean rooms. Entry also requires passing through a clean filtered-air blowoff station. Cosmetics are minimized, and most facial hair is covered. Wafers are protected from spittle spray at process, handling, and inspection stations. Unobtrusive spittle shields of clear plastic placed strategically to allow unobstructed visual observations can prevent skin, dust, perspiration, spittle, or sneeze spray contact to wafer surfaces. Instances of delayed effects from spittle contamination have been carefully documented in the detailed control procedures used for space program reliability. The best yield control for these subtle human contaminants is prevention through good equipment, training, and understanding of the need for clean wafers.

Wherever wafers are moved, they should be enclosed in clean boxes, with dust-free identification cards attached to the outside of the box. The boxes should be opened in clean laminar flow hoods, which blow clean filtered air over the product and out toward the operator to prevent external particles from approaching the exposed wafers. Slotted wafer carriers can be easily removed for wafer processing, then returned to the carrier box, which is kept sealed until the wafers return. The boxes and carriers must be replaced on a regular basis and cleaned to prevent particle buildup and chemical contaminants on the outside of boxes from getting onto wafer surfaces.

28.7.5 Fume Control

Severe deterioration of wafers exposed to chemical vapors is possible without adequate fume control. Furnaces using gaseous cleaning and doping processes produce by-products such as H_2O, HCl, and $POCl_3$. The traces of HCl that bypass the furnace hood systems can be quite damaging to metallized surfaces when the surfaces are exposed for several hours. Although such corrosion damage is microscopic, it can jeopardize yields and product reliability.

Wafer processes are generally better when the process steps are sequenced rapidly. A significant part of the improvement from fast throughput is the prevention of fume attack or absorption on the sensitive clean wafer surfaces. The most damaging contaminant vapors of this sort are phosphorous chemicals. The phosphorous will condense and cause a severe dopant effect known as *n skin* on any exposed wafer surface. Isolation of the furnaces, hoods, and handling equipment used for phosphorous processing is a necessary protection against this degenerating yield killer in the wafer fabrication areas.

A common vapor contaminant that is often associated with particle and fume control problems is relative humidity (RH) or water vapor. When RH is lower than 40 percent, solid particles tend to attach to wafer surfaces firmly as static electric charges begin to influence adhesion of such particles. When RH rises above 50 percent, chemical fumes tend to be absorbed with moisture on the wafer surfaces. These surface phenomena are not always identifiable except in the photo area, where consistent surface adhesion of organic resist films is a necessity for withstanding the action of acid etchants.

Just as photoresist adhesion effects are observed to be related to changes in RH, so the same surface absorption effects are surely occurring on all exposed wafer surfaces. This is a real concern where wafers entering or exiting from furnaces are not moved for some time. The loading and cooling vestibule areas of furnaces need carefully adjusted atmospheric balancing for even marginal fume and particulate control.

28.7.6 Static Electricity

Assembly process-handling problems are similar to those in wafer fabrication areas, but the grounding of operators and equipment to ensure a common static electric potential becomes necessary to prevent static electric damage to the thin oxide films in the circuit systems. One effective static electric control method is high RH, but this is not always practical, so ionized air blowers and mild radiation systems are often effectively used. Careful control of equipment-to-operator grounding and static charge monitoring in assembly handling operations is the most effective way to prevent unnecessary losses from static damage.

REFERENCES

1. W. Edward Deming, "Out of the Crises" (Compiled Notes), MIT Center for Advanced Engineering Study, 1982.
2. Margaret Wright Jenkins, "A New Preferential Etch for Defects in Silicon Crystals," *J. Electrochem. Soc.*, May 1977, Vol. 124, p. 757.
3. "Training Program—Product Engineers," Thomas Group, Inc., Irving, Texas, 1979.
4. E. L. Grant and R. S. Leavenworth, *Statistical Quality Control,* 5th ed., McGraw-Hill, New York, 1980.
5. K. Strater, H. Veloric, and N. Goldsmith, "Comparison of Yield-Defect Models for Monitoring IC Product Lines," RCA Coded Engineering Letter 672-16, March 3, 1981.
6. W. Price, "Wafer Electrical Testing in the Factory Environment," *RCA Engineer,* Vol. 30, March/April 1985, p. 47.
7. W. A. Bosenberg and N. Goldsmith, "Parametric Testing of Integrated Circuits—An Overview," *RCA Engineer,* Vol. 30, March/April 1985, p. 29.

P • A • R • T • 10

VLSI ANALOG CIRCUITS

CHAPTER 29
ANALOG CIRCUITS

Anton Mavretic
Department of Electrical, Computer and Systems Engineering
Boston University
Boston, Massachusetts

29.1 THE BASIC ANALOG CELL

29.1.1 Intuitive Design

The best circuit designers are those that have acquired a lot of intuition about circuit behavior and analysis. Analog circuit design is similar to other arts. We start designing with some basic blocks and eventually shape our ideas into a working model. A sculptor starts in the same manner. With a few basic materials and a feeling, he or she designs and creates an object that will be functional. Just as an artist can take many approaches to the solution of a problem, so too does the circuit designer. No matter how simple or complex the problem, there are a variety of approaches to its solution. The best of these solutions are based on simple analysis and intuition.

We will start our analysis of IC design with the biasing of bipolar basic cells. We will then proceed to MOS circuits and see that the dynamic equations of behavior of both technologies are very similar. If we take into account second- and third-order effects in our analysis, the technology will expose new problems and new challenges.

29.1.2 Bipolar Cell

We will treat the biasing arrangement from the point of view of IC design, not discrete design. Discrete design at low frequencies has been done well before and is covered in numerous electronics textbooks. The input base current to the transistor must be well controlled in order to temperature-stabilize the common-emitter amplifier. In discrete design the emitter resistor is often made large because of dc biasing considerations. The increased size of the emitter resistor will cause a degradation of the transconductance. The price we pay for stabilizing the base current is a reduction in gain. For improved gain the emitter resistor is then bypassed at the frequency of operation. In IC design it is not realistic to use large capacitors and resistors because of the amount of real estate they take up on the chip. We must therefore use other transistor configurations for active amplification, and by passing, pull-up, and load resistors and other functions that are accomplished by resistors and capacitors in discrete design.

Tight thermal coupling and close matching of components in ICs permit much more radical solutions. Figure 29.1 gives the basic cell of a common emitter amplifier. We will start with it and expand it into the full IC basic cell, which is the differential amplifier. We can use either *npn* or *pnp* devices in the basic cell configurations.

The biasing resistors R_1 and R_2 set the potential on the base V_b. Since $V_{be} = V_b - V_e = 0.6$ V, this sets the voltage on the emitter V_e. The current I_e is set by the selection of emitter resistance R_e. By assuming the collector current I_c and V_{ce}, we assign the power dissipation at the bias point and the unity-gain bandwidth f_T of the transistor. Unity-gain bandwidth will be discussed later, but it is important to note f_T when biasing, since it affects the speed of the performance of the transistor. We will see later that for high-speed operation (100 MHz to 1 GHz), large-bandwidth (f_T) transistors should be biased at a higher power dissipation ($P_d = I_c V_{ce}$). Unfortunately, the nonlinearities will affect the performance of the circuit, so optimization of biasing will be required.

Example 29.1: Basic Cell Biasing. Assume for Fig. 29.1, transistor Q_1 has $\beta = 50$, $V_{cc} = 15$ V and $V_{ee} = -15$ V. Let us also assume the bias points to be $I_c = 200$ μA and $V_{ce} = 5$ V with collector dissipation of $P_c = V_{ce} I_c = 1$ mW. With some anticipation in further development of the basic cell, we will assume $V_e = -0.6$ V. Therefore $V_b = 0$ V and $V_c = +4.4$ V (since $V_{ce} = 5$ V).

- Calculate the emitter resistance:

$$R_e = \frac{V_e - V_{ee}}{I_e} = \frac{-0.6 - (-15)}{200\ \mu\text{A}} = 72\ \text{k}\Omega$$

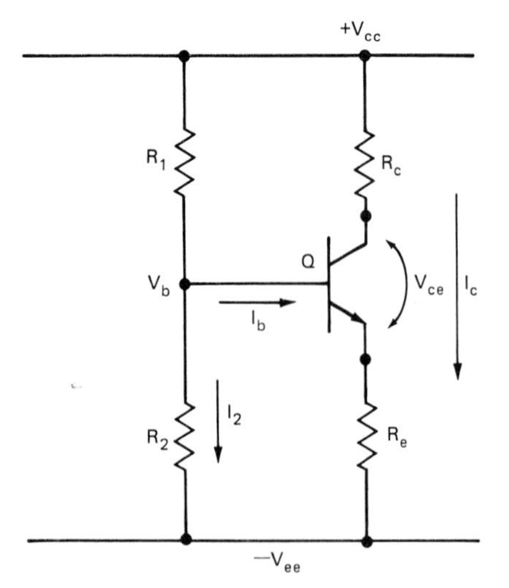

FIG. 29.1 Bipolar basic cell.

- Collector current is equal to emitter current for large dc $\beta = I_c/I_b = 50$
- Collector resistance:

$$R_c = \frac{V_{cc} - V_c}{I_c} = \frac{15 - 4.4}{200\ \mu\text{A}} = 53\ \text{k}\Omega$$

- Base biasing resistors R_1 and R_2

At this point we will make an assumption about the size of current that will flow through resistor R_2. The current through R_2 will be assumed to have the value $I_2 = I_c$. This assumption forces three consequences:

1. The large I_2 will lead to lower value of R_2 so that it may be possible to integrate.
2. Lower values of $R_1 \parallel R_2$ leads to better thermal stability; the best situation would be if the base were grounded.
3. Large I_2 increases total power dissipation of the circuit. This is not a desirable effect.

$$R_2 = \frac{V_b - V_{ee}}{I_2} = \frac{0 - (-15)}{0.2 \times 10^{-3}} = 75\ \text{k}\Omega$$

$$R_1 = \frac{V_{cc} - V_b}{I_2 + I_b} = \frac{\beta}{1 + \beta}\left(\frac{V_c - V_b}{I_c}\right)$$

$$R_1 = \frac{50}{51}\left(\frac{15 - 0}{0.2 \times 10^{-3}}\right) = 73.5\ \text{k}\Omega$$

$$R_b = R_1 \parallel R_2 = 37.1\ \text{k}\Omega$$

From discrete design we learned that if R_b/R_e $\beta/2$, then the device will be reasonably stable and collector current change ΔI_c due to temperature change will be small. In the above case this condition is obviously met at the expense of power dissipation of the bias resistors.

29.1.3 Constant-Current Sources and Voltage References

The electrical circuit element that provides a current to a load that is independent of the load resistance and voltage across the load is said to be an ideal current source. A constant-current source (CCS) can also be a controlled source whose value of the current is a function of some other voltage or current in the system. It should not be a function of the voltage across the load supplied by the CCS itself.

In ICs there are many applications of CCS's. Of primary importance are dc current sources for biasing applications. It is possible, as we will see, to design CCSs that approximate ideal conditions. A transistor to be used as CCS should be in the active mode. That is, $0.3\ \text{V} < V_{ce} < bV_{ce}$. A typical breakdown voltage for *npn* transistors used in ICs is about 40 V. Over the range of these collector voltages I_c will be relatively constant and independent of V_{ce}, provided base currents are low. We are assuming that early voltages for *npn*'s are about $V_A = 160$ V and $V_A = 80$ V for *pnp*.

Note that within the voltage compliance range the output current I_{out} increases slightly with V_{ce} or with the voltage across the current source. The conductance slope, however, is almost flat and constant over the voltage compliance range.

29.2 BIPOLAR OPERATIONAL AMPLIFIER DESIGN

29.2.1 Introduction

A good operational amplifier is essential for analog signal processing. An IC op amp design is based on (*a*) the characteristics of the signal to be amplified and conditioned and (*b*) the network characteristics, or the load where the signal will be delivered. Operational amplifiers can therefore be made to satisfy a particular requirement, but it will be highly specialized. The trend, however, is to design general-purpose amplifiers to satisfy many signal requirements. Several such circuit architectures have been synthesized in the past by manufacturers and tested in the marketplace.

The two major kinds are

1. Bipolar op amp, which has advantages of wide bandwidth, fast slew rate, and large current drive
2. CMOS op amp, with advantages of simplicity, small size, low power, and integratability on digital MOS processes

29.2.2 Analysis of the Bipolar Op Amp DC Biasing

A good op amp must be biased so that the bias currents are little affected by supply voltage V_{cc}, V_{ee}, and temperature. The circuit most often used for biasing the front end of the op amp is shown in Fig. 29.2. In this circuit, instead of connecting R_{ST} to the supply rail to get reference current, we make reference current dependent on output current source. This is done by sensing the output current by the current mirror and forcing the reference current to be equal to it. If we neglect collector resistances, r_{out}, and base currents in all transistors, we can explain the operation of the circuit as follows. Transistors Q_3 and Q_{11} and R contribute to calculate the current I_2:

$$I_2 \doteq \frac{V_{be}}{R} \tag{29.1}$$

$$I_2 \doteq \frac{V_t}{R} \ln \frac{I_1}{I_s} \tag{29.2}$$

which shows logarithmic dependence on current I_1. V_{be} referenced current sources have strong temperature dependence on V_{be}, typically in the order of −3000 ppm/°C. If resistor R is a base-diffused resistor, it would also produce an additional temperature coefficient of +2000 ppm/°C. From Eq. (29.1), we can see that total temperature dependence is about −5000 ppm/°C.

For example, if the circuit experiences $\Delta T = 100°C$ temperature change, the

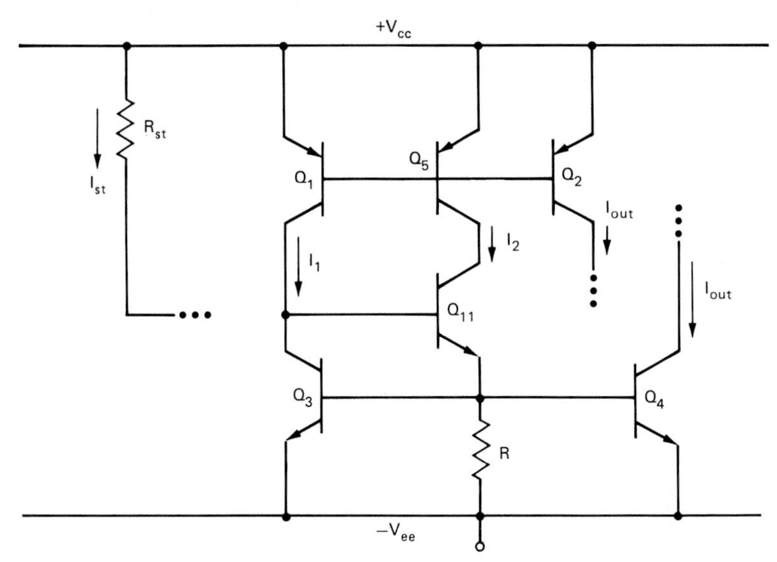

FIG. 29.2 Self-biasing V_{be} reference.

current I_2 would be reduced by at least 50 percent. In Fig. 29.2, the current is sensed by Q_1 and Q_5, and reference current I_1 is forced to be equal to I_2.

There are two constraints.

1. I_2 is logarithmically dependent on I_1, as shown in Eq. (29.2).
2. I_2 must be equal to I_1, as demanded by the current mirrors Q_1 and Q_5.

The operation of this condition is self-explanatory. Once the current level I_2 is set, it can then be reflected to other outputs through the current mirrors.

The circuit of Fig. 29.2 has a problem at start-ups. The circuit is obviously stable at $I_1 = I_2 = 0$. In order to avoid zero currents, we introduce high-impedance R_{ST} to V_{cc}, as shown in Fig. 29.2. The current I_{ST} should be very small, $I_1 \gg I_{ST}$, and it would have negligible effects on total circuit behavior.

The output circuits in Fig. 29.2, such as Q_4, can be greatly improved by the introduction of cascoded current mirrors. Several good circuits for op amp bias applications have been developed and can be read in the literature. Many of them are actually used in commercial applications. The rest of the op amp biasing is rather basic and is well covered elsewhere. We will, however, cover perhaps more salient points of thermal feedback that may mask out even the best-designed op amp at room temperature.

29.2.3 Thermal Feedback Effects on Gain

Op amps are capable of delivering powers of 50 to 100 mW to a load (special designs excluded). In the process of power delivery the output stage dissipates similar power levels. The temperature of the IC chip rises in proportion to the output dissipated power. Since the silicon chip and package are good thermal conductors, the whole chip rises to the same temperature as the output stage delivering the power. Despite this small size, a small temperature gradient 0.01 to 1°C develops across the chip. The output section is hotter than the rest. These temperature gradients appear across the input components of the op amp and induce an input voltage proportional to the dissipated power.

Thermal feedback is a single dominant source of self-heat. In order to model thermal feedback in an IC op amp, let the thermal constant be $\gamma_T = 0.6$ mV/W and P_d = power dissipated on the chip. Only a simplified analysis treatment in this section will be attempted. We can neglect the input drift due to uniform heating of the package. The effect could be large if the input stage drift is not kept low. This factor is layout-dependent also. In laying the floor plan, we must be careful in positioning the power-dissipating devices. The temperature difference across a pair of matched and closely spaced transistors is simply given by

$$T_2 - T_1 \approx \pm K_T P_d \qquad ^\circ\text{C} \tag{29.3}$$

where P_d is the power dissipated in watts and K_T is a constant in degrees Celsius per watt. The sign can be neglected since the direction of the gradient is unknown. The thermally induced input voltage V_{int} can be calculated as

$$V_{\text{int}} = \pm K_T P_d (2 \times 10^{-3}) = \pm \gamma_T P_d \tag{29.4}$$

where

$$\gamma_T = K_T (2 \times 10^{-3}) \qquad \text{V/W}$$
$$K_T \approx 0.3^\circ\text{C/W} \qquad \text{in TO} - 5 \text{ package}$$

An IC op amp is laid out to approximate either line or point source. Input components are put on isothermal lines. Stripe geometry on output transistors generates predictable isothermal lines. The output power dissipation in class AB is

$$P_d = \frac{V_{\text{out}} V_s - V^2}{R_L} \tag{29.5}$$

where

$$V_s = \begin{cases} +V_{cc} & \text{if } V_{\text{out}} > 0 \\ -V_{ee} & \text{if } V_{\text{out}} < 0 \end{cases}$$

Note that V_{int}, the thermally induced input voltage, has double-hump shape.

29.2.4 Improvements

In order to improve IC design and counteract thermal feedback deficiencies, we can make the following improvements:

1. Use a larger die.
2. Use clever CAD layout, personalized for good isothermal lines.
3. Use power package with heavier copper base, thus reducing γ_T to as low as 50 μV/W.

Some final things to note about thermal feedback are

- Thermal gain causes low-frequency distortion due to transfer characteristics.
- Differential thermal coupling falls off with 20 dB/dec at around 100 Hz. Therefore, an op amp operating at high frequencies is typically susceptible to the problems introduced by thermal feedback.

29.2.5 Small-Signal Frequency Analysis

The basic simplified bipolar IC amplifier is shown in Fig. 29.3. From the level-shifting second stage it follows that

$$V_{out}(S) = \frac{g_{m1}}{SC_c} V_{in}(S) \tag{29.6}$$

and

$$A_V(\omega) = \left| \frac{V_{out}}{V_{in}}(S) \right| = \left| \frac{g_{m1}}{SC_c} \right| = \frac{g_{m1}}{\omega C_c} \tag{29.7}$$

Here, dc and low-frequency gain analysis are not included since we have looked at it before. Therefore, from Eq. (29.7), the loop behavior is such that unity gain

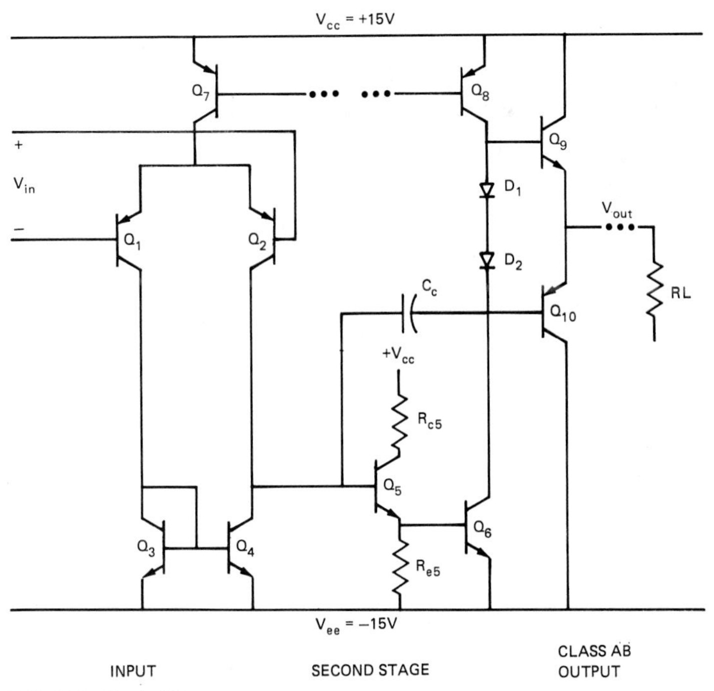

FIG. 29.3 Basic IC op amp.

ω_μ can be set by choosing C_c such that negligible excess phase would build up over a 90° phase shift.

- Low f_T in the *pnp* level-shift stage
- Stray capacitances
- Nondominant second-stage poles, etc.

Example 29.2: Calculating Unity-Gain Bandwidth. For a typical bipolar op amp, we have $I_1 = 10\ \mu\text{A}$ and $C_c = 30$ pF, and the unity-gain bandwidth is

$$f_\mu = \frac{g_{m1}}{2\pi C_c} = \frac{0.19\ \text{mA/V}}{2\pi \times 30 \times 10^{-12}} = 1\ \text{MHz} \tag{29.8}$$

29.2.6 Slew Rate

The slew rate (SR) of an op amp is the maximum rate of change of output voltage with time, $(dV_{\text{out}}/dt)_{\text{max}}$. Typical slew rates are from 0.5 to 100 V/μs. The slew rate varies, depending on type of input stage. The input step of about 50 mV typically will cause output to slew at 90 percent of maximum rate. For example, LM741 slews at this rate with input of 120 mV, whereas LM118 or FET Input op amps requires about 1 V of overdrive.

Since slew rate is a function of both frequency and signal levels, the bias current (current sources) in the design of the op amp must increase in value as we go from input to output. The signal level is growing from stage to stage within the op amp and it is desirable to preserve the SR. Therefore, I_2 (CCS for stage 2) must be larger than I_1 (CCS for stage 1) for an adequate slew rate.

Input voltage is large enough so that all CCS current $2I_1$ is diverted into the integrator:

$$\text{SR} = \left.\frac{dV_{\text{out}}}{dt}\right|_{\text{max}} = \frac{I_c(t)}{C_c} = \frac{2I_1}{C_c} \tag{29.9}$$

By combining Eqs. (29.9) and (29.8), we obtain

$$\left.\frac{dV_{\text{out}}}{dt}\right|_{\text{max}} = \left(\frac{2I_1}{g_{m1}}\right)\omega_\mu = 2\omega_\mu\left(\frac{I_1}{g_{m1}}\right) \tag{29.10}$$

$$g_{m1} = \frac{I_1}{V_T} \qquad V_T = \frac{kT}{q}$$

$$\text{SR}|_{\text{max}} = 2\omega_\mu V_T \tag{29.11}$$

which is the maximum slew rate for a bipolar device. The conclusion is that for a given op amp with fixed slew rate and maximum frequency of operation, the amplitude of the undistorted signal is limited.

Power Bandwidth. Power bandwidth is defined as the maximum frequency at which full output swing (10 V peak-to-peak) can be obtained without distortion. For a sinusoidal input,

$$\frac{dV_{out}}{dt} = \omega V_p \cos(\omega t)$$

$$SR = \frac{dV_{out}}{dt} = \omega V_p \qquad \text{at } \omega t = 1$$

$$\omega_{max} = \frac{SR}{V_p} \tag{29.12}$$

This is the maximum reproducible frequency without distortion. In LM741, for example, $f_{max} = 0.67(10 \times 2\pi) \approx 10$ kHz if the output swing is 10 V peak-to-peak.

Improving Slew Rate. From SR $= 2\omega_\mu(I_1 g_{m1})$, it follows that by reducing g_{m1}, the slew rate will be improved. Rather than reduce g_{m1} by itself, we will consider the ratio of g_{m1}/I_1 in transconductance degradation. It is given by

$$g_{m1(new)} = \frac{g_{m1}}{1 + g_{m1}R_e} \tag{29.13}$$

or

$$g_{m1(new)} = \frac{g_{m1}}{1 + I_1 R_e / V_T}$$

where R_e is emitter resistance in either leg of differential amplifier. If the quantity I_1/g_{m1} in the SR equation (29.10) increases rapidly, the voltage drop I_1R_e will become larger and could exceed V_T (25 mV at +25°C). Therefore, if I_1/g_{m1} increases, SR increases.

Typically an IC op amp will have a value of I_1R_e approximately equal to 500 mV. It may produce $f_\mu = 16$ MHz, and the slew rate will be improved to

$$\left.\frac{dV_{out}}{dt}\right|_{max} = SR_{max} = 2\omega_\mu \frac{I_1}{g_{m1}} = 100 \text{ V}/\mu\text{s} \tag{29.14}$$

If the emitter resistances are not matched well, the penalty will go to increased voltage offset, drift at the input, and noise. For a typical high-speed op amp, R_e is approximately 2 kΩ, and hence each 4-Ω mismatch causes 1-mV offset.

29.3 CMOS ANALOG CIRCUIT DESIGN

The circuits presented in this section include the active resistor or load, current sinks and sources, current mirrors, the inverter, the source follower, the differential amplifier, and the CMOS op amp.

29.3.1 Active Resistor

A resistor in MOS is achieved by simply connecting the gate to drain as shown in Fig. 29.4*a* and *b*. For the *n*-channel device the substrate should be placed at the most negative power supply voltage V_{ss} in order to eliminate body effect. Simi-

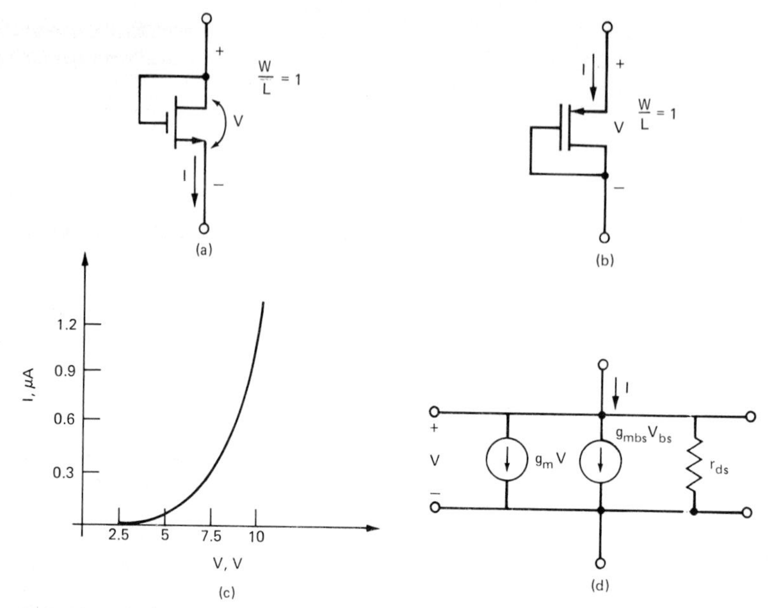

FIG. 29.4 Active resistor. (*a*) *n*-Channel; (*b*) *p*-channel; (*c*) *I-V* characteristics; (*d*) ac model.

larly, the substrate of the *p*-channel device should be connected to the most positive power supply voltage V_{dd}. Figure 29.4*c* illustrates the *V* versus *I* characteristics of both Fig. 29.4*a* or *b*. A connection of the gate to the drain forces operation in the saturation region; the *V-I* characteristics can be written as

$$I = I_d = \frac{K'W}{2L}(V_{gs} - V_T)^2 = \beta(V_{gs} - V_T)^2 \tag{29.15}$$

or

$$V = V_{gs} = V_{ds} = V_T + \frac{I_d}{\beta} \tag{29.16}$$

where β is defined as

$$\beta = \mu_o C_{ox} \frac{W}{2L} = \frac{K'W}{2L} \tag{29.17}$$

where W = width, L = length, and K' = constant. If either V or I is defined, then the remaining variable can be designed by using either Eq. (29.15) or (29.16), solving for the value of β, and setting appropriate values for W and L.

Connecting the gate to the drain forces V_{ds} to control I_d, and therefore the channel transconductance becomes a channel conductance. The small-signal model for either the *n*-channel or *p*-channel active resistor is shown in Fig. 29.4*d*. The small-signal resistance of either of these circuits is

$$r_{out} = \frac{r_{ds}}{1 + g_m r_{ds}} \approx g_m^{-1} \quad (29.18)$$

where $g_m r_{ds}$ is greater than unity.

Example 29.3. Let $V_{dd} = +10$ V, $V_{ss} = -10$ V, $V_{out} = 1$ V, and $I = 100$ μA. Then using the model parameters for the MOS transistors of Table 29.1, we get

$$\beta_1 = 1.0 \times 10^{-6} \text{ A/V}^2$$

$$\beta_2 = 1.57 \times 10^{-6} \text{A/V}^2$$

$$T_{ox} = 500 \text{ Å}$$

This gives a W/L ratio of 0.05 for M_1 and 0.2 for M_2.

29.3.2 Current Sinks and Sources

An implementation of a current sink is shown in Fig. 29.5*a*. The gate potential is set to whatever voltage is necessary to obtain the desired value of current. If the gate-source voltage is held constant, then the large-signal characteristics of the MOSFET in Fig. 29.5*a* are shown by Fig. 29.5*b*.

If the source and bulk are both connected to V_{ss}, then the small-signal output resistance is r_{ds}. If the source and bulk are not connected to the same potential,

TABLE 29.1 Model Parameters for a Typical CMOS Bulk Process Using 100 *N* Substrate with 4–6 Ω · cm Resistivity

Parameter symbol	Parameter description	Typical parameter value		Units
		NMOS	PMOS	
μ	Channel mobility ($N_{sub} = 2.3 \times 10^{15}$ cm^{-3})	550	225	cm^2/(V · s)
C_{ox}	Capacitance per unit area of the gate oxide	0.7×10^{-7}	0.7×10^{-7}	F/cm^2
λ	Channel length modulation parameter ($L \geq 25$ μm)	0.01	0.01	V^{-1}
V_{T0}	Threshold voltage ($V_{bs} = 0$)	1 ± 0.2	-1 ± 0.2	V
γ	Bulk threshold parameter	0.5	0.5	V$^{1/2}$
ψ	Surface potential at strong inversion	0.6	0.6	V
PB	Bulk junction potential	0.7	0.7	V
FC	Forward bias nonideal junction capacitance coefficient	0.5	0.5	
KF	Flicker noise coefficient	10^{-14}	10^{-14}	
AF	Flicker noise exponent	1.0	1.0	

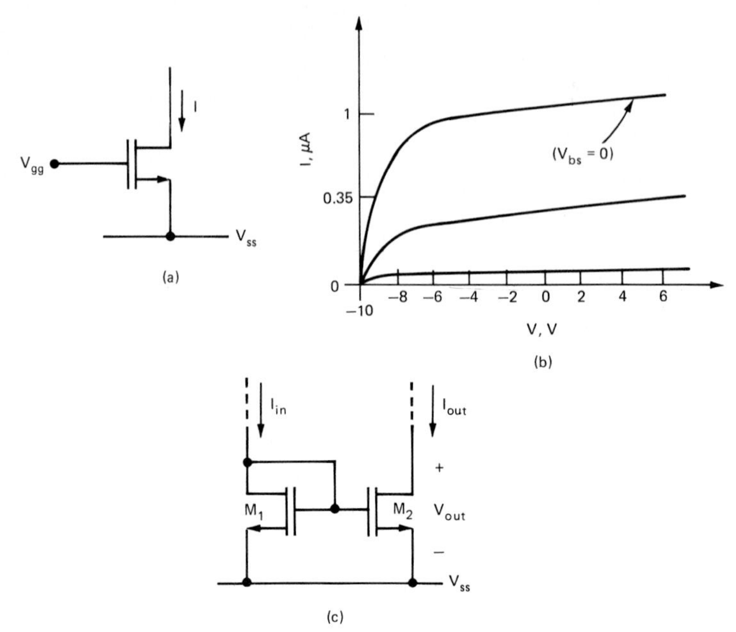

FIG. 29.5 (*a*) Current sink; (*b*) *I-V* characteristics of current sink; (*c*) Widlar current mirror.

the characteristics will not change as long as V_{bs} is a constant. This is an advantage of CMOS over NMOS or PMOS, since the current can be taken from the drain while the source can be kept at a constant potential.

29.3.3 Current Mirrors

The current mirror operates on the principle that if the gate-source potentials of two identical MOS transistors are equal, then the drain currents should be equal. Figure 29.5*c* shows the implementation of a Widlar *n*-channel current mirror. I_{in} is a current source, and I_{out} is the *mirrored* current. M_1 is in saturation since $V_{ds} = V_{gs}$, and $V_{ds2} > V_{T2}$. For this case we can use the equations in the saturation region of the MOS transistor. I_{out} may be solved in terms of I_{in}:

$$I_{\text{out}} = \left[\frac{(W_2L_1/W_1L_2)(1 + \lambda V_{ds2})}{(1 + \lambda V_{ds1})} \right] I_{\text{in}} = A_i I_{\text{in}} \tag{29.19}$$

where it is assumed that μ_o, C_{ox}, and λ are identical for both *n*-channel devices. Ideally, if $W_2/L_2 = W_1/L_1$, then $I_{\text{out}} = I_{\text{in}}$. Equation (29.19) shows that if V_{ds2} is not equal to V_{ds1}, then I_{out} is not equal to I_{in} unless λ is equal to zero. Since $V_{ds1} = V_{gs1}$ and $V_{ds2} = V_{\text{out}}$ is typically larger, this current mirror does not reproduce I_{in} precisely. Using the small-signal model for MOS, the output resistance can be expressed as

$$r_{out} = r_{ds2} = (\lambda I_d)^{-1} \tag{29.20}$$

In many cases, r_{out} of the Widlar current mirror is too low to be used for a specific application. One method of increasing the output resistance of a Widlar source is to use cascoded connection. The top half of such circuit has the effect of multiplying the output resistance by a factor of approximately $1 + g_m r_{ds}$. The overall effective output resistance of this circuit looking into the top drain is therefore

$$r_{out} = r_{ds}(1 + g_m r_{ds}) \tag{29.21}$$

29.3.4 Differential Amplifiers

A true differential amplifier must amplify the difference between two different potentials regardless of the common-mode value. It can be characterized by its common-mode rejection ratio, which is the ratio of the differential gain to the common-mode gain. The input common-mode range specifies over what range of common-mode signal levels a differential amplifier continues to sense the difference with the same gain. The performance characteristics of the differential amplifier are degraded by its offsets. In a CMOS differential amplifier, the voltage offset is the largest. Typical input voltage offsets of a CMOS differential amplifier are from 1 to 5 mV.

A CMOS differential amplifier that uses *n*-channel MOS devices is shown in Fig. 29.6*a*, where I_{ss} is a constant current source. The loads for M_1 and M_2 are simple *p*-channel current mirrors. If M_3 and M_4 are matched, then the current of M_1 will determine the current in M_3. This current will be mirrored in M_4. If $V_{gs1} = V_{gs2}$, then the currents in M_1 and M_2 are equal. Thus the current that M_4 sources to M_2 should be equal to the current that M_2 must take, causing I_{out} to be zero. If $V_{gs1} > V_{gs2}$, then I_{d1} increases with respect to I_{d2} since $I_{ss} = I_{d1} + I_{d2}$. This increase in I_{d1} implies an increase in I_{d3} and I_{d4}. When V_{gs1} becomes greater than V_{gs2}, I_{d2} will decrease. The only way that circuit equilibrium can be established is for I_{out} to become positive. It can be seen that if $V_{gs1} < V_{gs2}$, I_{out} becomes negative. This is a simple way in which the differential output signal of the differential amplifier can be converted back to a single-ended signal.

A CMOS differential amplifier that uses *p*-channel MOS devices is very similar to that of Fig. 29.6*a*. The circuit operation would be also identical to that of Fig. 29.6*a*, except in the bulk effect. The bulks of the *p*-channel devices can only be taken to the most positive potential V_{dd}. The *n*-channel bulks are not necessarily constrained to be connected to the most negative potential V_{ss}. This permits the *n*-channel differential amplifier of Fig. 29.6*a* to have the sources of M_1 and M_2 connected to their bulks. Thus, M_1 and M_2 would be made in a floating *p*-tub. A differential amplifier as shown is less sensitive to the bulk effects.

The small-signal analysis of the differential amplifier of Fig. 29.6*a* will be performed by the use of the model shown in Fig. 29.6*b*. This model is appropriate for differential analysis, where it is assumed that M_1 and M_2 are matched and the point of common sources is an ac ground. If we assume that the differential stage is not loaded, then the differential transconductance gain is

$$I'_{out} \approx g_{m1}V_{gs1} - g_{m2}V_{gs2} = g_m V_{id} \tag{29.22}$$

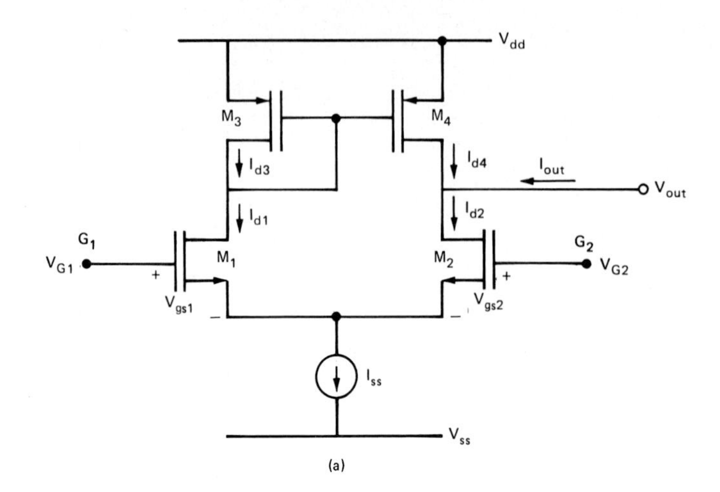

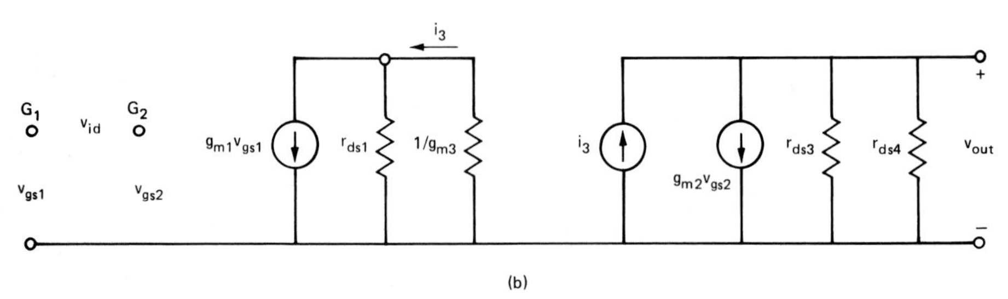

FIG. 29.6 (*a*) CMOS differential amplifier using *n*-channel input devices; (*b*) small-signal model for CMOS differential amplifier.

where $g_{m1} = g_{m2} = g_m$. The no-load differential voltage gain is

$$\frac{V_{\text{out}}}{V_{id}} = \frac{g_m}{g_{ds2} + g_{ds4}} \tag{29.23}$$

or

$$\frac{V_{\text{out}}}{V_{id}} = \frac{2}{\lambda_2 + \lambda_4} \sqrt{\frac{2\beta_1}{I_{ss}}} \tag{29.24}$$

Again we note the dependence of the small-signal gain on the inverse of $\sqrt{I_{ss}}$ similar to that of the inverter. This relationship is valid until I_{ss} approaches subthreshold values.

Example 29.4. Let $W_1/L_1 = 1$ and $I_{ss} = 10\ \mu\text{A}$ in Fig. 29.6*a*. Then the calculated small-signal gain of the NMOS differential amplifier is found to be 215. The small-signal gain of the PMOS differential amplifier under the same conditions is 138. It is lower due to smaller mobilities in *p* material.

The common-mode gain of the CMOS differential amplifiers as shown in Fig. 29.6*a* is ideally zero. The current mirror loads reject any common-mode signal. The common-mode response exists due to the mismatches in the differential amplifier. The parasitic effects are necessary if further analyses are carried out on SPICE.

29.3.5 Slew Rate

The slew rate performance of the CMOS differential amplifier depends upon the value of I_{ss} and the capacitance from the output node to ac ground. The slew rate in the CMOS differential amplifier is determined by the amount of current that can be sourced or sunk into the output capacitor. The slew rate of the CMOS differential amplifiers of Fig. 29.6 is

$$SR = \frac{I_{ss}}{C_L} \quad \text{V}/\mu\text{s} \tag{29.25}$$

Example 29.5. If $I_{ss} = 50\ \mu A$ and $C_L = 15$ pF in Fig. 29.6*a*, the slew rate is found to be 50 μA/15 pF = 3.33 V/μs. The value of I_{ss} must be increased to increase the slew rate capability of the differential amplifier.

The frequency response of the CMOS differential amplifier is due to the various parasitic capacitances at high-resistance nodes. The parasitic of constant-current source C_{ccs} is almost inactive since the node to which it is connected acts like an ac ground. C_M is the mirror capacitance, and C_{out} is the capacitance connected from the output node to the ground. The dominant pole will be due to the output capacitor because of the very high output resistance of the differential amplifier.

29.3.6 Design of CMOS Op Amps

CMOS op amps are very similar in their architecture and design to bipolar op amps. The important aspects of a CMOS op amp are shown in Fig. 29.7. The differential transconductance stage forms the input of the op amp and provides the differential to single-ended conversion. The level-shift high-gain stage is typically an inverter. The MOS bias circuits have been presented before. Thus, the only parts of the basic op amp diagram that have not been discussed are the compensation and the output buffer.

29.3.7 CMOS Op Amp Characteristics

An ideal op amp has infinite input resistance, infinite differential gain, and zero output resistance. The most important of these three values is the large gain. In practical cases, the output voltage V_{out} can be expressed as

$$V_{out} = A_v(V_1 - V_2) \tag{29.26}$$

V_1 and V_2 are the input voltages applied to the noninverting and inverting termi-

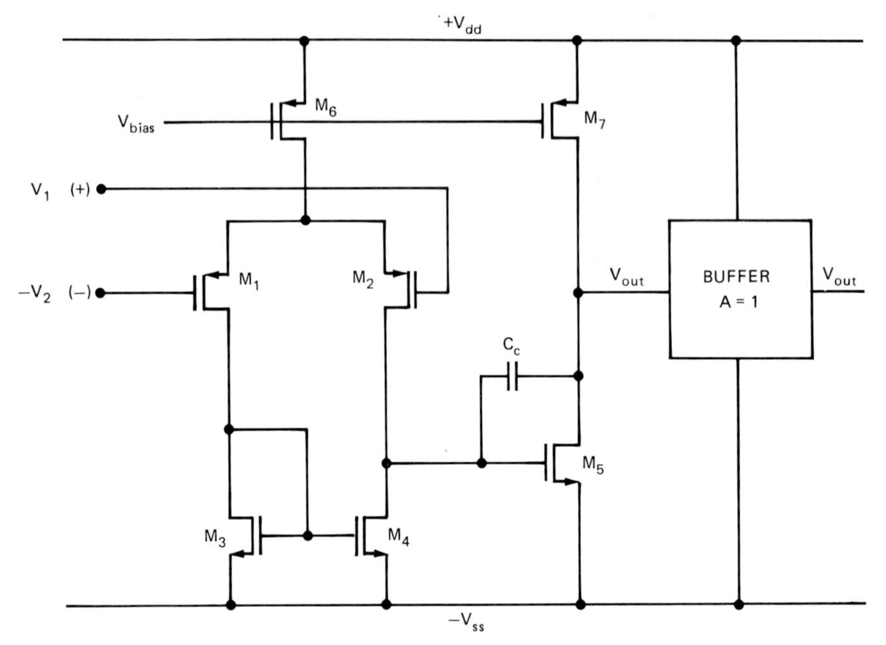

FIG. 29.7 Basic CMOS op amp.

nals, respectively. The difference between these two voltages is amplified by A_v and appears at the output of the op amp.

In analog-sampled data circuits the op amp is almost always used in an inverting configuration with the positive terminal connected to ac ground. The finite differential input impedance is modeled by R_{id} and C_{id}. The output resistance is modeled by R_{out}. The common-mode input resistances are given as resistances of R_{icm} connected from each input to ground. V_{os} is the input offset voltage that puts the output voltage to zero if both inputs are grounded.

The common-mode rejection ratio (CMRR) is modeled by the voltage-controlled voltage source indicated as V_1/CMRR. The other two sources are noise voltage $\overline{e_n^2}$ and noise current model $\overline{i_n^2}$. The op amp noises are called voltage and current noise spectral densities. Their units are mean square volts and mean square amperes, respectively. These noise sources have no polarity and are always assumed to add, since they are not correlated.

The output voltage of Fig. 29.7 can be defined as

$$V_{\text{out}}(s) = A_d(s)[V_1(s) - V_2(s)] + A_c(s)\left[\frac{V_1(s) + V_2(s)}{2}\right] \tag{29.27}$$

where the first term on the right is the differential portion of $V_{\text{out}}(s)$ and the second term is the common-mode portion of $V_{\text{out}}(s)$. The differential frequency response of the op amp is given as $A_d(j\omega)$ while the common-mode frequency response is given as $A_c(j\omega)$. A typical frequency response of an op amp is

$$A_d(s) = \frac{A_d\omega_1\omega_2\omega_3\cdots}{(s+\omega_1)(s+\omega_2)(s+\omega_3)\cdots} \tag{29.28}$$

where ω_1, ω_2 are poles of the operational amplifier. A_d is the low-frequency gain of the op amp. Figure 29.8*a* shows a typical frequency response of the magnitude of $A_d(j\omega)$. In this case, we see that dominant pole ω_1 is much lower than the rest of the poles. The intersection of the -20 dB/dec slope from the dominant pole and the 0-dB axis is designated as GB and is defined as the *unity-gain bandwidth* of the op amp.

Other nonideal characteristics of the op amp include the *power supply rejection ratio* (PSRR). The PSRR is defined as the product of the ratio of the change in the supply voltage to the change in the output voltage of the op amp caused by the ripple in the power supply bus and the open-loop gain of the op amp. PSRR $= (\Delta V_{dd}/\Delta V_{out})A_d$. An ideal op amp would have an infinite PSRR. The *common-mode input range* is the voltage range over which the input common-

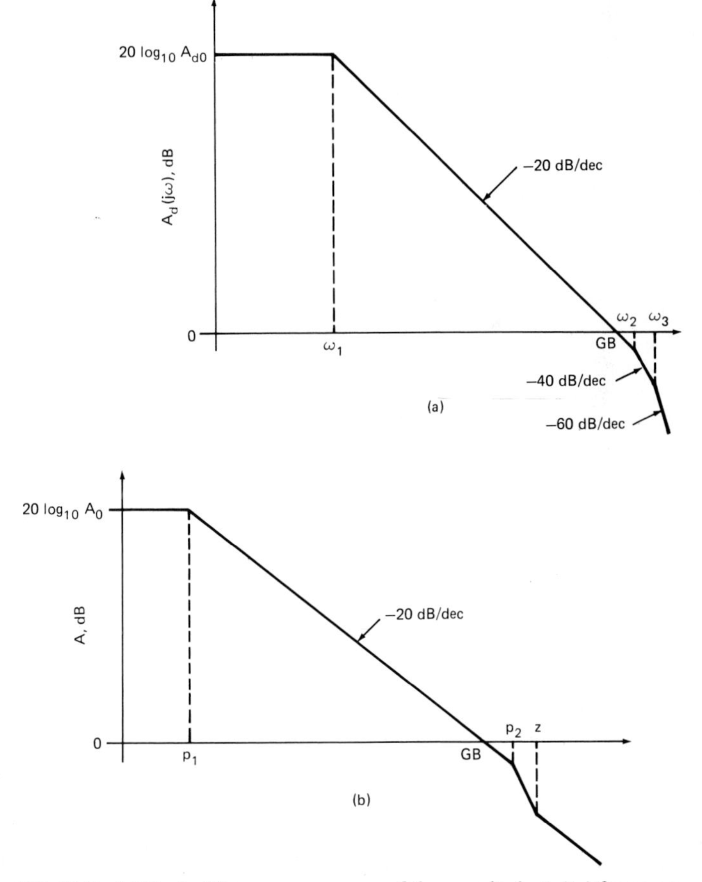

FIG. 29.8 (*a*) Typical frequency response of the magnitude $A_d(j\omega)$ for an op amp; (*b*) closed-loop frequency response of Fig. 29.9*b*.

mode signal can vary. Typically, this range is several volts less than the maximum and several volts more than the minimum power supply voltages. The output of the op amp has several limitations. One is the *maximum output current sourcing and sinking* capability, and another is the *output voltage swing*. Like the input common-mode range, the output voltage swing is typically limited to within several volts of the maximum and minimum power supply voltages.

29.3.8 General Approach to the Design of MOS Op Amps

The op amps are designed to meet certain specifications. In analog circuit applications, the following considerations may be pertinent, before amplifier design is attempted.

1. Open-loop gain
2. Output voltage swing
3. Power consumption
4. Power supply bus levels
5. Capacitance the op amp must drive
6. Power supply rejection ratio
7. Slew rate
8. Required unity-gain bandwidth
9. Common-mode input range
10. Substrate (back-gate) bias availability on chip
11. Settling time
12. Noise tolerance

The most important design criterion is also to be able to design an op amp that is insensitive to the process parameters.

In designing an op amp, one usually begins at the output, since the loading conditions are usually given. The design procedure must be iterative, since we cannot relate all specifications simultaneously. As a rule, all MOS transistors must operate in the saturation region, away from triode region, which is defined as

$$V_{ds} \geq V_{gs} - |V_T| \tag{29.29}$$

In the saturation region, the largest values of g_m and voltage swings are obtained. A typical CMOS op amp design starts with the following steps:

1. From the specifications, determine a block diagram that shows how the differential gains will be distributed throughout the op amp.
2. Input stage is differential and determines the common-mode input range, PSRR, and the noise performance. The output of this stage will generally have a low-frequency pole, which will be used by the Miller compensation technique to achieve the dominant pole.
3. The level-shifting gain stage should provide most of the gain of the op amp. This gain will permit Miller compensation techniques to be applied by using

small capacitances. The second and third poles of this stage should be moved outside of GB. See ω_1, ω_2 in Fig. 29.8*a*.

4. The output stage should have the desired current and voltage output capability. Since the output stage will consist of large devices, the frequency bandwidth will suffer. The poles should be moved above GB wherever possible. We must be aware of complex poles and pole-zero pairs, which lead to poor settling times.
5. The bias network should be designed to be power supply, temperature, and process independent.
6. Internal compensation is achieved by applying a capacitor around a high-gain, inverting stage with high input-output resistance. This will create a dominant pole. It may also be necessary to provide feedforward compensation in order to obtain a larger phase margin.

The specifications cannot be satisfied after going through the above steps for the first time. It is necessary to iterate, making changes that will meet the specifications. Once a preliminary design has been developed so that *W/L* ratios of the MOS transistors have been estimated, then the design is taken to the computer for further simulation on SPICE. It may also be necessary to computer simulate portions of the circuit before considering the entire circuit.

29.3.9 Compensation Methods

A good op amp should have a phase margin of 60° in the closed-loop unity-gain configuration. Smaller phase margins will cause the step response to "ring" and contribute to a longer settling time. The basic approach to compensation consists of two steps. The first step is to determine all the poles in the circuit including parasitic and loading capacitance. The parasitic capacitances include the moat bottom area to substrate capacitance, the moat sidewall area–to-substrate capacitance, the first poly-to-substrate capacitance, the metal-to-substrate capacitance and the second poly- (if used) to-substrate capacitance. We should also take into account the lateral diffusion in calculating the moat bottom area capacitance. The second step is to redesign.

An ac model of the operational amplifier suitable for compensation purposes is shown in Fig. 29.9.[1] This model represents two voltage gain stages with a compensation capacitor C_c around the second gain stage. If the gain is high, then this capacitor uses the Miller effect to accomplish compensation. The transconductance g_{m1} is of the first stage, and R_1 is the combined resistance to ground of the output of the first stage and the input of the second stage. The capacitance to ground at this node is C_1. The transconductance of the second stage is g_{m2}; R_2 and C_2 form the output load. If C_c is zero, then the roots of this case correspond to

$$p_1' = \frac{-1}{R_1C_1} \tag{29.30}$$

and

$$p_2' = \frac{-1}{R_2C_2} \tag{29.31}$$

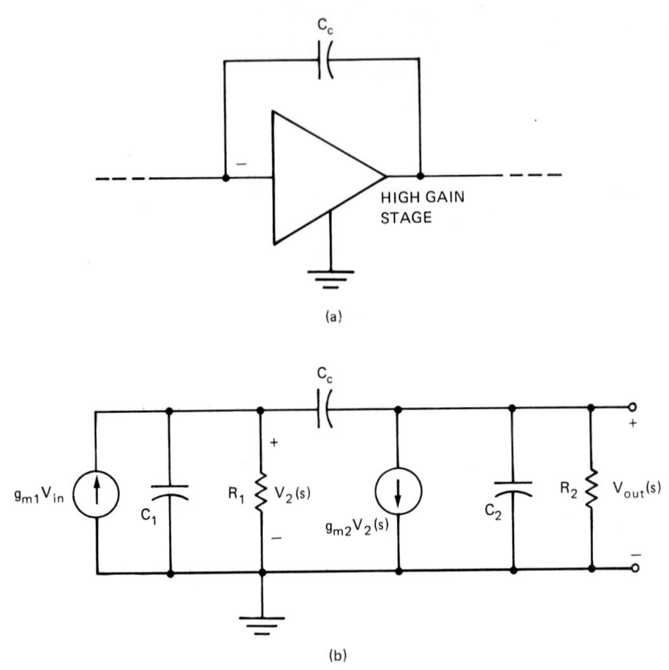

FIG. 29.9 Compensation scheme using Miller effect. (*a*) General circuit; (*b*) equivalent model.

These poles are on the negative real axis. When C_c is not zero, then the transfer function $V_{out}(s)/V_{in}(s)$ can be found as

$$\frac{V_{out}(s)}{V_{in}(s)} = \frac{g_{m1}g_{m2}R_1R_2(1 - sC_c/g_{m2})}{1 + s[R_1(C_1 + C_2) + R_2(C_2 + C_c) + g_{m2}R_1R_2C_c] + s^2R_1R_2[C_1C_2 + C_c(C_1 + C_2)]} \tag{29.32}$$

A second-order denominator may be expressed as

$$D(s) = \left(1 - \frac{s}{p_1}\right)\left(1 - \frac{s}{p_2}\right) = 1 - s\left(\frac{1}{p_1} + \frac{1}{p_2}\right) + \frac{s^2}{p_1p_2} \tag{29.33}$$

If $|p_2| \gg |p_1|$, then Eq. (29.33) becomes

$$D(s) \approx 1 - \frac{s}{p_1} + \frac{s^2}{p_1p_2} \tag{29.34}$$

Equating the denominators of Eqs. (29.32) and (29.34) gives

$$p_1 \approx \frac{-1}{g_{m2}R_1R_2C_c} \tag{29.35}$$

$$p_2 \simeq \frac{-g_{m2}C_c}{C_1C_2 + C_2C_c + C_lC_c} \tag{29.36}$$

and the zero is

$$z = \frac{g_{m2}}{C_c} \tag{29.37}$$

We can see that C_c has split p_1' and p_2'. In the process of applying the Miller compensation, a zero located on the positive real axis has been generated.

The frequency response of the compensated op amp is shown in Fig. 29.8*b*. The objective of the compensation is to have a single dominant pole p_1 and all other poles including p_2 the unity gain-bandwidth GB. The right-hand zero can destroy attempts to compensate the op amp if g_{m2} is not large. This zero increases the magnitude of the op amp gain, and causes phase lag. This phase lag can cause the op amp to become unstable when negative feedback is applied. The desired objective of compensation is to have sufficient phase margin in a given application. Typically, this value of phase margin is between 45° and 60°. This performance will be achieved when p_2 is equal to or greater than GB. It can be shown that GB is given as

$$\text{GB} = A_o p_1 \simeq g_{m1}R_1R_2g_{m2}\left(\frac{1}{g_{m2}R_1R_2C_c}\right) = \frac{g_{m1}}{C_c} \tag{29.38}$$

The ratio of p_2 of (29.36) to GB of (29.38) is expressed as

$$\frac{p_2}{\text{GB}} = \frac{g_{m2}C_c^2}{g_{m1}(C_1C_2 + C_1C_c + C_2C_c)} \tag{29.39}$$

If C_2 includes the load capacitor, then $C_2 > C_1$. If $C_c > C_1$, then Eq. (29.39) can be approximated as

$$\frac{p_2}{\text{GB}} \approx \frac{g_{m2}C_c}{g_{m1}C_2} \tag{29.40}$$

In the op amp design it is desired to have $|p_2| > \text{GB}$. Then Eq. (29.40) results in the following capacitor constraint for C_c and C_2 ($\approx C_{\text{LOAD}}$):

$$C_2 < \frac{g_{m2}}{g_{m1}}C_c \tag{29.41}$$

It is also necessary to put zero larger than GB to avoid the degradation of the phase margin. Hence,

$$\frac{g_{m2}}{g_{m1}} = \frac{z}{\text{GB}} > 1 \tag{29.42}$$

The MOS op amp currents are usually comparable in both the first and second stages. The transconductance varies as the square root of the bias current, so very large current differences are necessary to achieve g_{m2} greater than g_{m1}.

Since the zero is caused by feedforward path through C_c, the above situation can be avoided by the use of a buffer in the feedback path as shown in Fig. 29.10*a*. The network R_1, C_1, R_2, and C_2 is associated with the inverting high-gain stage. $AV_{out}(s)$ and R_{out} represent the buffer and its output resistance. If R_{out} is small, then the transfer function is

$$\frac{V_{out}(s)}{V_{in}(s)} = \frac{-g_{m1}g_{m2}R_1R_2}{1 + s[R_1C_1 + R_2C_2 + R_1C_c + g_{m2}R_1R_2C_c] + s^2[R_1R_2C_2(C_1 + C_c)]} \tag{29.43}$$

where the poles are

$$p_1 \approx \frac{-1}{R_1C_1 + R_2C_2 + R_1C_c + g_{m2}R_1R_2C_c} \approx \frac{-1}{g_{m2}R_1R_2C_c} \tag{29.44}$$

$$p_2 \approx \frac{-g_{m2}C_c}{C_2(C_1 + C_c)} \tag{29.45}$$

The zero has been eliminated, resulting in an improved phase margin.

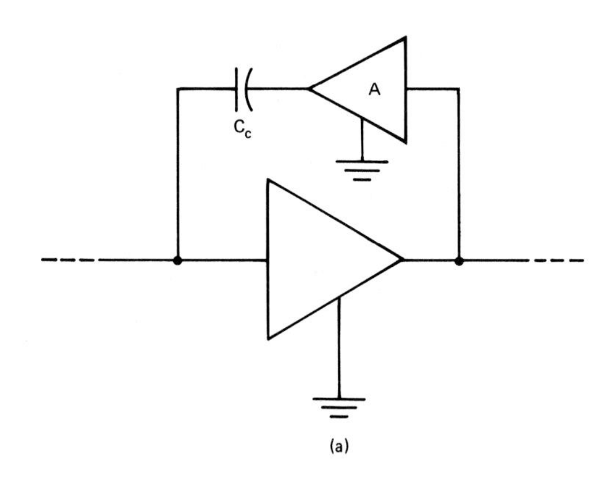

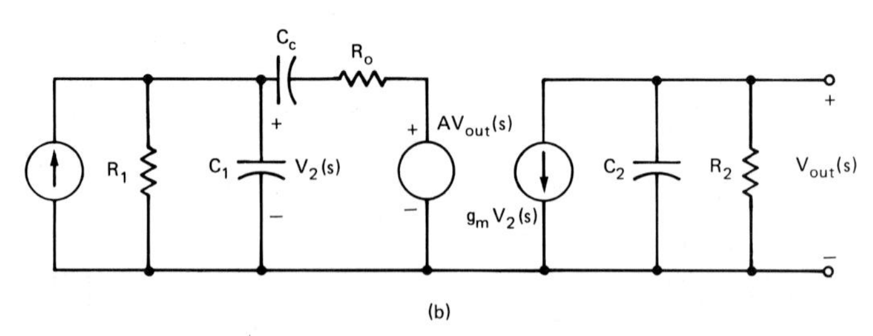

FIG. 29.10 (*a*) Modified Miller compensation scheme; (*b*) model for gain calculations.

Output resistance R_{out} should not be neglected and must be incorporated into the above analysis. The solution is

$$\frac{V_{out}(s)}{V_{in}(s)} = \frac{-g_{m1}g_{m2}R_1R_2(1 + sR_{out}C_c)}{1 + as + bs^2 + cs^3} \tag{29.46}$$

where
$$a = 1(R_1C_1 + R_2C_2 + R_1C_c + g_{m_2}R_1R_2C_c) + R_{out}C_c \tag{29.47}$$

$$b = 1(R_1R_2C_1C_2 + R_1R_2C_2C_c) + R_1R_{out}C_1C_c + R_{out}R_2C_2C_c \tag{29.48}$$

$$c = R_1R_2R_{out}C_1C_2C_c \tag{29.49}$$

After considerable operation on these equations, we get the poles

$$p_1 \approx \frac{-1}{g_{m2}1R_1R_2C_c} \tag{29.50}$$

$$p_2 \approx \frac{-g_{m2}C_c}{C_2(C_1 + C_c)} \tag{29.51}$$

$$p_3 \approx \frac{-1}{R_{out}[C_1C_2/(C_1 + C_c)]} \tag{29.52}$$

where we have assumed that R_1, $R_2 \gg R_{out}$. There is a zero due to R_{out} in Fig. 29.10*b*:

$$z = -\frac{1}{R_{out}C_c} \tag{29.53}$$

Example 29.6. Assume that $R_{out} = 30$ kΩ, $C_c = 30$ pF, $R_1 = 120$ kΩ, $R_2 = 100$ kΩ, $C_1 = 5$ pF, $C_2 = 32$ pF, and $g_{m2} = 400$ μA/V. The poles are located at

$$p_1 = -6.94 \text{ krad/s} \quad -1.017 \text{ kHz}$$
$$p_2 = -10.71 \text{ Mrad/s} \quad -1.57 \text{ MHz}$$
$$p_3 = -7.77 \text{ Mrad/s} \quad -1.14 \text{ MHz}$$

and the zero is at

$$z = -\ 1.11 \text{ Mrad/s} \quad -\ 163 \text{ kHz}$$

Note that the assumption that $|p_2| < |p_3|$ occurs in this example. We are interested primarily in the zero. We will not attempt to make a better solution for the pole locations. The zero at 1.11 Mrad/s (163 kHz) gives a leading phase shift that is desirable, but it also boosts the undesirable open-loop gain. If R_1, R_{out}, g_{m1}, g_{m2}, and R_2 are known, then C_c may be designed to realize the desired dominant pole, providing the zero does not deteriorate the compensation efforts. Since the zero is on the negative real axis, it could be incorporated into the compensation scheme using the concepts of lead compensation.[2]

Another means of eliminating the effect of the right-half plane zero of Eq. (29.37) is to insert a nulling resistor in series with C_c. Figure 29.10*b* can be used as an equivalent model for the resistor nulling method. This circuit has the following node voltage equations:

$$g_{m1}V_{in} + \frac{V_2}{R_1} + V_2 sC_1 + (V_2 - V_{out})\left(\frac{sC_c}{1 + sC_cR_z}\right) = 0 \qquad (29.54)$$

$$g_{m2}V_2 + \frac{V_{out}}{R_{out}} + (V_{out}sC_2 + (V_{out} - V_2)\left(\frac{sC_c}{1 + sC_cR_z}\right) = 0 \qquad (29.55)$$

These equations can be solved to give

$$\frac{V_{out}(s)}{V_{in}(s)} = \frac{a[1 - s(C_c/g_{m2} - R_zC_c)]}{1 + bs + cs^2 + ds^3} \qquad (29.56)$$

where

$$a = g_{m1}g_{m2}R_1R_2 \qquad (29.57)$$

$$b = (C_2 + C_c)R_2 + (C_1 + C_c)R_1 + g_{m2}R_1R_2C_c + R_zC_c \qquad (29.58)$$

$$c = R_1R_2(C_1C_2 + C_cC_1 + C_cC_2) + R_zC_c(R_1C_1 + R_2C_2) \qquad (29.59)$$

$$d = R_1R_2R_zC_1C_2C_c \qquad (29.60)$$

In practice, we assume that with R_1, $R_2 \gg R_z$ the poles will be widely spaced, and then we may approximate the roots as

$$p_1 \approx \frac{-1}{(1 + g_{m2}R_2)R_1C_c} \qquad (29.61)$$

$$p_2 \approx \frac{-g_{m2}C_c}{C_1C_2 + C_cC_1 + C_cC_2} \qquad (29.62)$$

$$z = \frac{1}{C_c(1/g_{m2} - R_z)} \qquad (29.63)$$

If we adjust the circuit so that $R_z = 1/g_{m2}$, then the zero is at infinity and it eliminates its influence on the circuit. One may wish to make R_z slightly larger than $1/g_{m2}$ to take advantage of lead compensation. Most often R_z is realized by MOS transistors; hence Eq. (29.38) will be satisfied over a wider range of operating conditions.

29.3.10 *n*-Channel CMOS Op Amp

An unbuffered CMOS op amp is shown in Fig. 29.11. The bias is provided by R_b and M_6. The differential amplifier consists of M_1, M_2 and the loads M_3, M_4 with M_7 acting as a current source. Transistors M_5 and M_8 form the gain stage and C_c is the Miller compensating capacitor. The advantage of this architecture is its simplicity and its excellent expected performance in applications with small capacitive loads, e.g., switched capacitor filter circuits. The biasing network R_b and

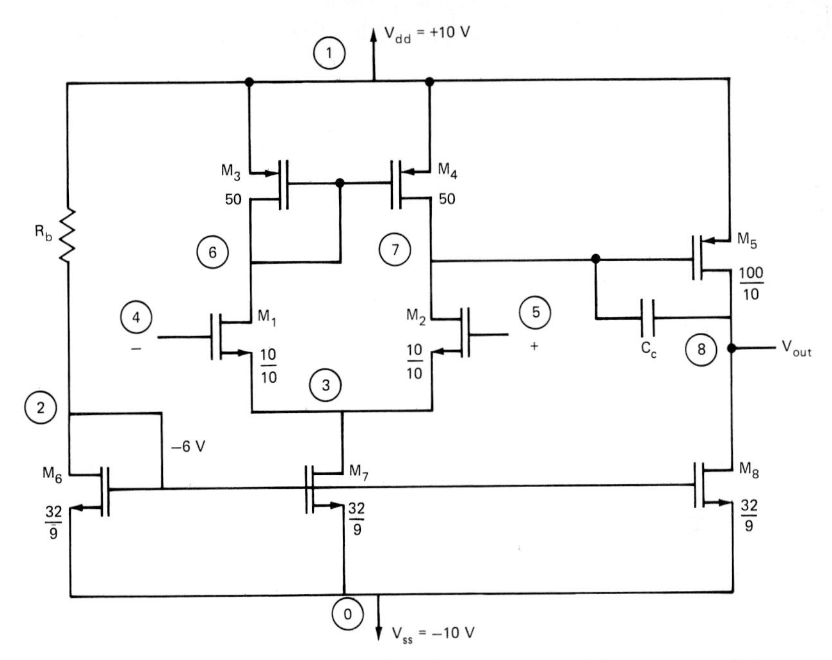

FIG. 29.11 Schematic of an unbuffered CMOS op amp. (For all devices: the bulk is connected to the source device; in the *W*/*L* ratios, *W* and *L* are given in micrometers.)

M_6 can be shared with other op amps. The first step in designing an op amp is to establish the dc considerations.

Example 29.7. Assume $V_{dd} = 10$ V and $V_{ss} = -10$ V, and since $|V_T| = 1$ V, choose the voltage at node 2 in Fig. 29.11 as -6 V and let $K_n' = 10\mu\text{A/V}^2$. This will provide a large enough current for moderately sized devices. Selecting R_b as 100 kΩ gives $I_6 = 160$ μA. For simplicity, we define S_i as the W_i/L_i ratio of the *i*th device. Now $S_6 = W_6/L_6$ is found from

$$S_6 = \frac{W_6}{L_6} = \frac{2I_6}{K_n'(V_{gs6} - V_{Tn6})^2} = \frac{32}{9} \tag{29.64}$$

Thus, select $S_6 = (32\ \mu\text{m})/(9\ \mu\text{m})$. In order to apply a current of 160 μA to the differential *n*-channel stage, select $S_7 = S_6$. Since the gate source voltages of M_8 and M_6 are equal, we have

$$S_8 = S_6 \frac{I_8}{I_6} \tag{29.65}$$

For M_4 to remain in saturation,

$$V_{gs5} = V_{gs3} \qquad \text{or} \qquad S_3 = S_5 \frac{I_3}{I_5} \tag{29.66}$$

Choosing $S_3 = S_4$ and $S_1 = S_2$ and since $I_5 = I_8$, Eq. (29.66) becomes

$$S_4 = S_5 \frac{I_4}{I_8}$$

but with

$$I_4 = \frac{1}{2} I_7 = \frac{1}{2} I_6$$

From Eq. (29.65), the condition for M_4 to remain in saturation becomes

$$S_5 = 2 \frac{S_4 S_8}{S_7} \tag{29.67}$$

The differential amplifier gain can be expressed as

$$A_d = \frac{\sqrt{K_n' S_1}}{\sqrt{I_1 \lambda^2}} \tag{29.68}$$

where $\lambda_n = \lambda_p = 0.01$ (all channel length modulation parameters are assumed equal since all lengths are about equal to 10 or 12 μm), $K_n' = 10\ \mu A/V^2$, and $I_1 = 80\ \mu A$. Therefore,

$$A_d = 35.36\sqrt{S_1} \tag{29.69}$$

The inverter gain A_I is given by

$$A_I = \frac{\sqrt{K_p' S_5}}{\sqrt{I_5 \lambda^2}} \tag{29.70}$$

where $K_p' = 5\ \mu A/V^2$, $I_5 = 160\ \mu A$, and $\lambda = 0.01$. Therefore,

$$A_I = 17.67\sqrt{S_5} \tag{29.71}$$

Choosing $S_1 = 1 = 10$ μm/10 μm, and $S_5 = 10 = 100$ μm/10 μm, we find $A_d = 35.36$ and $A_I = 55.88$. From Eq. (29.67), $S_3 = S_4 = 5$. Let $S_3 = S_4 = 50$ μm/10 μm. The overall gain becomes $A_d A_I = 1976$, and the output resistance is $= 1/2\lambda I_8 = 312$ kΩ. This resistance is large as expected for unbuffered op amps.

The ac considerations are the next step in the op amp design procedure. The transconductance of the first stage is given by

$$g_{m1} = \sqrt{2 S_1 K_n' I_1} = 40\ \mu s \tag{29.72}$$

and the second stage is

$$g_{m2} = \sqrt{2 S_5 K_p' I_5} = 126\ \mu s \tag{29.73}$$

If a minimum of 1.2-MHz unity-gain bandwidth is required, then from $\omega_\mu = g_m/C_c$, C_c is calculated to be 5.3 pF. Due to signal feedforward through C_c at high frequencies, a right half-plane zero appears at

$$\omega_0 = \frac{g_{m2}}{C_c} \tag{29.74}$$

and for $C_c = 5.3$ pF, the zero is at 3.8 MHz, and another pole is located approximately at

$$\omega_{p2} = \frac{g_{m2}}{C_L} \tag{29.75}$$

where C_L is the capacitive load at node 8. From Eqs. (29.74) and (29.75) it is seen that in order to minimize any resulting phase degradation, the g_{m2} stage should be designed to be larger than g_{m1}.

One way to reduce the effects of the right half-plane zero is by connecting a nulling resistance R_z in series with C_c. In this case, Eq. (29.50) gives the zero location as

$$\omega_0' = \frac{1}{C_c(1/g_{m2} - R_z)} \tag{29.76}$$

By proper selection of R_z, this zero can be moved away from the unity-gain bandwidth or even to the left half-plane where it would improve the phase margin of the amplifier.

Example 29.8. Let $g_{m2} = 126$ μs or $1/g_{m2} = 7.9$ kΩ. Then $R_z \approx 7.9$–8 kΩ, and the zero will be in the left half-plane with improvement in phase margin.

It is also possible to improve the ac performance of the CMOS op amp and hence its settling time by using external devices in series with the output of the op amp. These devices act as ON switches (pass transistors) and reduce the effective capacitive load seen at node 8.

The slew rate of the op amp in Example 29.7 is given by Eq. (29.25) as I_1/C_c or 16 V/μs. This slew rate limitation only occurs for input voltage changes greater than $|V_T|$. The use of an external biasing resistance for Fig. 29.11 allows the adjustment of the op amp performance.

29.3.11 *p*-Channel CMOS Op Amp

Another example of an unbuffered CMOS op amp uses a *p*-channel input pair. In bulk CMOS technology using an *n*-substrate, the substrates of the *p*-channel devices must be taken to V_{dd} so that the body effect can be minimized. The main issue of the body effect will be in limiting the negative potential swing of the common-mode input voltage range.

A design of this type of op amp follows very closely the design of the *n*-channel op amp of Fig. 29.11. M_1 and M_2 are the differential input pairs, M_3 and M_4 are the differential load, and M_7 is the current source of this stage. M_5 is the inverting amplifier with M_8 acting as its load. The C_c is the Miller compensation capacitor that is connected across the gain stage.

Example 29.9. At the first start, a dc design will determine the required gain. The transconductance of the second stage should be designed larger than the first stage in order to better drive the capacitive loads. Let $|V_T| = 1$ V and choose the voltage at node 2 to be 7 V. Let $V_{dd} = +10$ V and $V_{ss} = -10$ V. This will make $I_6 = 170$ μA for $R_b = 100$ kΩ. $S_6 = W_6/L_6$ is found from

$$\frac{W_6}{L_6} = \frac{2I_6}{K_p'(V_{gs6} - V_{T6})^2} = 1.7 \tag{29.77}$$

where $V_{gs6} = 3$ V and $k_p' = 5$ μA/V². Select $S_6 = 17$ μm/10 μm. For $I_7 = I_6$, select $S_7 = S_6 = 17$ μm/10 μm. Thus, $I_1 = I_2 = I_7/2 = 85$ μA. Since $V_{gs8} = V_{gs7}$, we have

$$S_8 = S_7 \frac{I_8}{I_7} \tag{29.78}$$

For $I_8 = 200$ μA, S_8 is found to be 20 μm/10 μm. The condition for M_4 to remain in saturation is

$$V_{gs5} = V_{gs3} \quad \text{or} \quad S_3 = S_5 \frac{I_3}{I_5} \tag{29.79}$$

Also select $S_1 = S_2 = 20/10$, $S_3 = S_4$, and $I_5 = I_8$; letting $S_5 = 100/10$, the above equation becomes $S_4 = S_5(I_4/I_8)$. However, $I_4 = \frac{1}{2}I_7$, and another condition for M_4 to remain in saturation requires an aspect ratio

$$S_4 = \frac{S_5 S_7}{2S_8} \quad \text{or} \quad \frac{S_4}{S_5} = \frac{1}{2}\frac{S_7}{S_8} \tag{29.80}$$

The differential amplifier gain is

$$A_d = \sqrt{\frac{K_p' S_1}{I_1 \lambda^2}} \tag{29.81}$$

where $\lambda_n = \lambda_p = 0.01$ (all channel length modulation parameters are assumed equal, since all lengths are 10 μm), $K_p' = 5$ μA/V², and $I_1 = 85$ μA. Therefore,

$$A_d = 24.25\sqrt{S_1} \tag{29.82}$$

The inverter gain A_I is given by

$$A_I = \sqrt{\frac{K_p' S_5}{I_5 \lambda^2}} \tag{29.83}$$

For $K_n' = 10$ μA/V², $I_5 = 100$ μA, the above equation becomes

$$A_I = 31.6\sqrt{S_5} \tag{29.84}$$

Selecting the aspect ratios as $S_1 = 2$ and $S_5 = 10$, we find $A_d = 34.3$ and $A_I = 100$. From Eq. (29.80), $S_4 = S_3$ is found to be 42 μm/10 μm. It is designed for the inverting stage to have the highest gain. The overall gain becomes $A_d A_I = 3430$ (81.6 dB). Based on the above ratio selection, the transconductances are

$$g_{m1} = \sqrt{2S_1 K_p' I_1} = 41.2\ \mu\text{s} \tag{29.85}$$

and

$$g_{m2} = \sqrt{2S_5 K_n' I_5} = 200\ \mu\text{s} \tag{29.86}$$

Assuming that the designed unity-gain bandwidth f_u is 1.2 MHz, then from $\omega_\mu = g_m/C_c$, $C_c = 5.5$ pF. Using a nulling resistor R_z in series with C_c will move the right half-plane zero to the left half-plane, thus improving the phase margin of the amplifier. The slew rate of this op amp from Eq. (29.25) is SR = 15 V/μS.

Another performance consideration is the power supply rejection ratio. The ratio PSRR can be defined to be

$$\text{PSRR}^+ = \frac{[V_{\text{out}}(\omega)/V_{\text{in}}(\omega)]V_s^+(\omega) = 0}{[V_{\text{out}}(\omega)/V_s^+(\omega)]V_i(\omega) = 0} \tag{29.87}$$

and

$$\text{PSRR}^- = \frac{[V_{\text{out}}(\omega)/V_{\text{in}}(\omega)]V_s^-(\omega) = 0}{[V_{\text{out}}(\omega)/V_s^-(\omega)]V_{\text{in}}(\omega) = 0} \tag{29.88}$$

Since the numerator of the above two equations is the compensated open-loop frequency response of the op amp, one must calculate the denominator gains in order to predict PSRRs.

$V_s^+(\omega)$ or $V_x^{-(\omega)}$ will cause ripple in the current sources in the op amp. The differential input stage will reject this common-mode input signal, since $V_{id} = 0$. The primary source of supply ripple is from the current source of M_5. Assuming that $V_{gs5}(\omega)$ 0 and that $I_8(\omega)$ is approximately calculated from Example 29.9, we get

$$I_8(\omega) = \frac{V_s^+(\omega) - V_s^-(\omega)}{R_b}\left[\frac{V_{SS} - V_{dd}}{(V_{dd} - V_{gs}) - V_{SS}}\right] \tag{29.89}$$

and the output voltage is

$$V_{\text{out}}(\omega) \frac{r_{ds5} \parallel r_{ds8}}{R_b} [V_s^+(\omega) - V_s^-(\omega)] \frac{V_{SS} - V_{dd}}{(V_{dd} - V_{gs}) - V_{SS}} \tag{29.90}$$

For the values from the above examples we get

$$V_{\text{out}}(\omega) \approx \frac{6}{2} [V_s^+(\omega) - V_s^-(\omega)] \tag{29.91}$$

We can see that the PSRR magnitude curve will have the shape of the open-loop gain $|A_v(\omega)|$, but it will be reduced by a factor of ⅙.

ACKNOWLEDGMENTS

Thanks are due to my graduate students who read this manuscript carefully and suggested improvements. In particular, I would like to thank Aung Thet Tu for typing the manuscript through several revisions with great patience. Thanks are also due to my wife, Darinka, who gave me great support during the preparation of this chapter.

REFERENCES

1. Y. P. Tsividis, "Design Considerations in Single-Channel MOS Analog Integrated Circuits—A Tutorial," *IEEE J. Solid State Circuits,* **SC-13**(3):383–391 (1978).
2. J. K. Roberge, *Operational Amplifier—Theory and Practice,* Wiley, New York, 1975, Chaps. 5 and 11.
3. P. R. Gray and R. G. Meyer, *Analysis and Design of Analog Integrated Circuits,* 2d ed, Wiley, New York, 1984.
4. J. G. Graeme, G. E. Tobey, and L. P. Huelsman, *Operational Amplifiers—Design and Applications,* McGraw-Hill, New York, 1971.
5. J. S. Brugler, "Silicon Transistor Biasing for Linear Collector Current Temperature Dependence," *IEEE J. Solid-State Circuits,* **SC-2**:57–58 (1967).
6. R. Gregorian and G. C. Temes, *Analog MOS Integrated Circuits for Signal Processing,* Wiley, New York, 1986.
7. Y. Tsividis and P. Antognetti, *Design of MOS VLSI Circuits for Telecommunications,* Prentice-Hall, Englewood Cliffs, N.J., 1985.
8. P. M. Chirlian, *Analysis and Design of Integrated Electronic Circuits,* Harper & Row, New York, 1987.
9. A. S. Sedra and K. C. Smith, *Microelectronic Circuits,* CBS College, New York, 1982.
10. J. Millman and A. Grabel, *Microelectronics,* McGraw-Hill, New York, 1987.
11. A. B. Grebene, *Bipolar & MOS Analog Integrated Circuit Design,* Wiley, New York, 1984.
12. S. Soclof, *Analog Integrated Circuits,* Prentice-Hall, Englewood Cliffs, N.J., 1985.

P • A • R • T • 11

SPECIAL TOPICS

CHAPTER 30
PHYSICAL DESIGN LIMITATIONS OF VLSI ECL ARRAYS

Andrew T. Jennings
Unisys Corporation
Paoli, Pennsylvania

The cost-effective use of VLSI ECL gate arrays requires that tradeoffs be made between circuit speed, power dissipation, packaging technology, and system architecture. The physical limitations and performance characteristics of these devices and some aspects of logic partitioning are discussed.

30.1 INTRODUCTION

High-performance processors are commonly constructed using semicustom bipolar devices. In this chapter we will examine some of the physical aspects of ECL VLSI gate arrays and the associated packaging with which the architects and logic designers must be familiar prior to beginning a new design. Our focus is on the effect, rather than the cause, of the lower-level physical phenomena that are described elsewhere in this and other reference works.

Designing a computer system is a complex task requiring expertise in many diverse disciplines. Some of the major areas of technical expertise are electronic circuits design, packaging, system architecture, and logic design. Prior to the development of LSI-VLSI, these disciplines could operate largely independently. Those involved in the architectural design needed little or no understanding of the components from which the system was to be constructed; the circuits and packaging engineers were largely unconcerned with the architecture of the system under design. With the introduction of VLSI devices, and the parallel development of more advanced packaging techniques, the range of options available to those involved in all aspects of the design has greatly expanded. The architect may now devise and employ strategies that would have been considered impractical only a few years ago. Similarly, given a reasonable understanding of the architectural elements of the system, the packaging engineer can adjust these aspects of the design to more closely support the higher-level concepts.

The development, design, test, and manufacture of VLSI devices and their associated packaging is an extremely, and increasingly, costly exercise. The failure on the part of the architects and logic designers to fully appreciate the limitations and performance characteristics of the technology being used is likely to lead to a system that meets neither the performance nor the cost goals that were placed upon it.

30.2 TECHNOLOGY LIMITATIONS AND TRENDS

The design process from conception to physical hardware requires many years to complete. It is safe to assume that in the intervening years, technological advances are likely to be made. Therefore, the initial stages of any design have to be based on a set of assumptions as to the state of the technology at the time that the system is to be actually realized in physical hardware. In this section, we will describe some of the current limitations of ECL technology and project what may be expected of the next generation of gate arrays.

Die Sizes and Gate Densities. The yields obtainable from the equipment on which the dice are fabricated limit the current maximum die size to approximately 400 × 400 mil. The fabrication of a die any larger than this on existing equipment would result in unacceptably low yields. The gate densities on such dice depend on the feature size. The feature size on present arrays is typically in the range 1.5 to 2.0 μm, yielding densities of approximately 10,000 gates per die. The introduction, in the future, of improved equipment capable of handling larger wafers can be expected to permit the maximum die size that can be produced at acceptable yields to increase to at least 600 × 600 mil. This increase in die size, along with a reduction in the feature size to below 1 μm, can be expected to produce gate densities in the range of 30,000 to 50,000 gates in the early 1990s. Figure 30.1 shows the basic trend in gate densities, and Fig. 30.2 shows the corresponding trend in unloaded gate speeds.

Metal Pitch and Layers. The metal interconnect between the IC devices and resistors on the chip may be defined by the linewidth plus the spacing between the lines (the metal pitch) and the number of layers. For a given number of gates there must be an adequate availability of interconnect; otherwise, it is impossible to place and route the chip.

The metal pitch tends to decrease in proportion to the reduction in feature size. Present technologies support three interconnection layers—two signal layers and a power layer—with metal pitches of 3, 6, and 15 μm, respectively: the smaller the metal pitch and the larger the number of layers, the greater the availability of routing channels to support higher-density interconnect. Currently under development are metal pitches of 2, 3, 4, and 15 μm, using four layers of metal. The introduction of this technology will significantly increase the routing channels on high-density dice, permitting gate densities of 40,000 to 50,000 gates per die to be supported.

I/O Pins. As the gate density rises, there is a corresponding increase in the demand for I/O pins to support the logic contained on the die. The number of I/O pins that may be connected to a chip is limited by the die size and the pad bonding technology used.

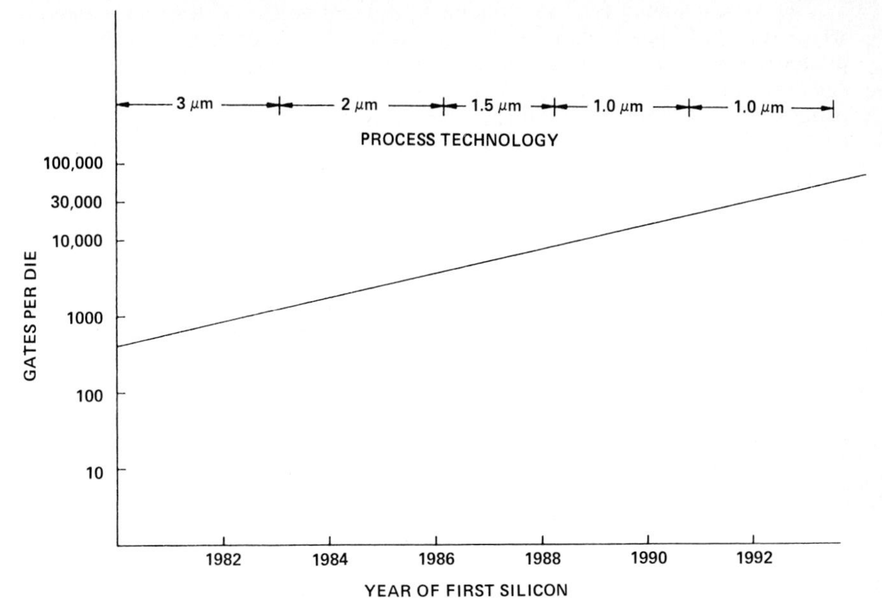

FIG. 30.1 Gates per die trend.

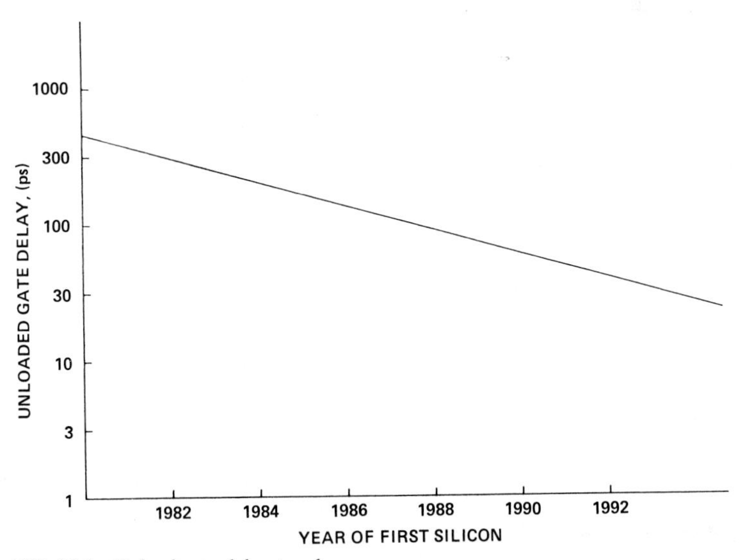

FIG. 30.2 Unload gate delay trend.

The most common form of bonding is through the use of peripheral pads, in which the I/O pads are placed on the perimeter of the die. If wire-bonding techniques are used, the bonding pads are typically 4 mils wide and placed on 6-mil centers. Equipment that is used currently limits the bipolar die size to approximately 400 × 400 mils, restricting the number of bonding pads per side to 65.

With the use of tape automated bonding (TAB) techniques, the bonding pad can be reduced to 2 mils in width and placed on 5-mil centers, allowing for up to 80 pads per side. Future developments can be expected to handle die sizes of at least 600 × 600 mil in area and a reduced pad pitch of 4 mil. Therefore with TAB bonding, the number of pads per side will approach 150, allowing for 600 pins on a die containing 50,000 gates. However, in order to meet the high current demands and provide low-inductance ground connections, 25 to 30 percent of these pins will be required for power.

An alternative to peripheral pads is the use of area array pads. This technique is sometimes referred to as *flip-chip technology*. In this case the pads are placed on the face of the die rather than on the perimeter. This technology permits a greater number of I/O pins to be interfaced to a die of a given size than is possible

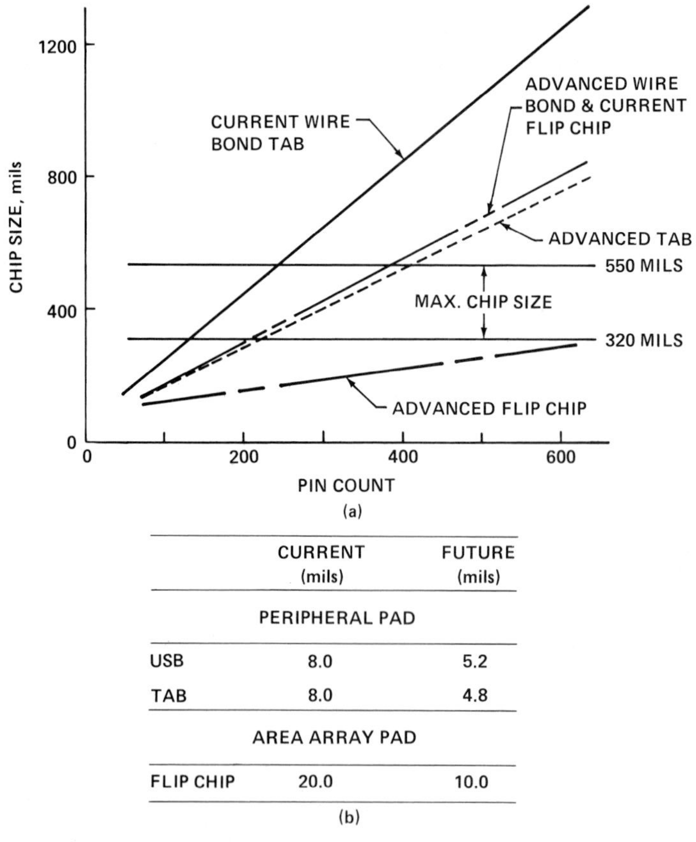

	CURRENT (mils)	FUTURE (mils)
	PERIPHERAL PAD	
USB	8.0	5.2
TAB	8.0	4.8
	AREA ARRAY PAD	
FLIP CHIP	20.0	10.0

(b)

FIG. 30.3 High pin count VLSI connection. (*a*) Chip size versus pin count; (*b*) pad pitch.

with the use of peripheral pads. The pad pitch is currently 10 mils. Even though this approach has the potential to accommodate a high number of I/O pins, problems with stress in the solder joints due to differential expansion increase as the die size increases. Figure 30.3 shows the number of I/O pins that can be interfaced to a die of a given size using these bonding techniques.

30.3 THERMAL CONSIDERATIONS

The system thermal design is a major factor, and in many cases the limiting factor, in the implementation of a VLSI system. The task of interconnecting and cooling a large number of high-pin-count, high-power VLSI chips is formidable. Junction temperatures must not exceed 100 to 120°C under worst-case conditions to provide proper functionality and typically should not exceed 60°C to satisfy system reliability requirements. The system designers must trade off gates per chip, speed per gate, and power per gate against the thermal and density constraints of the packaging system.

Advances in semiconductor technology have produced smaller device geometries with an improved speed-power product. Figure 30.4 shows the relationship between gate power and unloaded gate delay. On a very large VLSI device the

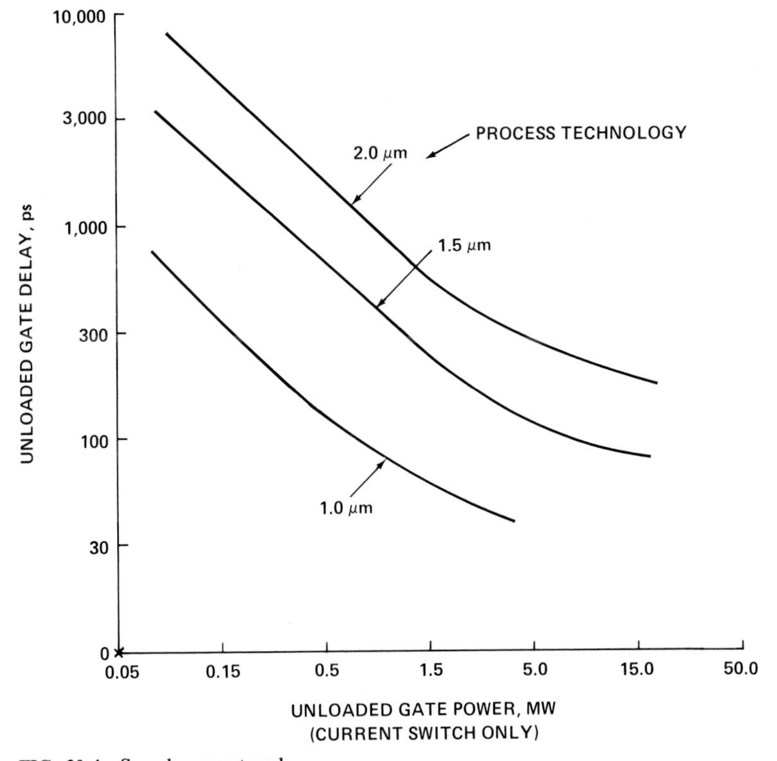

FIG. 30.4 Speed-power trend.

internal cell interconnection represents a significant delay factor. To minimize this interconnect delay, one must use a high-current, high-power emitter follower to drive the capacitive load of the interconnecting foil and the intrinsic capacitance of the gate input. These high-current emitter followers require a large junction size, thus reducing the overall gate density for a given chip size. For a defined chip size and chip power, the number of gates per chip can be traded off, within limits, against the loaded gate delay. For example, similar size chips fabricated with a 2-μm process and a 15-W dissipation provide 10,000 gates with a loaded gate delay of 300 ps, while a 7000-gate implementation, using higher-current internal drivers, may have a loaded gate delay of only 220 ps. The next generation of VLSI chips will provide 30,000 to 50,000 gates with loaded gate delays in the range of 110 ps at a power dissipation of 40 to 50 W.

The sophistication of the thermal design required to cool a high-performance processor is related to the power per unit area dissipated by the logic devices. Earlier systems fabricated with dual in-line packages (DIPs) had power densities in the range of 0.3 W/in^2 and were readily cooled by serial airflow. Higher-power single-chip packages, with power densities in the range of 0.6 W/in^2 required the addition of heat sinks on the logic package and specialized air-cooling systems. But where unpackaged dice are mounted in close proximity in a multichip module to provide high system performance, the power density normally exceeds the limits of air cooling. Existing multichip modules have power densities in the range of 2.2 W/in^2 and are liquid cooled. The power dissipation of the logic devices and the planar packaging densities must be considered in the system design. Present cooling systems could not accommodate a dense multichip packaging scheme that utilized large, high-power VLSI devices. For example, the highest-performing, liquid-cooled, multichip module currently in production (see Fig. 30.5) provides a thermal resistance of 5°C/W on an individual chip basis. Considering the junction temperature should not exceed 60°C for long-term reliability, and assuming a typical coolant fluid temperature of 20°C, the logic chip power dissipation in a multichip module is restricted to 8 W.

Work is in progress in the industry to design cooling systems for high-power, high-density modules. The performance and availability of such packaging systems must clearly be considered in the early stages of a VLSI system design.

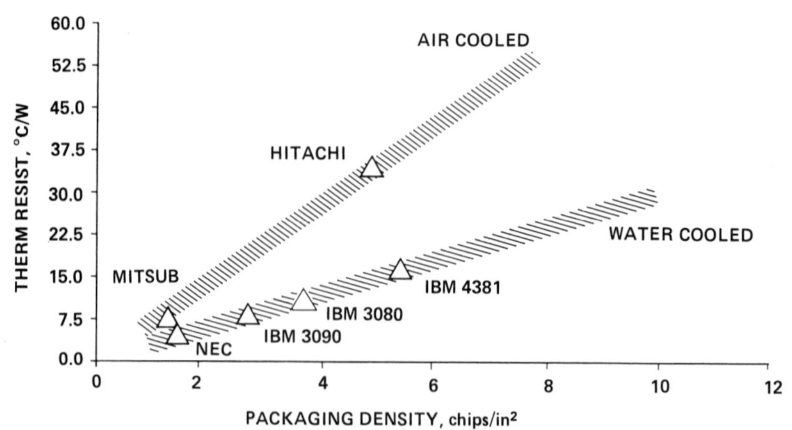

FIG. 30.5 Technology parameters for multichip modules.

30.4 PERFORMANCE

In this section we will describe a basic methodology that may be used in a performance evaluation study aimed at providing an estimate of the achievable frequency for the technology under consideration. We also describe the major parameters that must be quantified.

The performance of a large commercial processor is commonly rated in millions of instructions per second (MIPS). This measure may be expressed in terms of the clock frequency and the average number of instructions completed per clock, such that

$$\text{MIPS} = (\text{completed instructions per clock}) \times \text{frequency}$$

From this relationship, it is apparent that the required performance may be derived from two sources: the architecture, which defines the instruction completion rate, and the speed of the technology, which determines the clock frequency. Assuming that the performance goals of the system under development are fixed, the higher the clock frequency, the less sophisticated the architecture need be, and vice versa.

One of the earliest decisions that must be made in the design cycle of most systems is the frequency at which the logic is to be designed to operate. The performance characteristics of VLSI devices are usually stated in terms of the intrinsic, unloaded gate delays at a given power level. With the development of smaller device geometries, such statistics can be extremely misleading, since the contribution of the raw gate speeds to the overall performance of the technology is of increasingly minor significance.

Correct assessment of the performance of the technology requires an evaluation of all elements that contribute to the overall logic delay. Such an analysis is performed by considering the total delay incurred in a "register-to-register" path. These are the logic paths that both start and end at clocked storage devices, such as flip-flops.

The total delay incurred in a register-to-register path may be broken down into three major components:

1. The on-chip loaded gate delay and foil delay.
2. The delays incurred by the interconnect between the chips.
3. The delays incurred by the interconnect between the carriers on which the chips are mounted. These carriers are typically printed circuit boards or, increasingly frequently, multichip modules.

The total path delay may be defined as

$$Q_0D_0 + Q_1D_1 + Q_2D_2$$

where Q_0 = number of gate levels in typical path
D_0 = loaded gate delay plus the foil interconnect delay
Q_1 = number of chip crossings in the path
D_1 = chip crossing delay
Q_2 = number of board or module crossings
D_2 = board or module crossing delay

In the following sections we examine each of these elements in turn. Throughout this discussion we will assume that the point on the speed-power curve that the technology is to operate is fixed.

30.4.1 On-Chip Delays

On-chip delays are calculated in terms of the number of gate levels in the logic path and the average delay per level. This delay is a function of the loaded gate delay and the length of the foil interconnect between the gate levels.

Loaded Gate Delays. The loaded gate delay is influenced by the unloaded gate speed, the drive capability of the circuit, the number of loads, or fan-out, on the net, and foil length. The typical gate delay rises as the number of loads is increased and the foil length being driven increases. Figures 30.6 and 30.7 show the effect of these factors, for constant emitter follower and circuit switch currents, on the gate delay.

Interconnect. The foil length is a function of the number of elements to be connected, their pitch, the average fan-out, and the availability of routing channels with which to make the connections. The effect of these variables on the average interconnect length has been studied by a number of individuals.[1,2] These studies have led to the development of several different ways of predicting the foil length. Once the number of elements n, their pitch p, and the fan-out f are known, the following formula may be used to predict the average net length:

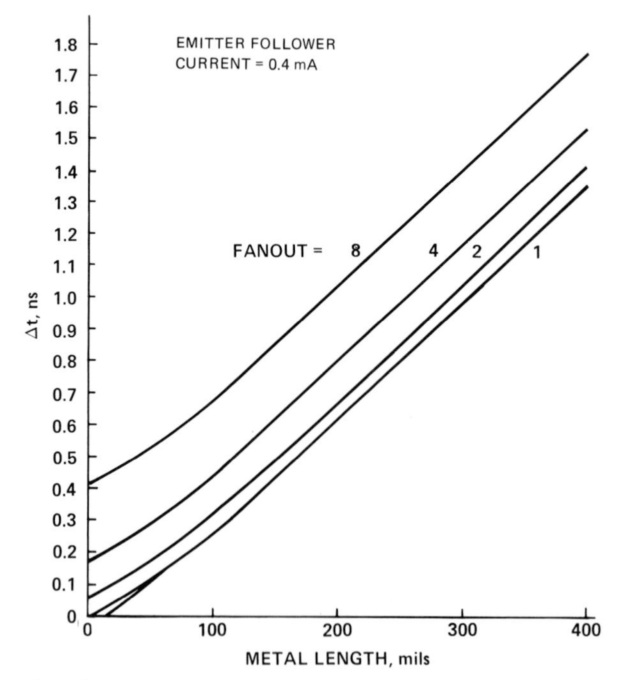

FIG. 30.6

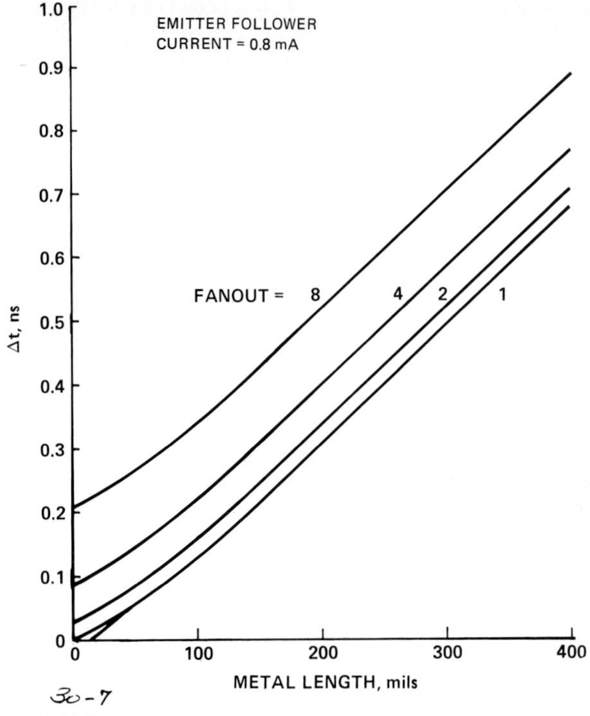

FIG. 30.7

$$\text{Length} = \frac{2}{9} pA(B - C)(0.2 + 0.8f)$$

where

$$A = \frac{1 - 4^{a-1}}{1 - n^{a-1}}$$

$$B = \frac{7(n^{a-0.5} - 1)}{4^{a-0.5} - 1}$$

$$C = \frac{1 - n^{a-1.5}}{1 - 4^{a-1.5}}$$

and the value of a, derived from the exponent in Rent's rule, is typically 0.65.

Empirical formulas of this nature are based upon the analysis of the net lengths found in existing designs, the majority of which are constructed from SSI-MSI and relatively low density LSI components. They should therefore be used with caution. Analysis of the actual foil lengths in LSI arrays indicate that studies based upon pre-LSI data tend to overestimate the foil lengths. For example, the interconnect delay in an aluminum conductor when the foil pitch is 3 μm is approximately 200 ps/in. Given the current gate densities of 10,000 gates/chip on a die size of 400 × 400 mil, the average foil length can be estimated at 50 mils/load. The data shown in Fig. 30.6 were measured from a chip that has an unloaded gate delay of 120 ps. It shows the additional loading penalty as function the number of fan-outs and the foil length. The on-chip fan-out is typically in the range of 2.5 to 3 loads per net. Given these factors, the loaded gate delay is approximately 350 ps, a degradation factor of nearly 3 to 1.

30.4.2 Package Crossing Delays

To assess the total delay incurred by package crossings, we must first establish the number of package crossings, the number of loads on each net, and the length and type of interconnect between the packages at each level of packaging. In addition, at the chip level, it is necessary to consider the penalties incurred in the output drivers.

Number of Package Crossings. The gate density that a given level of packaging can support, whether it is a chip, a printed circuit board, or a multichip module, affects the logic delays by influencing the number of times each level of packaging is likely to be crossed in a typical logic path.[3] Figure 30.8 shows the number of package crossings in a typical logic path as a function of the gates per package. As can be seen, the number of crossings may be reduced by increasing the packaging circuit density, provided that the logic can be partitioned in such a way that it can tolerate the changed relationship between gates and I/O pins.

Figure 30.8 is based upon data obtained from existing designs, which use numerous packages. However, the number of crossings is also a function of the total number of packages. For example, if the entire design is contained within a single package, the number of crossings will be zero regardless of the number of

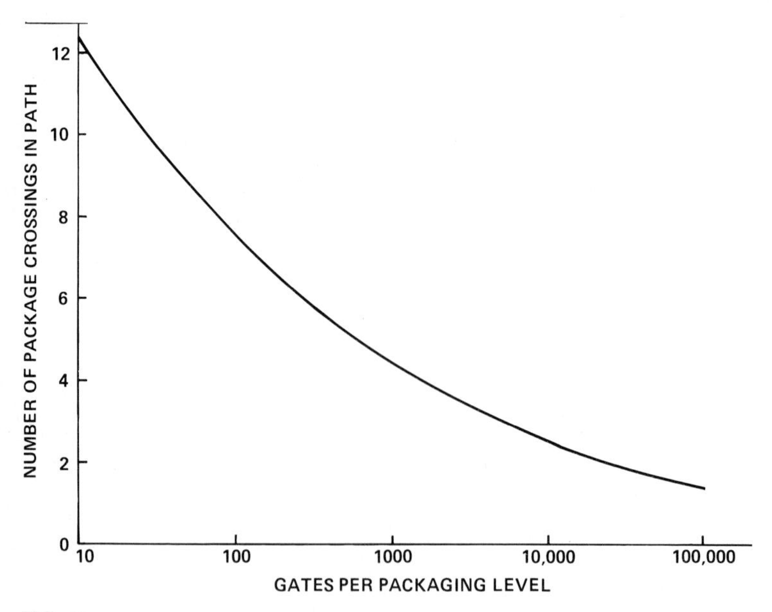

FIG. 30.8

gates that the package contains. It is not well understood, at present, how quickly the number of crossings approaches zero as the number of packages approaches unity.

Number of Loads. The number of loads on a net also decreases as the number of gates per package increases. As shown in Fig. 30.9, the average number of loads approaches unity fairly rapidly.

Output Cell Delay. Additional path delay is incurred in the output cell used to drive the external connections. This delay stems from the need to provide a higher drive capability on the external connection than is required on chip and from the foil interconnect between the die and the external pin. This latter delay is directly related to the size of the package being used.

Faster internal gate speeds are also reflected in the switching speed of the output driver. In many cases these output edges are intentionally slowed by adding on-chip capacitance to reduce the off-chip packaging complexity. Faster output edges cause increased crosstalk in the interconnects, high-amplitude reflections from impedence discontinuities and simultaneous switching noise on the die. Slowing the output driver edges adds further delay to the logic paths.

Loading Delay. For each additional load on a net, a stub is created between the line and the die within the package. The effect of these stubs is additive, and the delay penalty incurred is determined by their lengths. Once again, this length is determined by the package size. If multisource nets are used, buss contention can occur, in which case the total interconnect delay will double.

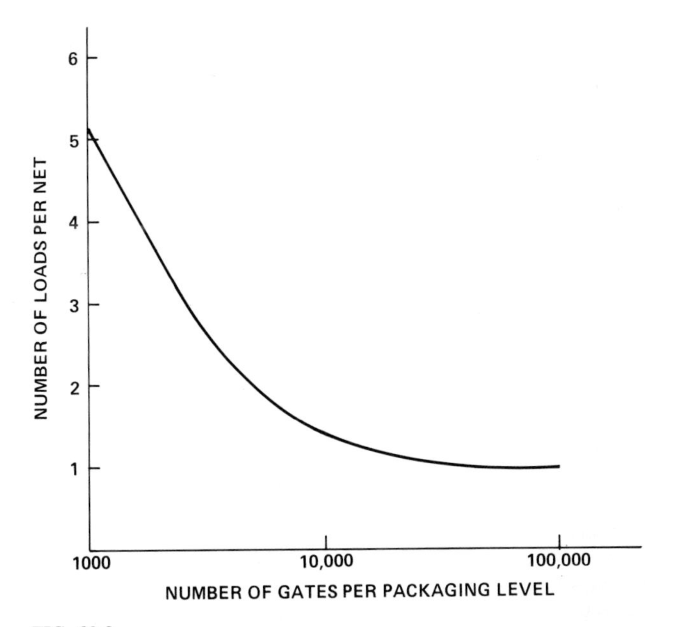

FIG. 30.9

Interconnect. The length of the interconnect between the chips can be estimated by using the formula in Sec. 30.4.1. The major difference is that the pitch of the elements to be interconnected is much greater.

Clock Skew. Unless the clock circuits are to be customized, the inherent variations in the processes used to fabricate the die, the packaging, and the interconnect will result in some degree of clock skew. Empirically, this typically amounts to 10 percent of the clock period.

30.4.3 Performance Analysis Examples

Considering all of the previously mentioned factors, it is now possible to make an estimate of the performance of the technology and the frequency at which it can be expected to operate. Assume that the processor under design is projected to require a total of 600,000 gates of logic, and the chip to be used is a 5000-gate array with an unloaded gate speed of 60 ps. The dice are to be packaged in pin-grid arrays, which in turn are to be mounted on a single printed circuit board at a pitch of 1.5 in. This will require that the board dimensions be approximately 20 × 24 in. The projected delay characteristics of such a system are summarized in Table 30.1. As can be seen, the unloaded gate delay contributes less than 15 percent to the overall delay, and the packaging and interchip interconnect delays amount to 75 percent of the total delay.

In Table 30.2 the results of a similar analysis are summarized. Exactly the same processor design is to be implemented by using a different packaging technology. In this case the same array as used in the previous example is to be mounted in liquid-cooled, multichip modules. Each module contains 25 dice,

TABLE 30.1 Performance of Single-Chip Processor Packaging

Unloaded gate delay	60 ps
Loaded gate delay (D_0)	130 ps
Number of levels (Q_0	20
Total on-chip delay	2600 ps
Chip-chip delay	
Off-chip driver	450 ps
Package	600 ps
Fan-out loading	200 ps
Interconnect	1150 ps
Total (D_1)	2400 ps
Number of Crossings (Q_1)	2.5
Total chip-chip delay	6000 ps
Total	8600 ps
+ 10% clock skew	9460 ps
Relative delay contributions	
Unloaded gate	12.7%
On-chip loading and foil	14.8%
Packaging and clock skew	72.5%

TABLE 30.2 Performance of Multichip Processor Packaging

Unloaded gate delay	60 ps
Loaded gate delay (D_0)	130 ps
Number of levels (Q_0)	20
Total on-chip delay	2600 ps
Chip-chip delay	
Off-chip driver	450 ps
Package	200 ps
Fan-out loading	75 ps
Interconnect	450 ps
Total (D_1)	1170 ps
Number of crossings (Q_1)	2.5
Total chip-chip delay	2925 ps
Module-module delay	
Interconnect	1680 ps
Package	1000 ps
Total (D_2)	2680 ps
Number of crossings (Q_2)	0.7
Total module-module delay	1876 ps
Total	7401 ps
+ 10% clock skew	8141 ps
Relative delay contributions	
Unloaded gate	14.7%
On-chip loading and foil	17.2%
Packaging and clock skew	68.1%

mounted face down using TAB bonding. This packaging sharply reduces the interconnect and the chip crossing penalties incurred between die contained within each multichip module. However, despite the use of this packaging, the performance in terms of average system delay per gate improved by only 15 percent, and the contribution of the packaging and interconnect to this delay remained largely unchanged. While a 15 percent increase in frequency may be significant, it should be remembered that improvements in frequency rarely result in corresponding improvements in the overall performance of the system.

30.5 LOGIC PARTITIONING

Logic partitioning is the assignment of both the logic to the gate arrays and the gate arrays to the chip carriers, if more than one carrier is to be used. The skill with which the logic is partitioned into the physical components has a direct bearing on the cost of implementing the design for two major reasons:

1. As the density of the components has increased, the volume to the chip vendor has tended to decrease, leading to both a reduced learning curve and fewer components over which to amortize the cost of developing the new process.

There are signs that these factors may be causing the historic trend of decreasing gate cost with increasing chip density to be leveling off, if not actually reversing.

2. The drive for higher speeds and greater densities demands that increasingly sophisticated and expensive packaging and cooling systems be developed.

Therefore, there is more pressure now to produce efficient designs when VLSI components are used. The partitioning of the logic in a complex, high-performance processor is, at present, more of an art than a science. Most of the published works on this subject tend to deal with the partitioning and construction of processors of a highly specialized nature. Unfortunately, the majority of designers do not have such freedom, being constrained by specifications such as IBM's S/370 principles of operation, which closely define the functions to be implemented.

Rent's Rule. One of the main areas of focus in any discussion on the partitioning of logic into VLSI arrays is the ratio of signal pins to gates on the array. In 1960, from a study of the random logic in the IBM 1401 processor, E. Rent of IBM's Endicott Development Laboratory, devised the formula $P = cG^a$, where P is the number of external I/O pins required to support a partition of G logic gates. From Rent's study, the constant c was 4.8 and the exponent a was 0.67. For current and future ECL gate arrays using peripheral pads, the constant c has a value of approximately 1.8 and the exponent p has a value of approximately 0.5. These values suggest that there are insufficient I/O pins on the array to support the number of gates contained within them. However, Rent's rule was based upon the study of the I/O requirements of relatively small blocks of *random* control logic and should not be applied to large blocks of logic that implement a complete set of functions. For example, consider an entire processor to be a single functional logic block; its I/O requirements, consisting primarily of the interfaces to its main memory storages, are independent of the number of gates used to implement the processor itself. In this case, the I/O requirements are defined by the data paths, not by the complexity of the function being performed upon them. What prevents the implementation of the processor on a single chip is not the availability of I/O pins (most VLSI arrays have sufficient pins to handle such an interface) but the fact that there are insufficient gates on the chip with which to implement the function.

Functional Partitioning. Most designs start with a high-level block diagram that delineates the major, conceptual, functional blocks and the data paths that interconnect them. In the past, any attempt to deduce the hardware packaging boundaries from such a diagram would have been unsuccessful. For example, Fig. 30.10 is a high-level block diagram of the central processor module (CPM) used in the Unisys A15 series systems. The conceptual design consisted of the five functional blocks shown in the diagram. The hardware realization of this conceptual design requires 22 logic boards containing several hundred LSI gate arrays and many supporting SSI-MSI components. It is not possible to determine the hardware packaging boundaries without examining the detailed board schematics. The lack of correspondence between the functional blocks and the hardware building blocks occurs because none of the components from which the A15 CPM is constructed is large enough to completely contain any meaningful function, let alone those implied by the blocks shown in Fig. 30.10.

Subsequent advances in both chip and packaging technologies have now made

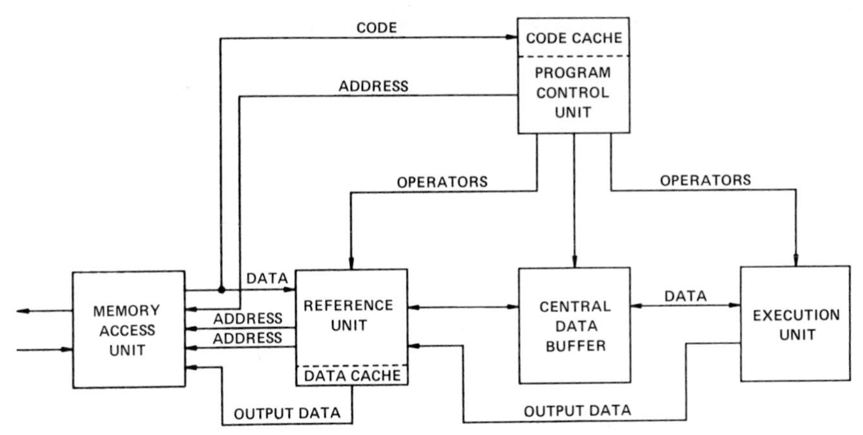

FIG. 30.10 Central processor module.

this untrue. It is now possible to restrict very large and complex functions to the confines of a single packaging entity. The key to successful logic partitioning lies in maintaining a strict relationship between the functional and physical partitioning at all levels of the design. Wherever this relationship is maintained, not only will Rent's rule not apply but the interconnect between the packages will be reduced and tend to be regular rather than random.

A natural consequence of attempting to force a direct correspondence between a functional block and a physical package is to concentrate the interconnection complexity at the lowest level of packaging, the die itself, where the wiring densities are greatest and the interconnect lengths are shortest. The methodology required to support such logic partitioning is summarized in the following rule:

> Never conceive of, or draw, a block diagram without first establishing a high degree of confidence that each block can be implemented within the confines of a single packaging entity.

If this rule were to be applied to Fig. 30.10, it would indicate that the architect responsible for conceiving this structure was confident that the chip carriers which were to be used to implement the final design had both sufficient I/O pins to support the interfaces that were drawn and could carry sufficient chips to implement the implied functionality of each block in its entirety. As was previously stated, it was not possible to satisfy this requirement at the time the A15 CPM was designed.

Changing Functional Partitioning. In Sec. 30.4.3 we compared the delay characteristics of a system using single-chip packages to one using multichip modules. Given the fact that the packaging delays account for more than half of the delays seen in a typical logic path, the improvement in performance was not as great as might have been anticipated.

To take full advantage of this change in packaging technology, designers must change the architecture and functional partitioning of the system such that it reflects the new physical boundaries. If the functional partitioning were to be ad-

TABLE 30.3 Performance of Multichip Processor Packaging with Functional Logic Partitioning

Unloaded gate delay	60 ps
Loaded gate delay (D_0)	130 ps
Number of levels (Q_0)	20
Total on-chip delay	2600 ps
Chip-chip delay	
Off-chip driver	450 ps
Package	200 ps
Fan-out loading	75 ps
Interconnect	450 ps
Total (D_1)	1170 ps
Number of crossings (Q_1)	2.5
Total chip-chip delay	2925 ps
Total	5525 ps
+ 10% clock skew	6078 ps
Relative delay contributions	
Unloaded gate	19.7%
On-chip loading and foil	23.0%
Packaging and clock skew	57.3%

justed to eliminate intermodule crossings in a typical logic path, the full potential of the technology would be realized. If such an architecture and implementation strategy were to be employed, the resulting delay characteristics of the system would be as shown in Table 30.3. Not only has the performance in terms of average delay per gate improved, but also the contribution of the packaging to the overall delay has decreased significantly.

The architecture required to obtain this optimal low-level partitioning is almost inevitably highly parallel. The MIP rate of the processor was previously defined in terms of the operator completion rate rather than the more traditional use of clocks per operator. Parallel structures can be used to increase the operator completion rate while maintaining a constant number of clocks per operator.

Logic Slicing. The A15 CPM was designed to use 1200 gate arrays with 56 I/O pins (signals only). The single most significant advance in the technology that has occurred since this design permits the functional and physical partitioning to merge is the availability of devices that have sufficient I/O pins to support complete data paths without requiring that they be sliced. For example, Fig. 30.11 represents a basic, chip-level block diagram of the A15 CPM's primary arithmetic adder. This adder performs basic functions such as floating-point and integer addition and comparisons. The complete function is implemented in 25 arrays and a substantial quantity of supporting SSI-MSI (for the sake of clarity, the SSI-MSI components are not shown in Fig. 30.11). This entire function can be implemented in a single 10,000-gate, 180 I/O-pin array. The twenty-five 1200-gate arrays are all pin bound. The single 10,000-gate array is not.

Sharing Logic. In this particular example, the reduction in both chips and logic used to implement the same function is not solely attributable to the fact that the

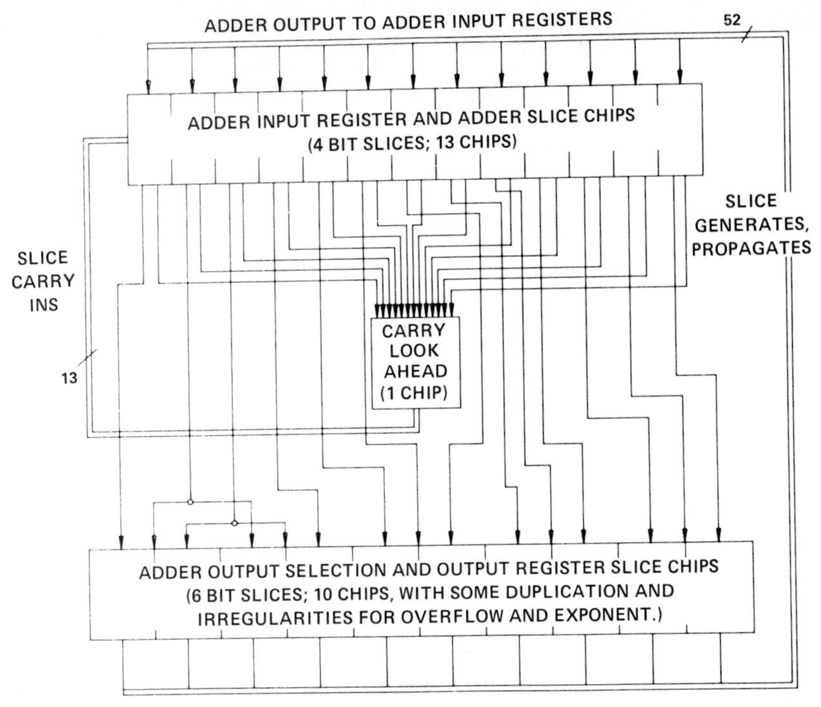

FIG. 30.11

data paths need no longer be sliced. The fact that using the LSI arrays requires a considerable investment in hardware led to sharing the same physical hardware with other functionally separate logic, in this particular example, with the logic used to perform multiplication. The sharing of logic in this manner further increases the random control logic required to support the function and, consequently, the I/O demands placed upon the arrays. When VLSI arrays that can support the full width of the data path are used, this form of logic sharing is not a wise strategy because it places the burden of the interconnect onto the packaging, resulting in both unacceptable delay and pin-limited arrays.

30.6 SUMMARY

We have examined some of the physical limitations of ECL gate arrays and how these limitations influence the system design. As the devices and packaging have become more dense, it has become increasingly important to adjust the functional partitioning of the design such that it matches the physical boundaries imposed by the technology. With the introduction of discrete, functional subblocks within an array, such as dense storage structures, the requirement to maintain a disciplined, functional partitioning to the design will be extended to considering the internal structure of the devices, in addition to the chip and packaging boundaries.

ACKNOWLEDGMENTS

I would like to thank my colleagues Bob Braun and Frank D'Ambra for their assistance and encouragement in preparing this paper.

REFERENCES

1. W. E. Donath, ''Placement and Average Interconnect Lengths for Computer Logic,'' *IEEE Trans. Circuits Systems,* **CAS-26**:272–276 (1979).
2. W. R. Heller et al., ''Wirability—Designing Wiring Space for Chips and Chip Packages,'' *IEEE Design and Test of Computers,* **1**(3):43–50 (1984).
3. D. Balderes and M. L. White, ''Package Effects on CPU Performance of Large Commercial Processors,'' *35th IEEE Elec. Comp. Conf.,* Washington, D.C., May 1985.

CHAPTER 31
VLSI IN COMPUTERS

R. W. Keyes
IBM T. J. Watson Research Center
Yorktown Heights, New York

The VLSI era began in 1970 with the introduction of the single-chip 1024-bit memory and the single-chip microprocessor, each containing well over 1000 transistors. Steady increases in the level of integration since have produced the well-known revolution in information processing. All information-handling system components have been affected: fast memory, low-cost memory, microprocessor chips, and gate arrays. Even disk files and displays have felt the impact of the techniques of microelectronics.

The great economic value of VLSI has given rise to a large investment in development of the technological base that will enable the past record of progress to persist well into the future. The innovation and invention that are needed to ensure the continuation of the historical trends is well supported; there is good reason to believe that progress will continue at fast rates, or even be accelerated by the growing economic base that depends on VLSI technologies.

This chapter examines the future of VLSI as it relates to large information-processing systems. The basic parameters of silicon technology are determined by extrapolation of history, tempered as appropriate by a view of what is physically feasible and of known inventions that have not yet entered the marketplace. The extrapolations provide a view of the physical nature of future computers. All aspects of computer technology must be included.

The application of projections of the capability of technology to the highest performance systems will be considered. It will be assumed that the development of large computers will occur by evolution, i.e., that no drastic revolutions will change the basic character of such machines.

31.1 EXTRAPOLATION

Extrapolations are broadly based on the assumptions that present levels of development effort will continue and produce results comparable to those achieved in the past. They also assume that such results will be economically desirable, that they will enable a user to exercise data-processing functions at reduced cost.

Figure 31.1 is an extrapolation of the basic capabilities of semiconductor technology as measured by the minimum dimension produced by the lithographic process with acceptable yield. The dates used in plotting the extrapolations of this

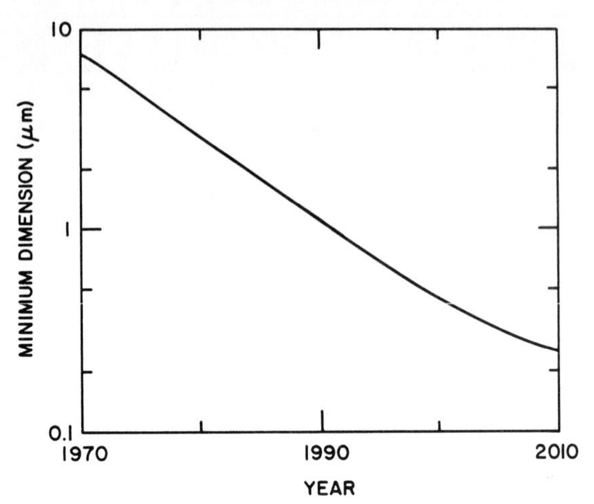

FIG. 31.1 History and projection of the minimum lithographic dimension.

section are intended to mean the dates of delivery of systems incorporating the technologies in question. There is no fundamental physical barrier to the achievement of the progress shown.[1]

Figure 31.2 shows how a chip may be regarded as composed of a large number of squares, each with side equal to the minimum lithographic dimension shown in Fig. 31.1. The character of each small square can be controlled independently of the other squares. Thus, the number of such small squares on the chip is the number of independently controllable or resolvable elements. Quite a few such resolvable elements are needed to fabricate, for example, the source, drain, and gate electrodes of a transistor and the insulating regions between circuits. Many features on a chip are larger than the minimum possible dimension. For example, wires may have greater than the minimum possible width to improve yields or to reduce resistance.

The number of resolvable elements per chip increases because of decreasing dimension and because of increasing chip area. Figure 31.3 plots the projected maximum number of resolvable elements per chip and also shows an extrapolation of the number of elements needed to construct a logic gate or a memory bit together with its associated wiring and off-chip communication devices. Ingenuity has reduced the complexity of circuits in the past, at least as measured by resolvable elements in the surface plane of the chip, and the continuation of the exercise of this kind of ingenuity is anticipated.[2,3] Increasing chip size, decreasing minimum dimension (miniaturization), and circuit inventiveness all contribute to the progress of integration.[2]

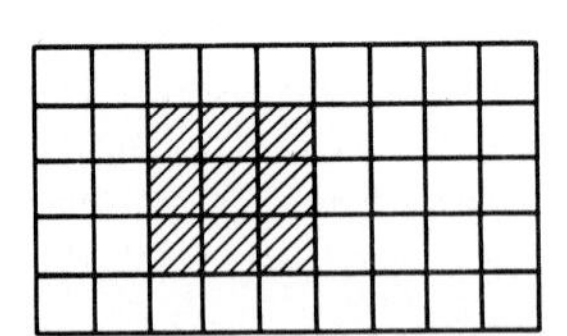

FIG. 31.2 Illustrating the concept of a resolvable element, a square with side equal to the minimum lithographic dimension. For example, the shaded area may be used to make an ohmic contact of sufficiently low resistance.

The main memories of large computers contain hundreds of millions of bits and are, consequently, very sensitive to cost per bit. Low cost per bit is

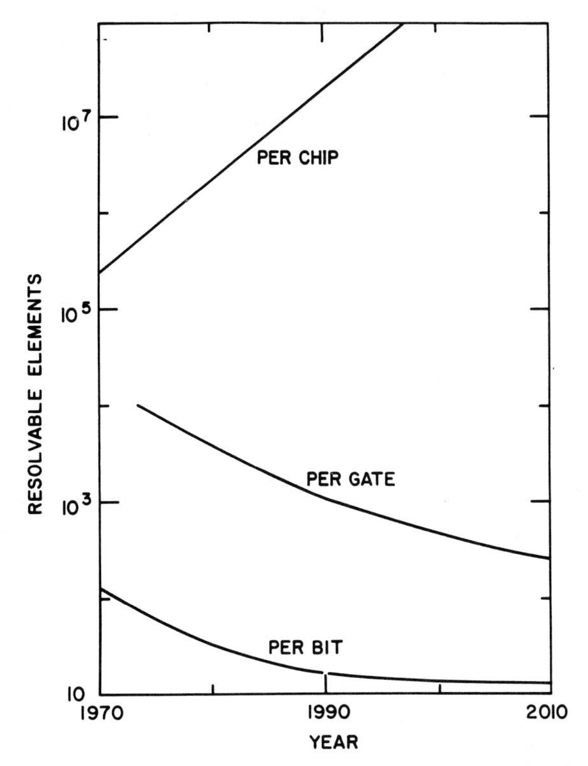

FIG. 31.3 The number of resolvable elements on a chip of maximum dimension and estimates of the number of such elements needed to construct a logic circuit and a memory bit plus the other elements, such as wires, decoders, and drivers that accompany them on a chip.

achieved through miniaturization and integration. The large volumes of chips and the relatively simple structure of field-effect transistor (FET) random-access memory enables memory technology to lead progress toward those objectives. The projection of bits per memory chip is shown in Fig. 31.4.

The transistors on chips for logic are organized into small circuits or *gates* that perform elementary logic operations. The logic chips for large computers are manufactured as gate arrays, a collection of gates that can be subsequently interconnected to form many functionally different logic chips. The wired gate-array chips are interconnected to form the processing unit of a computer. The growth of circuits per chip (Fig. 31.5) shows a slower growth for logic than for memory. (All projections relating to logic chips here refer to gate arrays and, in particular, do not apply to microprocessor chips.) These projections do not strain the ability of technology to provide larger chip sizes, and it is assumed that chip size grows at a rate that will accommodate the component growth shown in Figs. 31.4 and 31.5. For comparison with Fig. 31.1, the logic chip edge is 0.25 cm and the memory chip edge is 1.8 cm in the year 2000.

The power delay product of logic on a chip is largely fixed by the wire capacitance on the chip that must be charged and discharged at each logic operation.

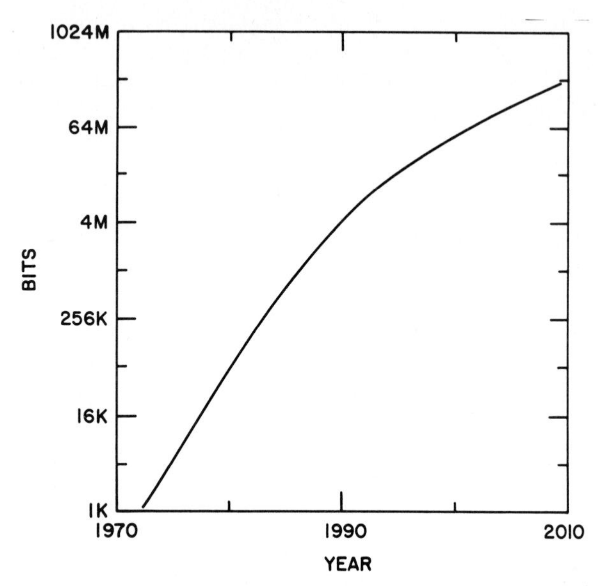

FIG. 31.4 History and projection of the maximum number of bits per chip for random-access memory.

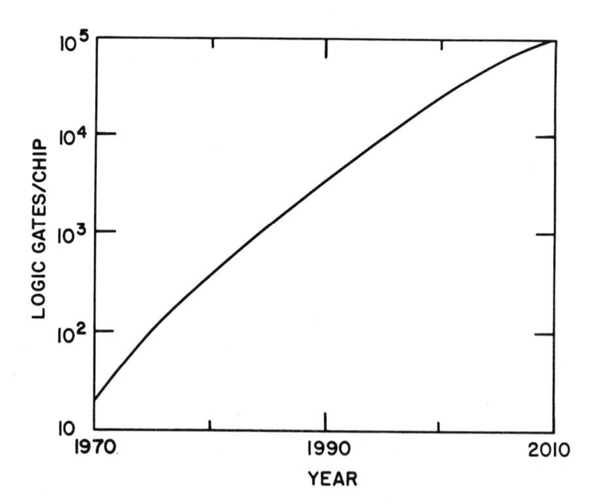

FIG. 31.5 Projected level of integration of logic chips for large machines measured as the number of gates or logic functions per chip.

The capacitance per unit length of wire is just a dimensionless geometric factor times the dielectric constant of an insulator and remains constant at about 2 pF/cm as technology advances. The logic chips in large machines are gate arrays, and the amount of wire needed to interconnect the gates has been analyzed statistically.[4] A number N_c of channels in which a wire can be placed must be provided in each cell of the array. The results of the statistical study can be roughly summarized by the expression $N_c = 6N^{1/6}$, where N is the number of logic gates on the chip. Each channel occupies an area $WA_c^{1/2}$, where W is the width of a channel and $A_c^{1/2}$ is the length of a channel that traverses a cell. Thus an area $N_c WA_c^{1/2}$ must be available for wiring in each cell. The wiring is placed in several, say K, layers in modern technology. Setting the available area equal to the required area, we have

$$KA_c = N_c WA_c{}^{1/2} + A_D \tag{31.1}$$

where A_D is the area occupied by devices in the cell, obtainable from the minimum dimension and the resolvable elements (Figs. 31.1, 31.3). A_c and the total length of wire channel per cell ($N_c A_c^{1/2}$) are found as the solution of Eq. (31.1). Thus the characteristics of the chip follow from the projections of level of integration (Fig. 31.5), dimension (Fig. 31.1), and circuit design (Fig. 31.3).

The energy per operation is determined as the charge drawn from the power supply in an operation multiplied by the power supply voltage. Summarizing, the energy per operation is (capacitance per unit length) × (average wire length per logic gate) × (signal voltage) × (power supply voltage). An energy dissipated in charging the capacitance of devices must be added to this. The energy dissipation per operation is shown in Fig. 31.6. According to these estimates the energy is primarily the energy dissipated in device capacitances in the early part of the period but will be increasingly dominated by dissipation in the wiring capacitances in the future.

Use of an energy per operation implicitly assumes that power and delay can be traded against one another and that it is their product that remains constant and characterizes a technology. The energy dissipated has decreased by about four orders of magnitude since 1960. The sources of this decrease may be very

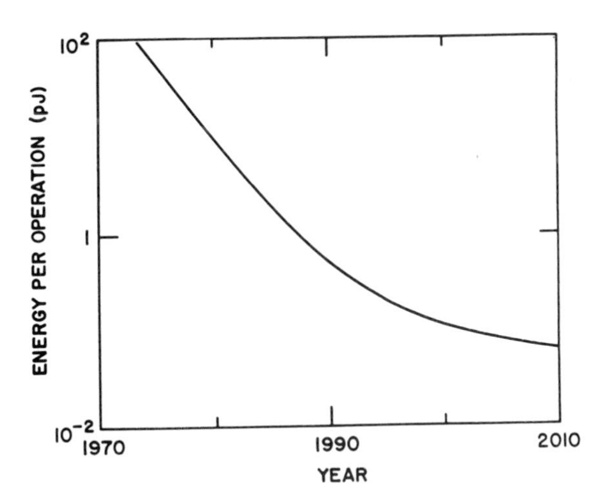

FIG. 31.6 The energy dissipated per logic operation on a chip, calculated as described in Sec. 31.2. The dissipation is dominated by device capacitance at low levels of integration and by wire capacitance at high levels of integration.

roughly assigned as follows: one order of magnitude to decrease in voltage (the voltage is approaching a lower limit and cannot be decreased much further); two orders of magnitude to miniaturization; one order of magnitude to improved designs, such as replacing *pn*-junction area with semiconductor-insulator interface.

A large amount of fast communication among chips is essential to the functioning of general-purpose computers made of many chips. Additional power is dissipated in the circuits that transmit signals from a chip to other chips. The circuits drive long connections that must be treated as transmission lines. The power required is determined by the impedance of the transmission line (and the signal amplitude) and is little affected by miniaturization. The communication can be described physically as a certain number of signal terminals on a chip and a representative length of chip-to-chip wire. The former is given by a relation known as Rent's rule, which relates the number of signal terminals that must be provided on any partition of a system to the number of logic gates contained in the partition.[5] Its form is

$$N_t = \alpha N^{\gamma} \tag{31.2}$$

Values of α and γ are determined by experience with logic designs; if an insufficient number of signal terminals are available, all of the circuits in the partition cannot be utilized. The best values of α and γ are given differently by different authors. Here $\alpha = 4$ and $\gamma = 0.5$ are used.

Some additional statistics relating to the logic chips, calculated as described in the preceding paragraphs, are shown in Fig. 31.7: the chip area, the total length of wire on the chip, and the number of external signal connections that will be needed according to Rent's rule.

Another factor that has contributed to the growing power of computational systems is the increasing number of logic gates that can be brought together to work as a unit. In fact the trend to larger system size seems to be accelerating, and our view of the future of this parameter is presented in Fig. 31.8.

The ability of packaging technology to remove the heat created by the dissipation in logic operations is another determinant of machine performance. Because of the projected rapid increase in number of circuits per chip, it is desirable to increase the density of power dissipation. Difficulties in removing heat directly degrade circuit speed because of the power-speed trade-off. However, advances in cooling technology also seem imminent. It is possible to quantify this limit as follows. The characteristic heat transfer parameter of a planar packaging technology is the maximum rate per unit area at which heat can be removed from the package.

The rate at which heat can be removed from dense assemblages of circuitry has been gradually increased by the introduction of new methods as shown in Fig. 31.9. Circuits are conventionally cooled by transfer of heat from a semiconductor or semiconductor module to air. However, air cooling is limited to a power density of about 1 W/cm^2, as shown in Fig. 31.9. Emerging technologies, in which heat is transferred directly to a liquid without the intermediary of air,[6,7] break through the 1-W/cm^2 limit.

31.2. DEVICES

The projections of logic technology in Figs. 31.1 through 31.6 are not based on specific assumptions about devices. The dominant device in high-speed logic for

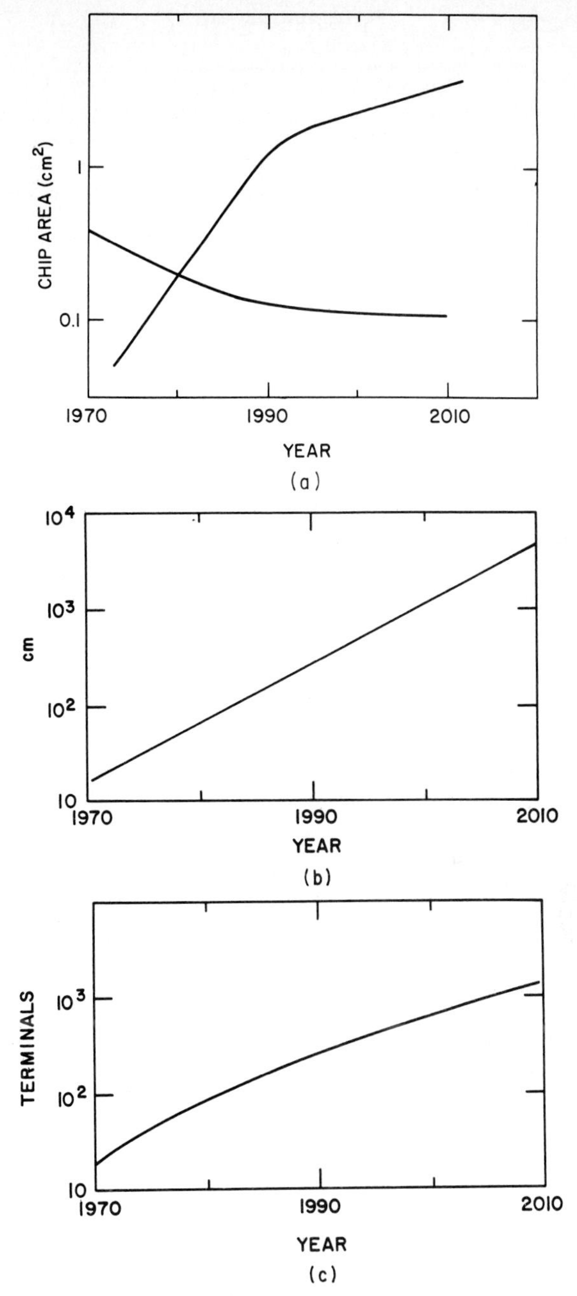

FIG. 31.7 Characteristics of the logic chip, estimated as described in Sec. 31.1. (*a*) Chip area. The area of logic chips decreases slowly. The projected area of memory chips is also shown as the curve of rapidly increasing area; (*b*) total length of wire connections on the chip; (*c*) numbers of signal connections between the chip and its substrate.

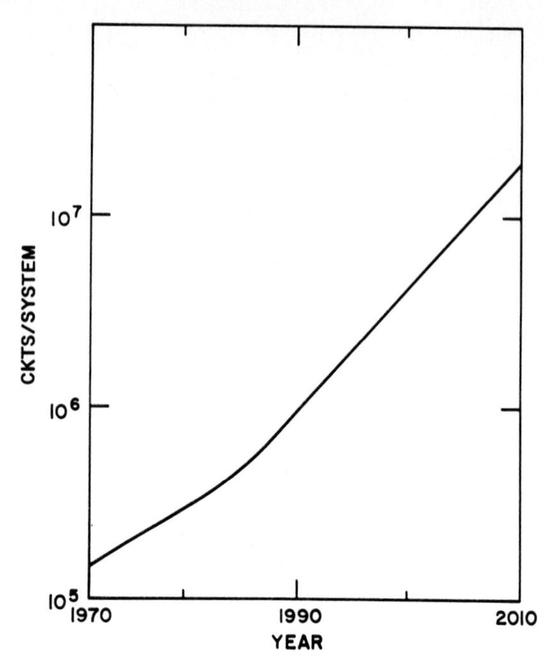

FIG. 31.8 The growth of the size in number of logic gates of large computing systems.

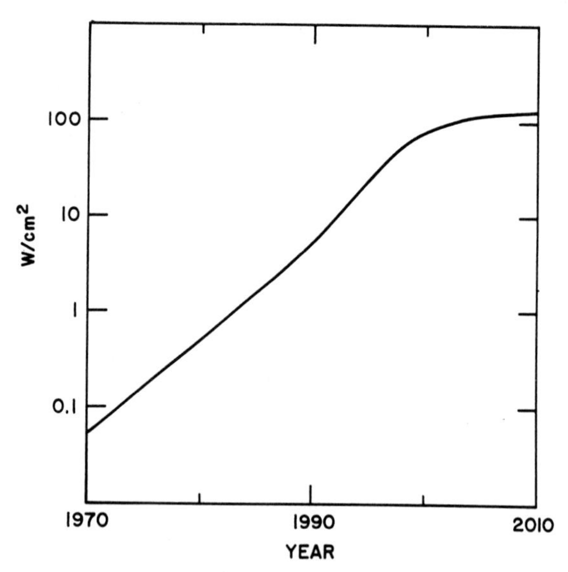

FIG. 31.9 History and projection of the rates of heat removal from packages.

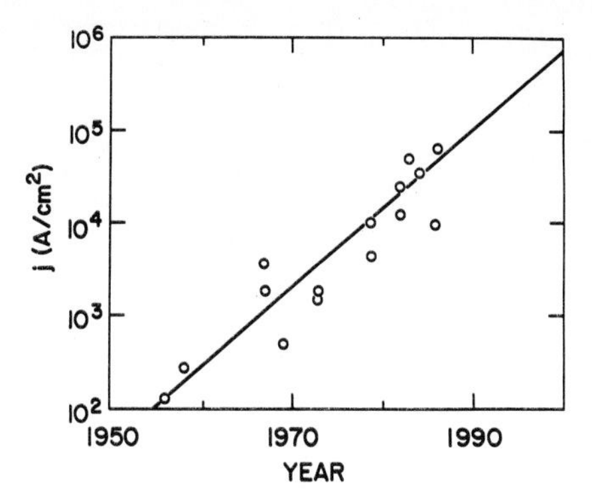

FIG. 31.10 Trends in bipolar transistors for logic. The current density at the emitter junction has increased rapidly with time.

large mainframe computers has been the *npn* silicon bipolar transistor. All large computers made since 1970 have used such transistors; the projections are based on the history of bipolar technology. Major changes in the design and fabrication of the bipolar transistor have taken place during the time period spanned in the figure. It is worthwhile, therefore to examine trends and limitations in bipolar transistor technology and possible alternative high-speed devices. Continuing advances in the technology of the bipolar transistors is essential to the more powerful machines. Figures 31.10 to 31.12 show the history and projection of the properties of the transistors.[8] (In contrast to the other figures in this chapter, the dates here refer to the publications of designs in the technical literature.) Ever-increasing speed through many stages of miniaturization has been obtained by designs that keep the current approximately constant, in other words, by steady increases in current density (Fig. 31.10). It is seen that the continuation of present trends will demand a further increase of a factor of 30 to 100 in current density. The high projected current densities will strain the technology of contacting the devices.

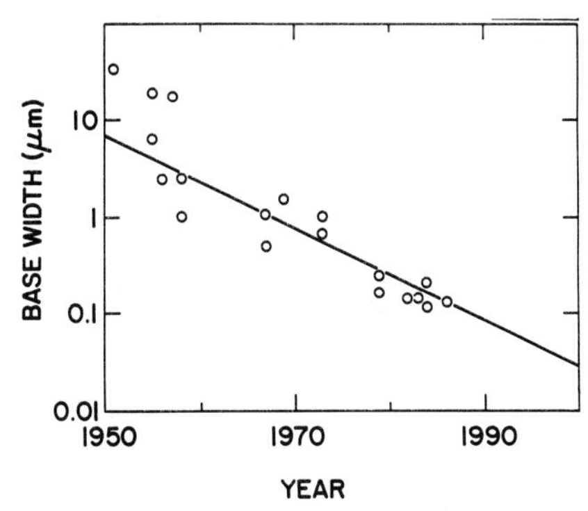

FIG. 31.11 Trends in bipolar transistors. Increasing current density and faster response demand reduction of base thickness.

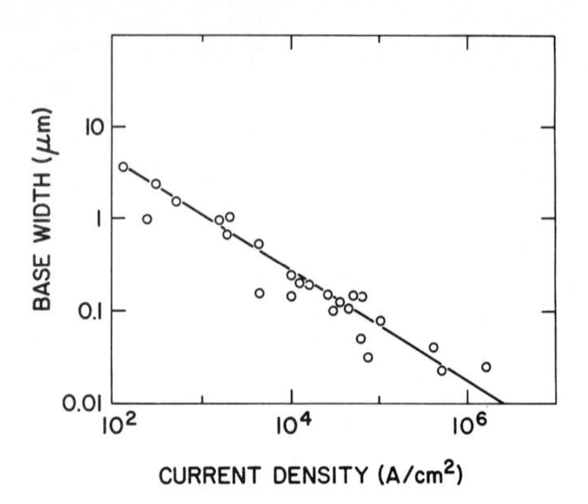

FIG. 31.12 Trends in bipolar transistors. Correlation between current density and base thickness.

One of the advances in transistor design that has made the high current densities possible is decreasing base thickness. Electrons carry current across the base by diffusion, and a large concentration gradient aids the flow. The evolution of base width through time is illustrated in Fig. 31.11. Basewidths less than 0.03 μm must be anticipated by the end of the century. Great demands will be made on fabrication technology.

The relation of the base width to the current density is shown in Fig. 31.12. This relation is conditioned by several constraints that must be simultaneously satisfied. The combination of basewidth and base doping must be such that the base is not completely depleted by the reverse bias applied to the collector and that the resistance to the flow of base current is not too large. The doping must not be so high as to lead to heavy doping effects such as tunneling and bandgap decreases at the emitter junction and in the emitter. But high performance requires that the charge stored in the base be low, and the time that an electron spends in crossing the base must be minimized. This last objective means that the base must be thin.

The problems emerging to retard the advances of bipolar transistor technology are the fabrication of the very thin base layers, the decrease in the energy gap of silicon at the high doping levels required in thin layers, and high current densities. Conventional methods define the base as the difference between the base-collector and the emitter-base junctions formed by introduction of impurities from the surface. As the difference decreases, control of it becomes increasingly difficult. Epitaxial emitters offer to eliminate the problem of controlling differences. If an epitaxial emitter can be made of a different semiconductor with a higher energy gap than the base, then the need for very heavy doping of the emitter is also relieved. A suitable alternative higher-gap material is not known for silicon. However, opportunities exist in the GaAs-AlAs alloy system, wherein increasing aluminum content increases the energy gap while remaining epitaxially compatible with a lower gap of GaAs. Difficult problems of materials science remain to be solved in the practical realization of this system. Handling the high current densities in future bipolar technologies will also require innovations in material systems for contacts and interconnections.

31.3. MODELING SYSTEM PERFORMANCE

A model that relates the physical description of technology to rates of operation can be used to forecast the performance of future systems.[9] The physical implementation of a system, however, offers a great many opportunities for choice, so that a wide variety of physical designs are actually encountered. In addition, physical implementations, especially the packaging, are extremely complex in detail. Thus any model that aims to be generally applicable must be simplified and can only hope to represent a limited number of important features in a semiquantitative way. In most cases neither theory nor experience suffices to provide unique relationships between the properties of chips and the characteristics of system components built from the chips. Such relationships must be regarded as deterministic in modeling; they are described by simple mathematical formulas fitted to the qualitative and semiquantitative information that is available.

31.4. MODEL OF PACKAGED DELAY

The forecasts of integration level, *pt* product, and cooling capability (Figs. 31.5, 31.6, and 31.9) can be used in a physical theory to predict the progress of packaged logic delay. The limiting delay is regarded as the sum of two parts, the circuit delay, which is obtained by dividing the *pt* product (Fig. 31.6) by power, and the chip-to-chip transit time. The latter, however, increases with power; since the power density is limited to values shown in Fig. 31.9, the more power expended per chip, the further apart the chips must be placed. The power per circuit can be chosen to minimize the total delay.

The contribution of the off-chip delays to the average depends on the frequency with which signals leave a chip and the length of the off-chip wiring on the package. Little accurate information that sheds light on either of these quantities is available.

As more gates are placed on a single chip, it becomes possible to perform more operations without going to another chip. Rent's rule, Eq. (31.1), is a reflection of this fact: the number of off-chip signal paths per gate on a chip decreases with increasing integration. As the off-chip connections are presumably needed, the probability of a signal leaving a chip must depend on the number of them, and we write it as a weighted ratio of off-chip to total destinations:

$$P_{\text{off}} = \frac{N_I}{N_I + \beta N} \tag{31.3}$$

Here β is a constant and N_I is given by Eq. (31.2). From Eq. (31.2) P_{off} may be written

$$P_{\text{off}} = (1 + \beta\alpha^{-1} N^{1-\gamma})^{-1} \tag{31.4}$$

The average length of a chip-to-chip connection is taken to be an appreciable fraction of the total system size. If n_c chips are mounted in a square array on a surface the size of the square measured in terms of the nearest chip-to-chip distance, the *chip pitch* is $n_c^{1/2}$. n_c is the total system size (Fig. 31.8) divided by N.

If connections were made randomly among chips, their average length would be ⅔ $n_c^{1/2}$. One can expect to do considerably better than this by proper placement of chips; however, the system performance is determined more by the worst cases than the best ones, and a conservative choice should be made in estimating performance. A value $n_c^{1/2}/4$ is used here. The average off-chip delay parameter used below is $m = P_{off} n_c^{1/2}/4$.

The desire to avoid the long off-chip delays is the main motivation for increasing the level of integration of logic chips. Advantages in cost and reliability are also realized. Let the chips, each containing N logic gates, be packaged on a planar substrate from which heat can be removed at a maximum density Q (Fig. 31.9) per unit area, with each chip occupying an area L^2. The time taken for a signal to propagate from one chip to another is mL/c_1, where c_1 is the velocity of electromagnetic waves on the interconnections. Here also it is necessary to be conservative; the velocity must take account of vias and other interconnections. Further, call the power dissipated in the high-current circuits that drive the transmission lines that connect one chip to another P_D and the power dissipated per logic gate p. Then heat removal requires

$$P_D + \mathrm{N}p \leq \mathrm{QL}^2 \tag{31.5}$$

It is also required that L^2 be larger than the chip area A. The total packaged delay t_p is the sum of the circuit delay t_c and the chip-to-chip propagation time. All of these effects are taken into account by expressing t_p as a function of the power as

$$t_p = \frac{U}{p} + \frac{m}{c_1}\left(\frac{Np + P_D}{Q} + A\right)^{1/2} \tag{31.6}$$

m is the factor that weights the chip-to-chip delays defined above. There is a value of Np that minimizes t_p, given by the equation

$$\frac{(Np)^2 m}{2NUc_1Q^{1/2}} = (Np + P_D + QA)^{1/2} \tag{31.7}$$

Equations (31.6) and (31.7) have been used to determine p and t_p. The resulting logic delay is shown in Fig. 31.13. The calculation predicts that it will decrease through more than another order of magnitude through the remaining part of the century.

31.5 INSTRUCTIONS

The average logic delay translates into millions of instructions per second in the following way. After about 10 delays sufficient asynchronism has developed among parts of the system that all results are temporarily stored and the system is resynchronized. The period of the synchronizing operation is called the *cycle*. The effect of the number of logic gates in the system (Fig. 31.8) is that as the number of gates is increased the number of cycles and, therefore, the number of delays needed to carry out an instruction decreases.[10] An estimate of the number of delays per instruction as a function of the number of logic gates in the system

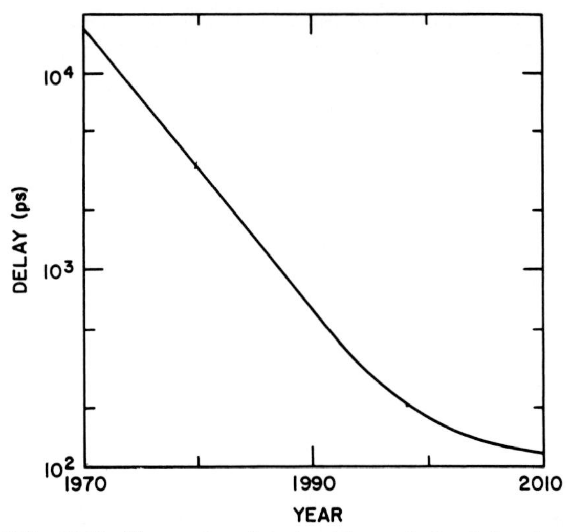

FIG. 31.13 The packaged logic delay calculated by the method of Eqs. (31.3) to (31.5).

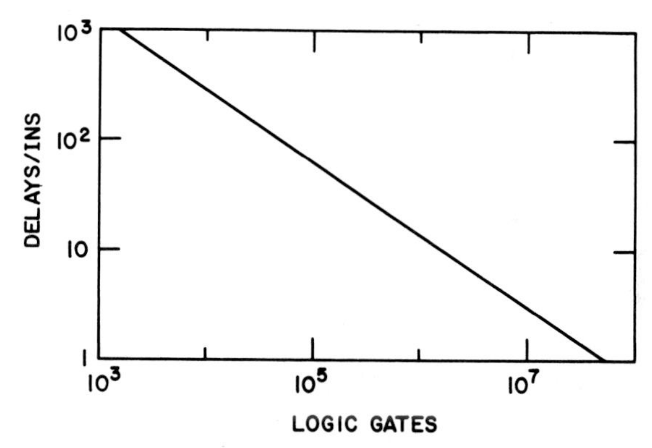

FIG. 31.14 The number of logic delays needed to execute an instruction as a function of the size of the system in number of logic gates.

is shown in Fig. 31.14. The number of delays per instruction decreases as the two-thirds power of the number of gates, a result in conformity with Hwang's limits on the advantages of parallelism.[11]

31.6 MEMORY ACCESS

Performance is also affected by the time needed to obtain access to information in the main memory. The access time of memory chips increases with the number of bits on the chip but decreases with miniaturization and increasing power applied to the chip. A rough fit to reports of the properties of random-access memory chips is

$$\tau_A = \frac{0.1B^{0.4}d^2}{P} \tag{31.8}$$

where τ_A = chip access time
B = number of bits on the chip
d = minimum dimension of features on the chip
P = power supplied

There is apparently quite a bit of room to trade time against level of integration in Eq. (31.8). The penalty that must be paid in gaining speed at the expense of B is cost; higher levels of integration translate to lower cost per bit. The largest computer memories are therefore almost always built from chips of the highest available level of integration. As discussed below, they usually also include a more expensive, consequently smaller, fast memory.

A certain amount of decoding to find the right chips and delays in transmission of signals over wires is encountered in accessing large memory systems. These delays are accounted for here by multiplying the chip access time by the one-sixth power of the number of chips in the memory. The memory access times in large machines are much larger than the machine cycle. As memory information is needed as often as every cycle, direct access to the main memory is not feasible. A small, fast buffer or *cache* memory is provided that is intended to hold the information relevant to the task of the moment. Occasionally, however, the needed information is not found in the buffer, and an access to the main be memory must be made. This event is known as a buffer *miss* and its frequency is described by an average miss ratio, the fraction of memory accesses that cannot be satisfied in the buffer. Miss ratios of a few percent are achieved in modern computers. The average time per instruction must be increased by the product of the miss ratio and the main memory access time.

31.7 SYSTEM CHARACTERISTICS

Figures 31.5, 31.8, and 31.9 contain the assumptions about future technology that are used to construct a picture of computers of the coming decades with the models of the preceding sections. Figure 31.15 shows the growth of the processing rate in millions of instructions per second calculated from Figs. 31.8, 31.13, and 31.14, that is, without attention to the limitations of memory access. The growing importance of the latter is illustrated in Fig. 31.16, which compares the time per instruction with the projected main memory access. The amount of time lost by each cache miss grows proportionately larger with time. The impact of this situation on processing power is shown in Fig. 31.17 for several values of the miss ratio. It is seen that the increasing disparity between processor times and main memory access times will seriously degrade the performance of systems for values of the miss ratio that are likely to be achieved. It seems likely that an additional stage will be needed in the memory hierarchy.

A motive for the use of VLSI in large systems mentioned above is illustrated in Fig. 31.18, where the number of wires on the chip and on the module are plot-

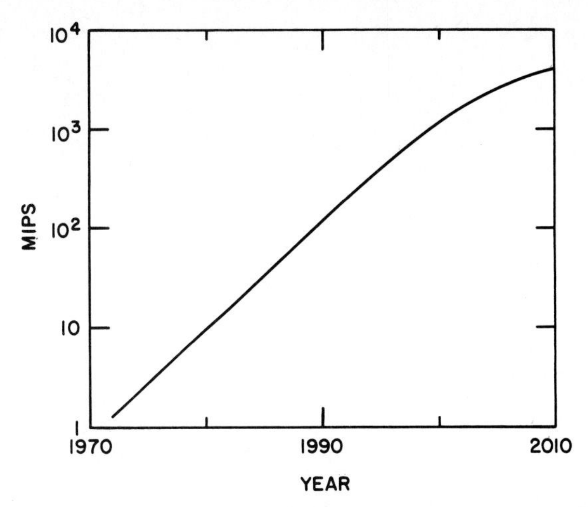

FIG. 31.15 Millions of instructions per second according to the results presented in Figs. 31.13 and 31.14.

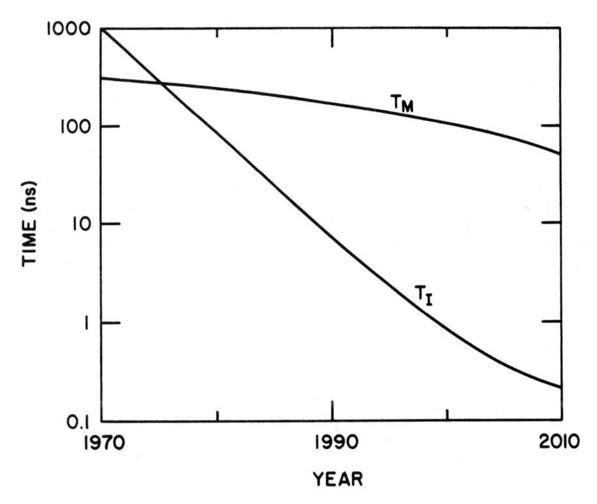

FIG. 31.16 Comparison of the processing time per instruction T_I with the main memory access time T_M.

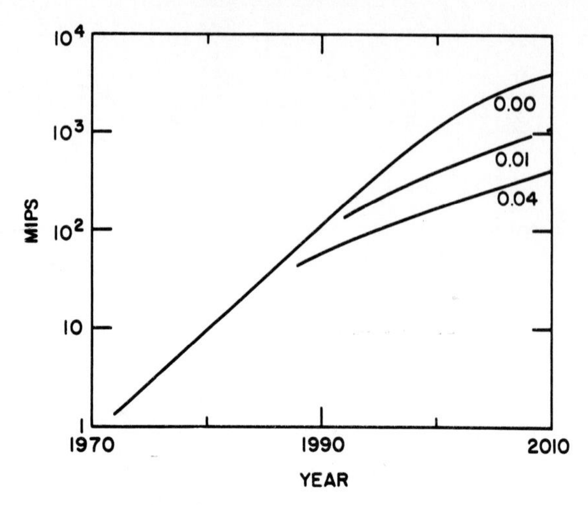

FIG. 31.17 The effect of cache miss ratio on the processing power of systems. The line for miss ratio 0 shows the result presented in Fig. 31.15, which does not take into account the need for memory access.

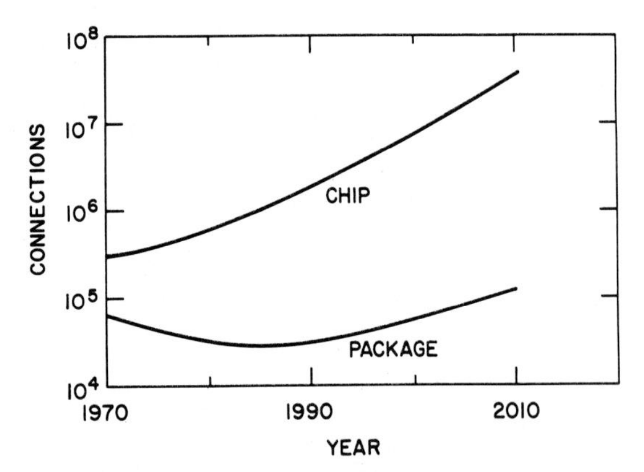

FIG. 31.18 Numbers of connections that are placed entirely on a chip and that must traverse the package. The increasing proportion of wires that are placed on a chip, where they are shorter and less expensive than wires on the package, is a major advantage of VLSI.

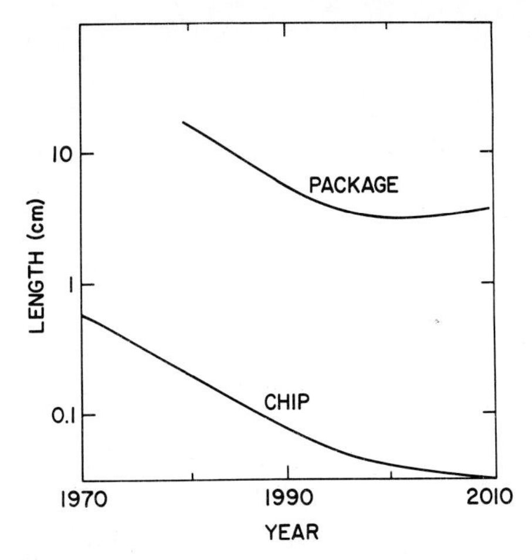

FIG. 31.19 The decreasing length of both on-chip (wire length per gate) and chip-to-chip interconnections is an important aspect of the decrease in delay.

ted. The fraction of wire interconnections that can be placed on the chip increases rapidly. This is an important economic factor, as the cost of a connection on a chip is much less than the cost of a connection on a higher package level.

Another aspect of the steady improvement in speed is the decreasing lengths of interconnections resulting from miniaturization and integration, shown in Fig. 31.19. Both the amount of wire associated with each gate and the average length of a connection between chips decrease.

Difficulties of the projected technologies are illustrated in Figs. 31.20 and 31.21. Figure 31.20 shows the density at which wire must be placed on the chip and on the package. Continuing miniaturization and development of techniques

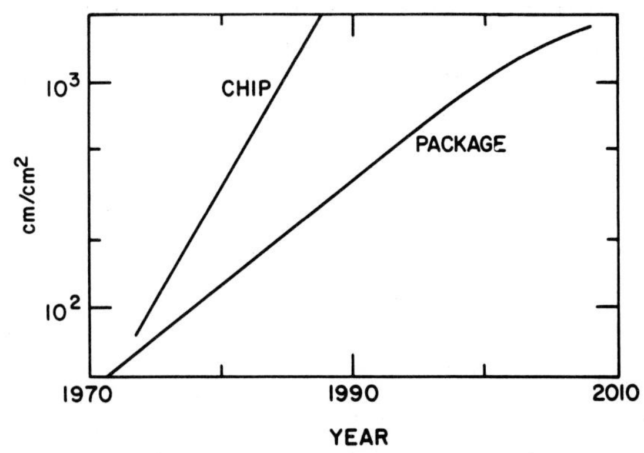

FIG. 31.20 The increasing density of wire on chips and packages.

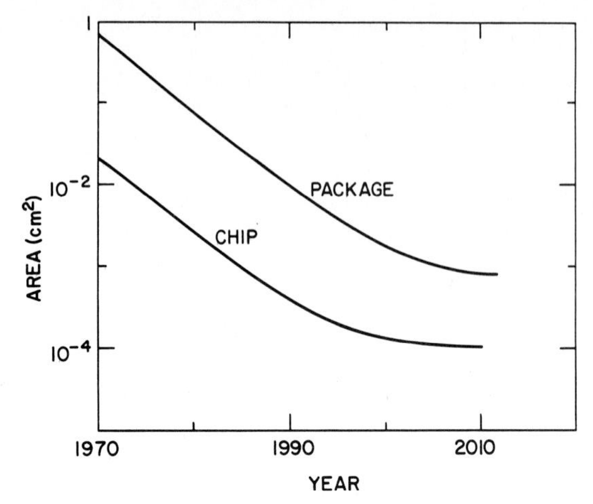

FIG. 31.21 The decreasing area per connection available for chip-to-package connections on both the chip and the package will require attention to miniaturization of this aspect of VLSI technology.

for placing wiring in increasing numbers of layers will be needed. Figure 31.21 shows the decreasing area on the chip and on the package that will be available for each interconnection between the chip and the package wiring.

31.8 CONCLUSIONS

Utilization of the advanced capabilities for the fabrication of integrated circuits that will become available for random logic chips will require solution of several difficult problems: Design and construction of the increasingly complex chip wiring, provision of many more external electrical connections to increasingly small chips, and devising circuits that can tolerate large interconnection resistances are among the needed advances. Assuming that some progress can be made in these areas, the evolutionary development of large machines should lead to single processors capable of handling over 300 million instructions per second by the end of the century.

REFERENCES

1. J. D. Meindl, "Ultra-Large Scale Integration," *IEEE Trans. Electron Devices,* **ED-31**:1555–1561 (1984).
2. G. E. Moore, "Future Directions of Silicon Device Technology," in *Solid State Devices,* E. A. Ash (ed.), Conference Series No. 40, Institute of Physics, London, 1977, pp. 1–6.
3. K. Kimura et al., *IEEE J. Solid State Circuits,* **SC-21**:(1986) 381–387.
4. W. R. Heller, W. F. Mikhail, and W. E. Donath, "Prediction of Wiring Space Require-

ments for LSI,'' *Proc. 14th Design Automation Conf.*, New Orleans, June 20–22, 1977. pp. 32–42.

5. B. S. Landman and R. L. Russo, ''On a Pin versus Block Relationship for Partitions of Logic Graphs,'' *IEEE Tran. on Computers,* **C-20**:1469–1479 (1971).
6. D. B. Tuckerman and R. F. W. Pease, ''High-Performance Heat Sinking for VLSI,'' *IEEE Electron Device Letters,* **EDL-2**.
7. A. J. Blodgett, ''Microelectronic Packaging,'' *Scientific American,* **121**(1):86–96 (1983).
8. R. W. Keyes, ''Trends in Bipolar Transistors,'' unpublished manuscript, 1986.
9. R. W. Keyes, ''The Evolution of Digital Electronics towards VLSI,'' *IEEE J. Solid State Circuits,* **SC-14**:193–201 (1979).
10. A. Masaki and T. Chiba, ''Design Aspects of VLSI for Computer Logic,'' *IEEE Trans. Electr. Dev.,* **ED-29**:751–756 (1981).
11. K. Hwang and F. A. Briggs, *Computer Architecture and Parallel Processing*, McGraw-Hill, New York, 1984.

CHAPTER 32
VLSI TECHNOLOGY APPLICATIONS IMPACT

Z. J. Delalic
Department of Electrical Engineering
Temple University
Philadelphia, Pennsylvania

The rapid progress in integrated circuit technology has dramatically changed the field of microelectronics. This microelectronic revolution has not only provided the means to put 100,000 or more circuit elements on a single chip, but also it has produced many new products used in many fields of science and everyday life. The new integrated circuits are characterized by low cost, small size, and/or low-power-consumption devices. Some of these new products, which have become feasible for the first time due to the technological breakthrough in LSI-VLSI technology, will be discussed in this chapter.

32.1 SIGNAL PROCESSOR

For a long time the implementation of digital signal processing techniques has been constrained by the limitations of component technology. Most of the signal processing has been done using the most flexible approach of a software implementation on a general-purpose computer.

The first attempt to change this was the utilization of specialized, high-speed, array processors attached to a conventional general-purpose host computer. This trend started in the early 1970s. During the early 1980s the evolution in component technology brought needed flexibility by use of programmable digital systems for signal processors. The bridge between the two fields of digital signal processing and VLSI system design is being built. Figure 32.1 presents the historical progress in these two fields as well as the necessary firm foundations for the successful and safe passage across this bridge.

In order to design a signal processor that takes advantage of advanced component technology, computational requirements must be understood and there must be an appreciation for the constraints imposed by technology.[1]

The algorithm for this type of processing device needs to include the time-domain processing as well as the transform functions. The time-domain process-

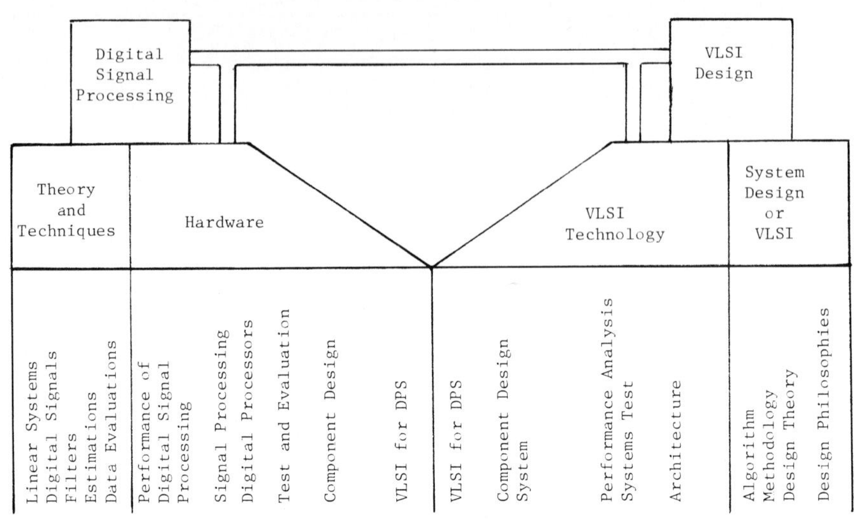

FIG. 32.1 Bridging VLSI technology and digital signal processing.

ing consists of digital filtering, correlation, digital encoding, and signal averaging. The transformed processing includes various transform algorithms. As far as signal processing architecture is concerned, there are classical general-purpose computer architectures, such as von Neumann and Harvard architectures, and modular signal processor architecture, which can be adopted for the new technological approach.[2]

The VLSI technology provides a decrease in the device delay times so that even conventional von Neumann–based architectures can be expected to provide a reduction in machine cycle time of perhaps an order of magnitude. Hopefully, in the future it will be easy to obtain 100 or 1000 subprocessor elements, all working simultaneously inside a given processor wafer. For this parallel activity it is necessary to reevaluate the processing functions required by each application and to evaluate algorithmic efficiency, as well as the regularity of the data flow and the needs for global versus local memory. When using VLSI technology to develop a signal processor chip, one has to be aware of the problems associated with processor interconnections and input-output considerations. In other words, minimum global communication is preferred. For this reason, two kinds of architectures are most common: (1) using many identical but relatively small look-up tables to achieve concurrency, and (2) addressing the issues of developing an optimum local processing element requiring only local nearest-neighbor interconnections. The second approach is more commonly used with NMOS technology and 5-μm design rules. Most of these types of processors can be programmed both at the local and array levels for performance of many algorithms.[3]

32.1.1 General-Purpose Signal Processor

General-purpose programmable signal processors are very useful for expanding digital signal processing (DSP) applications into a variety of different fields. Real-time simulation can be used to evaluate and improve an algorithm as well as incorporate a new idea into a given algorithm. This is a fast, cost-effective, compact means for product development. Any necessary modification in the system functions can be easily done by software. General-purpose signal processors using VLSI technology were developed in the late 1970s with NMOS and CMOS technology. The CMOS version decreases power consumption almost by one-tenth when compared to conventional design.[4] Every new chip development results in more functional capabilities of the processor and at the same time increases complexity and the number of transistors in the device. The number of transistors in a DPS chip has increased from approximately 10 in the late 1970s to 10^5 in 1980s, and is continuing to increase, as shown in Fig. 32.2.

By expanding general-purpose processors the programming becomes increasingly difficult. One of the older versions of the general-purpose signal processors is the uPD 7720 family.[5] These processors were originally used in a variety of fields ranging from voice-band data modems, low-bit-rate speech coders, vocoders, to speech recognizers, etc. The next generation of these processors

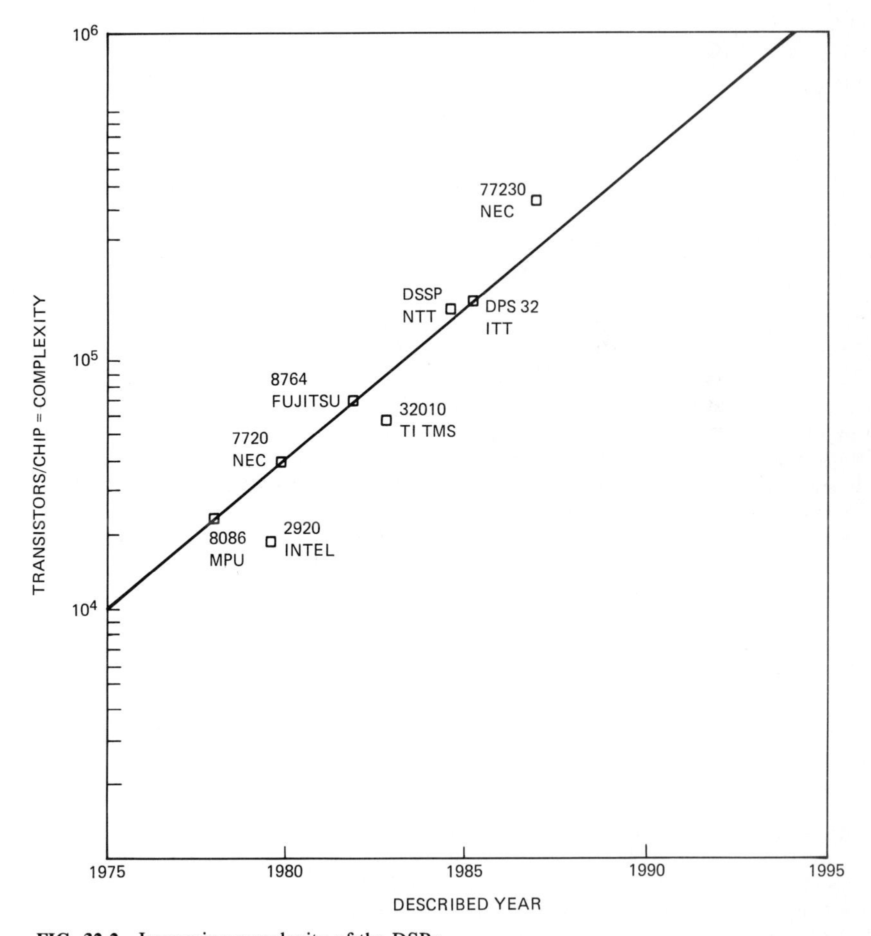

FIG. 32.2 Increasing complexity of the DSPs.

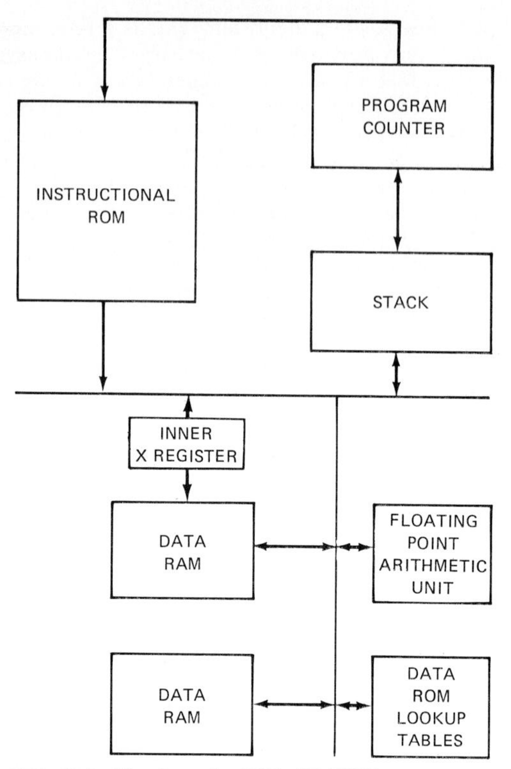

FIG. 32.3 Floating-point DSP uPD 77230.

with enhanced arithmetic operation capability were developed as 32-bit floating-point signal processors (uPD 77230). The uPD 77230 consists of two data ROMs, two data RAMs, a program counter, and a stack amplifier command sequence counter as shown in Fig. 32.3. One data ROM is for the look-up tables, and the other is an instruction ROM that provides the general-purpose subroutines and monitor. In one of the data RAMs the command sequence is stored, while the second data RAM is used for I/O data and delay inner-state data.

Although general-purpose processors are very useful for many applications, a definite shift to dedicated chip design is taking place. The architecture of the general-purpose processor can be optimized to be suitable for a number of specific applications so that most parts of the chip design can be based on the existing processor. One example is the programmable signal processor for a MODEM using the CCITT (Consultative Committee International Telephone and Telegraph) standard 32-kbits/s ADPCM algorithm.[6,7] The MODEM signal processor was developed to provide more enhanced capabilities necessary for implementation of the adoptive equalizer. Flexibility and modification capability, especially in the software, are important for many different variations of the data MODEM chip. With this in mind, the ADPCM processor was developed utilizing floating-point format operations while allowing algorithm modifications. Figure 32.4 represents a CCITT standard ADPCM processor.

The natural application for the programmable signal processor is in real-time

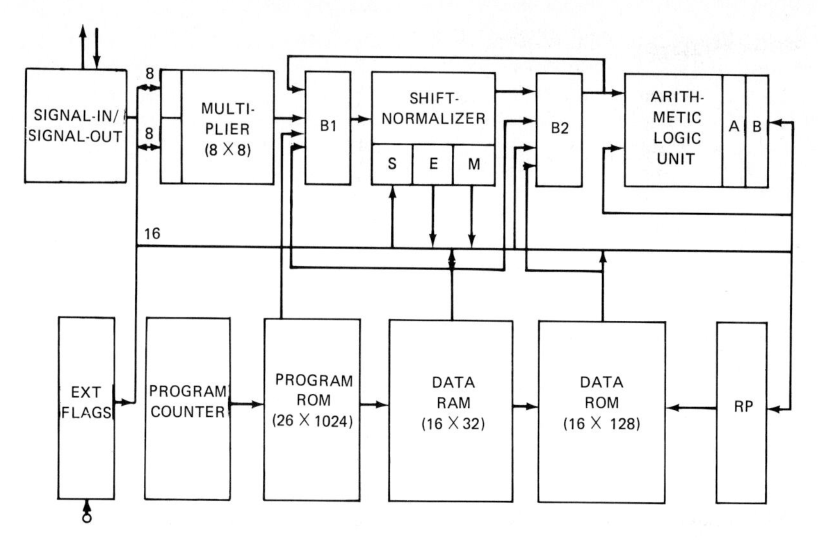

FIG. 32.4 CCITT standard ADPCM codec processor.

video processing. One picture frame can be divided into two-dimensional cells, and through programmable software the signal processor can process these cells in real time. The video signal processor (VSP) approach, dividing one frame picture into cells and processing them through the processor, is graphically shown in Fig. 32.5.

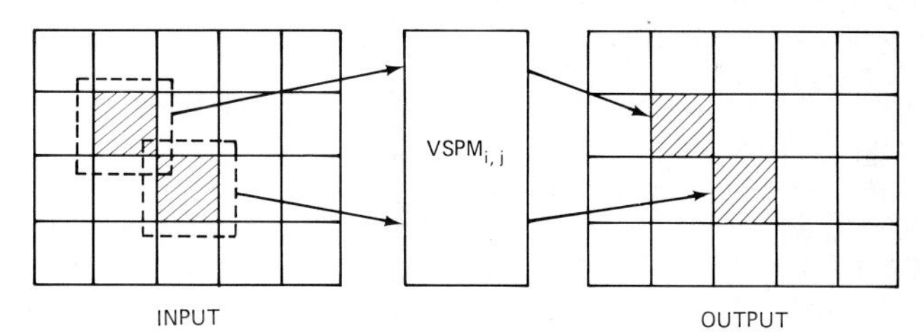

FIG. 32.5 VSPM (multi-VSP) processor approach.

32.1.2 Floating-Point Arithmetic DSPs

More and more newly developed DSPs are employing floating-point arithmetic. One such unit is the MSM6992 which consists of (*a*) the arithmetic block executing floating-point and fixed-point arithmetic; (*b*) the internal data memory block; (*c*) the instruction control block; and (*d*) the interface control block.[8] The DSP instructions consist of instructions for floating-point arithmetic, fixed-point arithmetic, logical operations, shift operations, transfers, input-output, jumps, and so on. The processor is capable of executing a maximum of eight different instructions to provide high processing capability.

Since these DSPs were used first in the field of telecommunication, the next step in their development was in the area of the speech processing and speech recognition. For this reason, the CODEC algorithms were used, which recently became very sophisticated, resulting in the acceleration of DSP processing speed. The data format for DSPs of this type consists of 22 bits/word—6 bits for the exponent and 16 bits for the mantissa. Both exponent and mantissa are expressed by the two's complement. This format allows realization of speech analysis processing. The second generation of DSPs could claim the MSM6992 as a successful processor that executes high-speed floating-point arithmetic and is expandable for future needs.

32.1.3 Programmable Array DSPs

High-performance signal processors using programmable array processing were successfully developed by Tetronix (M275 and TriStar).[9] The M275 consists of an arithmetic unit, an address computation unit, a memory control unit, and an instruction fetch unit with a built-in loop control. The data are processed simultaneously in a single clock cycle. The programmer can independently control all of the units for each clock cycle of the program execution. This processor allows communications with another processor through a bus request system that allows another machine to take control of its data memory. The other processor can write into the memory as well as interrupt. One of the important features is an overflow trap that allows testing from overflows during a high-speed algorithm without wasting any cycle.

Further advancement in this type of processing was made by developing TriStar, which is a 32-bit programmable DSP fabricated using 1.5-μm, double-metal CMOS technology. In this processor the word length is 12 to 16 bits. Both processors use fixed-point arithmetic. TriStar has a multiplier-accumulator that allows a single-cycle implementation. Both of these processors use single-cycle instruction, parallelism, and powerful data movement structure to overcome the inefficiency of single-bus architectures.

32.1.4 DSPs with Parallel Computation Algorithm

The Motorola DSP 56000 family is a user-programmable DSP in 15-μm, double-level metal, low-power CMOS technology.[10] The block diagram of the architecture is shown in Fig. 32.6. The core of the DSP consists of a data arithmetic logic unit (ALU), an address arithmetic unit (AAU), and a program controller connected by a multiple-bus architecture. The parallel operation of the execution units provides the resources needed to execute most instructions in a single 97.5-ns instruction cycle.

The arithmetic and logic operations are performed by the data ALU. A 24- by 24-bit fixed-point, hardware multiplier-accumulator ALU provides a 48-bit product and two 56-bit accumulators. The address arithmetic unit calculates addresses to locate data in the memory. This unit can provide two independent memory addresses at each instruction cycle and update them using two arithmetic units.

The instruction flow control, instruction decoding, and exception processing are provided by the program controller. The program controller consists of a program address generator, a program decode controller, and a program interrupt controller. For normal instruction a two-stage instruction pipeline is used.

The program interrupt controller reduces the overall timing associated with servicing the 18 interrupt sources. The peripheral devices and external interrupt pins can be programmed to one of three interrupt priority levels. The time-critical

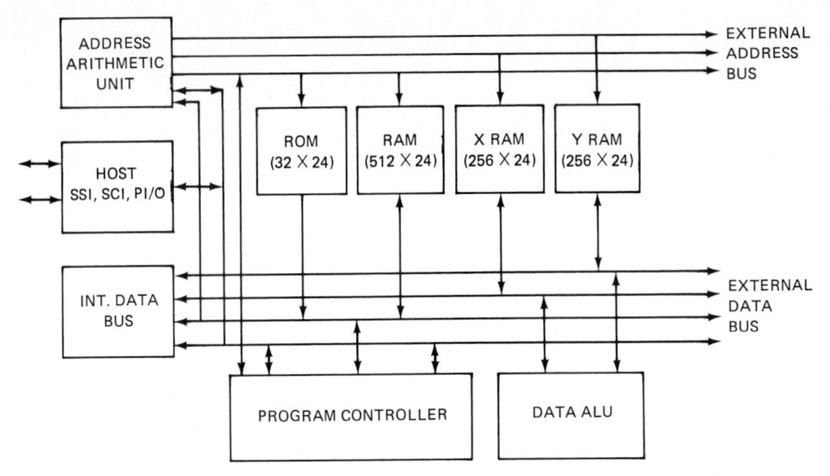

FIG. 32.6 DSP 56001 block diagram. (SSI = synchronous serial interface; SCI = serial communications interface; PI/O = program input-output.)

interrupts receive the highest priority. The fast interrupts can process up to 1.7 million interrupts per second.

This user-programmable DSP in RAM-based architecture with bootstrap mode has extensive I/O capability. The programs are developed for a finite impulse response (FIR) digital filter, a second-order infinite impulse response (IIR) digital filter, and a radix 2, complex, fast Fourier transform (FFT).[11,12]

32.1.5 DSP with Bit Slice Architecture

Most computations involved in digital signal processing can utilize matrix-vector multiplication. At present, most of the DSPs on the market implement a fast algorithm to compute a transform. These types of chips are dedicated to a particular transform and cannot be programmed for other signal processing tasks. On the other hand, the processor design to compute the transform by the matrix-vector product approach has the flexibility of being programmable. It will be much simpler to design the VLSI chip if a large number of identical processors are utilized in a specific implementation. A simple linear transform can be written as

$$Y = My$$

where y = input vector
M = transformation matrix
Y = output vector

For fixed-point representation, the input vector in binary form, with word length W bit planes and length L, is $y(i)$'s. The ith partial product of the kth transform coefficient with $y_i(l)$ is the ith bit of $y(l)$:

$$Y_i(k) = \sum_{l=0}^{L-1} y_i(l)M(k,l)$$

The total time for computation of a transform point is t_a ($\log_2 WL$), using ($WL - 1$) address. By using a bit-slice architecture, variable lengths of the input data can be processed with increased or decreased word length as required in different applications. This algorithm achieves the theoretical lower bound of the area-time squared measure (recall that the basic unit is an adder). The applications for this type of processor are for fixed-point computation. This approach allows selective computation of transformed parameters and is used when nonuniform sampling of the transform is needed.[13]

32.1.6 DSP Based on QRNS

The more complex applications of DPSs usually are based on the quadratic residue number system (QRNS).[14] With QRNS the interaction between the real and imaginary channels in complex arithmetic is eliminated. The processor computational operations usually are based on FFT, FIR filtering, inverse FFT (IFFT) convolution, correlation, and multiplication. This type of processor may be used by itself or as a coprocessor. The QRNS allows complex multiplication to be performed by two standard multipliers, leading to considerable computational savings, which is one of the most important features for signal processing. The real and imaginary components can be isolated from each other for the purpose of noninteractive parallel operations. The QRNS computation can be defined as

$$Z_1 = X_1 + jy_1 \qquad Z_1^* = X_1 - jy_1$$
$$Z_2 = X_2 + jy_2 \qquad Z_2^* = X_2 - jy_2$$

where $j = \;\; - 1$ and the operations are performed as

$$(Z_1,Z_1^*)(Z_2,Z_2^*) = (Z_1 \cdot Z_2, Z_1^* \cdot Z_2^*)$$

This design approach is quite suitable for VLSI implementation, and it has the reputation of having the appropriate simplicity, modularity, and hardware programmability. The main functions performed by the processor include FFT, I FFT, FIR filtering, convolution, correlation, and multiplication. The data flow is employed by microprogramming. Each operation has a single instruction. The instructions are loaded and stored in an instruction register. The programmable logic array (PLA) decodes the instructions and issues the appropriate control signals. The given configurations of the processor can be changed by receiving new instructions. The number of FFT passes and the number of data points to be processed have to be given. The synchronization of the different algorithms and the operation steps are performed by the control part of the processor. The arithmetic operations and the I/O operations for a particular function overlap.

The main advantage of this processor is in the parallelism on the computational level established by using QRNS. Since there is no interaction at all between real and imaginary channels, the degree of parallelism is doubled. The hardware for the multiplication complexity can be drastically reduced by using real multiplications in combination with lookup tables. It is estimated that the hardware reduction is about 50 percent by employing QRNS over using conventional arithmetic. The lookup tables have been used as building blocks modules, which are considered efficient for VLSI implementation for small and medium-sized moduli. Commonly, this processor is used in applications such as speech processing, in the frequency range of a few kilohertz to several megahertz.

32.1.7 DSP with Fault-Tolerant Capabilities

Bit-serial architecture plays a vital role in the design of VLSI DSPs. The best results can be achieved by the computational concurrency of many bit-serial structures operating in parallel.[15] There was substantial work done at Cornell University on the development of a bit-serial structure and fabrication of a high-performance signal processor. This type of design led to the development of a very high speed integrated circuit (VHSIC) class processor.[16–18] These processors operate on fixed-point and block floating-point data, which will expand into full floating-point and increase the computational capabilities of the next generation submicrometer VLSI technology.

The complex multiplier-accumulator (CMAC) developed at Cornell multiplies two complex numbers

$$M_1 = A + iB \qquad \text{and} \qquad M_2 = C + iD$$

and the product is accumulated by use of a third complex number $M_3 = E + iF$ to produce $M_4 = PR + iPC$, where $PR = AC - DB + E$ and $PC = AD + DB + F$. The floating-point format allows the multiplication to be processed as the product of mantissas and addition of exponents. The mantissas must be aligned and the exponents adjusted before any other operation will happen. By using a bit-serial environment, the aligning is done with linear shift registers. The normalization is performed after the multiply-accumulate process is completed and the elimination of leading zeros are performed.

The CMAC unit can communicate asynchronously with a host or other CMAC processors, so the central controller is not needed. Fault detection (the bit-scan) is built into the test for the computational components of the CMAC. This is similar to scan path and level-sensitive scan design (LSSD) techniques. A self-testing and self-isolation method is employed for each CMAC in the linear array that operates in a test and computation mode. If the fault is detected the isolate time of the CMAC goes high, and when the fault is clear the host processor is alert to erase information. This built-in fault tolerance is a very attractive feature for the next generation of VLSI DSPs, since the complexity of these devices is growing daily while controllability and observability of the internal parts of the processor are being diminished.

32.1.8 Adaptive Signal Processors for Different Functions

The partial-rank algorithm used in adoptive signal processing such as beam forming is perfect for utilizing VLSI technology. This processor uses the flexible systoliclike architecture. Higher-order algorithms are easily implemented by adding identical hardware.[19] The algorithms for beam forming cover the range from stochastic algorithms to full matrix methods. The stochastic techniques update the vector of adoptive coefficients with each input sample.[20] On the other hand, full matrix methods update the adoptive gains by inverting an $n \times n$ autocorrelation matrix to save the linear equations. The partial-rank algorithm is flexible and achieves fast convergence at lower cost.[21]

One of the interesting architectures for the same specialized DSPs is the multiformat discrete cosine transformation (DCT). This type of the transformation turns out to be the first step of most image coding algorithms.[22,23] This type of chip is in the developmental stage by Thompson Semiconducteurs in 1.25-μm CMOS technology, consisting of 70,400 transistors, in the area of 25 mm, with an internal clock of 13.5 MHz. This processor implements various sizes of DCTs (up to 16 × 16 with direct-reverse DCT) and has 8-bit spixels and 16-bit internal accuracy.

The VSP (vector signal processor) developed by Zoran Company performs high signal processing in time and frequency domains for numerous standard DSP operations. The chip consists of multipliers, accumulators, and sequencers as building blocks for more powerful configurations. It has a very flexible architecture. This means that the customization of the architecture as well as arithmetic precision requirements can be designed into the system while the system is in the design stage. The ZR34161 VSP operates using a vector or block of data.[24] The efficiency in this processor is achieved by automatically fetching a given vector of data from system memory into internal memory.

Figure 32.7 illustrates the general architecture of the VSP, consisting of the bus interface unit, execution unit, and memory and registers. The bus interface unit and execution unit are independent processors that can operate in parallel. The memory and registers are shared between both of these units. The register section and memory consist of over 128 complex-word-data RAM, instruction FIFO (first-in–first-out), floating-point-scale RAM, registers, and sinusoidal lookup tables. This processor is programmed with a "high-level" instruction set. The instructions for FFT, magnitude square, demodulate, etc., are strictly functional. It is also possible to add additional VSPs to a bus for increased throughput.

In conclusion, most of the signal processing systems need good optimization of data flow through a processing unit. The most important functional block in the processing unit is a multiplier accumulator. The data flow is also controlled by dedicated generators and/or sequencers, which add complexity to the system. In the conventional system there is the constantly increasing speed of the multiplier-accumulator or pipeline and busing architectures. The VLSI technology has the advantage of introducing new signal processing algorithms, and many so-called "not practical" for the old technology are now being exploited in high-performance DSP systems.

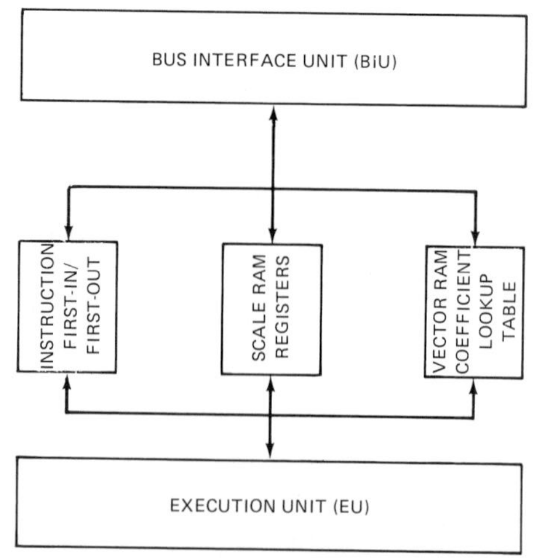

FIG. 32.7 General VSP architecture.

32.2 NAVIGATION, AUTOPILOTING, AND MACHINE CONTROL

With the rapid advances in VLSI, a growing number of complex digital signal processing applications are becoming economically feasible. For example, automobiles could use some of the processors to help automate navigation, map reading, autopiloting, engine and fuel control, etc.

32.2.1 Navigator

The number of proposed designs for automotive navigation and map display received a great deal of attention in recent years. The electronic map is capable of rotating to follow the movement of the vehicle, and the scale of the map can be changed by the vehicle operator. The operator is responsible for selecting destinations by entering a street address or two intersecting streets.[25] The direction, using an arrowhead, and the distance to the destination could be displayed, making the driver aware of those parameters at any time. Most of the navigator systems consist of a processor, a cassette tape drive, a display, a solid-state compass, and two wheel sensors. The future of automotive navigation depends on improvement of navigational algorithms and fabrication of more economical VLSI DSPs.

32.2.2 Self-Controlled Vehicle

The automatically controlled movement of the vehicle, with no fixed guide wires, has been of interest to designers for a long time. Present progress in technology is making this goal more realistic. This type of vehicle is now in the improvement stage but is not without unsolved problems. Some of the very demanding specifications are (1) high level of accuracy, (2) easy operation, (3) fast measurement of its own position, (4) easily understood traveling instructions and ability to change direction quickly, and (5) precise traveling along the desired track with high reliability. There are a number of proposed algorithms, such as the continuous path method and the point-to-point method.[26] The continuous path method makes a traveling vehicle easier to manipulate and easier to change traveling course. On the other hand, the steering and position of the vehicle are influenced by the traveling speed, and that makes speed changing an unstable operation. Neither method is totally satisfactory.

A somewhat different approach is to calculate the position of the vehicle from the rotations of the measuring wheels (not driving wheels) and correct its measurement error at a predetermined position, using optical methods. The traveling course instructions could be represented by a sequence of simple commands for stopping or traveling along a path. The instructions given to the vehicle are a sequence of elementary movements at a point or along the path. Each elementary command consists of an action, a path specification, a speed specification, and a heading specification. These specifications are further broken down to path type, start and/or stop position, radius, etc. The position is specified by coordinates or the angle. All these commands and sequences are stored in the memory unit of the chip. The movement control unit of the chip has the ability to determine the exact rotational speed of the driving wheels in accordance with the current command.

The state variables are used to express the vehicle's state. The relationship between these state variables and the control variables can be expressed as follows:

$$V_l(t + T) = V_l(t) + \int_t^{t+\tau} C_l \, d\tau$$

$$V_r(t + T) = V_r(t) + \int_t^{t+\tau} C_l \, d\tau$$

$$\theta(t + T) = \theta(t) + \int_t^{t+\tau} \frac{V_l(\tau) - V_r(\tau)}{2} \, d\tau$$

$$V_x(t + T) = \cos\,[\theta(t + T)] \frac{V_l(t + T) + V_r(t + T)}{2}$$

$$V_y(t + T) = \sin\,[\theta(t + T)] \frac{V_l(t + T) + V_r(t + T)}{2}$$

$$X(t + T) = X(t) + \int_t^{t+\tau} V_x(\epsilon) \, d\tau$$

$$Y(t + T) = Y(t) + \int_t^{t+\tau} V_y(\tau) \, d\tau$$

where C_l = acceleration of left wheel
C_r = acceleration of right wheel
X, Y = coordinates of a position
θ = heading angle
V_l = velocity of left wheel
V_r = velocity of right wheel
V_x, V_y = velocity of components
T_l = length between two wheels

The control error function will sum all the individual errors as

$$E = W_0 + e_0 + W_1 e_1 + W_2 e_2$$

where e_0 = velocity error
e_1 = position error
e_2 = direction error
W_0, W_1, W_2 = weighting factors

The relationship between this function and the control variables C_l and C_r at given time $t + T$ are

$$C_l = \frac{-\,d\overline{E}}{dC_l} \qquad C_r = \frac{-\,d\overline{E}}{dC_r}$$

The values of C_l and C_r are determined from the current values of the state variables and the specifications of the current command. The desired rotational speeds can be defined by integrating the calculated accelerations. This is one of the simplest ways to automatically guide vehicles. The complete control, which depends on the evaluation of control errors and the prediction of control effects, is reliable and possible by using a simple VLSI processor.

32.2.3 Traffic

To have good control of traffic one should have a stable detection of a vehicle. At the present time most of the traffic control and surveillance systems of highways use point sensors measuring the presence of a vehicle at the given point. However, to have good control of the flow of traffic one needs two-dimensional information and not limited point sensor information. The use of a TV camera and an image-processing technique provides more useful information and better accuracy.[27] Greater improvement can be accomplished by using VLSI technology to develop a high-performance specialized digital image processor. With this approach, accurate detection of the vehicle's size can be determined, and the two-dimensional movement of each vehicle within the field of view can be easily established.

32.2.4 Autopilot

VLSI technology has been extensively used in commercial systems that require extreme accuracy, low power dissipation, and small area. The autopilot development using CMOS VLSI technology started a few years ago. One application was the missile autopilot developed by using an analog LSI fabrication in double-poly *p*-well CMOS technology, 5μm design rules.[28] The LSI chip was designed to perform digital and analog functions required for a missile control system. The chip consisted of (*a*) a block that performs demodulation and filtering of the yaw, pitch, horizontal and vertical steering commands from the guidance system; (*b*) a block that provides demodulation, filtering, and rate-command limiting; (*c*) a block that sums the signals from the two previous blocks and provides a modulated output; (*d*) a crystal oscillator block, which provides all the necessary clocking for the filter's demodulators and digital-to-analog converts; and (*e*) the autopilot control block, which controls the timing of the other functional blocks (see Fig. 32.8).

The demodulation and filtering section of the chip consists of two full-wave demodulators, which are single-pole low-pass filters, the biquadratic filters based on a standard capacitive feedback design, and two six-input summers that produce two outputs.

Block 2 performs very similar functions, including filtering, demodulation, and summing. Summing block 3 transforms a linear input into a digital output whose pulse width varies continuously with time. Usually, the major disadvantage in designing analog MOS circuits is the large power and area required to drive an off-chip load. This requirement needs a great deal of study. The output of this functional block goes through the summer and is compared to a reference voltage. The output of the comparator is a pulse whose width varies with the amplitude of the input voltages. This type of system using the VLSI technology will eventually spread into other fields where autopiloting can be utilized.

32.2.5 Machine Control System

The custom-designed LSIs in combination with a microcomputer are presently in experimental use to form a computer numerical control system for specific ma-

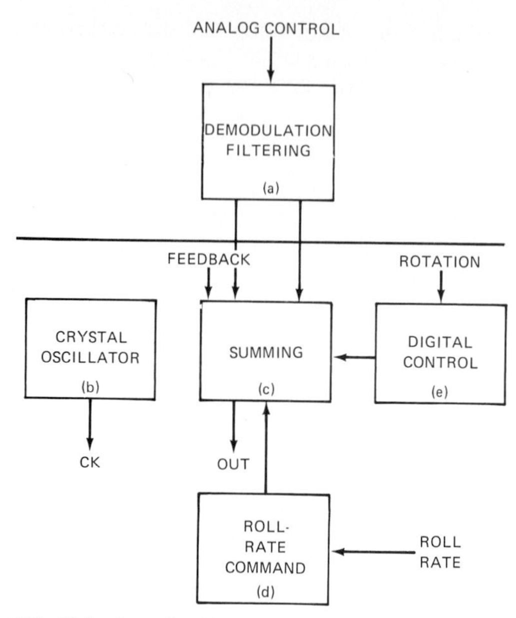

FIG. 32.8 Autopilot block diagram.

chine control. The VLSI custom-design processors are becoming more economical and attractive to designers who are planning to use VLSI technology to control stepping motor-actuated hydraulic servos to drive the machine. The advantage of using VLSI technology in comparison to the microprocessor is in programming both for general-purpose and custom VLSIs. Programming LSIs or VLSIs is easier than programming microprocessors.[29]

The general-purpose LSI can form the interface adopter between the microcomputer and the numerical machine control. One of the design LSI adopters consists of an 8-bit serial-to-parallel and parallel-to-serial shift register and input data latching on its peripheral parts. The adopter chip has the capabilities to control bidirectional data transfer between the LSI and machine. The built-in interval timer produces well-spaced and interpolated pulses to the stepping-motor-drive logic board. A slightly modified chip has been used for dc servomotor control. The objective for these two designs was to develop a cost-effective system suitable for a small firm.

32.3 COMMUNICATIONS

The VLSI technology is making significant strides in the field of communications. One example is a frequency synthesizer for a spread spectrum communications system. The important requirement for this type of design is the capability of tuning quickly, in small frequency steps, over a wide band. The best solution for the wide tuning bandwidth is through indirect digital synthesis, in which an integer multiple of a reference frequency may be produced, using a programmable divider in a phase-locked loop. Fine tuning is best achieved by direct digital synthesis.[30] One of the chips developed for this purpose consists of a 20-bit phase accumulator and a sine function ROM lookup table.[31] The rate-of-phase change is governed by the frequency, and because of that the output frequency is determined by the command word, which is added to the accumulator each clock period. The accumulator will store 2 rad as the largest values. The signal flows from the accumulator through the ROM table to the digital-to-analog converter (DAC). The process also involves filtering at the DAC stage.

To improve this approach and make it more flexible for other applications, we can modify the synthesizer to permit operation of two chips in parallel separated in phase by 90°. This CMOS digital frequency synthesizer is used in the fine resolution stage of the wideband synthesizer for spread satellite communication. The LSI chip was developed as a more general-purpose communication building block.

32.3.1 Telecommunications

In digital switching and transmission systems VLSI technology is widely used for development of new integrated circuits. One such circuit is an analog VLSI circuit used to make all its features programmable through software control from the central switching office. Some of the basic functions of the chip are (*a*) the codec filter for the voice channel, (*b*) the gain control for transmitting and receiving direction, (*c*) balance network, (*d*) two- to four-wire conversion, (*e*) three-party conferencing, and (*f*) a secondary analog channel. The input voice first goes through an antialiasing filter and then through a low-noise amplifier. There is a programmable gain stage in the voice path control to allow gain variation of +6 to −6 dB under software control. The signal has to satisfy the CCITT requirements, such as band limitation of 3.2 kHz. The other functional blocks are items such as the encoder and DAC. Usual telecommunication lines consist of two wires, and the transmission equipment uses four wires, which requires some type of a balanced network. The network is utilizing CMOS VLSI technology. This type of CMOS VLSI analog-digital interface circuit provides a high level of integration.[32]

The architecture for the newly standardized adoptive differential pulse-code modulation (ADPCM) for the coding and transmission of PCM has been developed and placed on a 16-pin CMOS VLSI digital signal processor.[33] This new approach shows a drastic improvement in channel bandwidth. The ADPCM is a digital function for encoding and decoding. The 8-bit PCM word represents an encoding operation that is formed in a 4-bit ADPCM word, received by the decoder, and then formed into an 8-bit PCM word. The design also includes adoptive filters.

The architecture is running in full-duplex mode since it is executing both functions in real time. The control section of the chip is a split control store structure with the macrocode and microcode controllers, and it is microprogrammable architecture. Individual controllers contain a latch, a program counter, a control store, and an instruction decoding logic. The macrocode controller determines the order in which the microcode routines are executed. The microcode basically controls the state of arithmetic elements of the chip and the data flow. The arithmetic part of the chip consists of the data memories and ALU.

One of the advantages of the split control structure is its die size. The other advantage is that the algorithms can be divided into a few functions that are repeated on different sets of data. This allows the microcode controller to execute the primitive functions while the macrocode controller passes information to the function through the address field. The split control structure allows the use of the address latch to indirect addressing of RAM or ROM and accomplishes the parallelism of the architecture. This chip is designed to run from a 20-MHz clock source and an instruction time of 100 ns.

32.3.2 Teleconferencing

The teleconferencing system in the near future will use wideband speech and will need acoustic echo control devices to achieve a quality. The classical echo cancellation algorithms are already on the chip using VLSI technology.[34]

The Motorola processor DSP 56200 is an algorithm-specific DSP developed to perform FIR and adoptive FIR filtering under the control of a host processor.[35] This device can be programmed for a number of filter storage elements (up to 256 per chip), and the DSP can be easily cascaded for longer filters. The FIR filters are common in DSPs, but this processor has the advantage of being able to cascade adoptive FIR filters when needed. For example, when a filter with a sharp cutoff is needed, the number of storage elements necessary to meet specification is very large, and this DSP will have noticeable advantages over other general-purpose DSPs. One of the applications of the adoptive FIR filter is in canceling echos in the telephone system. The echo produced in the telephone system acts as a delayed and attenuated version of the incoming signal. The FIR filter will match its impulse response to that of the echo path, and the output of the filter will be subtracted from the actual echo. The error signal, which is any difference between the actual echo and the synthesized echo from the filter, will be returned. Finally, the error will be small, and no echo will be heard.[36] Cascading more than one DSPs increases the capability of the processor to cancel a variety of the echo signals. More novel VLSI subband echo cancelers have been implemented and tested in two versions: one version is based on the two-band splitting of the speech spectrum by a quadrature mirror filter (QMF) pair, and the second uses pseudo-QMF banks to achieve the best reduction of the global computations. [In experimental work, subband echo cancelers exhibit better convergence properties than do full-band ones.] More development is needed in this area.

32.3.3 Telephone Radio Units

VLSI technology finds application in the land mobile radio unit. One of the single LSI chips FM modulator-demodulator for the 800-MHz band has already been developed.[37] This chip consists of digital and analog functions.[38]

The circuit consists of a digital phase-locked loop (PLL) block, the transmit block, and the receive analog baseband blocks. The digital PLL represents a high-speed programmable counter, consisting of a 32/35 dual-modulus prescaler, a swallow counter and main programmable counter, shift registers and latches, reference counter, and phase frequency comparator. The high-speed programmable counter operates up to 145 MHz. The shift registers and latches use external serial data to set up the high-speed programmable counter. The reference counter is a fixed divider for a reference oscillator.

The transmit analog baseband block consists of a transmit voice, data filters, antialiasing filters, smoothing filter, instantaneous deviation control, and input buffer amplifier. The amplitude of the transmitted voice signal is limited by the instantaneous deviation control. The input and reference voltages are compared in a voice limiter to keep the output signal amplitude below the reference level. The transmit voice filter is a fourth-order low-pass switched capacitor filter. The antialiasing and smoothing filters consist of a second-order *RC* active low-pass filter and cutoff frequency of 20 kHz.

The receive analog baseband block consists of receive voice and data filters, an antialiasing filter, an integrator, smoothing filters and input-output buffer amplifiers. Most of the analog baseband function blocks can be cut off during standby operation to reduce power dissipation. The input amplifier, bandgap reference, and transmit smoothing filter have to remain active even during standby operation to enable quick start-up operation. This type of circuit was implemented in the CMOS 2-μm technology. The chip consisted of about 3600 MOS devices, two bipolar devices, 1400 capacitors, and 15 resistors and occupied an area of 5.2 × 5.8 mm using a 44-pin package. The total power dissipation is 90 mW in the active mode. The performance of this single-chip FM modem baseband LSI was reported to be excellent. The chip is a very practical solution for the land mobile telephone radio unit.

32.3.4 Speech

One LSI developed chip for continuous speech recognition with an approximate vocabulary of 1000 words uses a design based on dynamic time warping (DTW).[39–41] It needs very high speed processing of a large volume of vector data. Ring array architecture[42] was used, which is flexible in terms of array size but maintains a high processor utilization and high throughput. The design utilizes processor elements (PE) as building blocks. The DTW algorithm for continuous speech recognition is

$$P_{ij}x = |u_i - z_j x|^2$$

$$v_{ij}x = \min \begin{cases} v_{i-2,j-1}x + a_1 p_{i-1,j}x + a_2 p_{i,j}x \cdots \\ v_{i-1,j-1}x + a_3 p_{i,j}x \cdots \\ v_{i-1,j-2}x + a_4 p_{i,j-1}x + a_5 p_{i,j}x \cdots \end{cases}$$

where the values a_k $(k = 1, 2, \ldots, 5)$ are the coefficients. The reference patterns are defined as $V = (u_1, u_2, u_i, \ldots)$ and $Z_x = (z_1x, z_2x, \ldots\ z_jx, \ldots)$ (x indicates each set of reference patterns). The above equations represent the vector distance and the cumulative distance calculations. The ring array DTW system consists of a host processor and a ring array processor that has several PEs. The system has two buses for host PE communications and local paths for local communications. The series of recursive equations given above are calculated by n PEs of the array in a parallel and pipelined way. The data flow can be accomplished by parallel computations in continuous and real-time processing.

The PE-LSI has been implemented using a 1.5μm CMOS and consists of about 270,000 transistors, 17,000 logic gates, and 31-Kbit RAM on an area of 13.8 × 14.3 mm. To accommodate continuous speech recognition for any size vocabulary, the chip can be built by utilizing the appropriate construction of integrated PE-LSIs.

One of the novel approaches for efficient processing of graphlike data structure has been used in a speech recognition system.[43] The architecture is based on the accumulate-minimum pipeline function, which complements the classical multiaccumulate structure found in many DSPs, and a difference-magnitude-accumulate function, which is a typical central operation in metric computations. These two operational approaches are merged into the graph search machine (GSM) architecture. The speech recognizer configuration consists of a codec analog-to-digital (ADC) conversion, which is necessary to be passed to the DSP device. In the DSP, signals are going through conditioning feature extraction (such as autocorrelation, filtering, and linear predictive coding) and distance calculation (Itakura-Saito distortion, rms distance, etc.), which are basically multiply-add operations.

The next step in the GSM architecture is add-minimum intensive operations, as shown in Fig. 32.9. The graphlike data structure is a good representation of spoken words and the grammatical structure of the language. The processor consists of eight principal circuit blocks: adder, minimum unit, argmin unit, program address generator, data address generator, I/O frame, program memory, and control unit. The block diagram is shown in Fig. 32.10. The chip is built using 1.5-μm CMOS VLSI technology. The architecture is flexible enough so that it can be used in different applications.

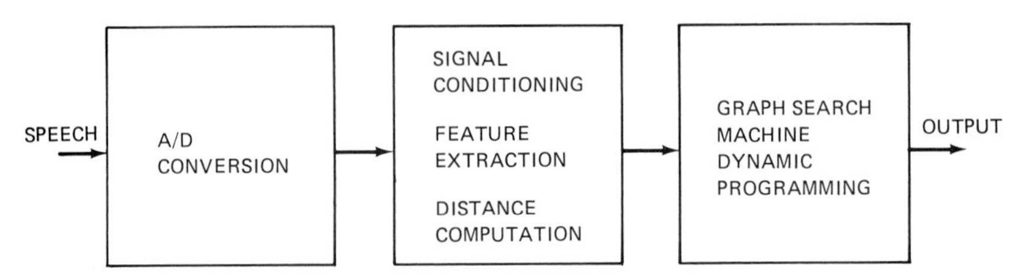

FIG. 32.9 Usual speech recognizer configuration.

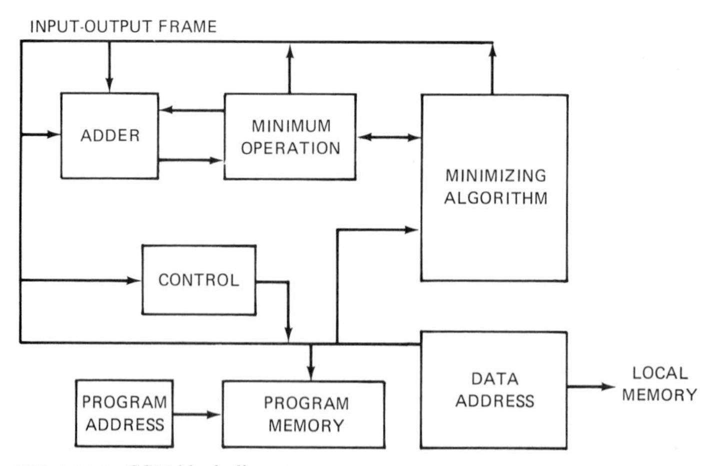

FIG. 32.10 GSM block diagram.

32.3.5 Music

With VLSI technology, an architecture can be developed for modeling the dynamics of musical instruments. It was implemented using CMOS 3-μm fabrication. Basically, the system design to generate musical sounds reproduces the harmonic content of sound. In nature, the sounds evolve in complex ways and cannot be modeled as simple oscillators with amplitude envelopes.[44]

One of the novel approaches to the musical instrument model is based on a number of inner-communicating elements responsible for computing a computation graph. The system can be organized in a ring structure with each element communicating to its nearest neighbors. They are controlled by a host that provides initial information, updates, and an interface to an external controlling device such as a keyboard or disk file. Functionally, each element consists of an array of indentical processing units, a buffer for holding coefficients supplied by the host, and a reconfigurable interconnection matrix, as shown in Fig. 32.11. The patterns of communication are defined by setting switches in the matrix prior to the computation.

The processor used is a serial-parallel multiplier structure capable of one multiply-add-delay step per work time. The programmable connecting matrix provides point-to-point communication between processors. The update buffer is a register bank to hold coefficients and is a parallel connection, the outputs of which are bit-serial lines from the connection matrix to the processing elements. The update buffer is two RAM structures: (*a*) writable from the parallel input bus with a decoder that selects one coefficient; and (*b*) readable 1 bit. The system implemented in CMOS technology generates realistic musical sounds in real time.

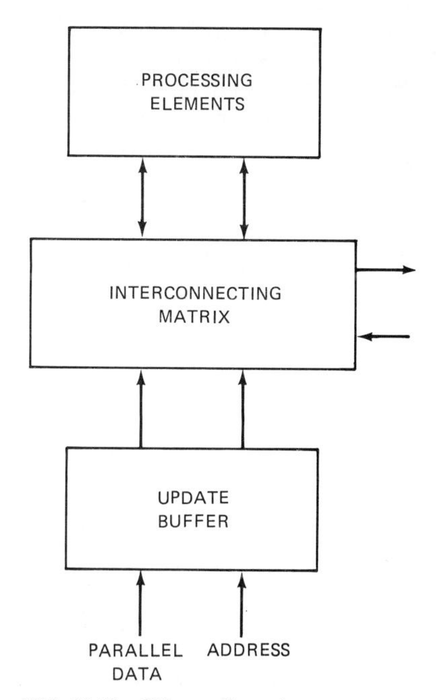

FIG. 32.11 Chip configuration.

32.4 IMAGE PROCESSING

Image processing is nothing more than one aspect of "information processing." Modern image processing relies to a great extent on research activity and improvements of computational algorithms and processor architectures, especially the architectures facilitated by VLSI technology. The key conceptual aspects of implementing image-processing systems is illustrated in Fig. 32.12.

FIG. 32.12 Implementation of VLSI image processing system.

The classical von Neumann architecture was one of the most widely used methods for image processing. However, VLSI designers are probing some of the nonconventional routes for implementing algorithms on the chips. The application of VLSI design and fabrication technology to the processing of image data is providing enormous success. The most significant advantages in using VLSI technology are in performance and sophistication, which are emerging as a consequence of using structured design techniques and the adoption of novel computational algorithms and processing architectures.

The systolic array is an excellent example of a specialized signal processing architecture which is highly amendable to VLSI design. It incorporates features that lead to a high computational efficiency and performance circuit. In many image-processing applications it is essential to perform local operations on every pixel in the image. The high performance obtained in the VLSI architecture is achieved by using numerous simple processors rather than a single large processor. This is achieved by pipelining the stages involved in computational or by multiprocessing independent computations in parallel. To balance the input-output bandwidth it is advisable to make multiple use of input data, which can be achieved, for example, by broadcasting the input data to all cells.

32.4.1 Dedicated Image Processor

One of the first commercial VLSI image processors, NEC uP07281, incorporates a novel internal circular pipeline architecture and powerful optimized image processing. This particular data flow architecture maximizes efficiency in a variety of image-specific multiprocessing applications.[45] The processors have internal circular pipeline architecture, which allows the processing unit to operate at a continuous rate of 5 MHz. The processor consists of the following units: link and functional tables (store object code) and data memory (temporarily stores the data), address generator and flow controller, queue (temporarily stores the information), processing unit (has multiplier and an ALU), input controller, output controller, and refresh controller. This processor is well suited for image processing involving two-dimensional convolution. It can rotate, for example, a 640- by 400-pixel image through an arbitrary input controller. The processor uPD7281 can also be used in cascade or in the ring scheme for a multiprocessor configuration.

One of the two-dimensional systolic-based devices is the geometric and arithmetic parallel processor (GAPP), which is capable of high-speed image-processing application (developed by NCR and Martin Marietta Aerospace). The device integrates 72 individual single-bit processors. The GAPP chip can perform 8-bit integer additions at a rate of 28 million operations per second running at a 10-MHz clock rate. Also, it can be cascaded together.

The AGILE (animated graphics imaging through list execution) developed by General Instrument Corporation, has graphics and a display system, which simulates full motion video. The AGILE can be used for displaying graphic programs as well as developing them. When the image enters manually (through a terminal), or from camera, the host system creates a linked list that is a chronological sequence of events, and it is described as three dimensional.

Another graphics-oriented special-purpose VLSI device (developed by Stanford) is the geometry engine, which is a four-component vector, floating-point processor. This processor can be cascaded into a pipeline configuration.

The new pipelined image processor developed at Tokyo University uses multiple-valued logic. The processor is without encoding and decoding since each pixel can be directly expressed by a single multiple-valued digit. The pattern-matching procedure is designed to perform four-valued image processing based on cellular logic operation. Also, two different templates can be processed simultaneously in a pipelined manner.

Multiple-valued image processing is essential in robotics, medical image processing, and video image processing.[46,47] The algorithm is based on cellular logic operations. An array of four-valued input data is transformed into a new data array. The images have four levels, and each pixel can be expressed by a single quaternary digit. The cellular logic operations are generalized by template or pattern matching, and two different templates can be processed simultaneously in a pipelined manner. The pattern-matching cells, which are building blocks of this type of design, operate with a 2-MHz data rate.

The algorithm, which uses multiple-valued logic, is based on near-neighbor logic operation. Figure 32.13 shows a configuration in the case of a 3 $\times$ 3 window with a center pixel.

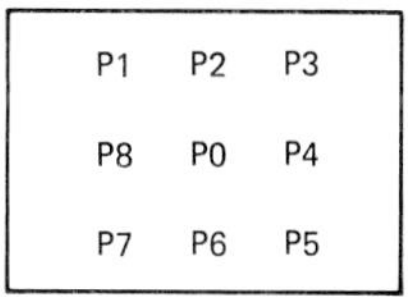

P1	P2	P3
P8	P0	P4
P7	P6	P5

FIG. 32.13 Near-neighbor logic operation.

P0 is a center pixel, and around it are the near-neighboring variables. The near-neighbor operations can be nonrecursive and recursive. The nonrecursive near-neighbor operation has a tendency to require a large number of templates for the state transition in a multiple-valued logic system. However, this problem can be solved by using the multiple-valued minimization technique. The recursive operation is effectively repeated until the output reaches a constant value.

The complete four-valued image processor consists of a three-stage linear array of the pattern matching cells, dual T gates, a pattern-matching cell, and the three-digit dynamic shift register. The important advantage of the four-valued image processor is that the number of the cells can be reduced to 50 percent of the corresponding binary implementation because of the direct processing of input pixels and the double matching procedures. This high-density image processor is an excellent start in VLSI multiple-valued logic.

32.4.2 The Image Processor for Finding Welding Defects

A very interesting application of the digital image processing and pattern recognition is in automatically identifying welding defects. Welding quality is very important. However, most of the checking is done by hand, subject to human error. Using the digital processor simplifies the process of recognizing welding defects in x-ray pictures.

The image algorithm is defined by

$$W(i, j) = Z(i, j)n_1(i, j) + n_2(i, j)$$

where $Z(i, j)$ = gray value of ideal image
$W(i, j)$ = gray value of observation image
$n_1(i, j)$ = photographic film grain noise
$n_2(i, j)$ = system noise

To make welding defects clear, we use the enhancement method of linear gray level transformation. The next concern is to find the projection function of W. The picture is segmented by the segment method (or region growing). The segment method sets the variable threshold in the row and/or column, and the two-valued image is obtained. It is a simple and very useful processor.[48]

32.4.3 Remote Sensing Radar

There is a growing need for faster operations in the area of remote sensing. This is due to the increased amount of data to be processed, resulting from the high tech sensor resolution and the amount of computations required for synthetic aperture radar image reproduction. The advances in VLSI technology allow large-sized hardware to be designed for specific purposes. The parallel and pipeline configurations are being used for ultrahigh-speed image processing.

One of the newly developed processors is a real-time VLSI video rate histogram.[49] The histogram of a digitized image provides a count of how often each pixel intensity occurs in the image. Integrated histograms approximate the probability distribution function for the intensities in the image. This function can be applied to each pixel of the original image. The other application of the histogram is image segmentation. For the histogram which has one sharp peak, a given segment can be assigned with a segment of similar intensity, and if the histogram has more than one peak, the region can be subdivided. The noise in an image will have less effect for a large group of pixels.

The histogram operation was implemented with the RAM and ALU on a single chip. A RAM location must be read, incremented, and rewritten for each sample period. This task is performed at a 10-MHz sample rate. The algorithm was developed so that two pipeline storages could be placed in the read-increment-write path. The RAM location is read into the memory output register after one simple period and then incremented and stored in the accumulator during the next period. The results are written to the RAM.

The programmable shifter is included for normalizing a histogram. A multiplexer follows the barrel shifter and selects whether data path or RAM address data is available on the chip output. The histogram chip is controlled by (*a*) a finite state machine that synchronizes the histogram calculation; (*b*) two counters, one of which is decoded by the PLA to generate signals for programming an external lookup table, and the second to generate the address for reading data from the RAM; and (*c*) a PLA. The histogram processor was fabricated in 3-μm NMOS.

32.4.4 Radar Image Processing

The latest synthetic aperture radar (SAR) uses Doppler processing to achieve high resolution in the direction of the radar platform motion.[50] Many radar returns are used for each pixel in the SAR image. The most frequently used radar processors are not capable of accomplishing very high image processing throughputs, and they do not have a large enough memory. The widely used optical techniques for processing SAR data do not have the accuracy and adaptability properties inherent in digital systems, which are essential for high-precision imaging. The advances in VLSI technology started the development of special-purpose systems for processing SAR data. Some of the algorithms presently used in SAR processing are range processing, azimuth processing, algorithms for computing the discrete Fourier transforms (DFT), convolution algorithms, and a parallel array processing.

Some of these algorithms are not suited for VLSI implementation, and others are very well suited. A class of architectures called a *systolic array* has been developed for implementing convolutions by the direct evaluation approach. These architectures are very regular, possess nearest-neighbor connectivity, and are well suited for VLSI implementation. The module arithmetic is difficult to perform with conventional hardware. However, with VLSI technology, fast parallel multiplication has become a relatively simple operation. One of the SAR signal processors most suitable for VLSI implementation is a processor array.

It is known that all operations required for generating high-precision SAR images can be very efficiently implemented with this type of processor. The processor configuration is well suited for implementing the basic operation in SAR processing, and this type of parallel processor is very efficient in processing large data blocks relative to the size of the PE array. SAR processing is carried out on large data blocks, since it is necessary to combine many radar returns. The main advantage of this processor is its regularity. The GRID (geometric rectangular image and data) processor array has processing power which can easily be increased by increasing the size of the array, and the data-processing word length is not constrained by the machine word length. The custom-designed VLSI GRID processor chip is economical and has compact implementation. The processor is fabricated in a two-level metal using 2.5-μm bulk CMOS technology and consists of 50,000 transistors with a maximum clock frequency of 10 MHz. The processor is arranged in a rectangular interconnected array of 64 bit-serial processing elements. Each PE consists of 64 registers. The block diagram of the GRID processor system is shown in Fig. 32.14.

The PE consists of a bit-serial architecture, which allows flexible data formats. A GRID processing element consists of two data buses, the register sets, the ALU, the nearest-neighbor switching network, the special-purpose 1-bit register, a 64-word × 1-bit dual-port RAM, the multiplier register, the enable regis-

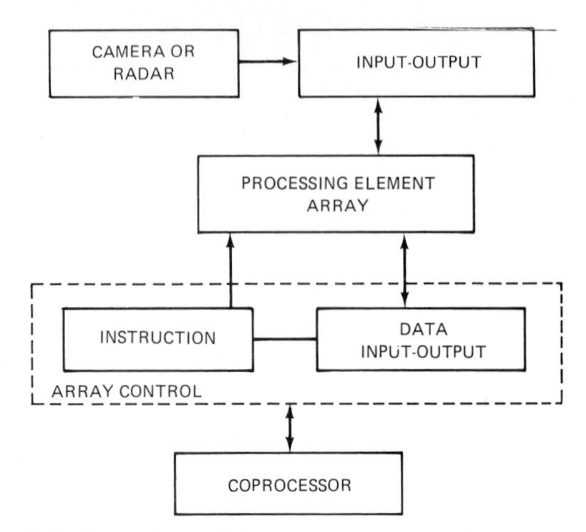

FIG. 32.14 A parallel processor—array architecture.

ter, and the external store register. The external store register allows the external memory expansion to suit a particular application. Much of the chip design effort goes into individual PEs to achieve an optimum trade-off between silicon area and efficiency. The trends are that VLSI technology and imaginative algorithm development are going to facilitate, perhaps drastically, the development of compact, high-performance, real-time SAR imaging processors.

32.5 MEDICAL APPLICATIONS OF VLSI TECHNOLOGY

A classical pioneer physiologist, Etienne Jules Marey, described the modern innovation in medicine as the use of the latest tools of physics and engineering to investigate, enhance, correct, alter, and improve phenomena of living organisms. In the biomedical field, instrumentation for the purpose of quantification in measurement has not pervaded all areas. One of the major tasks to be accomplished is the development of tools and technology for solving the problem of detection and quantitative measurement of living processes. Every new technological development, which improves and changes our external living medium, has definite application in our internal system. If one defines biological organisms as smaller subsystems of the larger universal system, then all the technological novelties can be applied to both systems. Certainly, technological devices have to be appropriately designed to be applied to the specific biological medium. VLSI technology is sweeping the engineering field and that automatically includes the biomedical field. Biomedical science uses areas such as measuring instrumentation, signal processing, image processing, and corrective functional devices (for external use and implementation, etc.). Clearly, progress in these areas will have impact on biomedical electronics.

32.5.1 Speech Processor–Cochlear Implant

VLSI technology has played an important role, and this technology is slowly being applied in medicine. Some of the examples of VLSI technology applied in the biomedical field will now be discussed.

A very important application is restoring hearing to the patients whose inner-ear hair cells have been destroyed but the auditory nerve remains in good condition. A low-cost DSP with a high computational rate (for example, TMS 320C10) and low power consumption could be used as a powerful, flexible, portable, cochlear implant speech processor.[51] The algorithm used in this processor was developed by Markel and Gray. The simplified inverse filter tracking algorithm (SIFT) may be chosen for pitch extraction.[52] The pitch extraction starts with collecting 200 to 300 speech samples at an 8-kHz sample rate and passing them through a digital low-pass filter with corner frequency around 1.1 kHz. The number of samples collected in each frame controls the range of the pitch. The algorithm has to consist of decision rules that will determine whether the samples collected are for voiced or unvoiced speech. Regardless of how attractive this programmable system is, it is only a start in exploring the utility of other fundamental speech features in addition to pitch and formants for excitation of a cochlear prosthesis.

32.5.2 Computer Vision

A major area where the signal processing chip is being employed is in computer vision or, simply, detection of abnormalities in bodies. Simple image processing can be used to determine the likely location of abnormalities. Some examples are detection of human hemoglobin fingerprints, tumors, storage tanks in aerial images, and abnormalities of the ribs, chest, and other areas.

For medical images, the major role of VLSI processors is in diagnostic screening. The automated detection of a specific abnormality in a given part of the body is fast and effective. One example is the automatic diagnosis of pneumoconiosis by texture analyses of the chest x-ray images. The dedicated processor, which includes the median filter, is capable of distinguishing between structural patterns, such as ribs, thick veins, and small capacities of pneumoconiosis. The proposed diagnostic system consists of preprocessing, feature extraction, and decision-making elements. Preprocessing includes a median filter to extract small opacities.[53] The output function is defined as the difference between the input function and the output image of a median filter. The flowchart of the process is shown in Fig. 32.15. Obviously, the system is capable of providing variable information about texture features in many other medical applications.

FIG. 32.15 Proposed diagnostic system for determining pneumoconiosis.

32.5.3 Image Processing in Cardiology

Image processing represents a great step forward in cardiovascular medicine. The reconstruction of the left ventricle from the biplane angiocardiograms has been a very useful medical tool. The functional ability and strength of the left ventricle plays an important role in therapeutic choices and successful prognosis in patients with cardiovascular diseases. The approach is to utilize a pair of orthogonal x-ray projection images taken from two perpendicular camera sets. The shape of the ventricle is reconstructed by dividing the projections into slices and processing slice by slice. The three-dimensional problem is transferred into a two-dimensional problem with the help of parallel cross sections. There are numerous proposed algorithms dealing with the shape of the cross section of the left ventrical. The discrete cross section defined by the two projection profiles is

$$\sum_{i=1}^{m} x_{ij} = P_j \qquad j = 1, \ldots, n$$

and

$$\sum_{j=i}^{n} x_{ij} = Q_i \qquad i = 1, \ldots, m$$

where P_j and Q_i are the two projections. Furthermore, the cross section is divided in half by the equal divisor (ED) curve. The cross section is always uniquely determined by one of its projections, and the problem comes down to estimating the real ED with respect to P_j or Q_i.

If i is the position of an estimated ED along the ruler (Q_i) coordinate, then an estimated cross section is obtained by assigning x_{ij} for each column through the construction. The x_{ij} is defined as

$$x_{ij} = \begin{cases} 1 & \text{for } |\, i - \hat{\imath} \,| \geq \frac{1}{2} P \\ 0 & \text{otherwise} \end{cases}$$

The estimated projection along the ruler is

$$\overline{Q}_i = \sum_{j=1}^{n} \overline{X}_{ij} \qquad i = 1, \ldots, m$$

The expected error index

$$E_Q = \sum_{i=1}^{m} |\, \overline{Q}_i - Q \,|$$

is hopefully close to zero.[54]

The same procedure is followed to find $\bar{P}_j$ and E_P if Q_i is assumed to be the constructor. The algorithm will obtain a unique estimate of ED for P_j or Q_i, since the shape of the ventricular cross section satisfies the regularity constraint.[55] This algorithm achieves better than 96 percent conformity for a regular case and better than 91 percent for an irregular case. The cross sections are reconstructed by this algorithm from the estimation of ED without imposing restrictive geometric conditions on the shape of the cross section. This is a reliable and feasible method for reconstruction not only of ventricular shapes but also of many other major internal organs.

32.5.4 Image Processing of the Stomach

One unpopular medical study is in the field of gastric emptying. The conventional methods suffer various problems, which are related to human interference, patient movement between images, and changes in the location of radioactive materials within the stomach. By using automated methods of moments that are invariant to scaling, rotation, and shifts, we can eliminate these problems. The automated analysis of gastric emptying (AGEA) is a completely automated method for finding the gastric output and generating gastric emptying rates for both solid- and liquid-phase materials. This particular study uses the moment invariants method to process and analyze the images of the stomach taken by the gamma camera. The moments are found as

$$m_{pq} = \int_{-\infty}^{\infty} \int_{-\infty}^{\infty} x^p y^p f(x, y)\, dx\, dy$$

where $p, q = 0,1,2,\ldots$ and the central moment is given by

$$u_{pq} = \int_{-\infty}^{\infty} \int_{-\infty}^{\infty} (x - \bar{x})^p (y - \bar{y})^q f(x, y)\, dx\, dy$$

where $\bar{x} = m_{10}/m_{\infty}$ and $y = m_{01}/m_{\infty}$

The discrete values of moments invariant under translation and rotation are

$$\phi_1 = u_{20} + u_{02}$$

$$\phi_2 = (u_{20} - u_{02})^2 + 4^{u}_{11}$$

$$\phi_3 = (u_{30} + u_{12})^2 + (u_{21} + u_{03})^2$$

$$\phi_4 = (u_{30} + u_{12})^2 + (u_{21} + u_{03})^2$$

$$\begin{aligned}\phi_5 = {} & (u_{30} - u_{12})(u_{30} + u_{12})((u_{30} + u_{12})^2 \\ & - 3(u_{21} + u_{03})^2) + (3u_{21} - u_{03})(u_{21} + u_{03}) \\ & \times (3(u_{30} + u_{12})^2 - (u_{21} + u_{03})^2)\end{aligned}$$

$$\begin{aligned}\phi_6 = {} & (u_{20} - u_{02})((u_{30} + u_{12})^2 - (u_{21} + u_{03})^2) \\ & + 4u_{11}(u_{36} + u_{12})(u_{22} + u_{03})\end{aligned}$$

$$\begin{aligned}\phi_7 = {} & (3u_{21} - u_{03})(u_{30} + u_{12})((u_{30} + u_{12})^2 - 3(u_{21} + u_{03})^2) \\ & - (u_{30} - 3u_{12})(u_{21} + u_{03}) \\ & \times (3(u_{30} + u_{12})^2 - (u_{21} + u_{03})^2)\end{aligned}$$

The "0" functions are normalized to make them invariant under a scale change by using normalized central moments:

$$n_{pq} = \frac{u_{pq}}{u_{00}^{\gamma}}$$

where

$$\gamma = \frac{p + q}{2} + 1$$

The steps used in determining the geometric properties of the stomach during ejection are as follows: (1) for images with adequate signal-to-noise ratio (SNR), automatic thresholding (to obtain a binary image) is achieved using a method based on an entropy criterion; (2) clustering analysis is performed, the biggest object (the stomach) is separated, and the contour is found; and (3) after finding the gastric contour a time activity curve is generated and a filling is performed to eliminate any gaps.

The first seven individual moments were computed (enough for excellent approximations) for normal and abnormal stomach geometry. The computation was performed in 15-min intervals from the start of the stomach emptying to the end of emptying. It was noticeable that the values of the moments for normal stomach geometry are smaller than those with abnormal stomach geometry. Plotting the moments of a number of normal and abnormal patients revealed that the normal moments follow the same pattern, which is different from the pattern formed by the abnormal curves. Apparently, there is a statistically significant difference in the geometry of the stomach between normal and abnormal samples. This is an excellent candidate for employing not very sophisticated DSP to produce automatic readings of the stomach's geometrical variation for diagnostic purposes. The DSP using VLSI technology is being developed and hopefully will be a useful tool in gastrointestinal medicine.[56]

32.5.5 Medical Implants

Manufactured insulin delivery systems have been available for some time. One well-developed system (Sandia Laboratories and the University of New Mexico), weighing 300 g, consists of an insulin pump, two lithium batteries, and electronic controls. The system is placed inside the abdominal wall. The silicon-rubber reservoir containing insulin represents an implantational part. An external control unit is built to control the external pumping rate. The system's operating frequency is 35 kHz. The amount of insulin delivered and the amount remaining within the reservoir is always known. The most acceptable technology for this type of control electronics is a low-power silicon integrated circuit using CMOS devices. The present system is built with CMOS ICs consisting of about 2000 transistors. The block diagram is shown in Fig. 32.16. The real bionic pancreas is with us to stay.[57]

This is one example where LSI and/or VLSI technology is immensely important not only in the size and weight reduction of the implant electronics but also in the medical status of the patient at any time. Incorporating the appropriate signal-image processor makes the task easy to accomplish.

Furthermore, a similar principle is applicable in monitoring and controlling other vital functions in the human body. Chemotherapy, as well as other hormonal controlled and druglike substances, can be administered in the body with minimum side effects. Obviously, VLSI technology opened the door to a number of different implantational devices.

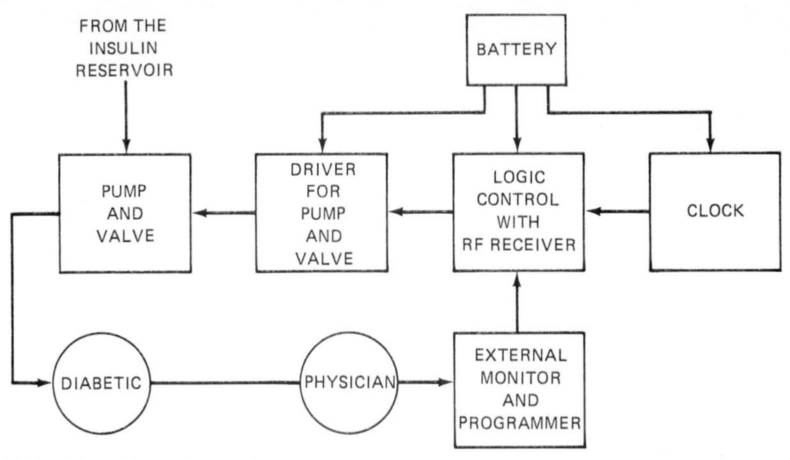

FIG. 32.16 Block diagram for an electronic pancreas.

32.5.6 Medical Telemetry Systems

Another area in the biomedical field that is very important in research is the telemetry system for neurophysiology. Until recent years, studies were conducted using external telemetry units to perform short- or long-time experiments on animals to learn more about different physiological functions. However, new technological advancements allow the use of implantable telemetry systems that can acquire and transmit neural firing patterns and other physiological parameters in moving unanesthetized animals.

One example of this type of study is in monitoring body thermoregulation by monitoring neural signals from microwave electrodes and correlating them with the muscle activity and body temperature while finding the algorithm for the neural thermostat. The already developed unit consists of filters, amplifiers, and multiplexers (which accept and process the neural action-potential waveforms), electrograms, and temperature signals.[58] The dedicated signal processor, utilizing the latest technology and with its low power and small volume, is able to perform long experiments using small batteries, which results in minimum disruption to the animal.

The system's multiplexing hierarchy samples each type of signal at a rate in accordance with its required low-bandwidth mode, which is based on the properties of the physiological signals. Spikes approximately 1 ms with a repetition rate of 500 Hz represent the neural signal. These signals are recorded, and the nerve cells are encoded solely in the time pattern of the spikes. The undistorted transmission of the waveform uses an 8-kHz channel, and a filtered waveform is transmitted by a 1-kHz channel. The electrogram signals are typically telemetered around a 300-Hz channel with variation for some specific investigation. For the study of temperature and some other body functions, the required bandwidth will be drastically reduced, to as low as 50 Hz. For this reason, multiplexing is implemented as a hierarchy of multiplexed channels, where the signals that require different bandwidths are sampled at different frequencies.

All the amplifiers are on the signal processor with varying ac coupled differential inputs, voltage gains greater than 120, and low-pass filtering. The signal processor includes an additional three drivers. Two drivers are connected to external thermistors, and one is for signal calibration. The control logic consists of cascaded Johnson counters that implement the multiplex hierarchy, set latches that turn the frequency modulation (FM) transmitter on or off, a ring oscillator, and a few different counters.

This is one of the IC-based telemetry systems necessary to be implanted for neurophysiological investigations. If progress can be made in analyzing thermoregulatory functions in the human body, hopefully one day these controls can be used to correct bodily functions by developing appropriate implantable control units. It is known that artificially producing the hot spots in the area of tumors will reduce the tumor size. However, the thermoregulatory system of the human body is opposing an increase in the temperature for a prolonged period of time and indirectly diminishing the hot spots. Progress has been made in this direction due to new technological advancements.

32.5.7 Specialized Implantable DSP

There is another area of important research being conducted in the biomedical area using implanted DSP. The hereditary disease sickle cell anemia is a disease of abnormal hemoglobin. The sickling of erythrocytes occurs only under low oxygen pressure, when hemoglobin molecules are deoxygenated. The process is reversed when normal oxygen pressure is restored and hemoglobin is reoxygenated.[59] The deformation of erythrocytes is caused by the formation of liquid crystalline tactoids of deoxyhemoglobin S in the cell. Observations have been made that high-intensity fields would decompose this abnormal cellular structure, and the erythrocytes will go back to the normal configuration. Developing the appropriate DSP to detect a low oxygen pressure in the erythrocytes and send a signal to induce a high-intensity field will decompose the abnormal cellular structure.

It can be seen that the new VLSI technology is making its mark in the field of biomedical physics and electronics. Different general-purpose DSPs can be used for diagnostic purposes, for implantation to control some abnormal body function, for implantation in animals to understand body functions, for improvement and/or replacement of the major senses in the human body, for medical imaging, for surgery, physical therapy, physical fitness, and so on.

32.6 DSPs FOR ENTERTAINMENT AND HOME

More and more video games are becoming sophisticated due to the new technology. Children's games and toys are becoming increasingly sophisticated, exciting, and very popular. The barking, giggling, and crying of the new breed of electronic pets are saturating the market. Some of the toys even respond to spoken commands. They purr or giggle when petted. They even dance. The toys produce a wide variety of voices and accents. They turn 90° at the command "turn around," at "stop-stop" they sit down and power themselves off. Some of the toys respond to human speech and attempt to repeat sounds they hear. One home electronic pet even has a security mode, with radio waves that direct any movement within 15 ft.

Home appliances, such as microwave ovens, refrigerators, televisions, home-controlled alarm systems, FM stereo tuners, super hi-fi by digital recording, and cordless infrared-transmitted headphones, are further examples. The more sophisticated cameras (regular, movie, and Polaroid) use LSI chips to replace some of the traditional mechanical parts. A variety of electronic watches on the market not only have remarkable accuracy but also are available with added functions such as heart rate measures, pager, and melody playing. Sophisticated electronic locks and hotel room keys are also taking advantage of the new technology. Industrial robots and other electronically controlled machines improve productivity, reliability, and design flexibility. To save energy, VLSI integrated circuits are used for the operation of electric motors, heating, and air conditioning units. To eliminate a number of mechanical parts, the Singer Athena sewing machine contains an NMOS VLSI chip. This machine can operate 36,000 operations in 1 h, and the controls contain around 1000 transistors. Physical fitness products are more available for home use due to the new technology. Computer printers are more reliable, quieter, and faster because mechanical parts have been replaced by VLSI processor control.

Many manufacturers that never used electronics are now incorporating LSI-VLSI chips in their products, which improve performance, reduce costs, and add new functions. The LSI-VLSI is bringing significant technological and social changes as the basis of a second industrial revolution. Often the users of LSI-VLSI chips require custom design, and since semiconductor manufacturers are usually not interested in custom design, users are forced to have an in-house design. For this reason users of custom design have been hiring chip designers. The future of the manufacturing processes for electronic products will experience drastic change. The conventional assembly line work force is being replaced by LSI-VLSI chip designers. The requirements for a large number of chip designers will create an enormous chip designer shortage. There is a clear message for educators to do something about change in the educational system to accommodate this trend.

32.7 EDUCATION

During the past several decades, technological developments have created industry needs and produced changes in engineering curricula to meet these needs. During the late 1970s custom design came into vogue. Many companies wanted proprietary chip designs that would create a competitive edge. Design and layout techniques for custom design were producing an inordinately high number of layout errors. While the errors would ultimately be resolved, the overall production schedules were slipping at a dramatic rate. Also, the complicating fact was that the product development cycle was lengthening and the product lifetime was getting shorter (see Fig. 32.17). As a response, a movement to incorporate computer-aided design techniques started.[60]

Upon entering the 1980s the development of CAE/CAD tools was somewhat disjointed. Each vendor had customized tools. Integrated circuit technology had matured with large-density chips at low cost per gate readily available (VLSI). However, the IC industry could no longer handle the demand by using internal talent. The customer became involved in CAD-type design. The concept of low-cost individual workstations was promoted. It should be pointed out that tailored systems imply that the computer software and terminals are purchased from separate vendors and supported by a group of in-house hardware-software experts.

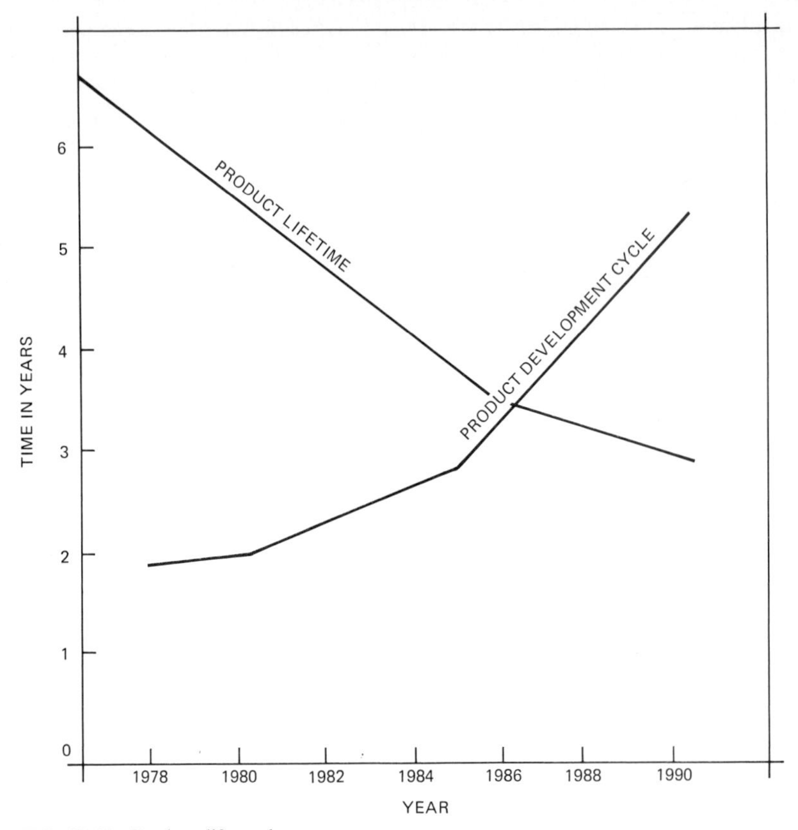

FIG. 32.17 Product life cycle.

Workstations are self-contained design vehicles that have integrated hardware and software tools.

Figure 32.18 shows that the demand for engineers with VLSI skills is high, but their availability is low. This demand will continue well into the 1990s. In the meantime, the VLSI design industry is trying to create CAD tools to make the design process opaque to the designer. The philosophy is that there are 600,000 logic and system designers while there are only around 3000 true integrated circuit designers in this country. Industry is now not necessarily looking for circuit process knowledge. Instead they are looking for a student who can understand the hierarchical design process. Through the use of ''silicon compilers'' the skills of IC design will supposedly no longer be required.

Many engineering schools have established intensive CAD programs to enhance instructions, attract high-quality students, and enhance the school's image. Most schools offer a course or two in NMOS and CMOS VLSI systems design. The VLSI course usually provides the students with essential materials in hierarchical design methodology for VLSI in an organized fashion. The emphasis in these courses is on fundamental design concepts. The students are introduced to the design rules and computations of circuit parameters for the layout. Great em-

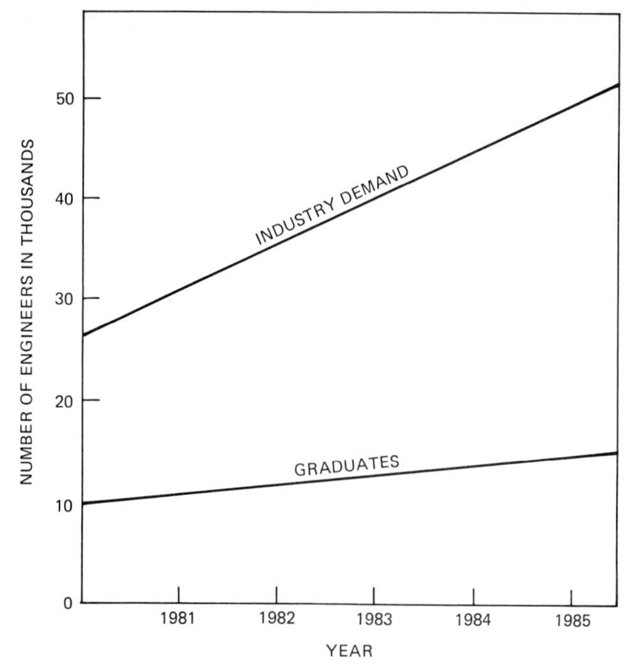

FIG. 32.18 Electronics engineering demand.

phasis is given in the area of power computations and system-level design. However, the new approach to education requires that each student be exposed to an organized lab on the VLSI workstations. They usually produce as a final project some sort of VLSI design. Some of these circuits are being fabricated. Students with this background are prepared for immediate productive design work. This education also closes the gap that existed between "theoretical education" and industry. More and more industries are utilizing the students as well as appropriate faculty to establish close working research relationships.[61] Ideally, the students involved in industrial research projects gain not only experience in the real design work but also a competitive edge in obtaining a position with a given company after graduation. The industry in this type of relationship has access to a pool of fully trained engineers.

REFERENCES

1. B. A. Bower and W. R. Brown, *VLSI Systems Design for Digital Signal Processing,* vol. 1. *Signal Processing and Signal Processors,* Prentice-Hall, Englewood Cliffs, N.J., 1982, pp. 5–64.
2. Earl E. Swartzlander, Jr., *VLSI Signal Processing Systems,* Kluwer Academic, Boston, 1986, pp. 67–116.
3. Graham R. Nudd and J. Greg Nash, "Application of Concurrent VLSI Systems to

Two-Dimensional Signal Processing", in *VLSI and Modern Signal Processing,* S. Y. Kung, H. J. Whitehouse, and T. Kailath (eds.), Prentice-Hall, Englewood Cliffs, N.J., 1985, pp. 307–325.

4. Sun-Yuan Kung, Robert E. Owen, and J. Greg Nash, *VLSI Signal Processing, II,* IEEE Press, New York, 1986.
5. Rikio Maruta and Takao Nishitani, "Signal Processing VLSI Developments: Prospects through Experience," in *VLSI Signal Processing,* II, Sun-Yuan Kung, Robert E. Owen and J. Greg Nash (eds.), IEEE Press, New York, 1986, pp. 223–230.
6. B. Hirosaki et al., "A CMOS-VLSI Rate Conversion Digital Filter for Digital Audio Signal Processing," *Proc. IEEE ICASSP'84,* vol. 3, 1984.
7. M. Yano, K. Inoue, and T. Senba, "An LSI Digital Signal Processor," *Proc. IEEE ICASSP'82,* Vol. 2, 1982.
8. Yoshikazu Mori et al., "Floating-Point Digital Signal Processor MSM6992 and Its Development Support System," in *VLSI Signal Processing, II,* Sun-Yuan Kung, Robert E. Owen and J. Greg Nash (eds.), IEEE Press, New York, 1986, pp. 248–256.
9. Charles L. Saxe and Arif Kareem, "Single Chip Digital Signal Processors," in *VLSI Signal Processing,* II, Sun-Yuan Kung, Robert E. Owen, and J. Greg Nash (eds.), IEEE Press, New York, 1986, pp. 257–266.
10. Kevin L. Kloker, "The Architecture and Applications of the Motorola DSP56000 Digital Signal Processor Family," *Proc. IEEE ICASSP'87,* vol. 1, 1987, pp. 523–526.
11. K. L. Kloker, "The Motorola DSP56000 Digital Signal Processor," *Proc. IEEE MICRO,* Dec. 1986.
12. K. L. Kloker, "Architectural Features of the Motorola DSP56000 Digital Signal Processor", in *VLSI Signal Processing,* vol. II, IEEE Press, New York, 1986, pp. 233–244.
13. N. Srinivasa, K. Rajgopay, and K. R. Ramakrishnan, "On a Programmable Signal Processor for VLSI," *Proc. ICASSP'87,* vol. 1, pp. 795–796, 1987.
14. Magdy A. Bayoumi, "A Quadratic Residue Processor for Complex DSP Applications," *Proc. ICASSP'87,* vol. 1, pp. 475–477, 1987.
15. Paul M. Chau, Kay C. Chew, and Walter H. Ku, "A Bit-Serial Floating-Point Complex Multiplier-Accumulator for Fault-Tolerant Digital Signal Processing Arrays," *Proc. ICASSP'87,* vol. 1, 1987, pp. 483–486.
16. R. W. Linderman, P. P. Reusens, P. M. Chau, and W. H. Ku, "Digital Signal Processing Capabilities of CUSP, A High Performance Bit-Serial VLSI Processor," *IEEE Conf. on Acoustics, Speech and Signal Processing,* March 1984, pp. 1611–1614.
17. R. W. Linderman, P. M. Chau, W. H. Ku, and P. Reusens, "CUSP: A 2-micron CMOS Digital Signal Processor," *IEEE J. Solid State Circuits,* **SC-20**(3):761–769 (1985).
18. P. M. Chau and W. H. Ku, "A Self-Testing VLSI/VHSIC-Class Signal Processor System," *Government Microcircuit Applications Conf. (GOMAC, 1986),* San Diego, Cal., November 1986.
19. Fathy F. Yassa and Steven G. Kratzer, "A VLSI Implementation of the Patial Rank Algorithm For Adaptive Signal Processing," *Proc. ICASSP'87,* vol. 1, 1987, pp. 507–510.
20. B. Widrow and S. Stearns, *Adaptive Signal Processing,* Prentice-Hall, Englewood Cliffs, N.J., 1985.
21. R. Manzingo and T. Miller, *Introduction to Adaptive Arrays,* Wiley, New York, 1980.
22. Demassieux Nicolas et al., "An Optimized VLSI Architecture for a Multiformat Discrete Cosine Transform," *Proc. ICASSP'87,* vol. 1, 1987, pp. 547–550.
23. F. Jutland et al., "A Single Chip Video Rate 16 · 16 Discrete Cosine Transform," *Proc. ICASSP'86,* 15.8.1, 1986.

24. D. M. Taylor and R. Retter, "A Novel VLSI Digital Signal Processor Architecture for High-Speed Vector and Transform Operations," *Proc. ICASSP'87,* vol. 1, 1987, pp. 531–535.
25. Stanley K. Honey and Walter B. Zavoli, "A Novel Approach to Automotive Navigation and Map Display," *IEEE Transactions on Industrial Electronics,* **IE-34**(1):40–43 (1987).
26. Takero Hongo et al., "An Automatic Guidance System of a Self-Controlled Vehicle," *IEEE Trans. Industrial Electronics,* **IE-34**(1):5–10 (1987).
27. Hidefumi Kobatake et al., "Measurement of Two-Dimensional Movement of Traffic by Image Processing," *Proc. ICASSP'87,* vol. 1, 1987, pp. 614–617.
28. Lawrence E. Larson et al., "An Analog CMOS Autopilot," *IEEE J. Solid-State Circuits,* **SC-20**(2):571–577 (1985).
29. Peter B. U. Achi, "A Low-Cost Computer Numerical Control Machine System Using LSI Interfaces," *IEEE Trans. Industrial Electronics,* **IE-34**(1):101–106 (1987).
30. David A. Sunderland et al., "CMOS/SOS Frequency Synthesizer LSI Circuit for Spread Spectrum Communications," *IEEE J. Solid-State Circuits,* **SC-19**(4):497–505 (1984).
31. J. Tierney, "A Digital Frequency Synthesizer," *IEEE Trans. Audio,* **AU-19**:48 (1971).
32. Bhupendra K. Ahuja et al., "A Programmable CMOS Dual Channel Interface Processor for Telecommunications Applications," *IEEE J. Solid-State Circuits,* **SC-19**(6):892–898 (1984).
33. Luis Bonet and Tim A. Williams, "A Split Control Store VLSI for 32 kbps ADPCM Transcoding," *Proc. ICASSP'87,* vol. 1 1987, pp. 511–514.
34. Andre Gilloire, "Experiments with Sub-Band Acoustic Echo Cancellers for Teleconferencing," *Proc. ICASSP'87,* vol. 3, 1987, pp. 2141–2144.
35. C. D. Thompson and J. P. Gergen, "A High-Performance VLSI/Echo Canceller Chip," *Proc. Speechtech,* **1**(3):31–34, 1986.
36. David E. Borth, Ira. A. Gerson, and John R. Haug, "A Cascadable Adaptive Fir Filter VLSIIC," *Proc. ICASSP'87,* vol. 1, 1987, pp. 515–516.
37. Tatsuji Habuka et al., "A Single-Chip FM Modem Baseband CMOS LSI for Land Mobile Telephone Radio Units," *IEEE J. Solid-State Circuits,* **SC-20**(2):617–622 (1985).
38. N. Kanmuri, "Major Techniques for High-Capacity Land Mobile Communication System—Mobile Subscriber Set," *Proc. NTT Int. Symp.,* July 1983, pp. 94–108.
39. Jun-ichi Takhashi et al., "A Flexible Linear Array Oriented VLSI Processor for Continuous Speech Recognition," *Proc. ICASSP'87,* vol. 1, 1987, pp. 499–502.
40. H. Sakoe et al., "Dynamic Programming Optimization for Spoken Word Recognition," *IEEE ASSP-26,* 1978, pp. 43–49.
41. N. Sugamura et al., "Japanese Text Input System Based on Continuous Speech Recognition," *Proc. ICASSP'86,* 1986, pp. 1125–1128.
42. J. Takhashi et al., "A Ring Array Processor Architecture for Highly Parallel Dynamic Time Warping," *IEEE ASSP-34,* 1986, pp. 1310–1318.
43. Stephen Glinski et al., "The Graph Search Machine (GSM): A Programmable Processor for Connected Word Speech Recognition and Other Applications," *Proc. ICASSP'87,* vol. 1, 1987, pp. 519–522.
44. John Wawrzynek, "A Reconfigurable Concurrent VLSI Architecture for Sound Synthesis," in *VLSI Signal Processing,* vol. II, Sun-Yuan Kung, Robert E. Owen, and J. Greg Nash, (eds.), IEEE Press, New York, 1986, pp. 385–395.
45. Raymond J. Offen, *VLSI Image Processing,* McGraw-Hill New York, 1985, pp. 1–47.
46. Michitaka Kameyama et al., "Design and Implementation of Quaternary NMOS Inte-

grated Circuits for Pipelined Image Processing," IEEE *J. Solid-State Circuits,* **SC-22**(1):20–27 (1987).

47. K. Preston, Jr. et al., "Basis of Cellular Logic with Some Applications in Medical Image Processing," *Proc. IEEE,* **67**:826–856 (1979).
48. Chen Su-xian and Li Xiao-song, "The Automatic Classification of the Welding Defects," *Proc. ICASSP'87,* vol. 2, 1987, pp. 618–621.
49. Brian Richards et al., "A Parameterized VLSI Video-Rate Histogram Processor," *Proc. ICASSP'87,* vol. 1, 1987, pp. 491–494.
50. B. Arambelopa and S. R. Brooks, "Algorithms and Architectures for Digital Synthetic Aperture Radar (SAR) Processing," in *VLSI Image Processing,* Raymond J. Offen (ed.), McGraw-Hill, New York, 1985, pp. 237–270.
51. L. Robert Morris et al., "Algorithm Selection and Software Time/Space Optimization for a DSP Micro-Based Speech Processor for a Multi-Electrode Cochlear Implant," *Proc. ICASSP'87,* vol. 2, 1987, pp. 972–975.
52. J. Market and A. Gray, *Linear Prediction of Speech,* Springer-Verlag, New York, 1976.
53. Hidefumi Kobatake et al., "Automatic Diagnosis of Pneumoconiosis by Texture Analysis of Chest X-Ray Images," *Proc. ICASSP'87,* vol. 2, 1987, pp. 610–613.
54. Z. D. Bai et al., "A New Approach for Reconstruction of the Left Ventricle from Biplant Angiocardiograms," *Proc. ICASSP'87,* vol. 2, 1987, pp. 1225–1228.
55. Z. D. Bai et al., "Reconstruction of the Left Ventricle from Two Orthogonal Projections," TR-86-33, Center for Multivariate Analysis, University of Pittsburgh, Pittsburgh, Pa., 1986.
56. R. Ech, Z. Delalic, and A. Abutaleb, "Automated Analysis of Gastric Emptying: Geometrical Properties," *Proc. 13th Northeast Bioengineering Conf.,* Philadelphia, Pa., March 1987.
57. W. J. Spencer, "For Diabetics: An Electronic Pancreas," *IEEE Spectrum,* June 1978, pp. 38–42.
58. Michael G. Dorman et al., "A Monolithic Signal Processor for a Neurophysiological Telemetry System," *IEEE J. Solid-State Circuits,* **SC-20**(6):1185–1192 (1985).
59. Z. Delalic et al., "Dielectric Constant of Sickle Cell Hemoglobin: Dielectric Properties of Sickle Cell Hemoglobin in Solution and Gel," *J. Mol. Biol.,* **168**:659–671 (1983).
60. Z. Joan Delalic and J. J. DiGiacomo, "VLSI Design Tools in Engineering Education," *1986 Frontiers in Education Conf. Proc.,* Austin, Tex., pp. 246–254.
61. Z. Joan Delalic, "Engineering Education in the High-Tech Era," *Int. J. Appl. Engineering Education,* forthcoming.

APPENDIX

LAWRENCE J. KOVACS
Unisys Corporation
Blue Bell, Pennsylvania

A.1 DEFINITIONS AND TERMS

acceptor An impurity capable of supplying a mobile hole.

angstrom, Å Lattice constant and measurement (1 Å = 10^{-8} cm).

band Energy level in a crystal where electrons can move freely.

conduction In metals, dependent upon density of unbound electrons; in semiconductors, dependent upon impurities, lattice defects, and temperature to move or scatter the holes and electrons.

crystal (lattice) Arrangement of atoms.

depletion layer A charge-free region in a semiconductor caused by the presence of an electric field.

diffusion Introduction of impurities in controlled amounts into a silicon substrate.

donor An impurity atom with a loose or free electron.

doping The act of supplying impurities to a silicon wafer or substrate.

electron A negatively charged particle or carrier.

hole A positively charged carrier that appears to move through the crystal structure.

impurity Boron, aluminum, gallium, indium, phosphorous, arsenic, and antimony are the most common impurities or dopants added to silicon.

junction The area surrounding the jointure of two dissimilar materials.

majority/minority carriers Greater than 50 percent or less than 50 percent, respectively, of the carrier concentration.

mask set Various patterns corresponding to progressive steps in the fabrication process, which combines to form the layout of a semiconductor.

monolithic Single silicon crystal.

photo resist Light-sensitive coating used to protect or eliminate portion of the surface wafer.

substrate The material upon which a semiconductor is fabricated.

A.2 SYMBOLS AND ABBREVIATIONS

A = Area, $L \times W$
a = Lattice constant, Å
BV = Breakdown voltage
C = Capacitance, F
c = Speed of light in a vacuum, cm/s
cm = Centimeter
D = Diffusion coefficient, cm^2/s
d = Distance
E = Energy, ev
f = Frequency, Hz
G = Conductance
h = Planck's constant, J · s
I = Current, A
k = Boltzmann's constant, J/K
L = Diffusion length
l = Length, m, cm, mm
m = Meter
mm = Millimeters
M = Mobility
N = Net impurity concentration
N_A = Acceptor impurity density
N_D = Donor impurity density
n = Electron concentration
P = Power
p = Hole concentration
ppm = Parts per million
Q = Surface impurity concentration
R = Resistance, Ω
r = Radius
s = Charge density
T = Absolute temperature
t = time
V = Voltage
v = Carrier velocity
w = Width
x = Distance
τ = Carrier lifetime (decay time), s
Ω = ohm

A.3 CONSTANTS

Angstrom unit, Å	10^{-8} cm
Avogadro's number	6.02204×10^{23} mol^{-1}
Bohr's radius	0.52917 Å
Boltzmann's constant k	1.38066×10^{-23} J/K
Elementary charge q	1.60218×10^{-19} C
Electron volt, eV	1.60218×10^{-19} J
Speed of light in a vacuum c	2.99792×10^{10} cm/s

A.4 LENGTHS AND CONVERSION FACTORS

Inches, in	Centimeters, cm	Millimeters, mm	Micrometers, μm	Angstroms, Å	Microinches, μin
1	2.54	25.4	2.54×10^4	2.54×10^8	10^6
0.3937	1	10	10^4	10^8	3.937×10^5
3.937×10^{-2}	0.1	1	10^3	10^7	3.937×10^4
3.937×10^{-5}	10^{-4}	10^{-3}	1	10^4	39.37
3.937×10^{-9}	10^{-8}	10^{-7}	10^{-4}	1	3.937×10^{-3}
10^{-6}	2.54×10^{-6}	2.54×10^{-4}	2.54×10^{-2}	2.54×10^2	1

A.5 SCALING FACTORS

Prefix	Multiplier	Abbreviation
exa	10^{18}	E
peta	10^{15}	P
tera	10^{12}	T
giga	10^9	G
mega	10^6	M
kilo	10^3	k
hecto	10^2	h
deka	10^1	da

Prefix	Multiplier	Abbreviation
deci	10^{-1}	d
centi	10^{-2}	c
milli	10^{-3}	m
micro	10^{-6}	μ
nano	10^{-9}	n
pico	10^{-12}	p
femto	10^{-15}	f
atto	10^{-18}	a

A.6 COMMON ACRONYMS INCLUDING LOGIC FAMILIES AND TECHNOLOGIES

BiMOS Bipolar metal-oxide silicon technology.

CAD Computer-aided design.

CMOS Complementary metal-oxide silicon technology.

CMOS/SOS Silicon on sapphire technology combined with complementary metal-oxide silicon technology.

DTL Diode transister logic.

DT²L Diode transistor—transistor logic.

ECL Emitter-coupled logic.

IC Integrated circuit.

I²L Integrated injection logic.

I/O Input-output.

ISL Integrated Schottky logic.

LSI Large-scale integration.

MOS Metal-oxide semiconductor technology.

MSI Medium-scale integration.

NMOS *n*-Channel metal-oxide semiconductor technology.

PLA Programmable logic array.

PMOS *p*-Channel metal-oxide semiconductor technology.

RAM Random-access memory.

ROM Read-only memory.

SSI Small-scale integration.

TTL Transistor—transistor logic.

VLSI Very-large-scale integration.

VMOS Short channel metal-oxide semiconductor technology.

74ALS Advanced low-power Schottky technology.

74AS Advanced Schottky technology.

74AC Advanced CMOS technology.

74ACT Advanced CMOS technology TTL compatible.

74F Advanced Schottky technology.

74HC High-speed CMOS technology.

74HCT High-speed CMOS technology TTL compatible.

A.7 INTERNATIONAL SYSTEM OF UNITS

Terms	Unit	Symbol
Capacitance	farad	F
Conductance	siemens	S
Current	ampere	A
Electric charge	coulomb	C
Electric flux density	coulombs/meter squared	C/m^2
Electric field strength	volt/meter	V/m
Energy	joule	J
Force	newton	N
Frequency	hertz	Hz
Inductance	henry	H
Length	meter	m
Magnetic field strength	ampere/meter	A/m
Magnetic flux	weber	Wb
Magnetic induction	tesla	T
Mass	kilogram	kg
Potential	volt	V
Power	watt	W
Pressure	pascal	Pa
Resistance	ohm	Ω
Temperature	kelvin	K
Time	second	s

INDEX